POWER AND COMMUNICATION CABLES

POWER AND COMMUNICATION CABLES
Theory and Applications

Edited by

R. Bartnikas, Editor
Institut de Recherche d'Hydro-Québec
Varennes, Québec, Canada

K. D. Srivastava, Coeditor
University of British Columbia
Vancouver, Canada

IEEE Dielectrics and Electrical Insulation Society, *Sponsor*

IEEE Industry Applications Society, *Sponsor*

IEEE Power Electronics Society, *Sponsor*

IEEE PRESS SERIES ON POWER ENGINEERING

P. M. Anderson, *Series Editor*

IEEE PRESS

The Institute of Electrical
and Electronics Engineers, Inc.,
New York

McGRAW-HILL
New York   San Francisco   Washington, D.C.
Auckland   Bogotá   Caracas   Lisbon   London
Madrid   Mexico City   Milan   Montreal
New Delhi   San Juan   Singapore
Sydney   Tokyo   Toronto

This book and other books may be purchased at a discount from the publisher when ordered in bulk quantities. Contact:

IEEE Press Marketing
Attn: Special Sales
Piscataway, NJ 08855-1331
Fax: (732) 981-9334

For more information about IEEE PRESS products, visit the
IEEE Press Home Page: http://www.ieee.org/press

Printed in the United States of America

10   9   8   7   6   5   4   3   2   1

IEEE ISBN 0-7803-1196-5
IEEE Order Number PC5665

McGraw-Hill ISBN 0-07-135385–2

Library of Congress Cataloging-in-Publication Data

Power and communication cables : theory and applications / edited by R.
   Bartnikas, K.D. Srivastava.
      p.   cm. — (IEEE Press series on power engineering)
   Includes bibliographical references and index.
   ISBN 0-7803-1196-5
   1. Electric cables.   2. Telecommunication cables.   I. Bartnikas,
R.   II. Srivastava, K. D.   III. Series.
TK3301.P66   1999
621.319′34–dc21   99-18442
                                                            CIP

In fond memory of the members of the IEEE Insulated Conductors Committee,
who have gone before us,
who through their diligent and dedicated labors
have contributed to the advancement of cable technology
in service of humankind.

BOOKS IN THE IEEE PRESS SERIES ON POWER ENGINEERING

ELECTRIC POWER APPLICATIONS OF FUZZY SYSTEMS
Mohamed E. El-Hawary, *Dalhousie University*
1998   Hardcover   384 pp   IEEE Order No. PC5666   ISBN 0-7803-1197-3

RATING OF ELECTRIC POWER CABLES: Ampacity Computations for Transmission,
Distribution, and Industrial Applications
George J. Anders, *Ontario Hydro Technologies*
1997   Hardcover   464 pp   IEEE Order No. PC5647   ISBN 0-7803-1177-9

ANALYSIS OF FAULTED POWER SYSTEMS, Revised Printing
P. M. Anderson, *Power Math Associates, Inc.*
1995   Hardcover   536 pp   IEEE Order No. PC5616   ISBN 0-7803-1145-0

ELECTRIC POWER SYSTEMS: Design and Analysis, Revised Printing
Mohamed E. El-Hawary, *Dalhousie University*
1995   Hardcover   808 pp   IEEE Order No. PC5606   ISBN 0-7803-1140-X

POWER SYSTEM STABILITY, Volumes I, II, III
An IEEE Press Classic Reissue Set
Edward Wilson Kimbark, *Iowa State University*
1995   Softcover   1008 pp   IEEE Order No. PP5600   ISBN 0-7803-1135-3

ANALYSIS OF ELECTRIC MACHINERY
Paul C. Krause and Oleg Wasynczuk, *Purdue University*
Scott D. Sudhoff, *University of Missouri at Rolla*
1994   Hardcover   480 pp   IEEE Order No. PC3789   ISBN 0-7803-1029-2

SUBSYNCHRONOUS RESONANCE IN POWER SYSTEMS
P. M. Anderson, *Power Math Associates, Inc.*
B. L. Agrawal, *Arizona Public Service Company*
J. E. Van Ness, *Northwestern University*
1990   Softcover   282 pp   IEEE Order No. PP2477   ISBN 0-7803-5350-1

POWER SYSTEM PROTECTION
P. M. Anderson, *Power Math Associates, Inc.*
1999   Hardcover   1344 pp   IEEE Order No. PC5389   ISBN 0-7803-3427-2

CONTENTS

**CHAPTER 3   DESIGN AND MANUFACTURE OF EXTRUDED
SOLID-DIELECTRIC POWER DISTRIBUTION CABLES   171**

H. D. Campbell and L. J. Hiivala

**CHAPTER 4   EXTRUDED SOLID-DIELECTRIC POWER
TRANSMISSION CABLES   208**

L. J. Hiivala

CHAPTER 5   DESIGN AND MANUFACTURE OF OIL-IMPREGNATED PAPER INSULATED POWER DISTRIBUTION CABLES   243
W. K. Rybczynski

CHAPTER 6　　**LOW-PRESSURE OIL-FILLED POWER TRANSMISSION CABLES** 276

W. K. Rybczynski

CHAPTER 7　　**HIGH-PRESSURE OIL-FILLED PIPE-TYPE POWER TRANSMISSION CABLES** 296

W. K. Rybczynski

CHAPTER 8   VOLTAGE BREAKDOWN AND OTHER ELECTRICAL TESTS ON POWER CABLES   313
H. D. Campbell

CHAPTER 12   CRYOGENIC AND COMPRESSED GAS INSULATED POWER CABLES   551
K. D. Srivastava

CHAPTER 13   UNDERWATER POWER CABLES   582
R. T. Traut

CHAPTER 14   HIGH-VOLTAGE DIRECT-CURRENT CABLES   624
C. Doench and K. D. Srivastava

CHAPTER 15   TELEPHONE CABLES   652
R. Bartnikas

CHAPTER 16   UNDERSEA COAXIAL COMMUNICATION CABLES   782
R. T. Traut

CHAPTER 17   TERRESTRIAL AND UNDERWATER
OPTICAL FIBER CABLES   818
W. F. Wright

PREFACE

The present book, as may be inferred from its title, deals with the subject of power and communication cables. With respect to the power cable aspects, the book may be regarded as a sequel to and an extension of our two antecedent volumes entitled *Elements of Cable Engineering* and *Power Cable Engineering*. These were based essentially on an intensive course that has been offered for nearly three decades on a more or less biennial basis initially at the University of Waterloo and later at the University of British Columbia. Both books, which were well received by the scientific and technical communities, attempted to present concisely the various attributes of power cable design, manufacture, testing, and use. The second book differed from the first in that it included considerable additional material and a more detailed description of the subject matter.

There were two principal reasons that impelled us to proceed with the preparation of the present book: We wished to incorporate addenda concerning new and more recent developments in the power cable field not covered in our two earlier volumes and to include, in addition, the subject matter of communication cables that would encompass both the metallic conductor twisted-wire pair and coaxial cables as well as optical fiber cables. Evidently, the combined treatment of power and communication cables did not present itself as a trivial task because it did involve divergently different concepts and applications of two types of generic cables. Yet as power and communication cables are frequently installed adjacent to each other, either aerially on poles or in ducts or directly buried, cable engineers should be fully cognizant of the attributes of the two different cable categories. Moreover, there is increasing usage being made of hybrid cables, containing both power conductors and communication lines; also many cable manufacturers produce both power and communication cables, sometimes even on the same premises. It has been always our conviction that knowledge of both the power and communication fields constitutes a decided asset for a cable engineer. We have adhered to this belief in our cable course, in which lectures were given both on power and communication cables, and we have carried it over to the present book, which we consider to be rather unique in its undertaking and intent.

This book treats in detail the various facets of power and communication cable design, manufacture, testing and evaluation, and installation. In the power transmission and distribution cable area, it covers solid-dielectric-extruded, oil-impregnated-paper, self-contained dielectric liquid-filled, pipe, dc, submarine, compressed gas, and super-conducting type cables. Therewith included are also the descriptions of the associated ac, dc, impulse, dissipation factor, accelerated aging, and partial-discharge test procedures as well as field tests. In addition, the thermal, mechanical, and electrical behaviors of cable materials are dealt with and the various cable aging mechanisms are examined. Cable system design methods are treated in detail and the different types of joints, terminations, and splices in current use are described. The portion of the book devoted to communication cables provides an extensive description of both the metallic conductor and optical fiber cable transmission characteristics and indicates their particular areas of application and deployment. Telephone cable design, construction, manufacture, and installation methods are considered. Detailed descriptions are provided on long-haul trunk, distribution, and drop cables; particular attention is given to long distance submarine and transoceanic cables. The signal transmission techniques, utilized with the various types of communication cables employed, are delineated. A significant amount of space is allotted to community antenna TV cables and to the data transmission cables deployed in the all pervasive, and yet still rapidly expanding, local and wide area computer networks.

Our treatise on cables is timely in the context of a recent historical event, which occurred at the beginning of 1998, that demonstrated the inherent reliability potential of underground cables under adverse weather conditions. The ice storm, which came in early January, plunged large sections of Western Quebec and the Maritime Provinces, Eastern Ontario, northeastern New York, and the New England States into darkness and cold as overhead transmission and distribution lines collapsed under the weight of an accumulated ice layer over the aerially suspended conductors. Telephone service was also interrupted as a result of damaged aerial telephone cables. It is estimated that a population in excess of 3 million was affected by the storm; in some areas it took longer than a month to restore power. It may be of passing interest to the reader to know that a part of the chapter on telephone cables was written by one of us (R.B.) under arduous and rather primitive conditions, working over an extended period of time in candlelight and incommunicado.

The interdisciplinary venture of preparing a comprehensive text—comprising power and communication cables—though highly demanding, was particularly gratifying to one of us (R. B.), who has spent a considerable amount of time in both the telecommunication and electrical power fields. We wish to take this opportunity to express our gratitude to the contributing authors for their specialist viewpoints. This book took a substantial number of years to plan and prepare; in its seminal stages of preparation, we received much support and encouragement from the late D. A. Costello, General Manager of the Cable Division of Northern Telecom. His profound knowledge of and commitment to both the communication and power cable fields few could equal. For his expert and perceptive advice, we shall be ever greatly indebted to him. We are also very much obliged to the Sandford Educational Press for their permission granted to republish previous material that had appeared in our two earlier volumes on power cables. Finally, both of us spent a very substantial amount of time working on manuscripts in our homes, and in this respect we wish to thank our wives Margaret (R. B.) and Gladys (K. D. S.) for their individual sacrifice and understanding.

R. Bartnikas
K. D. Srivastava
Varennes, Québec

LIST OF CONTRIBUTORS

R. Bartnikas
Institut de Recherche d'Hydro-Québec
1800 boul. Lionel Boulet
Varennes, Québec J3X 1S1
Canada

H. D. Campbell
Formerly with Northern Telecom Ltd.
Cable Division
Lachine, Québec
Canada

C. Doench
Pirelli Cables North America
Power Cable Division
Lexington, SC 29072

L. J. Hiivala
Director, Applied Engineering
Alcatel Canada Wire, Inc.
140 Allstate Parkway
Markham, Ontario L3R 0Z7
Canada

G. Ludasi*
Formerly with Transmission Engineering
Hydro-Québec
Montreal, Québec
Canada

W. K. Rybczynski†
Chief Engineer,
Pirelli Cables, Inc.
St-Jean-sur-Richelieu, Québec
Canada

K. D. Srivastava
University of British Columbia
Dept. of Electrical and Computer
Engineering
2356 Main Hall, Room 434
Vancouver, BC V6T 1Z4
Canada

W. T. Starr*
Formerly with Raychem Corporation
Menlo Park, CA 94025

R. T. Traut*
Formerly with Fiber Optic Engineering
and Development
Simplex Technologies, Inc.
2073 Woodbury Avenue, P.O. Box 479
Portsmouth, NH 03802-0479

W. F. Wright
Manager,
Fiber Optic Engineering and Development
Simplex Technologies, Inc.
2073 Woodbury Avenue, P.O. Box 479
Portsmouth, NH 03802-0479

---

* Presently an engineering consultant.
† Deceased.

CABLES: A CHRONOLOGICAL PERSPECTIVE

R. Bartnikas

He, who does not know his past,
remains forever a child.

Cicero

1.1 PRELIMINARY REMARKS

Metallic conductor cable technology is perhaps one of the oldest fields of endeavor in electrical engineering, whose origins can be traced back approximately 150 years. Not far behind on the chronological scale are situated the dielectric medium cables, which had their seminal beginnings more than 100 years ago in the early studies of light ray transmission along pipes that culminated in the 1960s with the development of optical fiber cables. Although current cables represent a vast improvement over their predecessors, they have changed very little in their fundamental design. Thus, in metallic conductor cables, the conductor along which the power and signals are transmitted must be kept isolated and insulated from its surroundings by a dielectric layer, and in an optical dielectric conductor cable, the guided light signal must remain confined within a fiber core or dielectric channel by means of a light-reflecting enclosure or cladding. The cable characteristics are determined by the geometrical parameters of the cable such as metallic conductor radius, insulation thickness, optical fiber core diameter, etc., as well as the electrical and physical properties of the materials constituting these and other component parts of the cables. Since the geometrical parameters of cables are generally optimized, amelioration of any given cable design is controlled by the improvements effected in either existing cable construction materials or the availability of new materials with outstanding properties and processability. Consequently, cable technology by its very nature is essentially a materials-based technology. In this chapter, we shall follow the chronological evolution of power and communication cables and delineate the past causes and events that have led to the mutations of the two main generic categories of cable currently in use.

The development of early power cables can be said to have followed closely the extension of the cable techniques, which had been developed on low-current cables, lines for telegraphy, and explosive mine cables. In fact, in the incipient stages of their development, both power and metallic conductor communication cables were evolving with considerable similarity. Steel conductors were being replaced with copper, and the early insulation consisted of gutta-percha, rubber, or various impregnated systems. The cable installation techniques employed were also alike in many respects.

As power transmission and distribution was carried out at direct current (dc) or low frequencies (25, 50, or 60 Hz) but at high voltages, while that of voice information was transmitted at low voltages but at high frequencies, it was a natural consequence that differences between the two systems would become increasingly more apparent in time. Even at the relatively lower values of transmission frequency employed in the early telephone cables, the cables were usually long in comparison to the wavelength of the signal transmitted in contrast to the power cables, which were electrically short lines vis-à-vis the wavelength of the power frequency. Since in power systems the power supplied to the load was obtained commonly from the same circuit, the tendency was, therefore, toward interconnection and development of large systems comprising a relatively small number of large power generating units. With telephone systems, it was, however, necessary to maintain an independent channel of communication for each pair of communicators or talkers. Evidently, this enabled a small amount of copper to provide a relatively large number of circuits either by the use of many small-diameter conductors or by the superposition of a number of independent channels of communication on a single pair of wires [1]. The transmitted acoustic power was minuscule at the relatively low voltage employed; accordingly, not only the conductor size but also the insulation thickness required were small. The important considerations in communication cables centered on the need for low attenuation and speech distortion and negligible cross talk between adjacent circuits. To meet these requirements, the cables were to have low losses at the operating frequencies, with low capacitance per unit length and low mutual capacitance between adjacent twisted-wire pairs. These constraints differed greatly from those imposed on power transmission and distribution cables. Power cables were to transmit sizable blocks of power and, as a consequence, had to be operated at high voltages with large current flows in their conductors. Hence, the size of the conductor was to be sufficiently large to carry the necessary load current without overheating and limiting the voltage drop per unit length; also, the insulation thickness and its dielectric strength were to be sufficiently great to withstand the high operating voltage gradients or stresses. The capacitance per unit length was to be low to limit the capacitive charging currents; the dielectric losses were also to be small, particularly if the cables were to operate at the more elevated voltages. Thus, voltage and temperature became the preponderant parameters in the rating of power cables.

Notwithstanding the salient differences between power and metallic conductor communication cables, certain inherent similarities did nevertheless prevail. Many of the materials used on the two types of cables were identical. Certain manufacturing processes and techniques were the same. For example, the same sheathing and armoring could be applied equally well on both telephone and power cables. The larger diameter conductors used on some power cables were obtained simply by stranding the smaller conductors that could be used on certain communication cables. It is thus not surprising that the basic similarities in the construction and manufacture of power and metallic conductor communication cables were sufficient to allow most cable manufacturers to produce both types of cable. It is only with the introduction of optical fiber communication cables, whereby information became transmitted along dielectric conductors in lieu of the metallic wires, that there occurred a major bifurcation in cable manufacturing techniques. The silica fibers had to be drawn under highly controlled conditions to prevent contamination effects and ensure accurate diameter control, which was in the order of a fraction of a micrometer. In addition, a well-controlled layer of cladding material was required over the optical fiber core to impart the desired lightwave guiding characteristics to the optical transmission line. However, the remaining manufacturing steps resembled closely or were identical to those common to conventional metallic conductor cables. The cladded optical fibers were protected by extruded polymeric coatings and placed loosely within tubes of polymeric material or in a polymeric buffered structure containing strength members (steel or polymer) to provide mechanical protection. The jackets on the optical fiber cables were similar to those on metallic conductor communication cables. Metallic shielding was employed primarily to avert rodent damage, as opposed to metallic conductor communication cables where metallic coverings were in addition desirable for electromagnetic shielding, particularly as concerns surges, arcing from lightning, and adjacent power lines or cables.

In some applications, composite or hybrid cables were used, necessitating the cable engineer to have some familiarity with both metallic and optical communication cables as well as power cables. For example, certain optical fiber cables contained metallic conductor twisted-wire pair communication lines that are required for control purposes. The immunity of optical fiber cables to electromagnetic interference rendered them particularly suitable for deployment in conjunction with power cables either along the grounded shield of power cables or incorporated as a part of grounded conductors. In addition, there has been a considerable increase in the use of the same rights-of-way for both power and communication cables. In such circumstances, when copper conductor communication cables were employed, they had to be adequately shielded by metallic shields to be protected against ground faults that may originate from adjacent power cables.

1.2 POWER CABLES

Power cable technology had its beginnings in the 1880s when the need for power distribution cables became pressing, following the introduction of incandescent lighting. With urban growth, it became moreover increasingly necessary to replace some of the overhead lines for power transmission and distribution with underground cables. The illumination of the larger cities proceeded at such a rapid pace that under some circumstances it was impossible to accommodate the number and size of feeders required for distribution, using the overhead line system approach. In fact, this situation deteriorated so notably in New York City that, in addition to the technical and aesthetic considerations, the overhead line system began to pose a safety hazard to the lineworkers themselves, the firemen, and the public. As a result, the city passed an ordinance law in 1884 requiring the removal of the overhead line structures and the replacement of these with underground cables [2]. Similar laws and public pressure were applied in other cities, with the consequence that by the early 1900s, underground electrification via insulated cables was on its way to becoming a well-established practice.

A practical lead press was invented in 1879 and subsequently employed to manufacture 2-kV cables for Vienna in 1885 [3]. During the same period vulcanized rubber was used to produce cables on a commercial scale, although use of gutta-percha had already been made as early as 1846. Impregnated-paper power cables were first put on the market in 1894 by Callender Cables of England, using impregnant mixtures of rosin oil, rosin, and castor oil; only in 1918 were these replaced by mineral oils [3]. In North America impregnated-paper cables were first supplied by the Norwich Wire Company. Varnished cambric cables were introduced by the General Electric Company in 1902;

the high-temperature behavior of these cables was subsequently improved in 1910 by the addition of black asphalt.

It is interesting to note that some of the earliest power cables consisted merely of ducts with the copper conductors insulated from ground by glass or porcelain insulators. In fact, in 1889 the entire city of Paris was electrified using this scheme with sewers serving as ducts. Some of the more common early solid and liquid insulating materials employed in various underground cable installations were natural rubber, gutta-percha, oil and wax, rosin and asphalt, jute, hemp, and cotton. In 1890 Ferranti developed the first oil-impregnated-paper power cable; following their manufacture, his cables were installed in London in 1891 for 10-kV operation. It is most noteworthy that the cables had to be made in 20-ft lengths; as the total circuit was 30 miles in length, about 8000 splicing joints were required. Nevertheless, these cables performed so well that the last cable length was removed from service only in 1933. In 1892, Buffalo, New York, was illuminated with arc lamps, and for this purpose 7.5-kV rubber insulated cables of the concentric design, using an overall insulation thickness of $\frac{1}{8}$ in. were placed in service [4].

Cable installation continued to proceed at a rapid pace, so that by the turn of the century many major cities throughout the world had many miles of underground power cable. For example, already by the end of 1909, the Commonwealth Edison Company in Chicago had 400 miles of underground cable operated in the voltage range from 9 to 20 kV. Montreal had some 4500-ft circuits of three-conductor cables installed in ducts under the Lachine canal for 25-kV operation; the same voltage was used for cables traversing the St. Lawrence River in 1906. With some experience behind them, cable manufacturers were increasingly gaining confidence and during the St. Louis Exposition in 1904 power cables developed for voltages as high as 50 kV were put on display [4].

1.2.1 Oil-Impregnated Paper Power Cables

During the period prior to World War I, extensive use was made of oil-impregnated paper cables of the three-conductor belted type for voltages up to 25 kV. This antiquated design is illustrated in Fig. 1.1 [5]. Due to the nonuniform stress distribution in the cable construction, the belted cable proved to be highly partial-discharge susceptible when attempts were made to extend the operating voltage range with larger wall thicknesses to approximately 35 kV, to meet the increased power demand following World War I [4]. This problem was resolved by shielding the individual conductors, using 3-mil-thick copper tapes. The outside of the shielded conductors was thus maintained at the same ground potential. In addition, the belt insulation was replaced with a

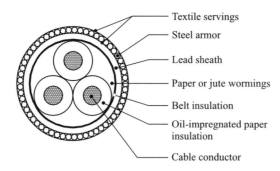

Textile servings
Steel armor
Lead sheath
Paper or jute wormings
Belt insulation
Oil-impregnated paper insulation
Cable conductor

Figure 1.1 Cross section of unshielded three-conductor belted-type high-viscosity oil-impregnated paper power cable.

binder consisting of fabric tapes and strands of interwoven copper wire. The purpose of the latter was again to maintain the shields of the three cables at the same potential (ground). Over the years, the conductor shapes of the three-conductor shielded paper insulated cables have evolved into three forms, namely circular, oval, and sectoral (cf. Fig. 1.2). In many utilities a substantial portion of the present-day distribution load is still carried at 35 kV via three-phase oil-impregnated paper belted cables, with the three conductors individually grounded. There is little inducement to replace these cables with solid extruded dielectric cables, whose outer diameter for an equivalent power rating would exceed that of the ducts accommodating the more compact three-phase oil-paper belted cables. Moreover, the oil-paper belted cables have been characterized by remarkably long in-service lifetimes that often exceed 65 years. Belted cables with unshielded conductors are still deployed but only for working voltages equal to or less than 15 kV.

With the individual conductors shielded, it was possible to extend the use of the three-phase belted cables for voltages as high as 69 kV, though on the average their application has been confined to voltages below 35 kV. The main reason for this upper limit has again been associated with the occurrence of partial discharges, which had in numerous instances led to the deterioration and failure of the dielectric at the elevated voltages. The partial discharges were found to take place in voids, which were formed either during the manufacturing process or during the load cycling while in service.

1.2.2 Oil-Pressurized Power Transmission Cables

The problem with void-associated discharges at the higher values of applied voltage, required for the transmission of electrical energy, was finally and effectively eliminated by the introduction of a low-viscosity oil-impregnated-paper insulating system; in this scheme, the formation of cavities was avoided by maintaining a pressure slightly above atmospheric on the insulating oil. The principle of this oil-filled-paper power cable was first expounded by Emanueli in 1917 and, in its original concept, the cable consisted of an oil-impregnated-paper insulation applied over a conductor strand, having a hollow center filled with oil (cf. Fig. 1.3). The first installation of a single conductor oil-filled cable rated at 66 kV was made in England in 1928; this was followed in 1931 with a 132-kV design installation. However, the first oil-filled cables installed on an actual system were those in Chicago and New York, in which use was made of a gravity feed of a minimum pressure of 1 lb/in.^{2}, which was just sufficient to prevent ingress of air and moisture.

Emanueli's design was based on the principle that the heated oil in the cable must be allowed to expand during the load cycle into an oil reservoir connected at its upper end; upon removal of the load, the oil is thus free to return or contract into the cable [6]. Present-day oil-filled cables differ very little in principle from Emanueli's early design, though higher oil or dielectric liquid pressures are employed

Figure 1.2 Configurations of shielded three-conductor belted-type high-viscosity oil-impregnated paper power cables: (a) round, (b) oval, and (c) sectoral conductors.

(a)   (b)   (c)

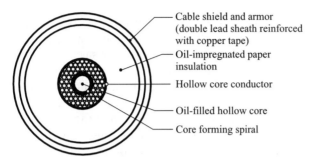

Cable shield and armor
(double lead sheath reinforced
with copper tape)

Oil-impregnated paper
insulation

Hollow core conductor

Oil-filled hollow core

Core forming spiral

Figure 1.3 Emanueli oil-filled hollow-conductor power cable.

and the reservoir tanks generally make use of pressurized air cells. The oil in the hollow-core oil-filled cables can be maintained under pressure by means of pressurized oil tanks at the ends of the cable. In other types of pressurized cables, pressure in the insulation may be maintained by gas pressure applied either to the lead sheath, which acts as a flexible membrane, or directly to the solid-liquid dielectric. Static oil pressures up to 75 lb/in.² are common. The minimum permissible oil pressure is about 10 lb/in.²; at the other extreme, pressures as high as 450 lb/in.² may be required due to vertical rises in the cable layout such as in the case of the Churchill Falls cables. With the majority of dielectric liquid or oil-filled cables, the maximum permissible stresses at the conductor may vary usually between 90 and 150 kV/cm; while with the so-called solid-type higher viscosity oil-impregnated paper insulated lead-covered (PILC) power cables, the equivalent stresses range below 40 kV/cm. Over the past decades, the operating voltages of oil-filled cables have been increasing steadily. In 1957, 14 circuit miles of 238-kV, oil-filled cable (insulation thickness of 0.835 in.) were installed in British Columbia; another 300-kV installation was made at Kitimat in the same province. An oil-filled cable system voltage of 230 kV is now common for underground power transmission in most of the larger Canadian cities. In the Montreal area, an oil-filled cable system voltage of 315 kV has been in effect since 1982. Throughout the world, a few oil-filled cable installations have been made for extra-high-voltage (EHV) levels up to 500 kV; most noteworthy is the 525-kV self-contained oil-filled cable system at the Grand Coulee Dam, Washington State [7]. The latter cable has a nominal insulation wall thickness of 30.5 mm, an aluminum sheath of 4.6 mm thickness, and a polyvinyl chloride (PVC) jacket of 3.1 mm thickness. Metallized carbon papers were used for shielding, and a copper conductor of 40.0 mm diameter formed the inner hollow core. The cable was designed to operate at a maximum oil pressure of $2.25 \times 10^3 \, \text{kN/m}^2$ and a dielectric loss of 17.4 W/m as compared to 12.8 and 7.2 W/m for the conductor and sheath, respectively. It is interesting to note that in recent years this particular cable had experienced failure. Attempts have been made to extend the self-contained dielectric liquid or oil-filled cable design up to voltages of 750 kV and beyond, using impregnated paper as the insulant. In such cable designs maximum stresses of 210 kV/cm are contemplated; as these cables are to carry loads greater than 2 GVA, the cable systems, due to their relatively high power losses, would require other than only natural cooling. A cross section of a possible 750-kV cable design, using an internal oil pressure of 15 atm, is delineated in Fig. 1.4 [8].

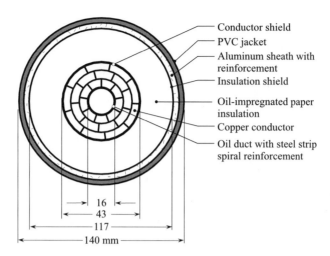

Figure 1.4 Cross section of 750-kV oil-impregnated paper self-contained oil-filled power cable with copper conductor cross-sectional area of 1100 mm^2 (after [8]).

In North America commencing with 1932, the first installation was made of an oil-filled pipe-type power cable. This cable differed from the self-contained oil-filled cable in that three individually shielded cable conductors were placed in an oil-filled pressurized tube as delineated in Fig. 1.5; the conductor and insulation geometry were similar to the high viscosity oil-impregnated paper-type cable, whereas the overall insulation thickness was roughly the same as that of the oil-filled cable of equivalent voltage rating. Also, the viscosity of the oil under pressure within the pipe was appreciably lower than that of the oil impregnant in the oil impregnated paper insulation of the cables themselves. The higher viscosity oil impregnant was required to prevent loss of impregnant from the paper insulation during the pulling procedure of the cables into the pipe. The first pipe-type cable, manufactured by the Okenite Company, was placed in an 8 in. diameter steel pipe with the insulation thickness of each of three-phase conductors, having an average value of 0.375 in. [9]. The experimental length was found to perform very well, and the first commercial installation of the pipe-type or Benett cable as it was initially called (after its inventor), took place in 1935. A total length of 17,000 ft of pipe-type cable rated at 132 kV was installed in Baltimore; an insulation thickness of 0.675 in. was used with the oil pressure set at 200 lb/in.^{2} and the operating temperature at 70°C. The pipe-type power cable proved to be more competitive economically, and by about 1954 the total circuit mileage of installed pipe-type power cable in the United States surpassed that of the self-contained oil-filled cable (cf. Fig. 1.6).

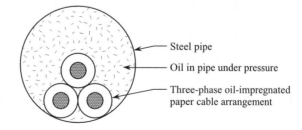

Figure 1.5 Oil-filled pipe-type power cable with three individually shielded oil-impregnated paper cables.

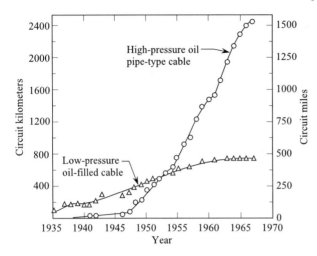

Figure 1.6 Early underground oil-type power transmission cable use in the United States in the voltage range from 69 to 345 kV (according to [10]).

The corrosion-resistant steel pipe enclosing the individual conductors of the pipe-type cable provides excellent mechanical protection to the three-phase system. It has been found to be particularly well adapted for servicing congested areas where usually the saving achieved in space is very significant as compared to self-contained oil-filled cables. Although pipe-type cables have been mainly used for underground installations, they can be installed as suspended systems in air or for submarine runs. However, in the latter case the corrosion-protective coating on the pipe must be supplemented by cathodic protection. Another marked advantage of pipe-type power cables lies with the fact that they normally are three-phase systems and thus have no problems associated with metallic sheath voltage rises or sheath losses as in the case of single-   conductor metallic sheath cables. The working pressures of pipe-type cables vary from 150 to 220 lb/in.^{2} and are thus substantially higher than those of hollow-core cables. The reason for this lies in the fact that the latter use a much lower viscosity oil (approximately 10 cP at room temperature). Since in pipe-type cables the occurrence of partial discharge is prevented by maintaining a sufficiently high pressure on any gas inclusions within the insulating system, the term high-pressure pipe-type cable is often used. As the individual conductors are not covered with a lead sheath, the relatively light weight of pipe-type cables allows pulling lengths up to 3500 ft during the installation process. To provide increased mechanical protection and to facilitate sliding movement, the individual cables are now constructed with a skid wire that runs spirally over the outside shield, thus minimizing the contact area between the inner surface of the pipe and the outside surface of the individual cable phases. The portion of the skid wire that is contiguous with the cable shield is flat. The skid wire arrangement represents an appreciable amelioration over the antecedent canvas wrapping used in the Benett cable [9]. Much the same as for the oil-filled cables, the operating voltages of pipe-type cables have been continuously on the increase over the first few decades following their introduction. The first 230-kV pipe-type cables in the U.S. were installed in 1956 at Garrison Dam, North Dakota. In 1964, the first 345-kV pipe-type cable was installed in New York City; its load rating was set at 484 MVA as compared to 188 MVA for an equivalent 138-kV cable [10].

The load-carrying capacities of cables of existing oil-paper designs may be further augmented by the use of forced cooling so as to remove the heat generated in both oil-filled and pipe-type cables. In pipe-type cables this is most easily accomplished by circulating and cooling the pipe filling oil, whereby the losses are usually reduced by about 50%. Circulation of the oil only already gives the decided advantage of maintaining a uniform temperature throughout the cable, thereby reducing the possibility of hot-spot zone instabilities. In present practice, forced cooling of self-contained oil-filled cables is most economically carried out by placing metal water-circulating tubes adjacent to the cables depicted schematically in Fig. 1.7. Other simple techniques of forced cooling involve submersion of cables in water troughs or enclosure of individual cables in cooled water pipes. A more sophisticated method of cooling being experimented with recently involves the heat pipe principle. Presently contemplated forced-cooling techniques could conceivably extend the capacity of dielectric liquid or oil-impregnated cables above 2000 MVA. However, for much greater loads neither increased cooling nor operating voltage (above 750 kV) can efficiently cope with the enhanced dielectric losses in the oil-impregnated papers themselves. Thus, for operating voltages in the range from 750 to 1000 kV, the oil-impregnated-paper dielectric must necessarily be replaced by low dielectric loss synthetic papers. In the 1980s composite paper-polypropylene-paper tapes, possessing inherently lower dielectric losses, became commercially available, and oil-pressurized cables with these types of composite tapes were produced. In Fig. 1.8 is portrayed a cross section of an 800-kV self-contained branched dodecyl-benzene dielectric liquid cable constructed with these tapes [11].

The composite polypropylene-paper tapes consist of a polypropylene film sandwiched between two kraft paper tapes. The dielectric loss and dielectric constant of this combination may be reduced by increasing in proportion the thickness of the polypropylene tape with respect to that of the two kraft paper tapes. For example, increasing the polypropylene-to-paper ratio from 42 to 60% reduces the dielectric constant and dissipation factor from 2.8 and 1×10^{-3} to 2.7 and 0.7×10^{-3}, respectively [9]. Table 1.1 provides additional data on the characteristics of the composite polypropylene–kraft paper tapes, with a polypropylene-to-paper ratio of 60%. Figure 1.9 illustrates the reduction in the overall losses of an 800-kV branched dodecyl-benzene-impregnated cable insulated with composite paper-polypropylene-paper (PPP) tapes as compared to equivalent kraft paper tape insulated self-contained dodecyl benzene filled cables for operation at 800 and 500 kV. The 800-kV PPP–dodecyl-benzene-filled cable is seen to have losses substantially lower than the 800-kV kraft paper–

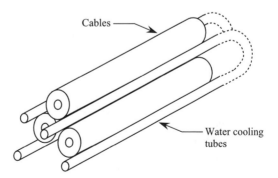

Figure 1.7 Forced-water-cooled cables [5].

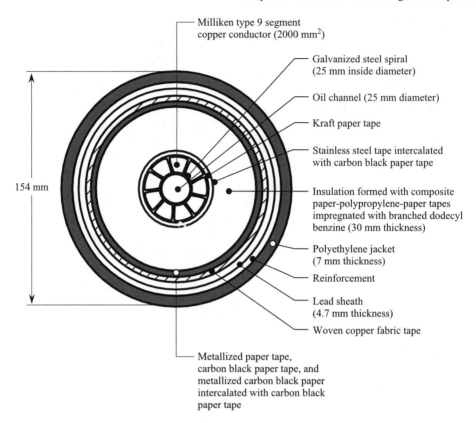

Figure 1.8 Cross section of 800-kV kraft paper–polypropylene–kraft paper (PPP) self-contained branched dodecyl-benzene–filled cable [11].

TABLE 1.1 Electrical Characteristics of Polypropylene–Kraft Paper Composite Tapes (PPP) and Kraft Paper Tapes Impregnated with Branched Dodecyl Benzene

Electrical Parameter	PPP Tape Thickness (μm)			Kraft Paper Tape Thickness (μm)
	120	170	220	200
Dielectric constant	2.57	2.58	2.61	3.4
Dissipation factor at 20 kV/mm, 80°C	5.4×10^{-4}	5.3×10^{-4}	5.9×10^{-4}	2.0×10^{-3}
ac breakdown strength, kV/mm	172	150	139	80
Impulse breakdown strength, kV/mm	284	256	245	150

Source: From [11].

dodecyl benzene filled cable and even slightly lower losses than a kraft paper–dodecyl-benzene-filled cable for operation at a reduced voltage of 500 kV. It is apparent from Fig. 1.9 that with the PPP self-contained dielectric liquid-filled cables one can readily obtain transmission capacities of the order of 2000 MVA even without a resort to forced cooling.

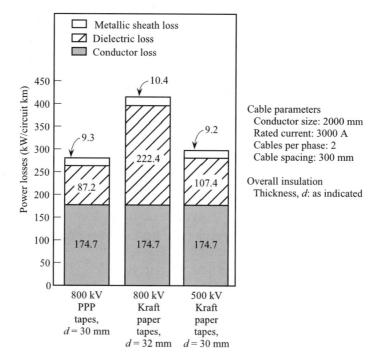

Figure 1.9 Comparison of power losses in 800-kV polypropylene–kraft paper self-contained branched dodecyl-benzene-filled cable and kraft paper self-contained branched-dodecyl benzene-filled cables (after [11]).

1.2.3 Solid-Dielectric-Extruded Power Transmission Cables

Following the introduction of solid-polyethylene-extruded power distribution cables in the 1950s, it became only a question of time until, with improved technology in materials and extrusion techniques, polyethylene would be applied to transmission cables. In these early days, polyethylene, because of its intrinsically low dielectric loss characteristics, was always viewed as an attractive substitute for the more traditional solid-liquid insulating systems. The last two decades have witnessed a marked increase in the number of solid-dielectric-extruded polyethylene transmission cable installations with increasingly higher operational voltages. Both from the environmental and maintenance considerations, the elimination of possible oil leaks and hydraulic oil reservoir tank systems have become attractive notwithstanding the current improvements in pressurized oil/dielectric liquid cable systems themselves. This tendency has continued to gain momentum in the face of the proven reliable long-term performance of the oil-paper cables with their termination and joint accessories. Presently, extruded cross-linked polyethylene (XLPE) cables are being designed for operation at 500 kV; joints and terminations of such cables form a very critical integral part of the overall cable system, and their design and continuing improvements are consuming much of the overall effort at this time.

Remarkable advances in the field of power transmission cables have been achieved initially by the French, who have made use of a thermoplastic-type low-density linear polyethylene (LDPE) on 225-kV cables. These cables have operated successfully since the time of their initial installation date in 1969, and by 1981 the total length of these cables used in the Électricité de France system had already reached 142 km [12–14]. The cables have been designed with a wall thickness of 22 mm to operate at a maximum voltage gradient of 100 kV/cm. The French cable design cross section is portrayed in Fig. 1.10. During the summer months, the cable depicted in Fig. 1.10, which has a copper (800 mm^2) conductor, is capable of carrying a load current of 750 A or a load of 290 MVA at a temperature of 70°C; the rated load-carrying capacity is approximately the same as that for an equivalent kraft paper self-contained oil-filled cable or slightly more as that of an oil-filled pipe-type cable. In 1985, the voltage of the subsequently manufactured LDPE cables was extended to 400 kV, corresponding to maximum operating stresses between 120 and 150 kV/cm. Finally, in 1995, LDPE cables were introduced for operation at 500 kV. Table 1.2 summarizes the French experience with LDPE power transmission cables [15]. Similar development work was undertaken in Germany in 1980 with projected cable insulation thicknesses between 25 and 28 mm [13, 16] where use was already being made of solid extruded dielectric cables for transmission voltages up to 100 kV.

The introduction of thermosetting-type solid extruded XLPE as the cable insulant on transmission cables resulted in an increase of 20°C in the operating temperature above that of 70°C allowable for the thermoplastic LDPE-insulated cables. The early work on XLPE insulation for power transmission cables has primarily been centered in Norway, Japan, Sweden, and the United States [13]. A lead-sheathed XLPE cable rated for 138 kV was commissioned for operation in 1976 at the Detroit Edison Company. The cable itself appears to have performed well initially, though difficulties were encountered with the joints. Subsequently, again in the United States, many studies have been carried out over the past several decades to perfect the 138-kV rated XLPE cables with efforts being directed toward improving cleanliness of the insulation and semiconducting shield interfaces, obtaining a more uniform distribution of the cross-linking agent, reducing cavity formation by increasing the extrusion pressure, and replacing the wet curing with a dry-curing process. This resulted in the production of 230-kV cables and the design of a 345-kV cable in 1980 [17]. In Norway four conductor lengths of 420 m of a 300-kV cable insulated with XLPE have been installed in 1980, and in Sweden 75 km of a 170-kV XLPE cable were put in service by the end of the same year [13, 18]. In Japan, it is now an established practice to employ XLPE-insulated power transmission cables at 275 kV along mountain slopes on power station reservoir sites [19, 20]. Work in Japan has progressed rapidly toward the development of 500-kV XLPE power transmission cables [21] with the first experimental length being actually placed in service in 1990 [22]. Similar work has been underway in France [15],

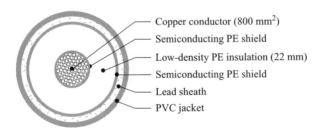

Copper conductor (800 mm^2)
Semiconducting PE shield
Low-density PE insulation (22 mm)
Semiconducting PE shield
Lead sheath
PVC jacket

Figure 1.10 Cross section of 225-kV low-density PE power transmission cable [12].

TABLE 1.2 In-Service Experience with LDPE Power Transmission Cables in France (up to 1995)

Voltage (kV)	Total Length of Cables (km)	Number of Terminations	Number of Joints	Date of First Installation
225	1195	3223	2055	1969
400	40	176	22	1965
500	<1	2	1	1990

Source: From [15].

England and Italy [23], and Germany [24]. Figure 1.11 depicts a typical cross section of an extra-high-voltage XLPE cable. Stresses in these cables usually range between 7 and 14 kV/mm, though maximum stresses as high as 27 kV/mm are contemplated. With the number of experimental and in-service installations of solid extruded dielectric transmission cables on the increase, meaningful data should begin finally to accumulate. It will be of great interest and practical significance to see whether extra-high-voltage polymeric transmission cables can achieve the same proven long-term performance reliability that has been so characteristic of oil-filled-type power transmission cables. The possibility of cavity occurrence and the propensity to electrical or water tree growth in polyethylene pose some uncertainties in this respect. On the other hand the low value of the dielectric constant and dielectric loss is an intrinsic property of plastic insulation that provides an ideal low-loss medium for underground alternating-current (ac) power transmission over relatively longer distances because of the lower associated capacitive charging currents. Moreover, plastic insulation is also attractive from the safety perspective because it does obviate the use of dielectric liquid or oil reservoirs and any possible oil leak difficulties that do arise occasionally with oil-filled cables.

In view of the relatively short experience, both in operational time as well as total cable length in operation, it is difficult to arrive at this time at a long-term prognosis on the expected performance of present and future 500-kV linear polyethylene and XLPE-insulated power transmission cables. A reliable long-term performance assessment would necessarily require answers to a number of pertinent

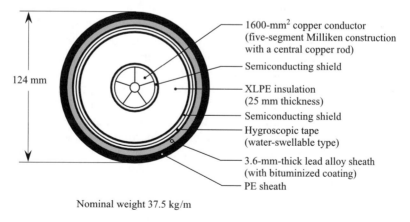

Nominal weight 37.5 kg/m

Figure 1.11 Cross section of 420-kV XLPE power transmission cable [22].

questions concerning the long-term thermal behavior of these cables [25]. For example, since crystalline regions in XLPE melt over a relatively wide range [26], the influence of oxidation on the partially crystalline polymer will differ substantially at the operating temperature (90°C) from that at the emergency temperature (130°C). Much of the needed oxygen for the oxidation of the polymer can be provided from the air entrapped at the corrugated metallic jackets of the cable as well as from the air dissolved in the solid components of the cable. The oxidation process will be retarded to some extent initially by a diffusion of the remnant antioxidants from the amorphous regions of the polymeric insulation [27] as well as from the semiconducting shields [28]. At elevated temperatures in addition to the more pronounced oxidation rates, the solubility of residual crosslinking agent by-products (acetophenone, cumyl alcohol, and dimethyl styrene) will be enhanced; although it is not known what influence this effect may pose to the long-term stability of the XLPE insulation in the extra-high-voltage cables. However, the very marked depression of the mechanical modulus at the emergency temperature of 130°C may result in thermomechanical damage in the cables. To circumvent this possibility, the emergency temperature would have to be reduced to the temperature of the polymer melting point of ca. 105°C or lower. Moreover, temperature changes, due to load cycling of the cable, may cause cavity formation arising from melting and subsequent recrystallization effects [29]. However, the diameter of these cavities may be too small to sustain partial discharge, unless coalescence of some of these cavities takes place. Other detrimental effects may arise with the in situ molding of splice joints that require temperatures as high as 150°C, which may lead to water formation from the decomposition of cumyl alcohol (a product of dicumyl peroxide) [30]. Subsequent water condensation may induce microcavity formation, thereby resulting in water tree growth that could conceivably lead to eventual failure of the high-voltage cable splice.

1.2.4 Solid Extruded Dielectric Power Distribution Cables

In terms of mileage, the amount of installed and manufactured volume of power distribution cables exceeds many times that of power transmission cables. Prior to and within the early 1950s various forms of rubber had been utilized in distribution-voltage cable-insulation applications, where flexibility and ease of handling were required as compared to the usually much more rigid solid-type high-viscosity oil-impregnated PILC cable systems. Butyl rubber compounds, because of their desirable electrical properties, were the most commonly employed materials for extruded solid-dielectric insulating systems. Following the discovery of the hydrocarbon thermoplastic polyethylene (PE) in England in 1933, polyethylene became rapidly the insulant of choice for RF (radio frequency) coaxial cables. It took a substantially longer time for polyethylene insulation to be introduced into the power cable area. However, since the time of its general acceptance in the 1950s, power distribution cables insulated with polyethylene have replaced virtually all of the butyl rubber cables and a significant portion of the oil-impregnated-paper insulated power cables used at operating voltages up to 35 kV [31]. But lower voltage PILC cables are still being manufactured, and there are many utilities in North America and overseas where PILC cables, due to their in-service longevity and reliability, continue to constitute a very substantial part of the overall underground distribution network.

Initially, the plastic cables manufactured in North America were made using low-density high-molecular-weight polyethylene (HMWPE); the latter is a thermoplastic material and, consequently, undergoes softening and flow at elevated temperatures. Its melting point is 105°C. The maximum normal operating temperature for HMWPE has been set at 75°C, with an emergency rating fixed at 90°C. It has been used since 1951 for distribution voltages up to 35 kV and sometimes higher [32]. Commencing with 1964, a changeover began toward the deployment of crosslinked polyethylene, which is a thermosetting compound with a higher melting temperature and exhibits as well a slightly better resistance to partial discharges [33]. With XLPE distribution cables, the conductor operating temperature is increased to 90°C and the emergency rating to 130°C. This improvement infers an equivalent ampacity increase of 12% over an HMWPE insulated cable or, alternatively, a proportional reduction in the conductor size for the same ampacity [33].

The early thermoplastic PE distribution cables, as well as their antecedent rubber cable counterparts, used carbon black tape conductor and insulation shields as opposed to the present-day cables, which utilize extruded semiconducting PE shields that provide better bonding between the insulation and conductor shields, thereby reducing the possibility of cavity occurrences at these interfaces and, hence, the incidence of partial discharge. Also the extruded semiconducting material has the advantage of smoothing out any sharp edges or protrusions on the conductor surface, thereby reducing the possibility of electrical and water tree growth at such asperity-like nucleation sites. The relatively low dielectric loss magnitude in the various PE-type insulations renders them exceptionally attractive for high-voltage applications; however, they are very susceptible to partial discharges and tree growth. Due to occasional imperfections in the extrusion technique, cavities tend to form in the extruded PE insulation wherein partial discharges can occur at sufficiently high voltage stresses; if the partial discharges are sustained for any appreciable length of time, failure of the PE insulation of the cable system follows inevitably. Though the exact nature of water tree growth mechanism is not fully understood, it has been observed to occur primarily in cables operating in high-moisture environments and to be much more prevalent in the earlier cable designs using linear or branched HMWPE cables constructed with carbon black tape shields. The incidence of water treeing was substantially reduced by the replacement of the carbon tape shields with extruded thermoplastic semiconducting shields and by the addition of tree-retardant compounds to the low-density HMWPE and eventually by the elimination of the linear low density HMWPE itself by XLPE. Finally, the 1980s saw further remarkable advances made to control water treeing by the introduction of the dry-curing process, supersmooth and clean semiconducting shields, and ultra clean, tree-retardant XLPE (TRXLPE).

In the late 1960s, ethylene-propylene-rubber (EPR) insulated cables, having clay filler contents as high as 50%, appeared on the market for voltage ratings up to 60 kV. These cables have a better flexibility than crosslinked PE, although they, very much like the thermoplastic PE and thermosetting XLPE cables, are still susceptible to partial discharges and tree growth. They are generally preferred over XLPE where mechanical flexibility is of prime concern. The operating and emergency temperature limits of EPR cables are identical to those of XLPE cables. Also the permissible short-circuit current value for both EPR and XLPE is set at 250°C as opposed to 150°C for HMWPE. The dielectric losses in EPR cables are much higher than those in either XLPE or HMWPE; these losses contribute to a significant energy loss at the higher operating voltages with the result that the use of EPR above 69 kV is less desirable unless for some reason cable flexibility is of paramount importance. It is noteworthy that it is only in North and South America and Italy that EPR cables are used to any great extent. In Italy already by the end of 1980, a total of 9000 km of EPR cable had been installed of which about 100 km were designed for 66-kV operation [13]. In North America in the distribution power cable area within the range from 5 to 35 kV, most utilities have now standardized on the use of thermosetting (XLPE or EPR) insulated cables. In this construction, the extruded shield over the conductor is most often semiconducting XLPE, though in some cases semiconducting EPR may be used. Over the insulation, the extruded shield is generally composed of semiconducting LDPE or XLPE. The jacketing materials normally employed consist of neoprene over EPR insulated cables and PVC or linear low-density polyethylene (LLDPE) over XLPE insulated cables. A TRXLPE cable construction, typical of the current practice followed in the 1990s, is depicted in Fig. 1.12; in parentheses are shown the common compounds employed earlier.

The cable portrayed schematically in Fig. 1.12 is of a concentric neutral construction in which the neutral is formed by a metallic tape shield. However, the shield may consist also of alternative constructions as, for example, a flat copper strap concentric neutral, a round copper wire concentric neutral, a small copper wire shield, or a corrugated metallic tape shield [32]. With strap and round copper wire concentric neutrals, it was a common earlier North American design practice to omit the jacket over the cable. In European primary distribution cable practice, it is normal to enclose the cable within a metallic sheath; although this adds considerable cost to the cable, it does provide protection against water ingress and water tree formation. The North American approach is to use metallic sheaths only on polymeric cables that are for special applications or that are to be used as transmission cables at EHV. It should be noted that with secondary distribution cables, i.e., those operating at 600 V or less, the neutral may consist of a separate single insulated conductor as in the triplexed construction or it may be applied in the form of coated solid copper wires around the two insulated conductors as in the parallel or two-phase arrangement [32]. As with all polymeric cables, either aluminum or copper conductors may be used. With larger conductor sizes in order to retain the necessary flexibility, the conductor must be stranded; while with smaller sized cables, the conductors may be either solid or stranded.

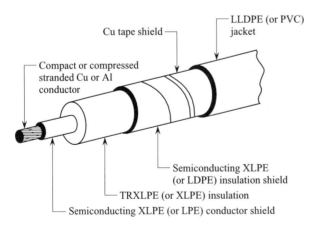

Figure 1.12 Typical TRXLPE (or XLPE) insulated power distribution cable.

There has always been considerable interest in the reliability of plastic insulated power distribution cables vis-à-vis the old conventional solid-type high-viscosity oil-impregnated PILC power cables. In the past, a great deal of statistical data on plastic insulated cables has been accumulated in the United States [17]. In this regard two rather revealing curves (released in 1978) are presented in Figs. 1.13 and 1.14, where failure rates are shown as a function of the year of installation for distribution cables rated between 5 and 36 kV. It can be discerned that whereas the failure rates up to 1978 of both HMWPE and XLPE were still on the increase with the year of installation, the failure rate of PILC cables, though higher, had reached a near asymptotic value. In examining the data further, one can also deduce the prognosis made by Thue et al. [17] that to attain a 10-fold cumulative failure rate, 6.2 years are required for HMWPE cables as opposed to 7.5 years for the XLPE cables. On the other hand for PILC cables, this value is reached in 100 years; although in the early years of PILC cable operation, the 10-fold increase had been manifest within 25 years (between the years 1925 and 1937). However, it was astutely predicted in 1980 by Eichhorn [34, 35] that as better materials appear and improved cable extrusion processes are developed, the performance of the plastic insulated cables should also improve just as it was the case with the early PILC cables. These predictions have indeed been borne out by the substantially improved performances of the XLPE and TRXLPE insulated distribution cables installed in the 1980s and 1990s. Furthermore, Eichhorn demonstrated that by analyzing statistically the results in [17] and comparing the situation where the total installed cable length of XLPE cables equals that of the PILC cables, the performance of XLPE in fact would appear better than that of PILC cables. Evidently, considerably more service data is necessary before definite trends in the failure rates and their full ramifications can be fully ascertained as concerns the comparative long-term performance of HMWPE, XLPE, and PILC distribution cables. For example, the type of soil for direct buried cables is an important parameter. Cables in wet soils may have to be rated higher than those in dry soils; but some discretion is required here since the soil conditions may vary appreciably with the seasons of the year.

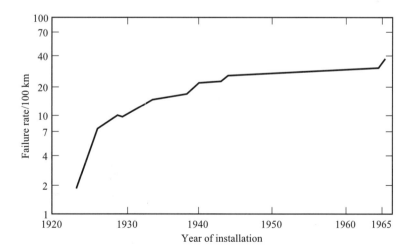

Figure 1.13 Cumulative rate of electrical failures of early PILC power distribution cables in the United States [17].

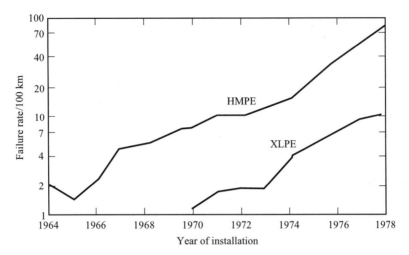

Figure 1.14 Cumulative rate of electrical failure of early HMPE and XLPE power distribution cables in the United States [17].

A rather revealing study on polymeric distribution cable failures was carried out by an Association of Edison Illuminating Companies (AEIC) Cable Engineering Section Task Group that collected HMWPE and XLPE distribution cable failure data between 1989 and 1991 [36]. The investigation examined the influence of electrical stress or cable insulation thickness and the type of cable jacket upon the in-service life of low density HMWPE and XLPE insulated distribution cables installed in ducts and directly buried. The findings are delineated schematically in Fig. 1.15 from which it is readily perceived that jacketed XLPE-insulated cables installed in ducts exhibit a low failure rate in contrast to the relatively high failure rates characterizing the directly buried unjacketed low-density HMWPE cables. Increased electrical stress or reduced cable insulation wall thickness is seen to affect adversely the failure rate of both the HMWPE and XLPE

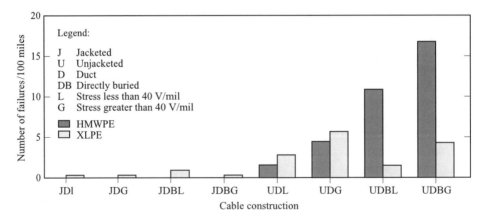

Figure 1.15 Low-density HMWPE and XLPE insulated power distribution cable failures in United States recorded between 1989 and 1991 (AEIC Cable Engineering Section Task Group [36]).

cables. Surprisingly, the unjacketed HMWPE insulated cables outperform the unjacketed XLPE insulated cables, when both are installed in ducts.

The intricate aging behavior and subsequent failure of solid extruded dielectric power distribution cables, under operating conditions in the field, are currently continuing to attract much attention. Considerable efforts have been expended toward developing suitable nondestructive diagnostic test procedures to determine a quantitative measure of the degree of aging of these cables while in service. One of the most popular diagnostic techniques in use currently is that of low-frequency testing, particularly as concerns the measurement of partial discharge and the dissipation factor [37–39]. Convenient lightweight low-frequency (0.1 Hz) high-voltage sources, having low kilo-volt-ampere ratings in comparison to equivalent high-voltage 60-Hz sources, have become commercially available, thus permitting partial-discharge site location as well as dissipation factor (tan δ) measurements to be made on distribution cables in the field. Since space charge effects predominate at low frequencies, particularly in aged cables, great care must be exercised when interpreting the tan δ measurement; in the presence of space charge effects, the tan δ value obtained at 0.1 Hz will generally exceed that at 60 Hz by an appreciable amount [40, 41].

A schematic circuit diagram of a typical 0.1-Hz test set is depicted in Fig. 1.16. The high-voltage dc is converted to a high voltage having a frequency of 0.1 Hz: A rotating rectifier in conjunction with a high-voltage choke convert the dc signal into an alternating polarity ac signal. The maximum 0.1-Hz alternating voltage output of the very low frequency (VLF) test set is typically 36 kV root mean square (rms) into a maximum capacitive load of 3 μF [42]. Low-frequency voltage withstand tests are also employed to assess aging of polymeric cables while in service [39]. They are frequently employed in lieu of the high-voltage dc tests, which are believed under some conditions to cause damage to aged polymeric cable insulating systems that may still have some useful life remaining under normal operating ac voltage conditions.

1.2.5 Underwater or Submarine Cables

Underwater or submarine cables, as their name implies, are used for traversing water bodies to interconnect systems, such as those between an island and an adjacent shore line. Submarine cables may generally be of the self-contained oil-filled type, pipe type, high-viscosity oil-impregnated paper, or solid extruded dielectric type. For short-distance transmission for large quantities of power, ac self-contained oil-filled cables are

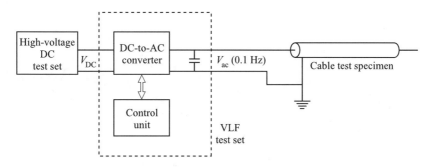

Figure 1.16 Very low frequency (VLF) high-voltage test set schematic with cable specimen under test (after [42]).

preferred. These cables make use of very low viscosity impregnating oils, which permit lower hollow conductor core diameters, guarantee positive oil feed to cable insulation under all load conditions, and prevent water ingress in case of external damage or failure. For power transmission over long distances low-cost solid-type high-viscosity oil-impregnated paper dc cables are deployed. Another application of dc cables concerns transmission system stability requirements, where isolation between two power networks must be provided; in the latter case, the distances may be short and the isolating dc intertie may involve either an underwater or a terrestrial crossing. For such short-distance dc cable applications, the self-contained oil-filled cable design is again the preferred choice. Submarine cables may traverse rugged land–water terrains and must, therefore, be well protected by steel wire armoring. Figure 1.17 depicts a profile of an ac submarine cable crossing at Long Island Sound [43] in which an oil-filled cable system, involving seven single-phase cables rated at 138 kV to supply a load of 300 MVA, was placed in service in 1970; the operating oil pressure in these cables is maintained between 179 (1170) and 270 lb/in.^{2} (1860 kN/m^2).

Perhaps one of the longest ac submarine cables installed was that in 1956 involving three single 138-kV gas pressure oil-preimpregnated-paper-type cables over a length of 80 miles across the Georgia Strait between Vancouver Island and the British Columbian mainland [5]. The cable system was designed to transmit a load of 100 MW. This transmitted load capacity exceeded substantially that of several important and almost concurrent earlier installations, namely the St. Lawrence River crossing in Québec in 1954 with a 75-kVA rated 69-kV ac single core oil-paper cable over a distance of 51 km; the Sweden–Gotland Island 100-kV dc oil-impregnated-paper cable over a distance of 100 km in the Baltic Sea for a load of 20 kVA in 1953; and, in 1951, the Sweden–Denmark 120-kV ac, flat three-core oil-filled cable rated for a transmitted load of 65 kVA over a distance of 5.5 km [44]. All of the foregoing power transmission capacities were again surpassed by that of the dc submarine cable link between England and France commissioned in 1961 [45]. The oil-impregnated-paper cables spanned a total distance of 52 km across the English Channel and were rated to deliver a maximum power of 160 MW at an operating voltage of ±100 kV dc. Over the last four decades the power transmission capacity of submarine cables has increased markedly with each rise in the ac or dc transmission voltage of the cables [46]. For example, the capacity of the ±450-kV dc submarine cable intertie across the St. Lawrence River between Québec and the New England states, commissioned in 1992, is 2000 MW [47]. The strong

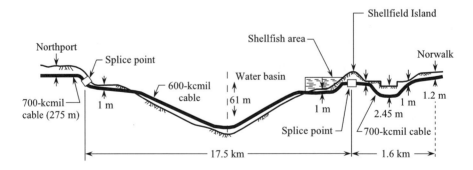

Figure 1.17 Profile of 138-kV oil-filled power cable crossing at the Long Island Sound (after [43]).

current, ship traffic, and patterns of ice flow in the river did not provide a benign environment to permit direct burial of the cables. A decision was thus made to install the cables in a 4.1 km long tunnel constructed in the limestone and shale strata beneath the riverbed. The tunnel has a 20-cm-thick circular shell of non-reinforced concrete as shown in Fig. 1.18, with the six 500-kV dc nominally rated self-contained oil-filled kraft paper tape cables placed on prefabricated reinforced-concrete troughs; a pair of cables act as spares, while four cables carry the load. Stability requirements dictated the deployment of dc cables in order to isolate the Québec system from the New England states systems. As mentioned already apart from system isolation applications, dc submarine cables are extensively used for longer underwater transmission distances where ac cables become uneconomical as a result of the excessive reactive power losses. For intermediate distances self-contained low-viscosity-type dielectric liquid-filled dc cables are preferred as higher voltages may be employed to transmit larger blocks of power. When the underwater distances become too large, at which time the required dielectric liquid pressures become too difficult to maintain, solid-type dielectric-liquid paper-impregnated dc cables must be employed with an attending compromise in the amount of transmitted power.

An interesting submarine cable design has been used in the crossing between New Brunswick and Prince Edward Island across the Northumberland Strait. Rated at 138 kV, two three-phase cables of the construction portrayed in Fig. 1.19 were laid to span a length of 13.5 miles. The cable construction is unique in the sense that it is basically a flexible oil-filled pipe-type cable. The rigid steel pipe of a normal pipe-type cable is replaced by a flexible lead sheath and armor, to provide a compact self-contained, pipe-type cable. Each cable is capable of carrying a 100-MW load under the present ac operating conditions; for optional dc operation with a ground return,

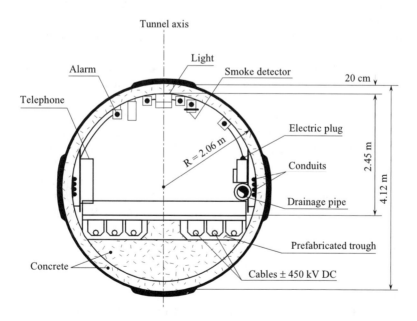

Figure 1.18 Disposition of six 500-kV dc nominally rated self-contained oil-filled cables in a submarine tunnel under the St. Lawrence River [47].

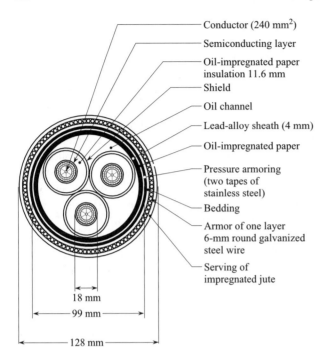

Conductor (240 mm^2)

Semiconducting layer

Oil-impregnated paper insulation 11.6 mm

Shield

Oil channel

Lead-alloy sheath (4 mm)

Oil-impregnated paper

Pressure armoring (two tapes of stainless steel)

Bedding

Armor of one layer 6-mm round galvanized steel wire

Serving of impregnated jute

18 mm

99 mm

128 mm

Figure 1.19 Cross section of Sieverts submarine power cable used on the Prince Edward Island–New Brunswick link. (Courtesy of the Maritime Electric Company.)

the capacity may be augmented to 300 MW per cable. For water depths in excess of 40 ft along the igneous crossing, the cables were plowed 2.5 ft below the ocean floor; whereas for depths less than 40 ft, specially dredged trenches were made for cable lays 6 ft below the ocean floor. Notwithstanding these burial precautions, the cable was damaged recently by a boat's anchor and had to undergo repairs with attending power interruptions on the island, which was fortunately equipped with standby thermal generator units.

Submarine cables have always been designed to withstand considerable mechanical stress and pressure, thus their construction is more robust than that of a normal type of underground cable. This is well borne out in the instance where a 72-kV oil-filled cable laid in Norway at a depth of 535 m in 1953 was subjected to excessive tensile stress during a storm at sea, resulting in appreciable radial deformation. The cable was nevertheless put in service and has remained in operation and as last reported on in 1980, without any interruption [48]. Though most submarine cables for operation under ac conditions are of the oil-filled or solid-type high viscosity oil-impregnated-paper design, an increasing number of XLPE-insulated cables are being installed. In Norway, a country that makes extensive use of submarine cables because of the geographical terrain of large numbers of fjords and islands, the first XLPE cable, rated at 12 kV, was installed in 1971. A total of 17 cables rated at 72 kV were already in operation in 1979 and a 145-kV rated cable was laid in 1980 [48]. The situation is rather interesting in the sense that the polyethylene cables placed for ac operation at or less than 24 kV are without a lead sheath; even some of the 72-kV cables have, in fact, been installed without an impervious lead sheath. In what manner this will affect water tree growth in the unsheathed submarine cables is not known, though it can be said that insofar as the reliability of performance of the lower voltage rated XLPE cables is concerned, no premature failures have yet been recorded. However, caution is to be exercised and the decision [48] in Norway in 1980 to apply lead sheaths on all future cables rated above 24 kV is certainly a very prudent one. At this juncture in our discussion, the appropriate question may be posed as to why terrestrial polyethylene power cables without hermetic sheaths when directly buried in moist soil are subject to tree growth and failure while submarine cables submerged in fresh or salt water are not. The only plausible explanation of the peculiarly divergent behavior of polyethylene cables in the submarine and terrestrial environments is the distinct possibility that the treeing mechanism thrives principally when the cables are being subjected to alternatively dry and wet conditions in the soil as is indeed the case with many terrestrial cables when water levels change during and with the seasons.

The application of XLPE insulation on dc submarine cables has met with considerable adversity; the excellent low dielectric loss characteristics of polyethelene result directly from its deep charge traps wherein free charge carriers are trapped and immobilized. But it is precisely these deep charge traps that render polyethylene unsuitable for dc power transmission. Polarity reversal under dc operating conditions can readily precipitate breakdown of the polyethylene due to the additive field of the space charge residing within the deep traps. It is well known that the incorporation of conductivity additives in the polyethylene can disperse some of the space charges, but the long-term effectiveness of these compounds is unknown. The long proven reliability of oil-paper systems for dc power transmission is difficult to improve upon. If it is desired to replace the oil-paper system for dc applications with substitutes, then perhaps it is more expedient to accomplish this with PPP tape composites [49] than with XLPE containing additives [50, 51].

1.2.6 Low-Loss Power Transmission Cable Systems

As rising urban population densities demand increasingly larger blocks of underground transmitted power, the present insulating systems of high-voltage power transmission cables are becoming subjected to more stringent performance requirements. With traditional dielectric liquid-impregnated kraft paper and solid extruded dielectric insulated cables, the rise in the demand of electrical energy can partially be met by raising the transmission voltage and adjusting accordingly the insulation wall thickness, and if necessary using forced cooling. However, even with adequate forced cooling techniques, the use of oil-impregnated kraft paper power cables cannot be efficiently extended much beyond 750 kV. Furthermore, even below 750 kV, when indirect forced cooling is employed, the required current ratings necessarily limit the lengths of the oil-impregnated kraft paper cables to a few thousand feet. It was demonstrated already in 1980 that higher load-carrying capacities could be achieved by direct cooling methods, such as placing the oil-filled cables in a water-filled cooling pipe [13]. This could conceivably result in cables being capable of carrying loads approaching those carried by overhead transmission lines. However, practical problems would arise as such procedures would require large quantities of water to be drained off whenever repairs were necessary. As discussed in Section 1.2.2, at voltages in excess of 750 kV, the dielectric losses may be maintained within an acceptable level by substituting the kraft paper with other synthetic polymeric materials such as PPP composites, having low dielectric loss characteristics. These systems when operated from 750 to 1000 kV can provide load ratings substantially in excess of 1000 MVA even without the aid of forced cooling.

The chemical and physical compatibility of polypropylene with mineral oils, polybutenes, and silicones was examined by Allam and his co-workers in a research project supported by the U.S. Department of Energy and the Electric Power Research Institute (EPRI) [52]. Various designs have been considered, and it was found that of the three evaluated dielectric liquids, polybutene appears to be most compatible chemically with the PPP composite material. Some swelling problems have been encountered with the polypropylene plastic, which, in addition, could not be fully impregnated to eliminate all of the minute cavities within its interstices. Moreover, cable bending problems were anticipated due to the inclusion of the polypropylene in a composite tape. In England work on the same laminate was also carried out by Arkell et al. [53] and prototype cables for operation at 132–275 kV were tested in 1980. The relative permittivity of these cables ranged from 2.6 to 2.8 as compared to 3.5 of the oil-impregnated kraft paper insulation, thus in effect reducing the capacitive charging current significantly in the synthetic insulation cables. In Section 1.2.2, an 800-kV PPP composite tape cable was described in which a dodecyl benzene impregnant was employed and for which no chemical compatibility problems were observed. The transmission capacity of the cable with natural cooling only was in the order of 1900 MVA.

Both LDPE and XLPE have excellent low dielectric loss characteristics as well as low dielectric constants, which should place their power transmission capabilities well in excess of 1000 MVA. However, some questions remain concerning their long-term upper operating temperature limits. Further marked reduction in the dielectric loss of power transmission cables necessitates the replacement of either the solid or solid-liquid dielectric insulant with a gas or vacuum medium. Vacuum and gases, with the proviso that the gases are utilized at stresses well below their ionization level, represent virtually perfect dielectrics of near-to-zero loss. Also the use of vacuum or gas insulants diminishes very significantly the magnitude of the capacitive charging current, since the real value of the relative permittivity or dielectric constant of these insulating media approaches unity if the effect of the solid insulator spacers is neglected. Although air fits the low dielectric loss category of requirement, there are other more suitable gas dielectrics, whose dielectric breakdown strength either at atmospheric pressure or above exceeds that of air by an appreciable margin [54]. A further reduction in cable losses may be achieved with gases at cryogenic temperatures at which both the dielectric losses as well as the resistive conductor losses are greatly decreased with the cable conductor operating in the superconducting mode.

1.2.6.1 Compressed SF_6 Gas Power Transmission Cables

Compressed SF_6 gas cables with epoxy support spacer insulation have been in use for more than two decades [55, 56]. In 1974, an early experimental high-power capacity compressed gas cable, rated at 550 kV for 3000-A load operation, was installed and tested by Cookson [57] at Westinghouse; these early feasibility test results proved to be very encouraging and stimulated much of the work toward the development of practical high-power transfer compressed SF_6 gas lines. As a result of their high transmission capacities, compressed SF_6 gas insulated cable systems have gained considerable popularity, particularly for short transmission distances where they can carry directly the power from an overhead line to a substation. For such short distances they present a lower cost as compared to conventional dielectric liquid or oil-filled cable systems with the associated terminal and auxiliary equipment [13]. Their use is also attractive for substations already having apparatus with compressed SF_6 gas insulation; in a number of ways, such cables may be regarded as extensions of existing compressed SF_6 gas bus systems.

A good example of the applicability of compressed SF_6 gas cables for connection between overhead lines and substations is the extensive use of these cables made in Japan. Commencing with 1979, the first compressed SF_6 line rated for 154 kV was installed; this was followed by a 275-kV line in 1980 and a 500-kV line in 1985. Table 1.3 tabulates the characteristic parameters of four compressed SF_6 gas cables installed between 1979 and 1990 [58]. It can be perceived from Table 1.3 that the transmitted power capacity of compressed SF_6 gas cables is substantially augmented with increasing transmission voltage. The increase in voltage rating is achieved by enlarging the spacing between the inner conductor and the outside tube enclosure of the line, i.e., the length of the epoxy insulator and its contour or geometric configuration. Increased current-carrying capacity requires inner conductors with greater outside diameters and support/spacer insulators with higher operating temperature limits. Hence, irrespective of whether increased power transfer is attained by an increase in the operating voltage or current, a larger outside cable diameter is the net result. Table 1.4 presents

TABLE 1.3   Characteristic Parameters of Compressed SF_6 Gas Power Transmission Lines

Commissioning Date	1979	1980	1985	Line 1, May 1988 Line 3, March 1998
Nominal voltage	154 kV	275 kV	500 kV	275 kV
Rated current	2000 A	4000A	6240A	8000A
Route length	Line 1, 164 m Line 2, 199 m	102 m	Line 1, 152 m Line 2, 140 m	Line 1, 153 m Line 2, 138 m
Cable cross section	Envelope Conductor Spacer/insulator			
Outer diameter of conductor	100 mm	160 mm	180 mm	180 mm
Inner diameter of enclosure	340 m	480 m	480 m	480 m
Type of insulators	Disk, cone, and post	Disk, post, and cone	Post, cone (with a rib)	Post, cone (with a rib)
Gas pressure	0.35 MPa at 20°C	0.35 MPa at 20°C	0.6 MPa at 20°C	0.5 MPa at 20°C
Conductor temperature	90°C	90°C	90°C	105°C
Construction details				
Conductor jointing	Welding	Plug-in contact	Plug-in contact	Plug-in contact
Enclosure jointing	Sleeve joint, and flange joint by welding	Sleeve joint and flange joint by welding	Plug and socket joint by welding	Plug and socket joint by welding
Welding method of enclosure joint	Manual	Manual	Automated	Automated

Source: After [58].

TABLE 1.4 Parametric Data for a Typical 275-kV Compressed SF$_6$ Gas Power Transmission Line with a Current Capacity of 8000 A

Number of conductors per tube enclosure	1
Conductor	
Outer diameter	180 mm
Thickness of hollow conductor	20 mm
Material	Aluminum alloy
Treatment of surface	Black alumite
Enclosure	
Inner diameter	480 mm
Thickness of tube conductor	10 mm
Material	Aluminum alloy
Treatment of surface	
Inner	None
Outer	Epoxy coating
Spacer	
Material	Epoxy resin
Geometric configuration	Post and Cone
SF$_6$ gas	
Pressure	0.5 MPa at 20°C

Source: After [58].

more detailed data on the dimensions and materials utilized in a 275-kV compressed SF$_6$ gas cable designed for a current-carrying capacity of 8000 A [58]. For both the inner hollow conductor and the outside cable tube enclosure, aluminum alloys are employed. Epoxy resin insulators, having a post-and-cone configuration are used with an SF$_6$ gas pressure of 0.5 MPa maintained at 20°C.

Compressed gas cables differ from conventional solid and solid-liquid insulated cables in that they are supplied by the manufacturer in sections and must be assembled in the field to complete their installation. The complete line assemblage consists of straight pipe units, which generally contain branched or T-connection sections and hinged aluminum bellows to allow for thermal expansion and contraction of the tube enclosures, as well as elbow sections and air insulator terminals. The configuration of some of these sections is illustrated in Fig. 1.20. The bellows of the straight sections also contain hinged conductors. The straight sections consist of extruded aluminum pipes, which may be in lengths of 12 or 14 m that are welded together at the plug and socket joints. The conductors within the pipes are joined by means of plug-in contacts.

The assembly of compressed gas cables in the field is frequently a cause and source of contamination, which may be in the form of different types of dust such as sand (SiO$_2$) and metallic particles [59]. The greater the number of sections to be assembled i.e., the longer the length of the required compressed SF$_6$ gas cable, the higher is the probability of contaminant particle incidence. This tacit constraint imposes in effect a practical limit on the length of compressed gas cables that can be assembled in sections. Metallic particles, which may comprise particle remnants from the manufacturing process of the pipes, or which may be produced by friction arising between the surfaces of movable parts in the system or due to other mechanical effects, affect adversely the breakdown voltage of the gas cables [60]. As the metallic particles execute motion under the influence of the existing electrical field between the conductor of the cable and the

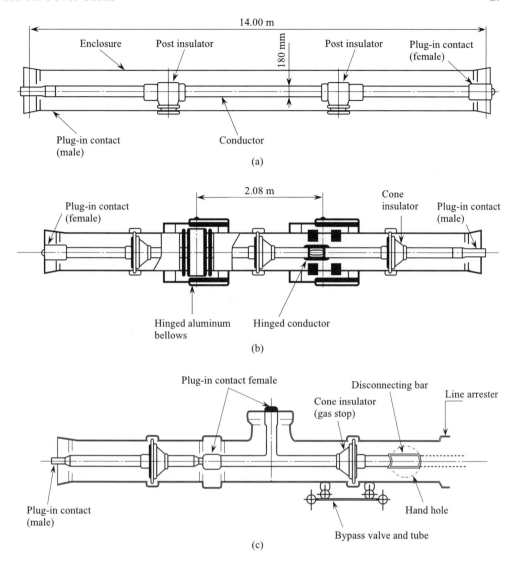

Figure 1.20 Typical sections of a compressed SF_6 gas power transmission cable: (a) straight section, (b) section with bellows, and (c) T-section (after [58]).

enclosure, partial discharges are initiated between the approaching particle and the site of its impending impact on the inner surface of cable pipe enclosure [61, 62]. In the presence of partial discharges, the SF_6 gas dissociates and reacts with oxygen to form some rather very toxic gases such as S_2F_2, SF_4, SO_2, SOF_2, and SO_2F_2. Although oxygen gas, which is required for some of the reactions to form the chemical radicals, may be available as a result of finite impurities in the SF_6 gas itself, it is much more likely to be provided by the traces of air adsorbed upon the surfaces of the components comprising the inner structure of the cable. The possibility of toxic gas production as well as the fact that SF_6 per se is a hot-house gas, coupled with data that indicates that gas leaks may and do indeed occur from the compressed cables themselves, have posed environmental concern. For this reason, there has been some effort directed toward possible replacement of the SF_6 gas by other gases, primarily gas mixtures such as nitrogen containing minute quantities of SF_6 gas. However, these gas mixtures have lower breakdown strengths than SF_6 and, therefore, entail the use of larger enclosure pipe diameters with attending increased costs.

The power transmission capacity of SF_6 cables is to a very appreciable extent also determined by their heat transfer characteristics [63]. The inner conductor operating temperature is usually within the range of 90–105°C. To prevent undue temperature rises within the gas cables, which may exceed the temperature limits of the epoxy resin spacer-insulators, the gas cable design must ensure an adequate heat dissipation rate by the cable envelope to its surroundings. The heat transfer from the inner conductor to the envelope is increased by application of black graphite paints over the surface of the conductor and the inner side of the tube enclosure. Since the heat transfer occurs by radiation as well as by convection within the SF_6 gas, it is much higher than for vacuum where heat transfer due to convection is nonexistent; consequently, the heat transfer rate is also increased with the pressure of the SF_6 gas. These considerations also indicate why vacuum insulated cables would experience difficulties in dissipating the heat generation at the surface of high current-carrying conductors. Since gas cables are normally installed outdoors and are unburied, the cooling of the external cable envelope itself occurs through radiation and external air convection. Thus, the wind direction and velocity have an important effect upon the cooling rate, and external heating of the envelope as a result of solar radiation must also be taken into account. However, installation of the gas cables at an angle to the horizontal ground plane appears to have only a minor influence [63].

Some work has been carried out toward developing nonrigid compressed SF_6 gas cables, since this would circumvent the incidence of particle contamination arising during cable part assembly in the field and provide the SF_6 cables with flexibility advantages inherent with conventional power transmission cables. In 1980, a flexible compressed SF_6 gas insulated cable design was described by Spencer et al. [64]. The cable, which is depicted in Fig. 1.21, is designed to operate at 362 kV and may have an overall length up to 80 m. The cable consists of a corrugated aluminum pipe as the outer conductor or shield and a multicore corrugated aluminium pipe as the inner conductor. It is expected that cables of this design may be operated at temperatures up to 140°C. Commercial equipment for the manufacture of flexible SF_6 gas cables has been made available in Germany. Since gas insulated cables have a higher characteristic impedance than oil-filled cables, they would be more susceptible to overvoltages when connected to the overhead transmission lines [12]. Accordingly, suitable protection must be provided. Although flexible SF_6 cables appear to have some practical advantages, no extensive use has been made of them thus far in the field.

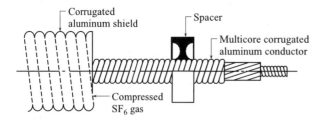

Figure 1.21 Flexible compressed SF_6 gas cable (after [13, 64]).

1.2.6.2 Superconducting Power Transmission Cables

Conceivably, compressed SF_6 gas cables should be capable of meeting any demand for larger capacity power transmission up to $\sim$10 GW. However, this projection is perhaps valid only for the short term in view of increased environmental concerns with SF_6 gas and the associated attempts to constrain its use in electrical apparatus. In the long term the high power transfers are more likely to be accommodated with superconducting cables, which are characterized by low dielectric losses and extremely low resistive losses in their conductors. Although low-temperature superconductivity in metals at liquid-helium temperatures was discovered in the early 1900s, serious efforts to develop low temperature superconducting and cryoresistive cables were only undertaken in the late 1960s. With the discovery of high temperature ceramic superconductors in 1986, most of the development work on low-temperature metallic superconductors was abruptly halted because with high-temperature superconductors the cable operating temperatures can now be raised to 77 K or higher—thereby reducing very substantially the refrigeration costs. Nevertheless, at the present even if a reliable high-temperature superconductor cable design were achievable, it would be economically quite uncompetitive with comparable conventional cable systems designed to carry the same power load.

There were principally two concerted projects carried out on the development of low-temperature metallic superconductive cables, that have achieved tangible practical results. One was pursued at the Brookhaven National Laboratory [65] and the other at Arnstein in Austria [66, 67]. The Brookhaven prototype was a three-phase cable rated for 1000 MVA at 80 kV; the cable was tape insulated, thereby permitting the construction of a longer cable specimen length of 115 m. The superconductor was made of niobium-tin (Nb_3-Sn) and was applied in tape form as illustrated in the profile schematic of the cable construction in Fig. 1.22 [68]; the coolant used was helium. Other pertinent details on the superconductor are provided in Table 1.5. The prototype cable underwent extensive testing and evaluation at Brookhaven between 1982 and 1986 but was never installed in the field for in-service operation. Additional parametric data on the Brookhaven helium-impregnated tape insulated superconducting cable is provided in Table 1.6, from which it can be seen that the dielectric losses in the cable were less

TABLE 1.5 Nb_3Sn Superconductor Characteristics of the Brookhaven Low-Temperature Superconducting Cable

Overall tape thickness, 110 μm
Nb_3Sn thickness (total per tape in two layers), $\sim$8 μm
Minimum bending radius (either direction), <10 mm
Cable inner conductor diameter, 29.5 mm
Cable outer conductor diameter, 52.7 mm
Inner conductor tape width, 6.5 mm
Outer conductor tape width, 11.7 mm
Rated rms current at 60 Hz, 4100 A
Surface field at rated current (inner conductor), 44.2 A/mm
Mean current density (inner conductor), $\sim$3700 A/mm^2
Critical current density at 4.2 K, $\sim$6 × 10^4 A/mm^2
Operating temperature range, 6.5–8.5 K

Source: After [68].

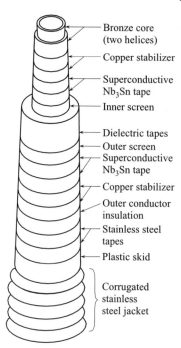

— Bronze core
(two helices)

— Copper stabilizer

— Superconductive
Nb$_3$Sn tape

— Inner screen

— Dielectric tapes
— Outer screen
— Superconductive
Nb$_3$Sn tape

— Copper stabilizer

— Outer conductor
insulation

— Stainless steel
tapes

— Plastic skid

Corrugated
stainless
steel jacket

Figure 1.22 Flexible low-temperature super-conducting prototype cable design at the Brookhaven National Laboratory (after [65]).

TABLE 1.6   Characteristics of Brookhaven Low-Temperature Superconducting Cable Prototype

Cable length installed (each), ~115 m
Cable outside diameter (over armor), 5.84 cm
Cryostat inner bore, 20 cm
Operating temperature, 7–8 K
Operating He density, 100 kg/m^3 minimum
Operating pressure, 1.55 MPa (225 psia) typical
Rated voltage (60 Hz), 138 kV, three phase
Rated current (60 Hz), 4100 A (continuous; 6000 A (60 min)
Operating voltage stress, 10 kV/mm
Operating surface current density, 44 A/mm
Maximum continuous power 1000 MVA (three phase)
Surge impedance load, 872 MVA (three phase)
Cable conductor loss at 4100 A (7 K), 0.2 W/m
Cable dielectric loss at 80 kV <0.06 W/m
Cable impedance, 25 Ω

Source: After [68].

than 0.06 W/m. In this design the dielectric performance characteristics are to a large extent determined by the coolant condition in the butt-gaps.

The first actual low-temperature superconducting cable installation was placed into service in Arnstein, Austria, in 1977. The cable operated in a superconducting mode at a voltage of 60 kV, employing a niobium film as the superconductor at liquid helium temperature with vacuum and liquid nitrogen employed for thermal insulation and helium-impregnated paper for electrical insulation. At the completion of the service trials in 1979, it was concluded that an upscaled cable of the same design could readily provide a load capacity between 2 and 3 GW [67]. Table 1.7 gives additional data on the Arnstein cable.

The only currently publicized low-temperature superconducting cable project still in progress is that in Japan [69]. The purpose of the study is to examine the behavior of solid extruded dielectric insulation (ethylene-propylene-rubber), which undergoes contraction as the cable is cooled down to cryogenic temperatures. Moreover, little is known on the degradation behavior of solid insulation at cryogenic temperatures, such as treeing.

The discovery of high-temperature ceramic superconductors has stimulated work to develop superconducting cables, which may operate at much higher cryogenic temperatures, hopefully with liquid nitrogen as the main coolant at 77 K. The first hurdle of producing a sufficiently pliable high-temperature superconductor for cables has been only recently surmounted. There are a number of high-temperature superconductors available, having a range of different transition temperatures, i.e., temperatures at which the conductor becomes superconducting [70]. Cable feasibility studies have been carried out using YBCO (yttrium-barium-copper-oxide) with a transition temperature of 90 K; however, this temperature is too close to the boiling temperature of liquid nitrogen (77 K), with the result that under overload or fault current conditions the temperature rise or swing of the cable conductor may be too large to restore the rated current of the high-temperature superconductor [71]. Thus, if YBCO is to be used as a cable conductor, liquid helium must still be employed to maintain the temperature rise of the conductor well below the temperature transition of 90°C for YBCO.

The high-temperature superconducting copper oxide ceramics per se do not have good flexibility characteristics and cannot carry high currents, nor can they withstand mechanical strains created by the Lorentz forces produced by the magnetic induction field and the current in the crystalline structure of the ceramic. High-temperature superconducting ceramic oxide wires are usually made by incorporating the superconducting filaments in a metal matrix, e.g., in the form of a tape having an approximate range of thickness that may extend from 200 μm to 0.2 mm with a width of 6 mm enclosed within a silver sheath [70]. The first step in the high-temperature superconductor wire-forming process is to synthesize a precursor powder, which may either contain the superconducting ceramic oxide initially or contain compounds from which the semiconductor can be produced by employing suitable mechanochemical processes. The precursor powder is then packed into a billet or tube, which is subsequently drawn into a wire. Following heat treatment, the wire is stacked into a bundle usually in the form of a tape in order to impart strength to the resulting multiwire ribbon. The high-temperature superconductor of bismuth-strontium-calcium-copper oxide ($Bi_2Sr_2Ca_2Cu_3O_{10}$), commonly referred to as BSCCO, has a transition temperature

TABLE 1.7 Arnstein Low-Temperature Superconducting Cable Characteristics

---

Voltage rating, 60 kV (three phase)
Cable length, 50 m (single phase)
Enclosure heat in-leak, 0.15 W/m (at 6 K)
Dielectric loss at 40 kV, 0.02 W/m (at 6 K)
Current rating, 1000 A

Source: From [67].

greater than 110 K and has been shown to be readily formed into longer wire lengths along with silver filaments for stabilization. The manufacturing procedure for the formation of BSCCO filaments is represented schematically in Fig. 1.23 [72]. In the process metal-oxide precursors of Bi_2O_3, $SrCo_3$, CaO, and CuO are reacted or calcinated at temperatures $\geq$ 800°C, followed by several cycles of grinding and reheating to form the BSCCO superconductor powder. The powder is subsequently packed into a silver billet, which is drawn through wire dies until its diameter is reduced to approximately 1 mm, whereupon a number of these wires are combined to form the desired tape conductor. The latter is then wound to the required conductor length and undergoes annealing in an oxygen furnace for a period of several days.

Although the BSCCO superconductor has exhibited great promise as concerns wire preparation, its performance at 77 K has been less impressive due to the low current densities that it can carry. Figure 1.24 depicts the critical surface for BSCCO formed by the temperature T, magnetic field, H, and the critical current density, J_c (the maximum current density, which the conductor can carry in the superconducting mode). As can be discerned from Fig. 1.24, the critical current density at the liquid-nitrogen temperature is only 5.4×10^2 A/mm^2 in the plane of the magnetic field at 77 K [73].

Notwithstanding the limitations of BSCCO superconductors, a prototype BSCCO superconductor cable was manufactured, using current oil-filled pipe-type cable technology and is depicted in Fig. 1.25 [72]. Liquid nitrogen is employed as the coolant. Detailed data on the cryogenic pipe-type cable is given in Table 1.8. A 120-kV SF_6 gas–liquid N_2 insulated cable termination for use in conjunction with a high-temperature superconducting cable for operation at 77 K has been recently described by Shimonosona et al. [73].

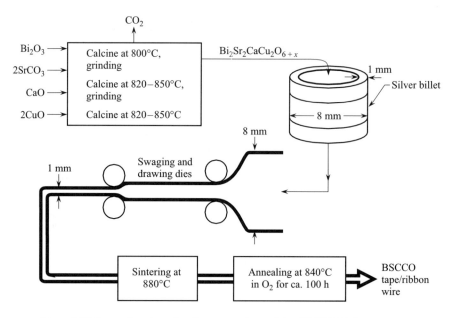

Figure 1.23 Processing steps for the preparation of the BSCCO high-temperature superconducting wire (after [72]).

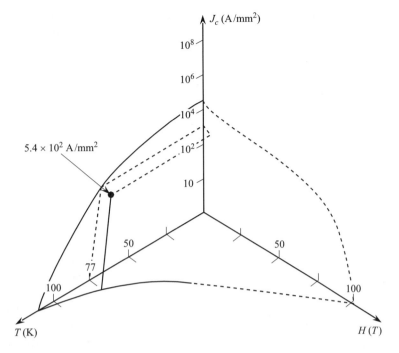

Figure 1.24 Critical surface of silver-sheathed bismuth-based high-temperature superconductor BSCCO (after [74]).

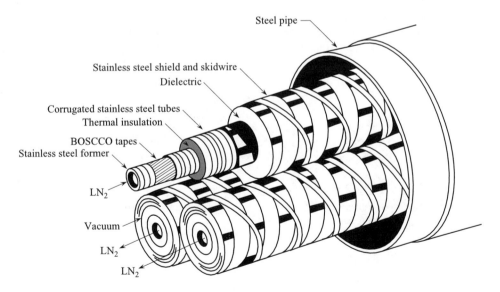

Figure 1.25 A 115-kV BSCCO three-phase high-temperature superconducting pipe-type cable (after [72]).

TABLE 1.8 Characteristics of 115-kV BSCCO Prototype Three-Phase High-Temperature Superconducting Pipe-Type Power Transmission Cable

Steel pipe diameter, 206 mm (8 in.)
Cable diameter in each phase, 85 mm
Cable length, 30 m
Power delivery capacity, 400 MVA
BSCCO silver-sheathed tape width, 4.5 mm
Number of BSCCO tapes in conductor, 150
Thermal insulation layer thickness, 12 mm
Total heat loss, 700 W/km MW

Source: From [69].

Figure 1.26 provides a schematic representation of possible power ratings either obtainable or anticipated with the different types of transmission cables available or contemplated in the future to meet the increasing electrical energy demand. It is apparent that much of the power transmission demand increase in the near future could conceivably be met by compressed SF_6 gas cables. However, for long-distance (≥ 100 km) underground transmission applications with power transfers in excess of 10,000 MVA, cryogenic cables offer the only possible solution [75–77]. But in the context of present-day economics and the technical difficulties associated with maintaining cable stability at extremely high current densities, this solution is far from being a trivial one. For superconducting cables to gain access into the underground power transmission area, their manufacture, installation, and maintenance must be shown to be cost effective. Notwithstanding the optimism that has accompanied the appearance of high-temperature superconductors, it must be observed that from the technical point of view it still remains to be demonstrated that high-temperature superconducting cables can equal or surpass the performance achieved with experimental low-temperature superconducting cables. The area of high-temperature superconductors is shrouded by the fact that there appears to be no generally agreed theory to explain the different characteristics and behaviors of the various high-temperature supercon-

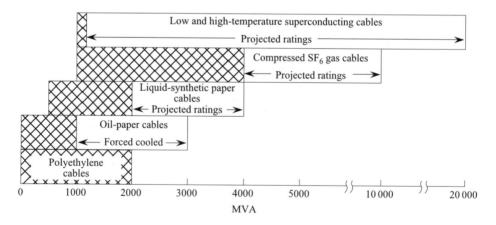

Figure 1.26 Ratings of present and possibly future power transmission cables.

ductors. It is only the YBCO that appears to fit well the accepted theory of super-conductivity of the high transition temperature cuprates [78]. However, YBCO is aty-pical of the class of high temperature superconducting compounds, whose transition temperature to the superconducting state is frequently found to be highly sensitive to the preparation technique [79]. This to a large extent reflects the complicated metallur-gical nature and chemical complexity of the high-temperature superconductors.

1.3 COMMUNICATION CABLES

1.3.1 Introduction

Communication cables trace their beginnings to telegraphy, shortly after the inven-tion of the telegraph by Morse. The first successful telegraph cable was laid between Dover and Calais in 1851. Its success in continuity of service was attributed to the use of a reliable insulation, namely gutta-percha, and a protective steel armor; the cable employed copper conductors [80]. The experience derived from this cable pointed the way to the laying of the first trans-atlantic telegraph cable in 1866. Following the invention of the telephone by Bell, terrestrial telephone cables began to appear. The first telephone cable was installed in 1882, consisting of wires insulated with gutta-percha and natural rubber, which were plowed underground over a 5-mile stretch along a railway line [81]. The earliest telephone cables were essentially single-wire grounded lines [82]. The transmission characteristics of the rather rudimentary early telephone cables were improved substantially by adequately insulating the individual wires; however, over long distances cross talk as a result of capacitive coupling between adjacent pairs became a serious problem. In 1887, the introduction of lead sheath shielding of twisted pairs and (in 1889) the insertion of capacitors between the series of two separate circuits, for the purpose of capacitance balance, were effective in reducing significantly the cross talk.

Long-distance telephone transmission became a reality in 1906, when a 145-km cable, with inserted loading coils in order to further diminish the attenuation of the line, was installed between New York and Philadelphia; this was followed in 1910 with a 730-km line between Boston and Washington in which use was made of the quad arrangement whereby three circuits were formed from four wires; this resulted in a reduction of the total number of pairs for the number of voice circuits to be carried over the cable. The 1920s witnessed a significant decrease in the overall telephone cable diameters due to a reduction in the copper wire diameters, a further decrease in cross talk with staggered twisting of the wire pairs, and replacement of the lead sheathing with tin-lead and antimony-lead compositions. Over this epoch of the terrestrial com-munication cable development, the wire insulation was formed with paper ribbons. The porous structure of paper provided a low dielectric constant medium, thereby minimiz-ing capacitive coupling between pairs.

Manufacture of paper-pulp-insulated wire cables began in the United States in the 1930s, which led to appreciable reductions in the space required per wire pair, thereby permitting a larger number of twisted-wire pairs per unit cable cross section. As in the case of paper ribbon cables, the wire gages employed were 22, 24, and 26 AWG (American Wire Gage), the notable exception being 19 AWG. Greater volumes of voice traffic during World War II and the postbellum period required larger cables with more twisted-wire pairs; for telephone cables containing 1800 pairs or more, the smaller diameter wires of 26 AWG size were employed. Concurrently, new cable sheaths of aluminum and different plastic materials were developed to replace the traditional but more expensive and heavier sheaths of lead alloys of tin, antimony, and cadmium. This allowed longer cable pulling lengths and resulted in greater ease of installation of aerial cables.

Plastic insulated twisted-wire multipair telephone cables first appeared in the 1950s and with rapidly decreasing cost of the plastic insulant (polyethylene or polyolefin) became quickly the twisted-wire multipair cables of choice. They have typically low installation and maintenance costs and are characterized by low dielectric losses. As a consequence of their lower attenuation characteristics, plastic insulated cables, commonly referred to as PIC cables, were employed initially for both long-distance analog transmission at carrier frequencies and subsequently for digital transmission, using pulse code modulation. The mutual capacitance between conductors in PIC cables, as for paper ribbon and pulp paper cables, was maintained at the accepted nominal value of 53 nF/km, and the PIC cables were manufactured with the same wire gage sizes as the antecedent paper-ribbon-insulated cables, i.e., including 19 AWG. At carrier frequencies to achieve lower attenuation, the 19 AWG larger diameter wire sizes were frequently employed; to obtain still lower attenuation, the wires were insulated with expanded polyethylene with which the mutual capacitance was reduced to 42 nF/km. Such lines were more economical in that fewer amplifiers or repeaters were needed to compensate for the attenuation losses.

The discovery of the thermoplastic polyethylene in the 1930s resulted not only in improved PIC twisted-wire cables but also provided an ideal flexible low dielectric loss medium for coaxial cables. Initial use of coaxial cables was made on electronic instruments and navigational applications as well as other military uses, principally in radar during World War II. It is interesting to note that the first coaxial submarine cable was installed between Havana and Key West in 1921; however, as this still occurred at a time prior to the discovery of polyethylene, the insulation consisted of gutta-percha. In Europe, a polyethylene insulated coaxial submarine cable was laid across the English Channel in 1947; its conductor was held in place by a solid polyethylene spiral tube, creating a partially air–solid dielectric insulating medium as depicted in Fig. 1.27 [80, 83]. It was shortly followed in 1950 by yet another polyethylene-insulated coaxial cable installation between Havana and Key West, along whose length repeaters were installed to amplify the transmitted signal. The first transatlantic coaxial submarine cable with repeaters and a rated capacity of 36 channels was installed shortly thereafter in 1956. With increased use of television, there developed in the 1960s a great demand for

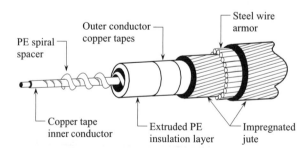

Figure 1.27 Early coaxial submarine cable with air-dielectric insulation, consisting of a polyethylene spiral spacer between the conductor and outer extruded polyethylene tube (ca. 1947) (after [83]).

coaxial cables with application to educational-TV, pay-TV, and community antenna TV (CATV) systems. For these uses low dielectric loss foamed or expanded polyethylene insulation was preferred. The CATV area has exhibited particular growth over the last few decades, and coaxial cable is still the preferred and most economic medium for transmission and distribution, though for long transmission distances increased deployment of optical fiber is now being made for backbone-trunk cables.

From the 1970s to the beginning of the 1980s, the air-dielectric coaxial-type cables dominated the long-haul telephone transmission field. The dielectric was principally air with intervening polyethylene disk spacers acting as supports for a solid copper inner conductor (7 or 10 AWG) within the coaxial cable. These cables replaced the twisted-wire multipair PIC cables, which could no longer meet efficiently the demands of the large volume of traffic, primarily due to the increased amount of data transmission along the cables whose original design had centered solely on voice transmission performance criteria. With digital transmission, using pulse code modulation and a repeater spacing of approximately 2 km, the maximum data rate transmission capacity achievable with air-dielectric coaxial cables was of the order of 400 Mbits/s. The mode of transmission in terrestrial coaxial cables was digital, using time division multiplexing (TDM) with pulse code modulation. This technique was introduced in the 1960s to replace the analog frequency division multiplexing (FDM) technique, which had dominated the telephone cable transmission field for approximately 50 years. The TDM technique was more effective in dealing with cross talk, signal distortion, and extraneous electrical noise. In contrast, however, the FDM technique remained the preponderant mode of transmission in long-distance underwater coaxial cables, including the last transatlantic coaxial submarine cable TAT-7 installed in 1983. It is important to emphasize that notwithstanding their structural differences in external armor or central strength member, all transatlantic underwater cables were insulated with solid polyethylene. A solid-polyethylene insulation was necessary to support a dc potential of several thousand volts, which was superimposed across the submarine coaxial cable to provide power to the signal repeaters placed along the cable.

The introduction of optical fiber telecommunication cables in the late 1970s revolutionized the communication cable field, since in these cables the transmission of voice and data information was accomplished via dielectric conductors in lieu of the traditional copper conductor wires. The eventual arrival of dielectric conductor communication cables was to be anticipated in view of Bell's photophone demonstration in 1880 [84] and the development of the theoretical basis underlying transmission along "dielectric wires" as described by Hondros and Debye in 1910 [85]. The realization of practical optical fiber cables required the development of low attenuation optical fibers and efficient light signal transmission sources and detectors. Reliable long lifetime semiconductor injection lasers for operation at room temperature were successfully designed in 1977 [86], though the early light sources used initially consisted of low power wide-frequency spectrum light-emitting diodes (LED); high-speed silicon *p-i-n* photodiode detectors became available in 1962 [87]. The development of silica fibers in 1970 led to what was at that time considered an acceptably low-loss transmission medium with an attenuation level of 20 dB/km at a transmission wavelength of 0.8 μm [88]. Even at these rather high attenuation levels, optical fibers represented an attractive transmission medium because of their instrinsically high transmission capacity. The use of optical fibers grew very rapidly following their introduction, and by the end of 1981 in Canada alone there were already 1500 km of installed optical fiber telephone cable, even though the attenuation losses in the early cables were as high as 4 dB/km as opposed to current monomode fiber cables with an attenuation of 0.2 dB/km at a wavelength of 1.55 μm.

The first transatlantic optical fiber telephone cable, having a data rate transmission capacity of 280 Mbits/s was laid in 1988. In this installation the transmission is carried out still with TDM, using electronic repeaters, which convert the optical signals to electrical signals for pulse shape regeneration purposes; after correcting for the attenuation loss, they are reconverted to optical signals before transmitting them along the cable. The vast majority of terrestrial optical fiber transmission systems are also operated in the TDM mode, and the latest long-haul lines have already achieved transmission capacities of 10 Gbits/s. With wave division multiplexing (WDM), the transmission capacity of optical fibers can be further augmented, and a commercial optical fiber transmission cable has already achieved a data transmission rate of 40 Gbits/s [89]. With erbium-doped optical fiber amplifiers, which are capable of simultaneous amplification of the transmitted optical signals using WDM transmission techniques, optical fiber transmission systems with a capacity of 0.1 Tbits/s are envisaged.

Following the introduction of optical fiber cables, there arose the perception and conviction that the rapid demise of copper conductor cables was imminent. Whereas this prediction was shown to be demonstrably correct as concerns the displacement of metallic conductor cables by optical fiber cables as the preferred transmission medium in the long-haul telephone cable area, it proved to be very much premature in the case of the local loop serving telephone subscribers and local area networks (LANs) such as Ethernet and IBM's Token Ring in which extensive use is still made of coaxial cables and twisted-wire pairs. Thus, at the present time, copper conductor communication cables are still being manufactured in large volumes to meet the demand for either new or replacement cables for short-distance voice and data transmission applications. The simple unshielded twisted-wire pair cable, which was originally designed for voice signal transmission, can now be operated over short distances at frequencies of several hundred megahertz with transmission rates up to 155 Mbits/s. Similarly, coaxial cables are now being made and used beyond their earlier 1-GHz limit. As long as metallic conductor communication cables remain cost-effective alternatives to optical fiber cables in meeting the necessary data transmission rate requirements, their use will continue for short-distance applications resulting in a continued proliferation of hybrid networks containing both metallic and optical fiber cable networks. In the sections that follow, we shall provide a cursory description of the types of communication cables currently in use in various application areas.

1.3.2 Twisted-Pair Communication Cables

In the field of telephone communications, it is customary to classify the cables into toll and exchange area cables. Toll area cables are employed for long-distance transmission, while exchange area cables comprise a category of cables that includes trunk, feeder, distribution, and video cables. Switching centers or central offices are connected by trunk cables, and feeder cables are used to connect the switching centers to the distribution areas from where the service drops to the subscribers are made with distribution cables. Video cables, as the name signifies, form the lines for picture signal links.

The early toll area or long-haul cables were of the multipair twisted-wire-type with either paper ribbon or paper pulp insulation. These were replaced with either regular

PIC cables of 19 AWG conductors with a mutual capacitance of 83 nF/mile or the low mutual capacitance (66 nF/mile) PIC cables. The latter also used twisted pairs with a conductor size of 19 AWG, but the twisted pairs were insulated with foamed polyolefin or polyethylene insulation, which required fewer repeaters as a result of their lower attenuation at carrier frequencies. The PIC toll area cables were replaced in the 1970s with composite multiple coaxial cables; the latter were subsequently themselves displaced from the long transmission link by optical fiber cables in the 1980s.

In North America, at the present, there are no multipair twisted-wire toll area cables in use; however, this is in contrast with the situation that prevails with exchange area cables serving the local subscriber loop. Trunk and feeder cables may consist of multipair twisted-wire polyolefin or polyethylene insulated PIC cables, having sizes of 900 pairs or less [90]. These cables may be filled with a jelly compound to prevent water ingress when buried or have an air-type core when used in aerial applications. The filling compound may be either petroleum jelly or an extended thermoplastic rubber (ETPR). Telephone cables, which are installed in ducts or directly buried, carry a protective polyethylene jacket and a metallic armor that may consist of aluminum, steel, and occasionally copper sheaths. In addition to providing mechanical protection, the metallic sheaths constitute effective electrostatic shields against induced voltages associated with lightning and, if properly sealed, prevent water ingress into the cable to avert increases in the mutual capacitance and possible shorts. At the distribution end, the service drop wire or distribution wire to the subscriber consists of two to six pairs; the buried distribution wire, in addition to a jacket, also carries a protective armor.

Figure 1.28 depicts schematically a twisted-wire pair telephone cable network serving a local loop. Since the late 1970s, the local loop networks were gradually converted from analog to digital carrier systems, which resulted in a substantial reduction in the number of twisted-wire pairs required in feeder cables. With digital transmission, the maximum length of a twisted-wire line usable without repeaters is contingent upon the

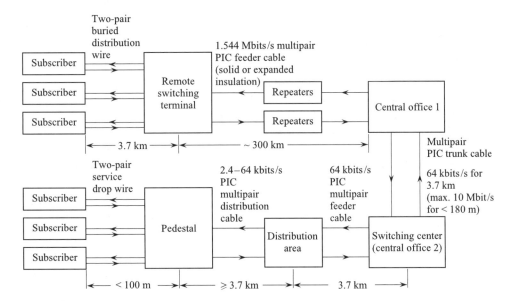

Figure 1.28 Multipair twisted-wire cables in the local loop.

gage of the copper conductor and the required data rate of transmission (cf. Chapter 15). For example, at a data rate of 2.4 kbits/s, the usable range of a 24 AWG wire pair is approximately 17 km without any repeaters; however, this value falls to approximately 2 km at a data rate of 1.544 Mbits/s. Most existing local loops provide data rate capacities up to 64 kbits/s without the use of repeaters for distances between 3.7 and 5.5 km between the central office and the distribution terminal. If repeaters are placed approximately 1.6 km apart along the feeder cable, then the distance between the central office and a remote switching terminal may be extended to several hundred kilometers. This antiquated long-distance PIC cable arrangement permits to increase the data rate to 1.544 Mbits/s and a corresponding increase in the number of subscriber lines at the remote switching terminal. As indicated schematically in Fig. 1.28, one twisted pair is employed for transmission and another for reception. The same data transmission capacity may be obtained, using high bit rate digital subscriber lines without the use of any repeaters with feeder cable lengths between the central office and the remote distribution terminal placed less than 3.7 km away. It is also interesting to note that the twisted-wire cable local loop network may be employed to transmit digital encoded video signals, using the asymmetrical digital subscriber line technique involving digital video signal compression methods at a rate of 1.5 Mbits/s within a closed loop less than 5.5 km.

Telephone cables, which are employed in the interior of buildings, e.g., central offices, commercial buildings, and subscriber homes, are classified as inside cables as opposed to all cables exterior to the building structures, which are denoted as outside cables. There are a number of salient differences between the two categories of cables. Station cable is a particular type of inside cable, since it is designed for both interior and exterior application [91]. It serves residences and consists of either two or four wire pairs with 22 or 24 AWG copper conductors as opposed to the normally larger inside cables, which are utilized in large commercial buildings and contain between 25 and 600 pairs of 24 AWG wire. Inside cables must comply with certain fire hazard regulations. Conductors of the wire pairs of inside plenum cables, which are installed in air circulating spaces (*plena*), are insulated with flame-retardant thermoplastics such as fluorinated ethylene propylene (FEP/Teflon) or ethylene chlorotrifluoroethylene (ECTFE/Halar). They may be extruded over the wire in either solid or expanded state as is also done with polyethylene or polyolefin insulation for cables for which the flame retardancy requirements are less stringent. Because of its low cost and flame-retardant characteristics PVC is also used as an insulant, notwithstanding its higher dielectric constant and dielectric loss. Table 1.9 compares the properties of the various

TABLE 1.9 Properties of Electrical Insulating Materials Used in Multipair Twisted-Wire Telephone Cables

Insulation	Operating Temperature (°C)	Dielectric Constant	Oxygen Index (%)	Volume Resistivity (Ω-cm)
PE	60	2.3	18	10^{15}
ECTFE	150	2.5	64	10^{15}
FEP	250	2.0	93	10^{18}
PVC	60	3.7–3.9	28–41	10^{14}

Source: After [91].

insulating materials utilized on twisted-pair telephone cables. For inside cables, flame retardant jackets must be used and, as the oxygen index values in Table 1.9 indicate, FEP is the best flame-resistant material available albeit the most expensive.

A very large volume of twisted-wire pair cable is now used in LANs. These networks comprise communication systems with a specific topology that do not form part of the publicly switched telephone systems, though they are usually interconnected with them for outside communication. They are found in commercial buildings, manufacturing sites, military installations, laboratories, campuses of universities, airports, etc. One such system, originally introduced between 1984 and 1989, is the IBM Token Ring network, whereby unidirectional data is transmitted either at a 4- or 16-Mbits/s rate, using the protocol of a token, i.e., digital code word passing between data stations, whereby the station with the token is the one that is permitted to transmit [92]. The transmission medium may consist of either shielded or unshielded twisted-wire pair cables. With shielded twisted-wire pair (STP) cables, the distance between workstations may extend up to 200 m; if unshielded twisted-wire pairs (UTP) are employed, then the distance between the workstations is decreased and the number of the interconnected workstations is reduced from 260 to 72 [92]. Moreover, the error rate in the transmitted data is adversely affected in the absence of shielding. Table 1.10 gives typical maximum attenuation and near end cross talk (NEXT) values as a function of frequency for a two-pair shielded twisted-pair cable with a flame-retardant PVC jacket, which is suitable for use in the IBM Token Ring network. Flame-retardant expanded PE or expanded FEP insulation is used over a 22 AWG bare copper conductor. Each pair is individually shielded with an aluminum/polyester foil, followed by an overall tinned copper braid shield. It should be mentioned here that these high-performance metallic conductor cables are also acceptable for use between the terminal boxes and the workstations with the 100-Mbit/s FDDI (fiber distributed data interface) Token Ring network. Uniform spacing of conductors and their twisting into pairs and prevention of

TABLE 1.10   Electrical Characteristics of High-Data-Rate STP and UTP5 Cables

Frequency (MHz)	STP 150 Ω, Two-Twisted-Pair Cable with Flame-Retardant Foam PE or FEP Insulation		UTP 100 Ω, Four-Twisted-Pair Cable with Polyolefin Insulation		
	Maximum Attenuation (dB/m)	Minimum NEXT (dB)	Maximum Attenuation (dB/m)	Minimum NEXT (dB)	Minimum Return Loss (dB)
4	2.2	58	3.7	65	23
8			5.3	61	23
16	4.4	50	7.6	56	23
25			9.5	53	22
62.5			15.5	47	21
100	12.3	38.5	17.9	44	21
155			25.3	41	21
200			29.1	40	18
300	21.4	31	36.6	37	18
310			37.3	37	17
350			40.0	36	17

Source: Belden Wire & Cable Company

looseness of twists leads to substantial improvement in the transmission line character-istics of twisted-wire pair cables, permitting their use at increasingly higher frequencies or data rates. This improvement is evident particularly with the more recently manu-factured high-grade unshielded twisted pair (UTP5) cables intended for high data rate cables, as classified in the ANSI/TIA/EIA 568-A, Category 5 standard. For compar-ison purposes, the electrical characteristics of a 350-MHz UTP5-type cable with a 24 AWG solid bare copper conductor insulated with polyolefin and covered with a low loss PVC jacket are tabulated in Table 1.10. The UTP5-type cables and their associated connector hardware are specified for data transmission rates up to 100 Mbits/s; with asynchronous transfer mode cell switching technology, their use may be extended to 155 Mbits/s.

From Table 1.10, it is apparent that the shielded twisted pairs, which are less susceptible to extraneous electrical noise and have typically a much more reduced radiated signal intensity, provide a significant improvement in the performance and signal transmission characteristics vis-à-vis the unshielded pairs. However, the lower cost of the latter presents the user with a compromising decision between economical and performance requirements. Shielded twisted-pair cables may also contain some unshielded pairs, but these are principally employed for voice communication circuits in the network. For the IBM Token Ring network, which operates either at 4 or 16 Mbits/s and which normally uses shielded twisted-pair cables, UTP5 should be more than adequate in view that even for the lower rated UTP3 and UTP4 cables the specified data transmission rate limits are already set at 16 and 20 Mbits/s. Figures 1.29 and 1.30 depict typical attenuation and NEXT characteristics of UTP3 and UTP5 cables, respectively.

The cable length between the multistation access unit is usually considerably less than the maximum permissible distance of 2 km in the IBM Token Ring network, which, as indicated, is a token-passing-type network; it consists of a closed-loop propagation medium in which hubs (automatic switches) or multistation access units (MAUs) are interconnected by means of a single 150-Ω shielded twisted-wire pair cable to form a ring as delineated in Fig. 1.31 [92]. To each hub are connected one or more workstations via shielded two-pair twisted-wire cables: one pair to receive the data and

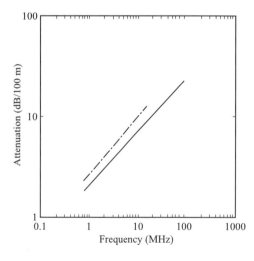

Figure 1.29 Typical attenuation characteris-tics of UTP3 and UTP5 type twisted-wire pair cables: — · —, UTP3 cable; —, UTP5 cable. (Courtesy of Belden Wire and Cable Company.)

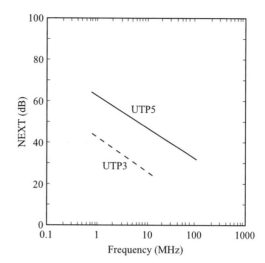

Figure 1.30 Typical near end crosstalk (NEXT) characteristics of UTP3 and UTP5 type of twisted-wire pair cables. (Courtesy of Belden Wire and Cable Company.)

another pair to transmit data. The hub, which acts as an automatic switch, senses when a station is on or off. If the workstation is off, then the switch shorts it out to maintain electrical continuity in the ring. If the workstation is on, the hub places the workstation in the ring circuit so that it can receive and transmit signals. As the token (digital code word) is circulated always in the same downstream direction within the ring, a station may capture it momentarily and then transmit the required data. Upon completion of the data transmittal, the workstation is obliged to pass the token along the loop whereby another workstation, which captures the token, may begin transmission. In this manner only one workstation can transmit data at a given point in time. In large cities with suburbs, in which commercial enterprises have a number of LANs operating over an extended area, the LANs are normally interconnected via optical fiber lines to form a metropolitan area network (MAN). Such networks may span across state, provincial, and international boundaries.

1.3.3 Coaxial Cables

The use of coaxial telephone cables attained its peak in the 1970s and early 1980s, when most of the long-haul telephone traffic in North America was carried via 75-Ω low-loss air-dielectric disk-spacer-type coaxial cables at a data rate of 274.176 Mbits/s. Thereafter their use rapidly declined as coaxial cables were replaced in the long-haul area by the less lossy optical fiber lines. Their 40-year dominance in the long-distance submarine telephone cable area came to a rather abrupt end in 1988 with the installation of the first transatlantic optical fiber telephone cable. However, the T4 (44.736 Mbits/s) telephone cable system functions with coaxial cables, and some coaxial cable may still be found in the local loop of the telephone system. But, at the present, the most extensive fields of coaxial cable usage continue to be those of CATV and certain LANs.

Community antenna television has become a well-established service that, shortly after World War II, has been distributing television programs via coaxial cables directly to the subscriber's premises. Initially, its operations were confined to rural regions where TV reception could only be obtained by means of antennas erected on high

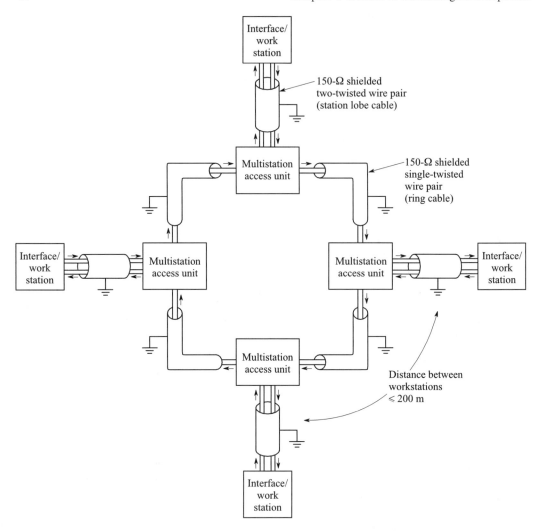

Figure 1.31 Simplified schematic of IBM Token Ring LAN network interconnected with 150-Ω shielded twisted-wire pair type cables [92].

aclivities or towers; eventually, however, in the 1960s, it began serving large urban populations, providing high-quality reception of channels otherwise not available in certain localities. Present CATV systems receive their television signals from geostationary communication satellites by means of approximately parabolically shaped dish-type antennas at the cable head (receiving end), from where the signals are transmitted and distributed to the subscribers.

The first CATV systems employed solid polyethylene insulated-type RG-11 coaxial cables with copper braid shielding obtained from military surplus, which became available after World War II [93]. The need to transmit video signals over relatively large areas within cities and their suburbs led to the development of low-loss coaxial cables; 75 Ω was selected for their characteristic impedance, because of its proximity to 77 Ω, which represents the minimum attenuation criteria in a coaxial cable (cf. Chapter 15).

Initially, CATV provided three channels from the low end of the very-high-frequency (VHF) band of 54 MHz up to 88 MHz. With the introduction of the ultra-high-frequency (UHF) TV broadcast band, the upper frequency limit was extended to 550 MHz and, finally, to 750 MHz as a result of further improvements in the electronic distribution equipment [93]. The upper bandwidth of 750 MHz, which already accommodates an analog video channel capacity of more than 100 channels, may be augmented by a further factor of 10 by means of digital video compression techniques. With the current wide-band analog equipment in the home, subscribers may receive 77 analog channels within the 54–550 MHz band; in the future, digital audio and video services will be offered within the 550–750 MHz band.

As already mentioned, the CATV system, because of the relatively longer transmission distances, required the development of low-loss coaxial cables. Low attenuation requirements necessitated the use of 75-Ω characteristic impedance cables in lieu of the traditional 50-Ω coaxial cables normally employed in instrumentation work; in addition, the losses are further decreased by utilizing foamed or expanded polyethylene insulation, having lower insulation conductance and permittivity than an equivalent solid-polyethylene coaxial cable. For trunk and feeder cables, a continuous aluminum sheath, which forms the outer conductor of the coaxial cable, is extruded over the foamed insulation. The diameter of the outside aluminum conductor of the cables is at least 0.5 in. and may even exceed 1.0 in., depending upon the size of the inner copper conductor, since the characteristic impedance is a function of the inner and outer conductor radii. However, the size of the inner conductor must be sufficiently large to carry the current to provide power to the amplifiers along the coaxial line. The aluminum sheath conductor is rigid and minimum bending radii are specified to prevent sharp bends, which may cause changes in the impedance along the coaxial line, e.g., an aluminum sheathed coaxial cable of 0.75 in. outside diameter has a minimum bending radius of approximately 7.5 in. The aluminum-sheathed coaxial cables, if placed in direct burial, must be protected against corrosion; this is accomplished by an extruded polyethylene covering over the Al sheath/shield.

The smaller diameter feeders on distribution cables, which branch off the larger diameter trunk cables, carry the TV signal to the local area of the cable TV subscribers. The coaxial drop cable, from the coaxial distribution cable to the service entrance of the dwelling, is a flexible cable and it may differ in construction depending upon the length of the drop distance. To maintain complete shielding of a flexible coaxial drop cable, aluminum-coated polymer tapes are frequently employed. Since the drop cable is most frequently aerially suspended, a copper-coated steel inner conductor is employed to provide built-in strength for the cable; also additional reinforcement may be obtained by means of a galvanized steel messenger wire. The latter is incorporated in the jacket such that the cross section of the overall cable assumes the form of a figure of 8. Coaxial service drop cables that are buried have foils applied over the shielding braid to prevent moisture ingress and protect against corrosion. These types of construction are illustrated in Fig. 1.32. All shown cable specimens have an identical construction, except that the coaxial cable, portrayed in Fig. 1.32(c), which is intended for direct burial applications, has a low-molecular-weight polyethylene grease film applied under its jacket to prevent water ingress and subsequent aluminum shield corrosion should mechanical damage of the jacket take place; a low-density polyethylene jacket is employed in lieu of a PVC jacket to provide a softer and more flexible cable at low temperatures. The laminated shielding tape, which consists of a polyester film

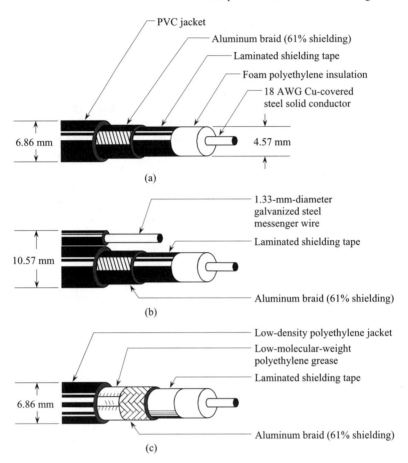

Figure 1.32 CATV type RG-6/U 75-Ω coaxial drop cables (a) aerial cable, (b) aerial
cable with reinforcing messenger wire, and (c) cable for direct burial.
(Courtesy of Belden Wire and Cable Company.)

sandwiched between two aluminum foils with an adhesive on one side of one of the
aluminum foils to ensure bonding to the insulation, is applied longitudinally over the
insulation. This ensures effectively 100% shielding over the coaxial cable, which renders
the cable less susceptible to extraneous interference and reduces as well the radiated
interference from the cable itself. In this respect, the drop cables are shielded as effectively as the trunk and feeder coaxial cables with the continuous but less flexible extruded
aluminum sheath. The coaxial drop cables have a minimum structural return loss (SRL)
of 20 dB; their velocity of propagation is equal to 82% of the speed of light, which is
essentially the same as that in the feeder and trunk cables, whose insulation also consists
of foam polyethylene.

The CATV tree topology network originates at the head end of the trunk cable,
where the encryption equipment is located and the TV signals that are received from a
satellite by means of a quasi-parabolic dish antenna are amplified prior to their transmission along the rigid foam polyethylene insulated aluminum-sheathed coaxial cable
(cf. Fig. 1.33). The outside diameter of the aluminum sheath is ≥ 0.75 in. and may in
some instances even exceed 1.0 in. If direct buried, the coaxial trunk cable is protected

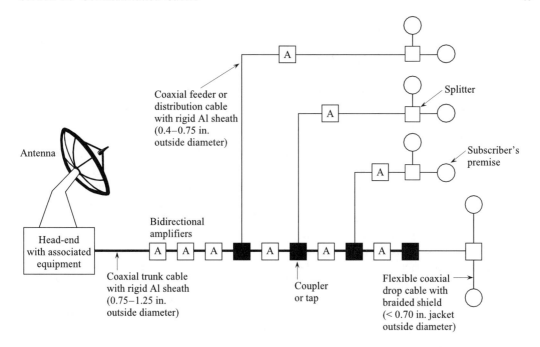

Figure 1.33 Simplified schematic diagram of CATV system.

against corrosion by means of a polyethylene jacket. The trunk cable has amplifiers inserted along its length at approximately 0.5-mile intervals, up to a maximum of 40 amplifiers [93]. By means of equalizers these amplifiers compensate for the increased attenuation at the more elevated frequencies (cf. Table 1.11). The signal levels are maintained using automatic gain control in the amplifiers; an ac current is supplied along the conductor to meet the power requirements of the amplifiers. The feeder or distribution cables include also amplification equipment, and as mentioned previously they are similar in construction to the trunk cables but have a substantially smaller aluminum conductor outside diameter, usually between 0.50 and 0.75 in. The flexible coaxial drop cables at the customer end are of the type shown in Fig. 1.32. The trunk cables may have lengths between a few miles up to approximately 30 miles with an attenuation range extending from 0.7 to 1.2 dB/100 ft at 300 MHz for the coaxial

TABLE 1.11 Attenuation Characteristics of CATV Drop Cables

Frequency (MHz)	Attenuation (dB/100 ft)	Frequency (MHz)	Attenuation (dB/100 ft)
5	0.45	400	4.21
55	1.53	450	4.47
211	3.02	550	4.95
270	3.44	750	5.83
300	3.64	870	6.30
350	3.93	1000	6.78

Source: Belden Wire and Cable Company.

cables aluminum sheath outside diameters ranging between 0.75 and 1.0 in. [93]. The shorter feeder cables, with outside diameters between 0.4 and 0.5 in., have attenuation values between 1.2 and 2.0 dB/100 ft; the still smaller 0.25-in.-outside-diameter drop cables are usually characterized by attenuation values in the range of ca. 3.5–6 dB/ 100 ft and a nominal pulse delay of 1.2 ns/ft. Table 1.11 gives the attenuation characteristics for frequencies between 5 MHz and 1 GHz of the CATV coaxial drop cables delineated in Fig. 1.32.

In a CATV system, two transmission paths are required: one for the inbound and another for the outbound signal. The outbound signal contains the TV programs for the subscriber, whereas the inbound signal (within the subband of 5–30 MHz) is used to obtain information from the subscriber end such as that on program billing [93]. This may be achieved by means of either a dual-cable configuration or a split configuration; the former technique [94], because it entails the use of two cables, presents an increased cost. The split configuration, delineated schematically in Fig. 1.33, necessitates the deployment of bidirectional amplifiers, which pass the lower frequency inbound signals and the higher frequency outbound signals with respect to the head end along a single coaxial cable. The frequency band of the outbound signals in most current CATV systems is within 50–550 MHz as opposed to 40–300 MHz, which characterized the earlier systems [94]. The splitters, inserted between the distribution and drop cables and shown in Fig. 1.33, divide the signal power in half, i.e., each subscriber receives half of the power from the distribution cable. Thus, a four-way splitter arrangement would permit the connection of four subscribers. The couplers or taps between the trunk and feeder or distribution cables simply divert a fixed amount of signal power from the trunk cable into the feeder cable.

The use of bidirectional amplifiers along a trunk cable in a CATV system adds to the maintenance costs of the overall system. Accordingly, in some CATV systems the coaxial trunk cables have been replaced with optical monomode fiber trunk cables, which can be operated without any amplifiers over distances up to 30–40 km due to their intrinsically low attenuation of ca. 0.4 dB/km at a wavelength of 1.31 μm. However, with such hybrid fiber–coaxial cable systems, the trunk portion of the cable must now have two separate optical fiber cables i.e., one for the inbound and another for the outbound signal [93].

A local area network in which extensive use of coaxial cable is being made is the Ethernet, which was introduced first by Xerox in the 1970s [95, 96]. Although recent versions of the Ethernet are now utilizing twisted-wire pairs, there are many coaxial cable connected Ethernets in service considering that the Ethernet is by far the most prevalent and popular LAN deployed because of its high speed, low cost, and widely available software and peripherals. The Ethernet topology is specified in the Institute of Electrical and Electronics Engineers (IEEE) 802.3 standard, with its protocol based on the carrier sense multiple-access with collision detection (CSMA/CD). The workstations in the Ethernet may be connected via interfaces to a coaxial cable bus as depicted in Fig. 1.34. The individual workstations monitor continuously the data transmission on the coaxial cable bus and transmit only when the cable is inactive. Should two or more workstations decide to transmit simultaneously, then this collision will be detected and the workstations will then attempt to retransmit subsequently their data at some randomly selected time interval [95].

The Ethernet employs coaxial cable that may be wired either with the so-called thick or thin coaxial cables, with both media capable of providing a maximum data

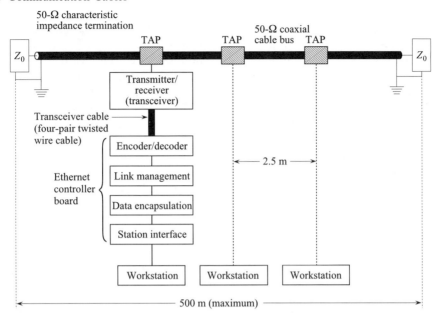

Figure 1.34 Ethernet topology with thick coaxial cable bus (IEEE 802.3, 10 Base 5) (after [94]).

transmission rate of 10 Mbits/s. The thicker or larger outside diameter coaxial cables, with a correspondingly larger conductor, can support a bus length of 500 m as opposed to the thinner or smaller outside diameter coaxial cables whose length is limited to 200 m [94]. The IEEE 802.3 standard refers to these two types of coaxial cable installations as 10 Base 5 and 10 Base 2, respectively; the term Base denotes a base-band system, i.e., because in base-band transmission the signal is sent without any modification, the system may be used either for digital or analog transmission. Both the thin and thick coaxial cables have a characteristic impedance of 50 Ω, since with digital signals the 50-Ω cable is preferred because it is less susceptible to signal reflection introduced by the insertion capacitance associated with the taps (interfaces); 50-Ω coaxial cables are also less affected by low-frequency electromagnetic interference than are the less lossy 75-Ω coaxial cables, which constitute the medium of choice for broadband applications in CATV systems [94].

The coaxial cable bus is terminated in its characteristic impedance of 50 Ω at both ends to prevent data interference due to reflections from the two cable ends. The workstations are connected to the coaxial cable bus via taps, with the separation between taps being fixed at 2.5 m to prevent phase addition of reflections between taps [94]. The maximum number of workstations is determined by the number of taps, which is usually set at approximately 100. The transceiver shown at the taps in Fig. 1.34 senses the presence of signals in the coaxial cable bus, recognizes whether there is a collision between two signals, and, respectively, transmits and receives signals to and from the coaxial cable bus in accordance with the CSMA/CD protocol. An important function of the transceiver is that of ground isolation between the signals of the workstation and those on the coaxial cable bus to avert differences in ground potential that may result in large cable shield currents, that may introduce noise as well as pose safety-related problems [94]. The transceiver cable is a specially designed four-pair shielded cable, which connects the Ethernet controller board to the transceiver and supplies power to the transceiver; the controller board's intelligence circuitry regulates the communication over the Ethernet. Although this section is devoted specifically to the discussion on coaxial cables, the importance of the transceiver or attachment unit interface drop cable in the Ethernet warrants its description here; it is depicted in schematic form in Fig. 1.35. It comprises four twisted-wire pairs with tinned copper conductors insulated with polypropylene, each pair shielded with a reinforced metallic foil, covered overall with a polyester tape, followed by a metallic foil–plastic composite tape and then a copper braid of 92% shielding effectiveness. The overall cable contains a 24 AWG stranded tinned copper drain wire, and the cable is protected by a PVC jacket. One twisted-wire pair in the transceiver cable has stranded tinned 24 AWG copper conductors and supplies power to the transceivers. The remaining three pairs with 28 AWG conductors are 78-Ω signal cables terminated in their characteristic impedance at both the transceiver and the network card of the personal computer (PC); one pair is employed for data transmission and another pair to receive data. The third pair is utilized either to transmit 10-MHz signals to the workstation whenever the transceiver detects a collision of data on the coaxial cable bus or a short burst of the 10-MHz signal to indicate the end of data transmission by a workstation. The transceiver itself is attached directly onto the coaxial cable bus by means of a sharp penetrating clamp that makes direct contact with the center conductor of the coaxial cable bus.

The so-called thin coaxial cable Ethernet uses smaller diameter coaxial cables, which are easier to manipulate and install. As a consequence, notwithstanding the reduced length of the thin coaxial cable bus (200 m) as compared to that of the thick coaxial cable bus (500 m), the use of the thin coaxial cable Ethernet is much more prevalent. Figure 1.36 depicts a typical thin Ethernet-type coaxial cable, which is specifically designed for use on the Ethernet coaxial cable bus. It consists of a 20 AWG tinned standard copper interior conductor insulated with foam polyethylene, shielded with a laminated Al-polyester-Al tape with a heat-sensitive adhesive applied on one of its sides, followed by a tinned copper braid of 93% shielding effectiveness and a PVC jacket. This particular cable is rated for a maximum temperature of 80°C. Cables rated for higher temperatures (150°C) use FEP (fluorinated ethylene-propylene, Teflon) insu-

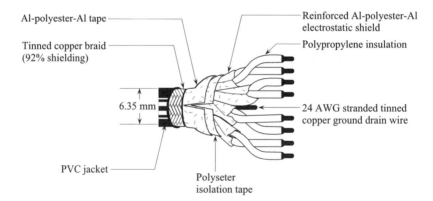

Figure 1.35 Four-pair twisted-wire transceiver cable; the stranded tinned copper conductors of one pair are 24 AWG, while the three remaining pairs of 78-Ω impedance have 28 AWG conductors. (Courtesy of Belden Wire and Cable Company.)

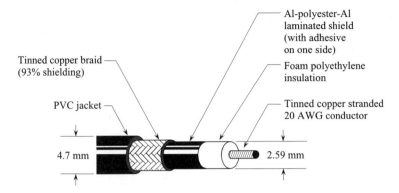

Figure 1.36 Ethernet thin coaxial bus cable, IEEE 802.3, 10 Base 2. (Courtesy of Belden Wire and Cable Company.)

lation and fluorocopolymer jackets. The attenuation characteristics of the Ethernet thin coaxial cable are given in Table 1.12. Comparison with Table 1.11 shows that the attenuation values of the Ethernet thin coaxial cable are significantly higher than those of the Ethernet thick coaxial cable, which accounts for the shorter permissible length of the Ethernet thin coaxial cable bus.

Over the many years of extensive use since its introduction in 1970, the Ethernet has become the most popular of the LANs; it has continued to mutate into forms, that have increasingly become more cost effective principally as a result of the replacement of the coaxial cables by twisted-wire pairs. The earliest Ethernet using twisted-wire pairs, which was introduced in the 1980s, was confined to a low data transmission rate of 1 Mbit/s, though it still maintained a maximum bus length capability of 500 m [95]. This antecedent technology is covered under the IEEE 1 Base 5 standard. The extension of the Ethernet to a data rate of 10 Mbits/s for twisted-wire pair connection entails the use of UTP3, 4, or 5 grade twisted-wire pair cables; this twisted-wire pair system is covered by the IEEE 10 Base T standard, which significantly reduces the distance capability of the network to a length of 100 m. The high data rate standard IEEE 802.3, 100 Base X defines a 100-Mbit/s network, which uses category UTP5 wiring and is limited to a distance of 150 m [96]. Another 100-Mbit/s capacity network is covered by the IEEE 802.12, 100 VG standard and uses HP (Hewlett-Packard)

TABLE 1.12 Attenuation Characteristics of a 50-Ω Ethernet Thin Coaxial Cable

Frequency (MHz)	Attenuation (dB/100 ft)	Frequency (MHz)	Attenuation (dB/100 ft)
1	0.43	100	4.2
2	0.60	200	6.1
5	0.90 (0.99 max.)	400	8.9
10	1.30 (1.40 max.)	700	12.1
20	2.00	900	13.9
50	3.00	1000	14.8

Source: Belden Wire and Cable Company.

technology; it is similar to the 100 Base X network in that a four-pair UTP3 wire system is employed in lieu of a two-pair UTP5 wire system in order to permit the higher data transmission rates with lower grade twisted-wire pairs. However, the maximum permissible bus length is consequently reduced from 150 to 100 m.

The 10 Base T, 1 Base 5, 100 Base X, and 100 VG networks, all of which use twisted-wire pairs, have a physical star-type topology; i.e., each workstation is connected directly via a twisted-wire pair cable to a hub or concentrator, which either simply connects the wiring together and passes the signals between the workstations (passive hub) or regenerates the signals prior to passing them (active hub) [95, 96]. Thus, if a problem develops with the cables of a given workstation, it only poses difficulties with the specific workstation involved and does not affect the performance of the other remaining workstations. However, the connections of the hub are configured logically so that each workstation forms part of a single logical bus as in the case of the coaxial Ethernet bus [92]. Thus, the maximum given twisted-wire cable lengths correspond to the cable lengths between the workstation and the hub as opposed to the actual length of the coaxial bus in the Ethernet wired with coaxial cable. A 10-Mbit/s optical fiber Ethernet (10 Base F) allows hub-to-workstation optical fiber cable lengths up to 4 km, thereby permitting the interconnection of workstations situated in different buildings spaced up to several kilometers apart.

Another important application of coaxial cables, dating from the 1950s has been in TV broadcast network studios. These cables bear considerable similarity to CATV cables in that they have a characteristic impedance of 75 Ω and use foam insulation, though where more flexibility is required, solid-polyethylene insulation is employed; in the solid-dielectric construction, the velocity of propagation is reduced to 66% of that of light in free space. The high-performance requirements for these studio cables usually necessitate the use of dual shields. Since the earlier mode of transmission of color video signals entailed the transmission of the picture information in analog form while the synchronizing signal information was in digital form, the design of video cables evolved into a form that was suited both for analog and digital signals. As a consequence, following the introduction of digital broadcasting in the early 1980s, coaxial cables have become an important transmission medium for digital video signals. Figure 1.37 portrays a typical coaxial video cable construction rated for a maximum temperature of 80°C which has a dual shield consisting of aluminum foil–polyester–aluminum foil

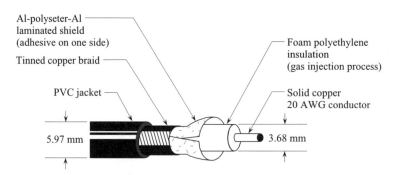

Figure 1.37 A 75-Ω precision coaxial video cable with double shield. (Courtesy of Belden Wire and Cable Company.)

laminates with a 100% shielding effectiveness followed by a copper braid of 95% shielding effectiveness. The attenuation characteristics of this cable are given in Table 1.13; its velocity of propagation is typically 83% of that of the speed of light, a value characteristic of foam-type polyethylene insulation.

1.3.4 Optical Fiber Cables

Optical fiber systems have several decisive advantages over the usual type of metallic conductor communication cables; these advantages reside primarily in their unusually high bandwidth capacity and their nonmetallic construction that renders them immune to electromagnetic interference and dangerous fault currents in the vicinity of electrical power lines. Although present coaxial cable systems may have bandwidths up to 1 GHz and additional bandwidth improvements resulting from the use of multiplexing and coding techniques, the bandwidths inherent with current commercial optical fiber cables, using modulated light transmission, have already achieved the 40-GHz level.

An optical fiber cable comprises a fiber core, which consists of silica, glass, or plastic material that may be clad with silica, glass, or plastic having a slightly lower index of refraction to ensure that the light waves are propagated along the core. Thus, the optical line acts essentially as a waveguide along which the optical signal is transmitted; the signal is launched by means of pulse-code-modulated LED device or laser sources and received by means of semiconductor light detectors (photodiodes). Attenuation and pulse dispersion are the two prime parameters affecting information data transmission rate in optical fiber cables. Several factors influence attenuation losses: (i) material absorption losses caused by a variation of the index of refraction of the material as a function of wavelength, e.g., these losses may arise in glass fibers from the presence of any transition metal ionic impurities and OH^- ions; (ii) material scattering losses as a result of density changes and structure variations within the fiber material itself; (iii) any fiber bending losses; which give rise to radiation losses; and finally (iv) waveguide scattering losses, which appear because of minute irregularities at the core and cladding interfaces. Protection against impurity ingress into the fiber's core and cladding is provided by a plastic coating applied over the cladding that also imparts mechanical strength to the fiber. The occurrence of pulse dispersion in optical fiber cables imposes limits on their data rate capacity and bandwidth. It is caused by intermodal, material, and waveguide dispersion mechanisms in the optical fibers [97].

TABLE 1.13 Attenuation Characteristics of a 75-Ω Precision Coaxial Video Cable, Whose Construction Is Described in Fig. 1.34

Frequency (MHz)	Attenuation (dB/100 m)	Frequency (MHz)	Attenuation (dB/100 m)
1	0.95	400	15.4
10	2.9	700	21.0
50	5.9		
100	7.9	900	23.9
200	10.8	1000	25.6

Source: Belden Wire and Cable Company.

As a result of the successful development of low-loss silica fibers at Dow Corning, optical fiber communication technology has become essentially a silica-based technology, and higher loss plastic materials are now only utilized for very short connection applications where high cable flexibility is required. Various forms of pure and doped silica are employed, with the silica cores having an index of refraction ≤ 1.48 and the cladding a correspondingly slightly lower value. The selected transmission wavelengths, within the infrared spectrum, in optical silica fiber cables are 0.85, 1.3, and 1.55 µm. The two latter are chosen because they are situated on the opposite sides of the major OH^- ion resonance absorption peak at two attenuation minima of the silica fiber. The value of 0.85 µm also evades some minor OH^- peaks, but falls in a relatively high attenuation region of the silica fiber over which the losses arise predominantly from Rayleigh scattering mechanisms. However, this transmission wavelength was used in the early multimode fiber telephone cables in the 1970s and early 1980s, because it coincided with the region of emission of the AlGeAs laser and LED light sources [98].

There are three types of silica optical fibers employed in communication cables, namely the multimode step index, multimode graded index, and single or monomode fibers. The multimode step index fiber was the fiber employed in the first optical fiber telephone cables installed in the 1970s (cf. Chapters 15 and 17). The relatively large core diameter of 50 µm in the early cables (with a cladding diameter of 125 µm) permitted easy termination and coupling of these fibers. Primarily broad spectrum modulated LEDs were utilized as light wave sources, and the cables consisted of relatively short spans operating at low data transmission rates at the wavelength of 0.85 µm. These cables were characterized by a relatively high attenuation of ≤ 4 dB/km, and their useful transmission distance was limited by their inherently high pulse dispersion caused principally by intermodal dispersion, which was intrinsically associated with the different length of paths of the various modes of light wave propagation within the multimode fiber as illustrated in Fig. 1.38. It is apparent from Fig. 1.38 that the number of modes within the multimode fiber is determined by the launch angle of the light signal with respect to the axis of the fiber; evidently, a light signal transmitted along the axis of the fiber will take the least time to reach the detector at the far end of the fiber. For a fiber of length ℓ_0, this minimum time of travel will be simply given by

$$t_{min} = \frac{\ell_0}{\upsilon}$$

$$= \frac{\ell_0 n_1}{c} \tag{1.1}$$

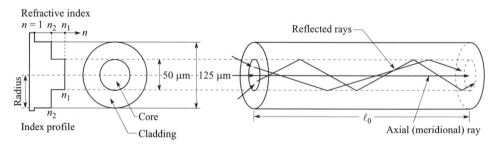

Figure 1.38 Transmission of light rays along a step-index multimode optical fiber.

where v is the velocity of the light signal in the fiber whose index of refraction is n_1 and c is the velocity of light in free space. The longest time of travel, t_{max} will occur when the light signal enters the optical fiber at the critical angle just sufficient for reflection to occur; if the length of the path of travel for this signal is equal to ℓ, then t_{max} is given by [99]

$$
\begin{aligned}
t_{max} &= \frac{\ell}{v} \\
&= \frac{\ell_0(n_1/n_2)}{(c/n_1)} \\
&= \frac{\ell_0 n_1^2}{cn_2}
\end{aligned}
\tag{1.2}
$$

where n_2 is the index of refraction of the cladding. Thus, a digitally transmitted square pulse along a step-index multimode fiber will be broadened by an amount

$$
\begin{aligned}
\Delta t &= t_{max} - t_{min} \\
&= \frac{\ell_0 n_1}{c}\left(\frac{n_1}{n_2} - 1\right)
\end{aligned}
\tag{1.3}
$$

Since pulse dispersion associated with intermodal dispersion is expressed in nanoseconds per kilometer length of cable, Eq. (1.3) may be rewritten in the form,

$$
\frac{\Delta t}{\ell_0} = \frac{n_1}{c}\left(\frac{n_1}{n_2} - 1\right)
\tag{1.4}
$$

for multimode step-index optical fibers; the quotient of $\Delta t/\ell_0$ is in the order of several tens of nanoseconds per kilometer, which imposes a limit on the useful transmission length of the fiber as well as the frequency of the input signal.

The idea of using a graded refractive index for the core to reduce intermodal dispersion in multimode optical fibers was first conceived in 1964 [100]. But more than a decade elapsed before such fibers were developed; this involved the doping of the silica core to produce a paraboloid-like distribution structure in the index of refraction in which the index diminished gradually from its maximum at the center of the core to a value equal to that of the cladding at the core-cladding interface. In this manner the speed of the light rays is reduced at and in the vicinity of the center of core as a result of the higher index along the shorter paths and increased toward the edge of the core due to the lower index of refraction along the longer paths. This results in all rays arriving almost simultaneously at the signal receiving end. Figure 1.39 illustrates schematically light wave propagation along a graded-index multimode fiber.

The grading of the core index in multimode fibers reduces the intermodal dispersion to well below 10 ns/km, which may even in some circumstances approach those of single-mode fibers. Both multimode and monomode fibers are subject to additional pulse dispersion as a result of material and waveguide dispersion. Waveguide dispersion is dependent upon the geometry of and the material comprising the waveguide as well as the frequency spectrum of the light source; that is the signal propagation velocity

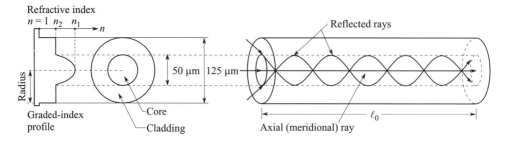

Figure 1.39 Transmission of light rays along a graded-index multimode fiber.

varies with the wavelength. The material dispersion, which arises as a result of the dependence of the refractive index upon wavelength (associated with atomic and molecular resonance absorption phenomena), is thus also dependent on the frequency spectrum of the light source. The effect of the light source upon the material and waveguide dispersions of the fiber is taken into account by expressing the dispersion caused by both phenomena in the units of picoseconds of pulse broadening per kilometer of fiber length per nanometer of light source spectral width (ps/km-nm) (cf. Chapters 15 and 17). Since the material and waveguide dispersion versus wavelength characteristics of doped silica fibers have slopes of different signs, a net zero dispersion point is found to occur on the wavelength scale and is evinced at 1.3 μm. It is for this reason that many of the early multimode fibers telephone cables, which have been initially placed in service for operation at 0.85 μm using the more economical but wider spectrum LED incoherent light sources, have been changed over to operate at the zero dispersion wavelength of 1.3 μm. Even for relatively short lengths of multimode fiber operating at 0.85 μm, the material and waveguide dispersions can give rise to pulse dispersions of nearly 100 ps/km-nm thereby imposing serious constraints on the data transmission rates that these cables are able to sustain.

The first installation of a long-distance multimode fiber telephone cable operating at the zero dispersion wavelength of 1.3 μm was effected in 1982 [98, 101]. In the application use was still made of an LED source to transmit over a cable distance of 205 km with repeaters placed a maximum of 10.3 km apart. The data transmission rate over this long-haul telephone cable was 34 Mbits/s. When the LED sources are substituted with coherent light-emitting laser sources having narrower spectral widths and more power output, longer lengths of multimode fiber cable may be used with correspondingly augmented repeater spacings. This approach was first demonstrated in 1987 with an inter-exchange trunk multimode fiber cable operating with a laser source at 1.3 μm over a distance of 42 km without repeaters [98, 102]. The system was capable of a transmission rate at 45 Mbits/s; the receiver consisted of a GaAs-FET preamplifier and *p-i-n* photodiode, which was characterized by a fast response time to the incident light energy.

Present graded-index multimode fibers, which are primarily employed in local area networks, have attenuation characteristics that in the limit may be made to approach very closely those of single mode fibers that have typically values of ~2.0 dB/km at 0.85 μm, <0.3 dB/km at 1.3 μm, and <0.2 dB/km at 1.55 μm. Commercially available graded-index multimode fibers have bandwidths up to 1 Gbit/s-km that may be extended to values as high as 6 Gbits/s-km when produced experimentally under highly controlled conditions in contrast to step-index multimode fibers whose bandwidth, as a result of severe intermodal dispersion, may be significantly below 50 Mbits/s-km. Note that the unit of bandwidth for optical fibers is expressed as a data rate or frequency-distance product, implicitly indicating that the bandwith is a quantity dependent upon the length of the fiber. For example, should a particular optical fiber of 1-km length have a bandwidth of 50 Mbits/s-km, then for a repeaterless distance of 10 km, its bandwidth would necessarily be reduced to 5 Mbits/s-km. Frequently bandwidth units for optical fibers are also stated in kilohertz per kilometer, megahertz per kilo-meter, gigahertz per kilometer, etc., with hertz substituting for the date rate unit of bits per second.

Commencing with 1981 concertive efforts were undertaken to evaluate the perfor-mance of high bit rate monomode optical fiber cables at the transmission wavelengths of 1.3 and 1.5 μm [98, 103–105]. The experimental work, which was directed toward the development of long-haul terrestrial and submarine monomode optical fiber cable systems, demonstrated that data transmission rates between 140 Mbit/s and 2 Gbit/s were readily attainable with monomode fibers with repeater spacings of 100 km [98, 103–109]. This led rapidly to the design and manufacture of long-haul monomode optical fibers, culminating in 1988 with the laying of the first transatlantic optical fiber submarine cable with a data transmission rate capacity of 280 Mbits/s at the wavelength of 1.31 μm (cf. Chapter 17).

Most of the early single-mode or graded-index monomode optical fiber telephone cables were designed for operation at the wavelength of minimum dispersion, i.e., 1.3 μm. Their operation at the attenuation minimum, occurring at the wavelength of 1.55 μm, was precluded because of the sizable material dispersion of 18 ps/km-nm manifest at the same wavelength. But with the development of dispersion-shifted single-mode optical fibers, whereby the refractive index profile of the fiber is modified in order to alter the waveguide dispersion behavior such that the wavelength of the material and waveguide dispersion cancellation is transposed from 1.3 to 1.55 μm, the dilemma has been resolved. There are still many older long-haul single-mode fiber cables operating at 1.3 μm, but the more recent installations of long-haul single-mode fibers are dispersion shifted (index profiled) for operation at 1.55 μm in conjunc-tion with narrow spectrum 1.55 μm laser sources and detectors (cf. Chapter 15).

In single-mode optical fibers, the prevailing intermodal dispersion that typifies the behavior of step-index multimode fibers, is eliminated by restricting the diameter of the core to ≤10 μm, so that the core size becomes too small to support a multimodal propagation process. As portrayed in Fig. 1.40, the single-mode fiber design also

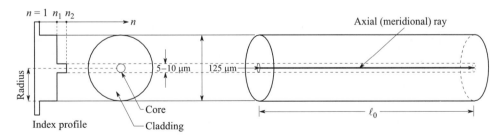

Figure 1.40 Transmission of light rays along axis of a single-mode fiber.

utilizes a step-index profile. The index of the single-mode fiber core may be typically ~ 1.460 (n_1) as compared with that of 1.457 (n_2) for the cladding [110]. By comparison, the corresponding values for both the step-index and graded-index multimode fibers are typically $n_1 = 1.470$ and $n_2 = 1.455$; i.e., the refractive index of the core, n_1, for both the step-index and graded-index multimode fibers is the same except that in the case of the graded-index multimode fiber n_1 represents only the maximum value of refractive index at the center of the core.

A single-mode or monomode fiber allows only one mode (the lowest order mode) to propagate at a given wavelength, the condition for this being that the radius of the core is to be less than twice the wavelength emitted by the laser source. Consequently, the core diameters of monomode fibers are by a factor of approximately 10–6 times smaller than those of multimode fibers. Optical fiber cables are characterized by their numerical aperture (NA), which is equal to $(n_1^2 - n_2^2)^{1/2}$; physically, it is a measure of the spread of a light beam emitted from the end of an optical fiber placed a distance d away from a screen mounted perpendicularly to the axis of the fiber. With this type of arrangement, it is defined by the term $2r/d$, where $2r$ is the diameter of the circular projection on the screen launched from the end of the fiber. The numerical apertures of multimode fibers are typically between 0.21 and 0.23, while those for single-mode fibers are at least a factor of two less. The value of the numerical aperture is important as it determines the coupling efficiency between the laser or LED light sources and the input end of the fiber. However, for long-distance transmission small values of NA such as those intrinsic to single-mode fibers are required to avert intermodal dispersion. Figure 1.41 depicts a typical single-mode optical fiber long-haul submarine cable. Note the extensive armor protection provided for the optical fibers, including the hermetically sealed copper tube to prevent moisture ingress into the cable. The cable is designed for operation at 1.550-1.565 μm and contains three single-mode fiber pairs (i.e., six fibers). Each pair is designed to carry eight channels, with each channel operating at 2.5 Gbits/s.

Virtually all of the long-distance multimode and monomode fiber telephone cables installed between 1980 and 1996 are operated using electronic repeaters in conjunction with electronic time division multiplexing (TDM). The repeaters optoelectronically regenerate and amplify the optical signals, which become attenuated and distorted as they travel along the optical fiber lines. The repeaters convert the optical signals into electronic signals, which are in turn amplified, reshaped, and retimed. The emerging signals, which have essentially a reconstructed form almost identical to the original electronic signals, are subsequently applied to drive a semiconductor laser whose output provides a train of optical pulses having the same magnitude and form as the original optical pulse at the input to the fiber line. Since the electronic repeaters add a very substantial expense to the optical transmission line, lower attenuation and pulse dispersion lines with longer repeater spacings are desirable. The improvement in the attenuation and pulse dispersion characteristics of optical silica fiber cables over the past 18 years has resulted in appreciable gains in the data transmission rate capacity of commercial optical fiber telephone cables as indicated in Fig. 1.42; the solid curve represents the gains made by cables operated using TDM, with a maximum capacity of 10 Gbits/s being attained in 1996 [89]. The isolated point of 40 Gbits/s for the year 1996 refers to cables operated using wave division multiplexing (WDM), which is analogous to the frequency division multiplexing (FDM) transmission method used with multipair twisted-wire cables, except that with optical fibers it is being carried

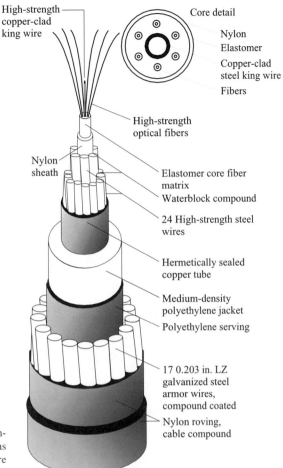

Figure 1.41 A 42-mm-outside-diameter single-mode optical fiber telecommunications submarine cable. (Courtesy of Simplex Wire and Cable Company.)

out in the optical frequency domain (cf. Chapter 15). The maximum data transfer rate capacity achieved thus far experimentally with WDM is placed at 2 Tbits/s [89]; the enormously large capacity potential with optical fibers using WDM is thus palpably apparent.

The early optical telephone transmission cables were characterized by their low transmission rate capacities, which in the case of step-index multimode fibers were limited by both their relatively high attenuation as well as their intermodal dispersion characteristics. With the development of graded-index multimode fiber cables, pulse broadening was reduced to an acceptable maximum range of values between 1.0 and 2.5 ns/km. These multimode cables, which had core and cladding diameters of 50 and 125 μm, respectively, were designed to operate in the temperature range between −40 and 60°C, being thus suitable for application in the Northern Hemisphere. Their attenuation values were still in the range of 4.0 dB/km at 0.85 μm and 2.5 dB/km at 1.3 μm, hence their useful length was essentially attenuation limited and relatively short repeater spacings of ∼12 km were required in pulse-code-modulated systems with laser sources. Increasingly higher transmission rate capacities were achieved following the

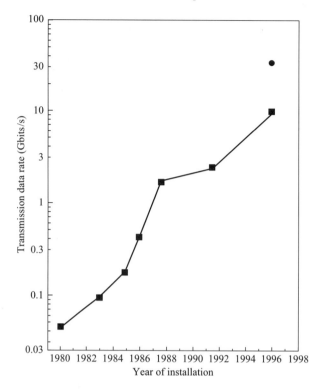

Figure 1.42 Growth of transmission capacity of commercial optical fiber telephone cables: ■, time division multiplexing (TDM) systems; ●, wave division multiplexing (WDM) systems (after [89]).

introduction of lower attenuation graded-index multimode silica fiber and single-mode silica fiber cables. The present single-mode optical fiber cables, which dominate the long-haul telephone cable field, operate at 1.3 μm and are characterized by attenuation values in the range of 0.3–0.4 dB/km. Since their pulse broadening due to material and waveguide dispersion at 1.3 μm is virtually equal to zero, their transmission capacity is still attenuation limited and repeaters must be employed approximately every 40 km. These systems, depending upon their length, can achieve transmission capacities of the order of 1 Gbit/s. It is interesting to observe that the capacity of the much longer transatlantic optical fiber cable, which also operates at 1.31 μm with a repeater separation of 70 km, is 280 Mbit/s per fiber pair (cf. Chapter 17). Nondispersion shifted single-mode fibers, which operate at a wavelength of 1.55 μm at which the reduced attenuation of the fiber is between 0.1 and 0.2 dB/km, may achieve data rates of 10 Gbits/s albeit at reduced repeater spacings. Since the chromatic (material and waveguide) dispersion at 1.55 μm is now finite and of the order of 18 ps/km-nm, the separation of the optoelectronic regenerative repeaters becomes, in addition to attenuation, also pulse dispersion or broadening limited at the wavelength of 1.55 μm. If the single-mode fibers are of the dispersion shifted-type (i.e., the chromatic dispersion at 1.55 μm is reduced nearly to zero), then the dispersion limitation becomes less stringent. Figure 1.43 compares the dispersion-shifted (index profiled) and the regular single-mode fibers, which are operated at wavelengths of 2 and 10 nm below 1.55 μm

[111–113]. It can be readily perceived from Fig. 1.43 that the transmission capacity per channel is greatly augmented in dispersion-shifted single-mode fibers, with further improvements affected the more proximate the fiber is operated to the dispersion-shifted wavelength of 1.55 μm. It is interesting to note that the 9000-km transpacific telephone cable installed in 1991, which operates at 1.55 μm, has a transmission capacity of 560 Mbits/s per fiber pair with repeaters placed at 110–120 km apart, which represents a substantial improvement over the transatlantic cable.

With the advent of erbium doped optical fiber amplifiers in the late 1980s [114–116], which operate in the wavelength region of 1.53–1.56 μm, large increases in the data rate transmission capacity of optical fibers are anticipated. The advantage of the optical amplifiers is that they can amplify simultaneously a large number of wavelengths in a wavelength division multiplexed system [116]. More detailed discussions on optical amplifiers are presented in Chapters 15 and 17, but a few cursory comments are in order here. The erbium-doped silica fiber amplifier essentially consists of a short length of monomode silica fiber (e.g. ~40 ft) in which the erbium ions are raised to a higher level by absorption of the light quanta from a powerful solid-state pump laser operating either at 0.98 or 1.40 μm. When a transmitted optical signal of wavelength 1.55 μm reaches the metastable erbium ions, the erbium ions when stimulated by a signal photon within the wavelength range of 1.53–1.56 μm fall to the ground state, emitting a coherent photon of identical wavelength, phase, and direction as that of the incident ray's photon, thus leading to amplification of the signal [89, 116]. A simplified schematic diagram of the erbium-doped fiber amplifier is depicted in Fig. 1.44 [111, 116]. The pump-signal multiplexer forms a low-loss wavelength selective coupler where from the pump light and signal light are introduced into the erbium-doped silica fiber. The purpose of the optical isolators is to prevent the occurrence of oscillations in the erbium-doped optical fiber amplifier in the presence of reflections. Although the erbium-doped silica amplifiers were developed in 1987, it was only in 1992 that the feasibility of transmitting a 5-Gbit/s channel over a distance of 9000 km without any optoelectronic signal regeneration along the line was demonstrated [89]. A problem, intrinsic to optical amplifiers, is that the transmitted signals are inevitably amplified along with all the incident noise by each successive optical amplifier inserted along the

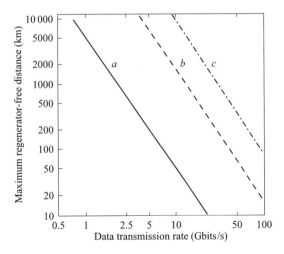

Figure 1.43 Chromatic dispersion-imposed maximum allowable nonregenerative cable distances with their corresponding single channel data rates for single-mode optical fibers: (a) single-mode fiber at the operating wavelength of $\lambda_0 = 1.55\,\mu m$ and dispersion-shifted single-mode fiber operated at (b) $(\lambda_0 - 10\,nm)$ and (c) $(\lambda_0 - 2\,nm)$ (after [111–113]).

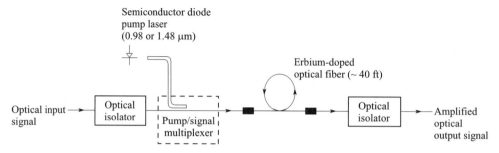

Figure 1.44 Simplified schematic diagram of erbium-doped silica fiber optical amplifier (after [111, 116]).

length of the fiber. In this respect the system behaves somewhat similarly to the wideband analog system used with copper wire cables. Here, however, the noise is generated in each optical amplifier due to spontaneous emission effects; thus each optical amplifier generates its own noise, which in turn is amplified by other optical amplifiers in tandem along the optical fiber lines. It is thus apparent that over a long optical transmission line, in addition to economic considerations, it is equally desirable to limit the number of optical amplifiers in order to mitigate noise accumulation effects.

The design, construction, and manufacturing procedures, which are utilized with optical fiber cables, have benefited greatly from the experiences acquired over more than a century with metallic conductor communication and power cables. The overall rap of the inner portion of the cable or core and the sheath construction of optical fiber telephone cables is either very similar or identical to that employed on metallic conductor telephone cables. The same situation prevails concerning armoring of the optical fiber cables (cf. Chapters 15, 16, and 17). However, in addition, optical fiber cables contain inner strength members of either steel or high-strength polymers for tension reinforcement of the cable. In many respects this approach is analogous to the use of steel-reinforced copper conductors in coaxial cables. As mentioned already, the fibers are further protected from compressive loads by placing them in the form of a bundle or ribbon matrix configurations within loose buffer tubes filled with a gel compound; alternatively, the individual fibers may be incorporated within a tight buffer coating, consisting of an extruded polymer layer. The tight buffered fibers, when stranded, result in a highly flexible cable construction of high strength and crush resistance. However, the typical long-distance multi-optical-fiber terrestrial telephone cables utilize a stranded loose tube (Siemens) or a stranded slotted core (Northern Telecom) or, alternatively, an enlarged central loose tube core (Lightpack cable, AT & T) design (cf. Chapters 15 and 17).

The use of optical fiber cable communication links by power utilities became popular in the late 1970s; this relatively early interest was primarily stimulated because of their immunity to electromagnetic interference. The optical fiber cables, used as communication links between power substations and as inner communication lines of ground wires along overhead power transmission lines, are usually of the four optical fiber core design—two optical fibers for the outgoing signals and two optical fibers for the incoming signals. An example concerns an installation carried out in 1981 at two transmission line substations [117]. In this installation, an ITT-developed optical cable, having the cross section shown in Fig. 1.45, was used to span a total distance of 1.6 km.

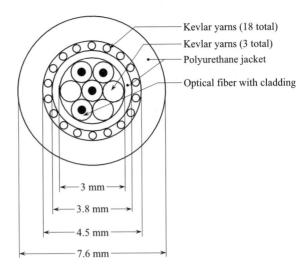

Figure 1.45 Multimode ITT optical fiber cable with protective sheaths [117].

The multimode optical fibers consisted of a doped silica core of 50-μm diameter with an NA of 0.23, enclosed in a borosilicate glass cladding giving a resulting overall diameter of 125 μm. To provide mechanical protection and to reduce microbending effects, a silicone color-coded coating was deployed over the four optical fibers within the cable. Additional mechanical protection was supplied by the use of Kevlar yarns and a double polyurethane jacket. The refractive index of the core was 1.48, and the optical fibers were designed for operation at a wavelength of 0.85 μm; the nominal attenuation of this early optical fiber cable was a relatively high value 6 dB/km.

Another interesting early application of graded-index multimode optical fiber cables occurred as inner communication links housed within bare aerial ground cables that are deployed as a lightning protection on overhead power transmission lines. This design developed by BICC in England is depicted in Fig. 1.46 [118, 119]. The overall

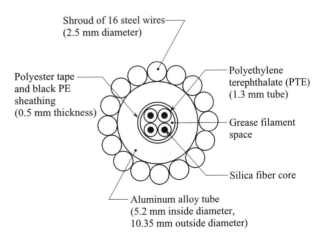

Figure 1.46 BICC optical fiber cable with a ground wire casing [118, 119].

cable shown was installed in 1981 between a substation and a measurement test station over a distance of 2.2 km as an overhead ground wire in a 735-kV transmission line. The lightning protection is attained via the grounded aluminum alloy tube conductor, and the cable support strength is provided by the steel wires. Thus the concentrically disposed optical fiber wires are completely protected from lightning. The cable is characterized by an attenuation value of approximately 2.7–3.7 dB/km for each of the four color-coded optical fiber conductors [118]; it was intended for operation at wavelengths between 0.80 and 0.90 μm [119].

In relation to LANs, we have seen from the previous discussions that twisted-wire pair cables can cost effectively support data rates as high as 100 Mbits/s; however, this can be accomplished only over relatively short lengths. When the required cable lengths are large, then the implementation of optical fiber cables becomes a necessity. For example, optical fiber cables may be used to connect copper-conductor-based LANs between buildings on a university campus or between different office sites of corporations in large cities. If the interconnecting distances are less than 10 km, then the lower cost graded-index multimode fiber cables are the medium of choice. Figure 1.47 portrays for illustrative purposes a typical multimode six-fiber cable that is commonly employed for indoor LAN applications to provide a high data transmission medium between mainframes and building distribution systems. The fibers have a standard 62.5-μm core diameter with 125 μm cladding covered by a coating of 900 μm outside diameter. The multimode fiber has an NA of 0.275; its bandwidth is, respectively, 160 and 500 MHz-km at the transmission wavelengths of 0.85 and 1.30 μm; the corresponding maximum attenuation values are 3.4 and 1.0 dB/km, respectively. The permissible minimum bend radii are 270 mm during installation and 180 mm over the long-term service conditions. It should be observed here that following recent standardization, most multimode fibers are currently manufactured with 62.5 μm core diameters in lieu of the 50-μm core diameters that characterized the earlier multimode optical fiber

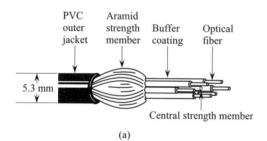

(a)

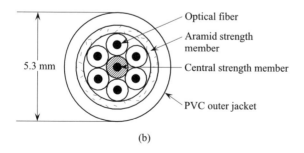

(b)

Figure 1.47 A 62.5-μm core six-fiber graded-index multimode optical fiber cable for indoor LAN connections between mainframes and building distribution systems: (a) profile view; (b) cross-sectional view. (Courtesy of Belden Wire and Cable Company.)

telephone cables. However, very recently, the 50-μm core multimode fiber has received increased attention, because its smaller value of NA vis-à-vis that of the 62.5-μm core fiber provides a greater transmission bandwidth [120, 121]. In addition the 50-μm core multimode fiber functions very well in conjunction with the recently developed vertical-cavity surface-emitting laser (VCSEL) at the wavelength of 0.85 μm.

Comparative bandwidth and corresponding maximum fiber cable length test data on 50 and 62.5-μm core multimode fibers has been recently compiled by the IEEE 802.3z Committee, concerning the backbone fiber interconnection of the newly planned Gigabit Ethernet Standard [121]. The results are presented in Table 1.14; it can be perceived that although at 1.3 μm both the 50- and 62.5-μm core multimode fibers have approximately equal bandwidths and maximum allowable interconnection fiber lengths, the 50-μm core multimode fiber is characterized at 0.85 μm by a larger bandwidth and, consequently, by a greater permissible maximum backbone link distance.

Since single-mode fiber cables with the associated laser or semiconductor diode light sources are more expensive, they are used in local area network applications only when greater data transmission rates and longer distances are involved. Figure 1.48 depicts a monomode optical fiber interbuilding cable containing six loose-type buffer tubes filled with a gel compound; there are six monomode fibers per tube. The cable is fully armored with a PE jacket and is suitable for direct burial; it is typically used to interconnect LAN networks for wide area network (WAN) applications. The buffer tubes are 2.5 mm in diameter, and the monomode fibers contained therein have respective nominal attenuation values of 0.45 and 0.35 dB at 1.31 and 1.55 μm, respectively; their corresponding maximum pulse dispersion values at the same wavelengths are, respectively, 2.8 and 18 ps/km-nm.

A local area network, which is specifically designed for optical fiber cables as the connecting media, is the FDDI (fiber-distributed data interface). It is essentially a high-speed Token Ring network defined by the American National Standards Institute in ANSI Standard X3T9.5, which specifies a ring with a data rate capacity of 100 Mbits/s that can support a maximum of 500 nodes or workstations with a separation of ≤2 km between the nodes [92, 122]. As in the Token Ring network, the workstations are connected in the form of a ring, but the connections are made with optical fibers, i.e., each station has an input fiber from the preceding station and an output fiber to the station in tandem. The FDDI loop contains in addition a secondary fiber in parallel with the primary fiber, whose purpose is that of a spare in the event that failure should occur in the primary fiber. In such circumstances, the primary and secondary fibers are joined to reconstitute the ring. The physical length of the ring itself may extend up to 200 km.

TABLE 1.14 Gigabit Ethernet Multimode Optical Fiber Bandwidth and Maximum Backbone Link Distance as a Function of Wavelength

Fiber Core Size (μm)	Modal Bandwidth (MHz-km)		Backbone Link Distance (m)	
	0.85 μm	1.30 μm	0.85 μm	1.30 μm
62.5	160	500	220	550
62.5	200	500	275	550
50	400	400	500	550
50	500	500	550	550

Source: From IEEE 802.3z Committee [121].

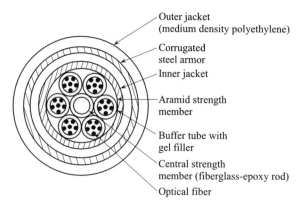

Outer jacket
(medium density polyethylene)

Corrugated
steel armor

Inner jacket

Aramid strength
member

Buffer tube with
gel filler

Central strength
member (fiberglass-epoxy rod)

Optical fiber

Figure 1.48 A 36-monomode optical fiber interbuilding trunk cable. (Courtesy of Belden Wire and Cable Company.)

Figure 1.49 delineates the topology of the FDDI network dual ring [92, 122]. The dual-attach station in the FDDI network dual ring is essentially a connection concentrator, which performs the function of the media or multistation access unit (MAU) in the Token Ring LAN network (cf. Fig. 1.31). It has two inputs and two outputs for the optical fiber cables, in contrast to the single input-output attach workstations, which are connected via the concentrators to the dual fiber ring. When the workstations are unactivated, they are bypassed by the concentrator. The dual-attach station remains always activated, since one of its prime purposes is to monitor continuously the performance of the optical fiber cables; should a fault be sensed in the primary optical fiber by the dual-attach station, the secondary fiber would be immediately placed into operation.

All single-attach stations comprising the ring have their inputs and outputs directly connected to a concentrator, which automatically maintains continuity of the ring path

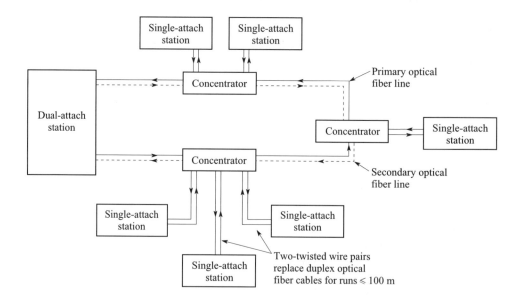

Figure 1.49 Simplified schematic of 100-Mbit/s dual-optical-fiber FDDI ring network (after [92, 122]).

irrespective of whether the machine is removed or turned off. Note that if the distance between the concentrator and a given single-attach station is less than 100 m, a lower cost two-twisted-copper-wire pair cable may be utilized in lieu of the duplex optical fiber cable. Adaptive circuitry would then form part of the concentrators, which would remain interconnected with the dual fiber backbone cable of the high-capacity FDDI long-distance ring. The concentrators also act as repeaters along the optical fiber ring over which data frames are transmitted in the form of a serial stream of bytes [92, 122]. The data is always transmitted in the same downstream direction, i.e., each station repeats the symbol stream, which it receives from an adjacent upstream station, and transmits it to its adjacent downstream station until the data frame reaches the station of origin and is removed from circulation. A new data frame is then initiated by the next station in possession of the token (digital code word) as in the case of the IBM Token Ring network.

Transmission along the FDDI dual ring is synchronized; accordingly, each node/workstation must synchronize its local clock to the received symbol stream from adjacent upstream node/workstation and decode the data. Thus, when transmitting the symbol stream to the adjacent downstream node/workstation, the node/workstation must combine its local clock signal with the data. The data in the FDDI dual ring is transmitted in the form of symbols, using an encoding process whereby four bits of data are transmitted along the fiber as a 5-bit symbol (e.g. 11160, 01001, etc.) corresponding to a cable data rate of 100 Mbits/s [92, 96]. The encoder employs a non-return-to-zero, invert on ones ((NRZI) waveform, whereby a 1 causes the LED source at the input of the optical fiber to switch states, that is from off to on or vice versa; while a 0 instructs the light-emitting source to remain fixed at its given state (on or off). However, the number of 0s in succession is limited to three in each of the five-bit symbols in order to limit the duration of the fixed-state period. The encoding/decoding procedure is illustrated schematically in Fig. 1.50; the flat cable shown is extensively employed in computer networks wherever parallel wire port connections are involved.

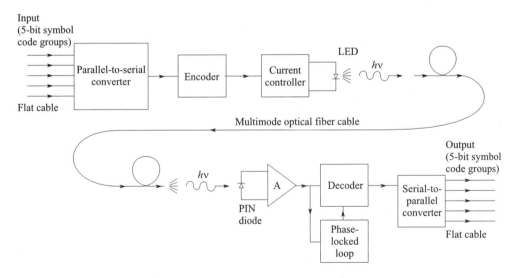

Figure 1.50 Schematic diagram of transmitter and receiver at two nodes of the FDDI network (after [92]).

The ANSI Standard X3T9.5 on the FDDI network specifies 62.5-μm core multi-mode fibers for distances ≤2 km and single-mode fibers for interconnection distances up to 60 km in length. A multimode duplex optical fiber cable depicted in Fig. 1.51 is suitable for short-distance applications; it is characterized by an NA value of 0.275 and has a core diameter of 62.5 μm with cladding and buffer diameters of 126 and 900 μm, respectively. It is designed for operation at wavelengths of 0.85 and 1.30 μm with corresponding maximum signal attenuation values of 3.4 and 1.0 dB/km respectively; its corresponding bandwidths at the same wavelengths are 160 and 500 MHz-km, respectively (cf. Table 1.14) It is intended for indoor LAN applications, but may be used also for outdoor installations when protection is provided in the form of ducts or conduits. When used as a cable between the concentrator and a single-attach station in the FDDI network, one fiber is employed for transmission and the other for the reception of data. For connections forming the actual dual ring in the FDDI network between the concentrators and the dual-attach station, one fiber acts as the primary fiber with the other fiber as the secondary or standby fiber (cf. Fig. 1.49). The optical fiber interconnections are made by means of a fixed shroud duplex (FSD) connector, which provides protection for the two signal transmitting and receiving fibers that are recessed within a shroud [92]. For long cable lengths dispersion-shifted monomode optical fiber is substituted for the multimode fiber, though the overall dual-fiber cable construction delineated in Fig. 1.51 remains essentially unaltered. Typical dimensions of the monomode silica fiber core, its cladding, and protective coatings are given in Fig. 1.52. At the transmission wavelengths of 1.31 and 1.55 μm, the single-mode fiber is characterized by attenuation values of 0.5 and 0.4 dB/km, respectively; the corresponding maximum chromatic dispersion values are 2.8 and 18 ps/km-nm, respectively.

In concluding our survey on communication cables, some general observations are in order, because over the last several decades the telecommunications transmission medium has undergone profound change. The changes were motivated by attempts to improve the quality of voice and video transmission and to accommodate increased

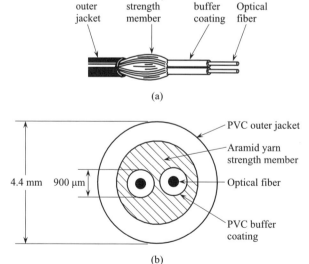

Figure 1.51 Duplex 62.5-μm core multimode optical fiber for use as a short-distance transmission medium in FDDI dual-fiber networks: (a) profile view; (b) cross-sectional view. (Courtesy of Belden Wire and Cable Company.)

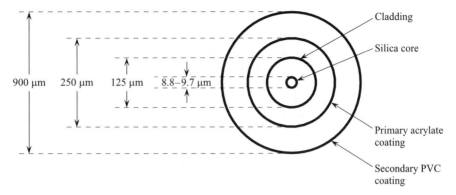

Figure 1.52 Dimensions of a monomode silica fiber with cladding and protective coatings. (Courtesy of Belden Wire and Cable Company.)

data transmission rates through digital transmission techniques and improved low-loss, wide-bandwidth cables. Long-distance communication and data transmission benefited greatly from the introduction of very low loss optical fiber cables, which rapidly displaced the metallic conductor cables used in long-distance transmission applications. However, the ever tenaciously cost-effective metallic conductor cables continue to be used for short-distance applications in local area networks and over a very substantial portion of the cable distribution network in the local telephone subscribers loop. The metallic cables in the local loop represent an enormous investment and various electronic techniques have been developed to enhance the signal transmission capacity of these in situ cables to meet the increased data requirements of the telephone subscribers' computer networks.

In 1984, the International Telegraph and Telephone Consultative Committee (CCITT) began work to establish standards for an integrated systems digital network (ISDN), which is to provide end-to-end digital connection over the existing telephone transmission and distribution network for voice and nonvoice services to which users may have access via standard multipurpose network interfaces. This will permit LANs to communicate directly over the telephone transmission system. Evidently, at present, the subscriber loop is an analog system and the digital signals from the subscribers' PCs must be transformed by means of a modem into analog signals for transmission over the local loop telephone cable network. The implementation of digital services in accordance with the ISDN plan is bound to bring the fiber cables closer to the telephone subscribers' premises. However, to exactly what degree this will affect the use of the traditional twisted-wire pair telephone cables in the subscriber loop is difficult to predict. The twisted-wire cable has proven to be very resilient, particularly for signal transmission over short distances. For those of us who were associated with the telecommunication industry in the 1950s, it may be recalled that in that epoch few would have conjectured that the twisted-wire pair cable, which was originally intended only for voice frequencies, would be tested at 600 MHz and be deployed for data transmission rates of 100 Mbits/s and, furthermore, in conjunction with data compression techniques for the transmission of high-quality video signals.

As local area networks proliferate, the demand for both traditional copper cable and optical fiber cable continues to increase. Most of these networks are copper wire based, though an increasing number are beginning to employ optical fiber cables for the network backbone and long-distance interconnections. Yet even in the all-fiber FDDI networks there is a place for copper cables involving short runs to the individual work-stations, because of the inherently lower expenses associated both as concerns the metallic cables themselves and their installation. In the present highly competitive environment, short-run installations tend to favor metallic conductor communication cables. Replacement of copper conductor cables with optical fiber cables will obviously not enhance the bit rate capacity of the LAN network itself, though the optical cables will augment the immunity of the network to extraneous electrical interference and obviate data security concerns by eliminating signal radiation.

REFERENCES

[1] H. S. Osborne, "Telephone transmission over long distances," *Electrical Commun.*, Vol. II, 1923, pp. 81–94

[2] E. B. Meyer, *Underground Transmission and Distribution*, McGraw-Hill, New York, 1916.

[3] W. A. Del Mar, *Electric Cables*, McGraw-Hill Book, New York, 1924.

[4] H. Foy, *High Tension Underground Electric Cables*, Electrical Publishing, New York, 1909.

[5] C. C. Barnes, *Power Cables: Their Design and Installation*, Chapman and Hall, London, 1966.

[6] L. Emanueli, *High Voltage Cables*, Chapman and Hall, London, 1929.

[7] J. J. Ray, C. A. Arkell, and H. W. Flack, "525 kV self-contained oil-filled systems for Grand Coulee Third Powerplant, design and development", *IEEE Trans. Power Apparatus Syst.*, Vol. PAS-93, 1974, pp. 633–39.

[8] G. M. Lanfranconi, G. Maskio, and E. Ochini, "Self contained oil-filled cables for high power transmission in the 750–1200 kV range," *IEEE Trans. Power Apparatus Syst.*, Vol. PAS-93, 1974, pp. 1535–1545.

[9] C. E. Benett and R. J. Wiseman, *Electrical World*, Vol. 100, 1935, p. 2891.

[10] C. F. Avila and A. F. Cory, "Underground transmission in the United States," *IEEE Spectrum*, Vol. 7, 1970, pp. 42–28.

[11] D. Couderc, Q. Bui-Van, A. Vallée, R. Hata, K. Murakami, and M. Mitani, "Development and testing of an 800 kV PPLP-insulated oil-filled cable and its accessories", Paper 21/22-04, Proc. CIGRÉ, Paris, 1996.

[12] L. Dechamps, R. Michel, and L. Lapers, "Development in France of high voltage cables with synthetic insulation," Report 21-06, Proc. CIGRÉ, Paris, 1980.

[13] G. Wauser, "CIGRÉ 1980—review and summary of high voltage cables," BP Chemicals Report, Geneva, July 1981.

[14] R. Jocteur and E. Favrie, "63 to 225 kV cables insulated with low density polyethylene. Operating characteristics, experience in production development envisaged," Report 21-06, Proc. CIGRÉ, Part II, Paris, 1980.

[15] P. Argaut and E. Favrie, "Recent developments in 400 and 500 kV XLPE cables," JICABLE, Versailles, 1995, pp. 1–5.

[16] U. Amerpohl, H. Koberh, C. van Hove, H. Schadlich, and G. Ziemeck, "Development of polyethylene insulated cables for 220 kV and higher voltages," Report 21-11, Proc. CIGRÉ, Paris, 1980.

[17] W. A. Thue, J. W. Bankoske, and R. R. Burghardt, "Operating and testing experience on solid dielectric cable," Report 21-11, Proc. CIGRÉ Paris, 1980.

[18] O. Mattison and S. Wretemark, "Development of XLPE cable systems and field test experience," Report 21-11, Proc. CIGRÉ, Paris, 1980.

[19] S. Nagasaki, N. Yoshida, M. Aihara, S. Fujiki, N. Kato, and M. Nakajama, "Philosophy of design and experience on high voltage XLPE cables and accessories in Japan," Report 21-01, Proc. CIGRÉ, Paris, 1988.

[20] K. Kaminaga, T. Asakura, Y. Ohashi, and Y. Mukaiyama, "Development and installation of long distance 275 kV XLPE cable lines in Japan", Report 21-102, Proc. CIGRÉ, Paris 1990.

[21] K. Kaminaga, M. Ichihara, M. Jinno, O. Fujii, S. Fukunaga, M. Kobayashi, and K. Watanake, "Development of 500 kV cables and accessories for long distance underground transmission lines; Part V; Long term performance for 500 kV XLPE cables and joints," *IEEE Trans. Power Delivery*, Vol. 11, 1996, pp. 1185–1194.

[22] K. Ogawa, T. Kosagi, N. Kato, and Y. Kawawata, "The world's first use of a 500 kV XLPE insulated aluminium sheathed power cables at Shimogo and Imaichi power stations," *IEEE Trans. Power Delivery*, 1990, Vol. 5, pp. 26–32.

[23] R. D. Rosevear, C. Larrive, B. Parmigiani, and P. Dejean, "Development of an XLPE cable system for 400 kV operation," JICABLE, Versailles, 1995, pp. 16–21.

[24] E. Peschke, R. Schroth, and R. Olshausen, "Extension of XLPE cables to 500 kV based on progress in technology," JICABLE, Versailles, 1995, pp. 6–10.

[25] R. J. Densley, R. Bartnikas, and B. Bernstein, "Multiple stress aging of solid-dielectric extruded dry-cured insulation systems for power transmission cables," *IEEE Trans. Power Delivery*, Vol. 9, 1994, pp. 559–571.

[26] S. Yan, K. Sheu, D. H. Damon, S. J. Huang, and J. F. Johnson, "Electric strength of XLPE containing acetophenone and other volatile substances," 1990 IEEE International Symposium on Electrical Insulation, IEEE Conf. Record 90CH2727-6, Toronto, 1990, pp. 305–308,

[27] B. Andress, P. Fischer, H. Repp and P. Pohl, "Diffusion losses of additives in polymeric cable insulations," 1984 IEEE International Symposium on Electrical Insulation, IEEE Conf. Record 84-CH1964-6-EI, Montreal, 1984, pp. 65–67.

[28] S. Haridoss, "Migration of vinyl acetate from semiconductive shields to insulation of power cables," IEEE International Symposium on Electrical Insulation, IEEE Conf. Record 90-CH2727-6, Toronto, 1990, pp. 281–285.

[29] C. W. Melton, D. Mangaraj, and M. Epstein, "Morphology of thermoplastic and cross-linked polyethylene cable insulation," 1981 Annual Report, Conf. on Electrical Insulation and Dielectric Phenomena, IEEE Conf. Record 81-CH1668-3, Whitehaven, 1981, pp. 299–305.

[30] Y. Miyashita, H. Itoh, and T. Shimomura, "Void formation in XLPE at high temperatures," 1985 IEEE International Conf. on Properties and Applications of Dielectric Materials, IEEE Conf. Record 85-CH2115-4, Xi'an, 1985, pp. 670–674.

[31] W. A. Del Mar and E. J. Merrell, "Polyethylene for power cables," *AIEE Special Publication S-48*, 1952, pp. 26–31.

[32] *Underground Cable Engineering Handbook*, Essex International, Power Conductor Division, Marion, Indiana, 1971.

[33] A. L. McKean, F. S. Oliver, and S. W. Trill, "Cross-linked polyethylene for higher voltages," *IEEE Trans. Power Apparatus Syst.*, Vol. PAS-86, 1967, pp. 1–10.

[34] R. M. Eichhorn, Discussion, Group 21-HV Insulated Cables, Proc. CIGRÉ, Paris, 1980, pp. 25–27.

[35] R. M. Eichhorn, in *Engineering Dielectrics, Vol. IIA, Properties of Solid Insulating Materials: Molecular Structure and Electrical Behavior*, R. Bartnikas and R. M. Eichhorn, Eds., STP-873, ASTM, Philadelphia, 1983, Chapter IV.

[36] J. H. Lawson, private communication: comments on the cable failure data collected between 1989 and 1991 by the AEIC Cable Engineering Section Task Group, March 3, 1997. (Also see IEEE Insulated Conductors Committee Minutes, April 1993.)

[37] J. Wonnay and H. Mathis, "Voltage tests and dissipation factor diagnosis of medium voltage cables with new high voltage function generator," Ninth International Symposium on High Voltage Engineering, Paper 4456-1, Graz, Aug. 28–Sept. 1, 1995.

[38] M. Baur, "Testing and diagnosis with dissipation factor measurement at 0.1 Hz on distribution cables," Minutes of the IEEE Insulated Conductors Committee, Appendix 12-A, St. Petersburg, Nov. 3–6, 1996.

[39] G. S. Eager, C. Katz, B. Fryszczyn, J. Densley, and B. S. Bernstein, "High Voltage VLF testing of power cables," *IEEE Trans. Power Delivery*, Vol. 12, 1997, pp. 565–570. (Also see N. N. Srinivas, B. S. Bernstein and R. A. Decker, "Effects of dc testing on ac breakdown strength of XLPE insulated cables subjected to laboratory accelerated aging," *IEEE Trans. Power Delivery*, Vol. 5, 1990, pp. 1643–1649).

[40] R. Bartnikas, in *Engineering Dielectrics, Vol. IIA, Properties of Solid Insulating Materials: Molecular Structure and Electrical Behavior*, R. Bartnikas and R. M. Eichhorn, Eds., STP873, ASTM, Philadelphia, 1983, Chapter I.

[41] R. Bartnikas, in *Engineering Dielectrics, Vol. IIB, Properties of Solid Insulating Materials: Measurement Techniques*, R. Bartnikas, Ed., STP 926, ASTM, Philadelphia 1987, Chapters I and II.

[42] IEEE Guide for Field Testing and Evaluation of the Insulation of Shielded Power Cable Systems, IEEE P-400 (Draft No. 12, 1997).

[43] P. Gassana-Priaroggia, J. H. Piscioneri, and S. W. Margolin, "The Long Island Sound submarine cable interconnection," *IEEE Spectrum*, Vol. 8, 1971, pp. 63–71.

[44] D. P. Sayers, M. E. Laborde, and F. J. Lane, "Possibilities of a cross-channel power link between the British and French supply systems," *Proc. IEE*, Vol. 101, Part. I, 1954.

[45] C. C. Barnes, J. C. E. Coomer, J. Rollin, and L. Clavreuil, "The British–French direct current submarine link," Paper No. 210, Proc. CIGRÉ, Paris, 1962.

[46] A. Nyman, "The Fenno-Skan HVDC commissioning," IEEE Power Engineering Society Winter Meeting, Feb. 3–7, 1991, New York.

[47] N. Bell, Q. Bui-Van, D. Couderc, G. Ludasi, P. Meyere, and C. Picard, "±450 kV dc underwater crossing of the St.-Lawrence River of a 1500 km overhead line with five terminals," Paper 21-301, Proc. CIGRÉ, Paris, 1992.

[48] J. N. Johnsen, T. A. Holte, K. Bjorlow-Larsen, and P. B. Larsen, "Submarine power cable development in Norway," Report 21-22, Proc. CIGRÉ, Paris, 1980.

[49] A. Fujimori, T. Tanaka, H. Takashina, T. Ienajo, R. Hata, T. Tanabe, S. Yoshida, and T. Kakikana, "Development of 500 kV dc PPLP-insulated oil-filled submarine cable," *IEEE Trans. Power Delivery*, Vol. 11, 1996, pp. 43–50.

[50] K. Terashima, H. Suzuki, M. Hara, and K. Watanabe, "Research and development of ±250 kV dc XLPE cables," *IEEE Trans. Power Delivery*, Vol. 13, 1998, pp. 7–15.

[51] T. Takeda, N. Hozumi, H. Suzuki, K. Fujii, K. Terashima, M. Hara, Y. Murata, K. Watanabe, and M. Yoshida, "Space charge behavior in full size 250 kV dc XLPE cables," *IEEE Trans. Power Delivery*, Vol. 13, 1998, pp. 28–39.

[52] E. M. Allam, W. H. Cortelyou, H. C. Doepken, Jr., A. L. McKeen, and F. A. Teti, "Development of low loss 765 kV pipe-type cable," EPRI Report EL 2196, Palo Alto, CA, Jan. 1982.

[53] C. A. Arkell, D. R. Edwards, D. J. Skipper, and A. W. Stannet, "Development of polypropylene/paper laminate (PPL) oil-filled cable for UHV systems," Report 21-04, Proc. CIGRE, Paris 1980.

[54] A. H. Cookson, "Electrical breakdown for uniform fields in compressed gases," *Proc. IEE*, Vol. 117, 1970, pp. 269–280.

[55] J. G. Trump, "New developments in high voltage technology," *IEEE Trans. Nuclear Science*, Vol. NS-14, 1967, p. 113.

[56] H. W. Grayhill and J. A. Williams, "Underground power transmission with isolated-phase gas insulated conductors," *IEEE Trans. Power Apparatus and Syst.*, Vol. PAS-89, 1970, pp. 17-23.

[57] Minutes of IEEE Insulated Conductors Committee Meeting, Kansas City, Nov. 11–13, 1974.

[58] Y. Kuroyanagi, A. Toya, T. Hayashi, and T. Raraki, "Construction of 8000 A class 275 kV gas insulated transmission line," *IEEE Trans. Power Delivery*, Vol. 5, 1990, pp. 14–20.

[59] K. Kaminaga and M. Koshiishi, "Development of compact 500 kV gas insulated transmission line-dust control during field jointing and method for detecting conductive particles," *IEEE Trans. Power Delivery*, Vol. 2, 1987, pp. 961–968.

[60] CIGRÉ Working Group 15.03 Report, "Effects of particles on GIS insulation and the evaluation of relevant diagnostic tools," Proc. CIGRÉ, Paris, 1994.

[61] A. H. Cookson and O. Farish, "Particle-initiated breakdown between coaxial electrodes in compressed SF_6," *IEEE Trans. Power Apparatus Syst.*, Vol. PAS-92, 1973, pp. 871–876.

[62] M. E. Holmberg and S. M. Gubanski, "Discharges from moving particles in GIS," *IEEE Trans. Power Delivery*, Vol. 13, 1998, pp. 17-22.

[63] D. Minaguchi, M. Ginno, K. Itaka, H. Furukawa, K. Ninomiya, and T. Hayashi, "Heat transfer characteristics of gas-insulated transmission lines," *IEEE Trans. Power Delivery*, Vol. 1, 1986, pp. 2–9.

[64] E. M. Spencer, R. W. Samm, J. Artbauer, and F. Schatz, "Research and development of a flexible 362 kV compressed gas insulated transmission cable," Report 21-02, Proc. CIGRÉ, Paris, 1980.

[65] E. B. Forsyth and R. A. Thomas, "Performance summary of the Brookhaven superconducting power transmission systems," *Cryogenics*, Vol. 26, 1986, pp. 599–614.

[66] P. A. Klaudy and F. Gerhold, "Practical conclusions from field trials of a superconducting cable," *IEEE Trans. Magnetics*, Vol. MAG-19, 1983, pp. 656–661.

[67] P. Klaudy, "In Arnstein war das erste supraleitende Kabel der Welt erfolgreich in Betrieb," *Energie*, Vol. 32, 1980, pp. 138–145.

[68] E. B. Forsyth, "Superconducting power transmission systems—the past and possibly the future," *Supercond. Sci. Technol.*, Vol. 6, 1993, pp. 699–714.

[69] Y. Mizuno, M. Nagas, M. Kosaki, M. Shimiza, and K. Horii, "Evaluation of ethylene-propylene rubber as insulating material for superconducting cable," *IEEE Trans. on Electrical Insulation*, Vol. 27, 1992, pp. 1108–1117.

[70] A. P. Malozenoff, "Superconducting wire gets hotter," *IEEE Spectrum*, 1993, pp. 26–30.

[71] T. Hara, K. Okaniwa, N. Ichiyamagi, and S. Tanaka, "Feasibility study of compact high-TC superconducting cables," *IEEE Trans. Power Delivery*, Vol. 7, 1992, pp. 1745–1753.

[72] P. M. Grant and T. R. Schneider, "The impact of superconductivity on electricity in the 21st century," Proc. World Energy Council 16th Congress, pp. 223–239, 1995.

[73] T. Shimonosona, S. Nagaye, T. Masada, and S. Iojima, "Development of a termination for the 77 K class high T_c superconducting power cable," *IEEE Trans. Power Delivery*, Vol. 12, 1997, pp. 33–38.

[74] K. Sato, N. Shibuta, N. Mukai, T. Hikata, M. Veyama, T. Kato, and J. Fujikama, "Bismuth superconducting wires and their applications," *Cryogenics*, Vol. 33, 1993, p. 243.

[75] P. Graneau, "A case for cryogenically cooled power cables," *Underground Engineering (J. Underground Power Technology)*, Vol. 2, 1971, pp. 41–43.

[76] D. C. Larbalestier and M. P. Maley, "Conductors from superconductors: conventional low temperature and new high-temperature superconducting conductors," *MRS Bull.*, 1993, pp. 50–56.

[77] M. Rahman and M. Nassi, "High capacity cable's role in once and future grids," *IEEE Spectrum*, 1997, pp. 31–35.

[78] P. Anderson, *The Theory of Superconductivity in the High-T_c Cuprates*, Princeton University Press, Princeton, 1997.

[79] T. Leggett, "Superconducting thoughts meet sceptical resistance," *Physics Today*, 1997, pp. 51–52.

[80] C. C. Barnes, *Submarine Telecommunication and Power Cables*, Peregrinius/IEE, London, 1977.

[81] F. W. Horn, *Lee's ABC of the Telephone, Vol. 5, Cable Inside and Out*, Training Manual, Geneva, II, 1976.

[82] W. E. Mongey, "The testing of long distance telephone cable during installation," *Electrical Communication*, Vol. II, 1923, pp. 219–223.

[83] F. W. Main, *Electric Cables*, Sir Isaac Pitman, London, 1949.

[84] A. G. Bell, "Apparatus for signaling and communicating, called photophone," U.S. Patent 235,199, 1880.

[85] D. Hondros and P. Debye, "Elektromagnetische Wellen in dielectrischen Drahten," *Annalen der Physik*, Vol. XXXII, 1910, pp. 465–476.

[86] R. L. Hartman, N. E. Schumaker, and R. W. Dixon, "Continuously operated (Al, Ga) as double-heterostructure lasers with 70° lifetimes as long as two years," *Appl. Phys. Lett.*, Vol. 31, 1977, pp. 756–759.

[87] R. P. Riesz, "High speed semiconductor photodiodes," *Rev. Sci. Instr.*, Vol. 33, 1962, pp. 994–998.

[88] F. P. Kapron, D. B. Beck, and R. D. Maurer, "Radiation losses in glass optical waveguides," *Appl. Phys. Lett.*, Vol. 17, 1970, pp. 423–425.

[89] A. E. Willner, "Mining the optical bandwidth for a terabit per second," *IEEE Spectrum*, 1997, pp. 32–41.

[90] *Northern Telecom Electrical Conductors Handbook*, 1977.

[91] W. D. Wilkens, "Telephone cable: overview and dielectric challenges," *IEEE Electrical Insulation Magazine*, Vol. 6, No. 2, 1990, pp. 23–28.

[92] J. K. Hardy, *Inside Networks*, Prentice-Hall, Upper Saddle River, NJ, 1995.

[93] J. Hamilton, M. Kolber, C. Schell, and L. Taupier, "Cable," in *The Communications Handbook*, J. D. Gibson, Ed., CRC/IEEE Press, Boca Raton, FL, 1997, Chapter 105.

[94] W. Stallings, *Local Networks*, Macmillan, New York, 1984.

[95] T. G. Robertazzi, "Computer networks," in *The Electrical Engineering Handbook*, R. C. Dorf, Ed., CRC/IEEE Press, Boca Raton, FL, 1997, Chapter 92.

[96] E. Tittel, *PC Networking Handbook*, AP Professional, Boston, 1996.

[97] G. R. Elion and H. A. Elion, *Fiber Optics in Communication Systems*, Marcel Dekker, New York, 1978.

[98] T. Li, "Advances in optical fiber communications: an historical perspective," *IEEE J. Selected Areas Commun*, Vol. SAC-1, 1983, pp. 356–372.

[99] E. Hecht, *Optics*, Addison-Wesley, Reading, MA., 1988.

[100] S. E. Miller, "Light propagation in generalized lens-like media," *Bell Syst. Tech. J.*, Vol. 43, 1964, pp. 1783–1809.

[101] R. David, D. Cleobury, and D. R. Cronin, "London–Birmingham: The first major-long-wavelength optical fiber installation," Proc. VIIIth European Conf. on Optical *Communications*, Paper A XIV, Cannes, Sept. 1982.

[102] D. Basch, D. Hanna and S. Vincent, "GTE long-wavelength fiber optic field trial," Technical Dig., 5th Topical Meeting on Optical Fiber Communications, Paper TUDD2, Phoenix, April 1982.

[103] M. Horiguchi, M. Nakahara, N. Inagaki, K. Kokura, and K. Yoshida, "Transmission characteristics of ultra-wide bandwidth VAD fibers," Proc. 8th European Conf. on Optical Communications, Paper A III-4, Cannes, Sept. 1982.

[104] K. H. Cameron, P. J. Chidgey, and K. R. Preston, "102 km optical fiber transmission experiments at 1.52 μm using an external cavity controlled laser transmission module," *Electronics Lett.*, Vol. 18, 1982, pp. 897–898.

[105] W. Albrecht, G. Elze, B. Enning, G. Walf, and G. Wenke, "Experiments with an optical long-haul 2.24 Gbit/s transmission system at a wavelength of 1.3 μm," *Electronics Lett.*, Vol. 18, 1982, pp. 746–748.

[106] J. E. Midwinter, "Studies of monomode long-wavelength fiber systems at the British Telecom Research Laboratories," *IEEE J. Quantum Electron.*, Vol. QE-17, 1981, pp. 911–918.

[107] P. Healey, P. Lindsey, D. M. Russ, and J. H. Stewart, "Field installation of a 31.5 km monomode optical fiber system operated at 140 Mbit/s and 650 Mbit/s," *Electronics Lett.*, Vol. 18, 1982, pp. 631–632.

[108] M. Chown and R. L. Williamson, "Development of optical submarine cable systems in the U.K.," International Conf. On Communications, Paper 7D.3, Philadephia, 1982.

[109] M. Washio, I. Kitazawa, H. Tsuji, and K. Takemoto, "400 Mbit/s submarine optical fiber cable transmission system field trial," Proc. VIIIth European Conf. on Optical Communications, Paper A XIV-6, Cannes, 1982.

[110] S. O. Agbo, A. H.Cherin, and B. K. Tariyal, "Lightwave" in *The Electrical Engineering Handbook*, R. C. Dorf, Ed., CRC/IEEE Press, Boca Raton, FL, 1997, Chapter 42.

[111] T. Li, "The impact of optical amplifiers on long-distance lightwave telecommunications," *Proc. IEEE*, Vol. 81, 1993, pp. 1568–1579.

[112] A. F. Elrefaie, R. E. Wagner, D. A. Atlas, and D. G. Daut, "Chromatic dispersion limitations in coherent lightwave transmission systems," *J. Lightwave Technol.*, Vol. 6, 1988, pp. 704–709.

[113] C. D. Poole, "Dispersion compensation in lightwave systems," *Tech. Dig. CLEO,* Paper CWFI, Washington, DC, 1993.

[114] R. J. Mears, L. Reekie, I. M. Jauncey, and D. N. Rayne, "Low noise erbium-doped fiber amplifier operating at 1.54 μm," *Electronics Lett.*, Vol. 23, 1987, pp. 1026–1028.

[115] E. Desurvire, J. R. Simpson, and P. C. Becker, "High gain erbium-doped travelling-wavefiber amplifier," *Opt. Lett.*, Vol. 12, 1987, pp. 888–890.

[116] E. Desurvire, *Erbium-Doped Fiber Amplifiers*, Wiley, New York, 1993.

[117] L. E. Jondry, "Fiber optic TI span line installations at a microwave site and two transmission substation," Canadian Electrical Association Meeting, Québec, October 1981.

[118] J. Péloquin and G. Missout, "Composite overhead ground wires with fiber optic lines," Canadian Electrical Association Meeting, Québec, October 1981.

[119] B. J. Maddock, N. J. Hazel, K. L. Calcut, B. Gaylard, P. Dey, R. J. Slaughter, and K. Lawton, "Optical fiber communication using overhead transmission lines," Paper 35-01, Proc. CIGRÉ, Paris, 1980.

[120] K. Brown, Discussion, IEEE-ICC Meeting of Working Group 14-2, Optical Fibers, Oct. 28, 1998, St. Petersburg.

[121] P. McLaughlin, "50-micron fiber gets a second glance," *Cabling Installation Maintenance*, Aug. 1998, pp. 33-38.

[122] W. H. Michael, W. J. Cronon, and K. F. Pieper, *Fiber Distributed Data Interface: An Introduction*, Digital, Burlington, MA, 1993.

Chapter 2

CHARACTERISTICS OF CABLE MATERIALS

R. Bartnikas

2.1 INTRODUCTION

Cable technology is by its very nature a materials-based technology. From the engineering systems point of view, cables are systems whose electrical behavior is very much dependent upon the materials comprising the cable construction. Since an electrical cable consists essentially of a metallic conductor surrounded by a dielectric, which is in turn wrapped in a metallic shield that is protected from the environment by a jacket, its electrical characteristics are thus largely determined by the metallic and dielectric materials utilized in the construction. In the design, installation, and use of cables, the cable engineer must therefore possess a sufficient background in cable construction materials in order to fully appreciate the limits of various cable constructions and systems. It is the intent of this chapter to provide a description of the construction materials employed in the past as well as those in use currently in the manufacture of electrical cables. While most of the data presented in this chapter pertain to power cables, some materials, such as copper and polyethylene, are also used in communication cables. Additional details on the materials employed in the construction of metallic conductor and fiber optic (dielectric conductor) communication cables are provided in Chapter 15.

2.2 METALLIC CONDUCTORS

Metallic cable conductors are selected on the basis of their electrical conductivity, coefficient of expansion, workability, breaking strength, heat resistance, and cost. Most of the communication and a large percentage of the power cable conductors are made of copper because of its low electrical resistivity, which is equal to $1.7241\,\mu\Omega$-cm at $20°C$; copper has also excellent mechanical properties that render it especially suitable for cable conductors. For example, although in comparison the electrical conductivity of silver exceeds that of copper, the unusual softness of the silver metal would make it entirely unsuitable for use as a cable conductor in addition to its relatively high cost. The electrical conductivity of copper is determined by its impurity content as well as the degree of annealing that it has undergone to change its crystalline structure. The copper commonly used for cable conductors is 99.9% pure, the remaining 0.1% comprising both metallic and nonmetallic impurities. The main nonmetallic impurity found in copper is that of cuprous oxide; since the latter occurs as a nonsoluble oxide in the copper conductors, it must be eliminated during the processing of copper. Accordingly, in the production process of copper for electrical uses, certain reducing agents such as calcium, lithium, magnesium, phosphorus, or silicon must be used so that the nonmetallic impurities may combine with the oxygen while the copper is still in a molten state [1]. It is precisely during this fabrication step that some metallic impurities are introduced into the copper metal. Perhaps the most deleterious of the metallic impurities is phosphorus. For instance, only a 0.04% content of phosphorus in copper may reduce its conductivity by as much as 80% whereas a relatively high content of arsenic of 0.4%, reduces the conductivity by 40% [2]. For this reason, it is well to refrain from using phosphorus as a reduction agent. It has been established that acceptable values of both electrical and thermal conductivity can be achieved when small quantities of deoxidant such as boron carbide, lithium, or zinc are employed [2].

The mechanical properties, in particular the tensile strength of copper, are greatly improved as the copper metal is subjected to work by rolling and drawing, both in the hot and cold state. The ultimate tensile strength of copper, in its annealed condition after cold working, may range from 28×10^3 to 32×10^3 psi. Cold working of copper metal, however, decreases its electrical conductivity up to several percent. Thus, in general the electrical resistance of copper conductors will be augmented with the tensile strength. For cold-worked copper having a tensile strength in the range from 40×10^3 to 60×10^3 psi, the percentage increase in resistance of the hard-drawn conductor over its resistance when annealed may be expressed empirically as $S/10$, where S is the tensile strength of the hard-drawn conductor in tons per square inch [2].

Copper rods received from rolling mills are covered with black scale, which must be removed by immersion in a dilute solution of sulfuric acid prior to drawing. The drawing process consists of pulling the rods through a succession of cooled steel dies of decreasing diameter of the opening of the dies, until the desired wire diameters are obtained. Depending upon the dies, a variety of conductor shapes may be obtained for the desired stranded conductor geometry e.g. circular, segmental, hollow core, etc. However, in the case of power cable conductors, before the copper wires emerging from the drawing process can be stranded, they must undergo annealing. Annealing is necessary to remove some of the hardness and stiffness from the copper wires and to return the conductivity of the copper wire to an acceptable value. Annealing is usually carried out in an inert atmosphere at temperatures of about 600°C for periods of one hour or more. It is recognized that annealing of work-hardened copper may proceed over extended periods of time even at temperatures as low as 100°C. However, for practical purposes where time is limited and to ensure that annealing will take place over the entire cross section of the wire coil, the annealing temperatures should not be less than 500°C. The density of high-conductivity annealed copper is on the order of 8.9 g/cm^3, with a corresponding melting temperature of 1083°C and thermal conductivity of 10.92 cal/cm^2-cm-s-°C. The volume resistivity temperature coefficient at a temperature of T degrees Celsius is equal to $1/(233.5 + T)/°C$ [2].

When cables are to be spliced or joined together, certain precautions should be taken in view of possible contact resistance problems. Copper exposed to the atmosphere will readily oxidize, so that the conductors must be thoroughly cleaned prior to joining. One redeeming feature is that the oxide itself is characterized by a negative temperature coefficient with the result that the resistance at the oxidized contact does fall with increasing temperature [3]. Moreover, the copper oxide is relatively thin and, therefore, as the joint is formed under mechanical pressure the film is crushed and contact is effected between the rough and protruding points of the copper metal conductors [4, 5]. While the contact resistance falls with the mechanical pressure applied, care must be exercised not to exceed the elastic limit of the copper conductors to be joined. To provide added protection against oxidation, corrosion, and high-temperature effects, copper cable conductors may be coated with metallic films. For example, tin is applied on conductors by an electroplating or hot-dip process; solderability is improved by the use of lead–tin alloy films. Although, the contact resistance is found to increase with the amount of lead in the alloy. Lately, it has been also found that tin will tend to diffuse into the copper conductor. If solderability is not the main concern, nickel coating may be used to obtain protection under high temperatures against oxidation and corrosion, as well as to provide a lower contact resistance for cable joining points. Nickel does, nevertheless, lack the good flexibility properties of tin.

Aluminum constitutes a direct alternative to copper for underground power cable application; in fact in North America, most present power distribution cables are manufactured with aluminum conductors. The aluminum used in cable applications is usually better than 99.5% pure with a direct-current (dc) resistivity of 2.8 $\mu\Omega$-cm at 20°C. Although the volume conductivity of aluminum is only 62% of that of copper, aluminum conductors are increasingly used in cables due to savings in weight and sometimes cost (depending upon current market fluctuations). If one considers the densities of aluminum (2.7 g/cm^3) and copper (8.9 g/cm^3), then for the same equivalent conductivity the weight of required aluminum is only about half that of copper, although this results in an aluminum conductor diameter of approximately 1.3 times that of copper [6]. Aluminum possesses a number of disadvantages vis-à-vis copper such as poor solderability, softness (which gives rise to cold-flow), and corrosion due to galvanic action when connected to or in contact with the more noble metals. In addition, as a result of oxidation, aluminum oxide films can form readily on the aluminum conductor surfaces. Because of their good dielectric properties, conductor splicing poses a problem; this can be overcome by using more expensive splicing techniques, such as argon welding to prevent oxide formation.

As in copper, the dc conductivity of aluminum is greatly affected by some metallic impurities. Small quantities of vanadium or titanium will have a very pronounced effect, whereas the presence of magnesium will decrease its conductivity only very slightly. The permissible metallic impurity contents in electrical-grade aluminum conductors have been specified in the American Society for Testing and Materials (ASTM) Standard B 233 [7]. These permissible metallic impurity levels are given in Table 2.1.

The minimum electrical-grade aluminum metal purity has been set at 99.5%. Aluminum has a thermal conductivity of 0.53 calories per second per square of thickness per 0°C and a coefficient of thermal expansion of 23.8×10^{-6}/°C. Its melting point occurs at about 660°C [8]. The tensile strength of annealed pure aluminum is 7×10^3 psi, though it may be increased appreciably to as much as 8.2×10^4 psi by alloying, heat treatment, and work hardening. All aluminum conductors used in electrical applications are annealed and tempered. The term temper is used to signify that the metal has undergone mechanical and thermal heat treatment to obtain the necessary physical properties. In the production of aluminum wire, a high-speed hot rolling process is employed to produce a wire with the necessary tolerances and properties. Aluminum used for electrical purposes then undergoes annealing (heat treatment) to impart ductility to the metal. In the annealing process, the aluminum metal is heated to a certain temperature and maintained at that temperature for a fixed time, whereby the fractured metal grains produced during the cold working process are allowed to recrystallize.

TABLE 2.1 Permissible Impurity Levels in Electrical-Grade Aluminum Conductors

Element	Composition (%)
Silicon (max.)	0.10
Iron (max.)	0.40
Copper (max.)	0.05
Manganese (max.)	0.1
Chromium (max.)	0.01
Zinc (max.)	0.05
Boron (max.)	0.5
Gallium (max.)	0.03
Vanadium + titanium (total max.)	0.02
Other elements (each max.)	0.03
Other elements (total max.)	0.10
Aluminum (min.)	99.50

Note: In accordance with ASTM B233 [7].

Annealing of commercially pure aluminum is normally carried out at a temperature of 350°C for a relatively short period [8]. With aluminum wires, the annealing process is more refined because care must be exercised to ensure proper tensile strength and elongation characteristics. For this reason the drawn aluminum wires must undergo what is termed a continuous annealing process, whereby the temperature is gradually raised and decreased to and from about 200 to 260°C, respectively. The annealing process generally lasts between 3 and 4 h.

The joining of aluminum conductors is more difficult than those made of copper because of the aluminum oxide film that forms on the aluminum conductor in a very short time following its exposure to the air atmosphere. Aluminum oxide constitutes a good insulator and must, therefore, be removed if two aluminum conductors are to be joined electrically. The oxide can be removed either mechanically by abrasion, or use of fluxes or with an ultrasonic method. Once the oxide is removed, the aluminum cable joint may be formed using a welding process whereby the metal surface is shielded with a flux or an inert atmosphere. Normally an argon gas welding process is used. A tungsten electrode is employed to form a high current arc within a shield of argon gas. The latter, upon ionization, assists in the removal of the oxide film and, by virtue of excluding the oxygen of the air, prevents its formation [8]. When aluminum conductors are to be joined to conductors of another metal, such as copper, great care must be exercised because of the position of aluminum in the electromotive series given in Table 2.2. Evidently, a direct contact between an aluminum and a copper conductor would result in the gradual corrosion of the aluminum metal. Although in the past tinned copper-aluminum joints have been used, the corrosion experience has not been satisfactory. Lately, there appears to have been considerable progress made in the use of transition washer-type connectors; simultaneous developments occurred in this area in both England and Germany. In England, a transition serrated brass-type washer was developed; in Germany, a transition contour aluminum alloy-type washer was put on the market. Both types of washer are used to form the interface between a bolted aluminum-copper joint. The sharp contour edges of the washers penetrate the oxide films on both metals to establish a good metal-to-metal contact. In addition, various types of greases usually containing AL_2O_3 powder are used to prevent further oxygen

TABLE 2.2 Electromotive Series

Element	Potential (V)
Sodium	2.71
Magnesium	2.37
Aluminum	1.66
Zinc	0.76
Iron	0.44
Cadmium	0.40
Nickel	0.25
Tin	0.14
Lead	0.13
Hydrogen	0
Copper	−0.34
Silver	−0.80
Gold	−1.50

contact with the metal. The role of the powder appears to be the provision of indurate particles to break the hard oxide film (on the aluminum surface) when pressure is applied. The experience with these type of joints has been encouraging thus far.

In North America the individual copper or aluminum wires are normally classified according to their size, using the American Wire Gage (AWG) system. The individual wires are usually stranded in a concentric lay to form power cable conductors; however, occasionally, solid cable conductors may be used for certain applications. For solid-copper conductors, AWG sizes generally range from 40 to 4/0, corresponding to 0.079 mm up to 11.68 mm in diameter in the metric measure, respectively. The dc resistance values for the lower and upper limit wire gages are 1.080×10^3 and $0.049 \, \Omega/1000 \, \text{ft}$ respectively. A 4/0 soft annealed conductor has an approximate strength of 5.98×10^3 psi; when the copper wire is hard drawn, this value is increased to 8.14×10^3 psi. For stranded conductors exceeding in diameter a size of 4/0 AWG, the diameter is measured in circular mils (cmil or CM) or thousands of circular mils (kcmil or MCM). One circular mil is defined as being equal in area to a circle having a diameter of 1 mil or 0.001 inch; thus, 1000 kcmil or $\text{MCM} = \pi/4 \, \text{in.}^2$. The size of concentric lay stranded copper conductors may attain a size of 2000 kcmil or MCM $(1.013 \times 10^3 \, \text{mm}^2)$ or a diameter of 41.45 mm, containing a total of 127 wires in the strand with a corresponding dc resistance of $0.00529 \, \Omega/1000 \, \text{ft}$

With solid-aluminum wires, the sizes used on power cables usually range from 35 to 4/0 AWG, corresponding to dc resistance values of 541.5–$0.080 \, \Omega/1000$ feet, respectively. Typically, the concentric lay stranded aluminum conductors begin at 6 AWG and proceed to 1000 kcmil or MCM, with 7–61 wires per conductor, respectively [9]. The reduced conductivity of aluminum vis-à-vis copper conductors, may lead to the use of very large aluminum conductor cables were it not for the economical limits imposed due to the additional amount of insulating material required to insulate and sheath oversized cables. It is difficult to generalize on the exact form of heat treatment that is given to the aluminum conductor metal, though it would seem that at least for most cables the conductor is either soft annealed or hard drawn.

Apart from aluminum and copper, no significant inroads have been made by other metals for conductor applications. There was a serious effort made to use sodium conductors; this was particularly attractive in view of the low cost of the sodium metal. The experimental sodium metal cables consisted essentially of polyethylene pipes filled with sodium. However, the corrosion and termination aspects together with the safety considerations, concerning containment of the sodium metal in case of accidental failure, created a rather formidable task to those engaged in the field. The efforts were thus gradually phased out. With the discovery of high-temperature superconductors, renewed interest has been stimulated on the development of superconducting cables. Since the subject matter of superconducting cables is dealt with in great detail in Chapters 1 and 12, we shall confine ourselves here only to a few cursory comments on the subject. The cable conductors for superconducting cables must be made of metals or copper-oxide-based ceramics that can operate in the superconducting mode. Metallic films, spirals, or strips may be used in such applications as extremely high currents are to be carried over small cross sections. Niobium (Nb) metal may be used as a superconductor at 4.4 K, while an alloy of niobium-tin (Nb_3Sn) performs well at 6.2 K. Superconducting metals or alloys constitute mechanically brittle systems and fabrication techniques must be developed for reinforcing their weak structures [10]. Especially fabricated flat wire strips of high-temperature superconducting copper-oxide-based ceramics, such as bismuth-strontium-calcium copper oxide (abbreviated as BSCCO), have been shown capable of carrying current densities on the order of 10 kA/cm^2 at 77 K. (cf. Chapters 1 and 12). With cryoresistive cables that are to operate in a reduced resistivity mode, aluminum or copper conductors may be used, however, only in a very pure state since impurities tend to reduce greatly any gains in the resistance reduction that can be achieved at temperatures down to 77 K. Both superconducting and cryoresistive cables have particular geometries for their conductor configurations. Since the operating temperature of high-temperature superconducting cables overlaps that of the cryoresistive cables, made with low-temperature superconductors, the interest in these cables was virtually extinguished with the discovery of high-temperature superconductors. In the case of conventional copper and aluminum conductor cables, certain configurations may also be used that require the individual wires to have other than circular geometry. For example, conductors having a nearly rectangular cross section may be used to obtain more compact strand structures as well as smoother overall conductor surfaces. Such cross-sectional wire shapes are required for self-contained hollow-core oil-filled cables (cf. Chapter 6).

The alternating current (ac) resistance of the cable conductors is higher than their dc value because of the skin effect, which arises because the center portions of the conductor are cut by a larger number of magnetic flux lines than are the outer portions. This causes the current to distribute more densely around the edges or the skin region of the conductor. As a result, the current density across the conductor becomes non-uniform, leading to an increase in the ac resistance of the conductor and, therefore, to increased energy losses. The skin effect is assessed in terms of the ratio of the measured ac resistance, R_{ac}, to the measured dc resistance, R_{dc}. If one defines the parameter x as [1]

$$x = \pi d \left[\frac{2 \times 10^{-9} f \mu}{\rho} \right]^{1/2} \tag{2.1}$$

where d is the conductor diameter in centimeters, f is the frequency in hertz, μ is the permeability of copper or aluminum and is equal to unity as neither metal is magnetic, and ρ is the resistivity of the copper or aluminum conductor in ohm-centimeters, then for $x < 3$

$$\frac{R_{ac}}{R_{dc}} = \frac{(1 + x^4/48)^{1/2} + 1}{2} \qquad (2.2)$$

and for $x > 3$

$$\frac{R_{ac}}{R_{dc}} = \left(\frac{x}{2\sqrt{2}} + 0.26\right) \qquad (2.3)$$

These empirical relations indicate that the resistance ratio increases with the frequency of the electric field and the conductor radius. In practice, it is also found that it increases with the load current carried by the cable. At power frequencies, the skin effect becomes significant with the larger conductor sizes, as is indeed apparent from the ac-to-dc resistance ratios given in Chapter 5. Readers interested in pursuing the subject further should consult the classical work of Dwight [11].

2.3 CONDUCTOR AND INSULATION SEMICONDUCTING SHIELDS

The purpose of a conductor shield or screen is to provide a uniform voltage stress over a relatively rough stranded conductor surface and in the case of extruded cables, in particular, to provide close bonding between the conductor and the adjacent insulation, so as to exclude any interspersed voids that may constitute sources of partial discharge. The purpose of the shielding over the outside insulation boundary is again twofold. First, with the outer shield grounded, the electrical field of the conductor attains radial symmetry and is confined to the insulation itself for safety considerations; second, intimate contact between the outer shield and insulation again serves the purpose of preventing partial discharges within air cavities between the insulation and the outer grounded metallic shield. In addition, the outer grounded shield provides also protection for the cable against any potentials that may be induced extraneously.

Certain differences exist between the type of screens utilized on the polymeric and oil-paper-insulated cables. For power cables using an oil-paper dielectric, the insulation and conductor shields consist of either copper tapes interlapped with kraft paper tapes or simply metallized kraft paper tapes. The copper tape thickness is usually on the order of 3 mils. The metallized kraft paper tapes may be metallized either with copper or aluminum, though it can be appreciated that an aluminum metallic film will necessarily be covered with a thin film of oxide having highly insulating properties. The thickness of the metallized tapes usually ranges from 5 to 7 mils, depending upon the thickness of the metallic film. Semiconducting carbon black tapes may also be used as conductor shields, but it has been found that the carbon black tends to contaminate the impregnating oil, giving rise initially to fairly high values of $\tan\delta$ at elevated voltage stresses. However, it has been established that these high values decrease eventually with time [12]. Present oil-filled cables have their shields often constructed using duplex paper that is semiconducting on one side and insulating on the other—analogous to the metallized tape design. With the duplex tape shielding, another layer of metallic shield is used. The construction of the latter is made of woven tinned copper wires within a synthetic fabric matrix.

Semiconducting carbon black tapes were also used as shields in the early linear polyethylene (PE) insulated cables. Due to poor adhesion between the tapes and the PE as well as occasional breaks or gaps between butting edges of carbon black tapes themselves, partial discharges often occurred within the voids formed at these faults. Polyethylene cables using carbon black shielding tapes were also found to be highly susceptible to tree growth [13]. The problem was ameliorated when the carbon black tapes were replaced by extruded semiconducting layers of PE, as delineated in Fig. 2.1.

The foregoing technique ensured good bonding between the shields and the insulation, thus effectively reducing the occurrence of air gaps and the associated partial discharges. The conductivity of the semiconducting polyethylene shields is varied by changing the carbon black content. For example, a crosslinked polyethylene (XLPE) semiconducting shield with a 25% carbon black has a resistivity of about 17 Ω-cm at 20°C, which increases to 45 Ω-cm at 75°C. Other materials such as polyvinyl chloride (PVC), ethylene-propylene-rubber (EPR), and irradiated polyolefins, containing semiconducting additives, can also be used. Either thermoplastic or thermosetting materials may be used for semi-conducting shields. It is not clear how the carbon black particles distribute themselves in the polymer matrix, but it is believed that they form chainlike paths along which current may be conducted. The particles in such paths need not be entirely contiguous so long as their separation is not greater than 100 Å so that electron tunneling may take place. The conductivity of the semiconducting shield increases with the carbon black content, and the permissible resistivities of the extruded shields are stated in the AEIC (Association of Edison Illuminating Companies) specifications on thermoplastic and XLPE and EPR cables [14, 15]. With both polyethylene and EPR cables, the thickness of the conductor shield is a function of the conductor size, whereas the insulation shield thickness is determined by the outer insulation diameter and also by whether the cable has an overall jacket. For cables with conductor sizes ranging from 8 AWG to 1000 MCM (kcmil), the average conductor shield thickness is between 15 and 25 mils, respectively. For insulation thickness diameters of 1.0–2.0 in. and larger, the average insulation shield thickness is fixed between 30 and 70 mils. When cables are constructed with an overall jacket, these insulation shield thicknesses may be decreased to 30 and 50 mils, respectively, because of the added protection of the jacket [14, 15]. For the thermosetting material insulated cables of XLPE and elastomeric EPR cables, volume resistivities for the conductor and insulation shields should not exceed 1000 and 500 Ω-cm, respectively at 90°C. For thermoplastic polyethylene insulation,

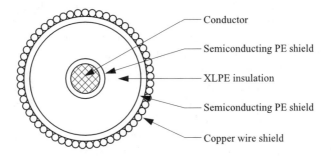

Figure 2.1 Crosslinked polyethylene (XLPE) cable with extruded semiconducting PE or XLPE shields with round wire concentric neutral applied spirally.

the specified values are identical but are with reference for a temperature of 75°C. Proper grounding of the extruded semiconducting insulation shield is affected by applying helically over it a layer of copper tapes, copper wires, or flat copper straps.

Little can be said on the actual chemical and physical composition of extruded semiconducting shielding or screening compounds because of their highly proprietary nature. The semiconducting materials are generally classified as thermoplastic or thermosetting semiconducting PE copolymers of the EEA (ethylene ethyl acrylate) or EEVA (ethylene ethyl vinyl acetate) type. Their chemical composition is adjusted to provide the desired properties for a given semiconducting shield, e.g. to provide either strongly adhering insulation shields or shields that can be readily stripped off the insulation surface to ensure ease of cable termination. In general, the EEVA copolymers are employed for strippable insulation and conductor shields—with the exception of copper conductors, where the shields react with the residual acetic acid content of the polymer [16]. The EEA polymers are well suited for conductor shields and as cross-linked insulation shields provide a strongly adhering semiconducting layer over the insulation. Because of the carbon content, their densities are greater than that of normal polyethylene, falling within the range of 1.12–1.14 g/cm^3. The tensile strength of semiconducting shields is on the order of 1800 psi or 12.5 MPa. The extrusion temperatures vary somewhat for the different semiconducting materials, ranging generally from about 120 to 200°C. Some semiconducting materials must undergo a drying period prior to their extrusion.

The electrical conductivity of carbon particle filled polymers depends on a number of factors, such as the filler content, the geometry and size of the filler particles, the manner in which the filler combines physically with the polymer, and the method of mixing. If the adhesive interaction between the carbon particles is high, then the carbon particles will form large clusters or aggregates and the conductivity of the carbon-polymer mixture will remain low because the formation of chainlike carbon conductive paths will be inhibited. If at the other possible extreme, arising from a high adhesion between the carbon particles and the polymer, the individual carbon particles become enveloped in the polymer, the conductivity again remains low. It is thus argued that the optimum compromise for conducting carbon chain formation occurs when the surface of the carbon particles is characterized by regions of varying degrees of adhesion such that a portion of the surface favors carbon-to-carbon aggregation necessary for conducting chain formation, while the remaining surface area develops a binding physical contact with the polymer [17]. The minimum value of the carbon particle concentration, having given surface properties at which conductive chain formation occurs, is termed the percolation threshold [18, 19]. At the percolation threshold, the conductivity of the carbon-polymer mixture increases very abruptly above the value characterizing the polymer insulate. This signifies that at the percolation threshold, the carbon-polymer structure comprises either continuous chains of carbon between the measuring electrodes or more likely sections of chains of carbon that are separated from each other at their end extremities by ≤ 100 Å. The distance of 100 Å represents the minimum distance between the chain ends required for electron tunneling to take place at the chain tips or junction points. As this distance is reduced, the conductivity will increase exponentially. The conductivity may be simply augmented by the addition of increasingly larger amounts of carbon to the polymer, but this will have serious adverse effects on the mechanical properties of the semiconducting shields.

Electronic conduction in polymers occurs primarily as a consequence of electron hopping over potential barriers between localized states or defect sites in the crystalline portion of the polymer structure. On the other hand in a true semiconductor, charge generation (comprising electrons) arises from a displacement of the charge carriers from the valence band into the conduction band due to the presence of impurities or dopants. In contradistinction, in the case of the so-called semiconducting polymers (or perhaps what should be better termed as polymers made partially conducting by the inclusion of carbon particles), the conductivity increase is produced either due to a normal ohmic conductivity when carbon particles form contiguous conducting chains or, alternatively, via the electron tunneling mechanisms when the individual ends of the carbon chains are separated by $\sim$100 Å or less.

If one considers two chains of carbon particles separated by a distance $\leq$ 100 Å at their tips, having an equivalent tip surface area A, then the conductivity at a given temperature T due to the tunneling effect between the two tips at the junction may be expressed as [20, 21]

$$\sigma = \sigma_0 \exp\left(-\frac{AdE_0^2/8\pi k}{T + AE_0^2/4\pi^2 xk}\right) \tag{2.4}$$

where the preexponential conductivity term σ_0, is assumed to exhibit only a weak temperature and field dependence, k is the Boltzmann constant, and the tunneling constant, x, is given by

$$x = \left(\frac{m\,\Delta H\,h^2}{2\pi^2}\right)^{1/2} \tag{2.5}$$

Here ΔH is the potential barrier (in electron volts) facing the electrons, m is the mass of the electron, and h is Planck's constant. The internal electrical field, E_0, is given by [21]

$$E_0 = \frac{4\Delta H}{ed} \tag{2.6}$$

Equation (2.4) indicates that for $T \ll AE_0^2/4\pi^2 xk$ the conductivity σ of the carbon containing polymer is temperature independent and, as long as the carbon chains are separated by less than 100 Å, the conductivity σ is governed by the tunneling mechanism such that

$$\sigma = \sigma_0 \exp\left(-\tfrac{1}{2}\pi xd\right) \tag{2.7}$$

At high temperatures when $T \gg AE_0^2/4\pi^2 xk$, the conductivity of the carbon-filled polymer becomes thermally activated such that

$$\sigma = \sigma_0 \exp\left(-\frac{AdE_0^2}{8\pi kT}\right) \tag{2.8}$$

The activation energy, ΔH, is now given by $AdE_0^2/8\pi$. Since the root-mean-square (rms) value of the thermally fluctuating voltage at the adjacent tips of the carbon chains is of the order of $(kT/C)^{1/2}$, where the capacitance between the tips $C = A/4\pi d$ [22], large fluctuations of the electric field arise between the individual carbon chain tips. As a consequence the current traversing the polymer film interposed between the carbon chain tips becomes thermally modulated. For high fields, the current behavior at the carbon chain junctions is similar in form to Eq. (2.4) but has the added field dependence given by [21]

$$J = J_0 \exp\left[\left(-\frac{AdE_0^2/8\pi k}{T + AE_0^2/4\pi^2 xk}\right)\left(\frac{E_a}{E_0} - 1\right)^2\right] \tag{2.9}$$

where E_a is the applied field. It is evident that the form of Eqs. (2.4) and (2.9) should at least qualitatively describe the situation for carbon-filled polyethylene employed for conductor and insulation shields. The shield materials used in practice should exhibit little temperature and field effects over the range of temperatures and operating voltages employed in order to maintain a reasonably constant conductivity. Some variation in the conductivity may take place due to internal molecular and carbon particulate rearrangements and density changes as the semiconducting materials expand and contract under load cycling.

The electrical behavior of semiconducting shield materials is assessed in terms of their volume and surface conductivities or resistivities as a function of temperature and electrical field. However, since surface resistivity is influenced very appreciably by absorbed moisture on the surface as a result of its exposure to ambient air during the sampling procedure, it is common practice to specify only the volume resistivity values. Nevertheless, surface resistivity measurements carried out at controlled low ambient humidity do supplement and complement the data obtained with volume resistivity measurements. Volume resistivity measurements normally are carried out using ASTM Standard D257 [23] and those of surface resistivity using ASTM Standard D991 [24] or D4496 [25] when the volume resistivity is in the range from 1 to 10^7 Ω-cm and surface resistivity ranges from 10^3 to 10^7 Ω (per square). Since the volume resistivity is a function of temperature, the measurement is carried out at applied voltages such that the heat dissipation is approximately ≤ 0.1 W and the loss current remains stabilized at some fixed value. If a guarded, three terminal parallel-plane electrode system is employed to measure the volume resistivity of a semiconducting sheet specimen of thickness d, with the voltage, V, being applied across this thickness, then the volume resistivity, ρ_v in ohm-centimeters is given by [23]

$$\rho_v = \frac{A}{d}\frac{V}{I} \tag{2.10}$$

where A represents the area of the cylindrical electrodes in centimeters squared and the current I is in amperes. If it is desired to measure the volume resistivity in the direction along the sheet i.e., in parallel to its surface, then a four-terminal electrode system, consisting of two current and two surface contacting potential electrodes separated by a distance, ℓ, as delineated in Fig. 2.2, may be used [24, 25]. With the foregoing arrangement, the volume resistivity is given by

$$\rho_v = \frac{wd}{\ell}\frac{V}{I} \tag{2.11}$$

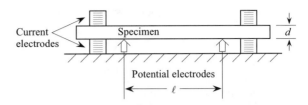

Figure 2.2 Four-terminal electrode system for the measurement of the surface resistivity and of the volume resistivity in a direction parallel to the surface of the sheet specimen [24, 25].

where w represents the width of the specimen or the rectangular current electrodes, such that the equivalent area of the specimen, A, is equal to wd. Here the quantity wd/ℓ is equivalent to the term A/d in the parallel-plane circular electrode case. With isotropic materials Eqs. (2.10) and (2.11) should give approximately equal values of ρ_v. Large variations in the test results would imply different preferential alignment of the carbon particle chains in the possible directions. The potential electrodes across which the voltage V is measured are displaced a certain distance from the current electrodes at which the current within the semiconducting specimen becomes more uniform. The arrangement depicted in Fig. 2.2 is also suitable for the measurement of the surface resistivity, ρ_s, which may be expressed as [25]

$$\rho_s = \frac{w}{\ell} \frac{V}{I} \tag{2.12}$$

where the surface resistivity ρ_s has the units of ohms per square or simply ohms.

As a result of some preferential orientation of the carbon particle chains, differences will be found in the volume resistivity values obtained between the volume resistivity measured in the radial and longitudinal direction of the extruded semiconducting shields. The AEIC specifications [14, 15, 26] state only the requirements for the volume resistivity of the semiconducting shields in the radial direction; these values are presented in Table 2.3. Examination of the tabulated values indicates that the specified values of maximum allowable resistivity for the insulation semiconducting shields are lower than those for the conductor. This reduced resistivity is to ensure more effective grounding of the cable, particularly to compensate at points where the concentric neutral wires may not make good contact with the insulation shield or become spread too far apart. The conductor shield will be subjected to a higher field gradient at which its resistivity will be reduced; however, some of this reduction will be compromised by an increase of resistivity due to the higher temperature in the vicinity of the surface of the conductor. It can be seen from Table 2.3 that at the emergency operating temperature of 130°C, only the conductor semiconducting shield resistivity limit is specified, since the semiconducting shield of the insulation is not expected to be subjected to as high a temperature. Also note that for thermoplastic compound type shields, the specified resistivity values are given for lower temperatures, which reflect their operating and emergency temperature limits of 75 and 90°C, respectively. Although Table 2.3 indicates that crosslinked polyethylene semiconducting shields may be used with EPR insulated cables, it is perhaps more common to employ especially formulated EPR semiconducting shields, since the oils and waxes utilized in the compounding of EPR affect adversely the conductivity of the crosslinked semiconducting shields. However, when EPDM (ethylene-propylene-diene terpolymer) is used in lieu of the EPR insula-

TABLE 2.3   Volume Resistivity Requirements for Semiconducting Shields

Cable Type	Shield Material	Radial Volume Resistivity (Ω-cm)				
		Room Temperature	75°C	90°C	110°C	130°C
Medium-voltage XLPE (AEIC-CS5, AEIC-CS7)	Thermoplastic conductor shield	—	$\leq 1 \times 10^5$	$\leq 1 \times 10^5$		
	Thermoplastic insulation shield	—	$\leq 5 \times 10^4$			
	Crosslinked conductor shield	$\leq 1 \times 10^5$	—	$\leq 1 \times 10^5$	—	$\leq 1 \times 10^5$
	Crosslinked insulation shield	$\leq 5 \times 10^4$	—	$\leq 5 \times 10^4$	$\leq 5 \times 10^4$	
Medium-voltage EPR (AEIC-CS6)	Crosslinked conductor shield	—	—	$\leq 1 \times 10^5$	—	$\leq 1 \times 10^5$
	Crosslinked insulation shield	—	—	$\leq 5 \times 10^4$	$\leq 5 \times 10^4$	
High-voltage XLPE (AEIC-CS7)	Crosslinked conductor shield	—	—	$1 \leq \times 10^5$	—	$\leq 1 \times 10^5$
	Crosslinked insulation shield	—	—	$\leq 5 \times 10^4$	$\leq 5 \times 10^4$	

Source: After [16].

tion, its partly crystalline structure is more compatible with the crosslinked semiconducting shields [16].

Among the types of carbon black available, acetylene black, which is a thermal form of black, has proven to be the most acceptable carbon black for semiconducting shields because of its lower impurity content as compared to other types of carbon blacks. It is produced by an exothermic decomposition of acetylene at $\sim 1000°C$ in water-cooled cylindrical containers [27]. Carbon black is essentially a polycrystalline solid, with the primary carbon particle containing between 10^5 and 10^9 atoms. The individual carbon atoms are arranged in an hexagonal structure [28, 29], and in the case of acetylene black, the crystallites consist of an average of seven planes [30, 31]; the average diameter of the parallel layers is 1.7 nm, and the average thickness perpendicular to the hexagonal plane is greater than 1.2 nm. The particle or grain size of various carbon blacks varies usually from 10 to 300 nm, but for semiconducting shield applications average particle sizes range from 24 to 70 nm as indicated in Table 2.4 [16].

In Table 2.4, the surface area of the carbon black aggregates is determined in terms of liquid nitrogen ($-196°C$) isotherms at a number of relative pressures. A high value of dibutyl absorption is indicative of increased porosity of the carbon black. This value

TABLE 2.4   Physical Properties of Semiconducting Shield-Grade Carbon Black

Parameter	Furnace Black	Acetylene Black	Extra Conductive Black
Primary particle size, nm	24–33	35	30
Nitrogen surface area value, m^2	70–250	85	800
Dibutyl phthalate absorption value, mL/100 g of carbon black	125–170	200	365

Source: After [16].

complements that of the area size, since a large area size may also characterize porous carbon black aggregates and should be distinguished from that obtained on large nonporous aggregates. Although the particle size of the three carbon blacks used for semiconducting shields is generally on the order of 30 nm, the conductivity of the semiconducting shields is not equal at equivalent amounts of the particular type of carbon blacks used. This is palpably evident from Fig. 2.3, from which it can be perceived that acetylene carbon particles when incorporated into a polymer matrix yield the lowest conductivity per concentration amount as a result of their exceptional dispersion characteristics. It is due to the dispersive properties of the acetylene carbon black that the semiconducting shields are characterized by their surface smoothness and are thus the preferred form of carbon black for this purpose. A concentration of 50 parts of carbon black per hundred parts of polymer constitutes the upper limit of carbon black concentration, since, beyond this limit, any increases in conductivity are offset by the practical constraint of reduced processability and poor mechanical properties of the extruded semiconducting shield layer. The use of acetylene blacks has resulted in the so-called supersmooth semiconducting shields, having appreciably lower numbers of protrusions. This is manifestly substantiated in Fig. 2.4(a), which compares the number and size of protrusions of regular and supersmooth insulation shields. The size or height of the protrusions was measured by means of a laser profilometer, with minimum protrusion height detection sensibility of 10 μm, which is appreciably better than the conventional optical microscope technique with a magnification of ×200 that is ineffective for measurements below protrusion heights less than 25 μm. The supersmooth semiconducting material intended for conductor shields must conform to higher standards than that of insulation semiconducting shields. Burns et al. [16] show that a 5 μm-long protrusion on the conductor shield of a 15-kV cable with an insulation thickness of 175 mils may lead to stress enhancement factors on the order of 30–210 times that of the normal stress at the conductor surface depending upon the geometry of the asperity point. Figure 2.4(b) presents typical protrusion density and height data obtained on regular and supersmooth conductor shields. Comparison of Figs 2.4(a) and 2.4(b) demonstrates the augmented smoothness of a supersmooth semi-

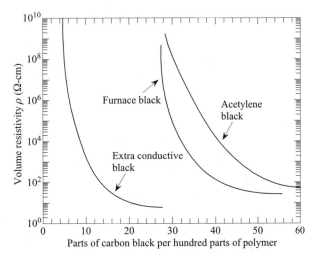

Figure 2.3 Volume resistivity vs. carbon black content with type of carbon black as a parameter (after [16]).

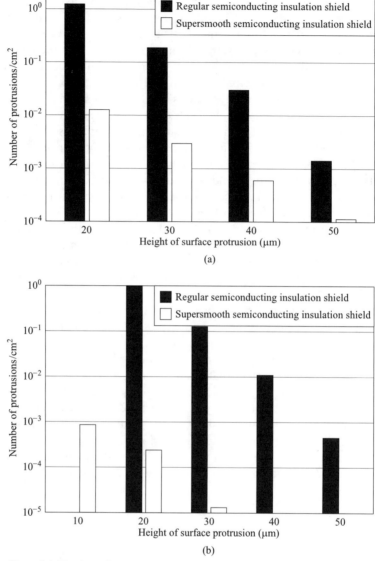

Figure 2.4 Number of protrusions per unit area vs. their height for regular and supersmooth semiconducting (a) insulation shields and (b) conductor shields (after [16]).

conducting conductor shield vis-à-vis that of an insulation shield for which the requirements are less stringent due to the lower electrical stresses employed.

There has been considerable evidence accumulated to indicate that ionic impurities contained in the carbon black particles within the semiconducting shield materials may be implicated in failures associated with water tree growth in distribution cables operating in wet environments. In their classical study Mashikian and Groeger [31] reported the detection of ionic minerals in the channels of water trees in polyethylene insulation adjacent to the semiconducting conductor shields, which infer the involvement of ions from the shields in the water tree growth process. At this time it is not obvious how these ions influence the water tree initiation and propagation rate process, though it is believed that the movement of the ions in the alternating electrical field and the resulting space charge field from the trapped charges may make a contribution to the water tree growth [32]. From the practical point of view, it is well recognized that reduction of the ionic content in the semiconducting shield decreases the incidence of water trees. This contention is supported by some early observations of water tree growth in cables installed in highly ionic contaminated environments [33]. The ionic impurities contained in the carbon blacks are both metallic (cations) and nonmetallic (anions). The latter generally originate from carbonates, halides, and sulfates; Burns et al., [16] used inductively coupled plasma emission spectroscopy on specimens of the ash of the different types of carbon black to quantify and identify the various cation species present. The results with some other pertinent data are presented in Table 2.5, from which it is apparent that acetylene carbon black is characterized by having the least cation concentration of the three carbon blacks normally utilized in semiconducting shield applications. Since some metallic ions such as sodium, potassium, calcium, and magnesium are anticipated to contribute significantly to the overall electrical conductivity of the carbon black incorporating polymers, this may explain (in part only) the lower conductivity per unit carbon black concentration in the polymer observed with acetylene black in Fig. 2.3. In terms of impurity effects the lower ash, sulfur, and volatile contents found in acetylene carbon black are significant in that the amount of ash reflects the inorganic content; the sulfur content is implicated in the aging rate of the cable; and the volatile content refers to the carbonyl and hydroxyl groups attached on the surface of the carbon particles under normal temperature operating conditions of the cable.

To improve the thermal and mechanical characteristics of semiconducting shield materials much effort has gone into the perfection of XLPE semiconducting shields, which may be utilized at higher temperatures [16, 34]. As a consequence the thermoplastic material shields that exhibit increased deformation at temperatures beyond $90°C$ have been discontinued and are no longer used on XLPE insulated cables. With the development of the triple extrusion process, the three layers comprising the conductor semiconducting shield layer and the insulation could be crosslinked simultaneously in

TABLE 2.5 Metallic Ion Contamination Levels in Ash Specimens of Typical Carbon Black Used for Semiconducting Shields

Property	Clean Furnace Carbon Black	Acetylene Carbon Black	Extra Conductive Carbon Black
Ash content, % by weight	0.02	< 0.01	0.5
Sulfur content, % by weight	0.05	0.001	0.30
Volatile content at 900°C, % by weight	1.0	0.3	0.5
Aluminum, ppm	3	< 1	100
Calcium, ppm	6	< 1	< 1
Copper, ppm	< 1	< 1	< 1
Iron, ppm	20	5	3
Magnesium, ppm	3	< 1	8
Potassium, ppm	50	< 1	2
Silicon, ppm	12	5	125
Sodium, ppm	10	< 1	30
Zinc, ppm	< 1	< 1	< 1

Source: After [16].

the vulcanizing tube. A further improvement in adhesion between the semiconducting shields and the insulation followed the development of a tree-retardant crosslinked polyethylene insulation and the deployment of radiant nitrogen vulcanizers [16]. All of the foregoing improvements led to some practical difficulties concerning the stripping of the insulation semiconducting shields for the purpose of cable termination and splicing. From the practical point of view the problem was adequately solved with the development of strippable semiconducting shields. However, it has never been satisfactorily described in what manufacturing process and which material additives impart the semiconducting materials the strippable properties. No doubt the concentration and geometry of the carbon black particles, semiconducting shield thickness, and temperature affect the strippability characteristics of the semiconducting insulation. Burns et al., [16] have carried out an empirical comparison of the adhesion characteristics of the early and current strippable XLPE insulation semiconducting shields; from Fig. 2.5 the improvement in insulation shield strippability over that of one decade ago is apparent; the current shields exhibit reduced adhesion and less variation of the adhesion with insulation shield thickness.

With increasing temperature, bonding or cohesive failure of the insulation semiconducting shields was often found to occur at higher ambient temperatures. When such bonding of the semiconducting shield occurred, removal of the film of the semiconducting shield from the surface of the insulation could only be effected by reducing the temperature of the environment. Moreover, Burns et al. [16] found that with the 1980s XLPE insulation semiconducting shields, the onset temperature of such bonding was on the order of 35°C (cf. Fig. 2.6). As the data of the obtained results indicate in Fig. 2.6, bonding appears to pose no problems for the XLPE insulation semiconducting shields that are of recent manufacture. Apart from the specific adhesion requirements for the particular semiconducting shields, the mechanical properties of both the semiconducting conductor and XLPE insulation shields are identical (cf. Table 2.6).

2.4 INSULATION

The vast majority of conventional power cables are insulated with either solid extruded dielectrics or liquid-impregnated papers. The use of the former now generally

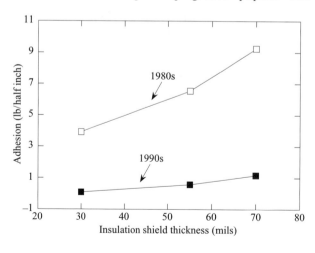

Figure 2.5 Adhesion characteristics of XLPE semiconducting insulation shields as a function of shield thickness for those manufactured in the 1980s and 1990s (after [16]).

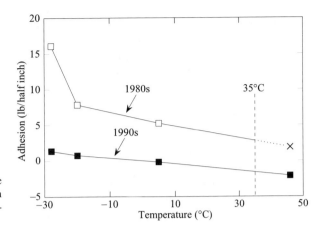

Figure 2.6 Effect of temperature on the adhesion characteristics of XLPE insulation semiconducting shields that were manufactured in the 1980s and 1990s (after [16]).

TABLE 2.6 Mechanical Properties of Semiconducting Conductor and Insulation XLPE Shields

Tensile strength	≥ 1800 psi
Tensile elongation	$\geq 150\%$
Brittleness at $-40°C$	$\leq 20\%$ failure rate

Source: After [16].

dominates the distribution power cable field, whereas the latter is still extensively used in high-voltage power transmission cables. Although the solid extruded insulating materials have undergone marked changes with time, the changes in the liquid-impregnated papers have been substantially less salient. In addition, considerable work has been carried out on other insulating systems, such as compressed gas-spacer systems and the more novel systems using cryogenic liquids; however, concerning the two later types of insulation, only the gas insulated systems have thus far attained a significant degree of acceptance in practical installations.

2.4.1 Dielectric Characteristics of Solid and Solid-Liquid Systems

Only a gas in its un-ionized state may be considered to be virtually free of dielectric losses. In solid and liquid insulation, dielectric losses arise due to the movement of electrons and ions, orientation of dipoles and interfacial polarization as a result of trapped space charges. The electrical conductivity σ and the real permittivity or dielectric constant, ε' constitute the two principal parameters that determine the electrical characteristics of an insulating material [35]. If in the presence of an external electrical field E the resulting leakage current density in the insulating material is J_l, then conductivity is given by

$$\sigma = \frac{J_l}{E} \tag{2.13}$$

If J_l is in amperes per centimeter squared (A/cm^2) and E in volts per centimeter (V/cm), then the units assumed by σ are siemens per centimeter (S/cm) or reciprocal ohms·centimeters ($1/\Omega \cdot$ cm). Should the leakage current density J_l arise entirely from the movement of free charge carries of one sign, then the conductivity may be alternatively expressed as

$$\sigma = e\mu n \tag{2.14}$$

where e is the charge of the mobile charge carriers, n is their concentration per cubic centimeter and μ is their mobility in centimeters squared per volt·second (cm^2/V · s). The conductivity σ is normally determined on material slabs in terms of their insulation resistance, R. For a measuring electrode area of A centimeters squared and a slab thickness of d centimeters, the conductivity is then given by

$$\sigma = \frac{d}{AR} \tag{2.15}$$

Good cable insulating materials have conductivities in the range of 10^{-16}–10^{-20} S/cm. Under dc conditions the conductivity is usually determined in terms of the dc insulation resistance R, by means of an electrometer, following a 1-min electrification in accordance with ASTM Standard Method D257 [23]. Often in lieu of the conductivity σ, the insulation resistivity ρ is determined. Since the resistivity is by definition the inverse of the conductivity (i.e., $\rho = 1/\sigma$), its units are in ohm-centimeters.

The real value of the permittivity or dielectric constant ε' is defined by

$$\varepsilon' = \frac{C}{C_0} \tag{2.16}$$

where C represents the measured capacitance in farads (usually at 60 or 1000 Hz) in accordance with ASTM Standard Method D150 [36] and C_0 is the capacitance in vacuo and is calculated from the geometry of the slab specimen using the expression

$$C_0 = \frac{\varepsilon_0 A}{d} \tag{2.17}$$

Here ε_0 denotes the permittivity in vacuo and is equal to 8.854×10^{-14} F/cm [8.854×10^{-12} F/m in the Système International (SI) units]. For air the real value of the permittivity ε' is equal to $1.000536\varepsilon_0$, while for most solid and liquid insulating systems employed for cables, the value of ε' at room temperature lies within the range of approximately $2\varepsilon_0$ to $3.5\varepsilon_0$.

When solid and liquid insulating materials are subjected to an alternating field E, the presence of losses causes the permittivity ε to assume a complex form

$$\varepsilon = \varepsilon' - j\varepsilon'' \tag{2.18}$$

where ε'' represents the imaginary part of the permittivity and is by definition given as

$$\varepsilon'' = \frac{\sigma}{\omega} \tag{2.19}$$

where ω is the radial frequency term. The value of σ in Eq. (2.19) must be distinguished from that obtained under dc conditions in Eqs. (2.13)–(2.15) because it now includes, in addition to the long-range charge carrier drifts in a direct field, the loss mechanisms arising from dipole orientation and interfacial or space charge polarization. Since the complex permittivity relates the dielectric displacement or flux density vector $\bar{\mathbf{D}}$ to the electric field vector $\bar{\mathbf{E}}$ in the relation

$$\bar{\mathbf{D}} = \varepsilon \bar{\mathbf{E}} \tag{2.20}$$

its complex value denotes a phase shift between the $\bar{\mathbf{D}}$ and $\bar{\mathbf{E}}$ vectors. If this phase angle difference is taken to be δ, then in complex rotation $\bar{\mathbf{D}}$ and $\bar{\mathbf{E}}$ become $D_0 \exp[j(\omega t - \delta)]$ and $E_0 \exp[j\omega t]$, respectively, where t represents time and D_0 and E_0 are the magnitudes of the two vectors. Expressing the exponential terms as trigonometric functions, it follows from Eqs. (2.18) and (2.20) that

$$\varepsilon' = \frac{D_0}{E_0} \cos \delta \tag{2.21a}$$

and

$$\varepsilon'' = \frac{D_0}{E_0} \sin \delta \tag{2.21b}$$

Thus from Eqs. (2.19), (2.21a), and (2.21b),

$$\tan \delta = \frac{\varepsilon''}{\varepsilon'}$$

$$= \frac{\sigma}{\omega \varepsilon'} \tag{2.22}$$

Here $\tan \delta$ is the dissipation factor of the insulating material. For power cables, the $\tan \delta$ value of the insulation should be as low as possible to minimize the dielectric power losses at the frequency of 60 Hz over the range of operating temperatures, particularly for the high-voltage transmission cables for which the dielectric loss component becomes a significant portion of the overall loss. For communication cables the $\tan \delta$ and ε' values should be low over the operating frequency spectrum to reduce the attenuation losses and maintain a high velocity of propagation. For power cables a low ε' is also desirable for the purpose of reducing their capacitive or reactive load on the power system. Some polymers such as polyethylene though appreciably susceptible to breakdown due to discharges and treeing are characterized by low real permittivity and $\tan \delta$ values, ~ 2.2 and $\sim 10^{-4}$, respectively, at low frequency and room temperature. The more lossy oil-impregnated kraft papers with corresponding values of $\varepsilon' \simeq 3.5$ and $\tan \delta \simeq 2 \times 10^{-3}$ are more resistant to treeing and the effects of discharges. When power cables are employed for dc applications, other considerations must be taken into account. For example, polyethylene insulated cables, in which charge carriers are held in both shallow and deep traps and which exhibit typically a low insulation conductivity, perform poorly under dc conditions. The deep electronic charge traps in the polyethylene insulation, trap and retain the electrons thereby greatly reducing their mobility in an electric field. When polarity reversal occurs across a dc cable, the space charge field of the trapped charge carriers adds to the external field; if the value of the two superimposed fields exceeds that of the breakdown field of the cable insulation, a catastrophic failure follows. In contrast a more lossy oil-paper insulation performs extremely well under dc conditions because charges from the shallow traps of ~0.5 eV (which are 2–3 times less than the deep electronic charge traps found with poly-ethylene insulation) [32, 37, 38] are more readily ejected, thus precluding a build-up of large space charge fields. Deep traps in the low-loss polyethylene insulation, which immobilize the charge carriers and thereby impart the polyethylene its excellent low dielectric loss properties, may at the same time be responsible for the propensity of the material toward tree growth. Incorporation of conductive dopants in polyethylene will tend to dissipate the trapped charge carriers and thus render it less susceptible to tree growth.

Over the power frequency regime, the dielectric losses are to a large extent determined by the movement of mobile charge carriers, which may be electrons, ions, or both. The most effective means of demonstrating that mobile charge carriers contribute to the overall dielectric loss within a given cable insulating material is to measure its dc conductivity as a function of temperature. Such a result is illustrated in Fig 2.7, which gives data obtained on EPR slab specimens representing several formulations and on filled and unfilled XLPE slab specimens, having thicknesses within the range of 404–770 μm. The decrease of dc conductivity σ with the inverse absolute temperature infers that the conduction process is governed approximately by a relation of the form

$$\sigma_{\text{dc}} = \sigma_0 \exp\left(-\frac{\Delta H}{kT}\right) \tag{2.23}$$

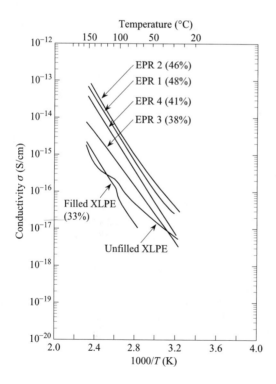

Figure 2.7 The dc conductivity of XLPE and EPR specimens as function of the inverse absolute temperature $1/T$, with filler content indicated in parentheses (after [39]).

which characterizes a thermally activated conduction mechanism. Here σ_0 denotes a preexponential conductivity factor, ΔH the activation energy of dc conduction, k the Boltzmann constant, and T the absolute temperature. The deviation of the characteristics in Fig. 2.7, from a straight-line relationship implies that Eq. (2.24) does not precisely describe the experimental conduction behavior of EPR and XLPE; however, over the more linear segments of the characteristics an activation energy of 1.0 eV is obtained with the exception of specimens EPR 3 and the unfilled XLPE for which ΔH is found to be 0.75 and 0.5 eV, respectively. The latter two values infer that the observed conduction mechanism in these two particular specimens is dominated by shallow traps; the current can thus be envisaged as consisting of charge carriers that hop between the shallow trap sites while any charge carriers that may be trapped in the deep traps remain immobilized and make no contribution to the conduction current. The range of the measured activation energies for the specimens evaluated in Fig. 2.7 is on the order that may be typical of either electronic or ionic conduction losses, and it is thus difficult to establish in terms of the ΔH values alone whether an electronic or ionic conduction process predominates. Nevertheless, other information may be employed to ascertain which process prevails. For example, the stepwise increase in conductivity, readily discernible in the proximity of the glass or melting transition of either filled or unfilled XLPE with rising temperature, suggests a stepwise contribution of the ionic charge carriers. (cf. Fig. 2.7 in the vicinity of ~100°C). With increasing temperature, some otherwise immobilized ions are rendered mobile due to structural changes arising from the melting of crystallites and softening of the amorphous phase as may occur in cables under prolonged higher temperatures under emergency operating conditions. In addition a thermosetting crosslinked polymer such as XLPE has various ion-containing impurities, which form chemical agents such as peroxides that are utilized to carry out the crosslinking. In addition, the water content of the various fillers employed for EPR or in filled XLPE facilitates ionic movement both at low and elevated temperatures.

At the lower temperatures, over which electronic conduction would be expected to play the predominant role, the highly reduced values of measured dc conductivity must be attributed to various trapping sites typically situated at the crystalline-amorphous interfaces (e.g. at the terminal vinyl groups in polyethylene and at various oxidation sites). As the temperature is increased, more of the thermally activated charge carriers are emitted from their traps, and the conductivity increases as both ions and electrons in increasing numbers hop between the various charge traps in the localized states of the polymer in the direction of the electrical field. Figure 2.7 was obtained at an electrical field, which was on the order of 20 kV/cm; at such field strengths, electrons are injected at the measuring electrode, causing the conduction process to deviate from the normally linear ohmic behavior. As the electrons are injected from the electrode surface, some become trapped in the vicinity of the electrodes, and the resulting space charge accumulation becomes a limiting factor on the conduction process. At this point in time, the electron trapping and conduction mechanism is said to become space charge limited. The manner in which the electrical potential influences the conductivity behavior in a XLPE specimen of ~430 μm thickness is portrayed in Fig. 2.8, which infers a power law dependence of the dc conduction current upon the applied potential over the more elevated voltages. Over the lower voltage, an ionic conduction behavior is clearly perceptible.

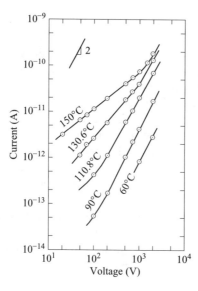

Figure 2.8 Current-voltage characteristics of a specimen of unfilled XLPE (after [39]).

Since both the electronic and ionic conduction losses are seen to be augmented by an increasing electrical field, the losses in the cable dielectric will be higher for cable insulations operated at the more elevated voltage stresses. Any dipole losses, if present in the cable insulation, will exhibit a loss that is invariant with voltage but that may change appreciably with temperature depending upon the value of the relaxation time characterizing the dipole orientation mechanism. This will be contingent upon the size of the dipoles, since in general only large molecular dipoles will give rise to losses over the power frequency regime. The effect of small molecular dipoles will manifest itself at frequencies in the radio frequency regime and will be of concern primarily for dielectrics employed in communication cables. If the dielectric loss in crosslinked polyethylene is measured at the fixed frequency of 60 Hz as a function of temperature, the so-called α absorption peak is observed at the crystalline melting point. It is believed to be associated with the relaxation of the carbonyl groups in the crystalline phase [40]. In elastomeric solids, such as EPR, it is conceivable that the rise of dielectric loss with crosslinking could result from the introduction of polar impurities produced by the chemical crosslinking reaction [41].

Figures 2.9 and 2.10 depict the real value of the dielectric constant and dissipation factor as a function of temperature at a frequency of 60 Hz, obtained on filled and unfilled XLPE as well as four formulations of EPR with the percentage of filler as indicated in the parentheses. It can be seen that the dielectric constant and dissipation factor of the filled materials are substantially higher than those of the unfilled XLPE over the entire temperature range. Both the unfilled and filled XLPE specimens exhibit a characteristic mutation in the dielectric constant in the vicinity of the crystalline melting transition temperature; the decrease of the dielectric constant with temperature for all the XLPE and EPR compounds reflects to an appreciable extent a diminution in their density with temperature. Nevertheless a certain amount of this decrease must be ascribed to dielectric absorption effects, as is evidenced by the unfilled XLPE over the crystalline melting region in Fig. 2.10. The observed α-type peak in tan δ is not apparent with the filled XLPE, where it may conceivably be obscured by the appreciably higher

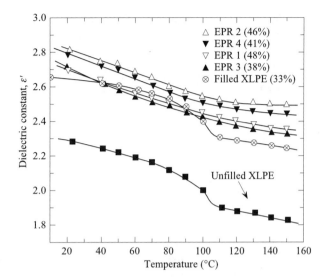

Figure 2.9 Dielectric constant of XLPE (filled and unfilled) and EPR as a function of temperature at 60 Hz, with percent filler content as indicated in parentheses (after [39]).

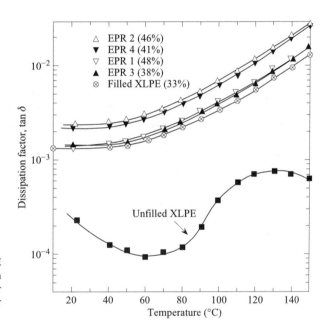

Figure 2.10 Dissipation factor of XLPE (filled and unfilled) and EPR as a function of temperature at 60 Hz, with percent filler content as indicated in parentheses (after [39]).

dielectric losses that occur within the filler phase of the dielectric. Since polymers containing fillers tend to absorb significant amounts of water when exposed to high-humidity conditions (c.f. Table 2.7), one would anticipate increased interfacial polarization losses, as well as enhanced ionic conductivity due to ionic impurities in particular at the higher temperatures. The rotational movement of various polar substances such as residual carbonyl groups and crosslinking by-products contributes directly to the dielectric losses; it influences also indirectly the dielectric loss by creating free volume spaces along which impurity ions can migrate. The involvement of ions in

TABLE 2.7 Water Content of Filled and Unfilled Polymer Insulation after Exposure to Steam and Water

Polymers[a]	Density at 25% (g/cm^3)	Filler Content (% by volume)	Water Content (ppm)
Unfilled XLPE	0.918	—	350
Filled XLPE (33%)	1.20	33	1150
EPR 1 (48%)	1.32	48	3200
EPR 2 (46%)	1.28	46	2900
EPR 3 (38%)	1.22	38	1150
EPR 4 (41%)	1.29	41	1750

[a]Filler content indicated as a percentage in parentheses.
Source: After [39].

the overall dielectric loss process is inferred in Fig. 2.11, which indicates increased losses with diminishing frequency at the elevated test temperatures. At still lower frequencies the ionic motion may become limited at the electrodes, giving rise to a space charge governed loss peak.

In contrast to XLPE, the dielectric losses in oil-impregnated paper are substantially higher; however, they are still less than those in EPR. The solid-liquid interfaces formed within the oil-impregnated-paper insulation provide boundaries where space charge forms, thus leading to appreciable interfacial polarization losses at both the paper tape–oil interfaces as well as on a smaller scale at the paper fiber–oil interfaces within the paper pores of the tapes themselves. The dielectric loss mechanisms in oil-paper systems are discussed in greater detail in the subsequent section and in Chapter 9, which deals in part with dissipation and ionization factor measurement on oil-impregnated-paper cables. Thus for comparison purposes, it will suffice to consider here in a cursory manner only the dissipation factor as a function of frequency with temperature as a parameter. The specimens employed here for illustrative purposes consist of cable models constructed with 5 mils kraft paper tapes of 0.73 g/m^3 density and impregnated with a low viscosity (9.2 cP) hollow-core-type cable oil and a high-viscosity (1260 cP) solid-type oil-impregnated-paper cable oil. The viscosities of the oils refer to a temperature of 27°C.

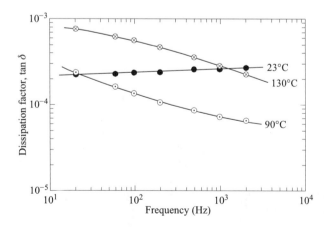

Figure 2.11 Dissipation factor vs. frequency with temperature as parameter for unfilled XLPE (after [39]).

Figure 2.12 shows that the dissipation factor in both cable models exhibits an increase with frequency over the higher frequency regime, which is a manifestation of the higher dipole losses in both the liquid and solid phases of the insulation system. At the elevated temperature the dipole absorption frequency shifts to a greater value and the tan δ values are lower over the range of frequency measured. The ionic conduction losses become predominant at the higher temperatures in the vicinity of 60 Hz and lower, as can be perceived from the increase of tan δ with falling frequency from the 70 and 71°C characteristics. Since ionic mobility increases with temperature or with falling viscosity of the oil, the higher tan δ values with diminishing frequency infer a significant ionic content in the low viscosity oil-impregnated-paper cable model and suggest also the existence of a space charge absorption peak at low frequencies ($\ll$10 Hz).

The dielectric breakdown strength of both oil-paper and solid-dielectric extruded polymer insulation is well above their normal accepted operating electrical stress values. The breakdown strength of oil-impregnated-paper cable insulation is typically on the order of 400 kV/cm, while that of polymeric insulation, such as polyethylene, is generally in excess of 700 kV/cm. The breakdown strength decreases with time that the cables are in service, and this is particularly perceptible for solid polymeric insulation for which it constitutes an effective measure of the rate of aging. The dielectric strength is determined using parallel-plane electrodes as described in ASTM Method D149 [43] or International Electrotechnical Commission (IEC) 243 [44]. Both of these standard test methods permit the use of either equal diameter aligned parallel-plane electrodes or unequal diameter electrodes for which exact alignment is not critical. Either type of test electrodes lead to approximately the same values of breakdown voltage, which are substantially below the true or intrinsic breakdown strength of the dielectric specimen tested. The latter value, which normally is on the order of $\sim$10^6–10^7 V/cm, may be obtained by means of recessed electrodes in order to yield more uniform electrical fields. However, as in practice the breakdown voltages of cables never approach those of the intrinsic strength for any insulation material, recessed electrodes are not used in routine breakdown tests. Even though breakdowns occur often at the edges of the equal

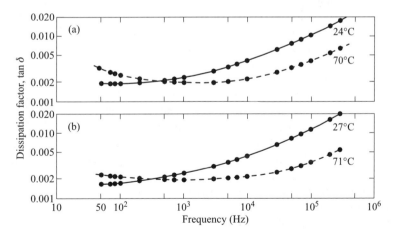

Figure 2.12 Dielectric relaxation spectra of kraft paper cable models impregnated with (a) low-viscosity (9.2 cP) and (b) high-viscosity (1260 cP) mineral oils (after [42]).

diameter electrodes or at the edge of the smaller diameter test electrode (for unequal diameter electrodes), the breakdown values obtained on slab specimens are of the same order as those obtained on actual cables.

There are a number of variables and parameters that determine the dielectric strength of an insulating material such as its molecular structure and geometry, geometry of test electrodes, as well as temperature. Since the actual breakdown event takes place in a time of $\sim 10^{-8}$–10^{-6} s, an electronic regulated mechanism is inferred. Also since under dc and impulse conditions the breakdown strength tends to be greater than under alternating fields, the overall ac breakdown process involves necessarily a thermal component. In addition, the repetitive occurrence of partial discharges in void occlusions under alternating fields contributes significantly to the reduction in the ac breakdown strength. If, for simplicity, one considers a crystalline polymer, then the thermal breakdown criterion can be more precisely postulated as a condition for which the average rate of energy gain by the free electrons accelerated in the direction of the field, $A(E, T, T_e, \xi)$, exceeds that lost in collisions with the crystal lattice $B(T, T_e, \xi)$; thus, the breakdown criterion may be expressed as [45]

$$A(E, T, T_e, \xi) = B(T, T_e, \xi) \tag{2.24}$$

where E denotes the applied electrical field, T the lattice temperature, T_e the electron temperature and ξ an energy distribution constant. Accordingly, the breakdown voltage continues to increase initially with temperature as a result of the augmented thermal vibrations of the lattice until, ultimately, a critical temperature is reached at which the electron-electron interactions surpass in importance the influence of the electron-lattice interactions, and the breakdown strength commences to decline. The forgoing behavior is principally observed with simple crystals and is not apparent in partially crystalline XLPE and the amorphous EPR as may be perceived from Figs 2.13(a) and (b), which delineate the ac breakdown strength as a function of temperature. The behavior of the unfilled XLPE indicates that as the temperature increases and the crystalline regions begin to undergo melting, the breakdown strength diminishes very rapidly and stabilizes after the melting transition temperature, approaching the level of the breakdown strengths observed with the EPR compounds, which exhibit little discernible systematic variation over the testing temperature range. The filled XLPE displays a fall in its breakdown strength in the vicinity of the melting transition temperature but which is much less pronounced than that of the unfilled XLPE, thus retaining a higher breakdown strength than the unfilled XLPE over the more elevated temperatures.

The variation of the impulse breakdown strength of XLPE, filled XLPE, and EPR compounds as a function of temperature is substantially similar to that observed with the ac breakdown strength. This is apparent from the impulse breakdown data portrayed in Figs. 2.14(a) and (b); however, it is noteworthy that the impulse breakdown values are considerably higher than the corresponding ones obtained under alternating voltage. This result together with the common observation made that the dc breakdown strength also exceeds that of the ac breakdown strength infer that under ac conditions the breakdown may be in part of a thermal nature. The thermal breakdown process involves the development of hot spots wherein heat is generated at a higher rate than the surrounding medium is capable of dissipating. The temperature rise at such sites may attain values at which fusion and vaporization can take place, thereby causing channel development along which breakdown eventually occurs. Thermally induced

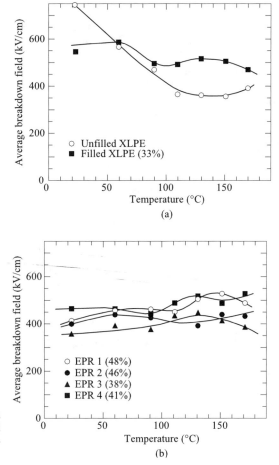

Figure 2.13 The 60-Hz breakdown strength vs. temperature for (a) filled and unfilled XLPE and (b) EPR, with filler content indicated in parentheses (after [39]).

breakdown is contingent upon the time of voltage application and is thus often implicated as a leading cause of breakdown in insulation systems under long-term operating conditions; nevertheless, it may also arise under short voltage pulse conditions. However, the probability of its occurrence in low-loss dielectrics, such as pure polyethylene, is relatively low. The condition for thermal breakdown is given by

$$\kappa A \; \Delta T/l = \omega \varepsilon' E^2 \tan \delta \tag{2.25}$$

where the left-hand side of the equation represents the heat transfer in joules per second (J/s) along a dielectric slab of length l in centimeters having a cross-sectional area A in centimeters squared in the direction of the temperature gradient due to a temperature difference ΔT; κ is the thermal conductivity constant in J/°C · cm · s. Note that κ is also referred to as the *coefficient of thermal conduction* or simply as the *thermal conductivity*. Its derived SI units are W/K · m or W/°C · m, but its CGS units of cal/°C · cm · s are still very much in common usage. The right-hand side of the equation comprises the dielectric loss dissipated in the dielectric in joules per second, where ω is radial frequency, E

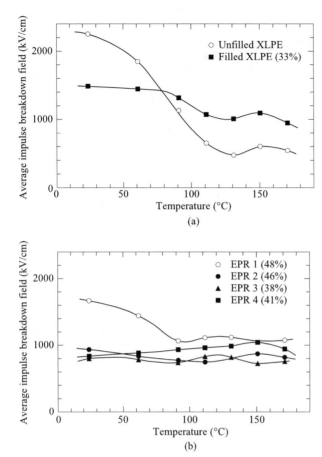

Figure 2.14 Impulse breakdown strength vs. temperature for (a) filled and unfilled XLPE and (b) EPR, with the filler content indicated in parentheses (after [39]).

the external electrical field gradient, ε' the real value of the permitivity, and $\tan\delta$ the dissipation factor.

It should be mentioned that an alternate theory for voltage breakdown is that based on the electron avalanche concept to explain the increase in the dielectric strength with decreasing thickness of the specimen under test. In the electron collision ionization theory, it is argued that positive ions, created along the path of the avalanche, drift toward the cathode where they accumulate forming a positive space charge. The latter enhances the electrical field at the cathode, which in turn further augments the electron emission until ultimately instability develops and breakdown ensues [45]. Support for the involvement of space charges in the dc breakdown process is provided by some earlier results obtained on polyethylene [46], which demonstrate the dc breakdown strength to be lower than the impulse breakdown strength at temperatures less than 30°C while above this temperature the converse situation prevails. The observed behavior suggests that over the lower temperature regime the dc breakdown process is space charge governed because there is ample time for the slow-moving ions to form a space charge region in the presence of a slowly rising dc field. Whereas at the more elevated temperature the accompanying increase in ionic mobility hinders charge trapping and space charge build-up is less likely to occur.

The breakdown strength of insulating liquids usually falls within the range from 80 to 300 kV/cm, yet when they are used as impregnants in conjunction with solid insulation such as suitably dried and degassed kraft papers or plastic tape substitutes, the resulting dielectric strength of the solid–liquid combination may attain values in the range from 400 to 800 kV/cm. Breakdown in the insulating liquid usually progresses from a field intensification site, such as an asperity on the test electrode surface where field emission causes the development of a gaseous phase wherein partial discharges form, ultimately leading to streamer channel formation and breakdown [47]. However, with the interposition of solid insulating tapes to form a barrier between the oil films in a cable, the possibility of this type of breakdown is greatly mitigated and completely eliminated in oil-pressurized cables. When plastic tapes are used either to replace the kraft paper tapes or alternatively are utilized in conjunction with the kraft paper tapes, as, for example, in the well-known composite paper-polypropylene-paper (PPP) tapes, great care must be taken to achieve full impregnation of the overall insulating system. The PPP tapes must be conditioned at a higher temperature ($\sim$80–100°C) to permit ingress of the impregnant into the amorphous-crystalline interfaces of the polypropylene insulant, i.e., to permit the filling of all the free volume within the plastic tapes. Completion of this filling process is indicated by a swelling of the polypropylene tapes to some limiting value of thickness, resulting in an improved breakdown strength of the solid-liquid insulation system. Figure 2.15 depicts the ac and impulse breakdown characteristics of a dodecyl-benzene-impregnated PPP insulating system; although the initial rise in both the impulse and the ac breakdown strengths is in part a consequence of the soaking or conditioning influence of the dodecyl benzene at the higher temperatures, the subsequent fall in both ac and impulse breakdown strengths at temperatures greater than 80°C must be attributed to another cause. One may conjecture that in view of the 55–65% crystallinity of the polypropylene tapes themselves, the reduction in breakdown strength at the higher temperatures may be ascribed to the effect of the crystalline region's matrices. However, it will be recalled that no such behavior is manifest in XLPE, which also has appreciable crystallinity. For the purpose of comparison, tests

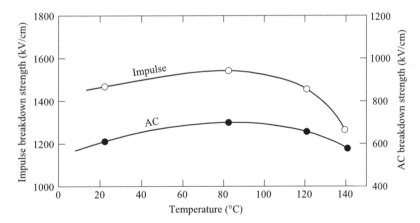

Figure 2.15 The ac and impulse breakdown strength as a function of temperature obtained on cable models constructed with PPP tapes impregnated with dodecyl benzene for an overall insulation thickness of 2.0 mm (after [48]).

on similar cable models constructed with kraft paper tapes and impregnated also with dodecyl benzene demonstrated the impulse and ac breakdown values to be substantially lower than those obtained with the PPP tapes [48]. In the case of kraft paper, the impulse breakdown strength of 1170 kV/cm at room temperature decreased to 1100 kV/cm at 80°C; ac dielectric strengths at the same two temperatures were 546 and 538 kV/cm, respectively.

2.4.2 Oil-Impregnated Paper

Oil-impregnated papers have been used since the earliest days of cable development and constitute even today one of the most extensively used cable insulants. Despite the recent advances made in the field of plastics for cable application, oil-paper insulation is still regarded as perhaps the most reliable composite insulation system for cable applications. Kraft papers consist of cellulose fibers felted together to form mechanically strong sheets. When properly dried and impregnated with dielectric liquids, electrical-grade kraft papers have good dielectric strength and loss characteristics suitable for both ac and dc cable applications. Kraft paper tapes possess a variety of other useful properties, such as lightness, flexibility, compatibility with insulating liquid impregnants, and may be manufactured to provide different grades of density, surface friction, strength, and thickness. The paper tapes employed on cables usually have a thickness of 5 mils, but thicknesses in the range from 2.5 to 7.5 mils are also used; the thinner tapes are commonly employed where high dielectric strength is required. The density of the kraft paper tapes is usually on the order of 0.75 g/cm^3. Consequently, as the density of the paper is a measure of the total cellulose fiber content, the density of the cellulose fibers themselves is 1.54 g/cm^3, so that the oil-impregnated-paper insulation contains roughly one half of the oil impregnant by volume. The porosity of the paper determines the ease with which the paper may be dried and impregnated. The mechanical strength of the paper is rigidly controlled to prevent tearing during the cable lapping operations; in the longitudinal and transverse directions, the strength of the kraft paper tapes is on the order of 11,000 and 5000 psi, respectively.

There are a number of parameters describing electrical insulating papers that are of considerable importance as they influence greatly both the electrical and mechanical properties of the kraft paper tapes [49]. One of these is the already alluded to paper density, which is expressed as a ratio of the weight to unit volume. Higher density paper tapes contain more water and thus tend to give rise to higher dielectric losses. On the other hand, they are less porous and thereby have a higher tensile strength and, when impregnated with oil, exhibit a higher dielectric breakdown strength. Since the higher density tapes, when impregnated with oil, are also characterized by a higher dielectric constant, the voltage gradient or stress in high-voltage ac cables can be more evenly graded or distributed by suitable manipulation of the paper density in the radial direction of the cable dielectric. To minimize the ionic content, and hence the associated dielectric loss in kraft papers, the papers are normally washed in deionized water. This process removes most of the chemical residues remaining after the manufacturing process; these residues are electrolytic in nature and may be present as ionizable bases, acids, salts, or as a combination. The presence of electrolytic ions in the papers can be established in terms of conductivity tests of the water extract and acidity or alkalinity (pH) determinations. The occurrence of pinholes and increased paper porosity can greatly weaken the oil-paper insulating system. Pinholes may lead to lower voltage breakdown values, whereas increased porosity may result in higher dielectric losses due to increased ionic mobility within the oil-filled papers. The arrangement and disposition of the individual fibers within kraft paper exerts considerable influence upon its electrical and mechanical properties. In the past, a great deal of research work has been carried out in this area; Fig. 2.16 shows the morphology of a typical kraft paper specimen. It is seen that in this particular specimen, the fibers are well intertwined to yield a densely packed structure.

When paper insulation tapes are applied helically on the cable conductor, great care must be exercised to ensure proper tension and overlay to provide constant-width butt gaps [50]. If the tension is too high, it will result in a tightly constructed cable with the result that upon bending, cracks and tears will develop in the tapes due to the high shear and tensile forces. If the tapes are applied at too low a tension, a loose cable construction will result and the cable upon bending will develop creases and even tears. In addition, near superposition of butt gaps may occur, thus resulting in the creation of dielectrically weak regions where breakdown may be initiated through the oil films within the butt gaps (cf. Fig. 2.17). Under visual observation, the created creases are discernible as hairline cracks in the kraft paper tapes. Schematically, the formed creases may be depicted as in Fig. 2.18. Creases form as the individual kraft paper tapes collapse in the butt gaps due to the end compression thrust caused by the bending forces when the radial tape pressure and the associated frictional forces are too high to allow free movement or sliding of the contiguous kraft paper tapes. As the tapes collapse along the butt gaps as a result of bending or flexing of a loosely constructed

Figure 2.16 Morphology of kraft paper (×168).

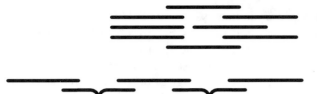

Figure 2.17 Near superposition of butt gaps over three layers of kraft paper tapes.

Figure 2.18 Creases depicted in schematic form.

cable, cracks appear along the line of collapse, assuming the sharp angular corners illustrated in Fig. 2.18.

In severe cases, the cracks may transgress into actual tears along the line of collapse. Examination of the cross-sectional area at the crease lines using a scanning electron microscope may provide striking evidence of serious tape weakening flaws. Figures 2.19(a) and (b) portray a typical top and profile view of a crease along a kraft paper tape. The profile view, taken at an angle of 10°, shows increased separation of the paper fibers at the crease line. A more attentive examination of the crease lines may reveal a rather pronounced separation of the paper fibers, indicating appreciable weakening of the kraft paper tapes.

Figure 2.20 provides a comparison of cross-sectional areas of creased and normal tapes. One can discern a partial tear at the crease line of the damaged kraft paper tape. It is apparent that any further development of the tear would lead to the creation of a butt gap having a thickness three times its normal value. Another inherent and common defect found in loosely taped cables is that of wrinkle formation. However, as oil tends to fill in the wrinkles for pressurized cables, partial discharges would be expected to take place only in solid impregnated paper cables, and even there the subsequent formation of waxes would eventually tend to extinguish these discharges. Thus, the formation of wrinkles poses less serious problems than does the formation of creases. Figure 2.21 shows an oil-impregnated cable specimen with wrinkle defects.

The rather peculiar behavior of electrical insulating papers is attributed not only to their unique physical structure, but also to a large extent to their molecular structure. The cellulose paper composition is defined by the general chemical formula $C_{12}H_{20}O_{10}$. When kraft papers are dried under vacuum and heat prior to impregnation, great care must be exercised that the cellulose paper does not lose some of its chemically bound water or water of hydration because irreparable damage would result as the paper loses physical strength. As the paper is heated beyond 200°C, the chemical bonds are readily broken and the chemical structure of the paper breaks down gradually to $H_2O + CO^2 + CO$ [51]. This process also takes place to some degree at temperatures below 100°C, but at a rate that is sufficiently slow not to give rise to serious problems. The above reaction occurs in the absence of external oxygen, so that it can progress readily within a cable insulation system. For this reason oil-paper cables should be operated at temperatures below 100°C.

The cellulose molecules of the paper fibers consist of a series of glucose repeating units arranged in celloboise pairs, as indicated in Fig. 2.22 [42]. The repeating unit is on the order of 10 Å, having an approximate width of 9 Å and thickness of 5 Å. The polymolecular character of the paper fibers prevents an exact determination of the molecular weight and chain length, but estimates range up to 10^5 Å for the former and 10^3 Å for the latter. Not shown in the molecular structure of cellulose are the carboxyl (–COOH) groups that occur at the rate of one per every hundred glucose units

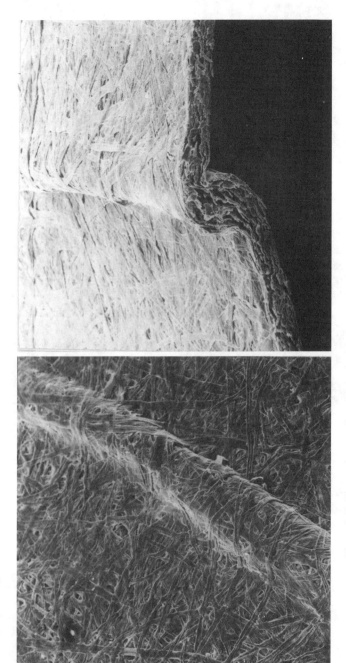

Figure 2.19 Views of crease line obtained with scanning electron microscope: (a) top view (50×); (b) profile view at 10° angle (90×).

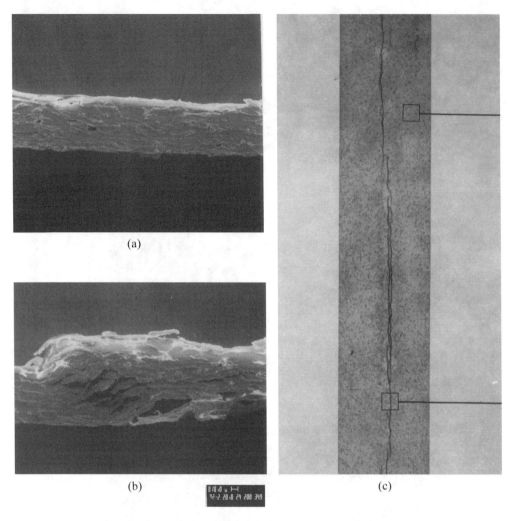

(a)

(b)

(c)

Figure 2.20 Evidence of crease-line-induced damage as viewed with scanning electron microscope: (a) profile view of normal tape; (b) profile view of tape at crease juncture; (c) top view of overall tape with crease markings. [Scale for views (a) and (b): 10 μm as indicated.]

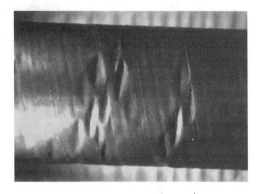

Figure 2.21 Wrinkling defects in an oil-impregnated paper cable.

1 cm

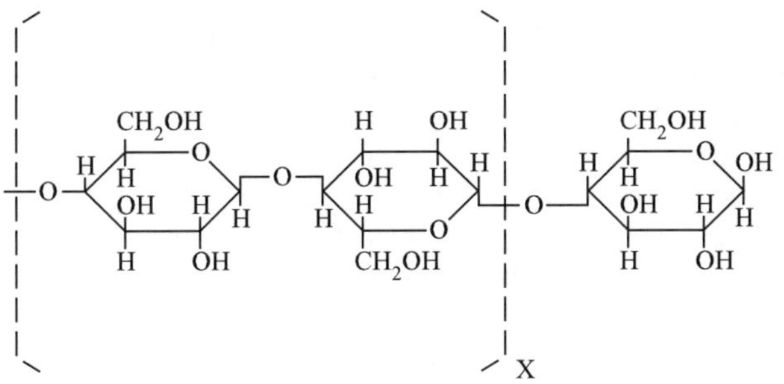

Figure 2.22 Molecular structure of cellulose (the carbon atom sites are represented by four bonds in each glucose unit).

and whose presence causes the cellulose to behave as a weak acid with an ionization constant of about 2×10^{-14}. Thus, if one views ionic conduction as a rate process, the carboxyl groups may account in some part for the existence of ionic losses within the paper at power frequencies. The paper pore size ranges from several hundred micrometers for an actual pinhole down to 10 Å for the smallest pore size within a single cellulose fiber (the pore sizes within the latter generally vary from 10 to 100 Å). The fibers form a complex interwoven channel system within the paper and, consequently, ample latitude exists for long-range ion excursions within oil-impregnated paper, provided the diameter of the ions is smaller than that of the pores [42, 52, 53].

Paper exposed to the atmosphere will absorb moisture from the air into the capillaries of the fibers and under normal conditions may contain as much as 12% water by weight; however, most of this water can be removed by heating the kraft paper under vacuum. In regions where the cellulose molecules exhibit an orderly arrangement (crystallites), water is retained by absorption forces. This colloidal or bound water is believed to be held by hydrogen bonds that exist between the H_2O dipoles and the hydroxyl groups of the cellulose molecules. The strength of these dipole bonds prevents the removal by vacuum of the colloidal water that will contribute to the overall ionic conduction losses if dissociated. It is generally found that vacuum-dried paper fibers retain between 0.5 and 1.0% of the bound water. At the power frequency dry paper has a dielectric constant of 2.2 and a $\tan \delta$ value of approximately 0.002. The ionic losses in the paper can be reduced by proper washing in deionized water to draw out some of the salts from the paper into solution.

In the past, oil-paper cable design and manufacture has largely centered on the use of mineral-type impregnating oils. However, the ability to procure mineral oils with suitable naphthenic and aromatic constituents to inhibit gas evolution at high voltages has greatly diminished with the result that presently extensive use is made of synthetic oils. For an extended period already, polytutene liquids, having viscosities of 10.6–11.1 cSt at 99°C, have been used in pipe-type cable applications. For the lower viscosity oil requirements in self-contained hollow-core cables, attention has been turned to alkyl benzenes or the newer low-viscosity oils having higher paraffinic contents but including alkyl benzene additives to inhibit gas evolution. In this regard it should be observed that the use of dodecyl and nonyl benzenes has been prevalent in European countries for already an extended period of time; in fact, these fluids have been also used for many years in North America but initially for specialized applications such as submarine cables (primarily the low-viscosity nonyl benzene). The difficulty with using alkyl benzenes in existing cable systems is associated with their action as detergents. They are easily contaminated as compared to the mineral oils already in the system and, in addition, tend to attack gasketting materials that have performed well in the presence of mineral oils. Consequently, alkyl benzenes must be either used in oil pumping systems designed specifically for their use or otherwise the gasketting materials, which are subject to degradations in their presence, and thus a source of ionizable impurities must be replaced. The gaskets that are most susceptible to degradation are those of nitrile rubber, and those that are least susceptible are of a viton rubber composition.

In selecting mineral oils for cable applications, a number of important impregnant attributes must be taken into account [54]. Table 2.8 provides a comparison of the properties of a few impregnating fluids in common use for oil-paper cables. The viscosity of the oil impregnant plays a preponderant role because during the cable manufacture it determines the impregnation rate and while in service under normal operating temperatures it determines the mobility of the oil within the cable insulating systems. As the load is removed from the cable and the insulation undergoes cooling, the initially expanded oil impregnant must be of sufficiently low viscosity at this temperature so as to prevent void formation that may lead to partial discharge. The viscosity of an insulating oil or fluid may be expressed in several units, depending upon one's preference and requirements. Basically, the viscosity is reported either in absolute (dynamic) viscosity or kinematic viscosity units. The unit of absolute viscosity is the poise and is defined as the ratio of force per unit area divided by the velocity gradient of the fluid. The dynamic viscosity is thus the proportionality factor, which relates the shear stress within the liquid to the velocity gradient (i.e., the rate of change of velocity normal to the flow of the liquid). One can write that [55]

$$
\begin{aligned}
1 \text{ poise} &= \left(\frac{\text{force}}{\text{area}}\right)\left(\frac{1}{\text{velocity/length}}\right) \\
&= \left(\frac{\text{dynes}}{\text{cm}^2}\right)\left(\frac{1}{\text{cm/s-cm}}\right) \\
&= \text{dyne-s/cm}^2 \\
&= \text{g/cm-s}
\end{aligned}
\tag{2.26}
$$

For reference purposes it is convenient to note that the viscosity of water at $20°C$ is 1 cP (0.01 P). Because the viscosity of most oils or impregnating fluids is also relatively low, it is customary to express their values of absolute viscosity in centipoises rather than poises. Note that in the SI units, 1 P is dimensionally equal to 0.1 Pa-s. A Brookfield viscometer is often used to obtain a measure of the absolute viscosity, which is expressed as a ratio of the shearing stress (dynes/cm^2) to the shearing rate strain (1/s). The measurement consists of rotating a flat conical surface with respect to a flat plate within the liquid in between and contiguous to the two surfaces; the conical axis must be set perpendicular to the surface of the flat plate. The fluid between the two surfaces is thus subjected to a uniform shear rate.

Although the absolute or dynamic viscosity is used extensively in research applied work, its usage is not prevalent in practice. Cable oils and impregnants are more

TABLE 2.8 Physical and Electrical Properties of Cable Oils

Property	Pipe Cable Oil	Heavy Cable Oil	Polybutene Cable Liquid	Self-Contained Cable Oil	Dodecyl-benzene
Viscosity, SUS				73.1 @ 20°C	12.0 @ 20°C
@ 37.8°C	763	2365	1200 @ 25°C	49.7	6.0
@ 99–100°C	60	101	63	32.5	4.5
Viscosity cSt	12,200 @ 25°C			13.9 @ 20°C	
@ 37.8°C	170	490		7.29	
@ 99–100°C	10	21	10.6	1.91	
Flash point open cup, °C	196.1	243.3	154	130	130
Acidity, mg KOH/gm	0	0	0.01	0.005	0.02
Pour point, °C	−26.1	−17.8	−34	−50.0	−45
Specific gravity					
@15.6°c	0.928	0.926	0.862	0.883	0.871
@25°C				0.876	
Coefficient of expansion, $cm^3/cm^3/°C$			0.00078	0.0007	
Thermal conductivity					
cal/°C · cm · s	0.00030	0.00030			
BTU/°F · ft · h	0.072	0.072		0.073	
Boiling point at 760 mm, °F			Highly volatile	~450	
Volatility, weight loss	~8%		>35%	4.7%	
Dielectric strength, kV/0.1 in. $(0.254\ cm^{-1})$	>30	>30		>30	>30
Dielectric constant					
@ 60 Hz	2.15	2.23	2.17	2.3	2.2
@ 10^3 Hz, 25°C	2.15	2.23			
@ 10^6 Hz			2.14		
Dissipation factor					
@ 60 Hz	0.001	0.001	0.0005	0.001	0.0011
@ 10^3 Hz, 100°C	0.0001	0.00005			
@ 10^6 Hz	<0.0001	<0.00005			
Volume resistivity, Ω-cm					
@100°C	2×10^{13} @ 25°C	7×10^{13} @ 25°C	1×10^{14}	5×10^{13}	2×10^{13}

commonly expressed in units of kinematic viscosity. The unit of kinematic viscosity is the stoke and by definition is equal to the absolute viscosity in poises divided by the density of the oil in grams per centimeter cubed. Thus, making use of Eq. (2.26) it follows that

$$1 \text{ stoke} = \frac{\text{dynamic viscosity}}{\text{density}}$$

$$= \frac{1 \text{ poise}}{\text{g/cm}^3}$$

$$= \frac{\text{g/cm-s}}{\text{g/cm}^3}$$

$$= \text{cm}^2/\text{s} \tag{2.27}$$

The kinematic viscosity expressed in stokes, or again more usually in centistokes (1 centistoke (cSt) equals 0.01 stoke), is a measure involving only the dimensions of area and time. A very convenient and quick measure of the kinematic viscosity can be performed by means of a Saybolt universal viscometer. Its principle of measurement rests on the number of seconds that are required for a 60-cm^3 volume of the fluid to pass through a tube at a given or specified temperature. Hence, the units of viscosity are termed Saybolt universal seconds (SUS). Most impregnating fluid specifications (cf. Table 2.8) state the kinematic cable-impregnant viscosity in both centistokes (cSt) and SUS units for specific values of temperature. Direct conversion tables exist between the units of cSt and SUS. If t is the time in seconds obtained with the Saybolt universal viscometer, then the kinematic viscosity in stokes or centimeters squared per second can be expressed by an approximate empirical relation [55]:

$$\text{cm}^2/\text{s} = \frac{0.00220t - 1.80}{t} \tag{2.28}$$

In addition to the viscosity parameter, the coefficient of thermal expansion of the oil must also be determined to specify the required size of the oil pressure tanks, as in the case of oil-filled and oil-pipe-type cables. The flash point of an oil provides a measure of the limit beyond which the impregnant cannot be heated to avert the fire hazard of the more volatile emitted vapors. At the other temperature extreme, the pour point of the oil is a useful quantity as it indicates the lower value of temperature at which partial wax separation within the oil will occur. Also at lower ambient temperatures, the pour point defines the temperature limit at which no free oil movement will occur within the cable. The gas content of oils is of particular importance when low-pressure oil-filled-type cables are involved where void formation and the associated partial-discharge effects are to be avoided. Tests for the gas content of impregnating oils are often carried out as a routine quality control procedure both by cable manufacturers and utilities. Another measure of the stability of insulating oils centers on their oxidation characteristics. The performance of the oil in service is determined by the extent to which it forms sludge and excessive acidity compounds in the presence of oxygen. To reduce the corrosivity of the oils, they should be free of sulfur and of inorganic chlorides and sulfates, usually introduced by either direct contamination or improper refining.

Particle analysis on oils or impregnants can be useful particularly if carried out on impregnants that have been in service for a period of time to determine whether they have been subject to particulate contamination in oil reservoirs, pumping systems, and cable pressure tanks. It has been generally established in practice that the presence of large quantities of particles, in particular of the larger size, does not tend to affect the dielectric loss in the oil. This is probably because the particles are not involved either in the transport of electric charge or, otherwise if they are, they do not transport large quantities of charge. However, the presence of particles does affect significantly the breakdown strength of the oil. The latter effect is attributed to the alignment of dipole chains formed by the particles to bridge the electrodes. Electrons can travel with relative ease along these chains by a hopping mechanism that may culminate in a breakdown of the bridged oil gap [56–58]. Table 2.9 provides typical particle count data obtained on a mineral oil specimen retrieved from a pressurized oil reservoir of a pipe-type cable operating at 138 kV. The particle count and size distribution is fairly representative of an oil that has not been subjected to significant particle contamination while in service. The breakdown strength of this particular oil was found to be 27.1 kV/mm, indicating that the number and size of particles have no marked effect on the dielectric strength.

From Table 2.9, it can be perceived that there are few large particles in the specimen. Large particles are arbitrarily designated to be those particles, that exceed a size of 20 μm. Perhaps the most convenient approach to particle count and size measurement involves the use of optical methods. The oil specimen to be analyzed is passed along a channel that intersects a beam of collimated light in line with a photodetector, as depicted in Fig. 2.23 [59].

The photodetector provides a pulse train output, that is proportional to the number of particles as well as their size. The particles may be either of a metallic or nonmetallic origin. Whether the particles are metallic or not can be determined by the use of a Ferrograph [60]. This technique is relatively new and involves the deposition of particles suspended in an oil specimen onto a so-called Ferrogram substrate whereby metallic particles with magnetic susceptibility are distributed close to the entry point of the insulating liquid and paramagnetic or particles of low magnetic susceptibility are deposited further along the Ferrogram substrate. Nonmetallic particles are found interspersed along the substrate. Optical methods are employed to distinguish between the metallic and nonmetallic particles. Perhaps the most effective technique is that of the bichromatic microscope, which makes use of transmitted red and reflected green light so that the nature of the particle is determined in terms of its color. Thus metallic particles appear in red color, while nonmetallic particles are green or yellow depending on the degree of light attenuation [59]. A typical replica of a Ferrogram is portrayed in

TABLE 2.9 Particle Size Distribution in a Used
Pipe-Type Cable Oil

Particle size, μm	>5	>10	>15	>25	>50
Number	117	25	11	3	1

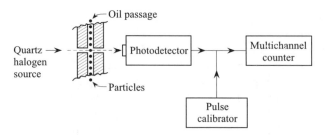

Figure 2.23 HIAC-type particle counter (after [59]).

Fig. 2.24, where a large particle in the center is easily discernible. In the original color photograph, the particle was visible in red color, thereby indicating it to be of a metallic nature; it was probably introduced into the oil by the pumping system. It appears superposed on a background of numerous small nonmetallic particles. As methods on particle analyses are being more developed and refined, they are becoming of increasing importance in studies relating to the contamination of cable oils.

The insulating oils—which are of pretroleum base, used as impregnants in conjunction with the kraft paper—comprise naphthenic, paraffinic, and aromatic constituents (cf. Fig. 2.25). Typically, a cable insulating oil may contain 12% aromatics, 38% naphthenics, and 50% paraffinic. Each given cable insulating oil comprises a variety of molecular sizes and structures. Its determined molecular weight represents thus only an average value of the molecular weight distribution of the oil. The length and overall structure of the oil molecules may assume a variety of forms, and Fig. 2.26 portrays one of many such possible arrangements. The aromatics tend to increase the dielectric losses of the oils, but their presence is necessary to inhibit gas evolution from the oils while under electric stress. The dielectric constant of the cable oils is on the order of 2.2 at

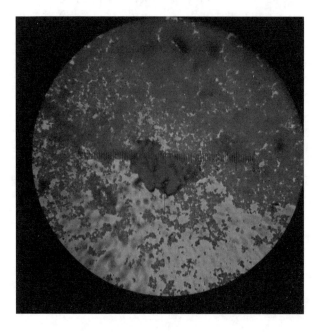

Figure 2.24 Ferrogram obtained on a used pipe-type cable oil (400×). (Courtesy of R. Olivier, Institut de Recherche d'Hydro-Québec.)

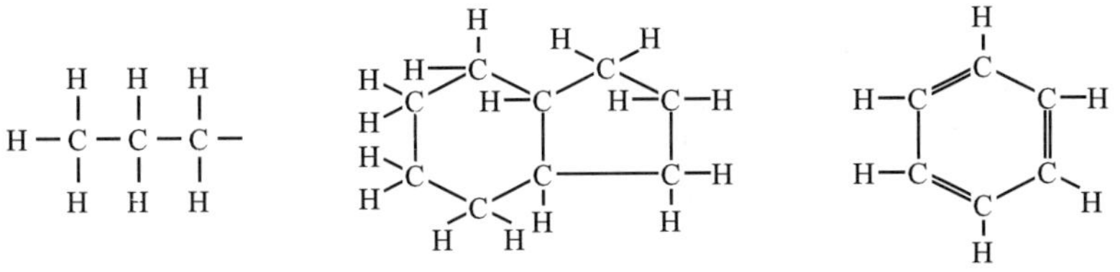

Figure 2.25 Molecular constituents of mineral cable insulating oil: (a) paraffinic structure; (b) naphthenic structure; (c) aromatic structure.

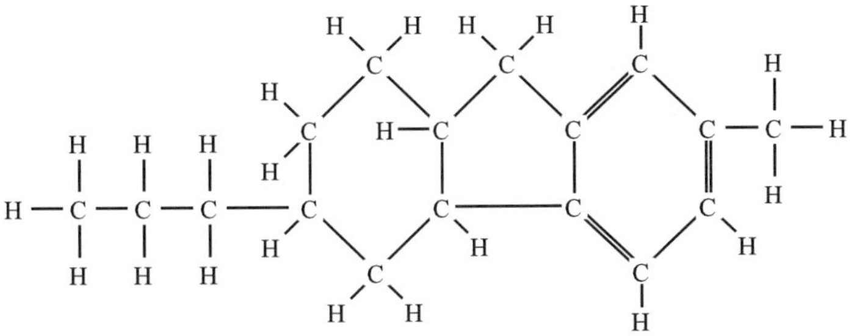

Figure 2.26 Possible molecular structure in mineral cable insulating oil.

room temperature and decreases with temperature due to a density decrease of the oils, whereas $\tan \delta$ is less than 10^{-4} at room temperature and exhibits a very marked increase with temperature due to increased ionic conductivity [57, 58]. The viscosity of the impregnating oils may vary anywhere from 10 to 1000 cP. The lower viscosity oils are used in oil-filled and pipe-type cables where it is important to maintain adequate oil flow throughout the cable (under pressure) to prevent void formation and the associated harmful partial-discharge effects, which could lead ultimately to failure at extra high voltages. With the lower voltage cables (solid-type oil-impregnated paper) the higher viscosity impregnants are used because the partial-discharge requirements are considerably less stringent. It should be emphasized that the lower viscosity oils are appreciably more susceptible to contamination and, consequently, greater handling precautions are required whenever low-viscosity impregnants are involved.

The magnitude of the dielectric loss in oil-impregnated papers is essentially determined by the paper and not by the liquid impregnants [37, 38, 42, 52, 53]. This can be seen from Figs 2.27 and 2.28, which compare the dielectric loss characteristics of an oil-filled cable oil obtained at room temperature and 85°C with those of a kraft paper insulating system impregnated with the same cable oil.

The dielectric constant of oil-impregnated paper is approximately 3.6 and exhibits very little decrease with temperature. The reason for the relatively high value of the dielectric constant is primarily the result of the relatively high dielectric constant of the cellulose fibers themselves, which is in the area of 6–10. Note that the reason this high value is not approached in the dielectric constant measured on the paper alone is due to the porosity of the paper, i.e., presence of air, which has a dielectric constant of unity. The dissipation factor of the oil-paper system is generally on the order of 0.002 and

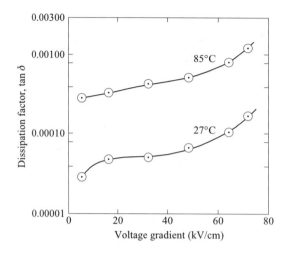

Figure 2.27 Typical dielectric loss characteristic of a mineral oil used in self-contained oil-filled power cables [57].

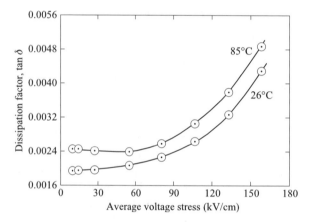

Figure 2.28 Typical dielectric loss characteristic of an oil–kraft paper insulating system impregnated with oil used in self-contained oil-filled power cables [42].

increases markedly with temperature due to increased ionic conduction [38, 42]. Extensive treatment of papers with deionized water may reduce the $\tan \delta$ value to as low as 0.0018 at 85°C. The dielectric losses resulting from this $\tan \delta$ value in the case of a 275-kV cable represent a 7% reduction in the current rating, whereas for a 400-kV cable they would constitute a 15% reduction. The relatively high dielectric losses in oil-paper systems prevent the use of oil-impregnated paper for voltages much beyond 750 kV, even when forced cooling techniques are employed. The dielectric strength of ordinary oil-impregnated paper is roughly 400–500 kV/cm at 60 Hz and between 1000 and 2000 kV/cm under impulse voltages. The dielectric strength improves with the density of the paper used; however, the dielectric losses are adversely affected. Moreover, the weakest breakdown point in the oil-paper system occurs not in the oil-paper proper but in the oil-filled butt gaps.

Polybutene oils have become increasingly popular for use in pipe-type oil-filled cables, both as higher viscosity impregnants and pipe filling oils. Some polybutene cables have been designed for operating voltages as high as 345 kV. The choice of polybutene oils centers primarily on their lower cost and good electrical and physical properties as compared to cable-type mineral oils. Polybutene oils are derived from the polymerization of olefins, consisting of very long chain molecules with isobutene as the base unit [61]. The long molecules are characterized by methyl group side chains and their molecular weights may range anywhere from about 300 to 1350, depending on their viscosity. Table 2.10 provides data on polybutenes manufactured by the Cosden Oil and Chemical Company. Examination of the table shows that polybutene oil 06SH is a typical pipe filling oil, whereas 015SH is a high-viscosity impregnating oil. Kraft paper insulation impregnated with polybutene oils exhibits an electrical behavior similar to that impregnated with equivalent viscosity mineral oils. Figure 2.29 compares typical data obtained on the dissipation factor behavior as a function of voltage stress with temperature as parameter for cable models impregnated with a mixture of 06SH and 015SH and an equivalent mixture of mineral pipe-cable filling and impregnating oils [62]. It can be perceived from the curves that at corresponding values of temperature, the losses in the mineral oil impregnated papers exceed somewhat those impregnated with the polybutenes. The cable model construction and impregnating apparatus used to obtain these results is described elsewhere [63]. The ac breakdown and impulse strengths of polybutene-impregnated papers have been found to be approximately 55.3 and 190 kV/mm, respectively; it is apparent that the values fall again close to those characteristic of mineral-oil-impregnated paper systems at room temperature [64]. However, with aging, the authors of the last reference did find some indication of thermal instability in their particular polybutene oils. This effect was manifest by a lower ac breakdown strength upon aging of the insulation. The same authors have studied mixtures of polybutene and alkyl benzene oils and found an improvement. This result supported the findings of Eich [65], who made use of a polybutene containing 10% of tridecyl benzene and was able to demonstrate an improvement in the dielectric strength of a pipe-type cable, which was characterized by breakdown values of 73.2, 96, and 50 kV/mm under the application of switching surge, impulse and ac voltages, respectively. In another study by McKean et al., [66] on a polybutene pipe-

TABLE 2.10 Physical and Electrical Properties of Polybutene Oils with Different Viscosities

Property	Oil type			
	OSH[a]	06SH[a]	015SH[a]	30SH[a]
Average molecular weight	340	450	570	1350
Viscosity, SUS				
100°F	110	575	3440	115,000
210°F	41	63	158	3070
Flash point, °F	270	310	320	450
Pour point, °F	−60	−40	−10	+40
Acidity, mg KOH/g	0.01	0.01	0.01	0.01
Water content, ppm	15	15	15	15
Dissipation factor (60 Hz, 20 V/mil, 100°C	0.0003	0.0003	0.0002	0.0002
Dielectric constant (1 MHz, 25°C)	2.14	2.16	2.17	2.24
Dielectric strength (kV, 0.1-in. gap, 80°C)	>35	>35	>35	>35
Volume resistivity, Ω-cm (100°C)	8×10^{14}	1×10^{15}	1.2×10^{15}	1.5×10^{15}

[a] Registered trade names of Cosden Oil and Chemical Company.

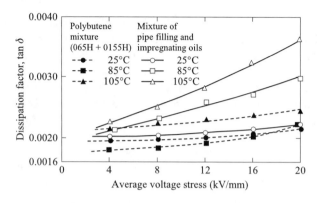

Figure 2.29 Dielectric loss characteristics of cable models impregnated with polybutenes and mineral oils. (Courtesy of J. Chan and S. Russek, Alcatel Canada Wire [62].)

type cable tested at the same Waltz Mill Testing Facility, it was shown that polybutenes yield low dissipation factor values and exhibit good stability against contamination and oxidation. Table 2.11 provides dissipation factor data on another polybutene pipe-type cable evaluated in the same project [67]. The $\tan \delta$ values indicate that the tested polybutene pipe-type cable is characterized by low dielectric losses. The excellent results obtained on the polybutene pipe-type cable at the Waltz Mill Test Facility were instrumental in ushering in more quickly the acceptance of polybutenes for pipe-type cables, both as impregnants and pipe filling oils.

Greater use is being made of alkyl benzenes because of the increased difficulty in obtaining mineral oils with acceptable gas inhibition properties (i.e., because commercial mineral oils are being provided with increasingly higher paraffinic contents). Alkyl benzenes perform exceptionally well because of their extremely low gassing or gas evolution tendencies when subjected to electrical stress using the Doble Engineering/ Pirelli measurement techniques as described in ASTM Method D2300. In the latter method, the oil to be evaluated for its gassing tendency is saturated with a gas and then subjected to a radial electrical field gradient. As the space above the oil is ionized due to the stress, the surface of the oil-gas interface is exposed to a bombardment of electrons and ions. This results in gases being evolved or absorbed from or by the oil; the volume of the evolved gas increase or decrease per unit time is determined in terms of the pressure changes with time [68]. Apart from their good electrical and gas inhibition properties, alkyl benzenes appear to have some other additional advantages. For

TABLE 2.11 Dissipation Factor as a Function of Voltage and Temperature of 500-kV Polybutene Pipe-Type Power Transmission Cable

Voltage to ground (kV)	$\tan \delta$		Operation temperature (°C)	$\tan \delta$
	27°C	90°C		
50	0.15×10^{-2}	0.14×10^{-2}	27	0.159×10^{-2}
100	0.15×10^{-2}	0.15×10^{-2}		
200	0.15×10^{-2}	0.16×10^{-2}		
300	0.16×10^{-2}	0.16×10^{-2}	71	0.160×10^{-2}
400	0.17×10^{-2}	0.17×10^{-2}	87	0.163×10^{-2}
460	0.17×10^{-2}	0.17×10^{-2}	105	0.185×10^{-2}

Source: From [32].

example, the presence of copper does not accelerate oxidation in alkyl benzenes to the same degree as it does in mineral oils. They exhibit excellent oxidation stability as well as low pour and high flash points.

The alkyl benzenes intended for cable oil applications are branched and have the general molecular structure depicted in Fig. 2.30. The value of x determines the length of the molecular chain. For a self-contained hollow-core cable alkyl benzene, the average molecular weight is on the order of 320 (cf. Table 2.12). However, due to a molecular weight distribution, lower molecular weight fractions are present though the removal of these during vacuum treatment prior to impregnation appears to be less marked than the removal of comparable fractions in mineral and polybutene oils. For hollow-core cable oils, pure alkyl benzenes or mixtures of alkyl benzene and mineral oils are a likely combination. The addition of alkyl benzenes to mineral oils improves the gas evolution properties of the latter. For pipe-type cables, polybutene–alkyl benzene mixtures are also employed. The alkyl benzenes have found considerable application as makeup oils for reservoirs of hollow-core cable systems, though great care must be exercised in such applications as concerns the chemical compatibility of gasket materials. As we have noted already, however, in contrast to North American practice, the use of dodecyl and nonyl benzenes is relatively widespread in Europe. Also in North America, there are now many self-contained hollow-core systems using the dodecyl and

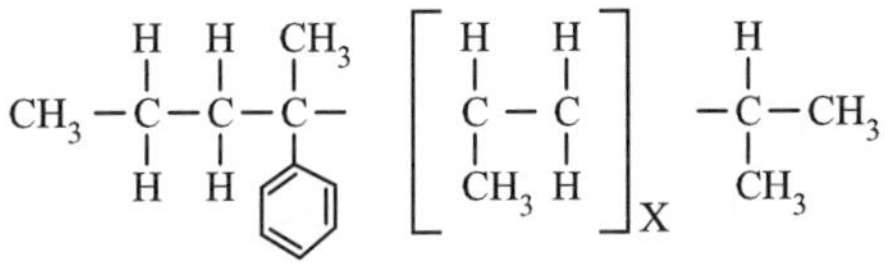

Figure 2.30 Molecular structure of branched alkyl benzene.

TABLE 2.12 Properties of Alkyl Benzenes and an Alkyl Benzene–Polybutene Mixture[a]

Property	Alkylate 21[b]	Dichevrol DF100[c]	Dichevrol DF500[d]
Average molecular weight	246	320	380
Specific gravity, 15.6°C	0.87	0.87	0.87
Viscosity, SUS			
@ 37.8°C		100	500
@ 98.9°C		38	59
Viscosity, cSt			
@ 37.8°C	6.0	20.6	108
@ 98.9°C		3.5	9.9
Flash point	127°C	350°F	350°F
Pour point, °C		−50	−20
Water content, ppm		25	25
Acidity, mg KOH/g		0.02	0.02
Dissipation factor (60 Hz, @ 210°F)		0.0002	0.0002
Dielectric constant (60 Hz, @ 210°F)		2.1	
Dielectric strength, (kV @ 77°F)		50	40
Volume resistivity (Ω-cm @ 210°F)		1×10^{15}	5×10^{14}
Gas evolution under stress, μL/min		−46 to −18	

[a] Data taken on dichevrol dielectric fluids provided by the Chevron Chemical Co.
[b] An alkyl benzene for use as additive in a hollow-core-type mineral oil.
[c] An alkyl benzene for use on self-contained hollow-core cables and forced cooled pipe cables.
[d] An alkyl benzene–polybutene mixture for pipe-type cables.

nonyl benzenes. Table 2.12 presents data on an alkyl benzene and an alkyl benzene–polybutene oil mixtures. The fluid DF100 is primarily intended for hollow-core cables and fluid DF500 is a pipe filling oil for pipe-type cables. For the actual cable impregnant in a pipe-type cable, a higher viscosity fluid would be necessary; it is understood that such alkyl benzene–polybutene fluids are in the development stage. The Alkylate 21 fluid given in Table 2.12 is likely to be employed as an additive for low viscosity mineral oils with application to hollow-core cables to inhibit gas evolution. Its molecular structure is given in Fig. 2.31.

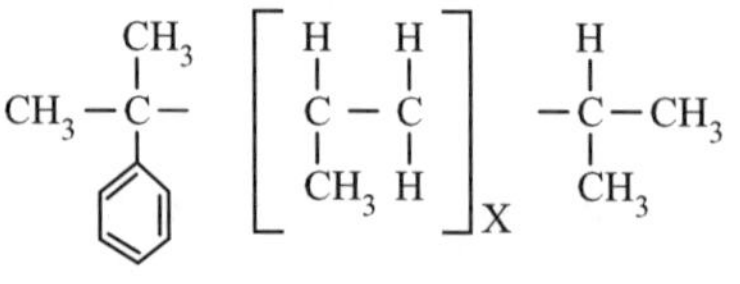

Figure 2.31 Molecular structure of Alkylate 21 (here $X = 2$).

2.4.3 Extruded Solid Dielectrics

Present-day solid-dielectric extruded power cables are primarily insulated with either linear PE, XLPE, or EPR. However, other types of solid dielectrics have been used in the past, while others offer still suitable and viable alternatives for special-purpose applications. In addition to dealing with present types of solid extruded insulations currently in use, this section will provide a description of antecedent types of solid insulations used for earlier cables not only because some of these types of cables are still in service but also because of their chronological importance in the field of cable development, design, and manufacture [69, 70].

2.4.3.1 Natural Rubber

In the early solid-dielectric extruded power cables extensive use was made of compounded and vulcanized rubber for insulations rated up to 25 kV. Vulcanization or crosslinking of the natural rubber was necessary to protect it against moisture absorption, to impart a higher tensile strength, and to prevent it from flowing freely while under pressure. This consisted primarily of adding sulfur to the natural rubber and heating the mixture to a temperature between 250 and 300°F and maintaining it under heat to allow both the physical and chemical changes to take place [69].

Natural rubber has the general formula (C_5H_8) and in this regard it is the same chemically as gutta-percha, also a natural gum, used in the earlier days of cable construction (cf. Figs. 2.32 and 2.33). Yet gutta-percha is a hard leathery substance because it is a trans-isomer of natural rubber [6]. Natural rubber is supplied from Brazil and in

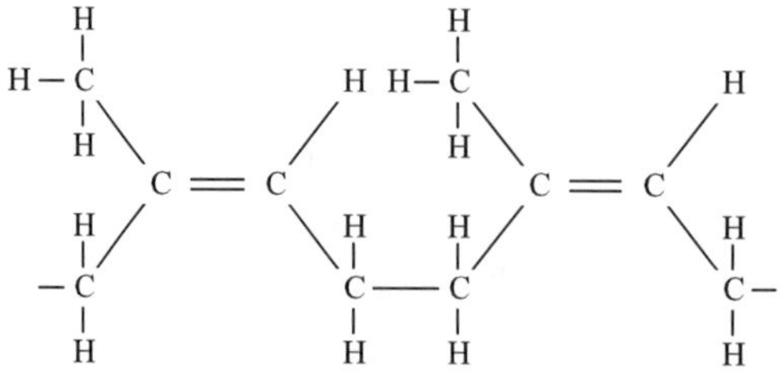

Figure 2.32 Molecular structure of natural rubber, cis-$(C_5H_8)_x$.

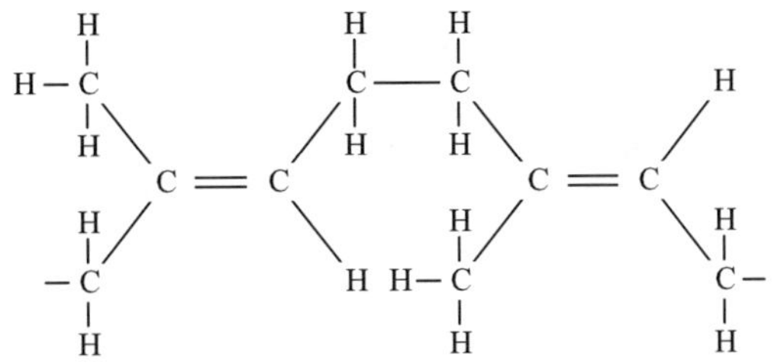

Figure 2.33 Molecular structure of gutta-percha, *trans* -$(C_5H_8)_x$.

its natural state must be macerated into spongelike sheets under running water whereby the solid organic and inorganic impurities are removed; therefore, these natural rubber sheets must be dried for moisture removal. The resulting rubber sheets thus become soft and elastic. Following the removal of extraneous solid matter and moisture, the sheets are placed in mixing rolls where fillers and softeners are added. In the early days, the fillers consisted of zinc oxide, litharge, chalk, whiting, paraffin wax, bitumen, and carbon black [69]. In more recent days, with the exception of the extensively used carbon black filler, other inorganic fillers such as mica, quartz, glass, clay, ceramics, as well as organic fillers such as cellulosic and synthetic fibers have become more popular. The final processing step for natural rubber involves mixing with vulcanizing agents (sulfur) and other accelerators. In the curing process a chemical reaction is induced to take place whereby crosslinks form between the individual polymer molecule chains, causing a change or improvement of the physical properties of the natural rubber; the acceleration agents merely enhance the rate of this reaction. Prior to extrusion onto the cable conductor, the natural rubber sheets may again be warmed and softened in the mixing rolls. Table 2.13 provides a comparison of the properties of natural rubber with other elastomers [51]. The term elastomer is commonly used to describe any material that at room temperature, while under tensile stress, will stretch to twice its normal length and then return to its original length following the removal of the externally applied stress.

2.4.3.2 Butyl Rubber

In the antecedent solid dielectric extruded power cables, considerable improvement in both the electrical and physical properties over those of natural rubber compounds was obtained with the development and subsequent introduction of butyl rubber. Butyl rubber is a long molecular chain compound that, although it superficially resembles rubber, is actually quite different in its chemical structure from rubber (cf. Fig 2.34). Butyl rubber is a polymer of isobutylene with a small amount of isoprene [71]. In the polymerization process sulfur is used as the crosslinking agent. Because of its low degree of unsaturation, butyl rubber must undergo vulcanization at temperatures higher than those used for natural rubber at comparable curing times [51]. Butyl rubber exhibits good resistance to the environment, partial discharge, ultraviolet radiation, oils, tearing, aging at elevated temperatures, and has good electrical properties. It was used extensively both for cable insulation and jacketing. Butyl rubber can be compounded with plasticizers to provide flexibility even down to −60°F (−51°C). Although butyl rubber is resistant to strong acids (sulfuric and nitric), it swells in petroleum solvents and is particularly sensitive to alkali substances at high temperature

TABLE 2.13 Physical and Electrical Properties of Elastomers

Property	Butyl Rubber	Ethylene-Propylene Copolymer (EPR or EPM)	Ethylene-Propylene-Diene Terpolymer (EPDM)	Neoprene	Natural Rubber	Silicone Rubber
Resistivity, Ω-cm	10^{17}	10^{15}–10^{17}	10^{15}–10^{17}	10^{11}	10^{15}–10^{17}	10^{11}–10^{17}
Dielectric strength (short time), V/mil	600	900	900–1050	150–600	—	100–655
Dielectric constant, 1 kHz	3.1–2.4	3.17–3.34	3.0–3.5	9.0	2.3–3.0	3.0–3.5
Dissipation factor, 1 kHz	0.0030	0.0066–0.0079	0.004 (60 Hz)	0.030	0.0023–0.0030	0.001–0.010
Tensile strength, gum, psi	2500–3000	500	200	3000–4000	2500–3000	400
Tensile strength, loaded stock, psi	2500–3000	800–3000	800–3500	3000–3500	3500–4500	600–1800
Elongation, % (gum)	750–950	200	200	800–900	750–850	200–800
Shore A hardness	15–75	30–90	30–90	20–95	20–100	30–80
Specific gravity, gum	0.91	0.86	0.86	1.23–1.25	0.92–0.96	0.97
Resilience	Low	Medium	Medium	High	Very High	Very low to high
Compression set resistance	F	G	G	F	G to E	G to E
Bondability to rigid materials	F to G	G	G	G	E	G
Processability	G	G	G	G	E	F to E
Ease of cure	G	E	G	G	E	F to E
Low-temperature brittle point, °F	−80	−60 to −100	−60 to −100	−40 to −70	−80	−90 to −200
Stiffening temperature, °F	0 to −20	−40	−40	+10 to −20	−20 to −50	−60 to −120
Continuous high-temperature limit, °F	300	300–350	300–350	225	300	500
Resistance to:						
Oxidation	G to E	E	E	E	G	E
Ozone	E	E	E	E	P to F	E
Tear	G	F to G	F to G	G	VG	F to G
Abrasion	G	G to E	G to E	E	E	P to G
Radiation	P	—	—	P	F	F to E
Dilute acids	E	E	E	E	F to G	E
Concentrated acids	P	E	E	G	F to G	E
Aliphatic hydrocarbons	P	P	P	G	P	P
Aromatic hydrocarbons	P	P	P	F	P	P
Chlorinated hydrocarbons	P	P	P	VP	P	P
Oil and gasoline	VP	P	P	G	VP	VP
Animal and vegetable oils	E	G to E	G to E	G	VP	P to G
Water absorption (swelling)	E	E	G	E	E	E
Sunlight aging	VP	E	E	VG	P	E
Heat aging (212°F)	G	E	E	G	G	E
Flame	P	P	P	G	P	F to E
Alkalies	VP	VG	VG	G	F to G	P to G

Abbreviations: E, excellent; VG, very good; G, good; F, fair; P, poor; VP, very poor.

Source: From [51].

124

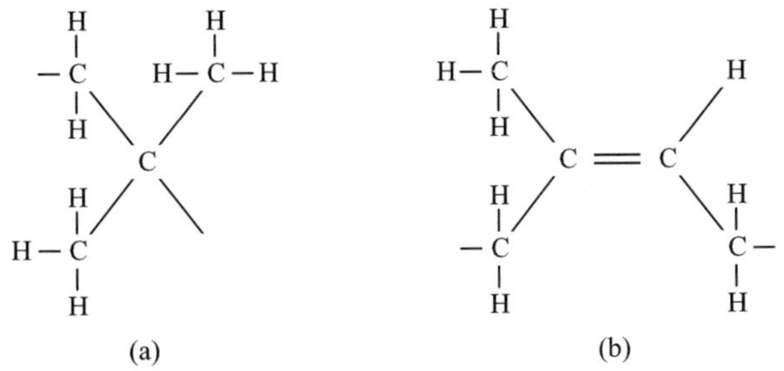

Figure 2.34 Building blocks (monomers) of butyl rubber: (a) isobutylene (95–99%); (b) isoprene (1–5%).

and humidity and has poor flame resistance as well. It has a volume resistivity of 10^{17} Ω-cm at room temperature, a dielectric strength of 600 V/mil; at 1000 Hz its dielectric constant is 2.9 and $\tan \delta$ has a value of 0.003. Butyl rubber may be used safely at temperatures up to 300°F (149°C); its low-temperature brittle point is about −80°F (−62°C) (c.f. Table 2.13) [51].

2.4.3.3 Ethylene-Propylene-Rubber (EPR)

Since its initial introduction into the commercial market during the late 1960s, ethylene-propylene-rubber has become perhaps the main current contender with the XLPE cable insulation for power distribution cables up to 35 kV, particularly where increased mechanical flexibility is required. Unlike the PE and XLPE used in cables, which may have approximately 50% crystallinity, most EPR formulations are almost completely amorphous; EPR is an elastomer synthesized from ethylene and propylene, using a ratio of 1:1. The molecular structure of EPR is shown in Fig. 2.35, where x is the repeating unit that may range anywhere from about 2×10^3 to 5×10^3. The average molecular weights of the EPR used in cables usually extend from about 1.5×10^5 to 2.5×10^5. In the structural form depicted in Fig. 2.35, EPR can be vulcanized only using peroxides, such as dicumyl peroxide and ditertiarybutyl peroxide [72]. But for vulcanization with conventional sulfur agents, the EPR molecule must contain 1,4 hexadiene units as delineated in the molecular structure in Fig. 2.36. The percentage of the third component or nonconjugated diene units generally does not exceed 6, for a ratio of 47 : 47% of ethylene to propylene, respectively; the other advantages of such terpolymers center on a more efficient use of curing agents, easier processing, and improved coloring. The filler content in EPR cable insulations is relatively high and may even exceed the 50% mark. Mostly clay, silicate and carbon black fillers are

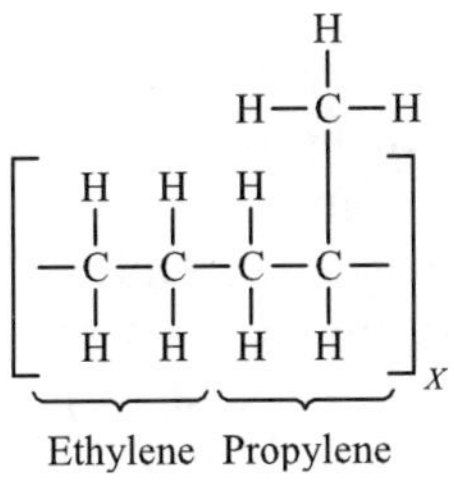

Figure 2.35 Molecular structure of EPR.   Ethylene Propylene

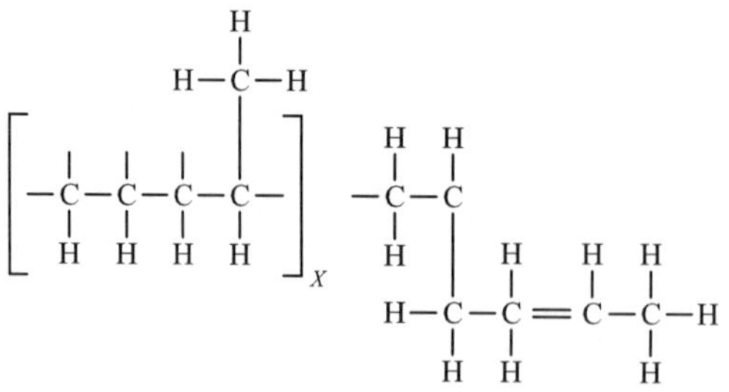

Figure 2.36 Molecular structure of EPR with 1,4 hexadiene unit.

employed. Clay is an inert filler and, therefore, does not improve the mechanical properties of EPR; it is used primarily because of its low cost and its ability to make the mixture easier to manipulate prior to crosslinking. The percent content of clay in EPR cable insulation may frequently range from 40 to 50 [72]. On the other hand, carbon black is utilized mainly as a reinforcing filler in EPR (as well as in all other rubbers). The resultant increase in the tensile strength is attributed to the action of carbon black particles blocking the short-range movement that is normally allowed by the crosslinking chains in the EPR [72]. The ideal carbon black particle size to effect improvement ranges from 200 to 500 Å. A carbon black content of several percent appears to provide the optimum content for best mechanical properties and does not cause too great an increase in the electrical conductivity and loss. Evidently, the dielectric losses in EPR are principally determined by the fillers.

The EPR copolymer is a low cost elastomer, and regular rubber processing equipment is used for processing it; EPR may be used for temperatures up to 300°F (149°C), has good resilience and stress–strain properties, good ozone and environmental resistance, good low-temperature characteristics becoming brittle only in the range from −60 (−51°C) to −100°F (−73°C). It has good electrical properties, having a volume resistivity between 10^{15} and 10^{17} Ω-cm, tan δ of 0.007, dielectric strength of 900 V/mil and dielectric constant of 3.2. It suffers from poor oil and flame resistance and, like PE and XLPE, is susceptible to degradation by partial discharges and treeing. Because of its somewhat higher dielectric losses, EPR cannot compete with XLPE for cable installations much above 138 kV. Its impulse strength is 400 kV/cm, which is appreciably below that of XLPE and oil-paper systems. For other physical and electrical data reference should be made to Table 2.13 [51].

As pointed out by Eichhorn [73] the optically opaque EPR compounds used in cable insulation may have different colors as a result of the various additives employed. In addition to the polymer itself and the filler, which is usually a finely divided surface-treated clay to provide the desired physical properties, the EPR insulating compound includes many other additives. The latter may comprise zinc and lead oxides (e.g. acid acceptors to stabilize the polymer in the presence of residual catalyst), paraffin wax and oils (to assist processing and for compatibility with the filler employed), a surfactant for the solid interfaces (usually organosilane), an accelerator (e.g. triallyl cyanurate, to accelerate the crosslinking reaction), an antioxidant (required for thermal stabilization), and possibly some carbon (to improve the mechanical properties). Traces of peroxides are also present, which are currently employed for chemical crosslinking in lieu of sulfur that was the common crosslinking agent in the earlier cable insulations. Although it is common practice in the cable field to refer to all cables insulated with ethylene-propylene-rubber as EPR cables, it may be well to emphasize some salient distinctions in the terminology concerning the ethylene-propylene-rubbers. The ethylene-propylene copolymer is commonly referred to as EPR or EPM, while EPDM denotes ethylene-propylene-diene terpolymers in which the third monomer may be ethylene norborn-diene or ethyldiene norbornene [73].

While EPR is tacitly assumed to be completely amorphous, it may in fact be partially crystalline. Crystallinity in the EPR polymers is essentially determined by the ethylene content, because it is the ethylene groups that tend to form organized repeated segments, while in contrast the propylene groups do not attain the lengths required to play an important role in the overall crystallinity of the polymers. When the ethylene content reaches 60% by weight, crystallinity is manifest in the copolymers; since most EPR formulations have ethylene contents in the range from 50 to 75% by weight, some EPR cables may exhibit some residual crystallinity [74]. The degree of crystallinity for a given ethylene content is also a function of the polymerization process and the nature of the catalyst employed. As a consequence of the crystallinity, the copolymer may exhibit melting points, usually in the range from 30 to 60°C, the higher temperature values reflecting increased ethylene segment lengths [74]. With increasing crystallinity, the processability of EPR compounds is reduced and the flexibility of the cable becomes poorer; however, the hardness of the extruded compound is augmented. Also, the capacity to accept higher filler contents is increased, thereby resulting in reduced production costs [75].

2.4.3.4 Silicone Rubber

Silicone rubber is classified as an organic-inorganic elastomer obtained from the polymerization of organic siloxanes. Silicone rubbers were first developed in 1943 and are employed where superior thermal stability is of prime importance. The basic structure of silicone rubber is given in Fig. 2.37. The properties of the silicone rubber are determined by the length of the molecular chain (value of x) and by the nature of the R substituents of the silicon atoms. Silicone rubbers are composed of the dimethyl-siloxane units, $(CH_3)_2SiO-$, the R substituents being simply methyl groups [76]. Silicone rubbers are prepared by the hydrolysis and subsequent condensation of lower molecular weight silicone fluids (also defined by the molecular structure in Fig. 2.37). To the resulting silicone gums are added fillers such as silica, calcium carbonate, titanium oxide, or iron oxide and catalysts to obtain the desired silicone rubber compounds. Crosslinking or vulcanization is carried out usually by the use of peroxides such as benzyl peroxide or dichlorobenzyl peroxide. Electron or gamma radiation sources may also be employed to obtain crosslinking, which becomes then a function of the radiation intensity. Silicone rubber compounds are processed on standard rubber mils; since silicone rubbers do not contain plasticizers or softeners, they are more resistant to embrittlement upon aging. To achieve improved mechanical properties, silicone rubbers like all elastomers are always reinforced with fillers, usually silica [72].

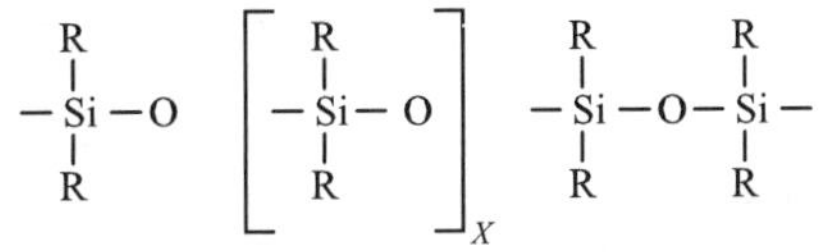

Figure 2.37 Basic building block of silicone rubber.

Silicone rubbers intended for extremely low temperature applications are flexible down to $-120°F$, with the brittle point occurring at $-180°F$ $(-118°C)$ [51]. For high-temperature applications, the silicone rubbers are designed to withstand continuous operation at $500°F$ $(260°C)$ and up to $700°F$ $(371°C)$ with intermittent operation [51] (cf. Fig. 2.38). They have low moisture absorption and have excellent resistance to partial discharges. Because of the retention of their desirable electrical characteristics under elevated temperatures, they are especially suited for extremely severe and demanding service applications. For a list of their properties, reference should be made to Table 2.13.

2.4.3.5 Polyethylene

Polyethylene is a long-chain hydrocarbon plastic produced by the polymerization of ethylene gas (C_2H_4) either under high or low pressures. The high-pressure process (1000–3000 atm at 200–250°C) yields low-density polyethylene, which has a waxy appearance and feels rubbery; its crystallinity varies from 40 to 60% and the molecular weight falls between 1×10^4 and 5×10^4 [72, 77]. The low density is necessarily the result of the branching introduced by the high-pressure process. The branched PE molecules give rise to an irregular matrix in which the molecules cannot be readily chain folded in a manner that is characteristic of the crystallites that are typical of large portions of the high-density polyethylene material. The high-density polyethylene produced at low pressures (60 atm) is stiffer, harder, and more brittle; its crystallinity varies from 60 to 90%, and the molecular weight may attain values as high as 20×10^6.

More than 10% of the world's total output of PE is being applied to electrical uses—largely as a result of its relatively low price, processability, resistance to chemicals and moisture, flexibility at low temperatures, and excellent electrical properties. The introduction of PE into the power cable field in the 1950s was largely responsible for initiating the gradual demise of natural and butyl rubber insulated cables. Polyethylene is produced in low (0.910–0.925 g/cm^3) and high (0.941–0.959 g/cm^3) density [78]. With a rise in density, the surface hardness, yield strength, stiffness, heat, and chemical resistance increase. The usual linear polyethylene consists of long chains of paraffin molecules of the form depicted in Fig. 2.39. If the molecules are very long and linear with very little branching, the molecules become closely packed: consequently, this type

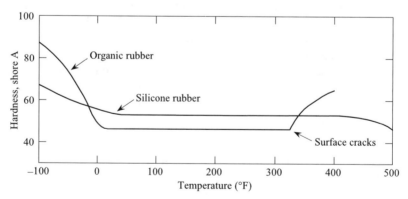

Figure 2.38 Effect of temperature on the hardness characteristics of silicone and organic rubbers (after [76]).

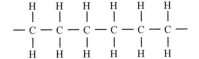

Figure 2.39 Molecular chain of linear PE.

of linear PE exhibits a high density. It is over these linear parts of the molecules that close packing occurs and ordered regions or crystallites result (i.e., PE becomes more crystalline). X-ray diffraction studies indicate the unit cell of the crystallite regions to be of a rectangular shape, having the dimensions of 7.42 by 4.94 Å and a depth of 2.54 Å in the direction of the paraffin chain axis [79, 80]. If the molecular structure of the PE becomes branched and less linear, its density decreases. For example, the low-density PE molecules are relatively highly branched and contain about 50–70 branches per molecule, thus resulting in a more amorphous or less crystalline structure.

It should be emphasized that the character of the branching that occurs in polyethylene is contingent upon the density of the polyethylene and the catalyst employed to initiate the polymeric reaction [81]. In low-density polyethylene the free radicals generated from the peroxide at elevated temperatures may either induce the ethylene molecules to add linearly in a normal fashion to each other or, alternatively, cause the polymer chain to grow short side chains. Long side chains may also propagate at sites where the radicals may remove a hydrogen atom. The appearance of numerous side chains of various lengths imparts the polymer a lower density. However, if catalysts containing transition metals are utilized, then long polyethylene chains result with negligible side branching; this approach is used to obtain long molecular chain lengths in both high-density as well as low-density linear polyethylene. In such circumstances branching is controlled by the addition of desired amounts of butene (C_4H_8) hexene (C_6H_{12}) and octene (C_8H_{16}); the latter may thus attach themselves to the side of the polyethylene molecule as delineated in Fig. 2.40 [81]. In the case of high-density

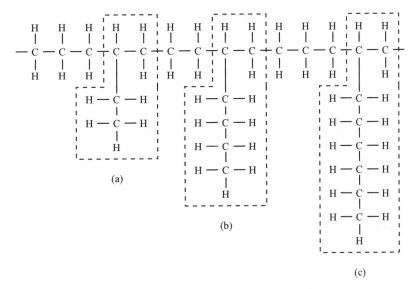

Figure 2.40 Polyethylene molecular chain with side branch formation due to the addition of (a) butene, (b) hexene, and (c) octene [81].

polyethylene, the addition of various amounts of butene, hexene, or octene either individually or in various proportions alter not only the extent of branching but also the crystallinity, since branching interferes with the chain folding mechanisms; the longer are the side chains, the lower is the crystallinity and, consequently, the density of the polymer.

In North America, the early manufactured PE power cables were of low-density, branched-type polyethylene. Because of the higher molecular weight of low-density polyethylene used, they are commonly referred to as high-molecular-weight polyethylene (HMWPE) cables. The amount of HMWPE cables manufactured began to decrease appreciably in 1972 with increased use and introduction of XLPE. For cable applications, low melt index PE is used, because in the extrusion process a lower viscosity material at the higher temperatures is desirable. Note that the melt index tends to decrease with the average molecular weight of the polymer; polyethylenes used for cable insulation should preferably have a melt index ranging from 0.2 to 3.0 g/ 10 min [81]. The HMWPE used on cables contains an antioxidant to prevent oxidation when exposed to heat. If polyethylene cables are exposed to light, a carbon black is added to screen out ultraviolet radiation, which, if not screened, degrades both the physical and electrical properties of the PE. The polyethylenes used on cables have a volume resistivity $>10^{16}$ Ω-cm, a dielectric constant and dissipation factor of 2.3 and 0.0002, respectively, at 60 Hz, and a dielectric strength of 500–700 V/mil; their water absorption is $< 0.02\%$. Table 2.14 provides data on the properties of polyethylene as a function of its density [51].

Polyethylene, like all plastics, is highly susceptible to degradation by partial discharges; furthermore, it may undergo additional degradation due to treeing when exposed to high-voltage stress [82, 83]. Treeing is generally initiated at high-voltage stress points; these points may either be voids, where intense partial discharges take place, or sharp metal or other irregular insulating impurity inclusions (e.g. ionic contaminants) that cause voltage stress enhancement. Trees may propagate in either dry or wet environments; in the former instance they are termed electrical trees and in the latter case they are denoted as water trees (since the stemlike channels contain water). Some water trees are also termed electrochemical trees in cases where certain minerals or ions cause permanent staining of the water trees. The conducting channels or branches of the electrical trees decrease the effective dielectric wall thickness between the cable conductor and insulation shield; hence electrical trees, once initiated, progress fairly rapidly until ultimately the remaining wall thickness is bridged and dielectric failure ensues. For water trees, the connection or link between the tree length and the actual failure event has thus far eluded a satisfactory explanation. Often it is found that cables may continue to operate without failure even when the water trees appear to bridge almost entirely the insulation wall thickness. Occasionally, it may transpire that an almost or completely bridging vented water tree initiates an electrical tree at one of its channel extremity points that begins to propagate rapidly and results in an abrupt electrical rupture of the dielectrically intact remaining wall thickness [84]. A prodigious amount of literature has been accumulated on research work carried out on water tree formation and growth under laboratory conditions, involving specialized procedures with needle electrodes and in studies on actual power distribution cables retrieved from service [83]. The work on PE cables is facilitated by the fact that the dielectric is translucent, and the tree growth and propagation processes may be readily discerned and observed. Although electrical trees are readily visible, water trees must be

TABLE 2.14 Properties of Polyethylenes

Properties	Low-Density PE	Medium-Density PE	High-Density PE	Irradiated PE (XLPE)	XLPE[a]
Volume resistivity, Ω-cm	>10^16	>10^16	>10^15	>10^15	>10^15
Dielectric strength					
Short time, V/mil	460–700	500–700	450–500	2500[b]	550
Step-by-step, V/mil	420–700	500–700	440–600	1800[b]	500
Dielectric constant					
60 Hz	2.3	2.3	2.35	2.3	2.30
10^3 Hz	2.3	2.3	2.35	2.3	—
10^6 Hz	2.3	2.3	2.35	2.3	2.28
Dissipation factor					
60 Hz	0.0002	0.0002	0.0002	0.0005	0.0003
10^3 Hz	0.0002	0.0002	0.0002	0.0005	—
10^6 Hz	0.0002	0.0002	0.0002	0.0005	0.0004
Arc resistance, s (ASTM D495)	Melts	Melts	>125	—	—
Density, g/cm^3	0.910–0.925	0.926–0.940	0.941–0.965	0.92	0.92
Modulus of elasticity in tension, psi × 10^5	0.17–0.35	0.25–0.55	0.8–1.5	—	—
Percent elongation, % (ult.)	20–650	100–600	15–700	>200	550
Tensile strength, yield, psi × 10^2	14–19	19–26	26–45	—	24
Compressive strength, psi × 10^3	—	—	2.4	—	—
Rockwell hardness	R10	R15	R30–R50	—	(Shore D)
Impact strength, ft-lb/in.	—	—	1–23	—	45
Heat distortion temperature (at 66 psi), °F	105–121	120–150	140–185	—	—
Thermal conductivity, cal/cm · s · °C × 10^{-4}	8	—	11–12	—	—
Thermal expansion, in./in. per °C × 10^{-5}	11–30	15–30	15–30	20	—
Water absorption, %	<0.02	<0.02	<0.01	Nil	—
Burning rate	Slow	Slow	Slow	—	—

[a] Union Carbide HFDE-4201 NT EC crosslinkable compound.
[b] A 5-mil film.
Source: From [51].

stained to render them permanently visible. The most common dye used for this purpose is methylene blue with a base, though acidic rhodamine dye has also provided acceptable results. One peculiar feature of water trees is that once the medium containing the water tree is removed from the water- or moisture-permeated environment, the water trees will become evanescent; yet they will reappear again when exposed to water. Under some circumstances a water tree may be rendered permanently visible even after all the water has been removed. This permanent staining is commonly believed to result from the presence of metallic ions in the water tree that may have been introduced by the impurities in the semiconducting shields or transported from the external environment via the water ingress. It is to be emphasized that the individual water trees are not branched and do not have permanent hollow channels that are electrically conducting as in the electrical trees.

Cables containing large numbers of water trees exhibit a lower ac breakdown strength [85], though some conflicting evidence has been obtained concerning this observation. It has been found that power cables containing moisture exhibit a high incidence of halo formation; these halos appear to have also a marked effect in lowering the ac breakdown strength of XLPE cables. This may account for the common observation that exposure of virgin polyethylene insulation to water will result in an approximately 10% decrement of initial breakdown strength even before the onset of water tree growth. In order to prevent completely water tree formation in polyethylene, the cables must be either installed in a dry environment or be protected from moisture ingress by an hermetic metallic sheath. The latter solution to the problem is considered to be too costly in terms of North American cable experience and practice, but it constitutes a standard procedure in some European countries where use is made of linear polyethylene as the extruded cable dielectric. Evidently, the foregoing European practice has avoided serious problems associated with water tree growth. The occurrence of electrical trees in solid extruded PE insulation is rare, because of the relatively low electrical operating stresses. Rapid electrical tree inception and formation requires high electrical stresses; thus electrical tree growth can be artifically initiated in PE by means of needle electrodes inserted in the cable insulation to simulate field enhancement points from asperities in the solid dielectric (c.f. Fig. 2.41).

Once an electrical tree is started at the stress enhancement point of a needle, it will propagate relatively rapidly until the remainder of the insulation is no longer capable of supporting the high stress and breakdown ensues; it is noteworthy that partial-discharge activity is observed to accompany electrical tree growth. The partial-discharge activity is only detected when the electrical tree channels attain a sufficient size to support electron avalanche formation that leads to partial-discharge inception. Space charge injected at the asperities under ac conditions consists of both electrons and holes; some of these charge carriers become trapped while others are detrapped and upon recombination emit ultraviolet (UV) radiation (electroluminescence) which causes chain scission in the polymer, thus inducing formation of the electrical tree channels. In contrast, water tree growth is considerably less rapid and occurs at appreciably lower electrical stresses with the water tree usually propagating very slowly either from the conductor shield toward the insulation shield as illustrated in Fig. 2.42 or vice versa.

Vented water trees that propagate from either the conductor or insulation semiconducting shields are considered to be more deleterious in their effects than those that originate from impurity inclusions or occluded voids, as depicted in Fig. 2.43. The latter type of trees, bearing the appellation of bow tie trees, seldom attain large

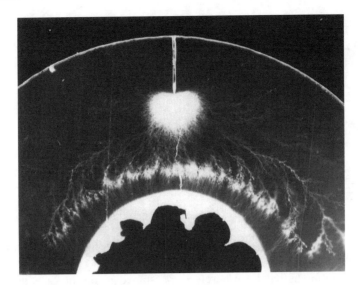

Figure 2.41 Typical electrical tree at a needle tip in early XLPE power cable insulation with extruded semiconducting shields (10×). (Courtesy of R.J. Densley, National Research Council.)

Figure 2.42 Typical water trees in early XLPE power cable insulation with extruded semiconducting shields (20×). (Courtesy of R.J. Densley, National Research Council.)

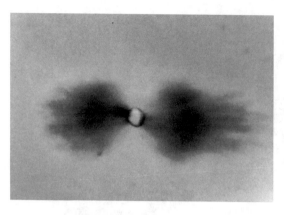

Figure 2.43 Bow tie water tree observed in 25-kV, 750 kcmil XLPE power cable; void diameter, 0.03 mm; tree length, 0.40 mm. (Courtesy of H. Orton, Powertech Laboratories.)

dimensions and their propagation appears to be of short duration. However, recently obtained evidence on XLPE indicates that the incidence of bow tie water trees may not be as innocuous as has been heretofore assumed to be the case, as we shall see in the subsequent section.

2.4.3.6 Crosslinked Polyethylene

In the late 1960s power distribution cables insulated with crosslinked polyethylene began making their appearance in Canada, following their earlier introduction in the United States in 1965. Cables insulated with XLPE presently dominate the distribution cable field in North America, Japan, and Northern Europe, and this situation is likely to persist for a long time unless a better dielectric can be found. In North America, XLPE transmission cables rated for 138-kV operation are common and a number of installations at 230 kV exist. In France, XLPE transmission cables rated at 225 and 400 kV are currently in use, while in Japan XLPE transmission cables are used at 275 and 500 kV levels.

Crosslinked polyethylene is extruded within the temperature range of 240–270°F (116–132°C). The crosslinked polyethylene is produced by compounding polyethylene with a radical at 240–260°F (116–127°C), whereby the free ethylene radicals in the PE molecules react with each other to form vulcanized or crosslinked material. The crosslinking process causes polyethylene to change over from a thermoplastic to thermosetting material with a marked improvement in both the physical and electrical properties. Because of its thermoset character, XLPE maintains its mechanical properties when exposed to temperatures that would cause linear PE to melt, lose shape, and flow. Impulse strengths as high as 700 kV/cm are readily obtained. For example, XLPE cables may be operated continuously at 90°C and intermittently at 130°C during fault conditions. It has good low-temperature properties, shows increased resistance to ozone and partial discharge as compared to linear PE, and has good impact, abrasion, and environmental stress cracking resistance characteristics. Limited flame resistance is incorporated by proper compounding or by the use of flame-resistant paints.

The addition of carbon black increases the tensile strength and hardness but affects adversely the electrical properties. Carbon black is added to guard against ultraviolet radiation, which due to its absorption by the carbonyl group (C=O) may induce degradation. Both PE and XLPE contain residual traces of C=O as well as some olefin groups with $C = C$ bonds that are particularly susceptible to photooxidation effects [6]. The reaction with oxygen leads to the formation of hydrogen peroxide groups ($-OOH$), that decompose to produce additional carbonyl groups, thereby perpetuating the degradation process. In cables containing carbon black that are exposed to ultraviolet radiation, antioxidants are added to provide further protection against oxygen-induced degradation.

In the crosslinking or curing process, crosslinks are induced to form between the individual polymer chains. The crosslinking agent is usually dicumyl peroxide when a chemical cross-linking agent is used. Alternatively, crosslinking may be induced with radiation, but this is usually confined to thin insulations because of the reduced effectiveness of radiation with increased insulation depth. As the dicumyl peroxide is terminally decomposed during the crosslinking process, acetophenone is formed. The presence of acetophenone in the XLPE cable insulation has generated considerable interest, though thus far no conclusive results have been obtained to ascertain its exact and precise effects on the XLPE insulation [86]. However, it should be pointed out that there are some indications that its presence does retard electrical [87] and water tree [88] growth. Recent evidence also indicates that acetophenone is actively involved in some chemical synthesis reactions in the degradation process occurring within discharging voids [89, 90]. However, it should be emphasized that acetophenone, $C_6H_5COCH_3$, has a molecular weight of only 120; it becomes a liquid above $20.5°C$ and diffuses fairly rapidly out of the insulation at elevated temperatures [91].

In the past XLPE power cables were steam cured, with the result that considerable moisture was retained in the insulation. To what degree this moisture remained in or diffused out of the insulation was not fully assessed, but it was believed that its residual presence may precipitate some of the water tree growth. Following the development of a dry curing process, most XLPE cables are now produced using this method. It was definitely established that the use of a dry curing process results in lower numbers of microvoids that may or may not be implicated in water tree growth. However, so long as the cables are exposed during their service to a moisture environment, the XLPE insulation should be able to absorb all the moisture necessary for water tree formation; this has been confirmed by Braun [92].

The prime advantage of dry-cured XLPE cables would appear to lie with cables that are covered with metallic sheaths to prevent moisture ingress such as would be the case with the higher voltage power transmission cables or distribution cables where metallic sheaths are used to provide additional mechanical protection. Nevertheless, dry-cured XLPE insulation, even when protected against external moisture ingress, may be subject to water tree growth; however, the water trees are of the bow tie type. Such trees have been demonstrated to occur in a dry cured 138-kV experimental transmission cable [93]. These are depicted in Fig. 2.44, and they have been generally attributed to residual water present as a by-product of the crosslinking reaction and possibly to some water introduced by the starting pellets. Many of the observed trees were microscopic in size; the largest dimensions were on the order of 30–70 μm and were observed to occur in the central portion of the insulation thickness. It should be emphasized that the bow tie tree counts or density is much greater in steam-cured XLPE as is evident from Fig. 2.45, which shows a typical average distribution curve of water bow tie tree count as a function of tree length or size. Note that $1 \, \text{mil} = 0.254 \times 10^{-6} \, \mu m$

In fact, the bow tie tree count numbers are so high that the manual tedium involved in their counting is simply overwhelming. Fortunately, recent developments

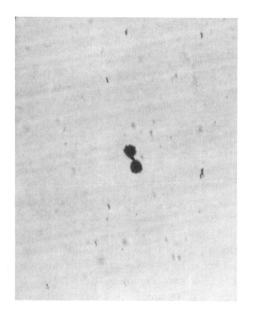 

Figure 2.44 Bow tie tree development in a dry-cured 138-kV XLPE metallic-sheathed power transmission cable: (a) 200×; (b) 400×. (Courtesy of H. St-Onge, Institut de Recherche d'Hydro-Québec.)

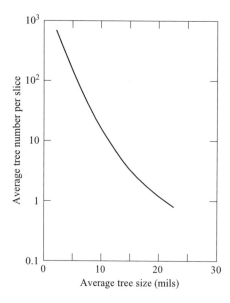

Figure 2.45 Typical bow tie tree distribution characteristic of a steam-cured XLPE power distribution power cable. (Courtesy of H. St-Onge, Institut de Recherche d'Hydro-Québec.)

in particle counting and image analysis techniques have eliminated this tedium. Computerized image editing instruments are now available that discriminate between different image shapes and forms, providing a direct readout of the image or particle size distribution sought after. As mentioned, in the past the occurrence of bow tie water trees was of questionable significance as they had not been directly implicated as a plausible cause of cable insulation failure. However, recent evidence, obtained by Pelissou [94], shows that the length of bow tie water trees may extend over an appreciable portion of the cable insulation wall and that, furthermore, bow tie water trees may be accompanied by electrical tree growth. Figure 2.46(a) depicts a bow tie water tree formed at a cavity in the vicinity of the conductor semiconducting shield in a 28-kV XLPE cable, which had failed after 14 years of operation following its installation in 1975. At its one extremity an electrical bow tie tree can be discerned to overlap the water tree portion. Figure 2.46(b) represents a bow tie water tree growth at a solid particle inclusion, with an electrical tree channel formation emanating from its two extremities. The occurrence of electrical trees would be expected to lead to short-term failure due to their rapid propagation across the insulation wall. In contrast, the deleterious effects of even large vented water trees that issue either from conductor or insulation semiconducting shields may be rendered innocuous by drying of the insulation, which results in their disappearance and thereby permits the insulation to recuperate a significant portion of its initial breakdown strength [91].

2.4.3.7 Comparison of EPR and XLPE Insulation

In view of the preeminence of the EPR and XLPE solid extruded dielectrics in the power distribution cable area and their growing use in transmission at high voltages, it is of considerable practical importance and technical interest to compare in some detail the properties of these two different types of insulating materials. Since we have already commented on their electrical properties and chemical structure, we shall primarily devote this section to a more detailed discussion of their mechanical and thermal properties.

The mechanical properties of EPR and XLPE compounds are contingent upon a number of factors. They are influenced by the macromolecular nature of the polymer as well as its chemical structure, molecular weight, crystallinity, and the copolymerization

 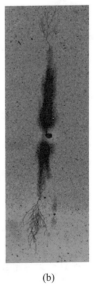

Figure 2.46 Bow tie water trees with contiguous electrical tree formations at their extremities at (a) a small cavity situated adjacent to the semiconducting conductor shield and (b) a solid particle inclusion in the bulk of the dielectric of 28-kV XLPE power distribution cable. (Courtesy of S. Pelissou, Institut de Recherche d'Hydro-Québec.)

(a)   (b)

and crosslinking mechanisms. External factors, principally thermal history, tempera-ture, magnitude of stress and strain, and time, exert pronounced effects on the behavior of the two polymers. The temperature to which the polymers are subjected constitutes perhaps one of the most preponderant factors that ultimately determines the suitability of a polymer for a particular cable application. It is a consequence of the viscoelastic nature of the polymers that it is the temperature to which the polymers are exposed over a given time period that determines the rate of deformation of the polymer or cable under extreme load current operating conditions. It is for this reason that stress-strain tests are commonly employed to determine the modulus of elasticity of polymers, which provides important data on the mechanical behavior of polymers. It will be recalled that, since stress is defined as the load per unit area and the strain as a dimensionless quantity representing the ratio of the change in length of the polymer specimen with respect to its original length prior to the application of the load, the modulus of elasticity E_e, given by

$$E_e = \frac{\text{stress}}{\text{strain}} \tag{2.29}$$

assumes the same units as stress, i.e., pounds per square inch (psi) or in SI units pascals (Pa) or newtons per square meter (N/m^2). Figure 2.47 depicts the modulus of elasticity as a function of temperature with the EPR and XLPE specimens subjected to a tensile stress [39]. Again as in the previous nomenclature, the values in the parentheses repre-sent the filler contents. The modulus of elasticity is plotted on a logarithmic scale in order to accentuate the melting point transition in XLPE and the differences between the two materials at higher temperatures. The modulus of elasticity scale represented in pounds per square inch (psi) may be changed to megapascals (MPa) by applying a multiplication factor of 6.9×10^{-3}. From the curves it is apparent that the modulus of

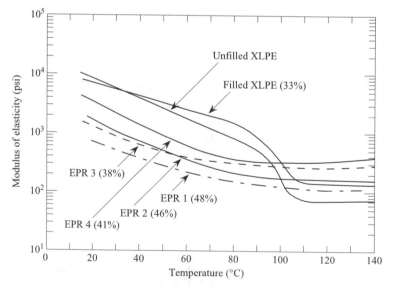

Figure 2.47 Modulus of elasticity in tension vs. temperature for EPR and XLPE compounds, with filler content indicated in parentheses [39].

elasticity in tension falls with temperature, becoming eventually nearly constant for both EPR and XLPE at the more elevated temperatures. Both the filled and unfilled XLPE display the characteristic mutation at the melting point, over which the modulus of elasticity exhibits a marked decrement. Both EPR and XLPE are typical viscoelastic polymers in that they tend to behave elastically at low temperatures where the strain is approximately proportional to the stress; whereas at the higher temperatures, over which the thermal agitation overcomes partially some of the bond forces, a quasi-viscous behavior is manifest.

The modulus of elasticity in compression as a function of temperature is shown in Fig. 2.48, and it is seen that its variation is similar to that in tension, except that the values obtained under a compressive stress exceed those in tension. This difference arises in part as a consequence of the tensile values being more affected by various possible defects in the physical structure of the compound, which exert negligible influence under compression stresses that tend to close them. The behavior implies that the modulus of elasticity under compression is a more representative characteristic of the compound itself, while the modulus of elasticity under tension characterizes the defect structure and distribution within the compound.

The bearing strength of plastic materials is a useful quantity when applied to the characterization of cable insulating materials in that it constitutes a measure of their ability to resist radial movement of the inner cable conductor, particularly when it is determined as a function of temperature. The bearing strength is obtained using ASTM Test Method D-953, which measures the bearing stress at which a 4% deformation is produced in the diameter of a bearing hole [95]. Hence, the bearing stress is equal to the applied load divided by the bearing area (product of the diameter of the bearing hole and the thickness of the specimen). Figure 2.49 indicates that the changes in the bearing strength as a function of temperature are analogous to those of the modulus of elasticity for tension and compression, with the exception that over the lower temperature regime its variations are more pronounced. It is apparent that the filled XLPE has the highest

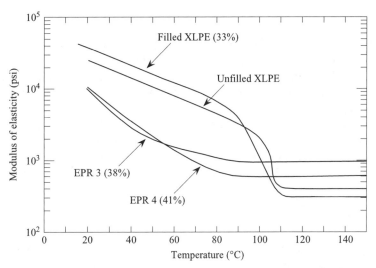

Figure 2.48 Modulus of elasticity in compression vs. temperature for EPR and XLPE compounds, with filler content indicated in parentheses [39].

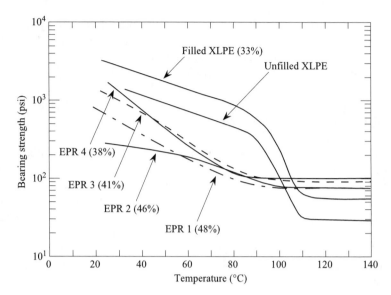

Figure 2.49 Bearing strength vs. temperature for EPR and XLPE compounds, with filler content indicated in parentheses [39].

bearing strength at the lower temperatures; also, it is interesting to note that two of the EPR compounds have relatively high bearing strength values at the low temperatures.

Hardness tests on plastics provide a measure of the material to withstand protrusion of the solid or stranded conductor into the extruded semiconducting shield. The hardness test is carried out using a diamond indenter with an applied load; the penetration of the indenter into the plastic constitutes a measure of hardness, which is proportional to the contact modulus of elasticity. The hardness H of the plastic (expressed in kg/mm^2) is calculated using the relation [39]

$$H = 37.8 \ W/h^2 \tag{2.30}$$

where W is the load in kilograms and h denotes the depth of penetration in micrometers of the diamond indenter. The hardness as a function of temperature of the different EPR and XLPE compounds is delineated in Fig. 2.50. Once again the results obtained are similar in behavior to the modulus of elasticity and bearing strength. It is apparent that the EPR compounds are harder than the XLPE compounds at the more elevated temperatures, while the converse is the case over the lower temperature régime.

Plastic insulation subjected to tension and compression forces at elevated temperatures over long periods of time may undergo permanent deformation (elongation or compression), which is termed creep. The creep modulus E_c is defined as

$$E_c = \frac{S_0}{\text{total creep strain}} \tag{2.31}$$

where S_0 denotes the initial value of the applied mechanical stress. Hence the value of the creep modulus characterizes the extent to which the plastic material is capable of resisting the externally applied mechanical forces.

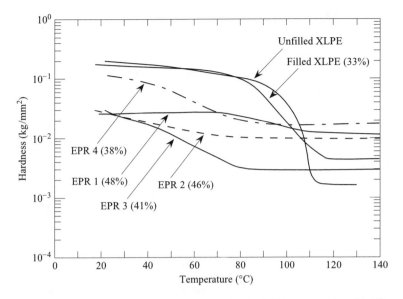

Figure 2.50 Hardness vs. temperature for EPR and XLPE compounds, with filler content indicated in parentheses [39].

Figures 2.51 and 2.52 portray the creep modulus in tension as a function of time at 70°C and temperature after 10 h respectively. Examination of the creep modulus in tension versus the time characteristic in Fig. 2.51 shows that at 70°C (below the melting point of XLPE), the filled and unfilled XLPE compounds exhibit a much greater creep modulus than the EPR materials over the shorter testing times; also the rate of creep in the EPR is much lower. It is apparent from Fig. 2.52 that the behavior of the creep

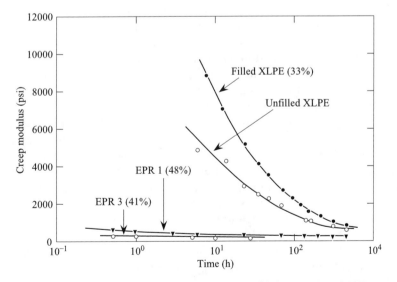

Figure 2.51 Creep modulus in tension vs. time at 70°C for XLPE and EPR compounds, with filler content indicated in parentheses [39].

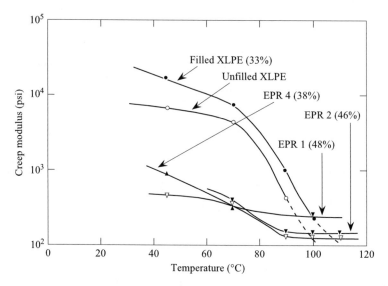

Figure 2.52 Creep modulus in tension vs. temperature after 10 h for XLPE and EPR
compounds, with filler content indicated in parentheses [39].

modulus in tension with temperature is similar to that observed with the modulus of
elasticity, bearing strength, and hardness tests. Thus the greatest values of creep modulus in tension over the lower temperature region are found to be associated with the
XLPE as compared to the EPR compounds, while the converse is true over the higher
temperature regime once the temperature attains values within the melting point of
XLPE. It is interesting to observe that relatively small variations in temperature can
cause rather large changes in creep modulus; in particular with XLPE, this may amount
to more than two orders of magnitude.

Figures 2.53 and 2.54 depict, respectively, the creep modulus in compression as a
function of time at 70°C and as a function of temperature after 10 h. Examination of
the characteristics demonstrates that the behavior of the creep modulus in compression
is essentially the same as that in tension, with the exception that the values of the creep
moduli in compression are substantially greater. Again this difference must be ascribed
to imperfections in the physical structure of the compounds, which tend to affect
adversely the creep modulus in tension.

Comparison of the mechanical test data on the XLPE and EPR compounds indicates that whereas for both of the materials the mechanical properties are appreciably
influenced by the temperature, the XLPE compounds have higher strength at lower
temperatures. However, as the softening temperature of XLPE is approached, its
mechanical strength falls rapidly in comparison to EPR. Since the rate of change of
the mechanical properties of XLPE with temperature is rapid, the rate at which the
XLPE will lose its mechanical strength will be much greater than that of the EPR
compounds. Nevertheless, with the proviso that a given XLPE cable is operated at
or below 90°C, the foregoing behavior is inconsequential in that its mechanical strength
exceeds that of the EPR compounds by an appreciable margin.

The heat transfer characteristics of the insulation constitute one the most important parameters that affects the current rating of power cable [96]. The temperature at

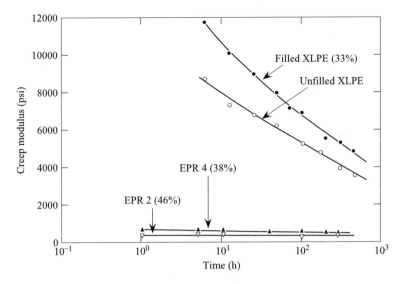

Figure 2.53 Creep modulus in compression vs. time at 70°C for XLPE and EPR
compounds, with filler content indicated in parentheses [39].

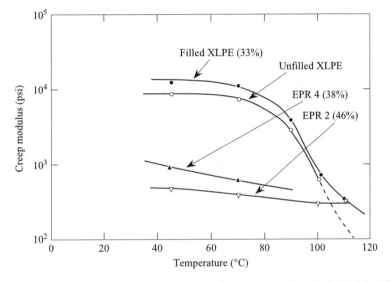

Figure 2.54 Creep modulus in compression vs. temperature after 10 h for XLPE and
EPR compounds, with filler content indicated in parentheses [39].

the conductor of the cable is contingent upon the thermal conductivity of the cable
components, their configuration as well as the medium into which the heat may be
dissipated. The thermal conductivity is measured using round specimen slabs of
12.7 cm diameter and 1.5 cm thickness. Figure 2.55 depicts typical thermal conductivity versus temperature curves obtained on XLPE and EPR compounds. It can be
perceived that the thermal conductivity of XLPE decreases with temperature, exhibiting
a very marked fall in the vicinity of 100°C, that is the melting point over which the

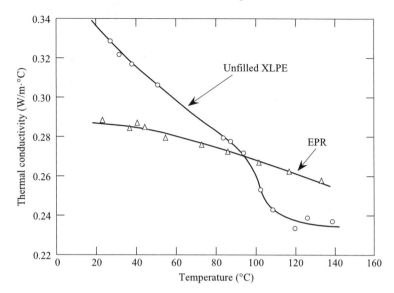

Figure 2.55 Thermal conductivity vs. temperature of XLPE and EPR compounds, obtained with ascending temperature [39].

XLPE changes from a partially crystalline to a completely amorphous compound. It should be emphasized that the abrupt decrement in the thermal conductivity at the melting temperature is not always observed [73]. In contrast to XLPE, the thermal conductivity of EPR decreases only slightly with temperature, demonstrating the beneficial effect of the more highly thermally conducting inorganic fillers employed.

Thermal expansion, defined as the ratio $\Delta V/V$, where ΔV denotes the change or increase in volume and V the reference volume at 25°C (cf., ASTM Method D 864 [97]), is delineated in Fig. 2.56 as a function of temperature. Both filled and unfilled XLPE exhibit a concave upward form over the temperature regime below the melting point, which is associated with the crystalline regions or crystallites whose lattice energies are proportional to their size [73]. As the temperature is augmented and the supplied thermal energy exceeds that of the small crystallites, the latter begin to melt. With further increase of temperature, the larger crystallites commence melting until the largest crystallites are ultimately melted at the so-called thermodynamic melting point. Thereafter the volume expansion continues at a lower but linear rate, which is essentially contingent upon the molecular structure of the XLPE compound. It can be observed from Fig. 2.56 that as the melting temperature is approached both the unfilled and filled XLPE compounds expand at a higher rate, which exceeds that of the EPR compounds. This difference in behavior is principally attributable to the crystallinity of the XLPE compounds and to the intrinsically lower thermal expansion coefficient of the inorganic fillers employed in the EPR compounds. It is interesting to note that the thermal expansion of XLPE and EPR is not linear over the entire temperature range. Consequently, we cannot speak of a thermal expansion coefficient that is valid over the entire temperature range. The latter situation is in direct contrast to copper and aluminum metals for which the expansion coefficient over the entire cable operating temperature range is linear.

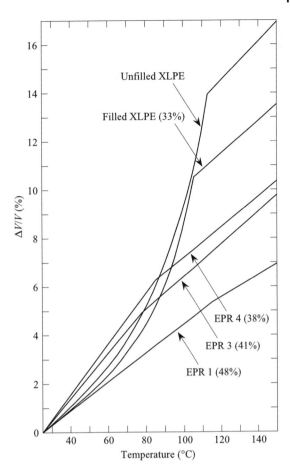

Figure 2.56 Volume thermal expansion, $\Delta V/V$, vs. temperature for XLPE and EPR compounds, with filler content indicated in parentheses [39].

Although the volume thermal expansion of EPR as a function of temperature does not manifest the concave upward variation of the XLPE compounds below the melting temperature, it does deviate somewhat from a true linear behavior over the test temperature range. The question arises as to whether some of the nonlinear variation manifest is in part due to some residual crystallinity in the EPR compounds. One of the most effective means of detecting crystallinity in the structure of polymers is by means of differential scanning calorimetry (DSC), which determines either endotherm or exotherm enthalpy peaks of the polymer specimen as heat energy is either absorbed or liberated to melt the crystallites or during recrystallization, respectively [98]. Figure 2.57 depicts the DSC traces obtained on the XLPE and EPR compounds, which were derived by subjecting the specimens to the same heating and cooling rate cycles of 20 K/min respectively [39]. A higher cooling rate is ordinarily utilized to obviate the effect of transformations resulting from recrystallization. All curves refer to the second run of the measurement, since the result thus obtained is independent of the previous unknown thermal history of the specimen.

The peaks of enthalpy in (a) and (b) of Fig. 2.57 indicate clearly the melting temperatures of the crystallite regions of the unfilled and filled XLPE. Examination of the characteristics indicates that the melting temperatures of the filled and unfilled

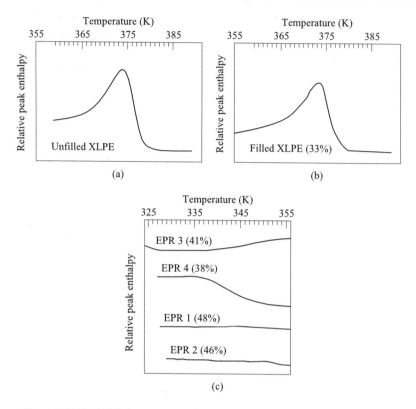

Figure 2.57 The DSC thermograms of XLPE and EPR compounds, with filler content indicated in parentheses [39].

XLPE are virtually identical. The relative flatness of the endotherms of EPR compounds 1, 2, and 3 infer that, neglecting the baseline variations, the compounds are essentially amorphous. However, the observed difference in the endotherm of EPR 4 reflects a 2% crystallinity content.

The melting of crystallites in XLPE takes place over a range of temperatures, extending up to ~110°C for the larger crystallite regions. The area under the enthalpy peaks is thus proportional to the crystallinity content of the polymer and can be employed by means of a suitable calibration to determine the crystallinity of the polymer. However, it must be emphasized that this constitutes only an approximate procedure, since it is also known that other variables such as the free peroxide content remaining after crosslinking may also affect the melting temperature range of the crystallites and thus the magnitude of the enthalpy peaks themselves. Another means of determining the crystallinity of XLPE is in terms of the specific volume $V(T)$, which may be derived from the volume thermal expansion curve; that is,

$$V(T) = V(25)\left(1 + \frac{\Delta V}{V}\right) \qquad (2.32)$$

where $V(25)$ represents the reciprocal of the density at 25°C, and $\Delta V/V$ is the volume expansion ratio at the temperature T. Then the degree of crystallinity is given by [99]

$$\text{Percent crystallinity} = 100 \times \frac{V_l - V(T)}{V_l - V_c} \qquad (2.33)$$

where V_l denotes the specific volume in the liquid state (above the melting temperature), $V(T)$ is the specific volume of XLPE, and V_c is the specific volume of the crystal, which is equal to 1.000 cm^3/g at 23°C [100].

2.4.3.8 Tree-Retardant XLPE

The incidence of water trees and their associated deleterious effects on XLPE power cable insulation have been greatly reduced by the introduction of cleaner XLPE compounds and cleaner and more smooth semiconducting conductor and insulation shields. Various attempts had been also made already in the past to develop additives with tree-retardant propensities to prevent or reduce appreciably water tree occurrence and propagation. In HMWPE considerable use was made of dodecanol $(CH_{12}H_{25}OH)$ as a tree inhibitor, though its effectiveness was found to diminish with time [84]. The early tree-retardant ethylene vinyl acetate (EVA), utilized in XLPE, tended also to lose its effectiveness over prolonged time periods under voltage stress. In 1983, a tree-inhibiting compound was made available by Union Carbide as an additive in their XLPE compounds. Little information is available on the molecular structure of this tree-retarding compound other than that it is commonly known to be polar. Accelerated electrical aging tests carried out on cables with smooth clean semiconducting shields and clean XLPE insulation containing the polar tree-retardant additive have indicated a greater resistance to water tree growth and that the size and numbers of water trees, which develop, are substantially smaller. This behavior is reflected in a longer cable life [101].

The introduction of the polar tree-retardant additive into the XLPE compound augments the dielectric loss markedly in an otherwise extremely low loss dielectric; nevertheless, the resulting dielectric loss magnitude is still much lower than that of an EPR compound. Figure 2.58 depicts the dissipation factor as a function of temperature at 20 kV on an aged and unaged cable having an XLPE insulation containing the polar tree retardant additive [102]. From the tan δ behavior it would appear that some residual products such as peroxides may inhibit molecular dipole orientation over the lower temperatures in the case of the unaged cable as compared to its aged state where the removal of these residuals permits freedom of dipole orientation leading to significant dipole losses at lower temperatures. The increase of tan δ with temperature both in the aged and unaged state suggests that the inclusion of the polar tree-retardant compound may have either reduced the effective charge trap depths in the polymer structure or introduced mobile ion constituents, whose mobility is enhanced at elevated temperatures due to thermal agitation. However, since charge trapping appears to be implicated with the mechanisms of both water and electrical tree growth, a change in the tree growth behavior in tree-retardant XLPE would imply an intrinsic reduction in the charge trap depths following the introduction of the polar tree-retardant compound [32].

Whether the introduction of the polar tree-retarded compound or the use of clean XLPE insulation with clean smooth semiconducting shields exert the predominant influence in inhibiting vented water tree growth in XLPE has not been fully ascertained. However, the inclusion of the polar tree-retardant compound is found to reduce

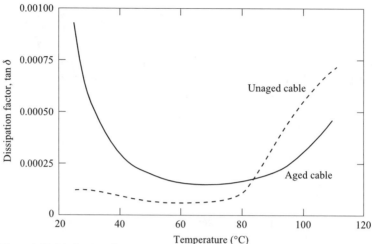

Figure 2.58 Dissipation factor vs. temperature at jacket surface of XLPE power distribution cable containing polar tree-retardant, before and after aging at 20 kV (after [102]).

definitely and very markedly the number of bow tie water trees. A behavioral trend in the incidence of vented and bow tie water trees is palpably evident from Fig. 2.59, which compares the incidence of water trees in a cable having a stranded aluminum conductor size of 120 mm^2, a conductor semiconducting shield of ≥ 0.3 mm in thickness, an insulation thickness of 5.5 mm, and insulation semiconducting shield between 0.3 and 0.6 mm in thickness [101]. It is apparent from Figs. 2.59(a) and (b) that, although the addition of a polar tree retardant to XLPE diminishes greatly the number and size of bow tie water trees when current state of the art semiconducting shields are employed, the effect on vented-type water trees is essentially insignificant. However, when higher purity carbon black particles are employed in the semiconducting shields, a decrease occurs in the number of vented-type trees in XLPE both with and without the tree-retardant additive, as substantiated in Figs. 2.59(c) and (d). Again a comparison of the two latter figures demonstrates that the addition of the tree retardant even with the cleaner carbon particle shields, decreases primarily the number of bow tie trees. Further examination of the results reveals that the tendencies in the tree growth are not always entirely predictable: for example, when cleaner carbon blacks are employed for the shields, the numbers of bow tie trees are appreciable greater than when less pure carbon blacks are employed, as becomes increasingly apparent when Figs. 2.59(a) and (b) are compared with Figs. 2.59(c) and (d), respectively.

Insulation that has been subjected to voltage stress and aged under wet conditions will form water trees, and the extent of this water tree formation will normally be detected in terms of reduction in its alternating breakdown strength as compared to that of unaged insulation [85]. Although it has been possible to relate the decrease in the ac breakdown strength to the length of vented water trees, no definite relationship has thus been established in the case of bow tie trees. Figure 2.60(a) delineates the effect of water tree length as a percent of total insulation thickness of XLPE cables upon the ac breakdown stress at 50 Hz. It can be seen that initially the breakdown stress falls rapidly with tree length, but as the tree approaches in length the insulation thickness,

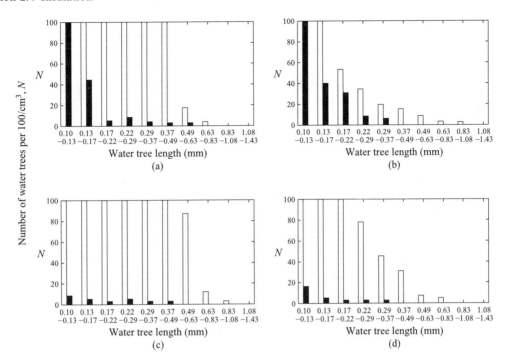

Figure 2.59 Effect of tree retardant and purity of carbon particles on length, type, and number of water trees in XLPE power distribution cables aged at 30 kV (50 Hz) in water tanks maintained at 40°C and containing 0.02 g NaCl/100 g H$_2$O (bow tie trees shown in light columns, vented trees in dark columns; upper and lower values of abscissa scale indicate range of recorded water tree lengths): XLPE cable with (a) standard semiconducting shields, (b) tree-retardant and standard semiconducting shields, (c) higher purity carbon particles in semiconducting shields, and (d) tree retardant and higher purity carbon particles in semiconducting shields (after [101] and [102].

the ac breakdown stress value approaches asymptotically the maximum tree length axis. This infers essentially that even though the water tree bridges the insulation, the insulation continues to support a low-voltage stress until such time when an electrical tree develops from the water tree and then propagates rapidly, precipitating breakdown once it reaches either one of the two semiconducting shields. It is noteworthy that the ac breakdown strengths of XLPE containing polar compound tree retardants are somewhat higher than those for the XLPE free of the tree-retardant additives when vented tree growths of approximately equal length are considered [cf. Fig. 2.60(a): points *b* and *d* are for cables with XLPE containing the tree retardant and *a* and *c* refer to XLPE cables without the tree retardant].

Figure 2.60(b) depicts aging test data obtained using a mean electrical aging stress of 5.5 kV/mm in water tanks containing the same NaCl concentration as employed for the results described in Fig. 2.60(a). The cables were not aged until failure, but their aging rate was rather evaluated in terms of their residual breakdown strength after 1 and 2 years of aging. The decrement in this value, when compared to the initial breakdown strength of the cable prior to the commencement of the aging test, is considered

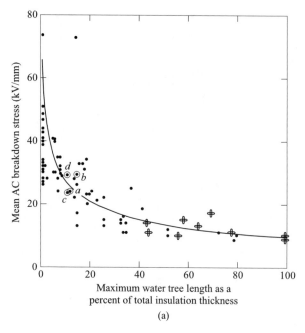

Maximum water tree length as a
percent of total insulation thickness

(a)

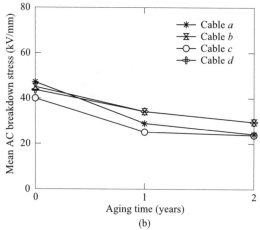

Aging time (years)

(b)

Figure 2.60 (a) Mean ac breakdown strength at 50 Hz vs. maximum vented water tree length obtained with XLPE power distribution cable specimens, both free and with tree-retardant additives: (dark circles indicate data obtained in laboratory and crosses are for field service data); points for *a*, *b*, *c*, and *d* refer to cables in Fig. 2.59. (b) Alternating voltage breakdown stress of 20-kV XLPE cable vs. aging time at 30 kV, 50 Hz. Cable *a*, a conventional XLPE cable; cable *b*, XLPE cable with tree retardant, but conventional semiconducting shields; cable *c*, XLPE cable with semiconducting shields containing carbon of higher purity; cable *d*, XLPE cable with tree retardant and higher purity carbon semiconducting shields (after [101, 102]).

to be an indicator or a measure of the degree of aging that has taken place. For Fig. 2.60(b), the ac breakdown stress values were determined using a voltage rise ramp of 2 kV/min [101]. It can be observed from Fig. 2.60(b) that the two cables containing the polar tree-retardant compound exhibit higher remnant breakdown strengths after 1 and 2 years of aging, thereby indicating a lower deterioration rate than XLPE cables without the tree retardant. Surprisingly, the XLPE cable without the tree retardant but with semiconducting shields of higher purity carbon black did not yield an exceptionally long-term performance; perhaps its poorer performance may have resulted from some other unknown constituents present in the insulation or the semiconducting shields. The behavior in Fig. 2.60(b) was also substantiated with additional results obtained by Katz [103].

2.4.4 Synthetic Solid-Liquid Insulations

Due to the relatively high values of $\tan\delta$ of oil-paper combinations, it is not possible to extend the use of oil-paper insulation for cable applications much beyond 750 kV even when forced cooling is utilized. Consequently, a great deal of effort has in the past been placed into finding suitable low-loss synthetic tapes as well as chemically compatible liquid impregnants, which would provide the same long-term performance reliability that has characterized oil-impregnated paper cable insulating systems.

An early study had been carried out to examine the effects of matched and mismatched dielectric parameters of the tapes and impregnants employed, using relatively high loss polycarbonate tapes ($\tan\delta = 2.5 \times 10^{-3}$) in conjunction with mineral oils and silicone liquids as the impregnants [104]. The results showed that the losses decreased with temperature as the dielectric parameters of the solid and liquid layers tended toward a better match at the more elevated temperatures. With oil-impregnated polycarbonate tapes, it was found that at room temperature and 85°C for a voltage stress of 180 kV/cm, the $\tan\delta$ values were, respectively, 3.0×10^{-3} and 2.2×10^{-3}. When the oil was substituted with a silicone liquid, the respective $\tan\delta$ values fell to 1.5×10^{-3} and 1.0×10^{-3}. The result thus demonstrated that the use of the particular silicone liquid, instead of the oil as the impregnant, could further reduce the losses. Had a lower loss polycarbonate polymer been available at the time, it is quite conceivable that the overall dielectric losses could have been lowered by at least a factor of 10. In a work dealing with laminated synthetic insulation cables, it was further demonstrated that cables using low-loss polycarbonate tapes impregnated with hydrogenated polybutene having a 5% additive of tri-isopropylbenzene were characterized by excellent stability at 345 kV/cm and $\tan\delta$ values of approximately 7.8×10^{-4}; at room temperature the latter value fell to 4.4×10^{-4} [105].

There have been other reported investigations on a number of solid-liquid systems, that appear to have considerable promise for low dielectric power loss cable applications. McMahon and Punderson [106] demonstrated that fluorocarbon polymers, when employed in conjunction with polybutene fluids, can yield excellent thermal and electrical characteristics, provided care is exercised to etch chemically the polymer surface and add small traces of silicone liquids to the polybutene. At 179 kV/cm and 100°C, with etched and unetched fluorocarbons, the $\tan\delta$ values were found to be 3×10^{-4} and 3×10^{-3}, respectively.

An alternate approach to synthetic solid-liquid insulation substitutes involves the use of porous plastic or synthetic paper tapes whereby greater ease in impregnation is achieved. At the same time it should be mentioned that with both the polycarbonates [104, 105] and the fluorocarbon [106] films, no impregnation difficulties were encountered using the normal low-viscosity impregnants. However, the porous plastics themselves may possess a disadvantage in terms of their lower breakdown voltage characteristics. Buehler et al. [107] have reported on cable models constructed with polyolefin synthetic paper tapes impregnated with polybutene liquids, for which $\tan\delta$ values on the order of 3×10^{-4} at 119 kV/cm and 80°C were obtained. Although polyolefin papers were found to dissolve in polybutene at 150°C, no detectable chemical effects could be observed at 80°C. Yet when submerged in a mineral oil at 100°C, the polyolefin papers exhibited appreciable swelling, suggesting some limited chemical instability. It is interesting to note that the morphology of polyolefin paper bears some very limited resemblance to that of kraft paper, as is readily discernible in Fig. 2.61.

Some work has been reported on cable models constructed with synthetic paper comprising polyester fibers and a porous polycarbonate phase [108]. The morphology of this type of paper bears considerable resemblance to the polyamide papers marketed in North America. Cables constructed with the composite synthetic paper and impregnated with dodecyl benzene tended to suffer greatly from buckling due to the bending as a result of the low tensile modulus of elasticity in the synthetic paper. The dielectric properties of the cables appeared to be satisfactory, exhibiting a $\tan\delta$ value of approximately 5×10^{-4} at 80°C and 100 kV; however, the occurrence of partial discharges at voltage stresses above 100 kV/cm suggested inadequate impregnation. The same authors [108] also provided data on an equivalent insulating system using mica-loaded papers for which the $\tan\delta$ value at 80°C was found to be 1.2×10^{-3} (the voltage stress value was not stated).

Many of the mechanical properties inherent with oil–kraft paper systems cannot be attained in their entirety with only synthetic material substitutes. Accordingly, research has been directed toward remedying this difficulty by using plastic–paper composites. One such popular composite insulation consists of polypropylene synthetic paper laminated in between two kraft paper layers. With this arrangement, one achieves a system that can approach the impregnation efficacy of a kraft paper system, in addition to providing some improvement in the dielectric loss and voltage breakdown characteristic

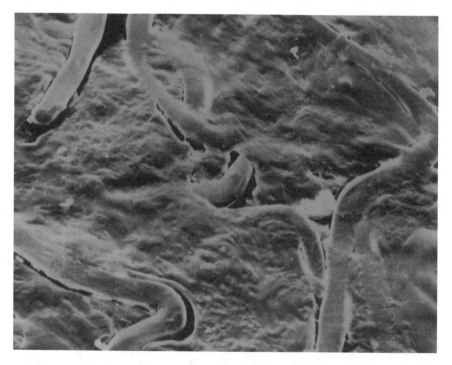

Figure 2.61 Surface morphology of a typical polyolefin paper, taken with electron scanning microscope (168×).

of a synthetic system. Evidently, the use of paper in the laminated system does introduce a certain degree of dielectric loss; however, it is a compromise between some additional dielectric loss vis-à-vis a gain in the bending characteristics of the cable. The polypropylene-paper laminate has been under extensive investigation in Europe, Japan, and United States. A number of Electric Power Research Institute (EPRI) supported studies on polypropylene–kraft paper composites have been carried out [48, 109, 110]. As mentioned in this chapter and in Chapter 1 some encouraging results have been obtained with PPP composites either of 125 or 155 μm thickness impregnated with dodecyl benzene [48]. It was found that the impregnated PPP system has an ac breakdown strength some 20–30% greater than that of the conventional oil–kraft paper systems. The foregoing result is ascribed to the relatively low dielectric constant of the PPP tapes that in effect lowers the stress distribution across the impregnant containing butt gaps (normally constituting the weakest link in the insulating system). Furthermore an increase in the impregnant pressure from 14.7 lb/in.^{2} (0.1 Mpa) to 284 lb/in.^{2} (2.0 Mpa) causes the ac breakdown strength to rise from 84.5 kV/mm (2113.5 V/mil) to 124 kV/mm (3112.5 V/mil), respectively. The impulse breakdown strength is also augmented by about 15% to a value of 229 kV/mm (5725 V/mil). In terms of these results, it was suggested that a cable insulated with PPP could conceivably be designed to operate at stresses 70% above those of conventional oil-paper cables. Table 2.15 provides pertinent electrical and physical property data on PPP tapes themselves and the dodecyl benzene impregnated systems. The impregnation process of the PPP tapes involves a conditioning step, which subjects the impregnated tapes to 100°C for 24 h. During this period the polypropylene undergoes swelling as the impregnant penetrates gradually its amorphous regions, forcing the molecular chains of the polymer to stretch. The conditioning procedure results in a marked reduction of the tan δ

TABLE 2.15 Physical and Electrical Properties of PPP Tape (Unimpregnated and Impregnated with Dodecyl Benzene)

	By PPP Tape Thickness	
Quantity	125 μm	155 μm
Polypropylene to total PPP thickness ratio	0.42	0.42
tan δ (unimpregnated)	0.001	0.001
tan δ (impregnated)		
18°C	0.00098	0.00093
100°C	0.00099	0.00096
Dielectric constant		
Unimpregnated	2.8	2.8
Impregnated	2.82	2.82
A-C breakdown, kV/mm	129.8	128.2
Impulse breakdown, kV/mm	225.2	217.8
Density, g/cm^3	0.89	0.89
Tensile strength, kg/mm^2		
Longitudinal	11.2	14.8
Transverse	5.9	8.1
Elongation, %		
Longitudinal	3.3	3.4
Transverse	9.9	10.4

Source: From [48].

value as is substantiated in Fig. 2.62(a). The variation of $\tan \delta$ with temperature suggests the involvement of ionic conduction in the dielectric loss mechanism; since the polypropylene tape barrier prevents free ionic movement between the adjacent paper layers, charge carriers accumulate at the plastic–paper interfaces, leading to a small but detectable interfacial polarization loss. It is interesting to note that other workers [110], who have carried similar studies on PPP tapes, using several impregnants did not observe any significant interfacial polarization effect with dodecyl benzene in conjunction with PPP [cf. Fig. 2.62(b)]. Pipe-type cables rated at 345 kV, using PPP-type insulation, have now been operating satisfactorily in New York City for more than several decades.

Perhaps the most extensive early investigations on solid-liquid systems under cryogenic conditions have been made by Jefferies and Mathes [111]. They have employed for their cable models spun bounded polyethylene fiber tapes and have carried out impregnation with liquid nitrogen and helium under a pressure of 75 lb/in.[2]; the $\tan \delta$ values at 197 kV/cm were found to be less than 10^{-5}. In a most surprising contrast, the systems, which made use of liquid helium, exhibited unusually high losses very much on the same order of magnitude as oil-impregnated paper. They could not

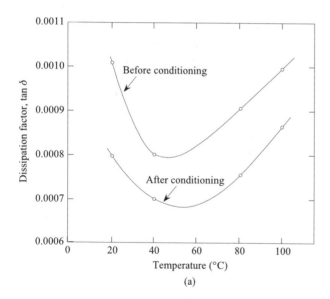

(a)

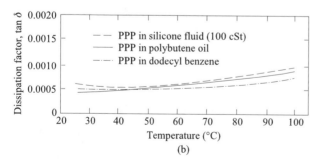

(b)

Figure 2.62 (a) Dissipation factor vs. temperature of 125-μm-thick kraft paper–polypropylene–kraft paper (PPP) composite tape impregnated with dodecyl benzene and subjected to voltage stress of 20 kV/mm (60 Hz), before and after being conditioned in dodecyl benzene at 100°C for 24 h (after [48]). (b) Power factor vs. temperature for kraft–polypropylene–kraft paper tape of 120 μm thickness impregnated with different dielectric liquids (after [110]).

advance a complete explanation for this phenomenon, although they believed that part of the reason may lie in the extremely low viscosity of the liquid helium ($36\,\mu P$). Although solid-liquid cryogenic systems appear to represent the ultimate in dielectric loss reduction, they nevertheless have a disadvantage in their high initial cost due to the associated cooling apparatus. However, this cost figure may be justifiably offset when large power transfers are involved.

As can be seen from the summary of the data in Tables 2.15 and 2.16, synthetic solid-liquid insulating systems represent viable low-loss substitutes for conventional oil-impregnated-paper cable systems. Use of plastic tapes in conjunction with certain organic impregnants can reduce the $\tan\delta$ value up to a factor of 10 as compared to that of oil-impregnated paper. With cryogenic liquid impregnants the corresponding reduction factor of $>10^2$ is most striking. For additional detailed discussions on cryogenic power cables, the reader is referred to Chapters 1 and 12.

2.4.5 Gas-Solid Spacer Insulating Systems

In terms of cable applications, sulfur hexafluoride, SF_6 gas is a serious contender for high-power-capacity gas-insulated power cables. Its breakdown strength is 2.3 times that of dry nitrogen at 60 Hz [51]. It is nontoxic and nonflammable, though under arcing or partial discharge conditions it will readily decompose into chemically active toxic gases, which can seriously degrade the adjacent insulation. Also, recently there has been considerable environmental concern expressed as regards its release into the atmosphere. Compressed SF_6 gas has been extensively used in electrical equipment (e.g., circuit breakers) since it became commercially available in 1947. For more than two decades, gas-insulated cables have been constructed using compressed SF_6 gas with the conductor in these cables being held in place by epoxy or PE spacers. The major technical problems associated with SF_6 cables hinge on the surface breakdown characteristics of the spacer insulation, which is greatly degraded in the presence of dust particles. The latter are usually introduced during the joining steps, as the individual pipe sections of the cable are being welded together. Presently SF_6 insulated cables are commercially available for voltages up to 500 kV, but their use is primarily restricted to short runs. For more detailed data on the characteristics of SF_6 and it uses in cables, the reader is referred to Chapters 1 and 12.

Filler containing epoxy resins is especially suited for gas-filled cable spacers, as the latter may be formed with metallic inserts necessary to keep the conductor in situ within the tube of the gas-filled cable [112]. In gas-filled cables, it is important to match the coefficient of thermal expansion of the epoxy to that of the metal, e.g. copper or aluminum, in order to avoid stress cracking or separation of the epoxy resin from the metal, which if not prevented, may lead to partial discharges [113]. This is accomplished by the use of fillers such as silica. Increase in the filler content also has the advantage of raising the thermal conductivity of the epoxy, thereby allowing a higher conductor operating temperature, which for gas insulated cables is usually above 120°C. There is a limit to the amount of fillers that can be used, since large filler contents ($>80\%$) cause difficulty in mixing and casting of the epoxy. Table 2.17 compares the properties of filled and unfilled epoxy resins [51].

There are numerous epoxy resin types available commercially; they are all derived from the basic unpolymerized epoxy molecule delineated in Fig. 2.63. Either organic acid or basic compounds are used as catalysts to open the oxygen bonds for the purpose

TABLE 2.16 Dielectric Data on Solid-Liquid Insulating Systems

Solid Tape	Impregnating Liquid	Average Voltage Stress kV/cm	tan δ at Room Temperature	tan δ at Operating Temperature	Reference of Result
Kraft paper tapes	Mineral oil	180	3.8×10^{-3}	5.7×10^{-3}, 85°C	42
Kraft paper tapes	Silicone liquid	180	2.7×10^{-3}	3.1×10^{-3}, 85°C	53
High-loss polycarbonate tapes	Mineral oil	180	3.0×10^{-3}	2.2×10^{-3}, 85°C	104
High-loss polycarbonate tapes	Silicone liquid	180	1.5×10^{-3}	1.0×10^{-3}, 85°C	104
Low-loss polycarbonate	Hydrogenated polybutene + 5% tri-isopropyl benzene	136	4.4×10^{-4}	7.8×10^{-4}, 100°C	105
Etched fluorocarbon	Polybutene liquid + 1% silicone liquid	179	—	3.0×10^{-4}, 100°C	106
Unetched fluorocarbon	Polybutene liquid	179	—	3.0×10^{-3}, 100°C	106
Polyolefin synthetic papers	Polybutene liquid	119	—	3.0×10^{-4}, 80°C	107
Polyester–polycarbonate synthetic paper	Dodecyl benzene	100	—	5.0×10^{-4}, 80°C	108
Mica-loaded papers	Dodecyl benzene	—	—	1.2×10^{-3}, 80°C	108
Spun-bonded polyethylene fiber synthetic papers	Liquid nitrogen	197	—	$<1 \times 10^{-5}$, 70 K	111
Spun-bonded polyethylene fiber synthetic papers	Liquid nitrogen	197	—	$<1 \times 10^{-5}$, 20 K	111
Spun-bonded polyethylene fiber synthetic papers	Liquid helium	197	—	Same as oil-paper, 4.2 K	111
Kraft paper–polypropylene–kraft paper (PPP)	Dodecyl benzene	200	9.3×10^{-4}	9.6×10^{-4}, 100°C	48

TABLE 2.17 Properties of Epoxy Resins

Property	Unfilled	Silica Filled
Volume resistivity, Ω-cm	10^{12}–10^{17}	10^{13}–10^{16}
Dielectric strength, V/mil		
Short time	400–550	400–550
Step-by-step	350	
Dielectric constant		
60 Hz	3.5–5.0	3.2–4.5
10^3 Hz	3.5–4.5	3.2–4.0
10^6 Hz	3.3–4.0	3.0–3.8
Dissipation factor		
60 Hz	0.002–0.010	0.008–0.03
10^3 Hz	0.002–0.020	0.008–0.03
10^6 Hz	0.030–0.050	0.02–0.04
Arc resistance, s	45–120	150–300
Specific gravity	1.11–1.14	1.6–2.0
Modulus of elasticity	3.0–3.5	3.5–4.0
in tension, psi $\times 10^5$		
Percent elongation, %	3–6	1–3
Flexural strength, psi $\times 10^3$	13–21	8–14
Compressive strength,		
psi $\times 10^3$	15–21	15–40
Rockwell hardness	M80–M110	M85–M120
Impact strength, ft-lb/in.	0.2–1.0	0.3–0.45
Heat distortion temperature, $^\circ$F		
at 264 psi	115–550	160–550
Thermal conductivity,	4–5	10–20
cal/$^\circ$C $\cdot$ cm $\cdot$ s $\times 10^4$		
Thermal expansion,		
in./in. per $^\circ$C $\times 10^5$	4.5–6.5	2.0–4.0
Water absorption, %	0.08–0.15	0.04–0.10
Burning rate	Slow to self extinguish	Self-extinguishing

Source: From [51].

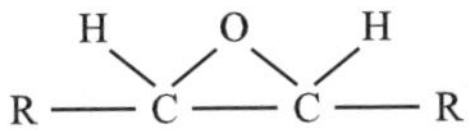

Figure 2.63 Epoxy molecule in its unpolymerized form.

of linking together the adjacent epoxy molecules into larger polymerized molecules. By controlling the rate and the time of the chemical reaction by the use of suitable accelerators and hardeners, as well as the nature of the catalyst used, a variety of epoxy formulations may be obtained. The density of epoxies is regulated by the amount of polymerization. As the polymerization proceeds, the density increases as a result of closer molecular packing as proximate interatomic bonds develop [114]. As the density increases, the epoxy in effect shrinks in size; this shrinking is easily offset by the addition of nonshrinkable fillers such as silica. Evidently, epoxies containing large amounts of filler have correspondingly lower densities. The question of shrinking must be carefully dealt with in the design of cable spacers in order to prevent separation from taking place between the metallic inserts and the epoxy.

2.5 MATERIALS FOR PROTECTIVE COVERINGS

Various materials are used for cable jackets or sheaths to provide protection for the insulation and shield it from the external environment. Such protective coverings enclosing the cable structure may be metallic or nonmetallic. The main function of the jacket or sheath material is to provide a mechanical outer protection for the cable and an impervious cover against moisture ingress. With oil-paper cables the metallic sheath must, in addition to the above two functions, prevent the escape of the liquid impregnant and act as a ground return for the neutral current in neutral grounded systems as well as a ground return for relay circuits.

2.5.1 Nonmetallic Sheaths

One of the older types of extruded thermosetting jacketing or sheathing materials used in the past, and still employed occasionally, is neoprene rubber. Neoprene is a synthetic rubber of polychloroprene (cf. Fig. 2.64), with oxides of divalent metals used as crosslinking agents. It has been in use as a jacketing material since 1933; neoprene rubber has the added advantages over natural rubbers in that it has better aging and temperature characteristics and exhibits greater resistance to ozone, UV radiation, oil and environment, and resists abrasion. As a jacketing material it possesses the important property of flame resistance.

Over the last few decades, cables with extruded jackets of polyvinyl chloride (PVC) have become popular, particularly in areas requiring high resistance to chemicals such as in chemical factories and oil refineries. However, it should be emphasized that due to its poor abrasion resistance, PVC does not pose any competition to neoprene rubber jackets that are mainly used in heavy duty applications such as on mine shaft cables. Since PVC has a tendency to creep under concentrated loads, PVC-jacketed cables should not be strung over sharp pointed supports.

Polyvinyl chloride is a thermoplastic compound, having a molecular structure similar to PE but with one of the hydrogen atoms in every second carbon atom replaced by chlorine; it is made by the polymerization of vinyl chloride (cf. Fig. 2.65). PVC exhibits good chemical and moisture resistance and, consequently, it is used as a cable jacketing material in the 60–105°C range. It is less costly than neoprene, and it is for this reason mainly that it has replaced neoprene in many cable jacketing applications. The properties of PVC may be applied to any particular cable application by the proper use of fillers, plasticizers, stabilizers, and pigments. Due to its polar structure, PVC has a highly temperature dependent dielectric loss behavior and, therefore, it is not used as a cable insulant in North America. In its manufacture and extrusion, there are some toxicity-related problems.

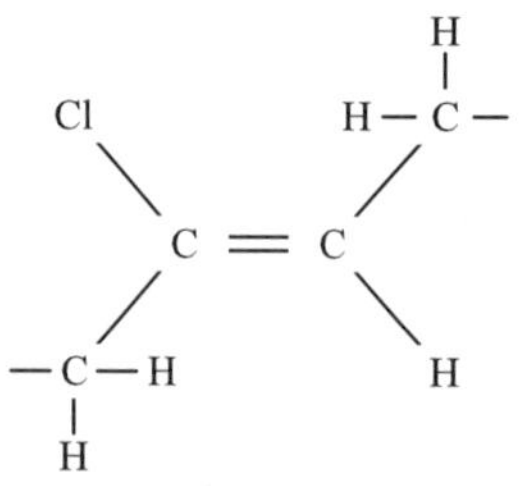

Figure 2.64 Building block (monomer) of polychloroprene and neoprene.

$$\begin{bmatrix} \begin{array}{cc} H & H \\ | & | \\ C & = & C \\ | & | \\ H & Cl \end{array} \end{bmatrix}$$

$$\begin{array}{cccc} H & H & H & H \\ | & | & | & | \\ -C-C-C-C- \\ | & | & | & | \\ H & Cl & H & Cl \end{array}$$

Figure 2.65 Polymerization of vinyl chloride:
(a) vinyl chloride; (b) polyvinyl chloride.

(a)

(b)

Polyethylene-type extruded jackets containing carbon black are also extensively used today. Due to their toughness and abrasion resistance as well as surface smoothness, PE-extruded jackets are especially suited for cables that must be drawn into ducts for installation. Also PE-jacketed cables are suitable for direct burial in dry locations, as the PE exhibits good resistance to the action of microorganisms and most soil chemicals. In wet environments, moisture ingress into the cables does occur via the PE jackets; the rate of the moisture ingress is contingent upon the type of polyethylene utilized. In North America both branched PE (thermoplastic) and XLPE (thermosetting) are used for the jacketing materials, depending on the temperature requirements.

In recent years the use of various density thermoplastic polyethylenes, namely low-density polyethylene (LDPE), linear medium-density polyethylene (LMDPE), and linear high-density polyethylene (LHDPE) has increased very markedly. For improved flexibility as a result of its lower density, LDPE is preferred, whereas greater mechanical protection is provided by the higher density and hence stiffer LMDPE and LHDPE compounds. Currently, for high flexibility applications LLDPE is preferred to LDPE primarily because of its improved water imperviousness and higher melting point [115]. The exposure of LLDPE to elevated temperatures does not appear to lead to any significant embrittlement [116]; this is supported by Fig. 2.66, which indicates little decrease in tensile elongation after prolonged subjection to a temperature of 100°C.

Carbon black has been used as a filler in polyethylene jackets for many years, principally to avert degradation such as cracking due to ultraviolet radiation. In the past efforts were made to utilize the same semiconducting carbon black containing compounds as those employed for semiconducting shields to provide the cables with semiconducting jackets. However, the efforts were unsuccessful because of the relatively high water permeation characteristics of these compounds. This difficulty appears to have been circumvented with the development of carbon containing (apparently LLDPE based) compounds, which have lower water permeability characteristics [116]. The volume resistivity of these semiconducting jackets is on the order of 10^4 Ω-cm at room temperature [16]. Semiconducting jackets effectively ground shorts

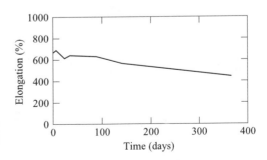

Figure 2.66 Tensile elongation vs. time of aging in air oven at 100°C for a carbon-filled LLDPE jacketing compound (after [116]).

and provide increased safety to workers by enclosing the power cable within an equipotential shield, which is maintained at ground potential by the concentric neutral wires. The semiconducting jacket may either overlap the concentric neutral wires or encapsulate them entirely.

Table 2.18 compares the physical properties of a number of thermoplastic polyethylene jacketing materials, including PVC and a semiconducting material presumably

TABLE 2.18 Physical Properties of PVC, Thermoplastic Insulating, and Semiconductive Polyethylene Jackets

Physical Property	Test Method	PVC	LDPE	LLDPE	LPDPE	LHDPE	Semiconducting Thermoplastic PE
Melt index (190°C, 2.1 kg), g/10 min	ASTM D1238	—	0.25	0.55	0.70	0.15	
Density (23°C) g/cm^3	ASTM D1505	—	0.93	0.93	0.95	0.96	1.12
Tensile strength							
psi	ASTM	≥1500	2100	2350	3500	4000	1600
MPa	D638	10.3	14.5	15.5	24.1	27.6	11.0
Tensile elongation percent	ASTM D638	≤100	650	650	900	800 ·	250
DSC melting point, °C		—	110	119	125	130	
Flexural modulus							
psi	ASTM	—	38,000	55,000	90,000	145,000	
MPa	D790		262	379	621	1000	
Dielectric constant (1 MHz)	ASTM D1531	—	2.60	2.50	2.52	2.56	
Dissipation factor (1 MHz) rad	ASTM D1531	—	0.005	0.0003	0.0003	0.0003	
Dielectric strength, 0.030 in. insulation thickness at 500 V/s rate of rise							
V/mil	ASTM	—	—	1230			
kV/mm	D149			46.6			
Heat distortion at							
115°C		—	90	22	7	0	
121°C		20					
135°C per cent		—	Melted	Melted	Melted	Melted	
Abrasion resistance, mg/100 revolutions	ASTM D3389	—	29	22	16	14	
Shore D hardness	ASTM D2240	—	54	56	61	66	
Cut-through resistance							
lb		—	2600	3000	3600	4000	
N × 10^4		—	1.20	1.3	1.6	1.8	
Water vapor transmission g/24 hr/m^2	ASTM E96	310.0	1.16	0.74	0.51	0.32	≤1.5
Brittleness temperature, °C	ASTM D3389	—	—	−60	—	—	−24

Note: LDPE, LLDPE, LMDPE, LHDPE, and semiconducting thermoplastic PE are Union Carbide compounds.
Source: From [116].

based on LLDPE [116]. It can be perceived from the table that the tensile strengths of the thermoplastic polyethylenes exceed appreciably that of the PVC compounds, with the exception of the semiconducting thermoplastic PE for which the tensile strength is approximately of the same magnitude due to the carbon content of the semiconducting PE. On the other hand, the water vapor transmission of PVC is higher by two orders of magnitude as compared to all the thermoplastic polyethylenes. It is also interesting to note that the tensile modulus of the thermoplastic compounds exceeds that of PVC.

2.5.2 Metallic Sheaths

Lead, with a density of 11.37 and a melting temperature of 319°C, represents perhaps one of the oldest sheathing materials used in the manufacture of power cables. Its initial use dates back to 1879, in Vienna, where the lead press was invented. In the sheathing operation, lead in its molten state is admitted into a lead press cylinder; upon solidification, a piston under hydraulic pressure forces the lead through an annular die, squeezing it tightly around the cable concentric with the die. Apart from its toxity, one of the main advantages of lead sheaths lies in the fact that the lead can be applied onto the cable at low temperature and pressure. The main disadvantages of lead-sheathed cables center on the high density of lead, which adds enormously to the overall weight factor of the finished cable. Copper bearing lead alloys are mainly used for lead sheath applications. For applications where the cables may be subjected to slow cyclic bending or other forms of vibrations, arsenic-lead or antimony-lead alloys are preferred.

Lead sheaths are subject to deformation under continuous load conditions due to their poor creep and elasticity characteristics. This is particularly manifested in the expansion of the lead sheaths following heating under elevated load conditions or due to the internal pressure buildup for cables installed under steep inclines. To constrain these effects, suitable reinforcement of the lead sheath is required. Another difficulty encountered with lead sheaths lies in their high susceptibility to failure arising from metal fatigue. Such failures often result from mechanical vibrations due to vehicle movements adjacent to the installation site or even during cable transport. Invariably, the fatigue failures are characterized by cracks or fissures along the crystalline grain boundaries of the lead.

The use of aluminum for the sheathing application of cables began in the 1940s in Germany. Aluminum sheathing was particularly attractive from the point of its low density (2.71) and good mechanical properties as compared to lead. Aluminum sheaths do not require the same mechanical reinforcement as do lead sheaths for pressures up to 200 lb/in.^{2}. Aluminum-sheathed cables, however, must have a larger bending diameter than those of comparable size with lead, if permanent sheath distortion and damage to the cable dielectric is to be avoided. Aluminum sheaths are extruded using a similar extrusion press to that of lead; but the extrusion temperatures range between 400 and 500°C as compared to 200°C for lead. The melting temperature of commercially available aluminum is 658°C, while that of lead is much lower at about 319°C [117]. Generally, an extra layer of paper or carbon black tape is applied over the outside of the insulation to provide protection against the higher extrusion temperature of the aluminum sheath. Additional information on metallic sheaths may be found in Chapter 5, particularly as regards their extrusion procedures.

Increasing use is presently made of copper sheaths, principally for the replacement of lead sheaths due to environmental concerns and for increased corrosion protection.

Copper has an advantage over aluminum since no extrusion press is required. Instead, the copper sheaths are applied over the cables in the form of a continuous weld or in corrugated form. For this application soft copper is employed. Copper sheaths are discussed in detail in Chapter 13.

2.6 ARMORING MATERIALS

Cable armor is employed for a variety of reasons to protect mechanically and add strength to the cable structures. The overall cable construction must be sufficiently strong to sustain any pulling or dredging and should be able to sustain crossing pressures by blunt objects or possible penetration by sharp objects. Underwater (submarine) cables usually require extra strong armor protection to provide tensile strength whenever unduly long cable lengths or rough terrain with steep acclivities are involved. Also, as mentioned earlier, oil-filled cables with lead sheaths require reinforcement of the latter to cope with the internal oil pressures.

Three conductor power cables are usually armored either with a double steel tape armor or a single or double steel wire armor. In three phase current systems, because of the small external field surrounding the conductor, the current losses in the steel armor are negligible. With single-conductor power cables attempts are usually made to use tapes or wires of nonmagnetic material, unless the overall armor cost considerations justify the losses within the steel armor employed. Where tensile strength is of prime consideration, a steel wire armor is used in lieu of the double metallic tape armor, which is normally applied helically onto the cable. There are a variety of stainless and galvanized low carbon steels available for armor wire application. Their tensile strengths usually range from about 2×10^5 to 3×10^5 lb/in.^{2}, having specific gravities in the order of about 8. Stainless steel is preferred where corrosion resistance is required. Also a number of high-strength nickel and copper-nickel alloys constitute suitable alternatives. Additional data on cable armor characteristics may be found in Chapter 13, which deals with underwater cables for which armor considerations constitute a very important aspect of the overall cable design.

If ungalvanized steel tapes are used, then a jute bedding saturated with a bituminous compound is applied over the cable prior and following the double tape armor. The tarred jute prevents indentation of the metallic sheathing by the armor wires; it also provides additional binding to the wires. The jute and the bituminous compound are omitted when galvanized steel tapes are used, and the cable is installed in dry locations where it may be necessary to avoid the fire hazard posed due to the presence of the jute and the bituminous compound. Both are flammable, since jute is a naturally occurring cellulose fiber derived from plants and bitumen is a dark-colored hydrocarbon mixture derived from petroleum products. Taped armor cables are primarily used for power feeder applications in factories and are usually suspended along walls or ceilings [9]. The metallic tape varies from 0.02 to 0.03 in. in thickness and from 0.75 to 3.00 in. in width, both dimensions increasing with the overall cable diameter including the jacketing material.

For buried power cables, where strength is required in the longitudinal direction, steel wire armors are used. The individual steel wire diameter varies from 0.083 to 0.203 in. for cable diameters ranging from 0.750 to 2.501 in. and over. For power cables used in vertical and submarine installations, a combination armor of helically applied steel tape and single steel wire armor is employed. For submarine cables, where excessive flexing is involved as a result of water movement, a double steel wire armor may be applied over which jute and bituminous compound are employed. The tarred jute prevents indentation of the metallic sheathing by the armor wires; it also provides additional binding to the wires. Note should be made of the fact that jute and the bituminous compound are only applied under the armor where metallic jackets are involved. With nonmetallic jackets, the armor is applied directly over the existing jackets. Often a film of whitewash is applied over the external jute and bituminous compound layer to prevent the cable turns from adhering to each other.

2.7 COVERINGS FOR CORROSION PROTECTION

Exposed metallic jackets and armors are subject to deterioration due to corrosion attack resulting from a number of possible causes. For cables buried directly in ground, corrosion may result from the chemical attack caused by weakly acidic or alkaline water, galvanic currents generated between dissimilar moist earths in contact with the metallic sheath or armor, development of a 60-Hz current component between the earth and the cable surface, and any possible stray direct currents. For cables installed directly in ducts, in addition to the above causes, corrosion may take place due to galvanic currents generated between dissimilar metallic surfaces in contact [118].

To prevent corrosion of the metallic sheaths and armors, the latter must be shielded from the environment by nonmetallic coverings. We have already mentioned in the previous section the use of jute in conjunction with bituminous compounds. Many applications, however, require more rigid and impervious coverings that resist the strong action of certain chemicals. At the present, a universal covering material that would function well in all corrosive environments is not available. However, by proper selection of any one of the previously described extruded jacketing or sheathing materials, adequate corrosion protection may be achieved. Table 2.19 shows the relative resistance of extruded PE, PVC, and neoprene coverings to various chemicals [9]. Table 2.20 lists some of the more salient properties of the three different coverings.

Corrosion may occur with primary-type distribution cables that have a bare concentric neutral of coated copper wire to ensure proper grounding [119]. Such cables are particularly susceptible in soils having high sulfur content; however, in many instances direct buried copper wires tend to resist corrosion remarkably well, since the copper neutral is cathodic to a large range of other metals (cf, Table 2.2). In cases where difficulties may arise, cathodic protection should be considered. In cathodic protection, a battery or a corrosion cell is formed between the cathode (the cable wire ground to be protected) and the anode. The anode may consist of metals high in the electromotive or galvanic series, such as zinc or magnesium. With the anode placed in close proximity to the possible or anticipated corrosion site on the copper grounding wires, a current will be established in the formed electrolytic cell between the anode and the cathode, leading to the gradual corrosion of the anode, thereby providing continued protection of the cathode.

TABLE 2.19   Chemical Resistance of PE, PVC, and Neoprene Coverings

Chemical	PE	PVC	Neoprene
Acids			
Sulfite liquors	E	VG	G
Sulfuric ($\leq 10\%$)	E	VG	VG
Hydrochloric ($\leq 10\%$)	E	VG	G
Nitric ($\leq 5\%$)	VG	G	NR
Acetic ($\leq 5\%$)	VG	F	NR
Alkalis			
Strong caustics (30%)	E	NR	E
Caustics ($<2\%$)	E	E	E
Sulfate liquors	E	NR	G
Bleaches			
Calcium perchlorate	G	NR	NR
Sodium hypochlorite	E	E	E
Hydrogen peroxide	G	G	P
Chlorine liquid	NR	NR	NR
Chlorine dioxide	NR	NR	NR
Esters			
Tributylphosphate	F below 60°C	NR	NR
Hydrocarbons			
Turpentine	VG	F	NR
Oil (splash)	E	VG	VG

Abbreviations: E, excellent; VG, very good; G, good; F, fair; P, poor; NR, not recommended.
Source: From [9].

TABLE 2.20   Properties of PE, PVC, and Neoprene Coverings

Protectives covering	Maximum Operating Temperature (°C)	Resistance to Moisture	Degree of Flammability	Minimum Temperature Installation °C (°F)
PE	80	Very good	Slow burning	−20 (−4)
PVC	75	Fair	Nonflammable	−10 (14)
Neoprene	75	Fair	Nonflammable	−20 (−4)

Source: From [9].

2.8 CONCLUSION

In this chapter a description has been presented on the construction materials utilized in the past that are currently in use and that have future applications in the manufacture of electrical cables. Space limitations have prevented us from delving too deeply into the science of cable materials; for greater detail of the subject matter, the more prodigious reader is referred elsewhere [120]. However, the chapter should provide the reader with sufficient background in materials-related topics to allow him to follow more effectively the ensuing chapters on cable design and manufacture.

2.9 GLOSSARY ON CABLE MATERIALS TECHNOLOGY

In the foregoing text we have used a number of terms that are peculiar to extruded cables. The definitions of these terms are listed:

- **Elastomer**   A material that at room temperature stretches under low stress to at least twice its length and returns to its original length with the stress removed.
- **Filler**   A substance (usually inert) added to a plastic to improve its physical properties or decrease cost.
- **Monomer**   The simple, unpolymerized form of a compound, which constitutes the building block of a polymer.
- **Plasticizer**   A chemical agent added to plastics to provide increased flexibility and softness.
- **Polymerization**   The chemical joining of two or more monomers and polymers of the same kind to form a molecule with higher molecular weight.
- **Stabilizer**   An additive used in some plastics to maintain the physical and chemical properties throughout processing and service.
- **Vulcanization**   A chemical reaction in which the physical properties of an elastomer are changed by reacting with sulfur or other cross-linking agents. It is now more common to use the term crosslinking in lieu of vulcanization, which was the preferred term with rubber-based solid extruded insulations.

REFERENCES

[1] Copper Development Association, *Copper for Bus-Bar Purposes*, CDA Publication No. 22, Thames House, London, 1936.

[2] Copper Development Association, *Copper Data,* CDA Publication No. 12, Thames House, London, 1935.

[3] G. E. Luke, "The resistance of electrical connections," *Electrical J.*, Vol. 21, 1924, pp. 66–69.

[4] F. H. Constable, "Growth of oxide films," *Proc. Royal Soc.*, Vol. 117, 1927, p. 385.

[5] C. L. Denault, "Electrical contact of bus-bar joints," *Electrical J.*, Vol. 30, 1933, pp. 281–282.

[6] A. King and V. H. Wentworth, *Raw Materials for Electric Cables*, Ernest Benn, London, 1954.

[7] *ASTM Standards on Metallic Electrical Conductors*, ASTM Standard B233. Permissible Impurity Levels in Electrical-grade Aluminum Conductors, Philadelphia, 1980.

[8] *Handbook of Aluminum*, Aluminum Company of Canada, Montreal, 1957.

[9] *Northern Telecom Electrical Conductors Handbook*, 14th ed., Lachine, Quebec, 1977.

[10] R. Roberge, "Alternative fabrication technologies for A15 multifilamentary superconductors," in *Superconductor Materials, Science*, S. Foner and B. B. Schwartz, Eds., Plenum, New York, 1981, Chapter 6.

[11] H. B. Dwight, "Skin effect in tubular and flat conductors," *Trans. AIEE*, Vol. 37, Part II, 1918, pp. 1379–1403.

[12] R. B. Blodgett and F. H. Gooding, "Parameters affecting the increase in dielectric loss, caused by carbon black screens, for oil-paper dielectric," *IEEE Trans. Power Apparatus Syst.*, Vol 82, 1964, pp. 121–130.

[13] W. Walstrom, "Investigation of insulation deterioration in 15 kV and 22 kV polyethylene cables removed from service," *IEEE Trans. Power Apparatus Syst.*, Vol. PAS-91, 1972, pp. 1023–1028.

[14] "Specifications for thermoplastic and cross-linked polyethylene insulated shielded power cables rated 5 through 46 kV," AEIC Specification CS5, New York, 1994.

[15] "Specifications for ethylene propylene rubber insulated shielded power cables rated 5 through 69 kV," AEIC Specification CS6, New York, 1996.

[16] N. M. Burn, R. M. Eichhorn, and C. G. Reid, "Stress controlling semiconductive shields in medium voltage power distribution cables," *IEEE Electrical Insulation Mag.*, Vol. 8, 1992, pp. 8–24.

[17] N. S. Enikolopyan, Ed., *Filled Polymers I*, Springer, Berlin, 1990.

[18] S. Kirkopatrik, "Percolation and conduction," *Rev. Mod. Phys.*, Vol. 45, 1973, pp. 574–588.

[19] J. C. Gurland and D. B. Tanner, *Electrical Transport and Optical Properties of Inhomogeneous Media*, AIP, New York, 1978.

[20] P. Sheng, E. K. Sichel, and J. L. Gittleman, "Fluctuation induced tunneling conduction in carbon-polyvinyl chloride composites," *Phys. Rev. Lett.*, Vol. 40, 1978, pp. 1197–1200.

[21] E. K. Sichel, J. I. Gittleman, and P. Sheng, "Transport properties of the composite material carbon-polyvinyl chloride," *Phys. Rev. B*, Vol. 18, 1978, pp. 5712–5716.

[22] N. G. van Kampen, "Non-linear thermal fluctuations in a diode," *Physica*, Vol. 26, 1960, pp. 585–604.

[23] ASTM Standard D257. Test methods for d-c resistance or conductance of insulating materials, *Annual Book of ASTM Standards,* Vol. 10.01, ASTM Conshohocken, 1998.

[24] ASTM Standard D991, Test method for rubber property volume resistivity of electrically conductive and antistatic products, *Annual Book of ASTM Standards*, Vol. 09.01, ASTM Conshohocken, 1998.

[25] ASTM Standard D4496, Test method for d-c resistance or conductance of moderately conductive materials, *Annual Book of ASTM Standards*, Vol. 10.01, ASTM, Conshohocken, 1998.

[26] "Specification for crosslinked polyethylene insulated shielded power cables rated 69 through 138 kV," AEIC Publication CS7, New York, 1993.

[27] K. Bosh, German Pat. 270, 199; 291, 901, 1913.

[28] B. E. Warren, "X-ray diffraction study of carbon black," *J. Chem. Phys.*, Vol. 2, 1934, pp. 551–555.

[29] B. E. Warren, "X-ray diffraction in random layer lattices," *Phys. Rev.* Vol. 59, 1941, pp. 693–698.

[30] J. B. Donnet and A. Voet, *Carbon Black*, Marcel Dekker, New York, 1976.

[31] M. S. Mashikian and J. H. Groeger, "Ionic impurities in extruded cable insulation: analytical detection techniques, sources, nature and effects," Second Int. Conf. on Insulated Paper Cables (JICABLE), Paper A6.1, Paris 1987.

[32] R. Bartnikas, "Performance characteristics of dielectrics in the presence of space charge," *IEEE Trans. Dielectrics Electrical Insulation*, Vol. 4, 1997, pp. 544–557.

[33] T. Tabata, H. Nagai, T. Fukuda and Z. Iwata, "Sulfide attack and treeing of polyethylene insulated cable—cause and prevention," *IEEE Trans. Power Apparatus Syst.* Vol. 91, 1972, pp. 1354–1360.

[34] N. M. Burns, "Characteristics of semiconducting shields," Minutes of the 86th Meeting of the Insulated Conductors Committee (ICC) of the IEEE Power Engineering Society, Dearborn, Michigan, April 29–May 2, 1990.

[35] R. Bartnikas, "Dielectrics and insulators," in *The Electrical Engineering Handbook*, R. C. Dorf, ed., CRC Press, Boca Raton, FL, 1993, Chapter 52.

[36] ASTM Standard D150, Test methods for a-c loss characteristics and permittivity (dielectric constant) of solid insulating materials, *Annual Book of ASTM Standards*, Vol. 10.01, ASTM, Conshohocken, 1996.

[37] R. Bartnikas, "Electrical conduction in medium viscosity oil-paper films—Part. I," *IEEE Trans. Electrical Insulation*, Vol. EI-9, 1974, pp. 57–63.

[38] R. Bartnikas, "Electrical conduction in medium viscosity oil-paper films—Part II," *IEEE Trans. Electrical Insulation*, Vol. EI-9, pp. 85–91.

[39] H. St-Onge, R. Bartnikas, M. Brannovic, C. de Tourreil, and M. Duval, "Research to determine the acceptable emergency operating temperature for extruded dielectric cables," Electrical Power Research Institute, Report EL-938, Palo Alto, 1978.

[40] A. L. McKean, "Investigation of mechanism of breakdown in XLPE cables," Electrical Power Research Institute, Report 7809-1, Palo Alto, 1976.

[41] E. O. Forster and W. C. Smith, "The effect of molecular structure on dielectric properties of polymers. II. Elastomers," 1974 Annual Report, Conf. on Electrical Insulation and Dielectric Phenomena, NAS/NRC, Publication 2416, Washington, DC, 1975, pp. 93–103.

[42] R. Bartnikas, "Dielectric losses in solid-liquid insulating systems—Part I," *IEEE Trans. Electrical Insulation*, Vol. EI-5, 1970, pp. 113–121.

[43] ASTM D149, Standard test method for dielectric breakdown voltage and dielectric strength of electrical insulating materials at commercial power frequencies, *Annual Book of ASTM Standards*, Vol. 10.01, ASTM, Conshohocken, 1998.

[44] Methods of test for electric strength of solid insulating materials, IEC Publication 243, Geneva, 1996.

[45] J. K. Nelson, "Breakdown strength of solids," in *Engineering Dielectrics, Vol. IIA, Electrical Properties of Solid Insulating Materials; Molecular Structure and Electrical Behavior*, R. Bartnikas and R. M. Eichhorn, Eds., STP 783, ASTM, Philadelphia, 1983.

[46] J. Artbauer and J. Griac, "Intrinsic electric strength of polyethylene in the high temperature region," *Proc IEE*, Vol. 112, 1965, p. 818.

[47] M. Pompili, C. Mazzetti, and R. Bartnikas, "Early stages of PD development in dielectric liquids," *IEEE Trans. Dielectrics Electrical Insulation*, Vol. 2, 1995, pp. 602–613.

[48] R. Hata, M. Hirose, and T. Nagai, "High pressure dielectric strength tests on PPP (PPLP) insulation," Electrical Power Research Institute, EPRI Report EL-3131, Palo Alto, 1983.

[49] ASTM Method D202, Sampling and testing untreated paper for electrical insulation, *ASTM Book of Standards*, Vol. 10.01, ASTM, Conshohocken, 1998.

[50] P. Gazzana Priaroggia, E. Occhini and N. Palmieri, *Fundamentals of the Theory of Paper Lapping of a Single Core High Voltage Cable*, Unwin Brothers, London, 1961.

[51] *Directory Encyclopedia Issue, Insulation/Circuits*, Lake Publishing, Libertyville, IL, June/July 1972.

[52] R. Bartnikas and R. Spielvogel, "Insulation resistance measurements at low frequencies," Proc. 11th Electrical/Electronics Insulation Conf, IEEE Publ. 73 CHO 777-3EI, Chicago, Sept. 30–Oct. 4, 1973.

[53] R. Bartnikas, "Dielectric losses in solid–liquid insulating systems—Part II," *IEEE Trans. Electrical Insulation*, Vol. EI-6, 1971, pp. 14–21.

[54] ASTM Method D2864, Standard definitions of terms relating to electrical insulating liquids and gasses, *ASTM Book of Standards*, Vol. 10.02, ASTM, Conshohocken, 1998.

[55] R. L. Daugherty and A. C. Ingersoll, *Fluid Mechanics*, McGraw-Hill, New York, 1954.

[56] J. A. Kok, *Electrical Breakdown of Insulating Liquids*, Philips Technical Library, Eindhoven, Netherlands, 1961.

[57] R. Bartnikas, "Dielectric loss in insulating liquids," *IEEE Trans. Electrical Insulation*, Vol. EI-2, 1967, pp. 33–54.

[58] R. Bartnikas, "Permittivity and loss of insulating liquids," in *Engineering Dielectrics, Vol. III, Electrical Insulating Liquids*, R. Bartnikas, Ed., Monograph 2, ASTM, Philadelphia, 1994, Chapter 1.

[59] T. Allen, *Particle Size Measurement*, Chapman and Hall, London, 1981.

[60] E. R. Bowen and V. C. Westcott, *Wear Particle Atlas*, Foxboro Analytical, Burlington, MA.

[61] G. Vincent, "Molecular structure and composition of liquid insulating materials," in *Engineering Dielectrics, Vol. III, Electrical Insulating Liquids*, R. Bartnikas, Ed., Monograph 2, ASTM, Philadelphia, 1994, Chapter 5.

[62] J. Chan and S. Russek, private communication.

[63] D. Couderc, "Dielectric response of oil-impregnated cable insulation at low temperature," 1980 IEEE International Symposium on Electrical Insulation, IEEE Conference Record 78CH1287-2-EI, Philadelphia, June 12–14, 1978, pp. 54–57.

[64] H. Fujita and H. Itoh, "Insulating oils for pipe-type oil-filled cables," *Dainichi-Nippon Cables Rev.*, No. 45, July 1970, p. 22.

[65] E. D. Eich, "EEI manufacturers 500/550 kV cable research projects, cable G—high pressure oil paper pipe type," *IEEE Trans. Power Apparatus Syst.*, Vol. PAS-90, 1971, pp. 212–223.

[66] A. L. McKean, E. J. Merrell, and J. A. Moran, Jr. "EEI-manufacturers 500/550 kV cable research project, cable C—high pressure oil paper pipe type," *IEEE Trans. Power Apparatus Syst.*, Vol. PAS-90, 1971, pp. 224–239.

[67] G. S. Eager, Jr., W. H. Cortelyou, G. Bahder, and S. E. Turner, "EEI-manufacturers 500/550 kV cable research project, cable D—high pressure oil paper pipe type," *IEEE Trans. Power Apparatus Syst.*, Vol. PAS-90, 1971, pp. 224–239.

[68] ASTM Method D2300 Gassing of insulating oils under electrical stress and ionization (Modified Pirelli method), *Annual Book of ASTM Standards*, Vol. 10.03., ASTM, Conshohocken, 1998.

[69] E. Molloy, *Cables and Wires*, Chemical Publishing, New York, 1941.

[70] W. A. DelMar, *Electric Cables*, McGraw-Hill,, New York, 1924.

[71] W. J. Greene and S. Verne, "Natural and Synthetic Rubbers," in *Modern Dielectric Materials*, J. B. Birks (Ed.), Heywood, London, 1960, Chapter 5.

[72] J. Tanaka and K. Wolter, "Composition and structure of dielectric solids" in *Engineering Dielectrics, Vol. IIA, Electrical Properties of Solid Insulating Materials: Molecular Structure and Electrical Behavior*, R. Bartnikas and R. M. Eichhorn, Eds., STP 783, ASTM, Philadelphia, 1983, Chapter 6.

[73] R. M. Eichhorn, "A critical comparison of XLPE and EPR for use as electrical insulation on underground power cables," *IEEE Trans. Electrical Insulation*, Vol. ES-16, 1981, pp. 469–482.

[74] R. J. Arhart, "The chemistry of ethylene propylene insulation, Part I," *Electrical Insulation Mag.*, Vol. 9, 1993, pp. 31–34.

[75] R. J. Arhart, "The chemistry of ethylene propylene insulation—Part II," *Electrical Insulation Mag.*, Vol. 9, 1993, pp. 11–14.

[76] W. Brenner, D. Lum and M. W. Riley, *High Temperature Plastics*, Reinhold, New York, 1964.

[77] J. B. Birks, "Synthetic high polymers" in *Modern Dielectric Materials*, J. B. Birks, Ed., Heywood, London. 1960, Chapter 6.

[78] ASTM Test Method D1248, Specification for polyethylene plastics molding and extrusion materials, *Annual Book of ASTM Standards*, Vol. 08.01, ASTM, Conshohocken, 1998.

[79] C. W. Bunn, "Molecular structure" in *Polyethylene*, A. Renfrew and P. Morgan, Eds., Iliffe, London, 1957, Chapter 7.

[80] P. J. Phillips, "Morphology and molecular structure of polymers," in *Engineering Dielectrics, Vol. II A, Electrical Properties of Solid Insulating Materials: Molecular*

Structure and Electrical Behavior, R. Bartnikas and R. M. Eichhorn, Eds., STP 783, ASTM, Philadelphia, 1983, Chapter 2.

[81] A. Barlow, "The chemistry of polyethylene insulation," *IEEE Electrical Insulation Mag.*, Vol. 7, 1991, pp. 8–19.

[82] G. Bahder, C. Katz, J. Lawson, and W. Vahlstrom, "Electrical and electrochemical treeing effect in polyethylene and cross-linked polyethylene cables," *IEEE Trans. Power Apparatus and Syst.*, Vol. PAS-93, 1974, pp. 977–986.

[83] R. M. Eichhorn, "Treeing in solid organic dielectric materials," in *Engineering Dielectrics Vol. IIA, Electrical Properties of Solid Insulating Materials: Molecular Structure and Electrical Behavior*, R. Bartnikas and R. M. Eichhorn, Eds. (Publication STP 783), ASTM, Philadelphia, 1983.

[84] E. J. McMahon, private communication.

[85] R. Bartnikas, S. Pélisson, and H. St-Onge, "A-C breakdown characteristics of in-service aged XLPE distribution cables," *IEEE Trans. Power Delivery*, Vol. 3, 1988, pp. 454–462.

[86] H. Wagner and J. Wartusch, "About the significance of peroxide decomposition products in XLPE cable insulations," *IEEE Trans. Electrical Insulation*, Vol. EI-12, 1977, pp. 395–401.

[87] T. Hayami and Y. Yamada, "Effect of liquid absorption on the treeing resistance of polyethylene," *1972 Annual Report, Conference on Electrical Insulation and Dielectric Phenomena*, Publ. 2112, NAS/NRC, Washington, DC, 1973, pp. 239–246.

[88] C. Katz and B. Bernstein, "Electrochemical treeing at contaminants in polyethylene and cross linked polyethylene insulation," *1973 Annual Report Conference on Electrical Insulation and Dielectric Phenomena*, NAS/NRC, Washington, DC 1974, pp. 207–216.

[89] M. Gamez-Garcia, R. Bartnikas, and R. M. Wertheimer, "Synthesis reactions involving XLPE subjected to partial discharges," *IEEE Trans. Electrical Insulation*, Vol. EI-22, 1987, pp. 199–205.

[90] M. Gamez-Garcia, R. Bartnikas, and R. M. Wertheimer, "Modification of XLPE exposed to partial discharges at elevated temperature," *IEEE Trans. Electrical Insulation*, Vol. 25, 1990, pp. 688–692.

[91] G. S. Eager, C. Katz, B. Fryszczyn, F. E. Fisher, and E. Thalmann, "Extending service life of installed 15–35 kV extruded dielectric cables," *IEEE Trans. Power Apparatus and Syst.*, Vol. PAS-103, 1984, pp. 1997–2005.

[92] J. M. Braun, "Comparison of water treeing rates in steam and nitrogen treated polyethylenes," *IEEE Trans. Electrical Insulation*, Vol. EI-15, 1980, pp. 120–123.

[93] J. E. Soden, R. T. Traut, H. St-Onge and D. Train, "Testing of a high voltage XLPE cable for dynamic submarine application," First Int. Conf. on Insulated Power Cables *(JICABLE)*, Paris, 1984, pp. 314–320.

[94] S. Pelisson, "Characteristics of field aged medium-voltage cables," Fourth Int. Conf. on Insulated Power Cables (JCABLE), June 25–29, 1995, Versailles, pp. 456–460.

[95] ASTM Method D953, Test for bearing strength of plastics, *Annual Book of ASTM Standards*, Vol. 8.01, ASTM, Conshohocken, 1998.

[96] ASTM Method C177, Test method for steady state heat flux measurements and thermal transmission properties by means of the guarded hot plate apparatus, *Annual Book of ASTM Standards*, Vol. 04.06, ASTM, Conshohocken, 1998.

[97] ASTM Method D864, Test method for coefficient of cubical thermal expansion of plastics, *Annual Book of ASTM Standards*, Vol. 08.01, ASTM, Conshohocken, 1998.

[98] K. D. Wolter, J. F. Johnson, and J. Tanaka, "Polymer degradation and measurement" in *Engineering Dielectrics, Vol. IIB, Electrical Properties of Solid Insulating Materials: Measurement Techniques*, R. Bartnikas, Ed., STP 926, ASTM, Philadelphia, 1987.

[99] R. Chiang and P. J. Flory, "Equilibrium between crystalline and amorphous phases in polyethylene," *J. Am. Chem. Soc.*, Vol. 83, 1961, p. 2857.

[100] M. G. Gubler and A. J. Kovacs, "La structure du polyethylene considéré comme un mélange de n-paraffines," *J. Polym Sci.*, Vol. XXXIV, 1959, pp. 551–568.

[101] R. M. Eichhorn, H. Schädlich and W. Boone, "Longer life cables by use of tree retardant insulation and super clean shields," Third Int. Conf. on Insulated Power Cables (JCABLE), Paris 1991, pp. 145–149.

[102] E. F. Steennis and A. M. F. J. van de Laar, "Characterization test and classification procedure for water tree aged medium voltage cables, *Electra*, No. 125, 1989, pp. 89–101.

[103] C. Katz, "Comparative evaluation by laboratory aging of 15 and 35 kV extruded dielectric cables," *IEEE Trans. Power Delivery*, Vol. 5, 1990, pp. 816–824.

[104] R. Bartnikas, "Silicone-polycarbonate cable insulations," IEEE Underground Transmission Conference, Proceedings Supplement, Pittsburgh, 1972, pp. 312–316.

[105] E. D. Eich, "Laminated synthetic insulation for extra high voltage cable," *Proc. CIGRE*, Part II, Report 203, Paris, 1966, pp. 1–9.

[106] E. J. McMahon and J. O. Punderson, "Dissipation factor of composite polymer and oil insulating structures on extended exposure to simultaneous thermal and voltage stresses," *1970 Annual Report, Conference on Electrical Insulation and Dielectric Phenomena*, NAS/NRC, Washington, DC, 1971, pp. 94–99.

[107] C. A. Buehler, R. W. Burvee, C. T. Doty, and J. V. Ugro, Jr., "Evaluation of a polyolefin paper-oil composite as an EHV cable insulation," *1970 Annual Report, Conference on Electrical Insulation and Dielectric Phenomena*, NAS/NRC, Washington, DC, 1971, pp. 70–77.

[108] T. Yamamoto, S. Isshiki, and S. Kakayama, "Synthetic paper for extra high voltage cable," *IEEE Trans. Power Apparatus Syst.*, Vol. PAS-91, 1972, pp. 2415–2426.

[109] G. Bahder, G. S. Eager, Jr., J. J. Walker, and A. F. Dima, "Development of 500 kV AC cable employing laminar insulation of other than conventional cellulosic paper," Electrical Power Research Institute, Report EL-1518, Palo Alto, 1980.

[110] E. M. Allam, W. H. Cortelyou, H. C. Doepken, Jr., A. L. McKean, and F.A. Teti, "Development of low-loss 765 kV pipe-type cable," Electrical Power Research Institute, Report EL-2196, Palo Alto, 1982.

[111] M. J. Jefferies and K. N. Mathes, "Insulation systems for cryogenic cable," *IEEE Trans. Power Apparatus Syst.*, Vol. PAS-89, 1970, pp. 2006–2014.

[112] R. Bartnikas, "EHV synthetic laminar cable insulation," *Underground Engineering (J. Underground Power Tech.)*, Vol. 4, 1973, pp. 17–20.

[113] C. H. de Tourreil and H. St-Onge, "Evaluation of epoxy resin formulations for SF_6 cable spacer materials," IEEE International Symposium on Electrical Insulation, IEEE Conference. Record 78CH1287-2-EI, June 1978, Philadelphia, pp. 167–171.

[114] T. W. Dakin, "Application of epoxy resins in electrical apparatus," *IEEE Trans. Electrical Insulation*, Vol. EI-9, 1974, pp. 121–128.

[115] J. H. Dudas, "Technical trends in utility specifications—1993 update," Minutes of the IEEE Insulated Conductors Committee (ICC) Meeting, St. Petersburg Beach, 1993.

[116] G. Graham and S. Szaniszlo, "Insulating and semiconducting jackets for medium and high voltage underground power cable applications," *IEEE Electrical Insulation Mag.*, Vol. 11, 1995, pp. 5–12.

[117] C. C. Barnes, *Power Cables: Their Design and Installation*, Chapman and Hall, London, 1966.

[118] E. Escalante, Ed., *Underground Corrosion* (A Symposium), (Publication STP741), ASTM, Philadelphia, 1979.

[119] *Underground Cable Engineering Handbook*, Essex International, Power Conductor Division, Marion, IN, 1971.

[120] R. Bartnikas and R. M. Eichhorn, Eds., *Engineering Dielectrics, Vol. IIA, Electrical Properties of Solid Insulating Materials: Molecular Structure and Electrical Behavior*, Publication STP 783, ASTM, Philadelphia, 1983.

Chapter 3

DESIGN AND MANUFACTURE OF EXTRUDED SOLID-DIELECTRIC POWER DISTRIBUTION CABLES

H. D. Campbell
L. J. Hiivala

3.1 INTRODUCTION

There is extensive use of solid-dielectric or polymeric insulated cables for the distribution of electrical energy in urban areas. The growth of solid dielectrics for medium-voltage power cable applications began in the early 1950s with introduction of butyl rubber and thermoplastic high-molecular-weight polyethylene (HMWPE), but oil-impregnated paper insulation continued to predominate until the mid-1960s when it was challenged by HMWPE, crosslinked polyethylene (XLPE), and ethylene-propylene-rubber (EPR). The paper insulated lead-covered (PILC) cable continues to be used, but it no longer holds its predominant position of decades ago.

The very rapid growth of underground power distribution in residential areas led to the use of polymeric insulted cables for medium-voltage single-phase applications. The underground residential distribution (URD) cable, because of its lower overall manufacturing and installation costs compared to PILC cables, was utilized to effect two important advantages of underground versus overhead distribution: improvement in the appearance of the residential environment and security from ice, wind, and other natural hazards. The introduction of XLPE and EPR has increased the capability of polymeric insulated cables because of their higher temperature ratings, so that now these cables have replaced the familiar PILC cable for most three-phase applications in industrial and commercial areas. This trend has extended to transmission circuits rated up to 138 kV in North America. For distribution circuits, XLPE is by far the most popular choice; EPR has a price disadvantage. Thermoplastic HMWPE is no longer used for new installations.

The purpose here is to provide a general outline of the design and manufacturing procedures in the production of solid-dielectric extruded power cables. In the discussion of manufacturing processes, we shall deal primarily with cable constructions of voltage ratings from 5 to 46 kV. It is to be noted that manufacturing processes often include proprietary techniques and they, therefore, must be treated in general terms. Typical cable designs are shown in Fig. 3.1.

3.2 DESIGN FUNDAMENTALS

It may be helpful and instructive to recall some basic theory and derive the expressions for cable capacitance, electric stress, insulation resistance, and circuit inductance.

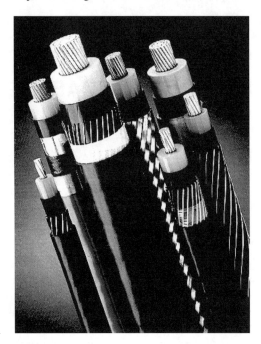

Figure 3.1 Selection of primary power distribution cable designs in the voltage range 5–46 kV. (Courtesy of BICC Cables.)

We begin by setting down some definitions. Units used in electrical engineering are those of the metric system. Although in everyday problems in North America imperial units are still used much of the time, the fundamentals and origins of the commonly used electrical formulas are more easily understood in terms of metric units. These are embodied in the International System of Units, designated SI units. There are seven SI base units of which the ampere is the electrical one. The other electrical units in this section are derived and can be completely defined in terms of the base unit. The following definitions are those of the *IEEE Standard Dictionary of Electrical and Electronic Terms* (Sixth Edition):

(i) An ampere is the constant current that, if maintained in two straight parallel conductors of infinite length, of negligible circular cross section, and placed one meter apart in vacuum, would produce between these conductors a force equal to 2×10^{-7} newtons per meter of length. It is useful to review the other derived units.

(ii) A volt is the potential difference between two points of a conducting wire carrying a constant current of one ampere, when the power dissipated between these two points is one watt.

(iii) A newton is the force that will impart an acceleration of one meter per second per second to a mass of one kilogram.

(iv) A joule is the work done by a force of one newton acting through a distance of one meter.

(v) A watt is the power required to perform work at the rate of one joule per second.

(vi) A coulomb is the quantity of electric charge that passes any cross section of a conductor in one second when the current is maintained at one ampere.

(vii) A farad is the capacitance of a capacitor in which a charge of one coulomb produces a one volt potential difference between its terminals.

(viii) A henry is the inductance for which the induced voltage in volts is numerically
equal to the rate of change of current in amperes per second.

To derive the expressions for capacitance, inductance, and electric stress, it is
necessary to recall the four quantities that describe electric and magnetic fields. They
are in the form of two vector pairs, the D and H and the E and B pair. The first pair is
determined from lines of force originating from charges and currents; the second pair
determines the forces acting on electric charges and current elements in space.

The relationship between the current I and the magnetic field strength H is according to Ampere's law:

$$\oint H \, dl = I \tag{3.1}$$

that is, the line integral of the magnetic field surrounding a current I is numerically
equal to that current. From Eq. (3.1) at a radius r,

$$H = \frac{I}{2\pi r} \tag{3.2}$$

The definition of the quantity I is illustrated in Fig. 3.2, where $\bar{\mathbf{B}}$ is the magnetic
flux density vector acting on the current $\bar{\mathbf{I}}$ in the same plane as the current going into the
page and the force vector acts in a plane perpendicular to the vectors $\bar{\mathbf{I}}$ and $\bar{\mathbf{B}}$. For
reasons of symmetry the field around only one wire is shown. In vector notation:

$$\bar{\mathbf{I}} \times \bar{\mathbf{B}} = 2 \times 10^{-7} \tag{3.3}$$

The units of this vector product are newtons per meter of circuit when the separation of
the one ampere current is equal to one meter. If in Eq. (3.3) the force on one ampere of
current were one newton per meter, the quantity $\bar{\mathbf{B}}$ would be unity, and its unit is called
a tesla or webers per square meter. According to Faraday's law the induced voltage in a
circuit is related to the rate of change of flux through the circuit. Figure 3.3 depicts a
closed circuit in space where the lines of flux pass through the loop perpendicular to the
plane of the loop. The induced voltage e is given by

$$e = -\frac{d}{dr} \int\!\!\int B_n \, dA \tag{3.4}$$

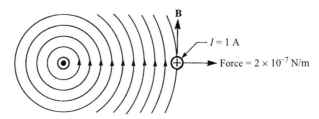

Figure 3.2 Force on two parallel currents.
(*Note*: For simplicity only the field around
one wire is shown.)

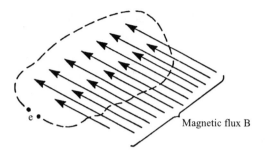

Magnetic flux B

Figure 3.3 Closed circuit in space.

where A represents the area. If $\int\int B_n \, dA$ is equal to 1 weber (Wb), then e by definition is equal to 1 Wb/s or 1 V. The minus sign merely indicates that the induced electromotive force (emf) tends to produce a current whose field would oppose the incident field.

Faraday's law may be obtained in another form where the magnetic flux is expressed in terms of the current I. This is accomplished by introducing the permeability of free space, μ_0, such that by definition

$$\mu_0 = \frac{B}{H}$$

or

$$B = \mu_0 H \tag{3.5}$$

Hence Eq. (3.3) may be written alternatively in vector notation as

$$\bar{I} \times \frac{\mu_0 \bar{I}}{2\pi r} = 2 \times 10^{-7} \, \text{N/m}$$

Both r and $\bar{I}$ are unity, and to determine the units of μ_0, Eq. (3.4) may be rewritten by considering the circuit in the configuration of Fig. 3.2:

$$e = \frac{\mu_0}{2\pi} \frac{dI}{dt} \int \frac{dr}{r} \tag{3.6}$$

from which the μ_0 units are seen to be in volts per ampere per second. But these are the units for the inductance in henries; therefore, $\mu_0 = 4\pi \times 10^{-7}$ H/m. The quantity μ_0 is a fundamental constant in the determination of the circuit inductance. In practical terms, it is the permeability of nonmagnetic materials. For magnetic materials the magnetic flux density B is increased and is, in general, not proportional to H. Thus, for magnetic materials

$$B = \mu_0 \mu_r H \tag{3.7}$$

where μ_r is the relative permeability.

In cable work electric fields are usually radial as shown in Fig. 3.4. The charge of q coulombs per meter on the conductor gives rise to an electric flux numerically equal to q. At a radius r, the electric flux density, according to Coulomb's law, is

$$D = \frac{q}{2\pi r} \quad \text{C/m}^2 \tag{3.8}$$

To make Eq. (3.8) a more useful expression, we need to have the charge in terms of the electric field. As in the case of magnetic fields a constant is introduced for free space such that

$$\varepsilon_0 E = D \tag{3.9}$$

where E is the electric field strength in volts per meter. From Eq. (3.8),

$$\varepsilon_0 = \frac{q}{2\pi r E} \tag{3.10}$$

Equation (3.10) defines the units of ε_0 in coulombs per volt-meter or farads per meter. The quantity ε_0 is called the *permittivity of free space*. Both Eqs. (3.6) and (3.10) deal with interrelationships of magnetic and electric fields; therefore, once the value μ_0 is defined, the value of ε_0 becomes fixed. The value of the latter may be determined from a fundamental expression often employed in electronic engineering:

$$c = \frac{1}{\sqrt{\varepsilon_0 \mu_0}} = 2.998 \times 10^8 \tag{3.11}$$

where c is the velocity of light in meters per second. From Eq. (3.11), ε_0 is found to be equal to 8.85×10^{-12} F/m. For practical dielectrics,

$$D = \varepsilon_0 \varepsilon_r' E \tag{3.12}$$

where ε_r' is the real value of the relative permittivity of the dielectric, usually called the *dielectric constant* or *SIC* (specific inductive capacitance).

The foregoing discussion illustrates two fundamental concepts: (i) the value of the two quantities permeability and permittivity of free space are determined from the definition of the ampere, the basic SI electric unit, and (ii) in radial fields, usually

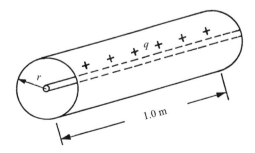

Figure 3.4 Radial electric field.

encountered in cable engineering, both the electric and magnetic field intensities vary inversely with the distance r from the source, as one would expect.

3.2.1 Inductance

By definition the self-inductance L of a circuit is

$$L = \frac{e}{di/dt} \tag{3.13}$$

where e is the instantaneous voltage; therefore, L has the units of volts per ampere per second. Also from Eq. (3.4),

$$e = \frac{d\phi}{dt} \tag{3.14}$$

where ϕ is magnetic flux in webers. Hence,

$$L = \frac{d\phi}{dt}\frac{dt}{di} = \frac{d\phi}{di} \qquad \text{Wb/A} \tag{3.15}$$

But from Eqs. (3.6) and (3.13),

$$L = \frac{\mu_0}{2\pi}\int \frac{dr}{r} \qquad \text{H/m} \tag{3.16}$$

The above expression is valid when nonmagnetic materials only are involved; this represents the usual case in cable engineering. The quantity under the integral sign is a measure of the total magnetic flux linking the current. The limits of integration can extend only to the position where the closed path around the circuit encloses zero current. In practice the flux external to the conductors can be calculated by assuming the currents to be line currents of negligible cross section positioned at the center of the conductors. This assumption does not change the external flux. The inductance of the conductor is calculated separately, assuming uniform current distribution.

Consider the case of two long parallel conductors of radius R separated by distance S as shown in Fig. 3.5 and carrying a go-and-return current I. The external flux linking conductor A is that of conductor A plus the flux of conductor B that links conductor A. Thus for conductor A

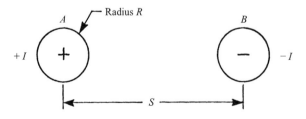

Figure 3.5 Inductance of parallel conductors.

$$L_{ae} = \frac{\mu_0}{2\pi}\left(\int_R^\infty \frac{dr}{r} - \int_S^\infty \frac{dr}{r}\right) \tag{3.17}$$

or

$$L_{ae} = \frac{\mu_0}{2\pi}\ln\left(\frac{S}{R}\right) \qquad H/m \tag{3.18}$$

For the two-conductor circuit in Fig. 3.5 the total external inductance L_e is twice that of conductor A, L_{ae}. Therefore,

$$L_e = \frac{\mu_0}{\pi}\ln\left(\frac{S}{R}\right) \qquad H/m \tag{3.19}$$

The internal inductance of each conductor can be determined with the assumption that the current density is uniform. This assumption is usually valid when the conductors are not large and not in close proximity and the frequency is 60 Hz (conditions normally experienced in distribution circuits). These factors will be discussed in more detail in a subsequent section.

If J is the current density in a conductor of radius R at radius r, then the current I is given either by $\pi R^2 J$ or $2\pi r H$ in terms of Eq. (3.2). Hence,

$$H = \frac{Jr}{2} = \frac{I'r}{2\pi R^2} \tag{3.20}$$

The portion of the total current I at radius r in the conductor is given by

$$I' = \frac{r^2}{R^2}I \tag{3.21}$$

so that

$$B_r\,dr = \frac{\mu_0 I r^3}{2\pi R^4}\,dr \tag{3.22}$$

where B is the flux density at radius r and

$$L_i = \int B_r\frac{dr}{I} = \frac{\mu_0}{2\pi}\int_0^R \frac{r^3}{R^4}\,dr = \frac{\mu_0}{8\pi} \tag{3.23}$$

Note that the internal inductance L_i of a cylindrical conductor is independent of the conductor diameter, provided the current density is uniform and has a value of 50 H/m. The total inductance L of the circuit in Fig. 3.5 is given as

$$L = L_i + L_e$$

$$= \frac{\mu_0}{4\pi} + \frac{\mu_0}{\pi} \ln\left(\frac{S}{R}\right)$$

$$= \frac{\mu_0}{\pi}\left(\frac{1}{4} + \ln\frac{S}{R}\right) \text{ H/m} \tag{3.24}$$

Expression (3.24) may be rewritten as

$$L = \frac{\mu_0}{\pi} \ln\left(\frac{S}{\text{GMR}}\right) \tag{3.25}$$

where GMR is the geometric mean radius. Alternatively,

$$L = \frac{\mu_0}{2\pi} \ln\left(\frac{S}{\text{GMR}}\right) \qquad \text{per phase of a three-phase circuit} \tag{3.26}$$

Also from Eq. (3.24)

$$L = \frac{\mu_0}{\pi}\left[\ln S + \ln\left(\frac{1}{R}\right) + \frac{1}{4}\right] \tag{3.27}$$

and

$$\ln\left(\frac{1}{\text{GMR}}\right) = \ln\left(\frac{1}{R}\right) + \frac{1}{4} \tag{3.28}$$

From which,

$$\text{GMR} = R\exp[-1/4] = 0.7788R \tag{3.29}$$

where R is the radius of a solid cylindrical conductor having uniform current density.

It may be observed from Eqs. (3.28) and (3.29) that GMR is a quantity that mathematically eliminates the internal inductance of a conductor and introduces a smaller conductor having an equivalent yet greater external inductance than the real conductor. This technique simplifies calculations. Conductors other than solid and round have GMR factors slightly different from those given by Eq. (3.29) depending on the size, number of wires, type of stranding, and cross-section shape. Values may be found in cable handbooks.

In a three-phase circuit where the cables are spaced equilaterally [Fig. 3.6(a)] the inductance of each phase conductor is the same and is given by Eq. (3.26); but, if the cables are not equilaterally spaced [Fig. 3.6(b)] the inductance of each phase will be different. In a long circuit where this unbalance may be important, the cables can be transposed at joints so that over three cable sections the average spacing will be equal. In such a case, it is convenient to have an equivalent space S for all three phases. This equivalent S is called the geometric mean distance (GMD). In the case just described it can be easily shown that

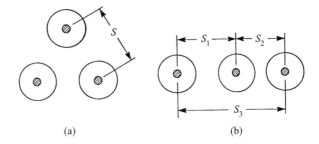

Figure 3.6 Different conductor spacing arrangements.

(a)   (b)

$$\text{GMD} = 3\sqrt{S_1 S_2 S_3} \qquad (3.30)$$

It is more convenient to deal with reactance than inductance. When the values of the constants are inserted, the inductive reactance of a three-phase circuit X_L is given by

$$X_L = (2\pi f)\frac{\mu_0}{2\pi}\ln\left(\frac{\text{GMD}}{\text{GMR}}\right)$$

$$= 0.0754\ln\left(\frac{\text{GMD}}{\text{GMR}}\right) \qquad \Omega/\text{km per phase at 60 Hz} \qquad (3.31)$$

3.2.2 Capacitance

From Fig. 3.4 and Eqs. (3.8) and (3.12) we may write for a radial electric field E,

$$E = \frac{q}{2\pi r \varepsilon_0 \varepsilon_0'} \qquad (3.32)$$

where E is the electric field gradient in volts per meter and q is the charge in coulombs per meter.

For a coaxial cable, which is the usual case for a shielded power cable, the voltage V between a conductor of radius a and a shield of radius b is given by

$$V = \int_a^b E\,dr = \frac{q}{2\pi\varepsilon_0\varepsilon_r'}\ln\left(\frac{b}{a}\right)$$

but since the capacitance C in farads per meter is equal to Q/V, then

$$C = \frac{2\pi\varepsilon_0\varepsilon_r'}{\ln(b/a)} \qquad \text{F/m} \qquad (3.33)$$

The value of C in Eq. (3.33) may be more conveniently expressed as

$$C = 0.0556\frac{\varepsilon_r'}{\ln(b/a)} \qquad \mu\text{F/km} \qquad (3.34)$$

where ε_r' is the dielectric constant or real relative permittivity of the insulation, and the capacitive reactance to neutral, X_c, is thus

$$X_c = \frac{1}{2\pi fC} = \frac{2653}{C} \qquad \Omega/\text{km at 60 Hz} \qquad (3.35)$$

where C is the capacitance to neutral in microfarads per kilometer.

3.2.3 Electric Stress

The electric stress or voltage gradient in a shielded cable of circular cross section from Eq. (3.32) is given by

$$E = \frac{q}{2\pi r \varepsilon_0 \varepsilon_r'} = \frac{CV}{2\pi r \varepsilon_0 \varepsilon_r'}$$

Or in terms of Eq. (3.33),

$$E = \frac{V}{r \ln(b/a)} \qquad \text{V/m} \qquad (3.36)$$

where the radius r is in meters.

3.2.4 Insulation Resistance

The resistance to the flow of direct current in a material may be expressed as

$$R_i = \rho \frac{l}{A} \qquad (3.37)$$

where ρ is the resistivity in ohm-meters, l is the length of the current path in meters, and A is the cross-sectional area in meters squared. It is noted that just as the electric stress and magnetic flux density vary inversely as the radius r from the conductor axis, so does the resistance to direct current flow in the insulation. It follows that the insulation resistance of a cable of circular cross section, length l meters, inner radius a, and outer radius b, is given by

$$R_i = \frac{\rho}{2\pi l} \int_a^b \frac{dr}{r}$$

$$= \frac{\rho}{2\pi l} \ln\left(\frac{b}{a}\right) \qquad \Omega \qquad (3.38)$$

A more common form of Eq. (3.38) is

$$R_{im} = k \log_{10}\left(\frac{D_2}{D_1}\right) \qquad \text{G}\Omega\text{-m} \qquad (3.39)$$

where R_{im} is the insulation resistance of one meter of cable in gigaohms, the outer and inner diameters of the insulation are D_2 and D_1, respectively, and k is a constant for the insulation. From Eq. (3.38), $k = 3.67 \times 10^{-10}$, where ρ is the resistivity in ohm-meters and k is in giga-ohm-meters.

3.2.5 Dissipation Factor

The term dissipation factor is a measure of the loss in the insulation as a consequence of the component of current in the insulation that is in phase with the applied voltage V. In Fig. 3.7(a) the in-phase component equals $I\cos\theta$, but the angle δ is normally so small that $\tan\delta = \cos\theta$ very nearly. An analog of a cable dielectric is shown in Fig. 3.7(b), where R is the equivalent alternating current (ac) resistance in parallel with a perfect dielectric, and the loss angle δ is small. In terms of this analog or equivalent circuit, the $\tan\delta$ of the cable dielectric is defined as the ratio of the equivalent parallel resistance R to the parallel reactance X_c:

$$\tan\delta = \frac{\text{real power}}{\text{reactive power}}$$

$$\frac{V^2/R}{V^2\omega C} = \frac{1}{\omega RC} \tag{3.40}$$

The value of R in Eq. (3.40) is not to be confused with the insulation resistance of the dielectric, which can also be represented as a resistance in parallel with the dielectric. The equivalent ac resistance R, a mathematical model, is a measure of the loss in the dielectric. A portion of this loss is a consequence of the applied alternating electric field at 60 Hz being out of phase with the resultant alignment of the molecular dipoles that tend to oscillate with the alternating field. But the most significant portion of this loss results directly from either electronic or ionic conduction processes that occur at 60 Hz. Because the field alternates, a loss is generated within the dielectric in the form of heat.

Most of the loss is manifested in a component of current, called the absorption current, or equivalent ac conduction current (cf. Chapter 9) that is in phase with the applied field. Under a constant direct voltage the absorption current generally decreases asymptotically toward a limiting value that may be 2 orders of magnitude less than the 1-min value [1, 2]. The limiting value is the true conduction current, and the time required to decay to within 1% of this minimum value depends on the characteristics of the material and may vary from a few seconds to many hours. Further details may be found in [1, 2]. The values of apparent insulation resistivity found in the literature are normally in reference to the 1-min value. The insulation resistance is strongly temperature dependent and may be perhaps 2 orders of magnitude less at rated temperature than at room temperature. Cable standards usually make reference to a temperature of 15°C.

Dissipation factor or $\tan\delta$ is a cable parameter of much more practical significance than insulation resistance. One obvious reason is that, unlike insulation resistance, it is

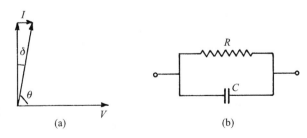

Figure 3.7 Current vector relationships and equivalent circuit of cable dielectric.

(a) (b)

measured in the normal operating mode of the cable. It is a measure of the total loss in the dielectric caused by polarization and ac conduction effects and is generally a sensitive indicator of abnormal cable insulation. In high-voltage power transmission cables the dielectric loss may be significant compared to the conductor losses, thereby reducing the cable ampacity. Dielectrics employed in solid-dielectric distribution cables exhibit low values of dissipation factor that are relatively stable over the whole operating temperature range (cf. Chapter 2).

The value of R, the equivalent ac resistance, is much smaller than the 1-min insulation resistance (R_i) value for polymeric insulations used in solid-dielectric cables. In fact, there is no mathematical relationship between the two quantities. The value of dissipation factor, obtained if the insulation resistance value were substituted into Eq. (3.40), would be much too small. Some idea of the insulation resistance required for a high-voltage insulation, as a limit, may be obtained if the quantities of capacitance and insulation resistance in Eqs. (3.33) and (3.38) are substituted into Eq. (3.40):

$$\tan \delta = \frac{1}{\omega \rho_i \varepsilon_0 \varepsilon_r'} = \frac{3.0 \times 10^8}{\rho_i \varepsilon_r'} \tag{3.41}$$

where ρ_i is an imaginary value of resistivity for the purpose of illustration in ohm-meters. Suppose a typical polymeric insulation is chosen that has a $\tan \delta$ of 0.005 and a dielectric constant ε_r' of 4.0, then one obtains a ρ_i value of 1.5×10^{12} Ω-cm. The actual value of ρ of such an insulation would be in the range of 10^{16} Ω-cm at 15°C. From this illustration one would not expect an insulation having a $\rho = 10^{12}$ Ω-cm to perform adequately on a high-voltage cable. Equally important, the resistivity of 10^{16} Ω-cm is not a sufficient condition to ensure a low dissipation factor.

3.3 DESIGN CONSIDERATIONS

The purpose of a cable system is to convey power from the source of energy to the load. The amount of power is, of course, proportional to the product of the circuit voltage and the conductor current (under given conditions of circuit length and load impedance). The designer must consider both the current-carrying capacity (ampacity) as well as the voltage rating for any given cable system. The ampacity is determined largely by the service environment and the installation conditions. Underground residential distributions systems serve many small loads over relatively large areas, so the choice of conductor size may well depend on voltage regulation rather than ampacity.

Underground distribution (UD) and transmission systems, on the other hand, require the determination of the appropriate conductor size, once the system voltage is decided. The procedure for the calculation of ampacities may be simple if the conditions are covered in standard tables [3, 4] or complex under some conditions. The concepts are fundamental. The difficulty arises from the many facets that determine the losses in the phase conductors and shields and the thermal resistances of the elements contributing to the total thermal path from the cables to ambient. The parameters are not always accurately known nor do they lend themselves entirely to analytical solutions. Procedures for the determination of ampacities are introduced in Chapters 5 and 11. Some additional facets are included here.

3.3.1 Ampacity

The advent of the digital computer has, of course, greatly reduced the number of tedious calculations, as well as made practical the investigations of many facets not possible only two decades or so ago [5–19]. Some elementary concepts are briefly described.

3.3.1.1 Shield Circulating Currents

Modern solid-dielectric cables for three-phase systems often have shields designed as the neutral conductor for unbalance currents. In such cases the shield is generally specified to be one third the conductance of the phase conductor. This feature, like single-conductor cables with a lead or aluminum sheath (cf. Chapter 11), permits the induced shield currents to become large enough to significantly increase the total $I^2 R$ losses in the cable, particularly if the cables are large and spaced (as opposed to triplexed). These losses can be eliminated by single-point grounding or by special cross-bonding techniques, procedures that may or may not be desirable options. Published Insulated Cable Engineers Association (ICEA) tables [3] deal with cases where the shields are grounded at multiple points. The losses are increased by increased cable spacings because of the more effective inductive coupling, conductor to shield. On the other hand increased spacing reduces the mutual heating, proximity, and eddy current losses. The relative effect of cable configuration when direct buried, in ducts, triangle, or flat formation, spaced or triplexed, can be ascertained from the tables.

3.3.1.2 Proximity, Skin Effect, and Eddy Currents

Proximity effect in cables is the result of the flow of alternating current in such a way as to reduce the magnetic field and, hence, the inductance of the circuit. It is a phenomenon similar to skin effect. In skin effect, the current tends to move away from the center of the conductor toward the periphery to reduce the internal inductance, while in proximity effect the current in one conductor tends to move off center toward the current in the adjacent phase or return conductor to reduce the effective spacing of the currents, hence the inductance of the circuit. Note that the effect is opposite to the force of repulsion between go-and-return currents that may become large during short circuits. The repulsion is the effect of moving charges (currents) in a magnetic field, while skin and proximity effects are due to the rate of change of the current. Proximity and skin effects become greater where the current flow can be easily distorted to cause relatively large decreases in the inductance, as in large closely spaced conductors. The nonuniform current density causes increased $I^2 R$ losses compared to that caused by direct current.

Eddy currents tend to occur in any conductor in which there is a changing magnetic field. If such a varying magnetic flux is passing through a plane surface in a direction normal to that plane, currents tend to flow around the flux lines in the plane. If the plane is conducting, the surface large and the field intense, then the losses tend to be large since they are proportional to $e^2 R_{\text{eff}}$, where e is the induced emf and R_{eff} the effective resistance. This will be the case when cables are in close proximity, the phase currents large and the magnetic flux normal to the concentric shields. Concentric wire and ribbon shields effectively reduce these losses to a negligible value. Tubular shields may have significant eddy current losses in large closely spaced cables. Eddy current losses may also be significant when cables are installed near metal beams, walls, and the like.

3.3.1.3 Emergency Overload Rating

Both EPR and XLPE cables have emergency temperature ratings of 130°C. There is evidence that these emergency ratings are satisfactory, provided the cable shield designs are adequate [20]. In the 90–130°C temperature range XLPE passes through a melting transition from a semicrystalline state to a rubbery state accompanied by a significant increase in its thermal expansion coefficient. If the metallic shielding structure cannot accommodate this increase in volume, the insulation wall may be severely distorted by the thermomechanical forces [21]. Inherently rubbery, EPR insulations do not experience the transitional thermal expansion of XLPE and exhibit significantly lower total volumetric expansion. A recent revision to ICEA Standard S-68-516 [22] has recognized the use of a class III EPR insulation with normal and emergency overload operating temperatures of 105 and 140°C, respectively. Tests are in progress to demonstrate the suitability of XLPE insulations at these same temperatures.

Both the Association of Edison Illuminating Companies (AEIC) and ICEA specifications put strict limits on the duration of emergency overloads and the number permitted during the life of the cable. Apparently, user practice deviates widely from the recommended practice [20]. One of the critical areas for thermomechanical forces at emergency temperature is in the cable joint. These accessories must be carefully designed to ensure that the relative movement of insulation, conductors, and shields are accommodated so that the design electric stress profiles are maintained.

3.3.1.4 Earth Interface Temperatures

The ampacity of a cable may be limited in some cases by the earth interface temperature [3]. The thermal resistance of the earth is sharply increased if the natural moisture is driven out by a heat source. It has been shown that the movement of moisture away from the cables is dependent on the temperature gradient in the soil, in addition to the actual temperature [7, 8]. There are many factors determining the soil thermal resistivity and moisture migration including soil composition, depth of burial, duration of heat flow, rainfall, and the like. It is especially important to note that if the heat flow is large, soils of initially low thermal resistance may be unstable due to moisture migration.

3.3.1.5 Short-Circuit Currents

The metallic shields of solid-dielectric cables in three-phase systems may not be required to act as a neutral. A good example is the case of industrial application. The shield must still be able to carry any fault or short-circuit currents impressed on them without damage to the cable structure. One method of calculation assumes that none of the heat generated in the shield will be conducted to the adjacent materials in the cable because of the short duration of the fault current. This assumption permits relative ease of calculation of the required cross-sectional area of the metallic shield but usually leads to conservative results [23–25]. It is the method used in ICEA P-45-482 [26] that recommends the withstand temperatures given in Table 3.1.

One difficulty with the ICEA method is the case of circuit breaker multiple reclosures in which the time durations are far too long for the assumption to hold. With the noted important assumption and limitation understood, the equation for the short-circuit current may be developed as follows:

TABLE 3.1 ICEA Maximum Withstand Temperatures

Cable Material in Contact with Shield or Sheath	T (°C)
Thermoset	350[a]
Thermoplastic	200

[a]For lead sheaths this temperature is limited to 200°C.

$$\Delta T(A l C') = I_8^2 \left(\frac{\rho_T}{A}\right)\Delta t \ l \tag{3.42}$$

where ΔT is the change in temperature (in degrees Celsius), C' is the specific heat (in J cm^3 °C^{-1}), A is the cross section area (in cm^2), l is the length of conductor in centimeters, I_8 is the short-circuit current in amperes, ΔT is the current duration in seconds, and ρ_T is the resistivity at temperature T in ohm-centimeters. From [26] the following constants for copper at 20°C are tabulated: specific gravity (8.93 g/cm^3), specific heat (0.092 cal/g °C), resistivity (1.72 µΩ-cm), and temperature coefficient (0.00393 or 1/254/ °C^{-1}).

Using these constants the value of C' becomes 3.44 J/cm^3 °C. Also since the resistivity at a temperature T is related to that at a temperature of 20°C (ρ_{20}) by

$$\rho_T = \rho_{20}\left(\frac{234 + T}{254}\right) \tag{3.43}$$

then inserting ρ_T from Eq. (3.43) into Eq. (3.42) permits the calculation of the short-circuit current, i.e.,

$$I_8^2 t = \frac{C' \times A^2 \times 254}{\rho_{20}} \int_{T_1}^{T_2} d\left(\frac{T}{234 + T}\right) \tag{3.44}$$

and finally,

$$I_8 = A\left[\frac{508 \times 10^6}{t}\ln\left(\frac{234 + T_2}{234 + T_1}\right)\right]^{1/2} \tag{3.45}$$

To conform to the units in [26], where A is in circular mils:

$$I_8 = 0.114A\left[\frac{1}{t}\ln\left(\frac{234 + T_2}{234 + T_1}\right)\right]^{1/2} \tag{3.46}$$

Values for aluminum and other commonly used conductor materials are given in [26].

3.3.2 Electric Stress

The electric stress for any given insulation thickness is always a maximum at the conductor shield as can be readily seen from Eq. (3.36). It might be thought that a maximum stress value for a given insulation would be determined and the insulation thickness for each conductor size determined from the value. This method would permit quite large variations in the average stress throughout the insulation wall that cannot be disregarded. In practice solid-dielectric cables are usually restricted to some minimum conductor diameter in industry standards and usually have the same wall thickness for the full range of conductor sizes. The maximum stresses at the conductor shield for HMWPE, XLPE, and EPR insulated cables are about 3.3 kV/mm and 4.3 kV/mm for the minimum conductor size of 25- and 35-kV cables, respectively. These values are conservative when compared to those for high-voltage XLPE cables. The most recent revision to AEIC Specification CS7-93 now allows the use of a reduced insulation thickness provided that the maximum stress does not exceed 6.0 kV/mm or 8.0 kV/mm for 69- and 138-kV rated cables, respectively.

3.3.3 Cable Insulation Levels

The cable insulation levels designated in industry standards as 100, 133, or 173% depend on the system neutral impedance to ground and the protective devices employed to deenergize the system in the event of a ground fault. If the system neutral is solidly grounded, then a ground fault will cause tripping of the circuit breakers. But if the neutral to earth impedance is high, a so-called ungrounded neutral, the neutral voltage to earth shifts and the phase to earth voltage on the unfaulted phases rises basically to the full phase to phase voltage. If the system is provided with relay protection such that ground faults will be cleared within 1 min, cables in the 100% category may be applied. Where the clearing time requirements of the 100% category cannot be met and yet there is adequate assurance that the faulted section will be deenergized in a time not exceeding 1 h, cables in the 133% category may be applied. Longer clearing times require the use of the 173% insulation level. A more detailed discussion is presented on this subject in Chapter 5.

3.3.4 Dielectric Loss

The dielectric loss in medium-voltage cables is normally very small compared to the I^2R loss in the conductor. This loss is a consequence of the in-phase components of current and voltage in the dielectric as shown in Fig. 3.7. The dielectric power loss P in a cable dielectric having a capacitance C is given by

$$P = IV \cos \theta \approx IV \tan \delta = V^2 \omega (C \tan \delta) \tag{3.47}$$

where C is the cable capacitance, and V is the operating voltage of the insulation. Equation (3.47) shows that the bracketed part is dependent on the cable insulation while the unbracketed part is dependent on the square of the applied voltage. Polyethylene (HMWPE) and unfilled crosslinked polyethylene are essentially nonpolar materials. All nonpolar dielectrics have a very low value of $\tan \delta$ and a dielectric constant (SIC or ε_r') of about 2.3 as a consequence of the molecular structure of the material. Tan δ of HMWPE and unfilled XLPE is about 0.05% at operating temperatures. Even at very high rated voltages, say 500 kV, the dielectric loss would be small compared to the conductor I^2R loss. Thus, from the point of view of dielectric loss these materials are very attractive for extra-high-voltage (EHV) power cables. The addition of mineral fillers in EPR increases both the permittivity and tan δ of the base material to typical values of 3.0 and 0.9%, respectively, at operating temperatures, resulting in a higher dielectric loss for this material than for XLPE. Since the dielectric loss is proportional to the square of the voltage, the losses are insignificant for medium-voltage cables. Continued development of EPR compounds has reduced the dielectric losses and extended the voltage range to at least 138 kV [27, 28]. Some XLPE insulations use filled materials with basically similar effects on $\tan \delta$ and permittivity as in EPR insulations.

3.4 DESIGN OBJECTIVES

Design requirements quite logically depend on cable application. Low-voltage cables do not require the use of low-loss, high-dielectric-strength insulations because much of the insulation wall is there simply for mechanical reasons. Portable cables require special considerations for good flexibility and the physical protection of cut- and impact-resistant coverings. Extra-high-voltage cables can employ only those insulations having low dielectric losses. The medium-voltage cable, used in modern distribution systems, requires a reasonable balance of physical and electrical characteristics. A modern underground power distribution cable is shown in Fig. 3.8.

3.4.1 Partial Discharge

Modern distribution cables, whether insulated with HMWPE, XLPE, or EPR, normally have insulation thicknesses that result in electric stresses about the same as those for the PILC cables they are replacing. Whereas the PILC cable has, from long experience, been known to provide troublefree service at these stresses, there is not nearly the same length of experience with similarly stressed polymeric cables. The oil-paper dielectric is known to be stable in the presence of partial discharge energies that are enormous compared to the levels permitted for polymeric cables in industry standards. Polymeric cables must be carefully designed to meet the partial discharge test values. The effect of partial discharges on cable life is very difficult to quantify. This is demonstrated by the fact that although there are substantial differences in the partial discharge resistance of different materials [27, 28], the test level is the same for them all. The partial discharge extinction level of any particular cable depends primarily upon the extrusion techniques, correct cooling rates of the extrudate, and shield application, shield design.

Solid-dielectric cables rated at 5 kV and higher are required to have a semiconducting (SC) shield at the conductor and over the insulation. This shielding technique is necessary to form an intimate interface between the cable insulation and its electrodes formed by the conductor and overall metallic shielding. Any gaseous void at an electrode insulation interface will be overstressed when subjected to electric fields normally encountered on medium-voltage cables. The resulting partial discharge may cause early failure.

Early cable designs that used semiconducting fabric tapes have been supplanted by extruded semiconducting material that provides a much improved electrode insulation interface. The conductor shield is bonded to the insulation, while the insulation shield must be either bonded to the underlying insulation or form a firm intimate contact. These shield interfaces must be free of voids and protrusions and compatible with the insulation over the operating temperature range of the cable. They must remain intact

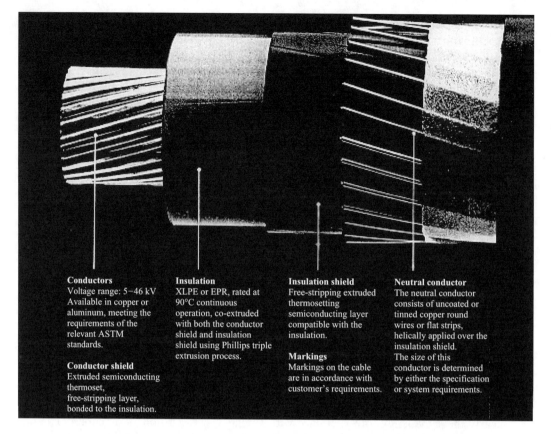

Conductors
Voltage range: 5–46 kV
Available in copper or
aluminum, meeting the
requirements of the
relevant ASTM
standards.

Conductor shield
Extruded semiconducting
thermoset,
free-stripping layer,
bonded to the insulation.

Insulation
XLPE or EPR, rated at
90°C continuous
operation, co-extruded
with both the conductor
shield and insulation
shield using Phillips triple
extrusion process.

Insulation shield
Free-stripping extruded
thermosetting
semiconducting layer
compatible with the
insulation.

Markings
Markings on the cable
are in accordance with
customer's requirements.

Neutral conductor
The neutral conductor
consists of uncoated or
tinned copper round
wires or flat strips,
helically applied over the
insulation shield.
The size of this
conductor is determined
by either the specification
or system requirements.

Figure 3.8 Typical utility underground primary 5 to 46-kV power distribution cable.
(Courtesy of BICC Cables.) *Note*: Overall jacket and binder tape is
optional. If required, the overall jacketing material may be one of the
following: insulating PVC or PE or semiconducting or insulating PE
with encapsulated neutral wires.

during expansion and contraction of the insulation caused by heating and cooling.
These compounds must not include constituents that may migrate into and degrade
the adjoining insulation. The insulation shield for commonly used utility distribution
cables is usually required to be easily removable for the installation of termination and
joints. These constraints require careful manufacture to meet the conflicting requirements of a smooth interface and ease of removal. Stripping tests specify a range of
suitable adhesion strengths [29–31]. For higher voltage cables, a firmly bonded insulation shield is specified [29, 30].

Initially, solid-dielectric cables used in URD systems were insulated with thermoplastic polyethylene (HMWPE) and used thermoplastic semiconducting compounds
that had adequate thermal performance. The introduction of XLPE and EPR insulations with emergency temperature ratings of 130°C or higher required the development of new materials. To ensure adequate thermal performance, especially at the
emergency operating temperature of the insulation, semiconducting deformation-resis-
tance thermoplastic polyethylenes (DRTP) or semiconducting thermoset materials are
employed.

The electrical resistivities of the semiconducting compounds must be stable over the operating temperature range and not degrade in service. Maximum resistivity values can be quite high before the loss becomes appreciable. Specified values are typically less than 1000 Ω-m. Exact values are given in industry standards and specifications [4, 22, 29–31].

3.4.2 Temperature Ratings

Both XLPE and EPR insulations perform well at elevated temperatures so that a normal operating temperature of 90°C has been established with emergency overload and short-circuit ratings of 130 and 250°C, respectively. Both the electrical and physical aging properties are acceptable at these temperatures. Permittivity, tan δ, short time ac, and impulse strengths are not degraded significantly at temperatures up to 90°C nor excessively even to temperatures of 130°C or more [27, 32]. Both polymers have a large coefficient of expansion compared to metal conductors and shields. In particular, XLPE experiences a relatively large increase in volume because of its gradual change from a mostly crystalline material at 90°C to a completely amorphous state at about 115°C. This characteristic requires the design of shielding systems to accommodate the significant increase in diameter at the emergency overload temperature, especially since the materials at this temperature become relatively weak. Industry specifications [29, 30] have a thermomechanical qualification test to demonstrate the ability of the cable to withstand temperature excursions up to 130°C without damage. The performance of modern solid-dielectric cables under emergency overload conditions has been investigated [20] and shown to conform well to the specified limits. Joints and terminations were not evaluated in the study, but the literature indicates that commonly used modular accessories for distribution cables are satisfactory for the 130°C emergency temperature [33].

3.4.3 Conductor Constructions

The choice of conductor construction is dictated by economics, cable application, and the type of insulation that will be used. The familiar sector-shaped conductor of the PILC cable was used to minimize the overall diameter of the cable. Until recently, this shape was only used with thermoplastic insulations. With the introduction of true triple extrusion, both EPR and XLPE cables are now available with this conductor type. The conductors used on extruded dielectric cables are normally circular in cross section, but the diameter can be reduced by about 10% by shaping the individual wires to minimize the void spaces. Compact stranded conductors, manufactured by compacting each layer of wires during the stranding process, are often utilized to gain most of the theoretical reduction and economy of reduced diameters.

Another construction is the compressed stranded conductor, which is effected by using a compression die over the outer layer only. This technique reduces the diameter slightly, but more importantly it provides a closely formed outer layer that is necessary for the application of an extruded semiconductive shield. If the outer layer of the strand is not tightly closed, the extrudate will tend to flow into the strand, causing defects in the shield layer. A further construction is the combination strand that utilizes wires of slightly larger diameter on the outer layer than the inner layers, thus creating the same effects as the compressed strand.

Underground primary distribution cables normally have aluminum conductors. Because of their cost advantage in the past, many utilities have standardized on their use. Solid aluminum conductors are often used up to the 1/0 AWG (American Wire Gage) size in order to prevent longitudinal water penetration. A sealant may be incorporated in the interstices of stranded conductors to provide the same benefit. Compatibility of this sealant with the extruded conductor shield can be determined in accordance with ICEA Publication T-32-645 [34]. The longitudinal water penetration resistance of the conductor itself can be determined in accordance with ICEA Publication T-31-610 [35]

3.4.4 Shields and Jackets

The metallic insulation shields of utility primary power distribution cables are usually constructed of copper wires or ribbon applied helically. Longitudinally folded and corrugated copper tapes may also be used. These constructions normally have a conductivity of one third the phase conductor on three-phase systems and a full neutral on single-phase systems. The wire and ribbon shields have been directly buried without a protective jacket. Experience [36], however, with corrosion of copper neutrals has led to the widespread use of jacketed cables. Among the jackets employed are polyvinyl chloride (PVC), polyethylene (PE), XLPE, or DRTP. The thermoset and DRTP jackets accommodate higher short-circuit and overload conditions than the thermoplastic designs. Semiconductive jackets are sometimes used to provide both corrosion protection and to maintain the shield at ground potential during faults, a construction that may be preferred for joint use random lay with telephone cables. (Random lay is a term describing telephone and power cable in the same trench with no controlled minimum spacing between them.) Jackets may encapsulate the shield elements or be applied over a separator of Mylar or other tape.

Industrial cables that are used in industrial plants for power distribution normally have metallic shields not designed to carry neutral current. These are electrostatic shields that act as drain conductors for the capacitive current. Such a cable is shown in Fig. 3.9. Fault currents will flow mostly in the separate neutral or equipment ground wires. The minimum cross-sectional area of the shield is 0.1 mm^2/mm of diameter over the insulation [4, 31]. Fault current capacity may be calculated by the method given in Section 3.3.1.5. Cables have jackets compounded to be fire resistant when used in buildings. Flame resistance has become a much more important cable characteristic in recent years. Experience shows that there is no absolute fire-resistant material. All

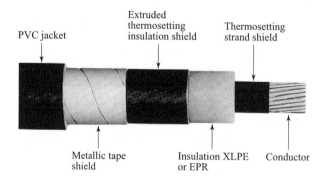

PVC jacket

Extruded
thermosetting
insulation shield

Thermosetting
strand shield

Metallic tape
shield

Insulation XLPE
or EPR

Conductor

Figure 3.9 Shielded power cable with semiconducting strand and insulation shields and overall copper tapes. (Courtesy of Alcatel Canada Wire.)

materials, even metals, burn if the ambient temperature reaches the ignition point. One of the chief requirements in the new generation of flammability tests is that the cable should not spread an industrial fire, a much more stringent requirement than the simple bunsen burner material flame tests. Industry standards, Institute of Electrical and Electronics Engineers (IEEE), and Underwriters Laboratories (UL) requirements [37, 38] incorporate full-scale vertical tray tests in which heat is applied by a large heat source strip burner to the lower end of a group of cables in the tray. The flame is required to be extinguished before spreading a specified maximum distance upwards along the cables. Cables for industrial applications often have an armor of interlocking aluminum alloy or sometimes galvanized steel (for multiconductor cables) applied over the inner jacket. An overall jacket may be applied for corrosion protection. These power cables are called Teck cables in Canada and are used in low- and medium-voltage applications for pulp mill, mining, and other severe environments. A low-voltage example is shown in Fig. 3.10.

In cold climates such as Canada, it is often required to have jackets with special low-temperature properties. The ability to withstand very cold temperatures ($-40°C$) is dependent both on the jacket material physical properties and total cable design. The underlying metallic components, anticorrosion flooding materials, and the compactness of the cable core all contribute to cold withstand capabilities of the jackets. Cold impact

Stranded copper or aluminum phase conductor

Mylar separator

Stranded copper for aluminum ground conductor

Thermosetting ozone, heat- and moisture-resisting Voltalene insulation

Inner jacket of high-temperature PVC

Interlocking aluminum alloy or galvanized steel armor

Outer jacket of PVC

Figure 3.10 Three-conductor 600-V XLPE Teck 90 cable. (Courtesy of Pirelli Cables and Systems.)

tests on finished cable samples [31] have been found to be a good criterion for cable installations and operation at very low temperatures.

3.4.5 Portable Cables

Modern heavy equipment used in mining operations, dredges, shovels, cranes, log handlers, and the like require power in the order of 10,000 hp or more. Such equipment may be supplied with cable rated up to 25 kV. Often the service environment is very severe. So the cables require very robust constructions to have a reasonable service life. Jackets are constructed from neoprene, nitrile butidiene, chlorosulfonated polyethylene (Hypalon), and sometimes partly synthetic and partly natural rubbers, depending on service factors, extreme cold (−55°C), oil resistance, flame resistance, color fastness, and other variables. The jackets are usually reinforced with a strong fabric braid or other means to provide increased tear resistance. Cable components, including the conductors and shields, must be capable of repeated bending, twisting, and relative movement. A prime requisite is flexibility. If one considers a cable bent into a circle, it is obvious that the outer periphery of the cable is greater than the inner by $\pi(D\text{-}d)$ where d and D are the inner and outer diameters, respectively. These changes in dimensions cause forces of tension and compression on the outer and inner peripheries, respectively. To accommodate these forces the metallic components of a cable are oriented in spiral or helical form. The shorter the helix, or "lay," the more flexible the cable tends to become. These factors are of great importance in portable cables. If the lay length (one complete revolution of the helix measured parallel to the cable axis) is short compared to the circumference of the bend in the cable, the average length of component does not change and flexibility is enhanced. Relative movement within the helix does take place and must be made as unrestricted as possible. It is also logical to construct the metallic components of many individual filaments, thus the conductors are constructed of many fine wires twisted together in a short lay and in a configuration to promote ease of relative movement while maintaining a firm conductor. The insulation is usually EPR because of its superior flexibility to the polyethylenes. Metallic shields are of braided or wrapped wires. Interstices of the phase conductors are filled with nonhygroscopic filaments to provide a round tightly integrated cable that provides a tough physical construction for a harsh environment. To provide ease of component relative movement, or slip, the conductor wires are normally stranded in rope format. Metallic shields should be capable of slip relative to their neighboring components. Lubricating powder, mica, for example, is applied over the fillers, completed conductors, and binder tapes to promote slippage relative to the extra-heavy-duty (EHD) jacket. Such a cable is shown in Fig. 3.11. Because these cables are not expected to have a service life equal to fixed installations, due to the arduous physical requirements, partial discharge tests are not specified.

3.4.6 Aging of Underground Cables

Exruded insulations for medium-voltage cables of circa 1950, butyl rubber, and other ozone-resistant compounds had the base polymer extended by fillers to provide optimum physical and electrical properties. Because it was recognized that these materials absorbed significant amounts of moisture in a wet environment, water soak tests were devised. The weight increase after prolonged water immersion at rated temperature was measured, the so-called gravimetric method of accelerated moisture

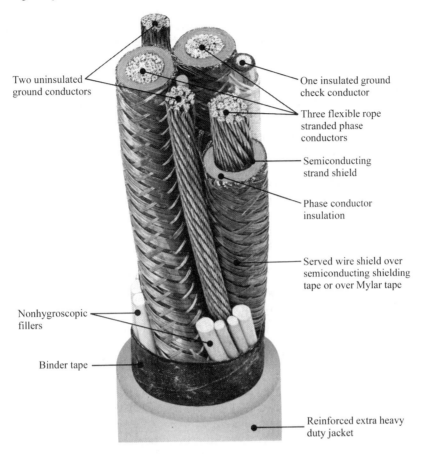

Two uninsulated
ground conductors

One insulated ground
check conductor

Three flexible rope
stranded phase
conductors

Semiconducting
strand shield

Phase conductor
insulation

Served wire shield over
semiconducting shielding
tape or over Mylar tape

Nonhygroscopic
fillers

Binder tape

Reinforced extra heavy
duty jacket

Figure 3.11 Medium-voltage type SHD-GC portable power cable. (Courtesy of
Pirelli Cables and Systems.)

absorption. These tests led to the elimination of hydrophilic materials. Later, electrical
tests were introduced for more accurate determination of moisture effects in the insulation. The EM60 test in industry standards [22, 31] provides for a maximum permittivity
(SIC) and increase in SIC after 14 days immersion at 75°C, and limits of power factor
(tan δ) stability. Both HMWPE and unfilled XLPE are hydrophobic materials. Since
they absorb very little moisture and long-term insulation resistance results of immersed
specimens were demonstrated to be stable over prolonged periods, it was naturally
expected that these materials could safety be used as medium-voltage cable insulation
in wet environments. In the early 1960s medium-voltage cables were in commercial use;
less than 10 years later premature failures of unexplained nature were being reported.

Perhaps the most well known of these reports was that of the Pacific Gas and
Electric Company of San Francisco [39] prepared in 1971. It reported numerous treelike
defects in HMWPE cable that had been removed from service because of unexplained
dielectric failures. Two further contributions to the original study have been made by
the original author and/or his associates [40, 41] in which it was concluded that all
polyethylene and crosslinked polyethylene in typical utility environments would eventually contain water trees. These trees sometimes develop into electrical trees and finally

lead to dielectric failure. Reports of cable failures suspected to be caused by water trees about the time of the original study by Vahlstrom [39] also originated in Japan [42]. These unexpected discoveries of possible unsatisfactory service life of polyethylene and crosslinked polyethylene power cables were disturbing to the power utilities and the cable industry and led to urgent studies of the problem in many countries including Canada [43–46]. Most of the work has been carried out on HMWPE and unfilled XLPE. The quantities of EPR and filled XLPE cables in underground distribution are relatively small; furthermore the opaque nature of these materials makes the detection of trees difficult. Trees can be observed easily in XLPE and HMWPE cables because these insulations are translucent. Water treeing of dielectrics is discussed and illustrated in Chapter 2. Water trees are apparent by their characteristic dark appearance when contrasted to the translucent, thin wafers cut from the cable insulation. Because of the lack of field evidence, it is not possible to conclude that EPR cables have failure rates comparable to HMWPE and XLPE. It has been reported that the treeing susceptibility of EPR cables is about equal to that in XLPE and less than that in HMWPE [47].

Water trees, sometimes called electrochemical trees, have basic characteristics different than electrical trees. Electrical trees are characterized by the occurrence of partial discharge, require high electric stress to initiate and rapidly lead to catastrophic dielectric failure. Water trees can be initiated at much lower dielectric stress, grow very slowly, are associated with no measurable partial discharge, and may completely bridge the insulation from conductor to shield without dielectric breakdown, although the dielectric strength is much reduced, in particlar the direct current (dc) breakdown value [48]. Electric treeing, a well-known phenomenon since the early days of electrical engineering, occurs in poorly designed or overstressed insulation systems. The mechanism of failure is known and understood. Water treeing is a process studied intensively since circa 1970, but the inception and growth of the treelike structure has no universally agreed theoretical basis. The literature on water treeing is large because the investigations, although only encompassing a short time period, are intensive. Bernstein [49] in his review of water treeing theory, gives the major requirements and factors influencing the growth but suggests that the mechanism of inception is not known. It is accepted that two fundamental conditions are required: (i) a polar liquid, usually water, must be present and (ii) voltage stress. Electrical trees require only voltage stress. Other factors, listed in no particular order of importance, have been enumerated by Bernstein [49]. These are aging time, material nature, contaminants/impurities, temperature, temperature gradient, cable design, magnitude of operating voltage stress, test frequency, antioxidant, voltage stabilizers, water nature, and semiconducting layer type.

It has been shown that water in the interstices of the stranded conductor greatly enhances the tree growth even when the cable is immersed in water, particularly when a temperature gradient exists in the insulation [50]. Badher et al. proposed a physical model of aging in polymeric cables [51] and later proposed short- and long-term electrical tests based on the model [52]. Lyle and Kirkland reported on an accelerated aging test procedure for the growth of water trees and determination of cable life when subjected to various test conditions [53]. The importance of water in the strand and increased temperatures were again demonstrated.

Although the problem of water treeing has not been eliminated, manufacturers are improving processing technology to improve the service life. Cables manufactured today are improved over those that exhibited early failures circa 1970. Tape strand shielding is now unacceptable. The number of voids, protrusions, and contaminants have been significantly reduced. Tree-retardant XLPE insulation compounds are now widely used. Some experts believe that voltage stress should be reduced even though this action would increase the first cost of the cable. Hermetically sealed sheaths have been developed [54], although these designs might be economically attractive only for transmission voltages. Experience indicates that such sheaths provide good service lives for cables rated at 138 kV and higher. The great importance of keeping water out of the strand is now accepted, and several manufacturers now offer a strand filler to prevent water ingress. Modern cable design has improved markedly over the last decade, but a final solution to water treeing and premature failure in wet environments has not yet been assured. Figure 3.12 illustrates the rising trend of failure rates. Recent field investigations [55] led to the conclusions that failures in XLPE begin to intensify after 10–15 years of service and that water treeing and internal defects (inclusions) were the most often identified problems. Unfortunately, there is an unavoidable time lag between possible effective corrective measures and corroborative evidence from service failure rates.

3.5 SOLID-DIELECTRIC INSULATION TECHNIQUES

Since the 1950s there have been great advances in the techniques of insulating medium-voltage solid-dielectric cables. Formerly the insulations were the thermosetting* compounds based on butyl rubber or the thermoplastic compound, polyethylene. The butyl rubber insulant was applied by an extruder, called a tuber. The vulcanization of the insulation was a separate operation in an open steam vessel or autoclave. Semiconducting fabric tapes were utilized for conductor and insulation shielding. In

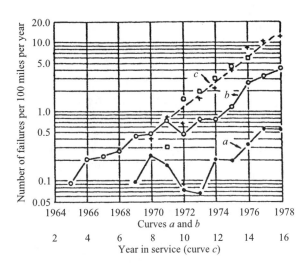

Figure 3.12 Failure rates of polymeric power cable rated 5–35 kV; after Thue [46]: (a) and (b). Average failure rate for all cables operating in a particular year: (a) XLPE and (b) PE. (c) Failure rate versus number of years in service: XLPE + PE.

*A thermoset is defined as a material, cured under heat, that does not soften when reheated. This definition is not strictly true for crosslinked polyethylene because it does soften and has lower physical strength than most vulcanized rubbers at high temperatures. However, for practical cable applications, the definition is suitable.

contrast, present insulating procedures employ continuous vulcanization (CV) machines for thermosetting insulations. Both the extruders for thermoplastic polyethylenes and CV machines usually apply the conductor shielding and insulation in the same operation. The higher permissible electric stress is in part achieved by the use of an extruded conductor shield that is bonded to the insulation. By the simultaneous extrusion of the two materials, a virtual chemical bond is achieved, and the inclusion of foreign material at the interface is minimized. Semiconductive shields are carbon-loaded copolymers or compatible polymers of the insulation compounds. Good processability of the semiconductive materials is an important characteristic, since a smooth conductor to insulation interface is essential to minimize highly localized electric stresses.

The three commonly used insulating materials for medium-voltage distribution cables are natural crosslinked polyethylene, tree-retardant crosslinked polyetylene, and EPR. These three insulants belong to the same group of polymers known as polyolefins. In general, this class of polymers are chemically saturated and cannot be crosslinked (vulcanized) with accelerators and sulfur as older rubbers such as butyl. The usual crosslinking agents are peroxides, and in particular, dicumyl peroxide. As a consequence of the replacing of sulfur with peroxide, it is no longer necessary to tin coat copper conductors to prevent sulfur copper reaction as in butyl or other rubber compounds.

The application of a polymeric material to a conductor for purposes of insulation in wire and cable manufacture includes three basic processes: compounding, extrusion, and crosslinking or vulcanization. Either one, two, or all three of the processes may be undertaken by the cable manufacturer. Thermoplastic polyethylene compound is purchased in compounded pellet form, i.e., the polymer is supplied ready for the extrusion process. Crosslinked polyethylene is also purchased in similar form but must be extruded and then vulcanized. On the other hand, it is common practice for the cable manufacturer to carry out all three processes for EPR and other elastomeric compounds.

3.5.1 Compounding

Crosslinked polyethylene for medium- and high-voltage applications is normally an unfilled insulation. Lower voltage applications may have fillers, finely divided clays, carbon blacks, and other compounds. Carbon black is required for application where the insulation must be shielded against ultraviolet radiation (direct sunlight) similar to thermoplastic compounds. A number of tree-retardant versions of XLPE are commercially available both in North America and Japan [56]. One version [57] uses an ethylene vinyl acetate (EVA) as the growth inhibitor. This additive was chosen in preference to organic fillers because of its better electrical performance, the $\tan \delta$ exhibiting a very small increase. It is believed that water tree growth is retarded by the barrier action of the additive. These additives tend to relieve the stress at treeing sites by field-dependent electrical conductivity [58]. A theory held by some experts, perhaps stated in over-simplified terms, is that very low loss (low $\tan \delta$) materials are subjected at discontinuities such as protrusions, voids, and contaminants to much enhanced electrical stresses compared to materials having appreciable loss like the rubber compounds of many years ago. The electrical losses tend to relieve the stress.

The base polymer of EPR is either a copolymer of ethylene and propylene (EDM) or a terpolymer EPDM. Both base polymers when compounded for cable insulation are normally called EPR, which like many other synthetic polymers represents a group of materials rather than an individual specific compound. The terpolymer is based on ethylene, propylene, and a diene monomer. Most conventional EPR insulation compounds contain 100–120 parts of filler and oil plasticizer per hundred parts of base polymer [27]. These additives are necessary to provide the desired physical properties and to aid processing. A new class of EPR elastomers have been developed that requires less loading (mostly clays and oil plasticizers) [27]. It is claimed [27, 28] that EPR compounds now commercially available are acceptable for transmission voltages because of the lower dielectric loss compared to compounds used for distribution cables; moreover, they are said to be suitable for wet locations because of their excellent water treeing resistance.

3.5.2 Extrusion

The purpose of the extruder is to apply the polymer to the conductor, after which it will be vulcanized if it is a thermoset material or cooled if it is a thermoplastic material. Figure 3.13 displays the essential features of thermoplastic extruder that is also, in general, representative of a thermoset extruder. Wire and cable applications require a crosshead extruder in which the die section is in a plane 90° to that shown. The conductor passes through a profiled cylinder, called a core tube, concentric with the die, as shown in Fig. 3.14. Figures 3.15 and 3.16 are examples of current extruder technology in use.

The correct application of the polymer that must be heated and forced into a concentric compact configuration around the conductor is dependent on a number of variables, principally, the geometry of the screw, barrel, and extrusion die, polymer temperature, screw and barrel cooling, and screw speed. The temperature of the compound moving along the barrel is affected by the mechanical working of the compound as well as the barrel external heating source; thus if the variables are not correctly controlled, the compound may be damaged from overheating. Also, diameter control

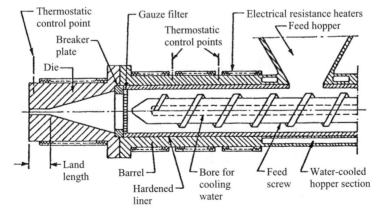

Figure 3.13 Typical extruder head for thermoplastic and thermoset insulation.

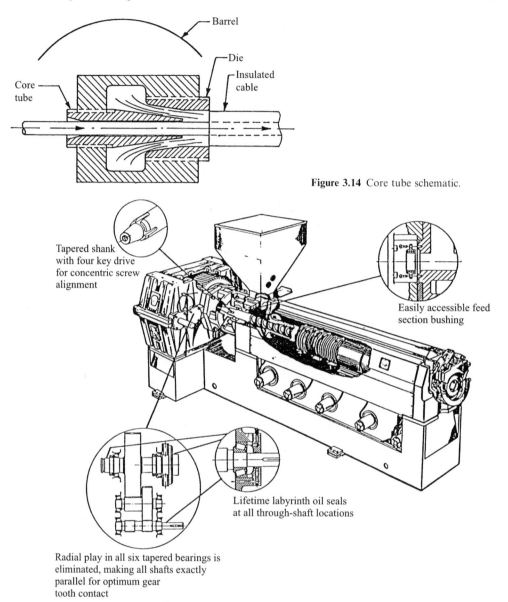

Figure 3.14 Core tube schematic.

Figure 3.15 Modern extruder for thermoplastic and thermoset insulations. (Courtesy of Davis Standard.)

may be unsatisfactory or void formation may result. The production speed for thermoplastic and tuber operation is limited only by the output of the equipment and the cable design, whereas in the CV operation the production rate is controlled both by the extruder and the rate of vulcanization.

3.5.3 Vulcanization

The continuous vulcanization CV process for small low-voltage conductors has been employed in Canada for more than several decades. The insulated conductor

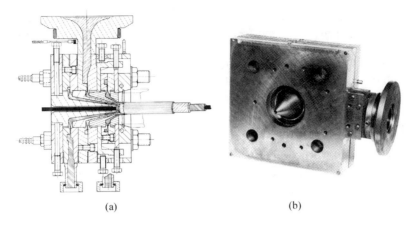

(a)   (b)

Figure 3.16 Triple-extrusion head: (a) cross-sectional view; (b) actual head. (Courtesy of Davis Standard.)

passes directly from the extruder into a long steel tube that contains steam at high pressure. At the far end of the tube a water-filled section cools the insulation that has been vulcanized in the steam section. Finally, the insulated cable passes from the high-pressure water-filled section through a water seal to the atmosphere and takeup equipment. In the earlier machines the high-pressure tube was set basically horizontal. These machines were suitable for small cables; however, it was found that cables weighing more than about 0.7 kg/m were dragged along the tube bottom with resultant damage to the partially cured insulation. To eliminate this restriction, vertical tubes were introduced. Since the length of the tube is of the order of 75 m or more, such vertical configuration required the construction of costly high towers. The evolution of the catenary CV (CCV) followed in which the first 12–30 m or more of the tube assumes a catenary curve. The cable follows its natural curve when passing through the first part of the tube and does not touch the bottom, at least until the compound has attained adequate physical strength. This machine configuration is an optimization of economics and performance for medium-voltage cable constructions. It can conform to most factory environments by having the extruder supported on an elevated platform near the roof. The far end of the tube, which continues to slope at a slight angle, may sometimes be placed below floor level.

It will be appreciated that such a machine that consists of a conductor pay-off, semiconductive compound extruder(s), an insulation compound extruder high-pressure tube, insulated conductor takeup equipment, conductor tension control, steam regulation, water pressure control, and other miscellaneous controls and equipment represents a large capital investment that easily exceeds $1 million. Production runs of many thousands of feet of any one particular cable size and type and infrequent shutdown times are required for economic utilization. The line speed depends on the extruder output and the time interval in the steam tubes to provide the required completeness of vulcanization. The extruder output is not difficult to establish, but the same cannot be said for the rate of vulcanization as the extrudate-covered conductor passes through the steam tube. Here one must know the thermal characteristics of the cable and the rate of vulcanization or crosslinking as a function of temperature and time, or heat history of the compound. It will be appreciated that conductor diameter, insulation compound, and wall thickness are all parameters of the heat history as well as steam pressure and line speed. Since it is of great importance to fully utilize the machine for economic reasons, it is essential that optimum line speed by established as a function of the many variables encountered in practical production. The task can be facilitated with the aid of a digital computer [59]. Line speeds of typical medium-voltage cables are of the order of 6–60 m/min.

3.5.4 Recent Developments

The crosslinking process has had a number of developments within the past decade. The three general processes are peroxide, silane, and irradiation. The first process has been outlined in Section 3.5.3. The second has become more commonly used recently, particularly in Europe. The third method is confined to limited use in communication and low-voltage thin-wall power and special-purpose cables.

A brief outline of silane crosslinking or curing is given in [60]. The process, which requires only atmospheric pressure, takes place in a water bath at elevated temperature. The chemical crosslinking process requires moisture. It follows that the rate of cure is dependent on the rate of moisture diffusion in the polyethylene. In Europe, production is normally linked to about 25-kV rated cables. It is used there extensively in filled flame-retardant building wires and control cables, instead of the PVC that is used in North America, to exclude the evolution of corrosive gases and dense smoke during fires. Silane cure has the advantages of low capital and energy costs, particularly as it can use existing thermoplastic extrusion lines. In practice wall thicknesses are limited to those of medium-voltage cable. There are two processes available: Monosil (or one step) and Sioplas (two step), each with its own advantages and disadvantages.

The peroxide CV technology, normally used in North America, has traditionally employed high-pressure steam to convey heat to the extrudate and effect cure. Steam cure tends to saturate the polymer with moisture and is one of the chief causes of microvoids in the insulation. To avoid this difficulty hot nitrogen gas curing has been introduced, often called dry cure. This process greatly reduces the number of microvoids having diameters of $1\,\mu m$ $(1\,\mu m = 4 \times 10^{-5}\,in.)$ or more [61–64]. Although there is not complete agreement on the beneficial effects of dry cure, substantial increases in short-term breakdown stresses of the insulation have been reported [61, 64]. It is not clear whether or not dry cure will significantly reduce the water-tree-induced failure rate. For example, one study [65] indicates that water tree growth rate will not be significantly affected. The new technique, however, is widely used in North America (cf. Fig. 3.17).

Another alternative to steam cure is the pressurized liquid salt (PLCV) system. This process utilizes eutectic salt with a melting point of about 150°C as the heat transfer medium. Because the heat transfer is rapid, relatively short curing tubes are required, and the pressure necessary is much lower than in the steam cure process. These features are said to have distinct advantages for the curing of jackets on cables, such as portable cables where a tough nonporous jacket is required without deformation of the enclosed insulated conductors. Such jackets have been traditionally cured under a metal mold (temporary lead sheath) in an autoclave; the PLCV technique has seen limited application, chiefly in Europe. Figure 3.18(a) depicts the steam cure and liquid salt types of CV or crosslinking systems. In the liquid salt CV system, the high-temperature eutectic salt is pumped over the cable surface along the slightly inclined

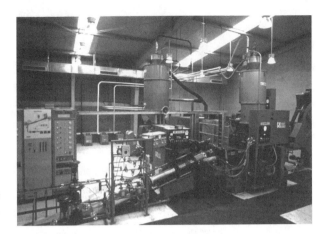

Figure 3.17 Arrangement of gas cure CV system in cable manufacturing facility. (Courtesy of Davis Standard.)

curing tube. This turbulent action improves the heat transfer to yield more rapid curing rates. While still under the system pressure, the cable is cooled in water; the salt in the cure tube is separated and returned to the reservoir for recirculation. The entire system is held under air or nitrogen pressure, independent of the curing pressure. The shorter curing tube length allows shorter lengths of cable to run. Figure 3.18(b) shows a dry curing system using nitrogen gas.

Because it has been shown that water trees often are initiated by contaminants in the insulation, much attention is given now to cleanliness of the raw materials. Modern methods include rigid inspection of raw materials entering the cable plant and hermetic sealing systems to exclude foreign materials during shipment from the polymer supplier to the storage area in the cable plant and finally during the transport to the extrusion process [47] (cf. Fig. 3.19). Devices are utilized to detect foreign material during insulation of the cable. For instance, laser inspection [66] can detect contaminates larger than about 50 μm. Triple extrusion (cf. Fig. 3.16) is widely used to ensure that the insulation SC (semiconductive) shield is applied over a surface free of foreign material and scratches or other defects that could occur if the unshielded cable is stored in cable plant atmosphere and conditions awaiting final processing.

3.6 RELATED TESTS

It is true, of course, that all tests on the completed cable represent some measure of both the quality of the materials and manufacturing techniques employed. It is the intent here to illustrate how some tests, whether they be routine industry tests or special tests employed by the manufacturers, are relevant to the control of manufacturing processes.

The partial discharge test result is indicative of the control of a number of processes. Foreign material either in the purchased raw material or introduced during manufacture can cause voids within the insulation. Incorrect extrusion of semiconducting conductor shields may result in a rough semiconducting shield to insulation interface. Again, incorrect application of the insulation semiconducting shield or incorrect cooling of the extrudate will permit the occurrence of occluded voids. Any of these defects may cause failure in the partial discharge test.

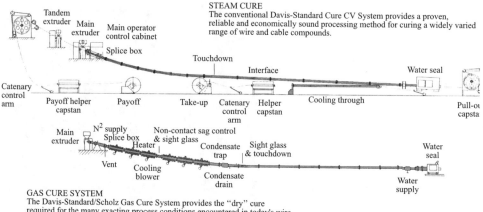

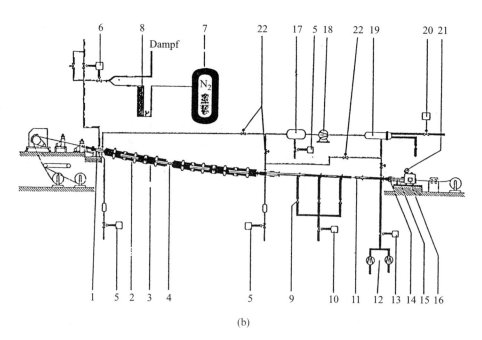

Figure 3.18 (a) Steam cure and liquid salt (PLCV) type continuous vulcanizing (CV) systems; (b) dry (gas) cure type continuous vulcanizing (CV) or cross-linking system: 1 splice box; 2, heating tube; 3, tube heating; 4, non-contact sag control; 5, condensate drain; 6, pressure regulation; 7, N_2 supply; 8, evaporator; 9, water overflows; 10, temperature regulation, water overflow; 11, cooling tube; 12, cooling water supply; 13, continuous cooling water regulation; 14, main sealing, double sealing; 15, lead wire sealing; 16, water or gas reservoir; 17, relief tank; 18, high-pressure circulating blower; 19, gas cooler; 20, cooling-water control; 21, gas excitation fan; 22, gas circulation. (Courtesy of Davis Standard.)

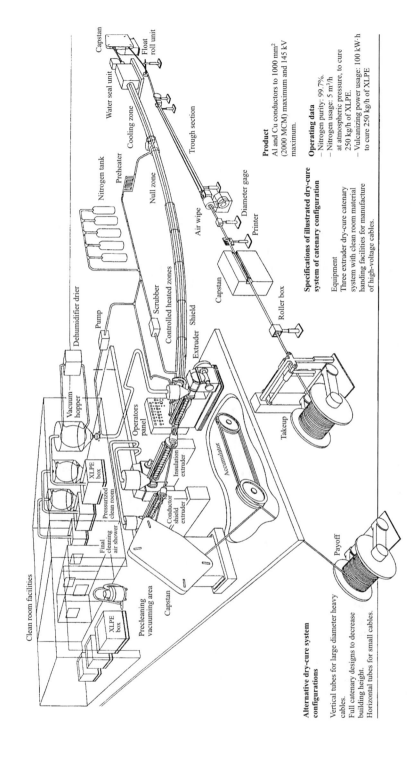

Figure 3.19 Dry curing system, including clean room facilities. (Courtesy of John Royle & Sons, Inc.)

203

Long-term water immersion tests are effective in detecting weaknesses in filled materials. Because of incorrect use and dispersal of fillers, moisture absorption may be high. Electrical deterioration as evidenced by high power factor and permittivity and low insulation resistance will occur. Load cycle testing in water has been shown to be effective in segregating inferior compounds [45].

The state of cure or crosslinking has significant effects on both physical and electrical properties of XLPE and EPR compounds. Deformation tests, solvent extraction tests, and physical properties both before and after aging are employed as controls for both material and manufacturing processes.

Other significant tests include the cyclic aging test, resistance stability test, and accelerated treeing test specified in AEIC documents [29, 30]. The latter test is for trial use, and engineering information and is, of course, related to the water treeing problem discussed in Section 3.4.6. Power factor, partial-discharge and accelerated aging (water treeing) tests are dealt with in detail in Chapter 9. The cyclic aging test is a significant test that simulates field conditions. The heavy current loading, for example, may indicate weaknesses in shielding designs. The resistance stability test (ICEA Publication T-25-425) is designed to ensure that SC shields are stable and will maintain their resistivities to acceptable levels over the life of the cable.

REFERENCES

[1] ASTM Standard D257-93 "Standard test methods for dc resistance or conductance of insulating materials," *Annual Book of ASTM Standards*, Vol. 10.01, ASTM, West Conshohocken, PA, August 1993.

[2] K. Yahagi, "Dielectric properties and morphology in polyethylene," *IEEE Trans. Electrical Insulation*, Vol. EI-5, 1980, pp. 241–250.

[3] NEMA Standards Publication No. WC50-1988/ICEA Publication No. P-53-426, "Ampacities including effect of shield losses for single-conductor solid-dielectric power cable, 15 kV through 69 kV (copper and aluminum conductors)," NEMA, Rosslyn, VA, September 1988.

[4] NEMA Standards Publication No. WC7-1992/ICEA Publication No. S-66-524, "Crosslinked thermosetting polyethylene insulated wire and cable for the transmission and distribution of electrical energy," NEMA, Rosslyn, VA, December 1996.

[5] W. A. Lewis and G. D. Allen, "Symmetrical-component circuit constants and neutral circulating currents for concentric neutral underground distribution cables," *IEEE Trans. Power Apparatus Syst.*, Vol. PAS-97, 1978, pp. 191–199.

[6] W. A. Lewis, G. D. Allen, and J. C. Wange, "Circuit constants for concentric neutral underground distribution cables on a phase basis, *IEEE Trans. Power Apparatus Syst.*, Vol. PAS-97, 1978, pp. 200–207.

[7] H. S. Radhakrishna, F. Y. Chu, and S. A. Boggs, "Thermal instability and its production of cable backfill soils," *IEEE Trans. Power Apparatus Syst.*, Vol. PAS-99, 1980, pp. 856–867.

[8] M. A. Martin, R. A. Bush, W. Z. Black, and J. G. Hartley, "Practical aspects of applying soil thermal stability measurements to the rating of underground power cables," *IEEE Trans. Power Apparatus Syst.*, Vol. PAS-100, 1981, pp. 4236–4249.

[9] M. Kellow, "Experimental investigation of the ampacity of distribution cables in a duct bank with and without forced cooling," *IEEE Trans. Power Apparatus Syst.*, Vol. PAS-100, 1981, pp. 3274–3283.

[10] M. A. Kellow, "A numerical procedure for the calculation of the temperature rise and ampacity of underground cables," *IEEE Trans. Power Apparatus Syst.*, Vol. PAS-100, 1981, pp. 3322–3330.

[11] C. W. Nemeth, G. B. Rackliffe, and J. R. Legro, "Ampacities for cables in trays with fire stops," *IEEE Trans. Power Apparatus Syst.*, Vol. PAS-100, 1981, pp. 3573–3579.

[12] E. Kuffel and J. Poltz, "Ac losses in crossbonded and bonded at both ends high voltages," *IEEE Trans. Power Apparatus Syst.*, Vol. PAS-100, 1981, pp. 369–374.

[13] M. A. El-Kadyk, "Optimization of power cable and thermal backfill configurations," *IEEE Trans. Power Apparatus Syst.*, Vol. PAS-101, 1982, pp. 4681–4688.

[14] A. Bernath and D. B. Olfe, "Optimization of ampacities for unequally loaded underground power cables," *IEEE Trans. Power Apparatus Syst.*, Vol. PAS-101, 1982, pp. 2348–2354.

[15] G. R. Engmann, "Low voltage cable ampacity with four or more conductors in rigid galvanized steel conduit," *IEEE Trans. Power Apparatus Syst.*, Vol. PAS-101, 1982, pp. 3201–3205.

[16] O. M. Esteves, "Derating cables in trays transversing firestops or wrapped in fireproofing." *IEEE Trans. Power Apparatus Syst.*, Vol. PAS-102, 1983, pp. 1478–1481.

[17] R. A. Hartlein and W. Z. Black, "Ampacity of electric power cables in vertical protective risers," *IEEE Trans. Power Apparatus Syst.*, Vol. PAS-102, 1983, pp. 1678–1686.

[18] W. Z. Black and S. Park, "Emergency ampacities of direct buried three-phase underground cable systems," *IEEE Trans. Power Apparatus Syst.*, Vol. PAS-102, 1983, pp. 2124–2132.

[19] G. Engmann, "Ampacity of cable in covered tray," *IEEE Trans. Power Apparatus Syst.*, Vol. PAS-103, 1984, pp. 345–352.

[20] C. Katz, A. Dima, A. Zidon, M. Ezrin, W. Zengel, and B. Bernstein, "Emergency overload characteristics of extruded dielectric cables operating at 130°C and above," Paper 84 T&D 332-3 presented at IEEE/PES 1984 Transmission and Distribution Conference, Kansas City, Missouri, April 29–May 4, 1984.

[21] M. Kellow and H. St. Onge, "Thermo-mechanical failure of distribution cables subjected to emergency loading," *IEEE Trans. Power Apparatus Syst.*, Vol. PAS-101, 1982, pp. 1914–1920.

[22] NEMA Standards Publication No. WC8-1992/ICEA Publication No. S-68-516, "Ethylene propylene rubber insulated wire and cable for the transmission and distribution of electrical energy," NEMA, Rosslyn, VA, December 1996.

[23] R. C. Mildner, C. B. Arenda, and P. C. Woodland, "The short circuit rating of thin metal tape cable shields," *IEEE Trans. Power Apparatus Syst.*, Vol. PAS-87, 1968, pp. 749–759.

[24] R. G. Lukaac, D. A. Silver, R. A. Hartlein, and W. Z. Black, "Optimization of metallic shields for extruded dielectric cables under fault conditions," Paper 84 T&D 339-8 presented at IEEE/PES 1984 Transmission and Distribution Conference, Kansas City, Missouri, April 29–May 4, 1984.

[25] M. A. Martin, D. A. Silver, R. G. Lukac, and R. Suarez, "Normal and short circuit operating characteristics of metallic shielded solid dielectric power cables," *IEEE Trans. Power Apparatus Syst.*, Vol. PAS-93, 1974, pp. 601–613.

[26] ICEA Publication No. P-45-482, "Short circuit performance of metallic shields and sheaths of insulated cable," ICEA, South Yarmouth, MA, March 1994.

[27] M. Brown, "Performance of ethylene-propylene rubber insulation in medium and high voltage power cable," *IEEE Trans. Power Apparatus Syst.*, Vol. PAS-102, 1983, pp. 373–381.

[28] E. Occhini, R. Metra, G. Portinari, and B. Vecellio, "Thermal, mechanical and electrical properties of EPR insulations in power cables," *IEEE Trans. Power Apparatus Syst.*, Vol. PAS-102, 1983, pp. 1942–1953.

[29] AEIC Publication CS6-82, "Specifications for ethylene propylene rubber insulated shielded power cables rated 5 through 69 kV," 4th edn., AEIC, New York, June 1982.

[30] AEIC Specification CS5-94, "Specification for crosslinked polyethylene insulated shielded power cables rated 5 through 46 kV," AEIC, Birmingham, AL, March 1994.

[31] C68.3-M1983, "Power cable with thermoset insulation," Canadian Standards Association, 178 Rexdale Boulevard, Rexdale, Ontario, Canada M4W 1R3.

[32] R. M. Eichorn, "A critical comparison of XLPE and EPR for use as electrical insulation on underground power cables," *IEEE Trans. Electrical Insulation*, Vol. E1-16, 1981, pp. 469–482.

[33] R. J. T. Clabburn and G. J. Clarke, "Jointing high voltage cables using heat-shrinkable components," Presented at the 7th IEEE/PES Conference and Exposition on Transmission and Distribution, April 1–6, 1979, Atlanta, Georgia.

[34] ICEA Publication No. T-32-645, "Guide for establishing compatibility of sealed conductor filler compounds with conducting stress control materials," ICEA, South Yarmouth, MA, February 1993.

[35] ICEA Publication No. T-31-610, "Guide for conducting a longitudinal water penetration resistance test for sealed conductor," ICEA, South Yarmouth, MA, July 1994.

[36] IEEE Publication 84TH0110-7-PWR. "Regional underground distribution practices," 9th IEEE/PES Transmission and Distribution Conference and Exposition, Kansas City, Missouri, April 29–May 4, 1984.

[37] IEEE Standard 383-1974, "Standards for type test of class IE electric cables," field splices, and connections for nuclear power generating stations," IEEE, New York, 1992.

[38] UL Publication 1277, "Standards for safety for electrical power, and control tray cables with optional optical-fiber members," Underwriters Laboratories, Northbrook, IL, 1998.

[39] W. Vahlstrom, "Investigation of insulation deterioration in 15 kV and 22 kV polyethylene cables removed from service", *IEEE Trans. Power Apparatus Syst.*, Vol. PAS-91, 1972, pp. 1023–1035.

[40] J. H. Lawson and W. Vahlstrom, "Investigation of insulation deterioration in 15 kV and 22 kV polyethylene cables removed from service. Part II," *IEEE Trans. Power Apparatus Syst.*, Vol. PAS-92, 1973, pp. 824–835.

[41] T. P. Lanctoe, J. H. Lawson, and W. L. McVey, "Investigation of insulation deterioration in 15 kV and 22 kV polyethylene cables removed from service. Part III," *IEEE Trans. Power Apparatus Syst.*, Vol. PAS-98, 1979, pp. 912–925.

[42] T. Fukuda, S. Suzuki, H. Goto, and Y. Nitta, "Water trees in 3.3 kV and 6.6 kV crosslinked polyethylene power cables," 1972 Annual Report, Conference on Electrical Insulation and Dielectric Phenomena, NAS-NRC, Washington, DC, 1972, pp. 66–68.

[43] M. Kurtz, "Treeing in polyethylene cable insulation," *Ontario Hydro Res. Q.*, first quarter, 1972, pp. 19–20.

[44] J. Densley, "The growth of trees in crosslinked polyethylene cable insulation," 1972 Annual Report, Conference on Electrical Insulation and Dielectric Phenomena, NAS-NRC, Washington, DC, 1972, pp. 43–45.

[45] E. J. Gouldson, A. S. Gahir, and M. Kurtz, "A contribution to the study of ageing of cable insulating materials," 11th Electrical Insulation Conference (IEEE Conference Record 73CH0777-3EI), Chicago, 1973, pp. 153–157.

[46] G. Bahder, C. Katz, G. S. Eager, Jr., E. Leber, S. M. Chalmers, W. H. Jones, and W. H. Mangrum, Jr., "Life expectancy of crosslinked polyethylene insulated cables rated 15 to 35 kV," *IEEE Trans. Power Apparatus Syst.*, Vol. PAS 100, 1981, pp. 1581–1590. (Also see W. A. Thue, J. W. Bankoske and R. R. Burghardt, "Operating and testing experience on solid dielectric cable," Proc. CIGRÉ, Report 21-11, Paris, 1980.)

[47] H. C. Doepken, Jr., A. L. McKean, and M. L. Singer, "Treeing, insulation material, and cable life," *IEEE Trans. Industry Applications*, Vol. PAS IA-17, 1981, pp. 205–210.

[48] A. Bulinski and R. J. Densley, "The voltage breakdown characteristics of miniature XLPE cables containing water trees," *IEEE Trans. Electrical Insulation*, Vol. E1-16, 1981, pp. 319–326.

[49] B. S. Bernstein, "Recent progress in understanding water treeing phenomena," 1984 IEEE International Symposium on Electrical Insulation, Conference Record 84CH 1964-6EI, Montreal, June 11–13, 1984, pp. 11–21.

[50] J. Sletbak, and E. Ildstad, "The effect of service and test conditions on water tree growth in XLPE cables," *IEEE Trans. Power Apparatus Syst.*, Vol. PAS-102, 1983, pp. 2069–2076.

[51] B. Bahder, T. Garrity, M. Sosnowski, R. Eaton, and C. Katz, "Physical model of electric ageing and breakdown of extruded polymeric insulated power cables," *IEEE Trans. Power Apparatus Syst.*, Vol. PAS-101, 1982, pp. 1379–1390.

[52] G. Bahder, M. Sosnowski, C. Kata, R. Eaton, and K. Klein, "Electrical breakdown characteristics and testing of high voltage XLPE and EPR insulated cables," *IEEE Trans. Power Apparatus Syst.*, Vol. PAS-102, 1983, pp. 2173–2185.

[53] R. Lyle and J. W. Kirkland, "An accelerated life test for evaluating power cable insulation," *IEEE Trans. Power Apparatus Syst.*, Vol. PAS-100, 1981, pp. 3765–3774.

[54] T. Nakabasami, K., Sasaki, K. Sogiyama, and M. Ishitobi, "Development of XLPE cable with new laminated waterproof layer," 1984 IEEE International Symposium on Electrical Insulation, Conf. Record 84CH 1964-6EI, Montréal, June 11–13, 1984, pp. 41–45.

[55] J. D. Mintz, "Failure analysis of polymeric-insulated power cable," IEEE Paper 84 T&D 345-5 presented at the IEEE/PES 1984 Transmission and Distribution Conference, Kansas City, Missouri, April 29–May 4, 1984.

[56] E. J. McMahon, "A tree growth inhibiting insulation for power cable," *IEEE Trans. Electrical Insulation*, Vol. EI-16, 1981, pp. 304–318.

[57] S. Nagasaki, H. Matsubara, S. Yamanouchi, M. Yamada, T. Matsuike, and S. Fuknaga, "Development of water-tree retardant XLPE cables," *IEEE Trans. Power Apparatus Syst.* Vol. PAS-103, 1984, pp. 536–544.

[58] D. M. Tu, L. H. Wu, X. Z. Wu, C. K. Cheng, and K. C. Kao, "On the mechanism of treeing inhibition by additives in polyethylene," *IEEE Trans. Electrical Insulation*, Vol. E1-17, 1982, pp. 539–545.

[59] V. L. Lenir, and R. C. Dodwell, "Optimization of CV cures by computer method," *Wire J.*, October, 1968.

[60] E. P. Marsden, Union Carbide Canada Limited, Montréal, private communication.

[61] T. Mizukami, K. Takahashi, C. Ikeda, N. Kato, and B. Yoda, "Insulation properties of cross-linked polyethylene cables cured in inert gas," *IEEE Trans. Power Apparatus Syst.*, Vol. PAS-94, 1975, pp. 467–472.

[62] S. Kageyama, M. Ono, and S. Chabata, "Micro voids in crosslinked polyethylene insulated cables," *IEEE Trans. Power Apparatus Syst.*, Vol. PAS-94, 1975, pp. 1258–1263.

[63] A. L. McKean, K. Tusuji, H. C. Doepken, and A. Zidon, "Breakdown mechanisms studies in XLPE cables—Part II," *IEEE Trans. Power Apparatus Syst.*, Vol. PAS-97, 1978, pp. 1167–1176.

[64] K. Kojimi, T. Fukoi, Y. Yamada, S. Kitai, and K. Yatsuki, "Development and commercial use of 275 kV XLPE power cable," *IEEE Trans. Power Apparatus Syst.*, Vol. PAS-100, 1981, pp. 203–210.

[65] J. M. Braun, "Comparison of water treeing rates in steam and nitrogen treated polyethylenes," *IEEE Trans. Electrical Insulation*, Vol. E1-15, 1980, pp. 120–123.

[66] A. Campus, G. Matey, and H. Legoupil, "Determination of the cleanliness of high voltage insulation compounds by a laser inspection technique," IEEE Paper 84 T&D 310-9 presented at IEEE/PES 1984 Transmission and Distribution Conference, Kansas City, Missouri, April 29–May 4, 1984.

EXTRUDED SOLID-DIELECTRIC POWER TRANSMISSION CABLES

L. J. Hiivala

4.1 INTRODUCTION

The use of extruded solid-dielectric power cables at transmission voltages of 60 kV and above began in the 1960s. When compared to oil-paper cables of either the low-pressure oil-filled or high-pressure pipe-type construction, polymeric insulated cables are easier to install and require less skill to splice and terminate. Unlike oil-impregnated paper insulated cables, they are less sensitive to atmospheric conditions. There is no need for an auxiliary oil pressurization system and the maintenance thereof. Together with the other advantages outlined above, this results in lower installation cost than for self-contained or pipe-type oil-filled systems. These are the main reasons why polymeric insulated cables have gained wide acceptance by electric power utilities around the world and are now used for underground transmission at voltages up to 500 kV.

4.1.1 Historical Overview

The excellent dielectric properties of polyethylene were first recognized during World War II. When it became apparent that polyethylene is very sensitive to partial discharges occurring within the insulation, much attention was paid to the elimination of microvoids and considerable effort was made to achieve a high level of cleanliness both in the compound and during extrusion.

Low-density polyethylene (LDPE) was first used at high voltages in the 1960s, and cables insulated with this material are now in operation at 500 kV. Because it has a normal operating temperature of only 70°C, some use has been made in France of high-density polyethylene (HDPE) because of its somewhat higher operating temperature of 80°C.

Initially, the development of polymeric cables for use at high voltages was made using thermoplastic polyethylene of either the low- or high-density type. When its low melting point became an obstacle, this was overcome by the development of crosslinked polyethylene (XLPE) in the 1960s. Because of its temperature stability, cables insulated with crosslinked polyethylene have a normal operating temperature of 90°C, compared with 70 and 80°C for low- and high-density polyethylene, respectively. This has a significant effect on the current-carrying capacity (ampacity) of such a cable.

4.1.2 Development Trends

Typically, polymeric insulated cables have been used initially in short sections without joints for the undergrounding of overhead transmission lines. As confidence grew in this new technology and joints became available, it was then used for long distance transmission. For example, Fig. 4.1 illustrates the history of XLPE cable development in Japan [1].

As can be seen from Fig. 4.1, the availability of reliable accessories, especially joints, has always retarded the acceptance of high-voltage polymeric insulated cables. For example, the first trial installation of 500-kV XLPE cable in Japan was made in 1990. This was more or less on an experimental basis, because a reliable joint had not yet been developed. Now that the extrusion-molded joint (EMJ) has been successfully developed, plans are underway for the installation of the world's first long-distance underground transmission circuits using 500-kV XLPE cable.

Because the installation time and skills required for such a joint are substantial, prefabricated and premolded joints are now being developed for use at 500 kV. Likewise, much work is being carried out around the world to increase the design stresses for polymeric cables, because the operating electrical stresses for such insulations are lower than for oil-paper cables, and thus the insulation thicknesses are greater.

4.2 DESIGN AND CONSTRUCTION

International Electrotechnical Commission (IEC) 60840 [2] is the international standard for power cables with extruded insulation rated up to 150 kV. It is a test standard in that it specifies routine, special, and type tests as well as electrical tests after installation but does not prescribe the design of the cable. While many countries have adopted IEC 60840 as their national standard, they have also issued specifications that define the constructional requirements for the cable.

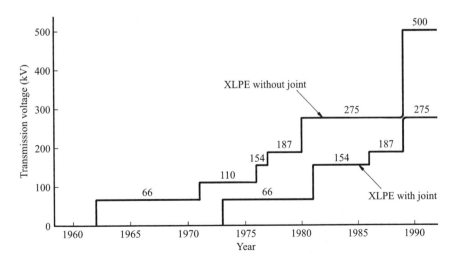

Figure 4.1 History of high-voltage XLPE cable development in Japan.

In the United States, for example, the Association of Edison Illuminating Companies (AEIC), a group of large investor-owned power utilities, has for many years issued specifications [3, 4] for such cables. Initially, the insulation thicknesses covered by these specifications were very conservative when compared to today's standards and specifications. In the most recent revision of the AEIC specifications, the use of reduced insulation thicknesses based on maximum stress designs was finally recognized. The maximum allowable stresses at the conductor shield interface are 6 kV/mm (150 V/mil) for 69-kV rated cables and 8 kV/mm (200 V/mil) for 115- and 138-kV cables.

Likewise, Electricite de France (EDF), the French national utility, has issued specifications [5, 6] for high- and extra-high-voltage polymeric cables. These specifications limit not only the internal but also the external gradient. The importance of the latter will be discussed in Section 4.3. Table 4.1 compares the maximum stresses at the conductor shield allowed by the AEIC and EDF specifications:

In order to increase the load transmission capability of extruded dielectric cables, lower loss designs must be considered. These losses are either current dependent or voltage dependent. While it is possible to reduce the former slightly by conductor design, it is necessary to use a conductor, such as a superconductor, that has a much lower resistance than the copper or aluminum used today in order to make a significant change. On the other hand, there is far more scope for a reduction of the voltage-dependent loss. The magnitude of this dielectric loss increases with the square of operating voltage, and does so very steeply at the higher voltages.

4.2.1. Conductors

As in the case of medium-voltage extruded dielectric cables, the choice of conductor construction is dictated by the cable application, choice of insulation and economics, including the cost of losses. The conductors used in high-voltage extruded dielectric cables are always circular in cross section, as shown in Fig. 4.2.

The simplest construction is the compressed stranded conductor which is effected by using a compression die only over the outermost layer of wires during the stranding operation. This technique not only reduces the overall diameter of the conductor slightly, but more importantly it ensures a tightly formed outer layer that is essential for the application of semiconducting conductor shield. If the wires in the outer layer

TABLE 4.1 Maximum Design Stress

Rated Voltage (kV)	Maximum Stress (kV/mm)	
	AEIC	EDF
63	—	6
69	6	—
90	—	6
115	8	—
138	8	—
150	—	6
225	—	11
400	—	16
500	—	16

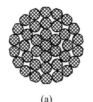 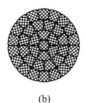 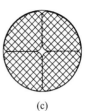

Figure 4.2 Types of conductors: (a) compressed; (b) compact round; (c) compact segmental type M[7].

(a)   (b)   (c)

are not tightly closed, the extrudate may flow into the interstitial spaces between the wires, causing defects in the conductor shield known as "fall-ins." For very large conductor cross sections, it may become necessary to use a semiconducting coated fabric tape over the conductor because of the larger wire sizes involved.

Another construction is the compact stranded conductor. During the stranding process, each layer of wires is compacted using either a combination of rolls and dies or dies alone. This technique allows for a reduction in the conductor diameter of approximately 8%.

One way of increasing the current-carrying capacity of a transmission circuit is by increasing the cross-sectional area of the conductor. However, for larger cross sections, the alternating current (ac) resistance of the conductor is increased due to the skin effect. For sizes larger than 1000 mm^2 (2000 kcmils) copper or 1600 mm^3 (3200 kcmils) aluminum, conductors of the well-known Milliken or type M construction are used. In such a design, the wires in each segment are transposed during stranding and compacting. The individual segments are lightly insulated from each other by a separator tape. It usually has four or more segments. In order to maintain a circular shape, an uneven number of segments may be used. For example, five such segments will prevent transverse sliding of the segments when the conductor is bent during manufacture, especially the extrusion process, or installation. Another measure that can be taken to maintain the tightness of a type M conductor is to prespiral the individual segments during stranding, so that they go together naturally when cabled.

4.2.2 Semiconducting Conductor and Insulation Shields

The extruded semiconducting conductor shield layer over the conductor serves two purposes:

1. It ensures a uniform electric stress distribution over a relatively rough conductor surface, thereby eliminating strand effects. Otherwise, protrusions at this critical interface can lead to a treelike breakdown path in the insulation wall, when the stress is high enough to cause failure of the cable.
2. This layer also ensures tight bonding between the conductor and the adjacent insulation, thus excluding any interspersed voids and the associated partial discharges.

Likewise, the purpose of the insulation shield layer over the outside of the insulation is again twofold:

1. With the outer metallic shield of the cable grounded, the electric stress distribution within the insulation attains radial symmetry and is confined to the insulation. This is an important safety consideration.

2. When this layer is in intimate contact with the underlying insulation, it again serves the purpose of preventing partial discharges in air spaces that would otherwise exist between the insulation and the metallic shield. For this reason, a bonded rather than a strippable insulation shield is used almost exclusively. Nevertheless, this shield layer should strip cleanly for the insulation during splicing and terminating operations. This can be accomplished by the moderate application of heat or the use of special tools.

The quality of the shield interfaces can be significantly improved by the use of a true triple extrusion process whereby all three layers are extruded simultaneously in a common crosshead. Techniques such as this will be discussed in Section 4.3.2.

As discussed in Section 2.3, another quality improvement entails the use of the so-called supersmooth semiconducting compounds. These incorporate acetylene carbon blacks to increase the smoothness of the shield interface (cf. Chapter 2). As shown in Fig. 2.4, a significant reduction in the number and height of shield protrusions can be realized through the use of such compounds. Specifications such as AEIC CS7 [3] limit shield protrusions to a maximum of 125 µm (5 mils) at the shield interfaces. In Japan, studies were carried out to determine the factors that affect the performance of cross-linked polyethylene cables manufactured with existing technology. Using an experimental technique called the prebreakdown partial discharge (PPD) detection method, it was possible to determine the types of defects from which insulation breakdown originated in model cables. In 26% of the cases, they were found to originate from shield protrusions and the remainder from contaminants in the insulation. Based on a tree initiation electrical stress of at least 300 kV/mm, an allowable defect level of 50 µm (2 mils) was determined for the design and manufacture of 500-kV XLPE cables [7]. Through a combination of three-layer simultaneous extrusion and filtering through a fine-wire mesh of no more than 30 µm, levels as low as 25 µm have been achieved [8].

As discussed in Section 2.3, the volume resistivity of the extruded semiconducting shields and their thermal stability are very important. IEC Standard 60840 only requires the volume resistivity of the shields to be measured as a type test to check that the materials used have satisfactory properties. These measurements are made on test pieces at the rated normal operating temperature for the particular cable design. This is done both without aging and after aging in an air oven for 7 days at 10°C above the rated conductor temperature. The maximum allowable values are 1000 and 500 Ω-m for the conductor and insulation shields, respectively.

On the other hand, AEIC Specification CS7 requires these measurements to be made not only at the normal but also the emergency operating temperature, which may be as high as 130°C. In addition, the volume resistivity test must be made on samples that have been taken on a periodic basis during each insulation extruder run.

4.2.3 Insulations

The manufacture of high-voltage power cables with an extruded insulation began with the use of thermoplastic (linear) polyethylene. Some use has been made of high-density polyethylene because of its slightly higher normal operating temperature. In order to improve upon this temperature and achieve thermal stability, thermoset compounds such as crosslinked polyethylene and ethylene-propylene-rubber (EPR) came into use. The permissible operating temperatures for these insulating compounds are shown in Table 4.2.

4.2.3.1 Thermoplastic Polyethylene

Because of its high dielectric breakdown strength and low dissipation factor and dielectric constant, low-density polyethylene was first used at 63 kV in France beginning in 1962. Its acceptance for use at increasingly higher voltages can be seen in Table 4.3 [9].

For high-voltage applications, EDF Technical Specification HN 33-S-52 limits the maximum stress for 63- and 90-kV rated cables to 6 kV/mm (150 V/mil) and 3 kV/mm (75 V/mil) at the conductor and insulation shield interfaces, respectively. In order to reduce the cable weight by up to 50% and decrease the overall diameter of the cable, these permissible limits have been raised to 7 kV/mm (175 V/mil) and 4 kV/mm (100 V/mil). This has enabled a reduction of the insulation thickness to 11 mm. By doing so and substituting an aluminum laminate moisture barrier for an extruded lead alloy sheath, this has enabled Electricite de France to deliver longer cable lengths and thus reduce the number of joints, utilize mechanized laying techniques, and reduce the total installation cost [10].

When it was recognized that polyethylene was susceptible to erosion by electrical discharges, considerable attention was paid to the elimination of voids within the insulation. Also, research indicated that its dielectric breakdown strength could be seriously affected by the presence of contaminants. These may be introduced on three different occasions: during the compounding process, during material handling and transfer, and during cable manufacture. With manufacturing improvements in these areas and the introduction of three-layer simultaneous extrusion, the use of low-density polyethylene has been extended to the 500-kV level. This has enabled a corresponding increase in the maximum allowable electrical stresses, as shown in Table 4.4 [11, 12].

Extrusion of all three layers is normally carried out in a triple head on a vertical line installed in a tower or shaft up to 100 m in height. Because it is generally recognized

TABLE 4.2 Maximum Conductor Temperatures

	Maximum Conductor Temperature (°C)		
Insulating Compound	Normal Operation	Emergency Operation	Short Circuit
Low-density polyethylene (LDPE)	70	80	150
High-density polyethylene (HDPE)	80	90	180
Crosslinked polyethylene (XLPE)	90	100–130	250
Ethylene-propylene-rubber (EPR)	90	100–130	250

TABLE 4.3 Use of LDPE Cables in French Network

Rated voltage, kV	63	90	225	400	500
Year	1962	1967	1969	1985	1990

TABLE 4.4 Maximum Stress by Rated Voltages Allowed by French National Standards

Stress Type	Maximum Stress (kV/mm)					
	63 kV	90 kV	150 kV	225 kV	400 kV	500 kV
Internal	7	7	7	11	16	16
External	4	4	4	5.5	7	7

that all polymeric-insulated transmission class cables should be protected against moisture ingress by a radial moisture barrier, an extruded lead alloy sheath is generally used.

For utility engineers, the generally accepted objective for failures is less than 0.2 faults per 100 k-y. If external mechanical damage and water penetration are excluded, the failure rate for low-density polyethylene insulated cables is significantly below this target. Thus, the use of polyethylene has not given way completely to crosslinked polyethylene, but the latter is beginning to make inroads.

4.2.3.2 Crosslinked Polyethylene

Although thermoplastic polyethylene is still used in France and some export markets, the use of crosslinked polyethylene is growing and is used almost exclusively in some countries.

As shown in Table 2.14, crosslinked polyethylene has similar electrical properties as do thermoplastic polyethylenes: high dielectric breakdown strength and volume resistivity and low dielectric constant and dissipation factor. Because of its thermoset character, crosslinked polyethylene maintains its mechanical properties when exposed to temperatures that would cause thermoplastic polyethylenes to melt, lose shape and flow. As a result, they can be operated continuously at a maximum conductor temperature of 90°C and occasionally at 105 or even 130°C [3] under emergency operating conditions. Most standards limit this temperature to 105°C. The ability of a cable to do so depends not only on its metallic shield design but may also be limited by the installation conditions and the accessories used with it. Nevertheless, this leads to a larger power transmission capacity for crosslinked than for noncrosslinked polyethylene cables.

In North America, XLPE transmission cables rated for operation at 138 kV are now quite common and a few installations have already been made at 230 kV. Although attempts were made in the early 1970s to qualify 138-kV XLPE cables employing higher operating stresses between 5 and 7 kV/mm, these were based on the use of the steam-curing process. It was not until cleaner insulating compounds and the dry-curing process came into use that such cables became generally accepted for high-voltage underground transmission.

In Europe, 400-kV XLPE transmission cables are now in use. Initially, these were trial installations of short lengths without joints. But more recently, a 22-km-long 400-kV line was installed in Denmark and commissioned in September 1997 [13], and a 7-km line is planned in Germany.

In Japan, the first 275-kV XLPE cable was installed in 1980 as shown in Fig. 4.1. With the successful qualification of an extrusion-molded joint in 1996, installation of a 40-km double-circuit 500-kV line is now underway in Tokyo with a scheduled completion date of November 2000 [14].

When compared to conventional oil-paper cables, the insulation thickness of a high-voltage XLPE cable is thicker, and thus it operates at a lower stress. For extra-high-voltage cables, which are generally rated above 150 kV, even higher stresses are required in order to limit the insulation thickness to a reasonable value.

Because extra-high-voltage cables often form the backbone of a transmission system, their reliability is of the utmost importance. Since such cables and their accessories are required to operate at higher electrical stresses than high-voltage cables rated below 150 kV, they have a smaller safety margin with respect to the inherent limits of the cable system. Likewise, such cables have a thicker insulation than high-voltage cables. Thus, thermomechanical effects become more important, especially when such cables are installed on bridges or in underground tunnels or duct and manhole systems. As for thermoplastic polyethylene cables, the reliability of crosslinked polyethylene cables is affected by the quality of the semiconducting shield interfaces and the cleanliness of the insulation. According to most experts, imperfections of more than 50 μm (2 mils) in size, which is the limit of current manufacturing technology, may be harmful, if the maximum electrical stress at the conductor shield interface exceeds 27 kV/mm (675 V/mil) [15]. In future it may be possible to exceed this value through the use of improved compounds and manufacturing processes.

In the same Japanese study [7] referred to in Section 4.2.2, it was shown that in 74% of all the cases the breakdowns originated from impurities in the insulation and in no cases did they originate from voids. Through the use of cleaner insulation compounds and improvements in manufacturing technology, it is now possible to design and manufacture XLPE cables for operation at increasingly higher operating stresses.

Table 4.5 will serve to illustrate the range of maximum electrical stresses now used around the world.

As will be discussed in Section 4.5, these values are only examples of current practice and should not be taken as recommended design levels. In consideration of accessory limitations, the maximum stress at the *conductor* shield interface may have to be reduced in order to limit the stress at the *insulation* shield interface to a more acceptable level. It should be noted that the maximum stress tends to decrease as the

TABLE 4.5 International Experience with Maximum Stress Designs

Rated Voltage kV	Maximum Stress					
	AEIC [3, 4]		NF [11, 12]		CIGRÉ [15]	
	kV/mm	V/mil	kV/mm	V/mil	kV/mm	V/mil
63			7	175		
69	6	150				
90			7	175		
115	8	200				
138	8	200				
150			7	175		
220					10	250
225			11	275		
275					12	300
345					13	325
400			16	400	14	350
500			16	400	15	375

ratio of the outside diameter D of the insulation to its inside diameter d gets smaller, that is, as the conductor size increases. The calculation of electric stress is explained in Section 3.2.3. Briefly,

$$E_{max} = \frac{V}{r \ln(R/r)} \quad \text{kV/mm} \tag{4.1}$$

where E_{max} is the maximum electric stress at the conductor shield interface in kilovolts per millimeter, V is the nominal line-to-ground voltage in kilovolts, R is the radius over the insulation in millimeters, and r is the radius over the conductor shield in millimeters.

In most countries, the design of (extra) high-voltage polymeric cables is usually based on maximum stress considerations. On the other hand, in Japan it is done on the basis of the average stress:

$$E_{mean} = \frac{V}{t} \quad \text{kV/mm} \tag{4.2}$$

where E_{mean} is the average electric stress across insulation in kilovolts per millimeter, V is the voltage in kilovolts, and t is the insulation thickness in millimeters.

The insulation thickness must be calculated on the basis of ac and impulse considerations using the following formula:

$$t = \frac{V k_1 k_2 k_3}{E_L} \tag{4.3}$$

where t is the required insulation thickness in millimeters, V is the voltage in kilovolts, k_1 is the temperature coefficient, k_2 is the degradation coefficient, k_3 is the allowance for uncertain factors, and E_L is the minimum breakdown strength for ac or impulse in kilovolts per millimeter.

Based on breakdown voltage tests carried out on model cables, the following were chosen as the design stresses for 500-kV XLPE cables [7]:

$$E_{Lac} = 40 \, \text{kV/mm} \qquad E_{Limp} = 80 \, \text{kV/mm}$$

The temperature and degradation coefficients were determined from an analysis of model and full-size cable test data after the application of ac and impulse voltages. The results of the various experiments lead to values [7] in Table 4.6.

By taking the higher of the insulation thicknesses determined under ac and impulse voltages and rounding up, a 25-mm insulation thickness was selected for 500-kV XLPE cables. The suitability of this insulation design was confirmed by manufacturing and

TABLE 4.6 Design Parameters for 500-kV XLPE Cable

Parameter	Alternating Current	Impulse
k_1	1.2	1.25
k_2	2.3	1.0
k_3	1.1	1.1

testing a 500-kV XLPE cable having a conductor size of 2500 mm^2 and an insulation thickness of 25 mm. As will be discussed in Section 4.5, the cable design must be coordinated with that of the accessories. When this was done, an insulation thickness of 27 mm was chosen for the cable [16].

4.2.3.3 Ethylene-Propylene-Rubber

Ethylene-propylene-rubber (EPR) is a term often used as the generic name for a broad family of compounds with very different formulations and, thus, different electrical and physical properties. Ethylene propylene copolymer (EPM) is a saturated polymer that exhibits excellent resistance to electrical discharges, water treeing, and deformation and also possesses good electrical and low-temperature properties. Because it can be crosslinked using peroxides, ethylene-propylene-diene (EPDM) terpolymer is now used. The typical physical and electrical properties of these compounds are shown in Table 2.13.

The use of EPR for high-voltage applications up to 150 kV evolved from its use at low and medium voltages. Because it can be formulated to give specific properties, it has been used in several applications because of its unique features.

In the early seventies, its excellent resistance to electrical discharges was discovered during the development of medium-voltage slip-on terminations. Today this has become an advantage when installing accessories on high-voltage EPR cables. As will be discussed in Section 4.5, the quality of the insulation shield interface may have a significant effect on the performance of accessories. However, EPR cables have proven to be resistant to discharges occurring along this interface as a result of minor surface imperfections.

The resistance of EPR to water treeing is well known and has been proven by extensive testing of medium-voltage cables [17]. Although so-called wet designs are used at lower voltages, metallic sheaths are generally the rule for high voltage. Because electric stress is known to be a contributing factor to the deterioration of any polymeric insulation in the presence of water, a radial moisture barrier of some form is desirable. Thus, lightweight shields such as copper or aluminum laminates can be used on EPR cables, even though they are not truly impervious.

Because of its rubberlike structure, EPR cables are more flexible than either LDPE or XLPE over a wide range of temperatures. This allows for a smaller bending radius during pulling and training into final position. This is especially important in generating stations and substations where space is often at a premium. Even in duct installations with several bends, longer lengths can be pulled, thus reducing the number of joints. Sections of 60- and 150-kV EPR cables have been pulled successfully in lengths up to 1500 and 1200 m, respectively [18].

At high temperatures, the mechanical properties of XLPE and EPR become quite different. As shown in Fig. 2.56, XLPE expands at a higher rate, which exceeds that of the EPR compounds, when its temperature approaches 110°C. Although the thermal expansion of XLPE and EPR is not linear over the entire temperature range, the coefficient of expansion for EPR compounds is approximately half that of XLPE. Thus, the thermal expansion of long lengths of EPR cable installed in ducted sections or on trays can be accommodated by simpler clamping arrangements.

Because the deformation resistance of XLPE drops sharply at temperatures above 100°C, caution must be used when establishing the emergency operating temperature [3]. On the other hand, high-voltage EPR cables can be operated at temperatures as high as 130°C. Long duration tests at elevated temperatures up to 150°C have shown no changes in the electrical or mechanical characteristics of 150-kV EPR cables [17].

Just as the thermomechanical properties of an EPR compound depend on its formulation, its electrical properties can be improved by the choice of the right formulation and proper compounding. All EPR compounds incorporate various types of fillers that lead to a noticeable increase in its dissipation factor, especially at elevated temperatures. Through the proper choice and treatment of fillers, optimization of the compounding process, and fine screening during extrusion, significant improvements have been made in the electrical properties of EPR. For example, the dissipation factor of a 132-kV EPR cable has been reduced from 0.0055 down to 0.0020 when measured at 110°C [17]. As a result, the dielectric losses amount to 0.5 W/m or 3% of the conductor losses at full load.

Because dielectric losses become increasingly important at higher voltages and are even more significant at voltages above 150 kV, EPR cables are not used for long feeders. Nevertheless, a trial installation of 220-kV EPR cables was made as early as 1984 in a Swiss substation.

As shown in Table 4.7, the dielectric constant and dissipation factor of EPR approaches that of impregnated paper. Thus, linear and crosslinked polyethylene are now preferred for extra-high-voltage transmission cables.

4.2.4 Metallic Shields and Sheaths

As discussed in Section 4.2.2, the extruded insulation shield layer must be grounded. This is most effectively done by the application of a metallic shield or sheath.

The simplest shield construction consists of one or more flat copper tapes wrapped helically around the cable core. In addition to fulfilling this shielding function, it is often used as the essential ground return path to carry the fault current in the event of a line-to-ground fault on the system. Thus, it must be able to carry this current for the time required to operate the overcurrent protective device without overheating and damaging the cable.

4.2.4.1 Tapes and Wires

Other constructions can be used in order to increase the short-circuit capability of the metallic shield. Concentric copper or aluminum wires applied helically is perhaps the simplest, as illustrated in Fig. 4.3. Another consists of a longitudinally folded and corrugated copper tape applied with an overlap under an extruded PVC or polyethylene jacket.

TABLE 4.7   Selected Properties of High-Voltage Insulating Materials

Material	Dielectric Constant	Dissipation Factor (%)	Thermal Resistivity (K-m/W)	Operating Temperature (°C)
LDPE	2.3	0.10	3.5	70
HDPE	2.3	0.10	3.5	80
XLPE	2.5	0.10	3.5	90
EPR	3.0	0.50	5.0	90
Oil-paper	3.5	0.30	5.0	85

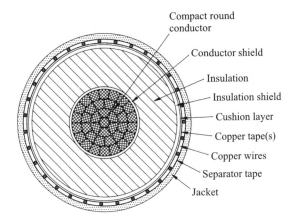

Figure 4.3 Cross section of solid-dielectric power transmission cable with concentric wire shield.

4.2.4.2 Laminate Shield

The shield constructions described above should only be used on high-voltage polymeric cables installed in locations known to be dry, such as the U.S. Southwest. Such cables must be either installed in a dry environment or protected against moisture ingress by a hermetic sheath.

Various countermeasures can be taken to protect the cable against water tree formation. Perhaps the most economical of these is a longitudinally folded copper, aluminum, or lead laminate shown in Fig. 4.4 with a sealed overlap and bonded to the polyvinyl chloride (PVC) or polyethylene jacket during extrusion. Where short-circuit requirements dictate, these may be used in combination with concentric copper or aluminum wires. If desired, longitudinal water blocking of the shield area can be achieved by the use of water-swellable powder or semiconductive tapes.

4.2.4.3 Lead Alloy Sheaths

For a truly impervious radial moisture barrier, an extruded lead alloy sheath represents perhaps one of the oldest sheathing materials used in the manufacture of power cables, as shown in Fig. 4.5. Because of its relatively low conductivity, concentric copper or aluminum wires or tapes may be used to obviate the need for an excessively thick sheath.

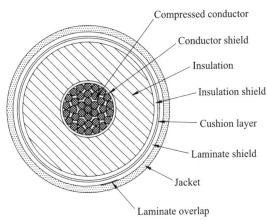

Figure 4.4 Cross section of solid-dielectric power transmission cable with laminate shield.

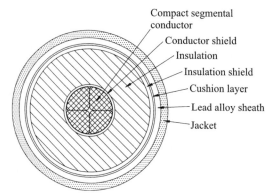

Figure 4.5 Cross section of solid-dielectric power transmission cable with lead sheath.

4.2.4.4 Corrugated Sheaths

As compared to lead alloys, aluminum sheathing is particularly attractive with respect to its lighter weight, greater short-circuit capability, and superior mechanical properties including fatigue resistance. Aluminum sheaths may be either extruded or longitudinally seam welded and then corrugated for enhanced flexibility, as shown in Fig. 4.6.

In some countries, corrugated copper [17] or stainless-steel [8, 19, 20] sheaths are used. Because of its conductivity, a copper sheath has the highest short-circuit capability when compared to any of the other sheathing materials. On the other hand, a stainless-steel sheath has the lowest eddy current losses. In a single-core cable carrying alternating current, the eddy current losses vary inversely as the sheath resistance, which in turn depends on its thickness and electrical resistivity. Even when compared to an aluminum sheath thickness of 3.2 mm, a 0.8-mm-thick stainless-steel sheath provides greater mechanical protection while reducing the eddy current losses of a 500-kV XLPE cable. It should be noted in passing that all three of the materials used as corrugated sheaths are environmentally friendly when compared to lead alloys for those applications that require a so-called green cable.

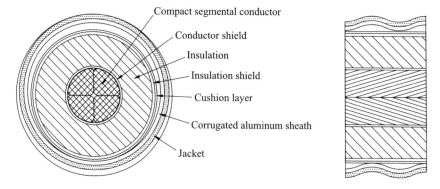

Figure 4.6 Cross and profile sections of solid-dielectric power transmission cable with corrugated aluminum sheath.

4.2.5 Protective Coverings

For all high-voltage power cables, the metallic shield or sheath is invariably protected by an extruded layer of either PVC or polyethylene (PE). This protective covering or jacket is required for a number of reasons.

In order to provide some degree of mechanical protection to the underlying shield or sheath, the jacket must be tough and resistant to deformation both during and after installation of the cable. For most installations, jackets are black in order to give them good resistance to sunlight.

Metallic shields and sheaths are susceptible to corrosion from a number of possible causes, as discussed in Section 2.7. In order to prevent such corrosion, an extruded covering must be applied. Although very few chemicals normally found in the ground along a cable route will have a significant effect on the jacket, care must be taken in the choice of jacket compound, when the cable is to be installed in a particularly corrosive environment. Table 2.19 compares the resistance of PE and PVC jackets to various chemicals.

Because of its susceptibility to corrosion, protection of the aluminum sheath on a transmission class cable is especially important. When the jacket is applied directly over an aluminum sheath, it is desirable to have a thin coating of bitumen or other flooding compound along this interface. If the jacket should be damaged for any reason, localized corrosion of the sheath may then occur. The flooding compound provides secondary protection by limiting penetration of the groundwater beyond the exposed area.

In order to verify the integrity of the jacket, a direct-current (dc) voltage test is normally carried out periodically to check that no damage has occurred. To facilitate application of the test voltage, a coating of colloidal graphite or some other conductive material is applied over the jacket during extrusion.

In a three-phase single-conductor cable system, sheath losses may be significant. They can be reduced by the use of a special bonding system. This involves grounding the cable sheaths at one point only and insulating all other points from earth. In this way, circulating sheath currents are prevented and losses are eliminated. As a result, the individual cables can be spaced further apart to reduce their mutual heating effect, and, thus, to increase their current rating.

Three basic variations of specially bonded systems are normally used: end-point bonding, mid-point bonding, and cross bonding. All such bonding schemes must of necessity be fully insulated. Thus, the cable sheath must be insulated from ground. This is done by having an extruded covering of PVC or PE on the cable sheath. Again, this jacket can be checked for damage by testing it regularly after the cable has been installed. In any of the special bonding arrangements, the jacket must have adequate dielectric strength to withstand not only the induced standing voltage on the sheath but also transient overvoltages arising from lightning or switching surges.

4.2.5.1 Polyvinyl Chloride

For several decades, PVC has been used as a jacketing material when extruded over the metallic shield or sheath of a cable. Because of its resistance to oils and most chemicals, it has found use in chemical plants and oil refineries.

The PVC compounds can be formulated for a particular application by the proper use of fillers, plasticizers, stabilizers, etc. Thus, there are a wide variety of such compounds. For cables installed in tunnels or on cable trays, flame-retardant compounds are available to limit flame spread. For cable installations in a cold environment, the low-temperature properties of a PVC compound can be enhanced by the incorporation of special plasticizers. Table 2.18 provides a comparison of the physical properties of a typical PVC jacketing compound with those of various polyethylenes.

4.2.5.2 Polyethylenes

Whereas PVC has been the established jacket compound in the United Kingdom, France, and Japan, polyethylene tends to be preferred in North America. Unlike PVC compounds, the only additions made to polyethylene are an antioxidant and carbon black. Because polyethylene does not suffer from the effects of brittleness at low temperatures, it has a particular advantage for installation in those countries where low temperatures exist for an extended period of time.

Thermoplastic polyethylenes of various densities are available for jacketing purposes as shown in Table 2.18. It is important to select one that has good resistance to environmental stress cracking, especially for HDPE, which is particularly notch sensitive.

For most applications, either LDPE or linear low-density polyethylene (LLDPE) is used for underground installations where flame retardancy is not a consideration. When cables are being installed at high ambient temperatures in tropical locations, HDPE is often used to minimize jacket damage during installation. Although greater mechanical protection is afforded by the use of HDPE, this comes with a trade-off in increased cable stiffness.

In those countries such as Australia, Malaysia, and Indonesia where ants and termites pose a threat to the cable, appropriate countermeasures must be taken. At one time, contact insecticides such as aldrin and/or dieldrin could be incorporated into a PVC jacket. While these proved to be very effective, increased concern about the toxicity of such additives precludes their use today.

Instead, very hard jackets such as HDPE are used either alone or in combination with a thin nylon coating. A variation on this design incorporates the nylon skin between an inner and outer HDPE jacket [20].

4.3 MANUFACTURING METHODS

All of the polymeric materials discussed in Section 4.2.3 have now been used in medium-voltage distribution cables for several years. The extension of their use to high- and extra-high-voltage transmission cables was made possible by the elimination of macrodefects occurring within the insulation. Such defects are now minimized by the use of cleaner insulating compounds, improved extrusion, and, in the case of crosslinked polymers, curing techniques. In order to reduce the insulation thickness of such cables and increase their reliability at higher voltages, microdefects must also be eliminated.

4.3.1 Compounding

For any polymeric material to perform satisfactorily in service, meticulous cleanliness is essential during both the compounding of the polymer and its extrusion. Thus, cable manufacturers have imposed high standards of cleanliness on their compound suppliers. Much greater care is now taken not only during compound manufacture but also during the material transfer process. It is transported from the compound supplier to the cable manufacturer in dedicated or nonreusable packages. As shown in Fig. 3.19, the compound is then unloaded from this package in a clean room having a controlled contaminant level such as Class 1000 of U.S. Federal Standard 209 [21, 22].

Other measures that may be taken include the use of special metal detectors to reject any pellets containing magnetic metals and extrusion of the compound through a fine-wire mesh. For example, in order to produce 500-kV XLPE cables with a permissible contaminant level of 50 μm, the insulation was extruded using a mesh of no more than 30 μm [23].

Unlike thermoplastic polyethylenes, crosslinked polyethylene and EPR require an additional compounding process. Instead of using precompounded insulating material from an outside supplier, some cable manufacturers use a totally closed system [24, 25]. The compounding plant is in-line with either a vertical continuous vulcanization (VCV) or Mitsubishi Dainichi continuous vulcanization (MDCV) extrusion line. In this way, the quality of the compound can be controlled from polymerization through to cable extrusion.

4.3.2 Extrusion

It is generally agreed that the most critical imperfections that may affect the life of a polymeric cable in decreasing order of importance are (1) protrusions at the conductor shield interface, (2) contaminants, and (3) voids. Protrusions at the conductor shield and even from the insulation shield of a high-voltage polymeric cable have been shown to be the point of initiation and propagation of electrical trees. When these trees bridge the insulation, cable breakdown occurs.

In addition to the use of supersmooth semiconducting compounds containing special carbon blacks, significant enhancements have been made in extrusion techniques. Initially a "pseudo-simultaneous" or dual-tandem triple extrusion process was used. The practice was to locate the first extruder applying the conductor shield on its own a few meters ahead of the dual head connected to the extruders applying the insulation and insulation shield. If the insulated conductor moved at all as it passed through the curing tube, the conductor shield surface was scraped by the dies of the dual head.

An improvement on this approach is the dual-dual configuration. The conductor shield and a thin layer of insulation approximately 1 mm in thickness are applied simultaneously in the first dual head. Then the main thickness of the insulation and the insulation shield are applied in the second dual head located a few meters down the line. In this way, any scraping that occurs at the entrance to the second head is immediately covered by the main insulation wall.

Although the dual-dual configuration has been used successfully in Denmark for the manufacture of 400-kV XLPE cables, the "perfectly simultaneous" or "true" triple extrusion technique is more commonly used elsewhere. As illustrated in Fig. 3.16, the conductor shield, insulation, and insulation shield are extruded simultaneously in a common crosshead. In order to prevent contamination while assembling and dismantling the extruders, these may also be totally enclosed within a clean room [14, 21].

4.3.3 Crosslinking (Curing) Methods

In the chemical process of crosslinking or vulcanization as it was called with the early rubber compounds, the modification of the polymer matrix in thermoset materials

such as XLPE and EPR is initiated by heating the core after extrusion. For the manufacture of high-voltage power cables, there are essentially three methods of continuous
vulcanization (CV).

At the time of extrusion, the thermoset materials are still thermoplastic, and the
only difference in technique is that of vulcanization. In the case of XLPE, for example,
the precompounded polyethylene, containing an antioxidant and a crosslinking agent
such as 2.5% dicumyl peroxide, is extruded onto the conductor at a temperature below
the peroxide decomposition temperature of 140°C. This is done in order to avoid
crosslinking during extrusion.

The insulated conductor then passes immediately into the heated curing zone of the
CV line where the heat required to raise the temperature of the insulation is provided.
Traditionally, saturated steam at high pressure was used. In this process, steam at high
temperature (200–210°C) and pressure (10–20 bars) diffuses into the insulation during
the crosslinking process. Because the solubility of water in polyethylene is so much
higher than at room temperature and under atmospheric pressure, the resulting insulation has a higher moisture content and many more microvoids.

The effects of these microdefects on the electrical performance of XLPE cables was
first recognized in the early 1970s and led to the development of various dry-curing
processes. As shown in Fig. 4.7, microvoids even smaller than those allowed by the
AEIC specifications are virtually eliminated.

Catenary continuous vulcanization (CCV) lines such as that shown in Fig. 3.19
have generally been preferred because the ratio of their output to capital cost is higher
than for vertical or horizontal lines. In any of these processes, the rate of crosslinking is
determined by the heating time and temperature. Since the surface temperature of the
conductor must be raised to 180°C to initiate crosslinking of the conductor shield layer,
CCVs with a catenary tube length in the order of 150 m have been used for the
manufacture of XLPE cables with large conductor sizes and thick insulations. By the
use of a low-melt index sag-resistant insulating compound and core rotation during
crosslinking, even 400-kV XLPE cables have been produced on a CCV line.

Whereas the curing and cooling tubes in a CCV line are arranged in a catenary
shape to match the natural sag of the cable core as it exits the crosshead at the top end
of the curing zone, the VCV process has the added advantage that sagging of the soft
extrudate due to gravity is precluded by housing the extruders at the top of a vertical
tower or shaft. In addition, no tension control system is needed to keep the cable core

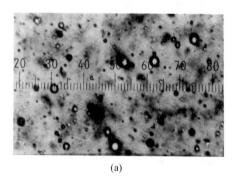
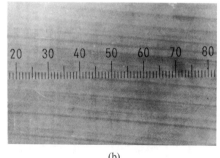

| (a) | (b) |

Figure 4.7 Photomicrographs of XLPE cable insulation: (a) conventional CCV
process; (b) MDCV process.

off the inner surface of a tube as is the case in a catenary line. On the other hand, VCVs as high as 100 m or more are required for the manufacture of XLPE cables rated 400 kV and above. The cost of erecting a VCV much higher than this may prove to be uneconomic.

For both the VCV and CCV, dry curing may be carried out in a number of ways. In the radiant curing process (RCP), crosslinking of the insulation takes place within a heated tube in a pressurized atmosphere of inert gas such as nitrogen [26]. Heat transfer is by thermal radiation from the walls of the tube onto the cable surface. The curing zone is divided into a number of sections that are heated either by a number of electrical elements attached to the outside of the tube or by circulating direct current through the tube itself. In a nitrogen atmosphere, the surface temperature of the core must be limited to 300°C to avoid decomposition of the insulation. At this higher curing temperature, the tube may be heated up to 450–500°C. Thus, it must be made from stainless steel or some other suitable metal.

Another approach to dry curing is the gas-curing system [27]. An inert gas such as nitrogen or sulfur hexafluoride is preheated to 300–450°C in a heat exchanger located outside the curing tube. Because the heat transfer coefficient by natural convection of an inert gas is so low, forced circulation of the high-temperature gas is utilized to overcome this shortcoming. It is also possible to use a fluid such as silicone oil as the heating and pressurizing medium [14] in order to improve the heat transfer.

A truly unique method of crosslinking is the long land die process depicted in Fig. 4.8 and often referred to as the MDCV process [24]. It is characterized by the use of a thin film of lubricant during extrusion to facilitate the smooth movement of the core inside the long land die, which is essentially an extension of the extruder die. Because the extrudate is in direct contact with the inside of the die, heat transfer is by thermal conduction rather than radiation or convection. As a result, the curing zone is relatively short and usually ranges from 12 to 18 m. Another characteristic of the MDCV process is the fact that it is a horizontal line. It is laid out to enable the conductor to travel forward without undergoing excessive bending. This makes it suitable for the manufacture of XLPE cables with large-sized conductors. The diameter of the core is precise, and it is always circular. Concentricity is ensured by the use of a low-melt-index XLPE compound and core rotation.

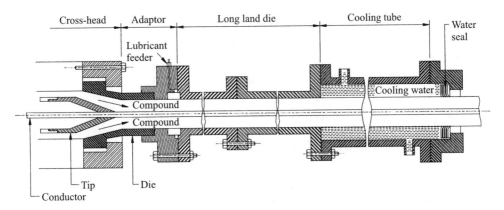

Figure 4.8 MDCV process.

Just as there are several methods for dry curing, some variation is also possible in the method of cooling. In a CCV line utilizing a silicone oil as the heat transfer medium in the curing zone, the same liquid is also used for cooling. For the other dry cure processes, cooling of the cable may be provided by the use of water in the cooling zone or by circulating nitrogen gas through an external heat exchanger. The latter is often referred to as a completely dry-curing and cooling (CDCC) process.

The use of water for cooling purposes is often questioned in conjunction with dry curing. As previously noted, the equilibrium moisture content of XLPE is highly temperature dependent. When compared to the temperatures normally employed for curing, the solubility of water in direct contact with polyethylene is approximately 2 orders of magnitude lower at room temperature. Thus, the moisture pickup during cooling is minimal.

Even when a CDCC process is used, the residual moisture content of the XLPE insulation is approximately 100–200 ppm. In addition to water vapor, the decomposition by-products of the crosslinking reaction include various gases including methane and ethane. For this reason, most manufacturers degas the cable prior to the application of an extruded metallic sheath or other radial moisture barrier. The core is left in an air oven or vacuum tank at elevated temperature for several hours in order to accelerate the diffusion of the volatiles out of the insulation.

4.4 TESTING

At the present time, extruded cable systems are covered by a range of tests from the initial development tests through to the final electrical tests after installation. Using the definitions proposed in the current extension of IEC Standard 60840 up to 500 kV [28], they may be classified as follows:

1. Prequalification tests are made on a complete cable system in order to demonstrate satisfactory long-term performance before supplying it on a commercial basis.
2. Type tests are made on a cable system in order to demonstrate satisfactory performance characteristics to meet the intended application before supplying it on a commercial basis. The term qualification test is often synonymous to type test.
3. Routine tests are made by the manufacturer on each manufactured component (length of cable or accessory) to verify that the component meets the specified requirements.
4. Sample tests are made by the manufacturer on samples of completed cable or components taken at a specified frequency in order to verify that the finished product meets the specified requirements.
5. Electrical tests after installation are made to demonstrate the integrity of the cable system as installed.

Each of these tests will be discussed in the following sections.

4.4.1 Development Tests

Development tests are normally carried out by the manufacturer before the prequalification tests. As the details and schedules of the development tests are not specified in industry standards, they are left to the discretion of the manufacturer and may, thus, differ widely between manufacturers. The CIGRÉ (Conference Internationales des Grands Réseaux Electriques) Working Group 21.03 of Study Committee 21 has suggested [29] that they should preferably include the following:

- Development tests are normally carried out by the manufacturer before the prequalification evaluation of materials and processes; for example, level of voids, contaminants, protrusions, etc.
- Evaluation of the Weibull parameters.
- Determination of the lifetime exponent.
- Tests on model and full-size cables.

4.4.2 Prequalification Tests

Once the above tests have been carried out, a prequalification test should be made as the final stage of the development process to gain some indication of the long-term reliability of the proposed cable system. This accelerated aging test only needs to be carried out once, unless there has been a significant change in the cable system with respect to materials, processes, design, and design levels.

For this test, approximately 100 m of full-size cable together with at least one of each type of accessory are tested in one loop. The test arrangement is meant to be representative of the actual site conditions. Thus, it may include design conditions such as duct, manholes, tunnels, and direct buried straight and in a bend. Figure 4.9 illustrates such a layout at the Institut de Recherché d'Hydro-Quebec (IREQ) [30].

The test voltage to be used for the prequalification test and the duration of its application must be adequate to confirm the life expectancy of the cable system and to reveal the effects of the relatively slow aging process, including insulation shrinkback, oxidation, corrosion, etc. Considering the possible existence of a threshold stress level, a

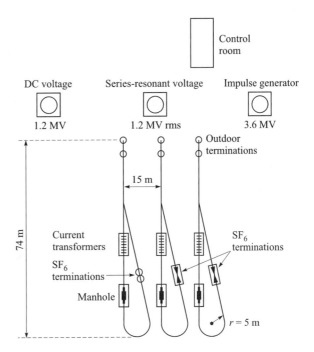

Figure 4.9 Installation scheme of cable test loops (after [30]).

line-to-ground test voltage of $1.7U_0$ is considered to be the upper limit with a corresponding test duration of at least one year. (U_0 is the rated rms power frequency voltage between each conductor and screen or sheath for which cables and accessories are designed.)

The test assembly is subjected to a number of daily heating cycles during which the conductor temperature is raised to its allowable maximum under normal operating conditions. This is usually done by circulating current through the conductor and, if need be, the sheath. The conductor is maintained at the specified temperature for 2–4 h and then allowed to cool naturally. For especially large-size conductors, it may not be practical to cool to ambient temperature within a reasonable time period.

Upon completion of the long-term ac test, a lightning impulse voltage test is then carried out on one or more cable samples to confirm that the ac test caused no harm. Because accessories may be damaged or disturbed during movement from the long-term test area to the laboratory, they are usually excluded.

Some test facilities such as those at IREQ and the Centro Elettrotechnico Sperimentale Italiano (CESI) are equipped to conduct such tests in situ [21, 30]. In this case, a switching impulse voltage test may also be carried out to qualify the entire cable system for those installations where, for reasons of insulation coordination, a demonstration of the switching impulse performance of both the cable and accessories is desired. This is usually only done for systems rated at 300 kV and above.

Table 4.8 summarizes the prequalification test protocol used for a typical 500-kV XLPE cable system [31].

4.4.3 Type Tests

Current industry standards specify the type, sample, and routine tests for cable systems rated up to 150 kV [2, 3, 32, 33]. However, the extension of IEC Standard 60840 up to 500 kV is now a new work item proposal [28] on the agenda of the International Electrotechnical Commission.

Type tests are made by the manufacturer before supplying on a general commercial basis a cable system to demonstrate satisfactory performance characteristics for its intended application. By their nature, these tests, once successfully completed, need not be repeated, unless changes are made to the materials, processes, or designs that might affect the performance characteristics.

TABLE 4.8 Parameters for 500-kV Prequalification Test

Test	Requirements
Heating cycle voltage test	500 kV ac phase-to-ground 8 h heating/16 h cooling Maximum conductor temperature 90–95°C One year duration (8760 h)
Lightning impulse voltage test	1550 kV ($1.2 \times 50\,\mu s$) Ten positive and negative impulses Maximum conductor temperature 90–95°C
Switching impulse voltage test	1175 kV ($250 \times 2500\,\mu s$) Ten positive and negative impulses Maximum conductor temperature 90–95°C

When the type tests have been successfully performed on one type of cable system with a specific value of rated voltage on one sample or two samples with different cross-sectional areas of conductors, the type approval is usually accepted as being valid also for the following:

(a) Cables with a design insulation thickness that gives a calculated maximum ac and impulse voltage stress at the conductor shield not higher than that type tested, and a calculated maximum stress at the insulation shield or in the main insulation part of the accessory not higher than that type tested.

(b) Cable systems in the same rated voltage class with all cross-sectional areas of conductor lying between the two on which the tests were made or cable systems according to (a). In this context, cable systems of the same rated voltage class are those of rated voltages having a common value of U_m, the maximum voltage that can be sustained under normal operating conditions at any time and at any point in a system, and therefore the same test voltage levels.

(c) Cable systems of similar construction, in the same rated voltage class and with cables having the same cross-sectional areas of conductor. Cable systems of similar construction are those with cables having the same type and manufacturing process of insulation material and the same forms of semiconducting shields and accessories of the same design. Repetition of the electrical type tests is not required on account of differences in the protective layers applied over the shielded cores, unless these are likely to have a significant effect on the test results.

The type tests consist of a combination of electrical tests on the completed cable system and nonelectrical tests on cable components and the completed cable. The electrical tests are normally carried out in the following sequence on one sample of the cable system:

(a) Bending test followed by partial discharge test at ambient temperature
(b) Dissipation factor measurement
(c) Heating cycle voltage test
(d) Partial discharge tests at ambient temperature and at elevated temperature
(e) Switching impulse voltage test
(f) Lightning impulse voltage test followed by a power frequency voltage test
(g) Tests of outer protection for buried joints

Table 4.9 summarizes the electrical type test requirements for a typical 400-kV XLPE cable system [34].

The nonelectrical type tests on cable components and the completed cable are summarized in Table 4.10, which indicates the tests applicable to each insulation and jacket material.

In addition to the tests shown in Table 4.10, a water penetration test is applicable to those cable designs that include barriers to longitudinal water penetration along the conductor and along the gap between the outer surface of the insulation shield and the radial moisture barrier. If the cable has a PVC jacket and the manufacturer wishes to claim that a particular cable complies with the requirements of IEC Standard 60332-1 [35], a test under fire conditions must be carried out on a sample of completed cable.

TABLE 4.9 Type Test Requirements for 400-kV XLPE Cable System

Test	Requirements
Bending test	Test cylinder diameter of 20 $(d + D) + 5\%$
	Three bending cycles
Partial-discharge test	Test voltage 460 kV
	Maximum discharge 5 pC
Dissipation factor measurement	Test voltage 230 kV
	Maximum conductor temperature 95–100°C
	Maximum dissipation factor 0.10%
Heating cycle voltage test	460 kV phase-to-ground
	8 h heating/16 h cooling
	Maximum conductor temperature 95–100°C
	20 heating/cooling cycles
Partial-discharge tests	Test voltage 460 kV
	Maximum conductor temperature 95–100°C
	Maximum discharge 5 pC
Switching impulse voltage test	1050 kV $(250 \times 2500\,\mu s)$
	Ten positive and negative
	Maximum conductor temperature 95–100°C
Lightning impulse voltage test	1425 kV $(1.2 \times 50\,\mu s)$
	Ten positive and negative
	Maximum conductor temperature 95–100°C
Joint protective covering	62.5 kV $(1.2 \times 50\,\mu s)$
	Ten positive and negative
	20 kV dc voltage
	One minute withstand
Crossbonding leads	125 kV $(1.2 \times 50\,\mu s)$ lightning impulse voltage test
	Ten positive and negative
	20 kV dc voltage
	One minute withstand

TABLE 4.10 Nonelectrical Type Tests for Insulating and Jacketing Compounds

Parameter	Insulation				Jacket	
	PE	HDPE	EPR	XLPE	PVC	PE
Mechanical properties						
(a) Without aging	×	×	×	×	×	×
(b) After aging in air oven	×	×	×	×	×	×
(c) After aging in air bomb	—	—	×	—	—	—
(d) After aging on complete cable	×	×	×	×	×	×
Pressure test at high temperature	—	—	—	—	×	×
Behavior at low temperature						
(a) Cold elongation test	—	—	—	—	×	—
(b) Cold impact test	—	—	—	—	×	—
Loss of mass in air oven	—	—	—	—	×	—
Heat shock test	—	—	—	—	×	—
Ozone resistance test	—	—	×	—	—	—
Hot set test	—	—	×	×	—	—
Measurement of density	—	×	—	—	—	—
Carbon black content	—	—	—	—	—	×
Shrinkage test	×	×	—	×	—	—

Note: × indicates that the test is applicable.

4.4.4 Sample Tests

In order to ensure compliance with specification requirements, sample tests are made on one reel length of cable from each manufacturing series of the same type and crosssection of cable. These tests include the following:

(a) Conductor examination
(b) Measurement of electrical resistance of conductor
(c) Measurement of thickness of insulation and jacket
(d) Measurement of thickness of metallic sheath
(e) Measurement of diameters
(f) Hot set test for XLPE and EPR insulations
(g) Measurement of capacitance
(h) Measurement of density of HDPE insulation
(i) Lightning impulse voltage test followed by a power frequency voltage test

The last test is only required for cables rated above 150 kV and is made once for each 20 km of cable core produced, or twice per year.

4.4.5 Routine Tests

Routine tests are carried out on each manufactured component (length of cable or accessory) to verify that it meets the specified requirements. Specifically, partial-discharge and voltage-withstand tests are carried out on each manufactured length of cable and on the main insulation of each prefabricated accessory to check that they comply with the requirements (Table 4.11).

For the voltage-withstand test on cables rated above 300 kV, the test voltage is lowered and the duration extended to avoid exceeding the threshold stress limit of 27–30 kV/mm [28]. This is done to avoid any possible weakening of the insulation that might later cause an in-service failure.

While the partial-discharge test is a well-established method of checking for voids within the main insulation of a cable or accessory, its effectiveness is limited for extra-high-voltage cables due to practical limitations of measurement sensitivity and

TABLE 4.11 Routine Electrical Tests

Rated Voltage (kV)	Withstand Test		Partial-Discharge Test (kV)
	Voltage (kV)	Duration (min)	
45–47	65	30	39
60–69	90	30	54
110–115	160	30	96
132–138	190	30	114
150–161	218	30	131
220–230	318	30	190
275–287	400	30	240
330–345	420	60	285
380–400	420	120	330
500	550	120	435

equipment availability. Because the partial-discharge test will only detect major defects, it is always used in conjunction with a voltage-withstand test.

For those installations where the protective covering or jacket serves to insulate the metallic shield or sheath from ground, a dc voltage withstand is normally carried out to verify its integrity. This is typically done by the application of 10 kV dc between the sheath and ground for jacket thicknesses of 2.5 mm or greater.

4.4.6 Electrical Tests after Installation

The purpose of this test is to check that the cable laying and accessory installation have been correctly made. The cable may, for example, have been accidentally damaged during shipping, handling, storing, pulling, and backfilling. Since it can be assumed that the cable insulation has not been damaged as long as the jacket is intact, it can be checked by a d.c. voltage-withstand test.

When it comes to the installation of the joints and terminations, an effective quality assurance program should always form the basis for the handover of the cable system to the customer. If a dielectric test of the main insulation is required, CIGRÉ Working Group 21.09 recommends [36] an ac voltage test in accordance with one of the following:

(a) Test for 1 h with the phase-to-phase voltage of the system applied between the conductor and the metallic shield or sheath.
(b) Test for 24 h without load at the normal operating voltage of the system.

The use of dc voltage tests is no longer recommended as they are both dangerous and ineffective. However, other tests such as oscillating voltage, very low frequency, and partial discharge are being evaluated for field use and may some day replace the ac voltage-withstand test.

4.5 ACCESSORIES

As shown in Fig. 4.1, polymeric-insulated cables were first used in short sections for the undergrounding of overhead transmission lines. At the ends of the cable circuit, terminations are required to make the connection to the other parts of the power system. As joints became available, such cables were then used for long-distance transmission.

The use of clean compounds, supersmooth shields, true-triple extrusion, and dry curing has enabled the use of polymeric cables up to 500 kV. But unlike the cable, the accessories must be assembled on site, often in inclement weather and in confined spaces such as tunnels and manholes. Thus, accessories have been developed to minimize the time required for their installation, to reduce costs, and to avoid contamination of the cable insulation during assembly.

Because accessories constitute an important part of the cable system, they must exhibit performance at least equal to that of the cable. Thus, they are always included in the prequalification and type tests on cable systems, as discussed in Section 4.4.

4.5.1 Preparations for Installation

Whereas the accessories used with oil-paper cables require the use of a fluid impregnant, polymeric cable accessories are generally much simpler. Nevertheless, the installation of such accessories still dictates a high standard of accuracy and cleanliness, even though the designs may incorporate various methods of eliminating imperfect shield interfaces, gaseous voids, and particulate contamination that may become the source of electrical discharges.

Despite the apparent simplicity of a polymeric cable, jointing and terminating such cables is a highly skilled process. Special precautions must be taken to ensure cleanliness and to enable visual examination of the prepared cable core. Depending on climatic conditions, the work area may also have to be heated or air conditioned. For polymeric cable accessories, visual examination upon completion is inadequate to confirm quality. In addition to an effective on-site quality assurance program, fitters are often required to complete a rigorous training program followed by a high-voltage test of the assembled joint or termination.

Because most polymeric transmission cables have a metallic sheath to exclude moisture and to provide a ground return path for the fault current, the cable sheath must first be vented to release any gaseous by-products of the crosslinking process, since some of these are flammable. If the cable core has been previously degassed at the factory prior to the application of the metallic sheath or radial moisture barrier, venting may not be required.

Extruded insulations such as LDPE, HDPE, and XLPE may exhibit a memory of the extrusion process. This is in the form of an elastic strain, which is released by cutting the cable, and a locked-in strain, which can be relieved by heating. To do so, the cable core is clamped and heated in a jig so as to straighten the cable and to encourage the insulation to shrink back. Even though the cable may have passed an insulation shrinkage test prior to shipment from the factory, this additional step is a worthwhile precaution, unless the joints and terminations have been especially designed to accommodate longitudinal retraction of the cable insulation.

For the preparation of the cable insulation, a set of special tools, which may be either hand-operated or motorized, is normally used. The bonded insulation shield is first removed with a stripping tool. In order to avoid leaving a step into the insulation at the end of the shield, which may become a point of stress concentration, the tool must compensate for any possible eccentricity of the cable core.

The insulation is then removed in order to expose the conductor. For prefabricated and premolded joints, a square cut may be used. Otherwise, the insulation is penciled for joints that are to be taped or field molded.

The semiconducting insulation shield is often removed by scraping it off with a piece of glass. The exposed insulation surface is then smoothed and polished by hand using successive grades of fine aluminum-oxide abrasive cloth. All traces of indentations and scratches along this interface must be removed, since they may lead to electrical discharges even under normal operating voltage. For high-stress cable designs, the performance of extrusion-molded joints (EMJ) has been shown [37] to be dependent on the smoothness of the insulation surface. To improve the electrical performance of the joint, a heat-shrink tube is applied over the insulation surface after finishing, and then it is heated and pressurized. Protrusion values in the treated portion of the insulation shield were found to be well below the quality control limit of 70 Tm [23].

The conductor connection is normally made by using a compression connector for copper conductors or metal inert gas (MIG) welding of aluminum conductors. The core is finally washed with a solvent and then visually examined prior to the application of the remaining components of the joint or termination.

4.5.2 Extruded Cable Terminations

Many of the termination designs used with high- and extra-high-voltage transmission cables evolved from those used on medium-voltage distribution cables. For example, heat-shrinkable terminations employing XLPE-based sleeves have been used up to 69 kV. Likewise, slip-on terminations using a factory-molded EPDM or silicone rubber stress control and modular sheds are in service up to 90 kV. Both of these dry designs are relatively inexpensive and do not require highly skilled fitters to install them.

The dry-type design for the termination of cable into gas-insulated switchgear (GIS) has been successfully type tested up to the 400-kV level [25]. It employs a pre-molded elastomeric stress control cone that is placed over the cable insulation and then inserted into an overall cast epoxy resin insulator. The stress cone is then kept under pressure by a mechanical spring arrangement to ensure good interfacial contact between the stress cone and both the cable insulation and the epoxy resin casting. Because no dielectric fluid or gas is required on the cable side of this termination, no pressure monitoring equipment or reservoirs are required.

The purpose of the stress cone in a cable termination is to control the electric field distribution beyond the end of the semiconducting insulation shield of the cable. It depresses the field into the cable insulation and thus significantly reduces both the radial and longitudinal components of the stress at the end of the insulation shield or the start of the stress cone as well as along the cable core interface. As illustrated by Table 4.4, the cable insulation thickness may be limited by the maximum stress not only at the conductor shield interface but also at the insulation shield. Thus, the use of a thinner cable insulation may be limited by the higher stress to which accessories would otherwise be subjected during operation.

For outdoor installations, the termination is housed in a porcelain insulator, as shown in Fig. 4.10 [31]. The insulation shield is terminated by a slip-on stress control cone, which is factory premolded from a rubber insulating compound. This may be based on chlorosulfonated polyethylene (CSP), EPDM, or silicone. For high-stress termination designs, a cylindrical capacitance-graded stress control cone may be used. It is factory preassembled into a polyethylene or impregnated paper roll laminated with imbedded aluminum electrodes to provide the necessary capacitive grading. This ensures a more uniform stress distribution along the length of the termination as compared to other designs. The porcelain insulator can be filled with either a high-viscosity silicone oil or SF_6 gas at low pressure (0.35 Mpa) [9, 38, 39]. If oil is used, an air space is provided above the filling oil to allow for thermal expansion. On the other hand, a means of maintaining and continuously monitoring the pressure is required, when a gas is employed.

For cable terminations into oil-filled or SF_6-insulated metal-enclosed equipment, a design such as that shown in Fig. 4.11 [31] is used. The cast epoxy resin insulator is the same as that employed for an oil-paper cable with an embedded electrode to prevent leakage of the SF_6 gas from the switchgear compartment into the cable termination. The insulator is filled with either a high-viscosity silicone oil or SF_6 gas. A method of pressurizing the filling medium must be provided.

Because of their large diameter, polymeric transmission cables are invariably of single-core construction. As discussed in Section 4.2.5, special sheath bonding arrangements are often employed to avoid induced sheath circulating currents and, thus, to increase the current-carrying capacity of the circuit. As with the cable sheath, the

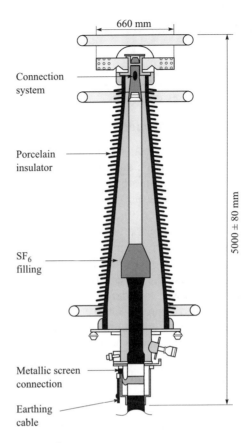

660 mm

5000 ± 80 mm

Connection
system

Porcelain
insulator

SF$_6$
filling

Metallic screen
connection

Earthing
cable

Figure 4.10 A 500-kV outdoor termination
of a solid-dielectric power transmission cable.

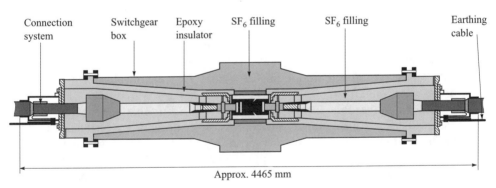

Connection system	Switchgear box	Epoxy insulator	SF$_6$ filling	SF$_6$ filling	Earthing cable

Approx. 4465 mm

Figure 4.11 Back-to-back GIS terminations.

termination must be insulated from ground. This is done by mounting outdoor terminations on support insulators and by incorporating a sheath sectionalizing insulator
into the entrance arrangement of GIS terminations.

As the cable conductor and sheath produce both tensile and compressive thermomechanical forces while in service, it is desirable to clamp the cable rigidly adjacent to
the terminations. Thus, fatigue failure of the metallic sheath is prevented and short-
circuit forces can be constrained.

4.5.3 Extruded Cable Joints (Splices)

As in the case of cable terminations, many of the joint designs used with power transmission cables evolved from those used on distribution cables. Heat-shrinkable joints using XLPE- based sleeves are relatively easy to install and provide an economical means of joining polymeric cables up to 69 kV.

4.5.3.1 Tape-Wrapped Joints

Taped splices have a long history of successful operation on distribution systems. Because of their versatile design, they have also found use at higher voltages. This type of joint can easily accommodate variations in cable dimensions and can thus be used to join two cables of different size.

Although a variety of splicing tapes are available, self-amalgamating EPR is generally used. By pulling the tape and stretching it to half of its original width, the self-amalgamating property required to bond one layer of tape to another is activated. This tensile force applied to the tape is translated into a cumulative radial pressure that serves to compress the splice inward. Thus, interfacial voids between successive tape layers are minimized, and deformation of the splice is avoided. Although joints can be hand wrapped, it is preferable to use a taping machine to improve speed and quality, because it is easier to control the tape tension and registration. This is especially true for 138 kV and higher voltage cables.

Such a splice is typically long (1.5–2.0 m) and up to twice the diameter of the cable itself. If the tapes are applied by hand, this will represent almost half the time required for splicing. Thus, it may take up to 24 h for a two- or three-man crew to complete a single 138-kV splice. The time required can be reduced and the quality much improved by machine taping. As a result, tape-wrapped splices are in service at 300 kV.

4.5.3.2 Field-Molded Joints

An alternative to the tape-wrapped joint is the field-molded joint in which the extrusion and crosslinking operations used during cable manufacture are reproduced in the field to reinstate and crosslink the joint insulation. Either crosslinkable XLPE or EPR tapes are wrapped around the cable ends. They are then heated and compressed in order to crosslink them.

This may be done in one of two ways. If a portable press with a split mold is used, additional insulation is injected by a small-screw extrusion press during the initial heating of the loosely wrapped tapes within the mold cavity. A simpler approach is to constrain the wrapped tapes in a mold. The mold is first heated to fuse the tapes to one another and to the cable. The temperature is then raised to initiate crosslinking of the tapes. Since gaseous by-products are released during this reaction, it is necessary to keep them in solution by accurately controlling the temperature and by the application of external pressure. This may be done by the use of a gas, liquid, or springs. To prevent the formation of contraction voids, the rate of cooling must also be carefully controlled.

As can be seen, this is a sophisticated process that requires a cycle of accurately controlled temperature and pressure for more than 12 h. As a result, the total time to make a 138-kV field-molded joint is almost 20 h. Because this technique is susceptible to contamination by airborne impurities during the taping operation, its use is generally limited to 150-kV class cables and lower.

Nevertheless, field-molded splices have several distinct advantages including the following:

1. They are essentially void free, since crosslinking takes place under pressure in a metal mold.
2. The length of insulation pencil can be relatively short, because there is a good bond developed between the cable and splice insulation.
3. As a result, they are quite compact and only slightly larger in diameter than the cable itself.
4. They represent only a small discontinuity in the cable system.

4.5.3.3 Prefabricated/Premolded Joints

Because of the limitations of the tape-wrapped and field-molded joints outlined above, prefabricated and premolded joints are well suited to installations in which working conditions are adverse, space is limited, installation time is important, and simplification of the splicing operation is required. Although the same degree of skill is required for the removal of the insulation shield and the polishing of the cable insulation, the steps of insulation penciling and taping can be eliminated. Instead, factory-molded and pretested insulation components are employed.

The distinction between the prefabricated and premolded joints is illustrated in Fig. 4.12 [30]. Both require a high degree of dimensional accuracy not only during manufacture of the cable and accessories but also while preparing the cable ends in the field.

The simplest design is a one-piece elastomeric molding, usually of EPDM or silicone rubber, which incorporates an internal electrode of semiconducting rubber embedded into the joint insulation as shown in Fig. 4.12(a). This electrode serves to shield the conductor connection and the square-cut cable ends. This design relies on the elasticity of the rubber to produce a radial compressive force onto the cable core. Because the large size of the molded component limits the radial stretch, a special tool is required. The tool stretches the joint body onto a hollow mandrel placed over the cable and then extracts the mandrel, so that the body exerts a compressive force onto the two cable ends. Because the cable sheath must be removed for at least twice the length of the molded unit, the overall length of the joint is increased.

A more complex design employs a rigid cast epoxy resin center portion with an embedded metal electrode as shown in Fig. 4.12(b). Flexible rubber stress control cones are used to reinforce the cable insulation at the end of the insulation shield. This permits stress cones with better range-taking capability to be fitted by hand without the need for special tools. The stress cones are maintained in intimate contact with the bore of the epoxy casting by a spring arrangement. Thus, thermal expansion and contraction of the cable insulation and compressive set of the rubber can be accommodated.

Both the premolded and prefabricated joints are installed in a metal casing as shown in Fig. 4.13 [31] to provide mechanical protection and prevent water ingress after installation. The casing is filled with a thermosetting resin or glass beads in order to limit movement of the joint and to improve radial heat transfer. The metal casing is insulated from earth by an outer protective covering, when special sheath bonding arrangements are employed.

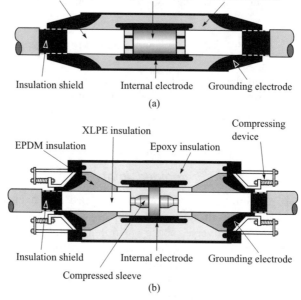

Figure 4.12 Types of joints: (a) slip-on premolded; (b) composite prefabricated (after [30]).

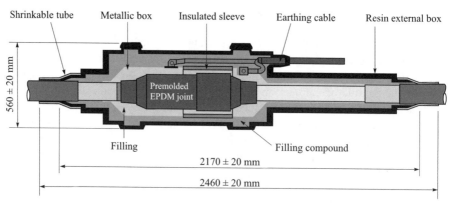

Figure 4.13 Sectionalized premolded joint for 500-kV XLPE power transmission cable.

Both joint designs have successfully passed the prequalification and type test requirements of the CIGRÉ recommendation for use with 400-kV XLPE power transmission cables [25, 38]. Additionally, the corresponding requirements for a 500-kV system have been met for the first time in the world by a premolded joint [31].

4.5.3.4 Extrusion-Molded Joints

The extrusion-molded joint is a variant of the field-molded joint discussed above. As shown in Fig. 4.14, the same type of resin as used for the cable is extruded on site into a mold using a small extruder, and the cable and joint reinforcing insulation are formed integrally. Thus the EMJ requires the use of cable manufacturing technology at the construction site.

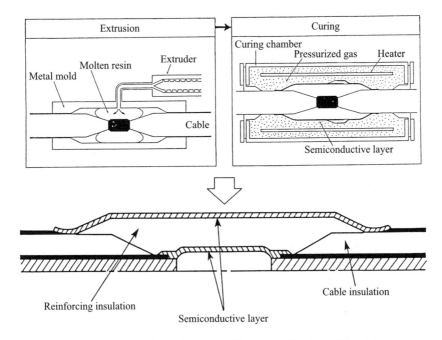

Figure 4.14 Extrusion-molded joint. (Courtesy of Sumitomo Electric.)

Following 2.5 years of basic research and 1.5 years of testing, the successful development of the extrusion-molded joint has allowed the application of 500-kV XLPE cable for long-distance power transmission lines in urban areas. It was necessary to have higher performance cable with reduced insulation thickness and to limit the overall diameter of the joint due to space constraints in tunnels and along bridges.

The factors governing the performance of an extrusion-molded joint were found to be voids and contaminants, especially in the joint reinforcing insulation [37]. This led to the development of the special techniques previously discussed for the treatment of the cable insulation surface following removal of the semiconducting insulation shield. As a result, the ac and impulse breakdown strength of the extrusion-molded joint were significantly improved. This allowed the adoption of a cable insulation thickness of 27 mm for 500-kV applications, identical to that previously used for 275-kV XLPE cables.

To ensure the high degree of reliability required for a 500-kV transmission line, a rigorous on-site quality control program must be followed during construction of the extrusion-molded joint. Preparation of the cable ends, extrusion of the joint reinforcing insulation, and finishing the outer semiconductive layer must all be carried out in a clean room. During the actual extrusion process, the compound is passed through a fine-wire mesh to screen out contaminants larger than 70 μm. In addition, a charge-coupled device (CCD) camera is used to detect any contaminants in the insulation resin that pass through a window section provided midway in the flow path of the resin. Finally, an X-ray examination is made after the crosslinking process [23]. As a result of the additional quality control checks required during the construction of an extrusion-molded joint to minimize contamination, the average total construction time for three

275-kV class extrusion-molded joints is 360 h, as compared to 150 h for the comparable prefabricated joints.

4.5.3.5 Transition Joints

For joining an extruded dielectric power transmission cable to an oil-paper cable, a transition joint is required. In principle, the design of such a joint is similar to that of the GIS termination shown in Fig. 4.11. A cast epoxy resin insulator with an embedded electrode is required to prevent leakage of the pressurizing medium from the oil-paper cable on one side into the polymeric cable on the other. Appropriate stress control cones must be used on both cable types. This back-to-back design has formed the basis for stop joints on oil-paper cables and early designs of 400- and 500-kV through-joints on polymeric cables [40].

4.6 CONCLUDING REMARKS

As illustrated by Fig. 4.1, the use of extruded dielectric cable for long-distance underground transmission lines was always limited by the lack of reliable joints. Thus, polymeric cables were initially used at increasingly higher voltages in power stations and for substation feeders, because suitable methods for terminating them were more readily available. Now that extrusion-molded and premolded joints have been successfully developed, the commercial application of 500-kV XLPE cable systems for long-distance transmission has been made possible. The material, design and manufacturing technology improvements that made this possible will also ensure the installation of more reliable and compact polymeric cable systems at lower transmission voltages.

REFERENCES

[1] L. H. Gross, "New high voltage XLPE insulation material," IEEE/PES Insulated Conductors Committee, Minutes of the 92nd Meeting, Appendix V-L, St. Pete Beach, FL, November 1992.

[2] IEC Publication 60840, "Tests for power cables with extruded insulation for rated voltages above 30 kV ($U_m = 36$ kV) up to 150 kV ($U_m = 170$ kV)," International Electrotechnical Commission, Geneva, Switzerland, 1993.

[3] AEIC Specification CS7-93, "Specifications for crosslinked polyethylene insulated shielded power cables rated 69 kV through 138 kV," AEIC, Birmingham, AL June 1993.

[4] AEIC Specification CS6-96, "Specifications for ethylene propylene rubber insulated shielded power cables rated 5 through 69 kV," AEIC, Birmingham, AL, April 1996.

[5] EDF Technical Specification HN 33-S-52, "Single-core cables with polymeric insulation for rated voltages of 36/63 (72.5) kV and 52/90 (100) kV (and up to 87/150 (170) kV)," Electricite de France, Paris, France, July 1988.

[6] EDF Technical Specification HN 33-S-53, "Single-core cables with polymeric insulation for rated voltages above 87/150 (170) kV and up to 290/500 (525) kV," Electricite de France, Paris, France, November 1992.

[7] T. Kubota, Y. Takahashi, S. Sakuma, M. Watanabe, M. Kanaoka, and H. Yamanouchi, "Development of 500 kV XLPE cables and accessories for long distance underground transmission line—Part I: Insulation design of cables," *IEEE Trans. Power Delivery*, Vol. 9, No. 4, 1994, pp. 1741–1749.

[8] M. Fukawa, T. Kawai, Y. Okano, S. Sakuma, S. Asai, M. Kanaoka, and H. Yamanouchi, "Development of 500 kV XLPE cables and accessories for long distance underground transmission line—Part III: Electrical properties of 500 kV cables," *IEEE Trans. Power Delivery*, Vol. 11, No. 2, 1996, pp. 627–634.

[9] E. Favrie, "225 kV and above XLPE transmission cables," IEEE/PES Insulated Conductors Committee, Minutes of the 98th Meeting, Appendix 15-D, St. Pete Beach, FL, November 1995.

[10] M. Pays, "Laminate sheaths for long life medium and high voltage cables," IEEE/PES Summer Meeting, Berlin, Germany, July 1997.

[11] French Standard C 33-252, "Single-core cables with polymeric insulation for rated voltages above 30 kV ($U_m = 36$ kV) up to 150 kV ($U_m = 170$ kV)," Union technique de l'Electricite, Paris, France, December 1993.

[12] French Standard C 33-253, "Single-core cables with polymeric insulation for rated voltages above 150 kV ($U_m = 170$) kV up to 500 kV ($U_m = 525$ kV)," Union technique de l'electricite, Paris, France, March 1995.

[13] B. S. Hansen, "420 kV XLPE cable project," IEEE/PES Insulated Conductors Committee, Minutes of the 100th Meeting, Appendix 13-D, St. Pete Beach, FL, November 1996.

[14] T. Kojima, "500 kV XLPE project for TEPCO," IEEE/PES Insulated Conductors Committee, Minutes of the 102nd Meeting, Appendix 13-D, St. Pete Beach, FL, November 1997.

[15] CIGRÉ, "Recommendations for electrical tests (type, special and routine) on extruded cables and accessories at voltages >150 (170) kV and ≤400 (420) kV," *Electra* No. 151, December 1993, pp. 21–29.

[16] T. Kubota, Y. Takahashi, T. Hasegawa, H. Noda, M. Yamaguchi, and M. Tan, "Development of 500 kV XLPE cables and accessories for long distance underground transmission line—Part II: Jointing techniques," *IEEE Trans. Power Delivery*, Vol. 9, No. 4, 1994, pp. 1750–1759.

[17] M. Laurent, "High voltage EPR cables up to 245 kV," IEEE/PES Insulated Conductors Committee, Minutes of the 99th Meeting, Appendix 5-A, Houston, TX, April 1996.

[18] M. Laurent, "15 years of experience with EPR cables rated 10 to 150 kV," Lecture presented at Nordic Wire & Cable Event, Göteborg, Sweden, September 1989.

[19] W. Torok, "Developments in Japan of up to 500 kV XLPE cables and joints," IEEE/PES Insulated Conductors Committee, Minutes of the 98th Meeting, Appendix 15-G, St. Pete Beach, FL, November 1995.

[20] H. A. Mayer, "Update on underground XLPE cable projects in Asia Pacific," IEEE/PES Insulated Conductors Committee, Minutes of the 102nd Meeting, Appendix 13-I, St. Pete Beach, FL, November 1997.

[21] S. La Pidus, "Berlin utility power company (BEWAG) 400 kV grid: status quo and future prospect," IEEE/PES Insulated Conductors Committee, Minutes of the 102nd Meeting, Appendix 13-H, St. Pete Beach, FL, November 1997.

[22] Federal Standard 209, Revision E, "Airborne particulate cleanliness classes in cleanrooms and clean zones," U.S. Government.

[23] K. Kaminaga, M. Ichihara, M. Jinno, O. Fujii, S. Fukunaga, M. Kobayashi, and K. Watanabe, "Development of 500 kV XLPE cables and accessories for long distance underground transmission line—Part V: Long-term performance for 500 kV XLPE cables and joints," *IEEE Trans. Power Delivery*, Vol. 11, No. 3, 1996, pp. 1185–1194.

[24] M. Okada, "MDCV process—a dry curing process for XLPE cable," Wire Association Technical Report, October 1976.

[25] M. D. Buckweitz, "Development and qualification of 400 kV XLPE insulated cable and accessories," IEEE/PES Insulated Conductors Committee, Minutes of the 98th Meeting, Appendix 15-B, St. Pete Beach, FL, November 1995.

[26] Y. Kawasaki, K. Otani, and H. Miyauchi, "Radiant curing process for manufacturing HV and EHV XLPE cables," Wire Association Technical Report, October 1976.

[27] T. Mizukami, N. Kato, N. Sato, K. Muraki, and K. Takahashi, "Inert gas curing system," Wire Association Technical Report, October 1976.

[28] IEC Committee Draft 20A/407/CD, "Tests for power cable systems. Cables with extruded insulation and their accessories for rated voltages above 150 kV ($U_m = 170$ kV) up to 500 kV ($U_m = 525$ kV)," International Electrotechnical Commission, Geneva, Switzerland, February 1999.

[29] CIGRÉ, "Recommendations for electrical tests (prequalification and development) on extruded cables and accessories at voltages >150(170) kV and ≤400(420) kV," *Electra*, No. 151, December 1993, pp. 15–19.

[30] J. -L. Parpal, "Prequalification tests of 345 kV XLPE cable system at IREQ," Conference Record of the 1996 IEEE International Symposium on Electrical Insulation, Montreal, Quebec, June 1996.

[31] J. -L. Parpal, "Prequalification testing of Alcatel 290/500 (525) kV extruded cable system at IREQ," IEEE/PES Insulated Conductors Committee, Minutes of the 102nd Meeting, Appendix 13-A, St. Pete Beach, FL, November 1997.

[32] IEEE Standard 48-1996, "IEEE standard test procedures and requirements for alternating-current cable terminations 2.5 kV through 765 kV," IEEE, New York, June 1996.

[33] IEEE Standard 404-1993, "IEEE standard for cable joints for use with extruded dielectric cable rated 5000-138 000 V and cable joints for use with laminated dielectric cable rated 2500–500 000 V," IEEE, New York, July 1994.

[34] H. T. F. Geene, "Development and qualification of 400 kV XLPE cable systems," IEEE/ PES Insulated Conductors Committee, Minutes of the 102nd Meeting, Appendix 13-E, St. Pete Beach, FL, November 1997.

[35] IEC Standard 60332-1, "Tests on electric cables under fire conditions. Part 1: Test on a single vertical insulated wire or cable," International Electrotechnical Commission, Geneva, Switzerland, 1993.

[36] CIGRÉ, "After laying tests on high voltage extruded insulation cable systems," *Electra*, No. 173, August 1997, pp. 33–41.

[37] N. Takeda, S. Izumi, K. Asari, A. Nakatani, H. Noda, M. Yamaguchi, and M. Tan, "Development of 500 kV XLPE cables and accessories for long distance underground transmission line—Part IV: Electrical properties of 500 kV extrusion molded joints," *IEEE Trans. Power Delivery*, Vol. 11, No. 2, 1996, pp. 635–643.

[38] J.-L. Parpal and J.-P. Crine, "Demonstration of the reliability of 345 kV XLPE cable and preloaded joints," IEEE/PES Insulated Conductors Committee, Minutes of the 98th Meeting, Appendix 15-C, St. Pete Beach, FL, November 1995.

[39] R. Rosevear, "XLPE insulated transmission cables 200 kV and above," IEEE/PES Insulated Conductors Committee, Minutes of the 98th Meeting, Appendix 15-F, St. Pete Beach, FL, November 1995.

[40] E. Favrie, "In service experience on prefabricated accessories in the field of 225 kV and higher voltages," IEEE/PES Insulated Conductors Committee, Minutes of the 98th Meeting, Appendix 13-D, St. Pete Beach, FL, November 1995.

Chapter 5

DESIGN AND MANUFACTURE OF OIL-IMPREGNATED PAPER INSULATED POWER DISTRIBUTION CABLES

W. K. Rybczynski[*]

5.1 BRIEF HISTORY OF DEVELOPMENT

Oil-impregnated paper insulated power cables, introduced many decades ago, still occupy a prominent place among cables for distribution and transmission of electrical energy up to the highest voltages. They have a longer service record than any other types available, and, if installed and maintained properly, they have a life expectancy much longer than that of more recently developed polymeric insulated types. Frequently, the mineral oil impregnants are replaced by other dielectric liquids, and for this reason oil-impregnated paper insulated power cables are commonly referred to also as dielectric-liquid-impregnated paper insulated power cables.

In their original form paper insulated power cables were composed of solid or stranded copper conductors, insulated with several layers of preimpregnated paper tapes, with two or three such insulated conductors assembled together with jute, or paper, fillers and fitted with layers of preimpregnated paper tapes applied over the cable assembly, forming an insulating belt. In order to protect the cable against the ingress of moisture, the assembly was then inserted into a lead pipe, which subsequently was swaged down, a method copied much later with the advent of aluminum sheaths. Such cables had to be manufactured in short lengths of about 7 m requiring an enormous amount of joints with the first major installation attributed to Ferranti in London in 1891, at a voltage of 10 kV to ground.

Later development of a lead press permitted cable manufacturers to produce longer lengths by a direct extrusion of the lead sheath over the cable assembly. Also the process of preimpregnating papers prior to their application has been replaced by an impregnating process of the entire cable assembly, which vastly improved insulating properties of the insulation, and permitted an increase of voltage rating of cables.

Belted-type cables, as described above, had an inherent weakness that came to light in cables designed for higher voltages. Poor performance of such cables operating with higher electrical stresses was attributed to the fact that in such cables not only was properly applied cable insulation on each conductor electrically stressed but so were also the spaces between convolutions in the multiconductor assembly, loosely filled

*Deceased.

either with crushed paper fillers or some other materials, by far inferior in electrical strength to the insulation proper.

This defect has been eliminated by the creation of shielded cables patented by M. Hochstadter in 1914, still known from its inventor to this day as type H cables. By the application of a conducting shield over each insulated core in the cable assembly and by providing a metallic binder over it, in close contact with the lead sheath, these loosely filled interstices were eliminated from the electric field and ionization of these spaces nullified. Later the same principle was applied to single-conductor cables in which larger conductor sizes and higher voltage ratings did not permit use of three-conductor design because of weight and overall diameter. Using modern manufacturing methods of the day, paper insulated lead-covered (PILC) cables were used for voltages up to 66 kV (phase-to-phase) as early as in 1932. Further increase in voltage rating of these cables known as solid-type PILC cables, required to satisfy ever growing demand for electrical energy, was not feasible because of the design limitations (low operating temperatures and limits in maximum electrical field gradients).

In order to increase dielectric strength of impregnated paper insulation of power cables, thus enabling them to be designed for higher voltage ratings, several new high-voltage cable types were developed and are given here in chronological order for future reference: (1920) low-pressure oil-filled cable by L. Emanueli in Italy; (1925) internal gas pressure cables by Atkinson and Fisher in the United States; (1931) gas compression cable by Hochstadter, Vogel and Bowden in England; (1931) oilostatic cable by C. E. Bennett in the United States; (1937) gas-filled cable by Beaver & Davey in England; (1936) impregnated pressure cable by Hunter & Brazier in England (developed in KEMA Laboratories in Holland); (1940) flat high-tension cable by J. S. Mollerhoj in Denmark; (1943) high-viscosity gas pressure pipe-type cables developed by General Cable for Detroit Edison; (1946) high-pressure oil-filled cable by M. L. Domenasch in France; (1950) Pirelli gas-filled cable by L. Emanueli in Italy [1–3]. Other gas-employing types, such as the gas-filled cable by Shanklin and the cushion gas cable by Dunsheath were simply derivitives of the gas-filled and impregnated pressure cable, respectively, and, as such are not included in the list.

As ever increasing transmission system voltages required high-voltage cables to be designed for higher and higher electric stresses, limitations imposed on gas-filled and gas compression cables because of corona discharge phenomena restricted further development of these cables, leaving the development field solely for oil-employing paper insulated cables. Only in some special cases, such as, for example, for very long submarine cables, gas-filled cables are still being considered because of an obvious attraction of an ease of feeding from compressed gas reservoirs.

5.2 ELEMENTS OF SOLID-TYPE OIL-PAPER CABLE DESIGN

In North America, the design procedures of solid-type oil-paper power cables are given in specifications issued by the Canadian Standards Association (CSA) and the Association of Edison Illuminating Companies (AEIC) [4, 5]. Although the two specifications differ in format, both give similar results regarding the cable design with an exception that the Canadian standard, because of the time lag, is based on a previous American edition already replaced by the present 10th edition, issued in 1926, in which insulation thicknesses were reduced by 4–6%, making paper-insulated cables slightly lighter and more competitive. A new edition of the CSA standard is based on this AEIC 10th edition; however, until its issuance, only cables made to the existing CSA standard were eligible for CSA certification and were thus specified by Canadian users.

5.2.1 Voltage Rating

Most, if not all, impregnated paper insulated power cables are used in related three-phase alternating-current (ac) systems, and therefore the cable rated voltage term always refers to phase-to-phase voltage in a three-phase system. This definition also applies to single-conductor cables, although in single-phase cables such voltage never appears and the cables are working at the phase-to-ground voltage. Cable rated voltages are related to the system voltage, as shown in Table 5.1, which is based on the system voltages encountered in North America. Only voltages above 5 kV have been included here as paper-insulated cables of lower voltage ratings are almost extinct in Canada.

5.2.2 Insulation Levels

Phase-to-phase cable rated voltage V does not define the power cable design voltage unless the related voltage V_0 between each conductor and sheath is also specified. In Europe and in international specifications the cable rated voltage is expressed by a fraction V_0/V, whereas in Canada this ratio is hidden in information whether the system is operating with solidly grounded, or ungrounded, neutral. Figures 5.1–5.3

TABLE 5.1 Maximum Continuous Voltage Versus System Voltage

System Voltage (Volts)		Cable Voltage (Volts)	
Nominal	Other Designations for Equivalent Systems	Recommended Rated Voltage	Maximum Continuous Voltage
7200	6600, 6900, or 75,000	8000	8300
4800/8320Y	8000	9000	9500
12,000	11,000 or 115,000	15,000	15,800
7200/12,470Y	11,450, 11,950, or 12,000	15,000	15,800
7620/13,200Y	—	15,000	15,800
13,200	13,500 or 13,800	15,000	15,800
14,400	13,200, 13,800, or 14,000	15,000	15,800
23,000	22,000 or 24,000	25,000	26,300
25,000	—	27,000	28,400
27,600	26,400	30,000	31,500
34,500	33,000 or 36,000	37,000	38,900
46,000	44,000 or 48,000	46,000	48,300
69,000	66,000 or 72,000	69,000	72,500

Notes:
1. Above table is taken from Appendix C to CSA Standard C68.1.
2. The data in the first and second columns are taken from ASA Standard C84.1.
3. Ratings in the third column are based on that report with the following considerations:
   - The design voltage be 5% greater than the rated voltage and
   - The maximum tolerable zone voltages specified for the system voltages should not exceed the maximum continuous operating cable voltages shown in the fourth column.

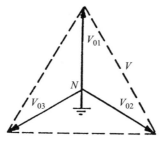

Figure 5.1 Solidly grounded neutral (GN) system; V_0 always equals $V/\sqrt{3}$.

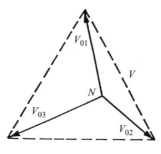

Figure 5.2 Ungrounded neutral (UN) system, delta or wye; V_0 may be greater than $V/\sqrt{3}$.

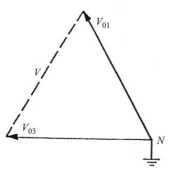

Figure 5.3 System operating with one phase grounded; $V_{0_1} = V_{0_3} = V$ and $V_{0_2} = 0$.

illustrate this point. In the systems with solidly grounded neutral (GN) (cf. Fig. 5.1), the phase-to-ground system voltage V_0 always equals $V/\sqrt{3}$; hence,

$$\frac{V_0}{V} = \frac{1}{\sqrt{3}} = 0.58 \tag{5.1}$$

Accidental grounding of one of the phases will trip the system momentarily. In delta or wye with ungrounded neutral (UN) (cf. Fig. 5.2), the phase-to-ground voltages V_0 may not be equal in each phase, exceeding $V/\sqrt{3}$. Besides such system may operate for a limited period of time with one phase accidentally grounded (cf. Fig. 5.3). In order to safeguard the cable against such occurrences, it has arbitrarily been decided to reinforce cable insulation, increasing the cable rated voltage by 33%. Therefore, for an ungrounded neutral system the ratio

$$\frac{V_0}{V} = \frac{1.33}{\sqrt{3}} = 0.77 \tag{5.2}$$

As the above differentiation still used in Canada presents some problems in cases when a neutral point is not solidly grounded but is connected to ground through a resistance, impedance, or Peterson coil, a concept of insulation levels has been introduced in American cable specifications; this also has gained acceptance in Canada.

Three insulation levels have been established replacing the orthodox grounded/ungrounded neutral concept. These are the 100, 133, and 173% levels. The 100% insulation level is designated by the nominal phase-to-phase system voltage. This is applicable only to systems where the normal voltage between the cable conductor and the insulation shielding tape, or metal sheath, will not exceed 58% of the phase-to-phase voltage. These cables may be applied, when the system is provided with relay protection, such that ground faults will be cleared as rapidly as possible, but in any case within 1 min. These cables are applicable to the great majority of cable installations, which are on grounded neutral systems; in addition, they may be used also on other systems for which the application of cables is acceptable, provided the above clearing requirements are met in completely deenergizing the faulted section. The 133% insulation level corresponds to that designated for undergrounded systems. Cables in this category may be applied when the clearing time requirements of the 100% level category section will be energized in a time not exceeding 1 h. Also they may be used when additional insulation strength over the 100% level is desirable. Cables of the 173% insulation level designation should be applied on systems where the time required to deenergize a grounded section is indefinite. Their use is recommended also for resonant grounded systems.

Table 5.2 shows the relationship between cable rated (phase-to-phase) voltage V and corresponding phase-to-ground voltage V_0 resulting from various insulation levels on which the cable design is based. Table 5.2 demonstrates that, for example, a 15-kV rated cable for a 173% insulation level is identical with a 25-kV rated cable for a 100%

TABLE 5.2 Cable Phase-to-Ground Voltages Expressed in Terms of Insulation Levels

Cable Rated Voltage V (Phase-to-Phase)	Corresponding Phase-to-Ground Voltage V_0 for Insulation Level of		
(kV)	100%	133%	173%
8	4.6	6.2	8.0
9	5.2	6.9	9.0
15	8.7	12	15
25	15	19	25
27	16	21	27
30	17	23	30
35	20	27	35
37	22	29	37
46	27	35	(46)
69	40	(53)	(69)

Notes:
1. Grounded neutral rating corresponds to 100% insulation level.
2. Ungrounded neutral rating corresponds to 133% insulation level.
3. Values in parentheses denote levels beyond the range for solid-type paper insulated cables.

insulation level, both designed for $V_0 = 15\,\text{kV}$. Also, a 35-kV (UN) rated cable is in fact a 46-kV (GN) rated cable. It follows that it is far more economical to provide a fast relay protection in the system and thus be able to use cables rated for 100% insulation level than to embark on the use of much more costly cables suitable for 133% insulation level applications. The 173% insulation level (one phase grounded indefinitely) is almost never used in practice and should be avoided wherever possible.

5.2.3 Selection of Conductor Size

The size of the conductor, or conductors, in a paper insulated power cable, as well as in any other type of power cable, depends on the required ampacity and may be obtained from the ampacity tables. In North America the main sources of information are the Insulated Cable Engineers Association [ICEA, formerly IPCEA (International Power Cable Engineers Associated)], the work of Neher and McGrath, and the International Electrotechnical Commission (IEC) [6, 7–10].

In order to enable the user to make necessary adjustments in case of parameters not covered by the cited ampacity tables, it may be of advantage to explain briefly basic physical laws on which the ampacity calculations are based [11]. The main factor governing cable ampacity is the temperature rise in the conductor(s), which is limited by the maximum conductor operational temperature, experimentally established in order to protect the dielectric against premature aging that would shorten the life of the cable. Table 5.3 lists these temperatures taken from the relevant North American standards covering solid-type paper insulated cables [4, 5].

It should be noted that the substantial difference in maximum temperature values in CSA standard and AEIC specification is only apparent. This may be seen from notes (a) and (b) to Table 5.3, which state that CSA values for normal operation may be increased by 10°C when the adequate knowledge of the overall thermal characteristic of the cable environment is available, whereas AEIC values are based on such a knowledge, otherwise advising reduction up to 10°C. Consequently, in theory, with minor exceptions, both standards give similar results regarding maximum temperature limits. However, in practice, the Canadian users tend to ignore the footnotes in the tables, accepting temperatures as shown and penalizing themselves by not applying the provision of the 10°C increase in temperature rating permitted by the CSA standard, even if such an increase is warranted. The new edition of CSA Standard C68.1 will eliminate this discrepancy.

The temperature rise permitted in the conductor is the difference between the maximum conductor temperature for normal operation T_c and the ambient temperature T_a, which in fact is the conductor temperature of a nonenergized cable in situ. In the determination of cable ampacity, limited by the temperature rise, the method of calculation is based on the fundamental theory of heat transfer in the steady state. In its simplest form this theory may be presented in the following way. The current passing through the cable generates heat due to conductor losses, dielectric losses, and the sheath losses. In medium-high-voltage cables dielectric losses may be neglected as having a minor effect, and the sheath losses only in large single-conductor cables are of an important magnitude. Therefore for simplicity only conductor losses will be considered.

The total heat generated in a cable per unit length H may be represented by the equation

TABLE 5.3 Maximum Conductor Temperatures

Rated Voltage (Phase-to-Phase), V (kV)	Maximum Conductor Temperature in Normal/Emergency Operation,[a] °C	
	CSA Standard C68.1[b]	AEIC Specification No. 1-68[c]
Multiple-Conductor Belted Cable		
1	85/105	95/115
2–9	80/100	90/110
10–15	75/95	80/100
Multiple-Conductor Shielded and Single-Conductor Cable		
1–9	85/105	95/115
10–17	80/100	90/110
18–29	75/95	90/110
30–39	70/90	80/100
40–49	65/85	80/100
50–59	60/75	65/80
60–69	55/70	65/80

[a] Emergency temperatures are based on the following assumptions: (1) their application will not exceed three emergency periods in any 12 consecutive months nor a maximum of one emergency period per year averaged over the life of the cable; (2) the maximum duration of each emergency period will not exceed 36 h.

[b] CSA standard recommends to use values for normal operation for ampacity calculation under assumption that hot spots may exist with temperatures exceeding those values by as much as 10°C. Temperatures 10°C above these tabulated values may be used for ampacity calculations when adequate knowledge of the overall thermal characteristics of the cable is available.

[c] AEIC specification advises that, if adequate information on the overall thermal characteristics is not known and the maximum temperatures shown in the table may be exceeded, these temperatures should be reduced by 10°C or in accordance with available data.

$$H = nI^2 R_{ac} \qquad (5.3)$$

where I is the current per conductor in amperes, n is the number of conductors in the cable, and R_{ac} is the conductor ac resistance in ohms per unit length at operational temperature T_c, including the skin and proximity effects. The generated heat must flow through the dielectric, sheath, outer protection, and surrounding medium, raising the temperature of these elements, until the steady state is attained. This flow of heat, caused by the difference of temperatures between conductor(s) and ambient, may be compared to an electric current flowing through a series of resistances under the influence of potential difference. Therefore, by analogy, the thermal resistance may conveniently be expressed in thermal ohms, this unit being the thermal resistance of a path through which a temperature difference of 1°C produces a heat flow of 1 W. Accordingly, the thermal Ohm's law may be written as

$$T_c - T_a = H \bar{R}_{ca} \qquad (5.4)$$

where T_c is the conductor temperature in degrees Celsius, T_a is the ambient temperature in degrees Celsius, H is the total heat dissipated per unit length of cable in watts, and $\bar{R}_{ca}$ is the thermal resistance per unit length between conductor and ambient in thermal ohms.

In the steady state the total heat generated in the cable must equal the total heat dissipated. Hence from Eqs. (5.3) and (5.4),

$$nI^2 R_{\mathrm{ac}} = \frac{T_c - T_a}{\bar{R}_{\mathrm{ac}}} \tag{5.5}$$

and the cable ampacity is thus derived as

$$I = \left[\frac{T_c - T_a}{n R_{\mathrm{ac}} \bar{R}_{ca}} \right]^{1/2} \tag{5.6}$$

where I is the cable ampacity (current-carrying capacity) in amperes, $T_c + T_a$ is the maximum permitted conductor temperature rise, and R_{ac} is the conductor ac resistance at temperature T_c per unit length of cable in ohms. It is not important which units are used for the length of cables in these equations, as long as same units are used for R_{ac} and $\bar{R}_{ca}$. For example, in North America 1 ft is used as a reference unit whereas in Europe 1 cm is employed. With the adoption of the SI (Système International) system of units, 1 m has become the measure of reference.

The thermal resistivity is expressed in thermal ohms, which by definition is the temperature drop in degrees Celsius produced by the heat flow of 1 W between opposite faces of a centimeter cubed of the material. Consequently, the thermal resistivity is expressed in degrees Celsius-centimeter per watt. If expressed in degrees Celsius-meters per watt, the value of thermal resistivity will be 100 times less. For example, for impregnated paper insulation the thermal resistivity ρ is equal to 600°C cm/W or 6.0°C m/W. Equation (5.6) demonstrates the basic relationship between the permissible temperature rise of the conductor, conductor ac resistance at operational temperature, thermal resistance of the heat path, and the resulting ampacity. From this equation the ac conductor resistance of the cable for a given ampacity, and from it the conductor size, may easily be found if the thermal resistance $\bar{R}_{ca}$ is known. This thermal resistance is composed of two basic components, namely the thermal resistance from the cable conductor(s) to the cable outer surface $(\bar{R}_{co})$ and the thermal resistance between the cable surface and the ambient in which the temperature T_a prevails, from where there is no further flow of heat $(\bar{R}_{oa})$. The thermal resistance R_{co} depends on cable geometry and on thermal resistivities of nonmetallic layers encountered by the heat path from the conductor to the cable surface, whereas the thermal resistance $\bar{R}_{oa}$ depends on the type of installation (duct, direct, buried, air), including such data as type of backfill and duct material, thermal soil resistivity, number of cables in a duct bank, or directly buried location, their physical separation, etc. In cables installed in air, it is acceptable to ignore mutual heating from other cables, if their separation is at least one cable diameter and the free air circulation is ensured. While the calculation of $\bar{R}_{co}$ is fairly simple, the problem of heat dissipation in the surrounding medium and method of calculation of $\bar{R}_{oa}$ is a complicated one.

Early theories (ca. 1900, Germany) considered the effect of earth surrounding the cable equivalent to that of a cylinder of material equal in resistivity to that of soil having a radius equal to the depth of the cable axis below the surface of the ground. More recent theories assume that the entire heat, dissipated from an underground cable, ultimately reaches the earth surface, and this theory has been exploited first by D. M. Simmons in 1932 and then fully developed by J. H. Neher and M. H. McGrath [8]. The work of the latter, published in 1957 on which the IPCEA power cable ampacities are based, is to this day the main basic work on which calculations of cable ampacities are based.

To select conductor size from Eq. (5.5), the ratio R_{ac}/R_{dc} must be known. This ratio depends on skin and proximity effects; hence, it changes with the conductor size and is more pronounced in three conductor cables than in single-conductor cables. This is demonstrated in Table 5.4 which gives the approximate ac/dc resistance ratio calculated for various conductor sizes at power frequency of 60 Hz. As in most specifications the dc resistances for various conductor sizes are given at ambient temperatures; it is essential to be able to convert the value R_{dc} at operational temperature to that at ambient temperature (usually at 20 or 25°C). The conversion calculation is based on the relationship

TABLE 5.4 Calculated ac/dc Resistance Ratios

Conductor Size		R_{ac}/R_{dc} Ratios for Single-Conductor Cables		R_{ac}/R_{dc} Ratios for Three-Conductor Cables (Concentric and Sector Strand)
AWG or MCM (kCM)*	Equivalent (mm²)	Concentric Strand	Segmented Strand	
4 or less	21.15	1.000	—	1.000
2	33.63	1.000	—	1.010
1	42.41	1.000	—	1.010
1/0	53.48	1.000	—	1.020
2/0	67.49	1.000	—	1.030
3/0	85.03	1.000	—	1.040
4/0	107.2	1.000	—	1.050
250	126.7	1.005	—	1.060
300	152.0	1.006	—	1.070
350	177.3	1.009	—	1.080
400	202.7	1.011	—	1.100
450	228.0	1.014	—	1.115
500	253.4	1.018	—	1.130
750	380.0	1.039	—	1.210
1000	506.7	1.067	1.018	
1250	633.4	1.102	1.028	
1500	760.1	1.142	1.039	
1750	886.7	1.185	1.053	
2000	1013	1.233	1.059	
2500	1267	1.326	1.102	
3000	1520	1.424	1.141	
3500	1773	1.513	1.182	
4000	2026	1.605	1.225	

*A MCM (kCM), thousand circular mils.

$$\frac{R_{T_1}}{R_{T_2}} = \frac{T_0 + T_1}{T_0 + T_2} \tag{5.7}$$

where for an annealed copper conductor $T_0 = 234.5°C$ and for an aluminum conductor $T_0 = 228.0°C$; R_{T_1} and R_{T_2} are conductor resistances at temperatures T_1 and T_2 respectively.

For very high voltage shielded cables, which usually are single conductor because of weight, it is necessary to add dielectric losses to the total heat generated in the cable. Hence Eq. (5.3) becomes

$$H = I^2 R_{\text{ac}} + P_d \tag{5.8}$$

where P_d is the dielectric loss per unit length in watts and is given by

$$P_d = \omega C V_0^2 \tan \delta \tag{5.9}$$

where ω is the angular frequency of the system, C is the cable capacitance in farads per unit length, V_0 is the system voltage to ground in volts, and $\tan \delta$ is the loss tangent of the insulation. The dielectric losses are distributed throughout the dielectric, but for simplicity they may be assumed to be concentrated at the conductor surface to be on the safe side. Hence from Eqs. (5.4) and (5.8) we have

$$I = \left(\frac{T_c - T_a - P_d \bar{R}_{ca}}{R_{\text{ac}} \bar{R}_{ca}}\right)^{1/2} \tag{5.10}$$

From Eq. (5.10) it may be noted that the influence of dielectric loss on ampacity is somewhat analogous to a reduction in permissible conductor temperature T_c.

As the dielectric loss is proportional to the cable capacitance, it therefore depends on the conductor size. Hence to obtain a correct answer, a trial-and-error method must be employed. Fortunately all exact calculations are essential only for very high voltage cables, beyond the voltage limits for solid-type paper insulated cables, for which the use of IPCEA ampacity tables would permit the selection of conductor size with adequate accuracy. It is not feasible within the scope of this chapter to fully exploit the problems involved in the calculation of cable ampacity, and anyone who would like to pursue this matter further should seek guidance in the literature [8, 9] and Chapter 11.

5.2.4 Selection of Conductor Material and Construction

Conductors of solid-type oil-paper insulated power cables may be either of annealed copper or aluminum. As the relative conductivities of annealed copper and aluminum are 100 and 61% respectively, it follows that for the same ampacity [i.e., same value of R_{ac} in Eq. (5.6)] the area of aluminum conductor must be increased to 161% of that of copper, equivalent to an increase in diameter by about 27%. Consequently, in general, for the same ampacity, aluminum conductor cables must be two sizes larger; hence, the economy in the cost of metal is offset by the increased cost of other components of the cable.

As nowadays splicing of aluminum conductors in power cables does not present any problems because of improved splicing technique with the use of suitable fluxes, the right choice of conductor material is a matter of economics. The choice of aluminum will always be a correct one in the case of high-voltage cables, which, for reasons of limitation in maximum electrical stresses, must have a certain minimum conductor size for a given voltage rating, irrespective of conductor material. For example, the minimum conductor size for 46 kV (GN) rated cable, permitted by the CSA relevant standard, is 4/0 AWG (American Wire Gage); hence, an aluminum conductor cable should be considered. On the other end of the scale, very large cables must be of copper because for manufacturing reasons it may not be practical to increase the conductor size.

Regarding stranded conductor construction of single-conductor cables, the choice may be made between concentric and compact round stranded conductors. Concentric stranded conductors are composed of several layers of round wires, each layer helically wound in opposite direction (see Fig. 5.4), whereas compact conductors are usually stranded in one direction, each layer subsequently rolled down to eliminate interstices and to decrease conductor diameter (see Fig. 5.5). In three-conductor cables, whenever sector-shaped conductors are permitted by the relevant standard, compact sector conductors are usually used (see Fig. 5.6). Very large single-conductor cables with cross-sectional areas $\geq 1013\,\mathrm{mm}^2$ (2000 circular mils, commonly abbreviated as kCM or MCM) have segmental stranded conductors in order to reduce skin effect and, consequently, the ac/dc resistance ratio (cf. Table 5.4). Such segmental conductors are usually composed of four sector segments, each separately stranded, and then assembled together to form a round conductor. Two opposite segments are lightly insulated with paper or carbon black paper tapes in order to interrupt the magnetic path. A copper tape intercalated with a paper tape is applied overall, which acts as a binder, also providing a smooth surface (cf. Fig. 5.7)

In spite of increased manufacturing cost, segmental conductor cables are found more economical because for the same ampacity such cables may be of smaller conductor size. For example, the ampacity of 2000-MCM (1013-mm^2) segmental conductor cable will roughly be equal to that of a 2250-MCM (1140-mm^2) cable with a

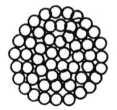

Figure 5.4 Concentric stranded conductor.

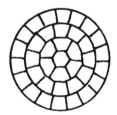

Figure 5.5 Compact conductor.

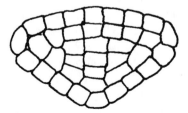

Figure 5.6 Sector conductor.

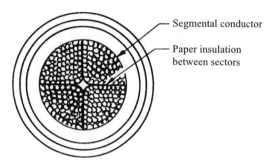

Segmental conductor

Paper insulation
between sectors

Figure 5.7 Segmental conductor power cable.

conventional conductor. Occasionally, segmental conductor cables are referred to as type M conductor cables in honor of their inventor Milliken. Lastly it has to be added that, whereas in North America copper is predominantly used in paper insulated power cables, in Europe (in particular in England) aluminum is almost exclusively used in three-phase cables with sector conductors. Ductility of aluminum permits these sector conductors to be made of solid aluminum, which offers not only cost benefits but also enables the use of economical splicing techniques, especially developed for that type of cable.

5.2.5 Selection of Insulation Thickness

Insulation thicknesses of oil or dielectric liquid impregnated paper insulated power cables, specified in current standards, are based more on tradition than on any scientific parameters. Because of the extremely good service record of these cables, any reduction of the insulation thickness in these cables always meets with a great reluctance by standardizing bodies, in spite of the fact that since the introduction of these type of cables many years ago, the manufacturing methods have greatly improved the quality of the insulation. Nevertheless, a modest reduction in the insulation thickness always takes place in any new revision of specifications, and so, for example, the current AEIC specification for solid-type impregnated PILC cable reduces the insulation thickness by 4–8%, also reducing by the same percentage the routine ac test voltage. The latter reduction is a result of North American philosophy that to be meaningful the test voltage must be high and must be a function of the insulation thickness, whereas in European specifications routine ac test voltages are directly related to the rated voltage of the cable and are much lower (35–40% lower for shielded cables).

The main parameter governing the thickness of insulation in the North American standards is the average stress value expressed in volts per mil, which is the ratio of the rated voltage to ground over the insulation thickness in mils (1 V/mil = 0.0394 MV/m = 0.0394 kV/mm). This parameter, however, is not constant across the entire voltage range but gradually increases with the voltage, for example, the insulation thickness for single-conductor cables in the current CSA Standard C68.1 is based on the average stress of about 1.5 MV/m for 8-kV (GN) cables, whereas that design value reaches 2.4 MV/m for 69-kV (GN) cables. Except for the lower voltage range, where the thickness of insulation also depends on the mechanical properties, the European philosophy is to base their design on the maximum stress that the insulation can sustain and, hence, on the maximum potential gradient prevailing in the insulation closest to the conductor surface.

In a single-conductor cable, as well as in a shielded cable with round conductors, the electric field is radial, and this gradient is fully calculable, considering such a cable, as a cylindrical capacitor with the inner electrode represented by the conductor surface and the outer electrode by the lead sheath or the insulation shield. Evidently for the purpose of calculation, irregularities on the conductor surface, caused by the conductor stranding, are to be ignored. In order to decrease the effect of these irregularities, conductor shielding, also called conductor screen, is introduced in higher voltage cables, composed either of carbon black or of metallized paper tapes. The use of compact stranding also improves the situation.

The potential gradient, or electrostatic stress, at radius r from the axis in such a condenser, may be expressed by a well-known formula

$$E = \frac{V}{r \ln(b/a)} \tag{5.11}$$

where E is the potential gradient (electrostatic stress) at radius r from the axis, V is the potential difference between both electrodes, b is the inner radius of the outer electrode, and a is the outer radius of the inner electrode. Considering that in a single-conductor cable, as well as in three-conductor shielded cable, the sheath, or shield, is at the ground potential, equal to zero, and that the conductor is as the rated voltage to ground V_0, the formula from which the maximum potential gradient or electric field could be calculated, derived from Eq. (5.11), becomes

$$E_{max} = \frac{V_0}{a \ln[(a + c)/a]} = \frac{V_0}{a \ln(D/d)} \tag{5.12}$$

where E_{max} is the maximum electric field appearing on the surface or on the surface of conductor shielding if such is provided [expressed in kV/mm ($=$ MV/m)], V_0 is the cable rated voltage to ground (in kV), a is the radius of conductor including shield, if any (in mm), c is the insulation thickness (in mm), and $D/d = (b/a)$ is the ratio of outer to inner diameter of insulation surface in the same units. According to recognized data, the maximum design gradient on which insulation thickness is based must not exceed 3.0 MV/m (3.0 kV/mm or 76 V/mil) for medium voltage rated, solid-type, impregnated paper insulated cables, and 4.0 MV/m (4.0 kV/mm or 102 V/mil) in the higher voltage range [25 kV (GN) and up]. Although, as it was mentioned before, insulation thicknesses in North American standards are based on different parameters, they more or less follow the above principle.

Except for old installations, one can hardly find in Canada paper insulated power cables in service, or in demand, with voltage rating below 8 kV (GN). Only belted cables are specified for such voltages in multiconductor cables. For the voltage range between 8 and 15 kV, according to the relevant CSA standard, the user has a choice of shielded or belted construction. No such possible choice exists for voltage ratings above 15 kV, for which only shielded cables may be specified. As opposed to shielded cables, in which the electrostatic field is radial and is fully calculable, even in sector construction (if the radii are known), belted-type cables have a nonradial and, therefore, a nonuniform field. The latter may be determined as a function of geometrical location only by the use of numerical methods. Therefore, in cases of belted cables, the cable dimensions are not determined from electric field gradient considerations; the electrostatic field may be calculated using computerized methods, that replaced the earlier techniques which made use of cable models in electrolytic tanks. Figure 5.8 shows the electrical stress distribution and equipotential lines in a typical three-phase cable at a particular instant when the top conductor is at maximum potential.

It should be noted that in a belted-type cable, the fillers are also highly stressed, and, as their dielectric strength is by far below that of the insulation, it is necessary in such cables to increase the insulation thickness on each conductor and on the belt, thereby lowering the average electrical stress to a safe value. Besides, in belted cables the insulation is also exposed to tangential stresses, to which the paper insulation is much less resistant, than to radial stresses in shielded cables. Cross sections of belted-type cables with round and sector conductors are presented in Fig. 5.9. For reasons explained above, belted cables cannot successfully be used for voltages higher than 15 kV. Even at that voltage level, shielded cables should be preferred, unless the cables are intended to be connected to already existing belted cables. Connecting shielded cables to belted type is more complicated because of stress cones, which must terminate the shields. Belted cables rated at 13 kV (GN) or higher must also have conductor shielding, usually of metallized paper, to improve cable performance by the elimination of the stranding effect. As shielded cables, in general, are more costly, there is no gain in specifying them for lower voltage ratings, for which the gradient considerations are of a lesser importance and the thickness of insulation is mainly selected from mechanical considerations.

Cross sections of shielded-type cables with round and sector conductors are represented in Fig. 5.10. Over each insulated conductor, the insulation shield is applied in the form of a thin copper tape intercalated with paper tape with an overall binder composed of similar materials. This represents a typical North American construction, whereas in Europe metallized paper insulation shield is preferred with copper woven fabric tape used as a binder. The role of a binder is not only to create a round cable assembly, compressing fillers in the interstices and keeping them in place, but also to provide a good electrical contact between individual insulation shields and the cable sheath, which must be solidly grounded to secure ground potential on the individual shields. To remove the effect of unevenness on the conductor surface, which would increase the maximum potential gradient in the insulation, all shielded cables must

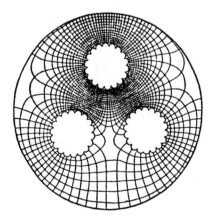

Figure 5.8 Electrostatic field in three-conductor belted cable.

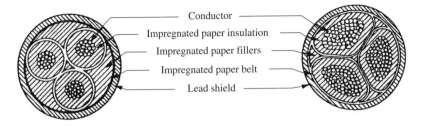

Figure 5.9 Three-conductor belted-type cable.

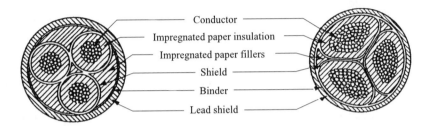

Figure 5.10 Three-conductor shielded-type cable.

also have conductor shield (screen), composed of layers of metallized paper. These layers are not indicated on Fig. 5.10.

According to CSA Standard C68.1 all single-conductor cables of a minimum voltage rating of 15 kV (GN), or 8 kV (UN), must have conductor and insulation shields. The reason for conductor shielding has already been explained. The requirement of insulation shield on single-conductor cables is warranted by the fact that such a shield, in frequent contact with the cable sheath at the ground potential (or very close to it because of sheath currents) removes from the electrostatic field spaces between the surface of the cable core and the cable sheath. Without the shield, these spaces or cavities, which could form themselves due to expansion and contraction of the cable core in service, caused by thermal load cycles, could lower the breakdown strength of the cable due to partial discharges.

5.2.6 Cable Impregnation

Pure wood pulp electrical-grade paper, applied on the cable conductor in a taped form, is only one component of the insulation in paper insulated power cables, the other being viscous mineral oil or high-viscosity oil/rosin compound applied degasified and hot to a previously dried and evacuated cable assembly. These insulating cable oils and compounds must have the following characteristics: a low power factor, an ability to absorb gas (which could develop due to ionization), a resistance to oxidation to prevent deterioration of insulation during cable manufacture and in service, an electrical stability (i.e., no change in electrical properties under storage conditions), and a set point, which should be as low as possible because of danger of voids formed during setting. In North America pure mineral oil is predominantly used, except for some manufacturers who adopted European practice and prefer using oil/rosin compounds instead. Oil/rosins (colophony) compound is more viscous at the cable operational temperature and has less tendency of displacement along the cable in service, a very important advantage. Another important advantage is greater resistance to oxidation and better electrical stability. Cables impregnated with oil/rosin compounds are called *mass impregnated* and, as they have less tendency of draining, they may be installed on a slope without a danger of hydrostatic pressure developing at the bottom end.

The reluctance to use oil/rosin compounds apparently originated from bad experience in the early stages of the development of paper insulated cables in North America. The difficulty of obtaining rosin with good electrical properties and the dramatic increase in power factor, when such rosin was added to their naphthenic oil, caused the North American cable manufacturers to abandon the use of rosin and to employ only highly refined naphthenic oil in their paper insulated cables. Although these problems have long ago been mastered by the use of rosin refined by distillation, and by proper compounding, the practice of using pure mineral oil in the cables on this continent still prevails in spite of their tendency to drainage. Such drainage in cables installed on a slope could cause bursting of joint sleeves at the lower end due to hydrostatic pressure, and the collapse of sleeves at the upper end of the run, a phenomenon which came to light in Canada some time ago and caused an alarm among the utilities. Later introduction of oil nonsoluble compounds in these joints eliminated this trouble.

For vertical riser cables, special nondraining impregnate has been developed and such cables are called mass impregnated nondraining (MIND) cables. Nondraining compounds are usually based on mineral oil to which polyisobutylene or high microcrystalline waxes are added. These compounds are fluid at the temperature of impregnation but become a gelatinous mass at the lower temperatures including the normal operating temperature. Care should be taken to avoid higher temperatures associated with overloads. Although very popular in Europe up to 30 kV, the MIND cables are nowadays very seldom used in North America, where for vertical risers polymeric insulated cables are almost exclusively employed. Because of the nature of the compound, the MIND cables must have electrical requirements relaxed with respect to conventional paper insulated cables; however, there is no such provision in the North American specifications. Therefore, the voltage rating of such cables here is restricted to a maximum of 15 kV (GN). As the MIND compounds at operational temperature are almost solid, they have very limited lubricating power and therefore belted MIND cables should be avoided, as paper tapes in the belt would particularly be exposed to tear stresses during bending of the cable.

5.2.7 Metallic Sheaths

As the interior of the dielectrical-liquid-impregnated paper insulated cables is highly susceptible to the ingress of moisture, the entire cable assembly must be covered with an impervious metallic sheath usually of lead, lead alloy, or aluminum. Lead is the oldest material used for cable sheathing that combines most desirable physical and chemical properties with the ease of application, because of a fairly low melting temperature (about 319°C) and the temperature of application of about 200°C, too low to damage the insulating material.

The lead sheath is applied over the cable core by a direct extrusion process, which forms a homogeneous sheath with a uniform wall thickness in practically unlimited lengths. It may be applied in hydraulic ram press or by a continuous extruder. In a ram press molten lead is fed into the chamber of the press, where it is cooled to the correct temperature and then extruded under hydraulic pressure through a die over the cable core. Extrusion must stop each time after the eligible quantity of lead from the chamber is spent and the chamber must thus be refilled again. This intermittent process results in the formation of charge marks in the sheath at each interruption, which undoubtedly are the weakest points of the sheath in spite of the precautions taken during extrusion. This inherent inconvenience of the ram press extrusion method has led to the development of a continuous extruder that has a screw-type feeder and conveys the metal to the cable through a die. Very long lengths of cable can thereby be sheathed without interruption. For example, for the well-known submarine cable link between the Italian mainland and Sardinia, executed by Pirelli S.p.A. in 1965, lengths of solid-type paper insulated cable up to 50 km (over 30 miles) were lead alloy sheathed with Pirelli continuous extruder in a special submarine cable plant near Naples.

The lead sheath not only provides an absolute barrier against an ingress of moisture, or chemicals, in liquid or vapor form but also provides some limited mechanical protection over the highly vulnerable cable interior, quite adequate in cables intended for a duct installation, even without outer covering. The bare lead-sheathed cables are seldom used nowadays and are usually provided with a thermoplastic covering. The main weaknesses of the lead sheath of cables are low resistance to creep under very low stresses, low fatigue resistance to high-frequency vibrations (from external sources), low bending resistance in daily load cycles, and low structural stability with tendency to form coarse grains. These deficiencies were particularly troublesome to North American users, who were installing their cables in ducts without any covering over lead, exposed to vibrations from the street traffic and to fatigue from manhole offsets. Therefore, in order to decrease the rate of cable failures caused by lead cracking due to fatigue, copper bearing lead was specified that offered a slight improvement in fatigue resistance and structural stability. Initially copper bearing lead was available as virgin lead from certain mines, however, later it became an alloy, artificially made with copper contents of about 0.06%.

In Canada copper bearing lead was used for a long time as a metallic sheath with an extruded thermoplastic outer covering. Its performance was, and still is, quite satisfactory. However, since it was found that, in a continuous extruder, there is a chance of copper separation in the form of copper crystals embedded in the cable sheath endangering its performance, the use of copper bearing lead alloys has been abandoned in Canada and replaced by a much superior, also easily extrudable, ternary alloy half-C containing 0.2% tin and 0.075% cadmium. These represent half of the quantities used in the composition of alloy C, which was popular in Britain some time ago, until it was removed from the list of recognized alloys by the British Standard Institution, because of the shortage of cadmium in United Kingdom, experienced right after World War II. Half-C (C/2) alloy is not in use by American industry.

Half-C is a further improvement over the copper bearing lead; however, if high hoop stresses are required from the sheath, then the F-3 alloy, developed by Anaconda is preferable. This arsenical alloy containing about 0.15% of arsenic, 0.10% of tin and 0.10% of bismuth is popular with some Canadian utilities, as it also displays extremely good creep resistance, adequate ductility, good resistance to fatigue, combined with high structural stability. Its drawback is that alloy F-3 can only be extruded in a ram press leaving charge marks in the cable sheaths which, as already mentioned, constitute weak points. Because of its extremely good fatigue resistance to vibrations, alloy F-3 is particularly suitable in cables exposed to long overland transit.

Another lead alloy with good fatigue resistance, gaining popularity in Canada, as it is extrudable in a continuous extruder, is a UK-developed lead alloy E, containing about 0.4% of tin and 0.2% of antimony. The addition of antimony is particularly effective in improving fatigue resistance to high-frequency vibrations encountered in long rail transit and in shipments by sea. The feasibility of continuous extrusion makes alloy E particularly suitable as a sheathing material for long submarine cable crossings. There are many other lead alloys developed for cable sheathing in North America and in Europe, however, as their use in Canada is minimal, they will not be discussed in this presentation.

Lead alloys are specified not only to protect the cables against vibration in transit, or to secure longer life of cables by increasing their fatigue resistance, but also in cases of cable risers, or cables intended for a run in which the difference in elevation could cause migration of compound from a higher to the lower portion of the cable, resulting in excessive hydrostatic pressure on the cable sheath. In order to protect the sheath against abnormal expansion caused by this pressure, the use of lead alloys would alleviate the necessity of excessive increase of lead sheath thickness, making the cable lighter and more economical. Allowable hoop stresses for the above types of lead and lead alloys used for maximum static head calculations are provided in Table 5.5.

To establish a suitable formula for the static head calculation, we must first derive the relationship between the maximum transversal pressure, exerted by the column of oil, or compound, within the cable sheath, and the hoop stress developed in the sheath, as a result of that pressure. Considering a cylindrical lead tube of inner diameter d and of uniform thickness Δd, the relationship between hydrostatic pressure exerted by the column of oil, or compound, inside the sheath and the hoop stress in the sheath may be expressed as

$$S = P\left(\frac{d}{2\,\Delta d}\right) \qquad (5.13a)$$

or

$$P = S\left(\frac{2\,\Delta d}{d}\right) \qquad (5.13b)$$

TABLE 5.5 Maximum Permissible Hoop Stresses on Lead Sheaths

Sheath Type	Pressure [psi (kPa)]
Pure lead	125 (862)
Cooper bearing lead and C/2 alloy	150 (1034)
Lead alloy E	175 (1207)
Lead alloy F/3	250 (1724)

where S is the hoop stress in the sheath (in psi or kPa) and P is the hydrostatic pressure (in psi or kPa). In SI units, pressure exerted by a column of 10 m of water at $4°C$ equals $98.07\,kPa$ ($= 1\,kgf/cm^2$). It follows that the hydrostatic pressure P (in kPa) may be expressed by a column h meters in height of liquid with specific gravity g in the following formula:

$$h = \frac{10P}{98.07g} = 0.102\frac{P}{g} \tag{5.14}$$

Substituting the expression for P from Eq. (5.13b) in the above formula we have

$$h = \frac{0.204S\Delta d}{gd} \tag{5.15}$$

where h is the height of the column of liquid in meters. Finally, Eq. (5.15), if used for the calculation of maximum allowable static head for the cable, expressed in SI units, may be rewritten as

$$h_{critical} = \frac{0.204S_{max}\,\Delta d}{gd} \tag{5.16}$$

where $h_{critical}$ is the maximum static head in meters and S_{max} is the maximum permissible hoop stress in kilopascals. Note the specific gravity g is equal 0.9 for oil and 0.95 for oil/rosin compounds. The equivalent formula developed in imperial units, still in use in Canada, is

$$h_{ft} = \frac{32.8P}{14.22g} = 2.3\left(\frac{P}{g}\right) \tag{5.16a}$$

where P is now in pounds per square inch and h in feet. Similarly, substituting the expression for P from Eq. (5.13) in the above formula, we obtain

$$h_{ft.critical} = \frac{4.6S_{max}\,\Delta d}{gd} \tag{5.16b}$$

If, for the required difference in elevation of a three-phase cable, the static head calculations prove that none of the alloys satisfies that condition without a prohibitive increase in the thickness of the sheath, then type SL paper insulated power cable may be considered in which each phase is treated as a separate single-conductor cable, individually lead sheathed. Three such cables are laid up in one cable assembly, over which a common outer covering is applied. Obviously, in such a three-conductor cable the ratio $\Delta d/d$ is greater than in a conventional type, resulting in a greater critical static head permitted for the cable [cf. Eq. (5.16)]. Another obvious solution is to specify MIND type cables, if the voltage rating is within the range for such cables. This solution is not used in Canada nowadays and polymeric insulated cables are preferred instead.

Lastly, aluminum sheaths have to be mentioned as the allowable hoop stress for aluminum is very high (6000 psi or 41,000 kPa), practically eliminating a necessity of static head calculations, except for some special cases. Although for reasons, explained later, aluminum-sheathed paper insulated power cables of the solid type are very seldom in demand in Canada, aluminum sheaths are offered here in the oil-filled version of paper insulated cables; and, as aluminum sheaths are also very popular in Europe, it may be in order to discuss them briefly in this presentation. The use of aluminum for cable sheathing always interested engineers, as aluminum is a much lighter material than lead (with a specific gravity of 2.71 versus 11.37 for lead), resulting in a possibility of producing much longer lengths, and, as it is much stronger mechanically, practically free from fatigue problems.

Aluminum sheathing was first developed in Germany during World War II due to a shortage of lead and abundant availability of aluminum. However, because of much higher temperatures and pressures required for extrusion of aluminum, new sheathing methods had to be found, and the Germans succeeded in the manufacture of a sheath from thin aluminum strip, which was welded and corrugated in one continuous process. The same process, although in much more sophisticated form is still in use at present times and is known as the Hacketal process, from the name of the company that first developed that method and is still supplying the equipment, although the Company name has been changed. After the war, the British took over the development of aluminum sheaths and introduced a novel method to draw the cable assembly through an oversized aluminum tube, afterwards swagging it down through a reducing die until a reasonably tight fit has been achieved. This process, also used in Canada, restricted cable lengths to those of the aluminum tube available and, therefore, was abandoned when direct extrusion by aluminum press became feasible. Finally, after a long and tedious development work carried out both in Germany and the United Kingdom, several designs of direct extrusion presses were tried out, from which only one, most ingenious press, patented by Hydraulik GmbH of Duisburg, Germany, emerged. It is a well-acknowledged fact that the Hydraulik press is the only reliable aluminum press, used by many cable manufacturers in the world, including one company in Canada. It has to be mentioned that, after an early bad experience with a press of another manufacturer, the Americans abandoned the use of aluminum for extruded cable sheaths and have not a single aluminum press in operation.

Because of the high temperature required for the extrusion of aluminum (450°C versus 250°C for lead) and a very high extrusion pressure, at least about twice as much as that required for lead, conventional designs of ram press had to be discarded and a new solution sought. Moreover, as the melting point of aluminum is also very high (658°C versus 319°C for lead), the cylinder of the press cannot be filled with molten aluminum, similarly to lead, and must be fed by preheated aluminum billets. Hence each time a new billet has to be introduced the extrusion must stop, leaving the cable assembly inside the die stationary and exposed to a prohibitive high temperature. Even with improved cooling, presses based on the above principle ran into difficulty associated with differences in sheath properties, when extrusion was stopped to recharge the billets.

As a result, on subsequent bendings, kinks in the sheath were formed near these stop marks, endangering sheath integrity. The Hydraulik Company resolved both of these problems by managing to maintain the extrusion, though at a very slow pace, while the main ram is withdrawn for the insertion of a fresh billet, preheated to 400°C in an induction oven. During this operation a subsidiary ram under the press takes over, raising the die block, thereby extruding metal from the bottom of the container. This auxiliary extrusion is maintained during a minute or so required for the insertion of a new billet. The volume of metal needed for this auxiliary extrusion is replaced as soon as the pressure builds up on the main ram. A special valve located at the bottom end of the container prevents back extrusion during the recharge. Therefore in that press the extrusion process never stops during the application of aluminum sheath onto the cable, eliminating the problem of charge marks as well as reducing the danger of insulation deterioration due to overheating.

In spite of the weight and cost advantage of aluminum-sheathed cables, Canadian utilities, the major users of paper insulated cables, are very reluctant to accept aluminum-sheathed cables in their systems. The main reason is that the aluminum-sheathed cables cannot be used in a duct and manhole installation designed for lead-sheathed cables because the aluminum sheaths are stiffer, even when corrugated. It follows that not only larger, nonstandard manholes have to be built for cables with aluminum sheath but also such cables cannot be used as spares for lead-sheathed cables to replace them in case of failure. Different and a more complicated splicing technique for aluminum sheaths, requiring retraining of personnel, may also be a determining factor. In Europe, where unlike on this continent power cables are mainly directly buried, eliminating manhole problems, aluminum-sheathed cables are generally used, with the aluminum sheath also replacing the neutral conductor in three-phase distribution, making such cables economically very attractive.

Lastly, it should be mentioned that aluminum occupies a high place in electrochemical series of metals and therefore there is a risk of corrosion of the aluminum sheath, due to galvanic action between aluminum and dissimilar metals in the presence of moisture, requiring the necessity of an effective anticorrosion protection.

5.2.8 Protective Coverings

The prime role of a protective covering, applied over the metallic sheath during the final stages of cable manufacture, is to protect the sheath against deterioration and damage in service as well as to decrease the possibility of damage in transit and during installation. The design of protective coverings largely depends on the type of installation, although other factors such as corrosive contamination of the environment, firmness of the ground, required mechanical strength, possibility of damage by rodents, teredos, and termites should also be considered. Distance and method of transportation from place of manufacture to the site could also influence the choice.

Protective coverings on large single-conductor cables in major three-phase installation, exposed to induced sheath voltages, are also expected to provide electrical insulation on cable sheaths, to insulate them from ground. In general, modern protective coverings may be divided into three groups, namely extruded nonmetallic coverings (anticorrosion coverings), fibrous nonmetallic coverings (auxiliary coverings in connection with metallic coverings), and metallic coverings (armor reinforcement).

5.2.8.1 Extruded Nonmetallic Coverings

Extruded nonmetallic coverings are composed of an extruded layer of thermoplastic material [polyvinyl chloride (PVC) or polyethylene] applied directly on the cable sheath (cf. Chapter 2). They not only efficiently protect the sheath against corrosion but also protect the sheath against abrasion during installation and should be used in all cases, when no other type of protection is specified. In case of the aluminum sheaths, such coverings are applied over a flooding compound covering the sheath. Because of the corrosive nature of aluminum, such anticorrosion protection is absolutely necessary for all types of installation. In the case of lead sheaths, this type of protection is not absolutely necessary, particularly when other types of protection are used over lead, as in the case of a metallic covering. Nevertheless, even if no outer protection is required, as for duct and manhole type of installation, extruded nonmetallic coverings applied directly over lead should be used, as they not only protect the sheath against electro-chemical corrosion from stray currents and offer additional protection against possible damage during installation but may also extend the cable life by the reduction of fatigue problems. During transit the coverings act as a damper, reducing vibrations that could lead to fatigue cracking of the sheath.

In Canada, the solid-type PILC power cables, intended for duct and manhole installation, are, in general, required to have an extruded nonmetallic covering, usually of a black PVC compound, as the latter is less susceptible to the deterioration caused by exposure to sunlight. The Canadian Standard C68.1 permits reduction of the lead sheath thickness by about 15% when extruded protective coverings are used, partly compensating for the additional cost of the coverings. Polyethylene, although a much better insulant and, as such, exclusively used on oil-filled cables, is seldom specified on PILC cables in Canada presumably because of ease of adhesion of the PVC splicing tapes used to extend corrosion protection over joint sleeves and pothead bases to polyethylene coverings. Lack of flame-retardant properties of the polyethylene coverings could also influence that particular choice. In large single-conductor cables in ac systems, the PVC jacket also acts as an insulation against sheath voltages.

5.2.8.2 Fibrous Nonmetallic Coverings

Fibrous nonmetallic coverings such as impregnated jute or hessian tape are used in modern PILC cables only in conjunction with metallic coverings, as bedding to protect the lead sheath against damage during the application of armor and as servings over the armor. They are generously covered with bituminous compound applied in hot condition during the application of the armor (cf. Chapters 2 and 13).

5.2.8.3 Metallic Coverings

There are various metallic coverings used to provide additional mechanical protection to the cables. Double steel tape armor (DSTA), offering limited transversal mechanical protection, is generally used in cables intended for direct earth burial. It is composed of two steel tapes, helically applied in two layers in the same direction with 50% registration, i.e., with the center line of upper tape coinciding with the gap between convolutions of the tape underneath. It is placed over an impregnated jute bedding with jute serving overall. Bituminous compound is generously used during the application of these layers with an antiadhesion agent applied overall. Because of its magnetic effects, DSTA cannot be used on single-conductor cables for ac operation and, consequently, another solution must be sought. Special metallic coverings in the form of a brass or bronze tape applied with complete coverage are used as protection against rodents and teredos.

Single-wire armor or, in special cases, double-wire armor are specified to provide PILC cables with adequate longitudinal strength required for cables intended for vertical installation in mine shafts or high-rise buildings as well as for submarine cables in which high longitudinal strength is also essential. Galvanized steel or aluminum alloy wires are used, although for single-conductor ac cables only aluminum alloy wires may be chosen. Wire armor must be applied over the bedding, usually of compounded jute, although a decay-proof cutched jute is occasionally specified for cable risers to avoid the use of compounds that could drip while in service. Also in such cases serving is omitted, and wire serving or stainless steel clamps are placed on the armor every 5 m or so to protect the armor against buckling and to increase the friction between the armor and the cable, thus safeguarding the cable lead sheath against longitudinal tension.

In some cases it is practical to replace jute bedding by an extruded protective covering applied over the lead sheath and, for further protection, to cover wires individually with an extruded layer of PVC or polyethylene. Obviously in this case, the number of wires must be reduced, as each covered wire will take additional space. The size and number of wires must be calculated to support the entire cable weight with a required factor of safety because cable lead sheaths cannot carry much weight, and the conductors should not be exposed to any longitudinal stresses. Finally, it has to be mentioned that to provide some anticorrosion protection in armored PILC cables, instead of an extruded covering, layers of compounded paper tapes are added on the lead sheath, underneath the jute bedding.

The above brief description by no means exploits all design possibilities of protective coverings, which in special cases must be custom designed in accordance with specific installation and service requirements.

5.3 CABLE MANUFACTURE

The PILC cable manufacturing process can be most effectively dealt with by describing the operations in the actual sequence. Since in Canada copper conductors are almost exclusively used in PILC cables, only copper conductor cables will be discussed.

5.3.1 Hot Rolling of Copper Wire Bars

Wire bars of electrolytic copper having the approximate dimensions of $54 \times 4 \times 4$ in. ($137 \times 10 \times 10$ cm) and weighing about 250 lbs (114 kg) each as supplied by copper producers are heated in an oil-fired furnace to about 900°C in a continuous operation and then guided individually to a roughing train in which the hot wire bars are passed between steel rolls back and forth, progressively increasing in length with a reduction in cross section. After a number of passes, the bar, now in the form of a long strip, is transferred to the intermediate and then to the finishing train in which after several more passes, now in one direction only, a round copper rod is obtained usually of a diameter of either $\frac{3}{8}$ in. (9.5 mm) or $\frac{5}{16}$ in. (8.0 mm) depending on its future use. At the end of the hot rolling operation, the copper rod is fed to the coilers in which it is wound into coils, subsequently cooled by a water spray. The entire hot rolling procedure is fully automatic with one man at the control desk guiding the entire operation. It should be pointed out that the hot rolling sequence is designed in such a way that after each pass, the overall copper cross section is not only reduced, but it becomes successively square, oval, diamond shaped, and ultimately round. In this way the entire material is worked out equally in the process, ensuring uniformity of properties.

The hot rolling operation results in the production of the so-called black rod, because of its characteristic dark surface coating of copper oxide, which must be removed before the rod can be used for further processing into wires. This is attained chemically by pickling, i.e., by placing the coils in a bath of diluted sulfuric acid for a certain time period. Afterwards the copper rod, which now assumes the characteristic reddish color of copper, must be washed in a neutralizing bath to remove any traces of acid.

5.3.2 Cold Drawing of Wires

From the rolling mill, the copper rod is transferred to the wire-drawing machines, in which the rod is pulled through a series of tungsten carbide or steel alloy dies, until the exact diameter, required for the production of cable conductor, is obtained. The drawing is performed in one continuous length by electrically butt-welding the rod before it is fed into the machine. Water-soluble lubricants are used to assist the drawing: They also act as coolants. The drawing of copper results in hardening of the material to the effect that the wire becomes of a hard-drawn (HD) grade with reduced ductility, elongation, and conductivity, thereby becoming completely unsuitable for further processing. Therefore, it must be transformed by annealing to attain characteristics of a soft-drawn (SD) grade copper.

5.3.3 Annealing

Annealing takes place in a bell-type lead furnace in which spools of copper wire are heated to temperatures in the range from 500 to 750°C for a period of time and then slowly cooled down to ambient temperature. During this entire cycle, the wire remains in a hermetically sealed furnace under an atmosphere of an inert gas in order to prevent oxidation.

5.3.4 Conductor Stranding

Stranding machines may be divided into two main types, known either as *tubular high-speed machines*, designed for fast stranding of small conductors or *cage machines*, used for the production of large, multilayer conductors. As in Canada paper insulated cables are predominantly used with large-sized conductors; cage-type machines are used for conductor stranding of these cables. Such machines are composed of several rotating frames, or cages, in which spools with wire are circumferentially placed on spindles. Each frame carries the spools for a single layer of wires. The rotation of frames is synchronized by a common drive with the axial pull of the conductor by a capstan placed at the end of the machines, resulting in a conductor composed of helically applied wires of a required lay. For standard stranded conductors the direction of rotation of each subsequent frame alternates, resulting in a multilayer conductor with each layer in reverse direction. This is done to neutralize any stresses in individual wires created during stranding operation. For compact stranded round and sector conductors, all layers are applied in the same direction; however, as each layer is rolled down after application, the danger of stress retention in the conductor becomes minimal.

Lastly, prespiralled sector stranding must be mentioned, as this method is very popular in Europe and is also used by one Canadian cable manufacturer. However, as this method requires a special manufacturing facility, it is not popular in the older

North American plants. The idea is to prespiral sector conductors during the stranding operation by rotating the sector forming rollers with a predetermined lay. As a result of prespiralling, the insulated conductors are assembled into round cable assembly without any twist, which otherwise must be created, tensioning paper tapes. This method also pays dividends in assembling segmental (type M) conductors in which the paper separators, dividing the sector segments, are very susceptible to damage during the conductor assembly using the conventional approach.

5.3.5 Insulating and Shielding

Insulating and shielding is performed in a paper lapping machine, composed of several lapping heads, which can rotate in both directions. In a modern paper lapping machine each lapping head may apply 10 or more tapes onto the cable from preslit paper pads through a very intricate arrangement of guides and springs. This arrangement permits accurate tensioning of each paper tape in accordance with a paper lapping chart programmed by means of a computer. Such a procedure is essential particularly in very high voltage cables having considerable insulation thickness because the selected tension must permit a slight transversal movement of tapes during cable bending; otherwise some tapes would become torn or crinkled and the quality of insulation decreased. A loose tape application should always be avoided as it would result in a soft cable core. It is important that the predetermined tension remains constant in the tapes, even when the machine must be stopped, which is a feat almost impossible to attain in old-fashioned lapping machines still in operation in many plants. In those machines any stoppage can result in a slight reversed movement of the lapping head, which, by removing tension from paper tapes, will tend to dislocate the tapes on the cable.

The relative speed of the cable provided in modern machines by a caterpillar capstan and the angular speed of lapping heads determine the length of lay of the tapes. This lay, the width of tapes, and the diameter underneath are interrelated, resulting in a small constant gap between convolutions in each layer of insulating tapes. This gap must not coincide with the similar gap in the next layer below or above. Although some North American manufacturers apply all paper tapes in the same direction, a better practice is to reverse the lay after a certain number of tapes in order to neutralize any mechanical stresses in the insulation. These stresses could create some problems during the winding of insulated conductors on the takeup reels located at the end of paper lapping machines. Conductor and insulation shielding tapes, if required, are applied in the same operation as insulting tapes, by placing pads of these tapes on an appropriate head in the paper lapping machine.

5.3.6 Laying-Up Operation

Multiconductor paper insulated power cables are assembled on laying-up machines with paper fillers placed in the interstices to form a round cable assembly. To allow production of cables in longer lengths, the laying-up machines are heavily built, as they must accommodate at least three large reels with insulated phase conductors on their rotating frames.

For assembling prespiralled sector conductors and segmental (type M) conductors (cf. Fig. 5.7) rotating frames with payoff reels must be equipped with motors, which during the operation must be able to slightly change the angle of the payoff reels to secure a correct assembling of prespiralled sector components into one round structure. This facility of laying-up machines is, however, generally not available in most North American cable plants. The laying-up machines are equipped with several paper lapping heads located after the closing die to allow in the same operation the application of a belt insulation in belted-type cables or a binder in shielded-type cables.

5.3.7 Impregnation

Because in paper insulated cables paper is only one component of the insulation, the other being impregnating oil, compound, or other dielectric liquid, the process of impregnation is extremely important. It takes place in the drying and impregnating tanks, where the cables are placed either on special processing metallic reels with removable flanges or on shallow trays. The impregnating tanks have hermetically sealed covers and are equipped with pipes for steam heating as well as auxiliary equipment to keep the tanks' interiors during the process either under vacuum or under pressure. The facility for electrical resistance heating by passing high current through the cable conductors is a useful requirement.

In the drying cycle the cable is exposed simultaneously to heat and vacuum for an extended period of time depending on the amount of insulation and its moisture content. Considerable amounts of moisture are usually drawn off the cable into a condenser incorporated in the vacuum line, and the amount of extracted water may be observed through inspection glass windows provided in the condenser. Evacuation is continued under high vacuum of about 1 mm Hg until a vacuum drop test and some electrical tests indicate the completion of the drying cycle. The cable is now ready for the impregnating cycle, which commences by the admission of the degasified impregnating compound into the tank. Impregnation proceeds under vacuum for a predetermined time, after which a period of impregnation under pressure follows in some cases. Finally, the cooling operation begins under closely controlled conditions to prevent the formation of voids due to contraction of the impregnant during the cooling cycle.

5.3.8 Application of Lead Sheath

After the process of impregnation is terminated and the cable attains a low enough temperature to reduce the possibility of oxidation of the impregnant, when exposed to the atmosphere, the cable is removed from the tank and transferred to a free rotating platform located in the rear of the lead press, or the lead extruder, when still on the reel, or used to reduce the exposure time of the still hot cable to cool air, while awaiting the leading operation. In the ram press method of extrusion, molten lead from a melting pot, located close to the press, is transferred to the container of the press by means of a removable chute. When the container is full, the dross is removed from the lead surface and then the press is operated until the surface of the lead ram is in contact with lead, sealing the opening of the container and exerting a certain pressure on the molten lead, which is left to cool down to a plastic state without exposure to air to prevent oxidation.

When a suitable temperature is reached extrusion commences as the lead from the container is pushed by the lead ram of the press through a die block, which contains the outer die and the point, a conical internal part of the die with a center hole for the cable core (cf. Fig. 5.11). The point, located at the tip of the point holder, together with the annular outer die control the diameter and the thickness of the lead tube. Graduated series of points are stored near the press, from which a proper one is selected whose

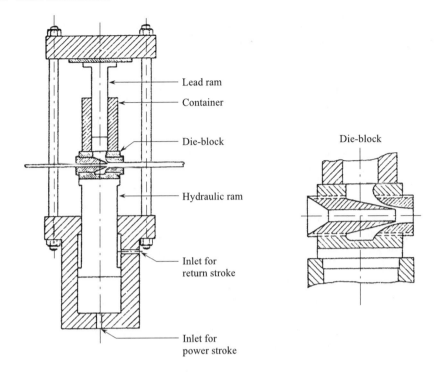

Lead ram

Container

Die-block

Hydraulic ram

Inlet for
return stroke

Inlet for
power stroke

Die-block

Figure 5.11 Diagram of ram press.

bore closely matches the diameter of the cable core. Overall diameter and the sheath thickness are controlled by changing the relative position between the outer die and the point. The diagrammatic sketch of Fig. 5.11 illustrates the arrangement of a Krupp upstroking press in which the lead ram remains stationary and the die block with the lead container move upwards during the extrusion process. In another type of lead press, called the down stroking press, the die block and lead container are stationary, and the lead ram moves downwards pushed by the hydraulic ram. As the method of controlling thickness and concentricity of lead sheath is rather crude, considerable skill is required from the lead press operator, who, before embarking on the actual sheathing, must first reach required dimensions of the empty lead tube extruded prior to the insertion of the cable core into the die block. Also considerable experience is required from the operator to eliminate a danger of defective charge marks.

After emerging from the press, the sheathed cable is cooled, usually by a water spray, and then it is automatically wound on a reel without tension by a slack control mechanism. Certain alloys, such as F-3, require some treatment after extrusion while the cable is still in hot condition. As pointed out before on the subject of metallic sheaths, ram presses are mainly used nowadays for extrusion of hard lead alloy sheaths, otherwise the continuous extruder is preferred. Without going into technicalities, it may be stated that the basic difference between the ram and continuous extrusion presses centers on the method of exerting pressure in pushing the plastic lead through the die arrangement. In the ram press this is done hydraulically; in a continuous extruder the same effect is achieved mechanically by turning a screw. In the Pirelli-General type extruder, this screw surrounds the cable core in the form of a steel cylinder with screw threads on its inner and outer orifices. The impelling screw rotates between these stationary inner and outer elements resulting in a double extrusion lead path leading to a forming chamber. The inner stationary tube through which the cable core passes is terminated with the point, which, with the die preceding it, controls the diameter and the thickness of the sheath.

Lead enters the machine at a temperature of about 360°C, and during the entire passage its temperature must be strictly controlled and kept at a predetermined level by induction heating and water cooling of various parts of the extruder. The independently developed Pirelli S.p.A. continuous extruder differs in certain features from the Pirelli-General machine, although its basic construction is similar. A differently designed screw provides only one lead path to the forming chamber, and air cooling replaces water cooling. Both these presses have one essential advantage over some other designs, i.e., they provide a flow of lead along the axis of the cable, without changing the direction of the flow in the forming chamber, thus securing more uniform application of lead onto the cable. Early extruders simply replaced the lead ram by a screw mechanism located horizontally, which propagated plastic lead into the die block at the right angle to the cable axis. Obviously in such an arrangement, also existing in ram presses, lead is supplied to the forming chamber perpendicularly to the axis of extrusion, hence the paths that the lead must traverse within this chamber are uneven, endangering the uniformity of properties of metal around the cable circumference.

5.3.9 Application of Protective Coverings and Armor

Extruded nonmetallic coverings are applied using plastic extruders that are somewhat similar in principle to lead extruders, except that because of dissimilarity of materials and lower pressures and temperatures required, their actual construction is completely different. In these machines the impeller screw is located at right angle to the cable axis and the thermoplastic to be extruded is fed in a granulated form. During its passage, the plastic is heated and worked out by the screw to attain a uniform mass, which is applied onto the cable in the die arrangement located in the head of the machine. After application, the covering is cooled in a long cooling through until the thermoplastic material becomes completely solid and the cable can safely be wound on a reel. Modern extruders are equipped with a facility called vacuum sheathing in which a slight vacuum is provided during extrusion to improve adhesion of the extruded covering to the metal sheath underneath.

It has to be added that, whereas in lead extruders the lead thickness is controlled by a die and a point, here the thickness depends on the speed of the cable through the extrusion head. Consequently, in order to obtain a required wall thickness with desired uniformity, the cable must be pulled during the extrusion step with a predetermined and uniform speed, secured by a caterpillar capstan located at the end of the cooling trough. Unlike the lead extruder, the plastic machine cannot extrude an empty tube and must have the cable inserted within the machine prior to the beginning of the operation.

Metallic coverings, such as double steel tape armor, or wire armor, are applied in the armoring machine, which is equipped with facilities to apply fibrous nonmetallic coverings, under and over the armor, with intermediate layers of bituminous compounds, in the same operation. The armoring machine consists of a number of rotating heads and cages, for each component covering, arranged in one line and driven from a common shaft through gears. Between each rotating flyer, steam-heated bitumen pots are located through which the cable passes between each subsequent application of coverings. After the last cage, the armored and jute served cable passes through the pot filled with a harder grade bitumen, after which an antiadhesion compound is applied to prevent adhesion on the bituminous surfaces of the cable. A capstan, located at the end of the machine, driven by the same shaft as the rotating flyers, provides a pulling force for the cable. Finally, the cable is automatically wound on the shipping reel without tension with the use of a slack controlling mechanism.

5.4 TESTS

After manufacturing each length of cable must undergo a series of routine tests. For example, CSA Standard C68.1 requires a number of routine tests to be made on each length of cable. Measurements of conductor resistance, thickness of insulation, and thickness of sheath are made. The required electrical tests include an ac high-voltage test and an ionization and power factor test [shielded cables rated 15 kV (GN) or 8 kV (UN) and higher]. Also, a number of routine tests have to be made on samples of finished cable, concerning mechanical integrity that consist of an examination and tests of nonmetallic coverings, insulation (thickness, tape wrinkling, registration), and lead bend. The lead bend test is performed on strips of lead taken from the cable, clamped in a hand-operated bending machine in which the strips have to withstand several 90° bends before failure. The protective coverings may be tested in accordance with the CSA Standard C170.

Besides the routine tests, which are mandatory, special optional tests listed in various standards may be specified by the purchaser and are carried out by special arrangement, at the purchaser's expense. Although seldom required, they are from time to time made by the manufacturer, at his discretion, in order to verify the quality of the product. These tests may include a bending test performed on cable specimens with, or without, nonmetallic coverings. Also for shielded cables rated at 15 kV (GN) or 8 kV (UN) or higher, a high-voltage time test, identical to the routine ac high-voltage test, but lasting 6 h, is performed. A dielectric power loss and power factor test is carried out at the ambient temperature $10°C$ higher than the maximum conductor temperature specified for normal operation. At each temperature the power factor must not exceed the specified value, when measured at the maximum phase-to-ground voltage. Measurements at the four different temperatures allow the researcher to draw a curve representing the dielectric power factor as a function of the conductor temperature, from which the cable thermal stability may be deduced. The significance of these tests is explained in the following section on cable electrical characteristics.

5.5 ELECTRICAL CHARACTERISTICS

The most important electrical characteristics of a high-voltage paper insulated power cable is the dielectric power factor and ionization factor, the latter being a power factor difference, when measured at a high and low electric stress (e.g., at 100 and 20 V/mil), as defined in the relevant standards. The values of power factor and ionization factor must be kept below a certain specified maximum value. However, as the power factor is a measure of the dielectric loss, it has no important bearing on the lower voltage cables; hence, normally it is required to be measured only in shielded cables rated 15 kV (GN), or 8 kV (UN) and higher. According to the CSA standards, single-conductor cables at these voltages must be shielded. As already pointed out in Chapter 3, the power factor of a dielectric is defined as a ratio of the real power loss to the reactive power loss. It can be also shown that for a good cable dielectric, the power factor, $\cos\theta$, is numerically equal to the dissipation factor, $\tan\delta$ of the dielectric (cf. Chapter 9).

A very important investigation concerns the relationship between power factor and temperature (carried out at the maximum operational phase-to-ground voltage), as the results may disclose, whether at the emergency conductor temperature there is a danger of runaway condition, which would create lack of balance between the heat generation in the cable and the heat dissipation, leading to cable destruction. It may be seen from the characteristics in Fig. 5.12 that an increase of temperature, much beyond 90°C, would cause a rapid increase in the $\tan\delta$ value and, hence, the dielectric loss magnitude. Heat generated by this increase would further augment the temperature of the cable insulation, causing in turn another increase of these losses in an avalanche effect (thermal runaway). It must be observed that these investigations are usually done by heating the cable sample externally. Hence, as the entire dielectric is reaching the recorded temperature, the results are far more critical than in an actual case, where only layers close to the conductor reach that temperature. In general, $\tan\delta$ vs. temperature tests are performed on solid-type paper insulated cables mainly for manufacturing information, as type tests.

Compulsory measurement of power factor is performed on each full length of manufactured shielded cable at the ambient temperature and the rated voltage-to-ground V_0. This is the most important test verifying the quality of the dielectric and is performed in conjunction with the ionization test, which requires that the power factor measurement be performed at two other voltage levels in order to check the power factor increase with the increase of stress. This increase, called ionization factor, is a measure of electrical discharge (ionization of voids) or other dielectric losses within the cable insulation under the influence of stresses higher than those in actual service. The ionization factor is defined as the difference in power factor at ambient temperature measured at two voltage levels, one very low, at which the cable insulation is assumed free from internal electrical discharges, with the power factor at a minimum value, and another at a higher than the V_0 value when corona discharge within the voids presumably takes place and the power factor is greater. These tests, required to be carried out on each cable length, provide an excellent check of quality of the entire production.

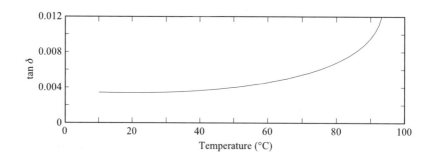

Figure 5.12 Typical $\tan\delta$ vs. temperature characteristics of shielded PILC cable.

Whereas the European standards specify these two voltage levels in terms of V_0, North American standards express them in volts per mil of insulation thickness. For example, the CSA standard requires that the ionization factor be established from two power factor measurements at 20 V/mil and 100 V/mil of insulation thickness. As power factor and ionization factor values are very small, they are usually multiplied by 100 and expressed in percent. For illustration purposes Table 5.6 gives actual results of ionization and power factor test performed on a three-conductor 500 MCM PILC shielded cable rated 15 kV (GN) ($V_0 = 8.7$ kV) having an insulation thickness of 175 mils (4.5 mm). Three phases were tested jointly to the CSA Standard C68.1.

The above results demonstrate that in a well-made cable, the ionization and power factor values are much below the specified maximum and the cable can hardly fail this test, unless some gross errors are committed during the cable manufacture. Nevertheless the manufacturer usually knows the order of power factor and ionization factor values expected from the test and will investigate any abnormality, even if the values are within the requirements of the standard.

According to North American practice, the specified ac test voltages are based on the average stress (about 200 V/mil) and in terms of rated phase-to-ground voltage are around $4V_0$. Even at such high test voltage levels, unrelated to real service requirements, paper insulated cables do not fail, as their withstand voltage levels are much higher. AEIC Specification No. 1-68 covering solid-type impregnated PILC cables specifies a high-voltage time test to be performed on shielded PILC cables, in which a specimen has to be subjected under a routine ac test voltage for 6 h. Thereafter the voltage is to be raised to an average stress of 400 V/mil and maintained for 7 h or to 350 V/mil for 14 h. The results have to be reported in equivalent volts per mil for a one-hour life test, using a rather complicated empirical formula quoted in that specification. The corresponding CSA standard lists only the first part of this test, which is to be performed only when requested. In general, routine ac high-voltage tests in any standard are of value in belted-type cables, whereas in shielded cables they are mainly performed for traditional reasons, as undoubtedly ionization and power factor test is a much more powerful tool to discover any defects in the cable. It has to be added that in Europe the routine ac test voltage levels are related to the cable ratings and are much lower than those in North America. For example, the International Electrotechnical Commission IEC Publication 55-1 specifies for shielded cables the routine ac test voltage equal to only $2.5V_0$ [12].

Impulse voltage tests are not included in North American standards for power cables. Nevertheless, in Canada the manufacturers check the impulse withstand voltage

TABLE 5.6 Power Factor and Ionization Factor of 15-kV PILC Cable

Test Voltage and Corresponding Electrical Stress	Power Factor	Power Factor Difference	Ionization Factor (%)
3.5 kV (20 V/mil)	0.00371		
8.7 kV (V_0)	0.00387	0.00029	0.029
17.5 kV (100 V/mil)	0.00400		

Note: CEA specified values: power factor at V_0 percent maximum; ionization factor 0.3% maximum.

levels of their shielded cables from time to time for manufacturing information. Results prove that these levels are much higher than the BIL (basic impulse level) of the system for which the cables are voltage rated. Practical experience shows that in terms of the average stress, the impulse withstand voltage for solid-type PILC cable ($V_0 = 23$ kV) size 250 MCM, with an insulation thickness of 390 mils (9.9 mm) has an impulse withstand voltage of 625 kV (crest), corresponding to the average stress of about 1600 V/mil and a maximum gradient of around 50 MV/m. The BIL value of the system for such cables is not higher than 200 kV.

Contrary to North American practice, Europeans include impulse tests in their standards, as type tests. For example, IEC Publication 55-1 includes such a test, with a minimum requirement of the crest voltage equal to $8V_0 + 40$ kV. Impulse tests are usually carried out with 10 positive and 10 negative impulses with a standard wave, which in Europe is 1×50 μs as opposed to 1.2×50 μs in North America. Lastly it has to be added that, whereas impulse tests for solid-type paper insulated cables are valuable for information purposes only, they are of prime importance in extra-high-voltage cables. For example in cables rated 500 kV or higher, it is not the maximum ac stress but the required impulse voltage level that forms the basic criterion on which thickness of insulation is based.

5.6 CONCLUSION

The most important limiting factor in the design of solid-type oil-paper power distribution cables is the impossibility to eliminate the danger of formation of voids in the insulation of these cables in service with no provision to suppress electrical discharges in these voids. Even though the gap spaces between the convolutions of lapped insulation are fully impregnated at the time of manufacture, the voids will inevitably form during the cable operation due to expansion and contraction of the impregnant caused by load cycles. When the electric stress in these voids exceeds the critical value, ionization of gas will follow leading to a slow deterioration of the dielectric. Therefore, to protect the cable performance, the design stresses in solid-type paper insulted cables must be kept within safe limits and insulation thicknesses selected accordingly. As, in general, the maximum gradient in these cables should not exceed 4 MV/m, which is an accepted critical value, it follows that for practical purposes solid-type cables cannot be safely designed for voltages higher than 69 kV (GN) ($V_0 = 40$ kV). This is the maximum voltage rating permitted in North American standards for solid-type oil-paper cables. Even at that voltage the insulation thickness specified is so large (0.650 in. = 16.5 mm) and permitted operational conductor temperature so low (55 and 65°C in Canadian and American standards respectively) that the 69-kV (GN) rated solid-type oil-paper cables are completely uneconomical and seldom required giving preference to more sophisticated paper insulated types, which will be discussed in subsequent chapters.

REFERENCES

[1] L. Emanueli, *High Voltage Cables,* Chapman & Hall, London, 1929.
[2] P. V. Hunter and J. Temple Hazell, *Development of Power Cables,* George Newnes, London, 1965.
[3] C. C. Barnes, *Power Cables, Their Design and Installation,* Chapman & Hall, London, 1966.

[4] "Paper-insulated power cable, solid-type," CSA Standard C68.1, Canadian Standards Association, Toronto.

[5] "Solid type impregnated-paper-insulated lead-covered cable specifications," AEIC Standard No. 1-68, 10th edn. Association of Edison Illuminating Companies, New York.

[6] *Copper Conductors*, Vol. I, "Power cable ampacities," IPCEA Publication No. P-46-426 International Power Cable Engineers Association, New York.

[7] *Aluminum Conductors*, Vol. II, "Power cable ampacities," IPCEA Publication No. P-46-426 International Power Cable Engineers Association, New York.

[8] J. H. Neher and M. H. McGrath, "The calculation of the temperature rise and load capability of cable systems," *AIEE Trans. Power Apparatus Syst.*, Vol. 76, Part III, 1957, pp. 752–772.

[9] "Electric cables—calculation of the current rating," IEC 287 Standard Series (new expanded number: 60287), International Electrotechnical Commission, Geneva.

[10] Power cable ampacities, impregnated paper insulation and copper conductors, IPCEA Publication No. P-48-426.

[11] H. Waddicor, *The Principles of Electric Power Transmission*, Chapman & Hall, London, 1964.

[12] "Test on impregnated paper insulated metalsheated cables," IEC Publication 55-1, International Electrotechnical Commission, Geneva, 1996.

Chapter 6

LOW-PRESSURE OIL-FILLED POWER TRANSMISSION CABLES

W. K. Rybczynski*

6.1 INTRODUCTION

In contrast to solid-type dielectric-liquid-impregnated power cables, low-pressure dielectric liquid or oil-filled (LPOF) cables, because of their principles of operation, must be designed in conjunction with the design of the entire installation. Therefore, it will not be possible within the scope of this presentation to embark on a detailed explanation of the design of an oil-filled cable system and only basic principles will be discussed. Low-pressure oil-filled cables, which constitute essentially the second generation of impregnated paper insulated cables, were developed by Emanueli in an effort to eliminate the problems of void formation and their associated ionization, which were the main drawbacks of solid-type cables, and to create a high-voltage power cable of a design suitable for voltage ratings much higher than those possible for cables of the solid type [1]. This has successfully been attained by replacing the traditional high viscosity cable impregnant by a highly degasified low-viscosity oil, completely filling the cable interior through longitudinal channels provided within the cable assembly. Through specially designed oil-filled cable accessories these channels are connected to feeding tanks, which accommodate oil surplus emitted from the cable during heating cycles, due to thermal expansion; they also return the oil to the cable during cooling periods to cover any oil deficiency caused by its thermal contraction.

In order to protect the highly degasified low-viscosity oil from gas contamination, the entire feeding system must be hermetically sealed and kept under pressure above atmospheric, provided by the feeding tanks of required characteristics. These reservoirs, the construction of which will be described later, must always be attached to the cables from the time of manufacture until the end of the cable service life. Therefore, in a well-made and properly installed oil-filled cable, no voids or pockets of gas are present after manufacturing and while in service. The cable is thus completely corona discharge-free at all times at any operating stresses, unless the voltage is raised to a level that causes breakdown of the oil. This would occur only at approximately 45 MV/m for an oil pressure of 1–2 atms (100–200 kPa) or at a gradient of 70 MV/m for an oil pressure of

* Deceased

15 atm (1500 kPa). It is important to emphasize here, however, that these enormous gradients are much higher than would actually be met in practice, as in general LPOF cables are designed for a maximum gradient of about 10 MV/m up to 15 MV/m for the higher voltage ranges where more sophisticated manufacturing methods are used. Although in a well-made oil-filled cable corona discharges do not occur, some variation in the tan δ value with the change in voltage is generally observed. This is caused by dielectric losses occurring with the solid-liquid dielectric of the paper-oil system itself.

Contrary to other types of cables, there is no concept of cable life in an oil-filled cable, which retains the quality of its performance throughout its useful service life except for the natural aging of cellulose. In fact, practical experience has proven that the power factor characteristics of some oil-filled cables may improve in service, presumably due to the fact that some of the charged particles or ions causing the dielectric loss become gradually immobilized and thus cease affecting the tan δ value [2]. Such a time conditioning method is sometimes used in manufacturing, with a conditioning voltage higher than rated for achieving faster results.

It has to be added that, although the principles of operation of an oil-filled cable, as may be seen from the above brief description, seem to be fairly simple, the practical application of these principles was extremely complex, as it required a great amount of development work not only on the cables themselves but also on accessories without which the cables could not function. Also as important was the development of a hydraulic design for an LPOF cable system, requiring a highly mathematical approach.

6.2 ELEMENTS OF OIL-FILLED CABLE DESIGN

In order to give the reader a means of better understanding the difference between the design of a low-pressure oil-filled cable and a solid type, the format of this section will resemble the format of similar sections on solid types in Chapter 5 with frequent cross-references, wherever necessary.

6.2.1 Voltage Ratings

Low-pressure oil-filled cables are used at voltages falling into the extra-high-voltage (EHV) category having ratings of 69 kV and higher and are not standardized in North America, as they are primarily employed for underground power transmission in big cities at voltages already in existence in their overhead systems. Therefore oil-filled cables of a great variety of voltage ratings may be found in service in Canada, the lowest being 69 kV [grounded neutral (GN)] and the highest 301 kV (GN), the latter being the rating of cables operating at 345 kV and installed in 1954 and in Kitimat, British Columbia, for the Aluminum Co. of Canada [3]. LPOF cables rated at 525 kV were installed in 1977 in the Grand Coulee Dam Hydro-electric plant on Columbia River (Washington) [4]. Presently, this represents the highest voltage rating of LPOF cables now in service. Development work to increase voltage rating of LPOF cables is still carried on, and 750-kV rated cables have recently been developed by Pirelli S.p.A. [5, 6]. All the above given voltages are phase-to-phase in three-phase alternating current (ac) systems, as the method of voltage ratings of EHV cables in North America is identical with that used for other types.

6.2.2 Insulation Levels

Extra-high-voltage cable systems are operated with the neutral point grounded and, therefore, the voltage ratings of EHV cables are specified in terms of the 100% insulation levels. For further details on the voltage insulation levels see Chapter 5.

6.2.3 Selection of Conductor Sizes

As explained in Chapter 5, the selection of conductor sizes is based on thermal calculations to ensure that in actual operation the maximum conductor temperature of cables, carrying required load at a specified maximum ambient temperature, will not exceed the maximum value permitted in oil-filled cables for normal operation. In Canada, in accordance with the manufacturers' recommendations, the value of 85°C is generally accepted for normal operation and 105°C for emergency loading.

The International Power Cable Engineers Association (IPCEA) Power Cables Ampacities Publication No. P-46-426 covering the ampacities of LPOF cables up to the voltage ratings of 161 kV [reprinted as Institute of Electrical and Electronics Engineers (IEEE) Publication No. S-135, SHO7096] may be used for preliminary selection of conductor sizes and subsequently confirmed by the actual step-by-step calculation. The Neher-McGrath method is used in North America, as pointed out in Chapter 5. In order to verify the conductor sizes, the designer must know beforehand all the relevant data, such as the cable ampacity for normal and emergency operation, load factor, number of circuits, geometry of the duct bank (or the cable trench), soil thermal resistivity and all pertinent ambient temperatures (soil, air, etc.). This data will assist in determining whether or not the preselected conductor sizes satisfy the requirements for the maximum conductor temperature (cf. Chapter 11).

The EHV underground transmission circuits generally involve large numbers of cables and, therefore, it is essential to carry out individual ampacity calculations in order to arrive at the most economical conductor size. Before the onset of the computer age, this constituted a very arduous operation; nowadays, however, computer programs are used and the correct solution is obtained with much less tedium.

6.2.4 Selection of Conductor Material and Construction

Except for 345-kV Alcan circuits in Kitimat, British Columbia, in which aluminum conductors were selected, annealed copper is universally used for LPOF cables in Canada; aluminum would not provide any economical advantage because of the necessity of increasing the corresponding conductor sizes and because of the more complicated joining procedures, which in the case of LPOF cables rely on compression sleeves that are not recommended for aluminum.

The conductor of a single-conductor LPOF cable consists of a central oil channel through which the oil is supplied to the insulation. In present cables this channel is formed by a self-supporting conductor composed of preformed segments as delineated in Fig. 6.1. The diameter of the oil channel must be sufficiently large to ensure a free movement of the surplus of oil to the feeding tanks during the heating cycle, without any prohibitive pressure rise, and, which is more critical, to ensure an adequate oil supply from the tanks during cooling cycle, without the danger of creating a negative pressure rise. This depends on the amount of oil present in the cable and is related to the cable size and its voltage rating. For manufacturing reasons the diameters of central oil

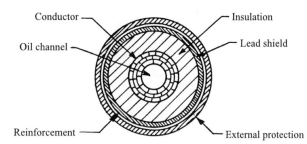

Figure 6.1 Cross section of single-conductor LPOF cable.

channels are standardized by individual manufacturers, ranging from 12 to 24 mm in Canada. For large conductors it pays occasionally to select larger than the necessary central channel size in order to decrease the ac/dc resistance ratio. Very large conductors must be of a specially adapted Milliken type, composed of six segments with the oil channel secured by a metallic spiral placed in the center.

Three-conductor cables employ conventionally stranded conductors in the assembly, and individual oil channels are provided in the interstices, as depicted in Fig. 6.2. Single-conductor LPOF cables permit the longest possible shipping lengths and thereby reduce to minimum the number of costly joints and manholes, or splicing bays. The use of three-conductor cables is rather restricted, and for this reason we shall confine our treatment here to single-conductor LPOF cables.

6.2.5 Selection of Insulation Thickness and Electrostatic Shields

All LPOF cables must be shielded. The traditional shielding materials consist of metallized paper. However, some LPOF cables have conductor shielding composed of several layers of carbon black semiconducting paper tapes, applied directly on the bare conductor. Some 30 years ago the cables permitted the increase of operating gradient in these cables without raising the pressure of oil, owing to higher ac breakdown strength of cables with conductor shielding; this allowed a reduction in the insulation thickness in such cables. In that respect the Association of Edison Illuminating Companies (AEIC) Specification No. CS4-79 [7] is very conservative, a fact generally recognized by North American users, who permit a reduction in the insulation thickness, provided that the electrical test levels of the specification remain unaffected.

The general criterion on which insulation thickness of LPOF cables is selected is based on the value of maximum voltage gradient, which in an ac cable appears on the

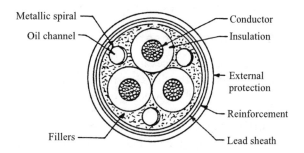

Figure 6.2 Cross section of typical three-conductor LPOF cable.

surface of conductor shield and is calculated from a simple formula given in Chapter 5. It has been stated before that this design gradient should not exceed 10 MV/m (10 kV/mm or 254 V/mil) in cables of the lower voltage range, but that it may be increased to about 13 MV/m (13 kV/mm or 330 V/mil) for higher voltage cables with the use of higher quality papers displaying very low dielectric losses, especially at higher temperatures.

Recently, the efficiency of shielding has been improved by the introduction of the so-called duplex paper, in which one side is semiconducting and the other insulating [8]. By applying this paper such that the insulating side is in contact with the insulation, a highly effective shield is obtained. In this case the overall metallic shielding consists of tinned copper wires woven into a synthetic fabric tape applied overall to secure a good electrical contact between the carbon black side of the duplex paper and the metallic sheath.

6.2.6 Dielectric Liquid Impregnants

The impregnating liquid required for LPOF cables must have a sufficiently low viscosity that is compatible with the acceptable volatility; the volatility of the liquid insulant must be sufficiently low to permit degasification under a high vacuum and not pose any flammability problems; the dielectric liquid must display the capability of gas absorption so that it can remove any traces of gases from the solid-liquid insulating system while in service; its power factor must be as low as possible over the entire range of temperatures in service; finally, it must display good chemical stability. It is not feasible to have all these properties optimized in a given mineral oil. For example, highly refined oils have a very low power factor but cannot meet the gas absorption requirements. Conversely, oils containing a high proportion of aromatic hydrocarbons have good gas absorption properties but would be unstable with regard to power factor. Therefore, a compromise must be made to develop oils having the suitable characteristics.

In the past only mineral oils were used for impregnation of LPOF cables, and such oils are still exclusively used in North America. Over the past years, a synthetic insulating oil (dodecyl benzene) has been developed in Europe, as a by-product of the detergent industry, and was found advantageous in LPOF cables because of its lower viscosity, high thermal and electrical stability, and extremely good gas absorption. However, due to its lower viscosity, it is appreciably more susceptible to contamination effects. As it was also more economical with availability on a large scale, it almost entirely eliminated the use of mineral oils in Europe. In North America the use of dodecyl benzene, though not locally available, has exhibited a marked increase over recent years; in many long-distance submarine cable crossings supplied by European manufacturers, employ either docyl, or the less viscous, nonyl benzene. Other suitable impregnants are described in Chapter 2.

6.2.7 Metallic Sheaths

For LPOF cables manufactured using the traditional techniques, metallic sheaths are applied on the dry core prior to impregnation. This is necessary, as the degasified low-viscosity oil must never be exposed to air, even for a short time, and once accepted by the cable must always be kept under slight pressure above atmospheric during further stages of manufacturing, storage, transport, and in the actual installation. More recently a new mass impregnation technique has been developed by Pirelli

S.p.A. in Italy, allowing cables to be impregnated prior to their sheathing. As this method has also been introduced in Canada, it will be briefly discussed in Section 6.3.

Metallic sheaths of LPOF cables are either of lead alloy or aluminum, depending on manufacturing facilities and preference of manufacturer. However, in some instances, as in the case of cables for duct and manhole type of installations, which with aluminum sheaths would require larger manholes, the user may insist on lead-sheathed cables, In Canada, following the European practice, lead sheaths in LPOF cables are made of half-C lead alloy. For improved fatigue resistance alloy E is some-times specified, as it may also be applied in a continuous extruder [9]. Ram presses are never used for lead extrusion of oil-filled cables.

A cable manufacturer, who has an aluminum press in his plant, can offer smooth or corrugated aluminum-sheathed oil-filled cables. Such cables are more economical, as they do not require reinforcement. However, the aluminum sheaths cannot be applied too tightly because of the very high extrusion temperature and also because (unlike cables with reinforced lead sheath) they cannot breathe, i.e., the sheath cannot expand and contract, following the oil pressure variations. Therefore, aluminum-sheathed cables require feeding tanks of somewhat greater capacity. For the same hoop stresses, aluminum cables are much lighter than cables with reinforced lead and, hence, are advantageous for vertical installations. For instance, the early 245-kV shaft cables supplied and installed in the Churchill Falls hydroelectric plant by the Canada Wire and Cable Company (now Alcatel Canada Wire) are covered with a smooth aluminum sheath. The thicknesses of the sheaths are based on the core diameter and are deter-mined by the relevant specifications. Because of its lower conductivity, the lead sheath thickness is sometimes increased to satisfy the short-circuit requirements.

6.2.8 Reinforcement of Lead-Alloy-Sheathed Cables

The name low-pressure oil-filled cable implies that the minimum oil or dielectric liquid pressure for operation of such a cable is required to be only slightly above atmospheric. However, in actual application the LPOF cables may be exposed to much higher oil pressures particularly when installed on a slope or in a vertical position. There are two basic types of reinforcements used on LPOF cables, which are identified by their function.

(a) *Transversal Reinforcement*: Here the reinforcement consists of nonmagnetic metallic tapes applied with a short lay that protect the lead sheath against the effect of transversal hoop stresses developed in the sheath by oil pressures that may exceed the safe limits. In modern oil-filled cables, fed by pressure tanks, a transversal reinforce-ment is always used, as it acts as an elastic component of the plastic lead sheath. Otherwise the sheath would always have a tendency to expand in a cumulative way after each heating transient. It is usually composed of a single layer of thin hard-drawn copper tapes, applied over the sheath protected by a nonmetallic tape underneath. If the sheath alone is not sufficient to carry the short-circuit current without overheating, transversal reinforcement of several layers of copper tape is used to lower the resistance of the short-circuit path.

(b) *Longitudinal Reinforcement*: In this case the reinforcement is composed of several nonmetallic narrow metal strips, applied with a long lay under a transversal reinforcement. It is used under special conditions where the oil pressure is so high that it creates longitudinal stresses in the sheath beyond the permitted limits. Because of the short lay of application, only transversal reinforcement is not capable of protecting the sheath against such stresses. The value of longitudinal stresses S_ℓ may be determined from Eq. (6.1):

$$P(\tfrac{1}{4}\pi d^2) = S_\ell \pi (d + \Delta d)\,\Delta d \tag{6.1}$$

where P is the hydrostatic pressure (in kg/cm^2 or in kPa), $\Delta d/d$ is the ratio of thickness/ inner diameter of the lead tube, both in the same units of length, and S_ℓ is the longitudinal stress in the same units as P. Considering that for all practical purposes $(d + \Delta d) \approx d$, the formula for the calculation of longitudinal stresses may be reduced to

$$S_\ell \approx \frac{Pd}{4\,\Delta d} \tag{6.2}$$

or

$$P_{\max} = S\left(\frac{4\,\Delta d}{d}\right) \tag{6.3}$$

where $P_{\max}$ is the maximum oil pressure permitted for the sheath with a transversal reinforcement only and S is the maximum permitted hoop stress in the sheath in the same units as $P_{\max}$. If the resultant value of $P_{\max}$ for Eq. (6.3) is lower than that actually required from the hydrostatic oil feeding calculations, either the value of Δd must be increased or, if such a solution is not feasible, a longitudinal reinforcement must be added.

By comparing Eq. (6.3) with Eq. (5.13) developed for bare lead-sheaths in Chapter 5, it can be seen that disregarding other factors, the use of transversal reinforcement doubles the permitted hydrostatic pressure of the oil in the cable. For calculations of the maximum oil pressure permitted for lead sheathed LPOF cables with transversal reinforcement only, the value for the maximum hoop stress for the half-C lead alloy of 21.5 kg/cm^2 (2108 kPa) is found to be acceptable for cables installed in air or ducts. For directly buried cable, this value may be increased by about 10%. It should be noted that because of the uniformity of stresses in the sheath, exerted from within by the oil pressure, the permissible hoop stress values for oil-filled cables are much higher than the comparable values for the solid-type impregnated paper insulated lead-covered (PILC) cables (cf. Chapter 5).

6.2.9 Protective Coverings

For normal land use in ducts or directly buried locations, LPOF cables have extruded nonmetallic coverings. Such coverings have the dual role of providing corrosion protection and an isolation for the induced sheath voltages, which in high-ampacity single-conductor cables in a three-phase system installed with separation may reach an appreciable amplitude under transient conditions. In Canada, high-molecular-weight black thermoplastic polyethylene is generally used because of its excellent resistance to cold temperatures, whereas in Europe polyvinyl chloride (PVC) is preferred.

In lead-sheathed cables, the coverings are usually applied over a fibrous tape, which covers the cable reinforcement; in aluminum-sheathed cables, it is directly applied over the aluminum sheath covered with a suitable flooding compound as an additional anticorrosion protection. Unless armor is required for special applications, such as submarine use, the cable is usually finished with a layer of conductive carbon black coating solely for the purpose of carrying out voltage tests on the jacket.

6.3 CABLE MANUFACTURE

As an LPOF cable is in fact a dielectric liquid or oil-impregnated paper insulated cable, its manufacture in many aspects resembles that of the solid-type cables, except that use is made of more sophisticated methods and controls, requiring specialized equipment in some phases. Therefore, it is proposed not to embark on the description of each operation, as was done in Chapter 5 on solid-type cables, but to concentrate on differences. Consequently, in this section only the manufacture as it pertains directly to single-conductor cables will be discussed.

6.3.1 Self-Supporting Conductor

The manufacture of the self-supporting conductor is carried out on a specially adapted stranding machine. The segments are drawn from rods on the bull-block machine after which they must be meticulously washed in a suitable fluid to remove any particles of copper, which, if left in the conductor, could be carried by oil into the dielectric and cause an increase in the power factor. Special closed-circuit washing equipment is used for that purpose. The importance of cleanliness during the manufacture of the conductor and the other subsequent operations, cannot be overstressed. Stranding of the self-supporting conductor requires considerable skill to prevent the conductor from collapsing due to the dislocation of segments, which must fall into place with absolute precision. Some carbon black shielding tapes are applied on the stranded conductor to impose further firmness in the segments in the conductor with a sacrificial paper tape applied overall as a temporary external protection to secure cleanliness of the surface of carbon black tapes until the time for the next operation.

6.3.2 Insulation and Shielding

Insulating and shielding is performed in a rather sophisticated paper lapping machine, similar to that described in Chapter 5 except that in this instance the entire machine is located in an enclosure in which the air ambient temperature and moisture content are kept within predetermined limits to reduce to minimum the shrinkage of the paper during the drying process of the cable core. The precut paper pads are also kept in this conditioned atmosphere. The application of paper tapes must be executed with a predetermined tension and a small gap between the convolutions so as to ensure that no paper tape will be creased during the subsequent bending on the process reel. These manufacturing precautions, described previously for solid-type cables, are very important in LPOF cable manufacture; paper lapping tension charts, calculated by a computer, are used. Additional carbon black paper tapes for conductor shielding and the insulation shielding tapes are also applied in this operation.

In the traditional method of production, the insulated and shielded cable core is placed in an evacuation tank for predrying and then is either lead alloy sheathed in a continuous extruder (cf. Chapter 5) or furnished with an aluminum sheath in an aluminum press. Lead ram presses are not used for LPOF cables because of stoppages and charge marks, which would endanger the sheath performance in service. After sheathing, the cable ends (with the core still in an unimpregnated and a semidry condition) are hermetically sealed with specially designed end caps equipped with oil fittings. The cable is then placed in an enclosure equipped with pipes for steam heating with the cable ends connected to an evacuating and treating circuit. Then the evacuation cycle commences, which may last several hundred hours, the time depending on cable length, insulation thickness, and the size of central channel. After the evacuation cycle is completed, usually verified by vacuum drop test, the cable is gradually filled with a highly degasified oil or dielectric liquid admitted in a preheated condition under vacuum from the storage tanks. This filling process lasts several hours.

To reduce evacuation time, which in higher voltage cables becomes very long [10], thus limiting the cable length as the entire evacuation is done through the central channel, a novel mass impregnation process has been developed in which the entire drying and impregnation takes place before sheathing. Therefore, the entire surface area of the cable core is exposed in the drying vessel and the desired level of evacuation is reached faster and more efficiently. Afterwards the cable is impregnated in the same vessel and the impregnated core sheathed with lead or aluminum without exposure to the atmosphere. As the length of cable to be treated depends solely on the capacity of the treating tank, much longer lengths can be produced by this method than by the traditional one.

From this moment of production, disregarding the method used for impregnation, the cable must always be kept at a slight pressure above atmospheric by means of a pressure tank connected to one end of the cable and usually located inside the cable reel. This overpressure is most essential, as it protects the interior of the cable from the ingress of moisture during any changes of the oil feeding in further manufacturing operations. In the next operation step, the cable is furnished with reinforcing tapes in a cage-type machine, also equipped with heads for the application of protective tapes under and over the reinforcement. Lastly, thermoplastic covering usually of a black polyethylene variety, is applied in a plastic extruder of a construction similar to that for use on solid-type cables, described in Chapter 5. Applications of a carbon black coating, performed in the same operation before the cable is placed on the reel, terminates the cable manufacture. During each of these operations the feeding of cable must be changed from the pressure tank located in the takeoff reel to another pressure tank in the takeup reel, always taking care that at no instance is the cable left without an oil feed.

6.4 TESTS

In North America the required tests on LPOF cables are usually specified by the customers themselves in their tendering documents. They are generally based on AEIC Specification No. CS4-79, dealing with lead-sheathed LPOF cables, with some additional tests from IEC Publication No. 141-1, such as, for example, the impulse tests, which are sometimes required as type tests but are not included in the American specification. Tests on aluminum sheaths are also based on IEC requirements. [7, 11]

6.4.1 Routine Tests

Routine tests usually follow AEIC Specification No. 4-69, which contains the following tests: (a) Conductor resistance test: the dc resistance of the conductor is measured. It is mainly used to verify the size of the conductor. (b) High-voltage test: performed at the ac voltage-to-ground corresponding to 300 V/mil of the nominal insulation thickness, as spelled out in the AEIC specification, disregarding whether or not the actual thickness has been reduced. These North American voltage test levels are about 5–10% below those specified in the relevant International Electrotechnical Commission (IEC) publication. This is a completely reversed situation from that in the solid-type cables, for which routine tests at much higher levels are required in North America (cf. Chapter 5). The duration of the voltage application is 15 min in both specifications. (c) Ionization factor test: in the IEC Publication No. 141-1, a similar test is called correctly the dielectric loss angle test so as to emphasize the fact that no partial discharges should take place in LPOF cables. The AEIC specified tests measure the power factor at voltage levels roughly corresponding to $0.13V_0$ and $1.23V_0$. Whereas, the IEC publication requires that these levels correspond to V_0 and $2V_0$ (reduced to $1.67V_0$ for cables having voltage ratings of $V_0 > 87\,\text{kV}$). However, since in LPOF cables the power factor versus voltage curve is relatively flat, this difference in voltage levels carries little significance. (d) Spark tests: an ac spark test at a level of about 90 V/mil is specified to check the overall polyethylene covering and is performed in conjunction with the extrusion operation of the covering. As the voltage application by the spark tester electrodes is very short, this test is usually not considered satisfactory for single-conductor cables exposed to any induced sheath voltages in service. In such cases, the covering usually must pass a proper ac high-voltage test with the cable on the shipping reel at the voltage level depending on the customer's requirement for the postinstallation test. Carbon black cable coating serves as an outer electrode in this test. (e) Thickness of sheath: average and minimum lead sheath thicknesses are measured to comply with the specification requirements. (f) Oil-check test: power factor and dielectric strength of the oil are checked in this test. These tests are carried out in accordance with the ASTM standards given in Book of Standards Vol. 10.03, published by the American Society for Testing and Materials (ASTM). (g) Expulsion test: a traditional North American test, that checks gas content in oil of a finished cable that could be admitted due to human error. This particular test is not recognized in Europe as a valid test and is not included in the IEC specifications. The procedure requires that the oil supply to the cable be interrupted during the test; this constitutes a dangerous practice and is contrary to the general principle of not leaving the cable without an oil feed at any given time.

6.4.2 Tests on Specimens

These tests are routine tests performed on specimens with certain specified frequency depending on the size or order. They are discussed below in the same order as listed in AEIC Specification No. CS4-79. (a) Mechanical integrity test: in this test a 1-ft. specimen is used to determine the minimum and average insulation thickness and to examine the quality of paper lapping after removal of the sheath. The lead sheath is also checked to determine its ability to withstand bending. (b) Bending test: after several consecutive bending cycles of a drained cable sample around a cylinder of a specified diameter, depending on the cable size, the deformation of the cable is checked by passing the straightened specimen through a tube having a diameter 0.25 in. larger than the overall cable diameter. A similar bending test is specified by the IEC; however, the test cylinder diameter is about twice that specified by the AEIC, and, instead of an outer deformation check, the specimen must subsequently pass a routine alternating voltage test. It is then dissected for examination of damage or displacement of all the components. (c) Test of corrosion protective covering: This AEIC test may be replaced by alternate requirements in CSA Standard C 170.2-1966. (d) High-voltage time test (ac): performed at room temperature by the application of a 430-V/mil stress for 6 h. This level is 43% higher than the routine high-voltage test level. There is no counterpart of this test in IEC Publication 141-1. The test appears to be of a little value. Since the withstand voltage of an LPOF cable is much higher than that specified, a defective cable would not pass the electrical routine tests. (e) Dielectric power loss and power factor test: carried out at the rated voltage and at not less than four ambient temperatures, namely room temperature, $80°C$, $90°C$ and $105°C$. At the first three temperatures, the power factor should not exceed the specified limits, which are given in Table 6.1. The measurement of power factor at $105°C$ is for engineering information only. A somewhat similar test is also included in IEC Publication 141-1; however, this test, classified as a type test, is required to be performed on only new designs. (f) IEC dielectric security test: although not included in AEIC specifications, it is sometimes required to be performed as a type test on a new design, usually in connection with only very large orders because it is complicated and costly. It is a specimen test composed of (i) a bending test, similar to that performed as a routine specimen test, (ii) a 24-h high-voltage test performed on the same specimen at a voltage level of $2.5V_0$ changed to $1.73V_0 + 100\,kV$ for cables rated $V_0 > 87\,kV$; (iii) an impulse voltage test performed after the bending test on another specimen at the normal operating temperature and the crest voltage level equal to the declared basic impulse level (BIL) value; and (iv) an alternating voltage test after the impulse test at a level specified for the routine high-voltage test.

6.5 ELECTRICAL CHARACTERISTICS

6.5.1 Dielectric Power Factor

Because there are no voids in the insulation of LPOF cables and, hence, no internal discharges in the dielectric, there is no significant increase of power factor with the increase of the applied voltage. However, a small increase in $\tan \delta$ with voltage, which is somewhat greater in cables equipped with carbon black paper conductor shields, does

TABLE 6.1   Power Factor Limits for LPOF Cables

Temperature (°C)	tan δ (%)		
	Cables with $V \leq 161$ kV	Cables with $V = 230$ kV	Cables with $V = 345$ or 500 kV
Ambient	0.35	0.30	0.25
70	0.30	0.25	0.23
90	0.35	0.28	0.25

occur. It has been proven that the more pronounced increase associated with the carbon tape constructions gradually disappears in service under the influence of a prolonged exposure to electric field. The power factor/temperature characteristics of LPOF cables are relatively flat. A typical curve obtained from actual measurements on a 230-kV LPOF cable at the rated voltage-to-ground is shown in Fig. 6.3.

6.5.2 Ionization Factor

AEIC Specification No. CS4-79 defines the ionization factor as a difference in power factor of the dielectric measured at two specified stresses, with a requirement that this difference must not exceed 0.10%. Actual measurements prove that in a well-made LPOF cable, the ionization is much less than the specified limit and rarely exceeds 0.025%, except in lower voltage cables, having greater capacitances, in which carbon black conductor shielding has a somewhat greater influence on the increase of $\tan \delta$.

6.5.3 Alternating-Current Withstand Voltage Level

Breakdown stresses of LPOF cables under alternating voltage conditions are very high. In cables with carbon black paper conductor shields, the values are on the order of 45 MW/m or higher. These breakdown stresses are many times greater than the operating stresses, as the maximum design gradient in LPOF cables is around 10–13 MV/m.

6.5.4 Impulse Withstand Voltage Level

The impulse strength of LPOF cables depends solely on the quality of paper, its thickness, the consistency of registration, and the quality of paper lapping. It is almost independent of the oil pressure. Contrary to cables with viscous impregnants, the impulse strength of oil-filled cables is only slightly reduced at higher temperatures and is not affected by the number of surges, as verified by extensive investigations in a number of leading laboratories. Actual impulse withstand voltage levels of LPOF cables correspond to the maximum gradient of about 100 MV/m. This value is sufficient to satisfy the actual service requirements without a necessity of increasing insulation thickness beyond that based on maximum design gradients at the power frequency. Only for very high voltage ratings, insulation thicknesses of LPOF cables are governed by their required impulse strength.

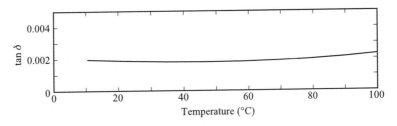

Figure 6.3 Typical power factor vs. temperature characteristics of 230-kV LPOF power transmission cable at rated voltage-to-ground.

6.6 PRINCIPLES OF OIL FEEDING

It is not possible within the scope of this presentation to explore fully methods of calculation of oil feeding design, so that only main principles of the design will be discussed. It has already been stated previously that the dielectric liquid or oil in the cable must be maintained in a degasified condition throughout the life of the cable, and therefore the whole oil system must be hermetically sealed. When the cable is placed in service and is put under load, its temperature gradually rises until it reaches a steady-state condition, at which the heat dissipated from the cable to the ambient becomes equal to the heat generated in the cable. During that period, the oil in the cable expands and must therefore be accommodated in the feeding tanks. Inversely, when the load is switched off, the cable temperature falls and the oil deficiency in the cable must be supplied from the tanks.

In both situations, before the steady state is reached, oil heating and cooling transients prevail in the cable, the former temporarily raising the pressure, with its maximum at the farthest point from the feeding accessory and the latter lowering it, with its minimum at the same point. This is because of the cumulative effect of oil viscosity and friction, which delays the flow of oil through the cable during these transients.

Pressure rises during heating transients are without much importance, as the rein-forced lead or aluminum sheath are strong enough to cope with them. However, a lowering of the pressure may cause dangerously low pressures (vacuum) in some parts of the cable, which must be avoided at all cost as ionization in these places will follow, endangering the cable life. Therefore, the distribution of the feeding tanks must be adequate and the selected pressures high enough to protect the cable against such an occurrence. Feeding the cable from the terminals does not present any problems; however, placing the feeding points along the line is rather costly, as it requires the installation of special feeding joints connected to additional tanks. In order to reduce the number of these joints in a major installation, and to eliminate them entirely in a minor one, the oil pressure at the feeding points must be increased to a predetermined level.

There are two basic systems of feeding the cable from reservoirs, namely a gravity system with the use of gravity tanks (now abandoned, except for older installations still in existence) and a variable pressure system that employs variable-pressure tanks. In a gravity system, the static oil pressure in the cable is assured solely by the elevation of the gravity tank location above the cable and is practically constant except for minor variations, as the oil in the tank is always under atmospheric pressure. However, physically, the oil is not exposed to atmosphere (as in a gravity tank) and the active oil is housed in hermetically sealed expandable cells inside the outer shell of the tank. Hence in a gravity system the only way to increase the static oil pressure in the cable is by placing the tanks on high, unsightly structures, a costly and often not feasible solution. Because of higher overall cost and limitations in design, this system is no longer in use. In a variable-pressure system, the static oil pressure in the cable is provided by variable-pressure tanks, and it changes considerably, depending on the cable temperature. With the temperature rise, oil expands in the cable and enters the tanks, causing the pressure to rise. Vice versa, when the temperature in the cable decreases, the oil is expelled from the tanks into the cable and the pressure falls.

Hence in a variable-pressure system, static pressures at full load are much higher than the static pressures at no load. As variable-pressure tanks have certain limits for maximum and minimum oil pressure and oil capacity, depending on their characteristics, a good feeding design requires a selection of tanks with suitable volume/pressure characteristics that are located suitably so as to ensure a maximum oil output.

A variable-pressure tank consists of a cylindrical steel cylinder, completely sealed, equipped with feeding outlets through which the tank may be connected to the cable feeding point. Inside the tank there is a stack of hermetically sealed cells filled with gas. The cells have diaphragm-type sides and, as the entire space between the cells and the tank wall is filled with active oil, the cells expand and contract following oil pressure changes, matched by gas pressure inside the cells. Consequently, the higher the oil pressure in the tank, the smaller the space occupied by cells, due to the increased gas pressure inside them, and the greater the oil volume inside the tank. As the cells offer a negligible resistance, acting only as a membrane separating gas from oil, the whole process follows Boyle's law, and the pressure/volume characteristics presented as a family of curves for various temperatures form the basis for the oil feeding design. For greater flexibility, variable pressure tanks of various sizes are available, as well as for various pressure ranges.

It may be proven mathematically that the maximum cooling transient occurs when a full load is switched off after a certain time of application while the heating transient is still in existence and oil is flowing toward the tank. Therefore, this maximum cooling transient, which for safety reasons is considered to appear in the cable as its temperature is still at the specified minimum, must be taken into account in oil feeding calculations. The easiest way to carry out these calculations is to represent the oil pressure in the cable by an equivalent column of oil, assuming that the pressure of 1 kg/cm^2 (about 100 kPa) corresponds roughly to an oil column of 11 m.

Except for the very short circuits fed from common tanks, in the larger installations every single-phase cable component of the same circuit has an identical and independent feeding arrangement. Consequently, the feeding calculations are carried out for one cable only, with the results applicable to other cables in the same run. In order to illustrate the various points raised in the discussion, a feeding diagram of a simple LPOF cable installation, with one variable-pressure tank per cable at each end, is presented in Fig. 6.4. The important points pertaining to a construction of such a diagram center on a number of aspects. The selected scale of elevations, representing various parts of the installation, is to be plotted with adequate accuracy. The evaluation of the lowest point of the cable is considered zero elevation. The scale of distances is without importance and is selected only to represent conveniently the profile of the cable route. The difference in elevation between various static and transient lines of the diagram and the profile of the cable route indicates the oil pressure in the cable at that particular point. Similarly, the difference in elevation between these lines and the tank elevation represents the oil pressure in the tank. The hypothetical maximum cooling transient based on the oil flow reversal, as already explained, plotted on the feeding diagram, must not at any point be closer to the line in elevation established for safety. Usually, the difference of 3 m is accepted, indicating a minimum pressure of about 0.3 kg/cm^2 (about 30 kPa). The cooling transient establishes the elevation of the so-called winter line (WL), corresponding to the minimum static pressure in the cable at no load and minimum ground temperature. The difference in elevation of WL and the tank denotes the pressure in the tank at

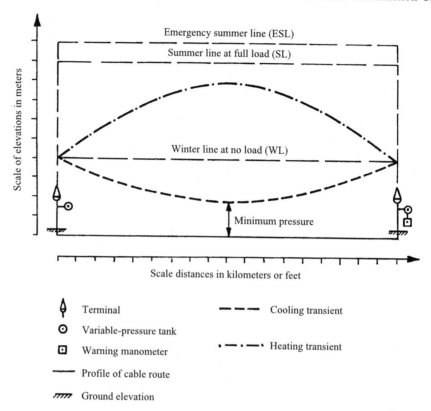

Figure 6.4 Variable-pressure feeding diagram of an LPOF power transmission cable.

the minimum air temperature corresponding to a certain volume of oil, which may be found from the tank pressure/volume characteristics. When the load is switched on, the heating transient appears and the pressure gradually increases in the system, as the surplus of oil flows into the tank, until the cable temperature and corresponding pressure become static again, with the new pressure line establishing itself at some elevation above WL.

In establishing the emergency summer line (ESL), representing the oil pressure at emergency load in summer, the designer must take care that the oil pressure in the cable does not exceed the permissible value for the transversally reinforced lead sheath and that the corresponding pressure in the feeding tanks is within limits of their volume/pressure characteristics. If, with the assumed feeding arrangements, the designer cannot satisfy the above requirements, he must lower the ESL value to safer limits by changing the number of feeding tanks or their type. With longer installations, particularly with a pronounced difference in elevation, the oil feeding system must be subdivided into several entities separated by stop joints.

Longitudinal reinforcements are used in lead-sheathed cables only in extreme cases, as such a solution is very costly. In that aspect aluminum-sheathed cables seem to be more advantageous. Note that the term summer line (SL) is used to indicate oil pressures at maximum normal operating temperature attained by the conductor at full load and under the ambient temperature during the summer months. As the actual pressures in service are expected to oscillate between those indicated by the WL and SL values, warning manometers with min/max electric contact set for these limits are installed, one per each feeding section, to give remote warning if these limits are exceeded.

6.7 NOTES ON SHEATH BONDING

Although sheath bonding in single conductor LPOF power transmission cables in general does not influence the cable design and, as a part of installation design problems, is outside the scope of this presentation, few remarks will be in order on this very important aspect of any major single-conductor oil-filled cable installation. In very short industrial installations, the matter is very simple, and sheaths are usually solidly bonded and grounded at one end, leaving the other end open. However, in underground high-ampacity transmission circuits with single-conductor LPOF cables installed with separation to reduce mutual heating, induced sheath voltages if left unchecked would reach dangerous levels.

Numerically, the induced voltage in the sheath of each cable is given by

$$V_{sh} = IX_m \quad \text{V} \tag{6.4}$$

where I is the current per conductor in A and X_m (ωM) is the mutual reactance between conductor and sheath. The mutual inductance may be determined with sufficient accuracy from

$$M = 2\ln\left(\frac{2y}{r_m}\right) \times 10^{-4} \quad \text{H/km} \tag{6.5}$$

where y is the interaxial cable separation and r_m is the mean radius of the cable sheath in the same units. As single-phase cables are never installed in a triangular formation, y may be taken as the geometrical mean distance between cable axes.

The aim of sheath bonding system is to limit the induced sheath voltage at any point of the line to a safe value, usually restricted to 100 V and to eliminate the sheath currents, or, if this is not possible, to reduce them to a level at which the resulting sheath losses would not visibly affect cable ampacity. The most commonly used method in this respect is the Kirke-Searing cross-bonding method in which the cable route is divided into sections, with each section divided into three equal subsections. At the end of each section the cable sheaths are solidly bonded and grounded, whereas at two intermediate points the sheaths are sectionalized by means of sectionalizing joints and are cross bonded, as shown in Fig. 6.5. As at each cross-bonding the induced sheath voltage changes its phase, it follows that at solidly bonded locations the resultant sheath voltage is zero and no circulating current would be present. Bonding conductors used in sheath bonding must have adequate short-circuit current-carrying capacity. In large systems, the individual grounding at the solidly bonded locations is not sufficient and a grounding conductor must be used along the cable route with a connection made to the substation ground. It should be noted that an ideal solution, as shown in Fig. 6.5, requires complete freedom in selecting joint location, whereas in practice the matter is far more complicated requiring numerous modifications in order to achieve the intended results.

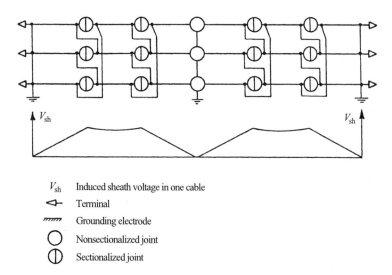

V_{sh}	Induced sheath voltage in one cable
◁	Terminal
⏚	Grounding electrode
○	Nonsectionalized joint
⊕	Sectionalized joint

Figure 6.5 Kirke-Searing sheath bonding diagram to limit the induced sheath voltage of LPOF power transmission cables.

6.8 LIMITATIONS OF LPOF CABLES

Long-lasting and worldwide service experience has proven that LPOF cables may successfully be used in ac operation up to the highest voltages in various types of installations, in very cold or tropical environments, as shaft cables with difference in elevation of several hundred meters, as well as in long submarine crossings in ac or dc operation, where the cables are fed from the ends by pumping plants [12, 13].

The cables may be built with a very large conductor and their ampacity further increased by water cooling. LPOF cables for 765-kV systems have been developed by Pirelli S.p.A. in Italy basically with the same design as their lower voltage counterpart but with very carefully selected components and more sophisticated methods of their application [5]. However, at such high-voltage ratings, the dielectric losses with respect to the conductor losses are high. Therefore, the only economical solution for a practical application of these cables is to increase their ampacity by forced cooling. Development of synthetic tapes, which would lower these losses, has been faced with some difficulty in the past because of strict impermeability requirements; however, as indicated in Chapters 1 and 2, these problems appear to have been well resolved by this use of composite polypropylene-paper-polypropylene (PPP) tapes.

6.9 SELF-CONTAINED HIGH-PRESSURE OIL-FILLED CABLES

It has been stated earlier that the ac withstand voltage of an oil-filled cable may substantially be raised if the oil pressure is increased. Hence, the design gradient in such cables may be increased and the insulation thickness reduced. Considering that obvious advantage, high-pressure oil-filled (HPOF) cables were first introduced in France with design gradients much higher than those in LPOF cables. However, as it was found that the impulse strength of these cables was not increased, it was necessary to obtain the required improvement by resorting to the use of very thin papers of high impermeability. Nowadays, HPOF cables with pressures on the order of 1200–1400 kPa have design gradients as high as 18 MV/m, resulting in significantly smaller insulation walls.

The HPOF cables are advantageous from the feeding point of view because they permit a greater distance between feeding points due to a smaller pressure drop under transient conditions. However, because of the high oil pressure, HPOF cables require stronger sheaths and very robust and costly accessories and are not economically competitive with LPOF cables. In addition, the voltage ratings of HPOF cables are limited by higher dielectric losses and higher capacitance due to the reduced wall thickness. Self-contained HPOF cables, completely unknown in North America should not be confused with HPOF pipe-type cables, which are very popular in North America and which will be discussed in Chapter 7.

6.10 SELF-CONTAINED OIL-FILLED CABLES FOR DC APPLICATION

Because of an ever-increasing number of dc power systems employing self-contained oil-filled cables in the world, a few words have to be said on basic differences between traditional design of self-contained oil-filled cables for three-phase power circuits and their design for dc bulk power transmission lines. In addition to Chapter 14, which is entirely devoted to dc cables, references included in this section will give the reader a possibility of obtaining more information on the subject.

The use of dc eliminates charging current and reduces overall losses unavoidable in ac transmission, such as dielectric losses and conductor skin and proximity losses. Hence dc oil-filled cables are particularly suitable for long EHV submarine cable crossing [14]. Another typical application is in a network requiring an increase in power without increasing the short-circuit current [15]. A dc cable can operate at dielectric stresses of more than double those in a similar cable in an ac system. In fact, the breakdown strength of the dielectric under dc stress approaches that of the impulse strength and is much higher than under ac stress. A transmission line consisting of two dc cables can transmit more than twice the power of a comparable ac line composed of three single-conductor cables. That advantage would offset the cost of additional ac/dc conversion equipment, provided that the transmission line is sufficiently long.

One of the basic differences in the design of ac and dc cables is the stress distribution across the cable insulator. In an ac cable this distribution depends on cable geometry with maximum gradient always appearing on the conductor shield decreasing logarithmically until it reaches the minimum value at the insulation shield. Contrary to this, in a dc cable the stress distribution is controlled by conductivity, depending on the temperature of the insulation and on electrical stress of the insulation. It follows that for each temperature drop across the insulation, the stress distribution curve is different and, under load, the maximum stress will appear at the outer surface of the insulation (cf. Fig. 6.6).

As the maximum stress must not exceed the maximum design value, the temperature drop across the insulation must accordingly be limited to a corresponding value. Hence, ampacity of a dc cable is governed by the maximum permissible temperature drop across the insulation provided that the maximum design conductor temperature is

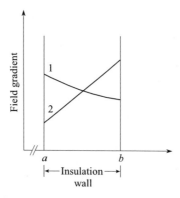

Figure 6.6 Schematic representation of stress distribution within insulation of dc cable.

also met. For a dc self-contained oil-filled cable, the following design parameters are used: (i) service working stress of 35 kV/mm (max) and (ii) maximum conductor temperature for continuous load of 90°C.

Because dc cable ampacity is limited by the maximum allowable electric stress at the insulation shield, it follows that the transmittable power would increase with the increase of the insulation thickness. This seemingly paradoxical feature may be explained by the fact that the increase in the thickness of insulation has a somewhat greater effect on increasing the cable ampacity than the negative effect of an increase in the thermal resistance of the insulation.

It is clear that dc self-contained oil-filled cables are particularly suitable for bulk power transmission links at highest voltages; at the present time technology for naturally cooled cables systems for +600 kV has been developed [16, 17]. It may be envisaged that such cable would be capable to transmit up to 3000 MW. Further extension of power could be achieved by forced cooling; however, in dc cable systems only internal conductor cooling would be efficient and such technology is now being developed [16]. Additional details on dc cables are provided in Chapter 14, which deals exclusively with dc cables.

REFERENCES

[1] L. Emanueli, *High Voltage Cables*, Chapman and Hall, 1929.

[2] R. B. Blodgett and P. H. Gooding, "Parameters affecting the increase in dielectric loss caused by carbon black paper screens for oil paper dielectrics," *IEEE Trans. Power Apparatus Syst.*, Vol. PAS-83, 1964, pp. 121–130.

[3] H. D. Short and J. T. Madill, "The 301 kV all-aluminum oil-filled cables at Kemano, British Columbia," *IEEE Trans. Power Apparatus Syst.*, Vol. 76, Dec. 1975, pp. 1329–1343.

[4] J. J. Ray, C. A. Arkell, and H. X. V. Flack, "525 kV self contained oil-filled cable systems for Grand Coulee Third Powerplant, design and development," *IEEE Trans. Power Apparatus Syst.*, Vol. PAS-93, 1974, pp. 630–639.

[5] G. M. Lanfranconi, G. Maschio, and E. Occhini, "Self contained oil-filled cables for high power transmission in the 750 to 1200 kV range," *IEEE Trans. Power Apparatus Syst.*, Vol. PAS-93, 1974, pp. 1535–1545.

[6] E. Occhini, G. M. Lanfranconi, and M. Tallarini, "Self contained oil-filled cable systems for 750 and 1100 kV. Design and tests". CIGRÉ, Paper No. 21-08, Paris, 1978.

[7] Specifications for Impregnated-Paper-Insulated Cable Low and Medium Pressure Self-Contained Liquid Filled Cable, AEIC No. CS4-79, Association of Edison Illuminating Companies, New York, 1994.

[8] F. J. Miranda and P. Gazzana-Priaroggia, "Self-contained oil-filled cables. A preview of progress," *IEE Proc.*, Vol. 123, 1976, pp. 229–238.

[9] D. G. Howard, "Selection of cable sheath lead alloys for fatigue resistance," *IEEE Trans. Power Apparatus Syst.*, Vol. PAS-96, 1977, pp. 80–87.

[10] R. M. Russek, "Drying and impregnating of power insulated power cables," *IEEE Trans. Power Apparatus Syst.*, Vol. PAS-86, 1967, pp. 34–52.

[11] Tests on oil-filled and gas-pressure cables and their accessories. Part 1: Oil-filled, paper-insulated, metal-sheathed cables and accessories for alternating voltages up to and including 400 kV, IEC Publication 141-1, International Electrotechnical Commission, Geneva (1996).

[12] G. M. Lanfranconi, G. Gualtieri, and M. Cavalli, "330 kV oil-filled cables laid in 1600 ft. Vertical shaft at Kafu Gorge Hydroelectric Plant," *IEEE Trans. Power Apparatus Syst.*, Vol. PAS-92, 1973, pp. 1992–1998.

[13] P. Gazzana-Priaroggia and G. Maschio, "Continuous long length AC and DC submarine HV power cables," *IEEE Trans. Power Apparatus Syst.*, Vol. PAS-92, 1973, pp. 1744–1751.

[14] O. Hauge, A. Berg, J. Norman Johnsen, G. Wettre, and K. Bjorlow-Larsen, "The Skagerrak HVDC cables," CIGRÉ, Paper No. 21-05, Paris, 1978.

[15] F. H. Last, P. Gazzana-Priaroggia, and F. J. Miranda, "The Underground HVDC link for the transmission of bulk power from the Thames Estuary to the centre of London," *IEEE Trans. Power Apparatus Syst.*, Vol. PAS-90, 1971, pp. 1893–1901.

[16] G. Bianchi, G. Luoni, and A. Morello, "High voltage dc cable for bulk power transmission," *IEEE Trans. Power Apparatus Syst.*, Vol. PAS-99, 1980, pp. 2311–2317.

[17] G. Bahder, G. S. Eager, G. W. Seman, F. E. Fischer, and H. Chu, "Development of ±400 kV/±600 kV high and medium-pressure oil-filled paper insulated DC power cable system," *IEEE Trans. Power Apparatus Syst.*, Vol. PAS-97, 1978, pp. 2045–2056.

Chapter 7

HIGH-PRESSURE OIL-FILLED PIPE-TYPE POWER TRANSMISSION CABLES

W. K. Rybczynski*

7.1 INTRODUCTION

High-pressure oil-filled pipe-type (HPOFPT) power transmission cables, also known as *oilostatic cables*, are a typical product of American inventiveness and practicality. If the mineral oils in these cables are replaced by other synthetic liquids, it is common to refer to them as high-pressure dielectric liquid-filled pipe-type cables. Such cables consist simply of three single core impregnated-paper-insulated solid-type cables, drawn into a previously installed steel pipe and filled with oil or another dielectric liquid under high pressure. During their first stages of development, the cores were protected by a thin lead sheath that was removed before insertion into the pipe, subsequently being replaced by a wrap of plastic tapes intercalated with metal tapes. Developed in the 1930s, the oilostatic cable soon became the number one extra-high-voltage (EHV) cable in the United States, for certain period of time almost completely eliminating any interest in low-pressure oil-filled (LPOF) cables, even though the latter were also first introduced in North America [1–7].

The rise in popularity of HPOFPT cables in the United States is attributed to a number of factors. First, there existed in the United States already a highly developed high-pressure pipe-laying technology based on crude oil pipelines. Second, the overall cable design was relatively simple with practically the same manufacturing processes as for solid-type paper insulated cables. Third, the absence of sheaths provided a light cable construction. Fourth, a reduction in splicing locations was effected because of feasibility of producing and installing the cable in much longer lengths than in the case of self-contained cables. Fifth, overall economy was achieved because in many cases existing city bylaws required that LPOF cables be installed in costly ducts and man-holes, whereas pipe-type cables could be laid directly in the ground. In Europe, where all LPOF cables are directly buried, even in the centers of large cities, the cost situation is entirely different and, consequently, with the exception of France, European manu-facturers have shown very little interest in this type of cable. In Canada HPOFPT cables were introduced first in 1957 and since that time have made a great impact, particularly in rapidly growing cities in the Canadian west. However, the LPOF cable

*Deceased

network has also grown lately and, for example, large cities like Montreal and Vancouver rely entirely on LPOF cables in their underground transmission systems [3].

As HPOFPT cables are in fact assembled in the field, their installation is of major importance. However, as the installation practices are outside the scope of this chapter, the reader is referred to Chapter 11.

7.2 PRINCIPLES OF OPERATION

The HPOFPT power transmission cables have been designed to extend the voltage rating of viscous oil-paper-impregnated cables beyond the limits imposed by ionization on conventional paper insulated lead-covered (PILC) cables. This has been achieved by inserting three single-conductor shielded cable assemblies, similar to those of solid-type PLC cables, into a steel pipe subsequently filled with low-viscosity oil and then pressurizing it to about 200 psi (1400 kPa). Due to the exerted pressure, the impregnating oil gradually penetrates the cable insulation dissolving any remaining gas occluded in it and thereby inhibiting all ionization. However, this is a lengthy process and may last several thermal cycles. Therefore, there is no similarity with HPOF cables in which the purpose of the pressure is to increase their dielectric strength. Consequently, the service gradients in HPOFPT cables are correspondingly much lower and are generally of the order of 9 MV/m. During the first months in operation, HPOFPT cables resemble more a compression cable than an oil-filled cable.

Expansion and contraction of the impregnating oil during thermal cycles is dealt with by an adequately designed and automatically operated pumping plant, capable of keeping the pressure of the oil in the pipe within preset limits, thereby supplying oil deficiency from a storage tank and removing the surplus back to the tank through relief valves.

7.3 ELEMENTS OF CABLE DESIGN

As the design of single-conductor cables used in a HPOFPT cable system is very similar to a design of solid-type single-conductor shielded cable assembly, modified by an addition of a moisture barrier and skid wires, frequent references will be made to the sections in Chapter 5 pertaining to solid type cables.

7.3.1 Voltage Rating

The HPOFPT cables have been developed in the United States to cover EHV ratings up to 500 kV and a large quantity of 345-kV cables have been in operation since 1963. In Canada the highest voltage rating of HPOFPT cables is 240 kV, this being the highest underground transmission voltage in most parts of the country.

7.3.2 Insulation Levels

In Canada all EHV cable systems are operating with the grounded neutral point requiring cables rated at the 100% insulation level, however in the United States, over the lower voltage ranges, cable ratings at the 133% insulation levels are also used.

7.3.3 Selection of Conductor Sizes

The Insulated Cable Engineers Association (ICEA) power cable ampacities covering ampacities of HPOFPT cables up to a voltage rating of 161 kV may be used as a guide for preliminary selection of conductor sizes (Publication No. P-45426; see [12] in Chapter 5). Later, the values may be confirmed by the actual step-by-step calculation following the Neher-McGrath method (see [8] in Chapter 5). As specified in the relevant Association of Edison Illuminating Companies (AEIC) specification, ampacity calculations must be based on the maximum conductor temperature of 85°C for normal operation and of 105°C for emergency operation lasting 100 h, reduced to 100°C if the emergency period is extended to 300 h. According to the AEIC definition, the emergency temperatures are based on the assumption that their application will not exceed a maximum of one emergency period in any 12 months, nor a maximum of 0.2 emergency periods per year averaged over the entire cable service life [8].

Final ampacity calculations must be carried out after the cable design is completed and after the formulation of component cables in the pipe is determined because of the magnetic pipe losses. This will be explained subsequently in Section 7.3.9 on the jam ratio considerations (cf. Chapter 11).

7.3.4 Selection of Conductor Material and Construction

Conductors of HPOFPT cables are of bare copper or aluminum and are of concentric round construction, except that for large cross sections, segmental conductors composed of four segments are selected to decrease the ac/dc resistance ratio. Compact conductors do not seem to offer any advantage except in cases where the decrease in conductor diameter is essential from jam ratio aspects.

The AEIC specifications permit using soft-drawn (SD), medium-hard-drawn (MHD) or hard drawn (HD) copper wires before stranding, depending on the tensile strength required to maximize pulling lengths, to reduce the number of splicing locations. Most common, however, is to design conductors from soft-drawn copper wires without annealing the conductor after stranding, thus hardening copper to the MHD grade having greater tensile strength. Aluminum of electrical grade is preferred in cases, where the pipe size is not affected by the increase of conductor diameter. Otherwise the gain in the lower cost of aluminum conductor would be offset by the higher cost of the carrier pipe. Also splicing of aluminum conductors is more troublesome as they must be welded, whereas compression connectors can be used for copper.

7.3.5 Selection of Insulation Thickness and Electrostatic Shield

Conductor shielding consists of conducting or semiconducting tapes applied immediately over the conductor. Metallized or carbon black paper tapes are used. Insulation thickness is specified in the relevant AEIC specification. However, as for each voltage rating the same thickness covers a range of sizes, it follows that to attain a most economical solution, the insulation thickness shown in the AEIC specification must in some cases be either reduced or increased in order to keep the maximum design gradient at the conductor within safe limits (around 9–10 MV/m).

In the very high voltage ranges, the required basic impulse level (BIL) values for the cable may affect the required insulation thickness. The AEIC specifications indicate that the maximum impulse stress at the required BIL should be limited to 2006 V/mil (79 MV/m) for cables rated at 115–151 kV, and 2260 V/mil (89 MV/m) for higher voltage rated cables. The reduction of insulation thickness, below that specified in AEIC specification, is fully permitted in Canada, provided that all testing requirements of that specification are met. Another important factor that could affect the selection of insulation thickness is the so-called jam ratio, defined as a ratio of internal diameter of the pipe to the overall diameter of the finished cable. The insulation shielding applied directly over the insulation surface may consist of conducting and semiconducting tapes such as, for example, metallized paper intercalated with carbon black paper tapes.

7.3.6 Impregnating Oil

High viscosity pure mineral oil or an equivalent synthetic dielectric liquid (cf. Chapter 2) very similar to that in solid-type cables is used for impregnation of insulated and shielded cable assemblies. Even though regarding power factor, particularly at higher temperatures, the high-viscosity oil is inferior to low-viscosity oil, it is used in HPOFPT cables for two important reasons. First, the impregnating oil must have a high resistance to oxidation because during the application of the moisture seal it is completely exposed to the atmosphere. Second, following manufacture, in storage, transport, and on site prior to the pipe oil filling, the impregnating oil must be prevented from leaking through the moisture seal; hence, in cold conditions its viscosity must be very high. As during the operation the impregnating oil will gradually mix with the low-viscosity pipe filling oil, both types of oils or dielectric liquids must be compatible.

7.3.7 Moisture Seal and Skid Wires

The moisture seal is applied directly over the insulation shield. It is a temporary barrier protecting the cable interior against the ingress of moisture, acting as a substitute sheath, and is very essential until the cable is installed and the pipe filled with oil. In modern HPOFPT cables, moisture seals are made of nonmagnetic metal tapes intercalated with metallized Mylar tapes. Over the moisture seal, skid wires are applied with a short lay to decrease the friction during the pulling operation into the pipe, at the same time protecting the moisture seal against damage during that operation. D-shaped skid wires, usually of dimensions 0.200 × 0.100 in. (5 × 2.5 mm), with their flat side toward the moisture seal must be designed and applied in such a way as to provide an electric contact between insulation shield and skid wires.

7.3.8 Carrier Pipe and Pipe Coating

Although fabricated and installed by other than the manufacturer, the carrier pipe is in fact an integral part of an HPOFPT cable replacing the cable sheath and, as such, must be considered by cable manufacturers as an element of cable design. It is a black, seamless, low-carbon steel pipe, usually about 0.25 in. (6.4 mm) thick, supplied in random or double random lengths of 40 or 80 ft (12 or 24 m). The type with flared ends and chill rings for welding is commonly used. Another type with bell and spigot may be slightly more economical, but it permits safe pulling in one direction only, thus restricting the installation procedures.

Although carrier pipes are always cathodically protected to prevent corrosion, the prime anticorrosion protection is an efficient pipe coating. The function of the pipe coating is twofold; it must protect the pipe from chemical corrosion and electrolysis and it must provide some degree of insulation for the cathodic protection, installed as a second line of defense in case of deterioration or mechanical damage of the coating. In the United States, somatic coating, applied hot to the pipe by direct extrusion in a specialized plant, is commonly used; in Canada taped coverings composed of a combination of woven glass fabric tapes and pipeline felts, sandwiched with layers of coal tar enamel, are preferred. After application of the entire coating, usually about 0.5 in. (13 mm) thick, tests are carried out for defects at 15 kV with a suitable detector equipped with a ring electrode. The same tests are repeated after the pipe is welded and is placed in the trench just before backfilling. After grit blasting to remove all dirt, rust, and mill scale, the interior of the pipe is coated with a special compound, that protects the cable interior against oxidation during storage, also decreasing friction during the pulling of cables. This interior pipe coating must not cause any contamination of pipe filling oil.

7.3.9 Coordination of Pipe and Cable Sizes

After the cable design is completed and overall diameter established, a suitable pipe size must be selected. Only standard pipes should be considered to avoid prohibitive cost, and therefore their sizes may not necessarily match calculated internal diameters. Hence the configuration of the cables inside the pipe may vary.

It is obvious that from the cost aspect, minimum pipe size should be selected. However, it must be such as to secure a minimum clearance between the cables and the pipe wall and an adequate relationship between inside diameter of the pipe and the diameters of each component cable to prevent jamming of conductors during pulling operation. Regarding the first condition, a clearance of 0.50 in. (13 mm) is usually acceptable and could be slightly less, if the section does not contain bends in which the pipe might develop ovality. The second condition is secured by examining the so-called jam ratio, expressed as D/d, in which D is the internal pipe diameter, and d is the nominal overall diameter of each component cable plus 1.5 times the thickness of skid wires, which also must be considered. If the calculated value of D/d is equal to or less than 2.5, the cables are most apt to fall into triangular formation, which is most advantageous from the point of view of the electric losses (see Fig. 7.1) In this case minimum clearance must be checked; however, no jamming of the cables during pulling is possible. This is the best solution but not always feasible because of restriction in availability of pipe of the right size.

If the value of D/d is greater than 2.5, a cradled configuration is likely to result in which one cable is at the bottom of the pipe with the other two resting symmetrically on both sides (see Fig. 7.2). However, if the jam ratio D/d is slightly under 3.0, jamming

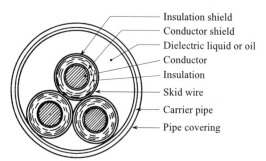

         — Insulation shield
         — Conductor shield
         — Dielectric liquid or oil
         — Conductor
         — Insulation
         — Skid wire
         — Carrier pipe
         — Pipe covering

Figure 7.1 Cross section of HPOFPT power transmission cable.

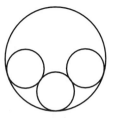

Fig. 7.2 Diagram of cradle formation.

may occur, as the bottom cable will tend to move upwards due to the pulling tension, resulting in serious damage to the cable. Considering possible ovality of the pipe and other factors, it is generally accepted that the value D/d must not be greater than 2.8. Therefore, in practice it is possible that to refrain from that condition, the designer, being unable to choose a suitable pipe diameter, would be obliged to increase d by increasing the insulation thickness. A situation may also arise in which the correct clearance value for triangular formation could be secured by slightly decreasing d by a reduction of insulation thickness or by using compact conductors. Of course, the probability of jamming could be completely eliminated by considering jam ratios much larger than 3.0; such a solution, however, would hardly be economical.

7.3.10 Pipe Filling Oil

Pipe filling oil (or dielectric liquid) is a low-viscosity, low-dielectric loss fluid specially developed for HPOFPT cables. As the oil in the pipe is mainly used as a pressurizing medium over a large volume, its characteristics are different from those used in self-contained oil-filled cables. The main differences in the oil characteristics are given in Table 7.1. It should be noted that, contrary to the situation in LPOF cables, the pour point for the pipe filling oil cannot be very low in order to permit the large volume of oil in the pipe to be solidified by freezing in case of repairs. As pointed out in Chapter 2, polybutene oils constitute suitable substitutes for the mineral oils normally used in pipe filling applications.

7.4 CABLE MANUFACTURE

It has been mentioned already that the insulated and shielded cable assembly closely resembles that of a single-conductor solid-type paper insulated cable. Hence the manufacturing is also similar to that described in Chapter 5 with some exceptions, which may be seen from a brief description of the operations. The cable conductors—whether of concentric, round, or segmental form—are identical to those used in solid-type

TABLE 7.1   Characteristics of Different Types of Cable Oils

	Impregnating Oil	Pipe Filling Oil	LPOF Cable Oil
Viscosity at 100°F (37.8°C), SUS	2365	763	46
Viscosity at 210°F (99°C), SUS	101	60	32
Flash point °C (Cleveland open cup)	249	202	132
Pour point, °C	−18	−26	< −46

cables. Except for different tensions of application of paper tapes and different type of grading of paper used to obtain a very hard core, essential in HPOFPT cables to prevent damage during pulling, the insulating and shielding operation is identical to that of an LPOF cable described in Chapter 6. Shielding is applied over the insulation in the same operation usually composed of metallized paper tape intercalated with carbon black paper tape. The impregnation process is identical to that described in Chapter 5 for solid-type cables. However, because of the much greater demands regarding electrical performance from HPOFPT cables, the entire process must be done in a much more sophisticated way with different parameters and better equipment. The drying and impregnating tanks are also much bigger because of the longer lengths involved. Although the impregnated cable core may be for a short time exposed to the atmosphere without any ill effects, because of a built-in resistance to oxidation of the viscous oil, this exposure, unavoidable during the moisture seal application, must still be as short as possible. To achieve this, the impregnation plant must be designed in such a way as to allow the cooled cable to remain completely immersed in oil while waiting for the application of the moisture seal. Therefore, impregnation with the cable placed in a pan or tray, and not on a reel, is preferred. The cable in the pan, completely covered with the oil, is then placed on a turntable, from which it is transferred for the moisture seal application.

Application of moisture seal and skid wires is performed in the same operation in a specially designed machine equipped with suitable heads and located close to the impregnating plant to shorten the exposure time of the cable during the transfer. This skid wire line is directly wound on a shipping reel, constructed in such a way that the cable may be completely sealed inside and kept under an atmosphere of dry nitrogen during storage and shipment. After the routine tests (see below), the reel with the cable is hermetically sealed by fixing a plastic-coated metallized cloth across the traverse and by purging the interior with dry nitrogen through a special valve placed in the flange of the reel. The dew point of the gas emitted from another outlet valve is checked and the purging stopped after the required dew point of about $-25°C$ is reached. Bags with silica gel are placed inside to absorb any traces of moisture. If after a short interval a second dew point check discloses no dramatic increase in moisture content of gas inside the reel, confirming the effectiveness of the seal, an aluminum sheath is placed over the cloth around the reel to provide mechanical protection. Subsequently, the reel is lagged and made ready for shipment or storage. To counteract the downwards tendency of oil migration due to gravity in the cable, the reels in storage or during a long transport should be rotated from time to time.

7.5 TESTS

Tests to be performed on HPOFPT cables are to satisfy the requirements of AEIC Specification No. 2.73, which also covers cable ratings up to 500 kV, although there is no practical service experience with HPOFPT cables at these higher voltages. In addition, major North American utilities may also specify an impulse test to be carried out on a specimen, as a type test, with the procedure in accordance with the International Electrotechnical Commission (IEC) Publication 230.

7.5.1 Routine Tests

Since routine tests are made on the cable as shipped from the factory, i.e., without any pressurization, the routine high-voltage tests must be made at a reduced voltage; hence they cannot fully establish operational quality of the cable. Only in conjunction with high-voltage specimen tests, performed on cables under full operational pressure, do these routine tests become meaningful. It has to be added that no tests carried out in the field on the pipe itself, prior or after the cable pulling, are included in this section as such tests are considered installation tests and are dealt with in Chapter 11.

The conductor resistance test is performed to verify the conductor size. However, as the maximum dc resistance values given in the AEIC specifications pertain to copper with 100% conductivity and 2% tolerance, then any HD conductors, if such are used, will be slightly larger to comply with this requirement. The high-voltage test is performed at the alternating voltage-to-ground specified in the relevant AEIC specifications. Compared to the routine test voltage levels for LPOF cables, the required test voltages are very much reduced and in terms of the rated voltage-to-ground these test voltages may be expressed as equal to about V_0 for 345-kV cables. The duration of the voltage application is 15 min. The ionization test is very similar to the test specified by AEIC for LPOF cables where the power factor is specified at two voltage levels, corresponding to 100 V/mil (3.9 kV/mm) and 20 V/mil (0.8 kV/mm). In terms of the rated voltage between each conductor and sheath, V_0 these values correspond to $0.71V_0$ and $0.14V_0$ for 69-kV rated cables gradually reduced with the increase of voltage rating to the value of about $0.5V_0$ and $0.1V_0$, respectively, for 345-kV cables. Under these conditions the ionization factor (the increase in power factor) should not be greater than 0.1% (0.001).

7.5.2 Tests on Specimens

These tests are routine tests performed on cable specimens with certain specified frequency. High-voltage tests are performed with a specimen of the single-conductor cable core inserted in a pipe filled with pipe filling oil and fully pressurized to the operational pressure not exceeding 200 psi (1400 kPa). A 30-ft specimen in the pipe is terminated with regular HPOFPT cable terminals.

The mechanical integrity test is used to determine the minimum and average insulation thicknesses and the condition of shielding tapes and the moisture seal. The insulation is also checked for the quality of paper lapping and for any wrinkles and tears of the paper tapes. Poor paper lapping, verified by the number of registrations in excess of those permitted in the specification, is considered a failure of the test; however, a certain amount of wrinkling requires that another specimen with the same degree of wrinkling has to pass a routine high-voltage test with the time of voltage application extended to 30 min.

The bending test is performed on a 10- to 20-ft specimen of the cable core with the skid wires still in situ. Before the bending test, the specimen is subjected to a temperature of $-10°C$ for not less than 2 h. After several consecutive bendings around a cylinder of a specified diameter, three 1-ft pieces, cut from the center of the specimen are examined for damage with the center piece also measured for deformation. If the total number of torn tapes in any 10 consecutive tapes is more than 2, or if minimum insulation thickness is less than 80% of either the specified or measured average thickness, the specimen is considered to have failed the test.

The dielectric power loss or power factor test and the next two specimen tests are usually carried out on the same fully pressurized 30-ft-long specimen. It is performed at the rated voltage and three temperatures, i.e., ambient, 70°C, and 90°C approximately. The power factor established from a graph should not exceed the limits specified in Table 7.2. The more severe requirements for HPOFPT cables of voltage ratings $V \leq 161\,\mathrm{kV}$, as compared to those of the LPOF cables, should not be interpreted as indicating that the LPOF cables have higher power factors but rather that the values in AEIC Specification No. 2-73 are more realistic than those in AEIC Specification No. 4-69, although in both specifications these limits seem to be too liberal considering the present state of the art. The ionization factor test is similar to the routine ionization test, except that this test is made on a fully pressurized specimen and therefore, the higher voltage level is based on much higher average voltage stresses of 180 V/mil (7.1 kV/mm) for 69-kV rated cables, reaching the value of 230 V/mil (9.1 kV/mm) for cables rated over 161 kV. In terms of V_0 these voltage levels correspond roughly to about 1.2–$1.3V_0$, depending on the voltage rating of cables. Between these values and the lower average stress levels corresponding to about $0.1V_0$, the increase of power factor (ionization factor) must not exceed 0.1% (0.001) in cables rated up to 161 kV and 0.05% (0.0005) for higher voltage rated cables.

The high-voltage time test is performed at room temperature and with the oil pressure reduced to 100 psi (700 kPa). The specimen must withstand the test voltage corresponding to 430 V/mil (16.9 kV/mm) for 6 h, which in terms of V_0 equals $3V_0$ for 69-kV rated cables gradually reduced to $2.2V_0$ for 345-kV cables.

The impulse test is not included in the relevant AEIC specifications and is only sometimes required as a type test. It is performed in accordance with the IEC Publication 230, with the crest voltage equal to the BIL of the system. Ten negative and 10 positive impulses are applied to a fully pressurized single conductor specimen with a standard wave of $1.2 \times 50\,\mu s$. Information regarding BIL values for various system voltages may be obtained from the IEEE Standard 48-1995.

7.6 ELECTRICAL CHARACTERISTICS

The electrical characteristics of HPOFPT cables are more or less in line with those described in a similar section in the chapter on LPOF cables (cf. Section 6.5). We shall discuss them briefly here in the same order.

7.6.1 Dielectric Power Factor

The power factor of HPOFPT cables, at the rated voltage and ambient temperature, is only slightly higher than in the case of LPOF cables, because of the suppression

TABLE 7.2 Limits of tan δ Values for HPOFPT Cables

Cable Temperature (°C)	tan δ (%)		
	Cables with $V \leq 161$ kV	Cables with $V = 230$ kV	Cables with $V = 345$ and 500 kV
20–40	0.40	0.30	0.25
80	0.35	0.25	0.23
90	0.40	0.28	0.25

of ionization of voids by external oil pressure. However, at higher than the rated voltages, the suppression of ionization in HPOFPT cables becomes less effective due to the higher voltage gradients. Accordingly, the increase in tan δ with applied voltage becomes quite pronounced as compared to that of the LPOF cables in which no internal discharges can take place under similar circumstances. However, it must be emphasized that in oil-pressurized cables either of the LPOF or HPOFPT type, the variation of tan δ with temperature is more likely to be associated with dielectric-loss temperature dependence effects in the solid–liquid dielectric system itself rather than with any highly improbable onset of partial-discharge. Regarding the power factor/temperature characteristics of HPOFPT cables at the rated voltage, one can perceive that they are somewhat similar to those obtained from LPOF cables. It should be noted, however, that with the increase of temperature, and a corresponding decrease in viscosity of the impregnating oil, there is a significant decrease in tan δ until the temperature in the region of operational temperature is reached. Only after further increase of temperature is this trend reversed and tan δ increases. A typical power factor/temperature curve obtained from actual measurements on a 230-kV cable at the rated voltage-to-ground is shown in Fig. 7.3.

7.6.2 Ionization Factor

The value of the ionization factor obtained during a routine ionization test is not representative of the actual working conditions; in this regard only the specimen test value is meaningful. In well-made cables, the results from both tests should be far below the AEIC specification requirements with the ionization factor values rarely exceeding 0.025% (0.00025).

7.6.3 Alternating-Current Withstand Voltage Level

In HPOFPT cables with carbon black paper conductor shields, the ac breakdown strength is very high. This can be verified by further extension of the high-voltage time test for an additional 3 h at the test voltage corresponding to the average stress of 520 V/mil (20.5 kV/mm), which equals $3.7V_0$ for 69-kV rated cables gradually reduced to $2.7V_0$ for 345-kV cables. The actual breakdown level is difficult to investigate because of the limitations in testing equipment and costly implications of such an occurrence.

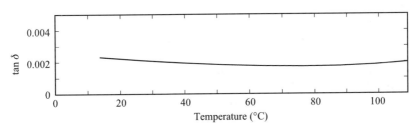

Fig. 7.3 Typical power factor vs. temperature characteristics of a 230-kV HPOFPT power transmission cable at rated voltage-to-ground.

7.6.4 Impulse Withstand Voltage Levels

Because of the type of paper and high-viscosity compound used in the HPOFPT cables, their impulse strength is extremely good. As stated in Section 7.3.5 the design impulse stress should not be lower than 79 MV/m (79 kV/mm or 2006 V/mil) for cables rated up to 161 kV and 89 MV/m (89 kV/mm or 2260 V/mil) for higher voltage ratings. The actual impulse breakdown stress is much higher than these values, but in general is not checked because of limitations of the testing equipment and other implications.

7.7 PRINCIPLES OF OIL FEEDING

In comparison with LPOF-type cables, oil feeding in the HPOFPT type cable is very simple, although the equipment is costly and complicated. Correct pressure of a nominal value of 200 psi (1400 kPa) is provided by a pumping plant equipped with two independent sets of high-pressure oil pumps fed from a large oil reservoir designed to hold at least three times the maximum calculated amount of oil expected to be expelled from the cable, when its temperature changes from a minimum value in winter at no load to the maximum in summer at full load. In such a way, $\frac{1}{3}$ of the tank capacity is filled with spare oil, $\frac{1}{3}$ with active oil moving back and forth from and into the cable during the load cycles, leaving the upper $\frac{1}{3}$ space empty, which is filled with dry nitrogen gas under slight pressure to avoid oxidation. During the heating cycle oil expelled from the cable returns to the tank through the relief valves, set to operate if the oil pressure in the cable exceeds the preset limits. When during the cooling cycle the oil pressure drops below the preset limits, the oil pumps automatically cut in, restoring the pressure in the cable. Each set of pumps must have sufficient capacity to perform that function, with the second set used as a standby. The whole process is completely automatic, although the plant may also be switched to a manual operation mode if required [4].

7.8 CATHODIC PROTECTION

Cathodic protection is a very important element in any modern HPOFPT cable installation to protect the carrier pipe against electrochemical corrosion in addition to the protective coating, which may be damaged or may deteriorate in service. This is achieved by the application of a small negative potential to the pipe with respect to earth. Steel is generally considered protected, if this potential is at least 0.85 V. Consequently, the design of a cathodic protection system must be such that the above condition is met in any portion of the pipe with the resulting minimal leakage current to other structures. Also the cathodic protection should not jeopardize continuity of the path for pipe fault current resulting from line-to-ground fault [5]. Practical experience shows that both these conditions may only be satisfied by the use of cathodic protection equipment, the design of which is usually based on an extensive corrosion survey.

The most classical method of meeting grounding and negative potential requirements, is by means of the so-called series battery system. In this system the cathodic protection equipment is composed of a 2-V lead–acid storage battery connected to the pipe and the substation ground and a selenium rectifier circuit for keeping this battery fully charged. The battery must be capable of carrying the maximum ground fault current. For additional assurance a spark-gap may be placed in parallel. At the remote end a so-called polarization cell is placed, which has a property of adjusting its potential to the current flow over the pipe, reducing it quickly, thus reducing the load of the rectifier. A specially developed low-impedance cell, also operating with a rectifier, has lately replaced the lead–acid battery. The output of this cell is usually set for 15 A at 2.0 V. Another method worth mentioning is to insert a resistor in the pipe grounding connection and to secure a desired negative potential by maintaining a large circulating current flowing through this resistor from a rectifier. Simplicity and ruggedness of this arrangement makes it particularly suitable in remote places, where maintenance is a problem.

7.9 LIMITATIONS OF HPOFPT CABLES

The most important limitations came to light some time ago, when the large network of 345-kV HPOFPT cables in the New York City area developed electrical breakdowns at an alarming rate. Investigations attributed these breakdowns, which always were occurring in the vicinity of joints, to the uncontrolled snaking of the cable during heating cycles. Apparently, the cumulative effect of the expansion caused dislocation of the insulating tapes near the joints, lowering the cable withstand voltage below the service requirements. Although a partial remedy was found in a form of spacers inserted into the pipes near the joints, a conclusion must be drawn that there must be some limitations in the insulation thickness of HPOFPT cables beyond which the behavior of these cables in service may become erratic. Therefore, it is unlikely that with all the precautions HPOFPT cables could safely be designed for voltages higher than 500 or 550 kV. Also, the manufacturing methods and the fact that HPOFPT cables are assembled in the field, seem to restrict the development of these cables for higher voltages, giving preference to self-contained oil-filled cables.

Magnetic pipe losses and the necessity of placing three single-phase cables in the pipe restricts their transmitting power capabilities. The above limitation and the danger of large oil leaks through a blow-out in the pipe, possible during the cable breakdown due to a high magnetic field caused by the short-circuit current, is also a factor that explains the reluctance of Europeans to consider this type of cable in their installations. On the positive side HPOFPT cable has some distinct cost advantages in North America, as the carrier pipe is considered a duct by North American utilities and is allowed to be directly buried, whereas in the same locations self-contained cables must be installed in ducts. Also HPOFPT cables may be shipped and installed in much longer lengths, sometimes exceeding 2 km, thus reducing the number of joint locations. An HPOFPT cable may easily be adapted for a short underwater crossing, simply by gunniting a layer of concrete over the protective coating of that portion of the pipe, for weight, and for additional mechanical protection. Many such underwater crossings have been executed in Canada, with one entirely submarine cable installation supplying l'Île d'Orléans across the St. Lawrence River, close to Québec City [6].

7.10 DEVELOPMENT OF HPOFPT CABLE FOR HIGHER VOLTAGES IN THE UNITED STATES

It was previously stated that an HPOFPT cable with a conventional design seems to reach safe operational limits at a voltage rating on the order of 500 kV. With the installation of new overhead transmission lines, rated 765 kV, it became necessary to study the possibility of development of HPOFPT cables to match that voltage rating and to be ready for a further extension to a next transmission level of 1100 kV in the future.

Since the use of present technology employing cellulose paper/oil dielectric as the cable insulation was not possible because of prohibitive dielectric losses, a need for a new insulant triggered an extensive research, sponsored by the U.S. Energy Research and Development Administration and by the Electric Power Research Institute (EPRI). The possibility of employing synthetic papers, polymer films and composite laminates impregnated with insulating liquids of characteristics compatible with the insulants was examined. From various combinations polybutene-oil-impregnated laminar insulation consisting of cellulose paper with a polypropylene paper (PPP) film was selected as most promising, and a 765-kV low-loss HPOFPT cable employing that insulation has been developed with satisfactory test results [10, 11]. Also a prototype of an HPOFPT cable rated ±600 kV has been developed by another U.S. manufacturer [12]. Practical application of the above developments will show whether or not the actual service experience would confirm the expectations.

7.11 GAS-TYPE CABLES

As gas pressurized power transmission cables are essentially oil-impregnated-paper cables, which are under high gas pressure in lieu of high oil pressure, it is logical to consider these cables as an appendage to high-pressure oil-filled pipe-type cables. Because of ease of feeding from compressed gas cylinders, in comparison to oil, the use of inert gas in extra-high-voltage cables was attracting various inventors, who in the second quarter of the twentieth century employed this pressurizing medium in several new EHV cable designs. The gas-employing types of cables may be divided into two main groups, both impregnated paper insulated, namely, gas-filled cables, also called internal gas-pressure cables in which the inert gas under pressure is in contact with the insulation and gas compression cables (external gas-pressure cables) in which gas is used solely as a pressurizing medium and is not in contact with the insulation. In both groups nitrogen gas is used as it is readily available.

Although originally these cables had been developed as self-contained cables some of them were later modified as pipe-type cables. In Canada gas-filled and gas-compression-type cables were introduced in the early 1950s mainly by overseas manufacturers; however, only a few of these installations are still in operation, including one major submarine cable installation across the Strait of Georgia in British Columbia, using 138-kV gas-filled cables. No new gas-employing cables were installed in Canada since 1961 and, because of voltage rating limitations and other factors, it is unlikely that the interest in these type of cables would ever be revitalized. Therefore, there is no purpose in including gas-utilizing types of cables in this presentation in the same manner as other paper insulated types, and only a brief description of gas-filled and gas compression types is offered for general information.

7.11.1 Gas-Filled EHV Cables

Gas-filled cables, also called internal gas pressure cables, were first developed by Atkinson and Fisher in the United States ca. 1925 and since that time underwent many modifications by various inventors, who created several new types of cables based on the same principle. In general, all these cables are insulated with paper, impregnated to various degrees, with the insulation in contact with dry nitrogen under a pressure of about 15 atm. (1500 kPa). In these cables, ionization is suppressed up to a certain critical voltage; however, the level of the incipient ionization is difficult to establish, as it depends on the distribution of voids, which undergoes change during cable operation. Conventional power factor measurements would give dependable results only if these voids were numerous and uniformly distributed; this may occur only in cables with a low degree of impregnation. Partial-discharge measurements as carried out in case of polymeric insulated high-voltage cables would be more sensitive, although because of the long lengths involved the results again may not be very accurate. Hence, with this type of cable there is an element of uncertainty as to when the ionization takes place, and, consequently, the designer must exert extreme caution in selecting the insulation thickness. Generally, it is estimated that a maximum working gradient of 8–10 MV/m (8–10 kV/mm or 203–254 V/mil) would represent 50–70% of the incipient ionization gradient. Although such margins of safety with respect to the incipient ionization level are rather small, gas-filled cables in 120- to 130-kV rated circuits operating with a maximum gradient of 8.5 MV/m (8.5 kV/mm or 216 V/mil) have enjoyed a good service record for several decades. Several types of gas-filled cables, based on Atkinson-Fisher principles, were developed. The most significant of these are briefly described below.

7.11.1.1 Beaver and Davy Gas-Filled Cables

The Beaver and Davy type of cable was developed in England in 1937. Longitudinal channels for gas distribution are provided in the interstices between the phases in three-phase cables or by a central channel in the conductor of single-conductor cables in analogy to the LPOF cable oil ducts. Because of the preimpregnated paper used in the insulation, the distribution of gaseous voids tends to be more uniform and the ionization level well defined. As there is no impregnation following the insulating step, this type of cable is found particularly suitable for long submarine crossings.

7.11.1.2 Hunter and Brazier Impregnated Pressure Cables

This fully impregnated, self-contained cable, developed by British engineers before World War II, was first installed in 1943. High viscosity compound impregnated insulation was used in the cable with the longitudinal transmission of gas provided by the space between the cable assembly and the lead sheath. However, in service, this space could gradually be blocked by the cable impregnate, slowing down the longitudinal penetration of gas.

7.11.1.3 Pirelli Gas-Filled Cable

Developed by L. Emanueli in 1950, as a three-phase 33-kV cable, the Pirelli gas-filled cable also uses insulation fully impregnated with a high-viscosity compound; however, unlike in Hunter-Brazier cable, the transmission of gas is secured by perforated synthetic rubber tubes, placed in the interstices of the cable assembly. The perforations allow the compressed gas to penetrate into the cable but do not allow the compound to enter the tube.

7.11.1.4 High-Viscosity, High-Pressure, Gas-Filled Pipe-Type Cable

This successful cable [12], still in use by Detroit Edison Co. in its vast 138-kV network, has been developed during World War II by General Cable Co. It is a pipe-type version of a gas-filled cable with a construction identical to that of HPOFPT cables, except that an extra-high-viscosity compound is used for cable impregnation instead of oil and that the 200-psi (1400-kPA) pressure is exerted by dry nitrogen. Because of the constructional similarity, this cable follows AEIC Specification No. 2.73, which covers high-pressure oil-filled pipe-type cables. To reduce the possibility of the impregnate leaving the cable via the moisture seal, a very high viscosity compound is used in the mass impregnation of cables. Any surplus of the high-viscosity compound is removed after impregnation, while the hot cable still rests in the impregnating tank.

It should be noted that, although after a certain length of time the gas will definitely penetrate the moisture seal and enter the cable core, the high-viscosity gas pressure cable at the beginning of its operation will behave similarly to a compression cable. The high-viscosity, high-pressure, gas-filled pipe-type cable, which since its introduction has undergone considerable development, is rather of a conservative design with an insulation thickness considerably increased with respect to the requirements of the relevant AEIC specifications, resulting in a maximum operating gradient of about 6.9 MV/m (6.9 kV/mm or 175 V/mil).

7.12 GAS COMPRESSION EHV CABLES

Gas compression power transmission cables, developed by the Enfield Cables Company in England, were introduced in Canada in the form of a self-contained cable (Montreal, circa 1952) and as a pipe-type cable (Toronto, circa 1951). Its construction follows the pattern of a conventional shielded PILC cable having an elliptical conductor with oil-impregnated-paper insulation and a reinforced elliptical lead sheath. The latter acts as a flexible diaphragm through which gas pressure is transmitted to the cable core. The noncircular construction is required to accommodate oil expansion and contraction due to load cycles.

A self-contained version of this single-conductor cable has been provided with an oversized external lead sheath with a reinforcement and a rubber sandwich-type protective covering. Gas under pressure was introduced between the external sheath and the diaphragm. A cable rated for 120 kV operation installed on the Jacques Cartier bridge in Montreal developed many failures in service due to lead fatigue cracks caused by severe vibrations; following 12 years of operation, the cable had to be dismantled. In a pipe-type cable version, in Toronto, three single-conductor cables of similar cable construction, with a reinforced lead sheath diaphragm and skid wires, were installed in a steel pipe and subsequently pressurized with nitrogen. After about 16 years of service, one of the 115-kV cable circuits developed frequent electrical failures and had to be dismantled. These failures apparently were caused by contaminated gas entering the cable interior through cracks in the lead diaphragm weakened by the corrosion of the reinforcement cover. It is interesting to note that in order to save the cost of replacement, the pipe, vacated by these cables, was utilized as a duct for three single-conductor

LPOF-type cables constructed without an outer protective covering, and pulled into the pipe, which was subsequently filled with oil to prevent corrosion.

It is worthy of note that the last gas compression pipe type cable installed in Canada (Ottawa, circa 1965) had a reinforced polyethylene sheath used as a diaphragm. This cable is still in operation; it is now functioning most probably as a gas-filled cable, as it is unlikely that a polyethylene sheath could act as an efficient impervious barrier for a long time period; it is thus highly likely that the gas is now in contact with the insulation. However, in this case the gas, pressurizing the pipe was not contaminated and did not therefore affect the cable operation.

7.13 CONCLUDING REMARKS

In conclusion it may be stated that whereas gas-filled power transmission cables may be considered as a good design with the limitations imposed by the voltage rating and minimum design gradient, gas compression power transmission cables cannot enjoy the same status. Disregarding the problems encountered in Canada with a gas compression pipe-type cable that originated from poor installation procedures, it appears that the basic difficulty with these types of cables centers on the lack of certainty of suppressing effectively the ionization close to the applied gas pressure. Even with a polyethylene diaphragm, the soundness of operation of these cables may be attained only when, after several years of service, these cables commence functioning as gas-filled-type cables. Therefore, it is unlikely that gas compression cables will ever enjoy great popularity. In contrast, gas-filled-type cables may still undergo further development with the use of SF_6 gas, which would tend to effectively increase the dielectric strength of the cables at power frequency, allowing them to be designed for much higher voltage ratings.

The current deployment of gas insulated transmission lines (GITL) for bulk transportation of electric power, employing SF_6 gas under pressure, has certainly spurred a new interest in the various forms of gas-filled-type cables. The subject matter of compressed SF_6 gas cables is treated in Chapter 12.

ACKNOWLEDGMENT

The author wishes to thank the management of Pirelli Cables Limited for the privilege of using the Company's Technical Library, and the engineers of that company for their valuable assistance.

REFERENCES

[1] J. S. Englehardt and R. B. Blodgett, "Preliminary qualifications of a HPOF pipe type cable system for service at 765 kV," *IEEE Trans. Power Apparatus Syst.*, Vol. PAS-94, 1975, pp. 1569–1583.

[2] G. Bahder, G. S. Eager Jr., D. A. Silver, and S. E. Turner, "550 kV and 765 kV high pressure oil-filled pipe type system," *IEEE Trans. Power Apparatus Syst.* Vol. PAS-95, 1970, 1976, pp. 478–488.

[3] Canadian Electrical Association Proceedings, "Summary of cable installations in Canada 69 kV and above," Toronto 1974, pp. 20–28.

[4] J. Jerome, "Cable pressurizing systems," *IEEE Trans. Power Apparatus Syst.*, Vol. 82, 1963, pp. 365–379.

[5] W. D. Lawson, "Corrosion control for pipe-type cable systems," *IEEE Trans. Power Apparatus Syst.*, Vol. PAS-97, 1978, pp. 1202–1207.

[6] Canadian Electrical Association Proceedings, "69 kV submarine pipe test cable installation across St. Lawrence River between North Shore and l'Ile d'Orleans," 1975, p. 24.

[7] P. Graneau, *Underground Power Transmission*, Wiley, New York, 1979, pp. 347–401.

[8] Specifications for Impregnated-Paper-Insulated Cable, High Pressure Pipe Type, AEIC No. 2-73, Association of Edison Illuminating Companies, New York, 1993.

[9] E. E. McIlveen, R. C. Waldron, J. W. Bankoske, F. M. Dietrich, and J. F. Shimshok, "Mechanical effects of load cycling on pipe-type power cable," *IEEE Trans. Power Apparatus Syst.*, Vol. PAS-97, 1978, pp. 903–913.

[10] E. M. Allam, W. H. Cortelyou, and H. C. Doepken, Jr., "Low-loss 765 kV pipe-type power cable," *IEEE Trans. Power Apparatus Syst.*, Vol. PAS-97, 1978, pp. 2019–2030.

[11] E. M. Allam, W. H. Cortelyou, and F. A. Teti, "Low-loss 765 kV pipe-type power cable — Part II," *IEEE Trans. Power Apparatus Syst.*, Vol. PAS-98, 1979, pp. 2083–2088.

[12] T. H. Roughley, J. T. Corbett, G. L. Winkler, G. S. Eager, and S. E. Turner, "Design and insulation of a 139 kV high pressure gas-filled pipe cable utilizing segmental aluminium conductors," *IEEE Trans. Power Apparatus Syst.*, Vol. PAS-93 1974, pp. 658–668.

Chapter 8

VOLTAGE BREAKDOWN AND OTHER ELECTRICAL TESTS ON POWER CABLES

H. D. Campbell

8.1 INTRODUCTION

Power cables are in one sense less complex than most other high-voltage electrical equipment. They transmit (or distribute) electrical energy but do not transform it, as a transformer does, or convert it, as an electric motor does. It is not surprising, then, that electrical tests of power cable are primarily tests of the insulation system. We shall consider first and primarily the alternating- and direct-current (ac and dc) tests on finished shipping lengths. A complete description of impulse testing techniques will be provided; other more specialized tests will be discussed briefly. References [1–7] give details of generating high test voltages and testing techniques. Specific standards, however, apply to the different types of power cables.

8.2 ALTERNATING-CURRENT OVERVOLTAGE TEST

Undoubtedly, the simplest and most fundamental of all electrical tests on electrical insulation is the routine high-voltage test. For some elemental low-voltage cable, and for some other simple electrical products, it may be the only routine electrical test on the completed product. This test is an overvoltage test since the applied voltage is higher than the rated voltage. It is sometimes called a proof test, alluding to the theory that it proves the insulation. A little thought leads to the conclusion that such over-voltage tests are a compromise at best, and, in some cases the question arises to the possibility that they should be eliminated. One can use a mechanical analog. A bridge, for example, is not loaded to three or four times its rated capacity to determine if it will remain intact. In the early days of electrical engineering, the overvoltage test was about the only electrical test available. Insulation resistance was another—but that is another story.

A compromise is required. The overvoltage must be such that the dielectric is subjected to stresses low enough not to cause significant damage yet high enough to detect defects such that a substandard product does not reach the user. Products that approach this ideal are low voltage wires and cables. Building wire, the type used in homes and factories is an example. This wire may be subjected to a spark test, passed at high speed through a mesh electrode, beads, or brushes, so that several thousand volts are impressed on the insulation for less than a second. The duration is so short that the insulation does not deteriorate, yet pin holes, inclusions, and other flaws are detected. The standard insulation, without flaws, can easily withstand a voltage several times higher than the test voltage without failure. Even if some significant defects are not detected, the operating stress of only a few volts per mil is so low that failure in service is unlikely to occur.

The ideal compromise is simply not possible for medium- or high-voltage cables. These cables operate at higher electrical stresses such that a different life mechanism is brought into play. Partial discharges may be present and significant in all high-voltage (HV) cables with the possible exception of pressure-type cables of the HV and extra-high-voltage (EHV) classes. All solid insulation, including solid-type paper insulated lead-covered (PILC) cables, contain to a varying degree gaseous voids that ionize under electrical stress. At the operating voltage these partial discharges have very low energy contents so that the aging mechanism in good-quality cables is very slow indeed. At high stresses these energies are enormously increased even at values less than one half the breakdown value. The problem of the test voltage value becomes difficult. It is difficult to establish the level at which the cable insulation is not damaged and yet defects can be detected. It is clear that such a simple go–no-go test cannot be definitive. Other tests such as partial-discharge level, power factor (tan δ) and the like are employed to provide quantitative information of the dielectric condition; however, an overvoltage test is a universal requirement and a reasonable value must be used. The duration of the test is usually either 5 or 15 min. Failure, of course, can occur at any time during the test, at 4.9 min for example. The test is arbitrary in the sense that any particular cable that meets the 5-min test may still fail at, say 5.1 min.

The routine test voltage levels applied to shipping lengths have been evolved over many years of experience by manufacturers all over the world. Although there may be different empirical formulas or rules in different specifications, and none in others, in general, the test stress is about three to four times the operating stress. These test levels would have to be increased by a further factor of 3 or 4 before approaching breakdown level. It is obvious, then, that the test levels are modest and are unlikely to damage significantly good-quality cable. It is also obvious that quite serious defects could exist in the insulation and not be detected. The routine high-voltage test is not a criterion of quality but rather evidence that the product has not been mishandled during manufacture. Failures that occur are almost always due to mechanical damage. Routine voltage test failures on high-voltage cables today are rare, probably not more than one in a thousand.

There are two different philosophies or rationales for routine ac voltage test levels on high-voltage cable: (i) the volts per mil philosophy as exemplified in paper insulated cables and (ii) the use of the rated cable voltage to which a factor, according to an empirical formula, is applied; in short, the volts per mil basis and the rated voltage basis. The volts per mil philosophy was general in the North American cable industry until circa 1970 when the Insulated Cable Engineers Association (ICEA) introduced the rated voltage basis on all polymeric insulated cables rated over 2.0 kV. Paper insulated cables continue to be tested on the volts per mil or kilovolt per centimeter basis. The two approaches to the problem are quite different. While the rated voltage idea is used in Europe and is being accepted in North America, there are still strong proponents of the opposing view. This rationale holds that the test voltage is used only to detect gross flaws in the insulation and provides no indication of satisfactory service at the operating level. A reduction of insulation thickness (by higher quality material or improved manufacturing processes) should be accompanied by a decrease in test voltage, otherwise the insulation may be too highly stressed. If the insulation wall is increased for any reason for the same rated voltage, then the test voltage should be increased so that defects will be detected as effectively as before. The rated voltage rationale holds that the insulation thickness is secondary. If the product is designed to operate at a given stress, the overvoltage test should be some factor of that stress regardless of the wall thickness. It is not logical in this view to test a product, having reduced wall thicknesses but of proven equal service capabilities, to a lower voltage level than the inferior replaced product. This entire syndrome may be discouraging to those that believe that engineering is an exact science and that all procedures are based on unassailable logic. The volts per mil logic is more suitable to the oil-paper insulation, since these insulations are of one general type; whereas polymeric insulations may have rather different characteristics all performing the same task. New and modified compounds will continue to appear requiring possibly differing thicknesses depending on their various characteristics for the same application. It would not be logical or practical to have a myriad of test voltage schedules differing relatively slightly one from the other when it is not known, as we have seen, which level is, in fact, the best compromise.

8.3 DIRECT-CURRENT OVERVOLTAGE TEST

The dc overvoltage test has been universally employed in factory testing for many years in the final test of rubber and plastic insulated high-voltage cables. This test has one fundamental advantage over the ac test. There are very small partial-discharge energies introduced into the dielectric even at voltages near the breakdown value, whereas the stress reversals (the rate being twice the power frequency) cause continuous bursts of discharges near each voltage peak during the application of the alternating voltage test. The direct overvoltage is much less time dependent than the alternating overvoltage test. Quite high test levels are not considered to have significant aging effect on the dielectric. The dc test level usually specified is about 2.5–3 times the ac level. In North America both the ac and dc tests are mandatory for routine tests on medium-voltage polymeric cables. It might be thought that the dc test eliminates the compromise of the test level but as we shall see in a later chapter on cable tests in the field, the dc voltage level is currently subject to much controversy. When the ICEA changed to the rated voltage concept for ac testing, it had at the same time decided to make the dc test level a function of the basic impulse level (BIL) of the polymeric insulated cables. In effect, this is also a rated voltage concept.

8.4 VOLTAGE TESTING OF PRODUCTION LENGTHS

Until now we have discussed in generalities high electrical stresses in cable insulation. However, these are not really different from any high-voltage equipment: transformers, generators, bushings, and the like. There are some facets of cable overvoltage testing that are different. A cable is often tested in lengths of 700 m or more. Electrically at power frequencies a cable is a large lumped capacitor—on the order of 300 pF/m. Suppose a shipping length of 115-kV cable, 700 m long is required to be ac tested at 130 kV. The reactive current I_c is easily calculated if C, the capacitance per meter, is assumed to be 300 pF. The reactive current is given by

$$l_c = V\omega C_s \tag{8.1}$$

where C_s is in farads and refers to the total specimen capacitance. In this case it is equal to 300×700 pF, which at a test voltage of V equal to 1.3×10^5 volts gives an I_c value of approximately 10 A. A transformer rated at least 1.3 MVA is required to test this cable of rather modest voltage rating. Not only is such a transformer large in terms of cost and physical dimensions, but the demand on the factory distribution system is also high.

8.4.1 Test Sets for Routine Measurements

There are two basic methods of reducing the cost of the testing transformer and the reactive load: (i) the use of a parallel resonant facility or (ii) the use of a series resonant facility. The parallel resonant system is illustrated in Fig. 8.1 [1, 2].

If a variable reactor is switched in parallel to the cable load either on the secondary side as in Fig. 8.1(a) or in the primary side as in Fig. 8.1(b), then the system may be tuned to resonance, and only the resistive losses are required to be supplied by the factory distribution system. In practice, several fixed reactors may be switched in various parallel or series-parallel combinations to approximate the ideal resonance condition. In the case of Fig. 8.1(a), the transformer current rating can be greatly reduced, but the high-voltage switching must be done manually. In the case of Fig. 8.1(b), motor-driven switches are practical, but the transformer must have the full rating.

The second more modern method to accommodate large reactive loads is the series-resonant system [8], used by many high-voltage cable manufacturers throughout the world. The basic series-resonant circuit is shown in Fig. 8.2. The moving-coil reactor is tuned to resonance with the cable capacitive load so that the feed transformer that supplies only the system losses may have quite a low-voltage rating. The moving-coil reactor (MCR) is adjusted to resonance by suitable controls and can be tuned to resonance automatically.

For high-voltage tests above 10 kV, the MCR is connected to the load through a high-voltage transformer, as in Fig. 8.3. This arrangement permits flexibility in design. Only the transformer operates at high voltage, and it can have series and parallel windings to extend the reactance range of the single reactor from 40:1 to 120:1 or more. The reactor and transformer are usually contained in the same tank and are of

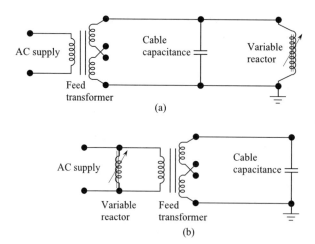

Figure 8.1 Parallel-resonance high-voltage test circuit: (a) secondary side; (b) primary side.

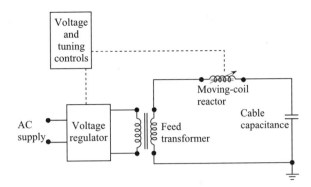

Figure 8.2 Basic series-resonant circuit.

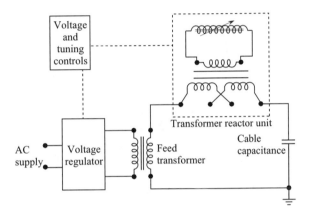

Figure 8.3 Series-resonant circuit for high-test voltages.

special design. The reactance of the MCR is varied by having two windings change their positions relative to each other so that no electrical switching is required. The ratio of reactance to loss resistance, $\omega L/R$ or the value of Q, is selected to provide optimum tuning and stability conditions. The transformer is designed for low winding distributed capacitance and leakage reactance to maintain the tuning range and provide good noise rejection properties. If such a system has a Q of 10, the required kilo-volt-amperes (kVA) (at unity power factor) is only one tenth of the test kilo-volt-amperes. If a fault occurs in the cable under test, the state of resonance is lost, thus the fault current is only one tenth the normal current. These features are very important in high voltage cable testing where routine tests may require 10 mega-volt-amperes (MVA) or more. The limited fault current has a particular advantage in development work where the faulted specimen is to be examined, since the fault damage is much reduced.

8.4.2 Routine Test Terminations

The second feature of routine factory testing of high-voltage cables to be noted is that a cable shipping length is not in itself a complete operating system. It is shipped in units to be joined and terminated in the field. Field terminations on cables rated above 5 kV or so include an end reinforcement to reduce stress concentration. For routine ac tests in the factory, the same requirement has to be met on cables rated above about

25 kV, but it is obviously not practical to place potheads on a 25-kV cable just to make a 15-min test.

A brief examination of the related theory will demonstrate the need for stress relief at the cable ends. In Fig. 8.4, a cross section of a stripped cable test end is shown to which a test voltage V is being applied. At point A on the insulation surface at the end of the grounded shield, the potential is zero. At point B, say 2 cm axially along the insulation surface, the potential is fixed by the relative capacitance to the conductor and ground. The capacitances to the grounded shield and remote ground are obviously small compared to the capacitance to the conductor; therefore, the potential at point B approaches that of the conductor, and the axial voltage stress along AB is high. The axial stress between B and C, a point near the end of the conductor is very low. Rather arbitrary points have been chosen, but the explanation is suitably illustrative of the facts that (i) most of the longitudinal stress is concentrated at the end of the shield and (ii) increasing the length of the bared insulation does not significantly change the stress concentration. If the test voltage is on the order of 40 kV (ac) or more, partial discharges in air at the end of the shield will be of sufficient intensity to initiate tracking along the surface of the insulation within a very short time. This conductive tracking simply advances the high stress concentration toward the end of the cable. Increasing the length of the bared end will only delay the eventual flashover. For cables rated above common distribution voltages, some form of stress relief is required in the routine ac test. Since we are dealing with repetitive production testing, the method should be quick and inexpensive.

It is well to be reminded here of dc test termination characteristics. Under dc voltage there is also a high stress concentration at the shield end; however, much higher rated voltage cables can be accommodated without any special precautions other than possibly increasing the clearance between the terminated shield and the end of the cable. The partial-discharge mechanism is much less severe than in the ac condition, since there are no polarity reversals. Second, the stress tends to be graded by the surface resistance of the insulation. Our discussion will continue to apply to ac testing. Routine factory dc tests are not, in general, required for EHV oil-paper cables and do not require stress grading at other voltage ratings.

There are four general methods available for alleviating the nonuniform stress distribution: changing the dielectric surrounding the cable end, field shields, capacitive grading, and resistive grading. The first method can be accomplished by immersing the cable end in oil. Electrical insulating oils have much higher dielectric strength than air, thus partial discharges are suppressed. The method is not nearly as effective as one

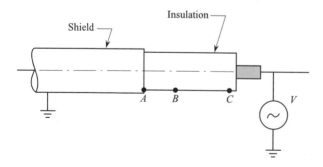

Figure 8.4 Cross section of a stripped polymeric power cable test end.

might suppose because the stress is likely to be high enough to cause some partial discharges in the oil, creating gas bubbles that rapidly propagate. Second, cylinders of oil are not easily or quickly fixed to the cable end. Third, the method is not readily adaptable to polymeric cables. The field shield method is really a form of capacitive grading. Normally a stress cone is built with tape or shaped rolls, as shown in Fig. 8.5. The shield is extended to the peak of the cone and the whole assembly immersed in oil. The stress is graded by the gradual increase in the diameter. At the end of the shield a new boundary condition exists (cf. Fig. 8.5) where the stress concentration is less than before; point B is now farther from the conductor than in the plain end condition. An insight into the order of decrease can be obtained if it is remembered that the electric field surrounding an isolated conductor in space varies as $1/r$, where r is the distance to the conductor center. A further improvement can be made by the use of a field shield that extends beyond the built-up insulation, as depicted in Fig. 8.6. The cable shield has now a larger capacitance to point B than before (cf. Fig. 8.5), thus effecting a more favorable stress distribution. The difficulty is that this method is expensive and tedious.

A different method of capacitive grading can be used in conjunction with a liquid dielectric, as shown in Fig. 8.7. One form is the condenser cone that consists of concentric tubular capacitors displaced axially that can effect a linear stress distribution along the cable length l. Units of this type are light, easily handled and installed, and require a surprisingly short test end. They can be used repetitively, which does offset their rather high initial cost. Nevertheless there are some technical difficulties. Normally a small reinforcement or stress cone has to be applied at the high stress region. They are very effective for certain applications and can be installed quite quickly. A resistive stress grading has long been used as an aid in termination design and is shown in Fig. 8.8. If the high stress region is coated with a resistive material, the capacitive coupling will result in a current flow i along the resistance. If the resistance is linear, the longitudinal surface stress will obviously be greater as the shield is approached. In practice a linear voltage drop is not attainable, but test voltages with the end in air can be obtained up to 100 kV or more. The method is simple and quick, though a limitation exists due to heat dissipation. This is by far the most useful termination method for

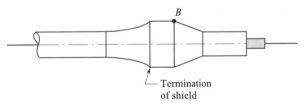

Figure 8.5 Stress cone termination for high-voltage power cables.

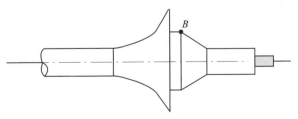

Figure 8.6 Field shield-type termination for high-voltage power cables.

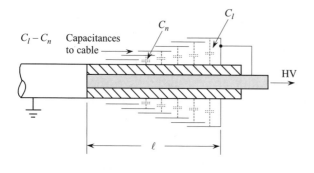

Figure 8.7 Capacitive grading-type termination for high-voltage solid–liquid dielectric power cables.

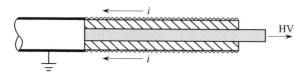

Figure 8.8 Resistive stress grading-type termination for high-voltage polymeric power cables.

routine tests on cables when the test's voltage does not exceed 100 kV. It is commonly employed in the routine full-reel high-voltage test and partial-discharge or corona test. High permittivity materials may also be used to effectively grade the impedance, as discussed in Section 10.13.3.

8.5 TESTS ON SPECIMENS

The routine testing of cable shipping lengths has been discussed in some detail. Cable development and quality control utilize many different overvoltage ac tests, for example: (i) quick rise test, (ii) high-voltage time test, and (iii) accelerated aging tests. On the other hand, the common insulation resistance test, a form of dc test, is utilized both for routine testing on shipping lengths and more specialized tests on specimens, e.g. long-term insulation resistance (IR) in water. Since one of the chief purposes in this section is to discuss the more common overvoltage test as an inspection tool of the product being shipped to the customer, specimen tests will be treated very briefly to illustrate one important concept.

Short-time dielectric strength tests are often used to compare materials. The material is usually in sheet or slab form and held between electrodes of prescribed geometry. The applied voltage, normally 50 or 60 Hz, is raised at certain standardized rates so that breakdown occurs, depending on specimen thickness and other factors, on the order of 1 min. Such a test method is useful in comparing new materials and in evaluating the effects of electrical, chemical, or physical stresses to which the specimen has been previously subjected. The dielectric strength of insulating materials, as listed in reference books and the like, is normally the short-time dielectric strength result. The high-voltage time test is specified in some industry standards for sample tests on high voltage cables. The voltage is applied in steps, usually of several hours each, so that a complete test to breakdown requires perhaps 12 h to complete. The breakdown mechanism is normally one of tree growth or partial discharge. The results provide some insight into the projected service life of the product. The minimum requirement to constitute the specification acceptance is based on no failure at a prescribed voltage level.

The testing of EHV cable specimens may require a 60-Hz source in the megavolt range. A schematic circuit diagram of a commonly used configuration is depicted in Fig. 8.9(a), where (1) is the voltage regulating transformer, (2) represents the compensating reactor [see also Fig. 8.1(b)], (3) is the high-voltage transformer (grounded unit with tertiary winding), and (4) is the line unit high-voltage transformer. The cable specimen undergoing test is represented by the capacitance (5), while the capacitance

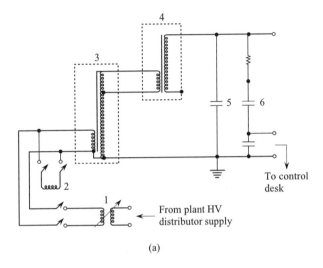

(a)

Figure 8.9 (a) Circuit diagram of a cascade testing transformer composed of two transformer units. (b) 2.1-MV cascade testing transformer (max. test voltage achieved with four units in series and two in parallel connected at the midpoint). (Courtesy of D. Train, Institut de Recherche d'Hydro-Québec.)

(b)

(6) constitutes part of the voltage divider. The foregoing cascade arrangement permits the use of similar high-voltage transformers insulated for only part of the total voltage to ground. The line transformer rests on an insulating base support. It is possible to extend the cascade configuration to more than the two units shown in Fig. 8.9(a). An extra-high-voltage testing system (comprising several transformer units) suitable for the evaluation of cable test specimens rated to 1000 kV is depicted in Fig. 8.9(b). Accelerated aging tests are designed to simulate the service conditions of the cable, but the aging factors of influence are deliberately intensified to make the test of practical duration. Voltage is applied continuously, while the load or conductor current is cycled to simulate field conditions. Typically, the conductor-to-shield applied voltage is two to four times the rated voltage. Cable performance is measured by the stability of such characteristics as tan δ and partial discharges rather than insulation breakdown. Often aging of the insulation is ascertained in terms of the ac breakdown strength of the aged specimen; recent results show it to be a very sensitive measure of the aging process [9]. The latter two tests, including accelerated electrical aging, are dealt with more extensively in Chapter 9.

It should be noted that the three tests have been discussed in ascending order of test duration. The first test has a duration of minutes, the second in hours, and the third in weeks, months-, and sometimes years. The third test has also a much higher level of sophistication and cost; however, one rather basic point is evident, i.e., the time factor. The effects of alternating voltage are highly time dependent. The expected service life is sometimes defined on the basis of an inverse power law of the applied voltage. It is not surprising that the nearer the test voltage approaches the operating voltage and the nearer the test conditions approach the service conditions, the more representative the predicted life will be. For these reasons the accelerated aging test or load cycle test is by far the most definitive test of the three described.

8.6 IMPULSE TESTS

Impulse test procedures for cables are inherently quite straightforward. The length of the test specimen is restricted so that it can be treated as a lumped capacitor. There are no interturn stresses or partial breakdown mechanisms as there are in a high-voltage transformer. This does not mean that the test is easy or inexpensive to perform on high-voltage cables. Here we shall briefly review the theory of the impulse generator circuit, some features peculiar to cable testing, and the applicability of such tests to high-voltage cables [2, 3]. One of the world's highest voltage impulse generators is shown in Fig. 8.10. Its rating is well above any of the required impulse levels for the evaluation of cable systems presently in use or envisaged in future applications.

8.6.1 Necessity of Impulse Tests

It is important to understand that the requirement for the verification of adequate impulse strength is not simply a function of voltage rating. For example, it is not correct to assume that EHV cables must be impulse tested simply because they will be used in important transmission systems where outages are very costly, while medium cables need not be tested because of their use on less critical systems. The impulse test requirement is a function more of the cable type of construction than the voltage rating.

Figure 8.10 A 6.4-MV impulse generator with capacitive resistive voltage divider, trigatron gap arrangement, and HV bushing in foreground. (Courtesy of D. Train, Institut de Recherche d'Hydro-Québec.)

It will be recalled that lightning protective devices must withstand continuously the rated power voltage of the system but be capable to discharge the lightning transient voltage such that the BIL of the system is not exceeded. Such protective devices have, in practice, a minimum ratio between the rated ac voltage (phase-to-phase) and the disruptive or discharge voltage, in order that the device (lightning arrestor) will discharge the energy of the transient but will valve off the follow through power current. This ratio is about 1:4 or 1:5. If the impulse strength of a high-voltage cable is, say, one order of magnitude greater than its rated voltage, no difficulty arises in lightning protection. A 15-kV oil-impregnated-paper insulated (solid-type PILC) cable can be readily protected by a lightning arrestor having a disruptive impulse voltage level of 110 kV, since the impulse breakdown value is probably more than double this value. If now by some means the ac strength is upgraded without a comparable upgrading of impulse strength, a different situation exists wherein the ratio of impulse strength to rated voltage is not much greater than the equivalent ratio of the lightning arrestor. Consider a high-pressure oil-filled pipe cable (HPOF pipe type). Here the design ac stress of the insulation has been greatly increased above the PILC cable because of the pressurized system and attendant elimination of partial discharges, but the impulse strength has not been substantially enhanced. The impulse strength has become the limiting design parameter. The designer and the user must be assured that the product meets the BIL requirement.

Thus, the pressurized cables (normally in North America oil-filled or oil-filled pipe-type cables) have an impulse to 60-Hz strength ratio that is much lower than unpressurized cables (PILC or polymeric) because of partial-discharge suppression by the pressure medium. It is the pressurized cable wherein the impulse test is normally mandatory. Impulse tests of the extruded polymeric (solid-dielectric) cables are normally only for engineering information. Their impulse levels are well above the required BIL of the system.

8.6.2 Lightning Impulse Waveform

The wave shape used in impulse tests is designed to be representative in a general way of transient voltages impressed on electrical equipment due to the influence of lightning strokes. There are two standard wave shapes: the Institute of Electrical and Electronic Engineers (IEEE) $1.4/40\,\mu s$ and the International Electrotechnical Commission (IEC) $1.2/50\,\mu s$.

The first dimension in the wave shape expression is the front time t_1 (see Fig. 8.11), and the second dimension is the time to half value t_2. The front time is measured by measuring the time difference τ at points A and B and then applying a factor. The two points A and B lie on the linear part of the curve; the straight line through points A and B is extended upwards and downwards to intersect the time axis and a line drawn parallel to the time axis through the maximum value of the wave. A perpendicular projected from the later point to the t axis determines the value of t_1. Clearly in the curve $t_1 = 1.67\,\tau$, the factor is that used by IEC. The IEEE uses a slightly different method. The front time as defined in Fig. 8.11 is a more representative value of the rate of rise than if the actual time to the peak value were used. Because of the slightly different definitions and the permitted tolerances, the two standard waves can normally be met by using the same circuit constants in the impulse generator for both standards. In cable testing the front time is not so critical as for (say) transformers; values from 1 to $4\,\mu s$ are permitted and one normally attempts to keep below $2.5\,\mu s$

8.6.3 Impulse Generator

It is not necessary here to discuss the impulse generator in detail, but a review of the fundamentals is useful [1, 2]. In Fig. 8.12 one stage of an impulse generator is shown, and for our purposes any generator can be reduced to this form. Extra stages merely increase the output voltage.

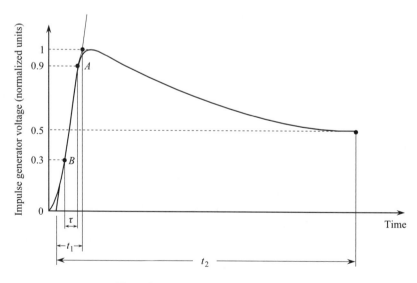

Figure 8.11 Impulse generator waveform.

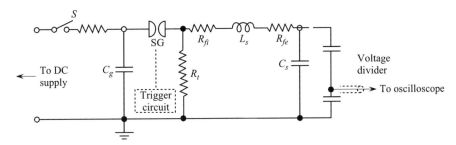

Figure 8.12 Schematic of a one-stage impulse generator.

When the sphere gap (SG) is triggered from the control room, it is seen that the generator capacitor C_g will discharge through R_{fi} and R_{fe}, the internal and external front resistors, into C_s, the test specimen. Since R_t, the wave tail resistor, is larger than the front resistor, it has little effect on the front time but determines the decay time or wave tail. Associated with the circuit is the unavoidable stray inductance L_s. Since R_t has little effect on the front time, an expression can be derived for the transient between time zero and the peak value by neglecting R_t. The circuit in Fig. 8.12 can be simplified to that in Fig. 8.13.

When the switch S is closed, it is clear that R_f must be of sufficient value to prevent underdamping, otherwise there will be an overshot on the wave and the transient will appear as shown in Fig. 8.14.

In a practical test R_f and C_s are so chosen that approximate critical damping is obtained, i.e.,

$$R_f \approx 2\left(\frac{L_s}{C}\right)^{1/2} \tag{8.2}$$

where

$$C = \frac{C_g \, C_s}{C_g + C_s} \tag{8.3}$$

This condition is not difficult to obtain in cable testing, since the stray inductance can be estimated to a fairly good accuracy. It is about $1.0 \, \mu H/m$ of circuit external to the

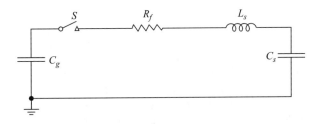

Figure 8.13 Simplified equivalent circuit of impulse generator.

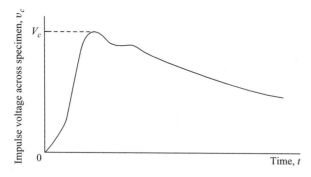

Figure 8.14 Overshoot effect on impulse waveform.

generator. The cable specimen can be treated as a lumped capacitor and adds very little to the circuit inductance. From Eq. (8.2) we have

$$L_s = \tfrac{1}{4} R_f^2 C \qquad (8.4)$$

With these assumptions, it is fairly easy to show that

$$v_c = \frac{CV}{C_s}\left[1 - \left(1 + \frac{2t}{R_f C}\right)\exp\left(\frac{-2t}{R_f \, C}\right)\right] \qquad (8.5)$$

where v_c is the voltage across the cable specimen when $t_0 \le t \le t_f$, C_s is the test specimen capacitance in microfarads, $C = C_g C_s/(C_g + C_s)$ in microfarads, V is the impulse generator terminal voltage at time, $t = 0$, t is the time in microseconds, R_f is the front resistance in ohms, t_0 is the time zero, and t_f is the time to peak value. Note that here the time to peak value is not identical to the front time, as defined in Fig. 8.11. Expression (8.5) appears complicated, but in fact we are normally interested only in one value of t, the front time. It turns out that the front time is given by

$$t_1 \approx 2.7 R_f C_s \qquad (8.6)$$

where C_s is the test specimen capacitance in microfarads, R_f is the front resistance in ohms, and t_1 is the front time in microseconds.

If in Eq. (8.5) we let $t \to \infty$, then the steady-state value or peak value becomes

$$V_c = \left(\frac{C}{C_2}\right)V \qquad (8.7)$$

If appropriate values of t_1 from Eq. (8.6) are substituted into Eq. (8.5), it is found that $v_c = 0.97 V_c$, verifying that the approximate expression agrees fairly closely with the derived Eq. (8.5). In a practical cable test, when R_f is large enough to prevent oscillation, the front time is normally greater than $1.5\,\mu$s.

In our simplifying assumptions the wave tail resistor was neglected because very little energy is lost in this resistor during the transient rise time. This resistor discharges the cable and generator capacitance in parallel and is sufficiently large that the circuit inductance has little effect. The discharge circuit is essentially an *RC* circuit with an exponential decay.

8.6.4 Test Specimens

Basically, the impulse test on a cable specimen involves the discharging of one capacitance (the generator) into another (the cable). It follows that the cable capacitance should be much smaller than the generator capacitance, otherwise the efficiency of the generator would be poor. Sometimes the question is asked whether entire reel lengths of cable are tested. A little arithmetic demonstrates that tests of such long lengths of cable are not feasible. Suppose we consider a 115-kV oil-filled pipe cable. Such a cable has a capacitance of say 400 pF/m. In our theoretical treatment only a single-stage generator has been examined, but a practical high-voltage impulse generator is composed of a number of stages, often 10 or more, that are charged in parallel but are discharged in series. Suppose the generator has a capacitance of about $0.5\,\mu F$ and a voltage rating of 200 kV per stage. The impulse strength of the cable is probably more than 600 kV, therefore at least four stages are required. The equivalent series capacitance is $0.5/4 = 0.125\,\mu F$. A cable length of approximately 300 m has a capacitance of $0.125\,\mu F$. The generator efficiency from expression (8.7) is 50% and the maximum crest voltage is 400 kV, well below the required test value. Further calculations would show that when practical values of stray inductance are used, the resulting front time would be too long. Another perhaps more fundamental reason for short specimen lengths will be discussed later.

In practice, the test load is usually between 4000 and $8000\,\mu F$. If the test load is too small, then difficulty with underdamping can be experienced, unless a wide range of front resistors is available. In this context it is necessary to place the test specimen as near the generator as clearance will permit. Another question that sometimes arises is whether there are reflections from the far end of the specimen. It is, of course, essential that the transient measured at the cable terminations is representative of the entire length of cable, and that there will be no perturbances on the transient. A heuristic approach will be used to indicate that reflections are not normally a problem.

The frequency spectrum of a 1.5/40-µs wave contains mostly components less than 50 kHz [6, 10]. The front of the wave, of course, gives rise to the higher components. Suppose the front time is 2.0 µs. A commonly used approximate relation between the maximum frequency component f_m and rise time t_r is

$$f_m \approx \frac{1}{2\pi t_r} \tag{8.8}$$

and substitution of 2.0×10^{-6} s for t_r in Eq. (8.8) yields an f_m value of ~ 80 kHz. From transmission line theory it can be demonstrated that the transmission line, the cable in this case, is equivalent to a lumped capacitance if the ratio of the length of the line, l, to the wavelength λ is small; i.e.,

$$\frac{l}{\lambda} < 0.04 \tag{8.9}$$

also

$$v = c\sqrt{\frac{1}{\varepsilon'_r}} \qquad (8.10)$$

and

$$f\lambda = v \qquad (8.11)$$

where ε'_r is the relative permittivity of the line dielectric, v is the velocity of propagation in meters per second, f is the frequency in hertz and c is the velocity of light (3×10^8 m/s). In an oil-paper cable the upper value of ε'_r may be taken as 4.0; therefore from Eq. (8.10) $v = 1.5 \times 10^8$ m/s. Also from Eq. (8.11) $\lambda = 1875$ m Finally from Eq. (8.9) the cable length l is found to be less than 75 m. From the foregoing, one would not expect difficulties from reflections for specimen lengths less than about 75 m. On the other hand, it becomes clear that cable lengths of the order of 500 m would be approximately equal to one-quarter wavelength for the higher frequency components. Hence, even if the impulse generator had the required energy, it would not be practical to test such long lengths of cable because of pronounced reflection effects.

8.6.5 Test Specimen Preparation

Cables having a BIL above about 300 kV require carefully constructed terminations to prevent flashovers. In addition, oil-pressure-type cables are often required to include a joint in the test specimen. Such procedures, particularly in EHV cables, make the cost of the impulse tests high. It may require two or three technicians a week to prepare the test specimen in the laboratory. Solid-dielectric cables, having an impulse breakdown level up to about 350 kV, are much easier to test. A simple resistive graded termination can be employed.

8.6.6 Calibration Procedures

After the test object is fully prepared and connected to the generator, it is often desirable to calibrate the measuring system against some standard, particularly if the test is required as an acceptance test by the user. A recognized standard for measuring the crest voltage, the most critical parameter in cable testing, consists of a sphere gap. For this purpose 50-cm spheres are adequate for rated voltages up to at least 138 kV. With the spheres set to a spacing of about 80% of the test specimen BIL, a series of impulses of both polarities is applied to the sphere gap and the cable in parallel. By a recognized standard procedure, impulses are recorded that are just sufficient to cause flashover of the sphere gap. The crest values of these impulses can be determined, after suitable correction for atmospheric pressure and temperature, from standard sphere gap tables. The average value should be within 3% of the average value obtained by reading the photographed transients recorded on the cathode ray oscilloscope (CRO). A useful instrument as an adjunct to the oscilloscope is the impulse peak voltmeter. This instrument records the crest voltage on a conventional meter scale. Each crest value can be read immediately after it occurs without the use of a camera. New digital recorders for impulse voltages are now extensively used in high-voltage testing and new international standards have been accepted for their use [6].

8.6.7 Polarization Effects

An impulse test for a customer usually begins at the BIL level with 10 impulses applied at each polarity. If it is desired to test the specimen to breakdown, the test voltage may be increased by 5% and the 20 pulses repeated. Breakdown will often occur, if one proceeds in this mode, at the first pulse after the polarity is reversed.

Such a breakdown is due to polarization effects. Although the specimen is short circuited between pulses by the impulse generator circuitry, the internal stress of the dielectric will not be completely relieved before the next pulse is applied. If the next pulse is of reversed polarity, the stress within the dielectric is enhanced and premature failure may occur. For this reason a pulse or series of pulses at reduced levels is applied after the polarity is reversed, before application of the full valued pulse.

8.6.8 Impulse Testing as a Development Tool

Impulse testing is an important technique in cable development. This fact is self-evident in the case of pressurized oil-paper systems such as those of oil-filled cables and oil-filled pipe cables; here the impulse strength, as previously discussed, is a mandatory design parameter. It may determine the insulation thickness to be used and, as a corollary, has a direct effect on the insulating paper to be chosen—tape thickness, apparent density, Gurley density, permittivity, and the like. The other insulation component, the impregnating oil, has characteristics, notably cleanliness and viscosity, that may significantly affect the impulse test.

The rationale for testing polymeric cables to impulse breakdown is not so obvious. Surely the impulse strength is comfortably in excess of that required, particularly for distribution cables. This is true but other information can be deduced from impulse test results. Let us suppose that the impulse strength of a polymer used for distribution cables is 40 kV/cm. A test schedule might include 10 specimens of such a cable rated at 15 kV. The standard wall thickness is about 175 mils (or 4.5 mm), so an average break-down value of 175–180 kV is to be expected. If the test values were found to vary between 150 and 225 kV, the investigator is likely to be disappointed, although the values are well above the BIL requirement of 110 kV. From experience he will know that such a variation in results would never occur in oil-filled cable. The large variation indicates poor quality control either by the polymer supplier or the cable manufacturer or both.

Impulse tests are not necessarily more complicated than the routine ac or dc tests. If there is an impulse facility available, this test is often no more difficult than the ac test. Indeed, specimen preparation is often simpler for medium-voltage ratings because resistive graded terminations can be utilized for impulse tests, whereas more sophisti-cated terminations are required if the specimen is to be subjected to an ac breakdown test. Furthermore, the energy released in an impulse failure is small compared to that in a conventional transformer-induced breakdown so that there is very little fault damage to mask the cause of a possible discrete insulation weakness. The dc test is also rather simple to perform, but it leaves the investigator with a question concerning its practical significance. Cables (at least the vast majority of cables) do not operate under a dc stress. They do get subjected to surge voltages, time-variant phenomena like 60 Hz, although admittedly in a higher frequency spectrum. In an ac test and an impulse test, the cable acts as a capacitor; the stress distribution is determined by permittivity. In the dc test, the stress distribution is determined by the insulation resistivity and may be quite different from the ac capacitive stress distribution.

REFERENCES

[1] T. J. Gallagher and A. J. Pearmain, *High Voltage Measurement Testing and Design*, Wiley, New York, 1983.

[2] E. Kuffel and W. S. Zaengel, *High Voltage Engineering Fundaments*, Pergamon, London, 1984.

[3] N. Hylten-Cavalins, *High Voltage Laboratory Planning*, ASEA-Haefley HV Test Systems, Haefley Instruments, Zürich, 1986.

[4] IEC, Publication IEC-60. High Voltage Test Techniques: Part 1, General Definitions and Test Requirements, 1989; Part 2, Measuring Systems, IEC, Geneva, 1994.

[5] Publication IEC-270, Partial Discharge Measurements, IEC, Geneva, 1981.

[6] Publication IEC-1083 Digital Recorders for Measurements in High Voltage Impulse Tests: Part 1, Requirements for Digital Recorders, 1991; Part 2, Evaluation of Software Used for the Determinations of Parameters of Impulse Waveforms, IEC, Geneva, NJ, 1996.

[7] IEEE Standard 4, Techniques for High Voltage Testing, Piscataway, NJ, 1995.

[8] S. J. Gladys, "Series resonance testing of ac cables using a moving coil reactor and transformer", *Insulation*, July 1969, pp. 45–47.

[9] M. M. Epstein, W. H. Stember, B. S. Bernstein, R. Bartnikas, and S. Pelissou, "Service aged XLPE cables: insulation characteristics and breakdown strength," *IEEE International Symposium on Elec. Insulation*, Pub. CH2196-4, Washington, DC, 1986, pp. 23-27.

[10] W. J. McKeen, "The frequency spectrum in impulse testing," IEEE Conference Paper CP. 58-56, Winter Power Meeting, New York, 1958.

Chapter 9

DISSIPATION FACTOR, PARTIAL-DISCHARGE, AND ELECTRICAL AGING TESTS ON POWER CABLES

R. Bartnikas

9.1 INTRODUCTION

In this chapter we shall consider three specialized testing methods and procedures on cables, involving the measurement of dissipation factor or power factor, partial discharge, and accelerated electrical aging. Each of these measurements requires a more detailed description than was provided in Chapter 8, which dealt principally with the more routine voltage tests under alternating-current (ac), direct-curent (dc) and impulse conditions.

Power factor tests are normally carried out on power cables insulated using liquid-impregnated systems. As pointed out earlier, despite its high reliability while in service, the oil-paper combination is nevertheless characterized by a relatively high dielectric loss magnitude. Since the power factor value is an indicator of dielectric loss (which is a function of the ionizable contaminants present in the solid-liquid insulating system), the power factor measurement is a very useful tool in assessing the quality of the impregnated cable insulation at high voltage. Partial-discharge measurements are in general not performed on solid-type oil-paper cables since the oil-paper system if properly dried and impregnated will tend to be free of corona discharges. Although it is possible that discharges may occur intermittently in the transient voids formed during certain periods of load cycling, their intensity and duration are usually not sufficiently high and long to give rise to any deleterious degradation effects. In this regard, it is now well recognized that although exposure of the solid-type oil-paper insulation to partial discharges may cause the evolution of gases such as hydrogen and methane accompanied by solid wax formation from the oil undergoing chemical changes, the wax thus formed because of its larger volume will tend to fill in the void or cavity space and thereby extinguish the discharges. With low dielectric loss polymeric power distribution cables, the power factor measurement is considerably less important; however, it is routinely performed on higher voltage polymeric power transmission cables in which the dielectric losses attain significant values because of their direct dependence upon the square of the applied voltage. The distribution power cable field, as mentioned previously, is essentially dominated by polyethylene (PE) or crosslinked polyethylene (XLPE) cables, both of which have dielectric systems with intrinsically low power factors. There the problem

331

is not so much dielectric loss and the associated dielectric heating as it is the occurrence of cavities and the associated partial discharges. For this reason, lower voltage solid extruded dielectric insulated cables—including even the higher loss ethylene-propylene-rubber (EPR) type of cables—are not evaluated in terms of the power factor measurement, but rather in terms of their partial-discharge characteristics.

9.2 DISSIPATION FACTOR OF A CABLE

Although the term power factor is widely used, it constitutes an incorrect usage; the term that ought to be used to describe the losses in the dielectric is the dissipation factor, tan δ, value. The distinction between the terms power factor and dissipation factor becomes more palpable if one considers the definition of tan δ. It is common to represent the dielectric of a cable by the parallel equivalent circuit shown in Fig. 9.1. The resistance R describes the lossy part of the dielectric, while C represents the lumped-circuit capacitance of the cable. When a voltage V is applied across the cable, the total current flowing in the cable, I, is given by the vector sum of the charging current I_c and the leakage current I_l:

$$I = I_l + I_c = (G + j\omega C)V \tag{9.1}$$

where G is the conductance and is defined as the reciprocal of R. The vector relationship in Eq. (9.1) can be depicted by the phasor diagram in Fig. 9.2. Hence, the dissipation factor tan δ of the dielectric in terms of Fig. 9.2 becomes

$$\tan \delta = \frac{I_l}{I_c} = \frac{1}{\omega RC} \tag{9.2}$$

There exists a definite relationship between the electrical parameters of the equivalent circuit and the dielectric parameters of the cable insulant. The dielectric parameters characterizing the cable insulation are the real permittivity ε', the imaginary permittivity ε'', and the conductivity σ. The value of the cable capacitance C is thus defined as

$$C = \varepsilon_r' C_0 \tag{9.3}$$

where C_0 is the capacitance in vacuo and ε_r' is the relative value of the real permittivity or dielectric constant defined by $\varepsilon'/\varepsilon_0$; here ε_0 refers to permittivity value of the

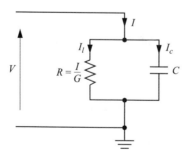

Figure 9.1 Equivalent lumped-parameter circuit of a cable.

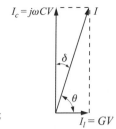

Figure 9.2 Phasor relationship of charging and leakage currents in a cable dielectric.

vacuum. In analogy with this definition, the imaginary permittivity is introduced to describe the lossy part of the dielectric so that Eq. (9.1) may be expressed by

$$I = j\omega\varepsilon_r^* C_0 V \tag{9.4}$$

where ε_r^* is the complex value of the relative permittivity and is by definition

$$\varepsilon_r^* = \varepsilon_r' - j\varepsilon_r'' \tag{9.5}$$

where ε_r'' is the relative value of the imaginary permittivity given by $\varepsilon''/\varepsilon_0$; so that the total current assumes the form

$$I = j\omega\varepsilon_r' C_0 V + \omega\varepsilon_r'' C_0 V$$
$$= (\omega\varepsilon_r'' + j\omega\varepsilon_r')C_0 V \tag{9.6}$$

It can be perceived now that in terms of the dielectric parameters, the dissipation factor value becomes

$$\tan\delta = \frac{\varepsilon''}{\varepsilon'} = \frac{\varepsilon_r''}{\varepsilon_r'} \tag{9.7}$$

Normally, the imaginary permittivity ε'', referred to occasionally as the loss factor, cannot be directly measured; instead the dielectric conductivity σ is measured. The conductivity is defined as the ratio of the current density J to the electric field gradient E. If we consider an elemental dielectric thickness d having a surface area A, then the capacitance in vacuo C_0, is given by

$$C_0 = \frac{A}{d}\varepsilon_0 \tag{9.8}$$

where ε_0 is the value of the permittivity in free space or vacuum. Substituting the expression for C_0 in Eq. (9.6) and dividing by the unit area yields the current density expression

$$J = (j\omega\varepsilon_r' + \omega\varepsilon_r'')\varepsilon_0 V/d$$
$$= (j\omega\varepsilon' + \omega\varepsilon'')E \tag{9.9}$$

The real part of this expression is simply the conductivity term; that is,

$$\sigma = \omega\varepsilon'' \tag{9.10}$$

Thus Eq. (9.7) becomes

$$\tan \delta = \frac{\sigma}{\omega \varepsilon'} \qquad (9.11)$$

Since the conductivity term of dielectrics, which are used as cable insulating materials, is very small, $\tan \delta$ is also small with the result that the loss angle δ is very small. For this reason it can be seen in terms of Fig. 9.1 that $\cos \theta \approx \tan \delta$. In antecedent cable engineering terminology, it had been customary to use the term power factor ($\cos \theta$); however, present usage prefers $\tan \delta$ as the more appropriate term.

An alternative means for deriving the expression for $\tan \delta$ is to consider the relationship between the electrical field, $\bar{E}$, and the flux density, $\bar{D}$, vectors [1] for the case of a perfect dielectric:

$$\bar{D} = \varepsilon' \bar{E} \qquad (9.12)$$

where the real permittivity is a real quantity such that no phase shift exists between the $\bar{D}$ and $\bar{E}$ vectors. Under the appearance of a loss in the dielectric, the $\bar{D}$ vector exhibits a lag with respect to the electric field vector $\bar{E}$ as the permittivity becomes complex and Eq. (9.12) must be rewritten as

$$\bar{D} = \varepsilon^* \bar{E} \qquad (9.13)$$

in complex notation $\bar{D}$, assumes the form $D_0 \exp[\omega t - \delta]$ and correspondingly $\bar{E}$ becomes $E_0 \exp[\omega t]$. Here D_0 and E_0 represent the respective magnitudes of the vectors $\bar{D}$ and $\bar{E}$. Hence, since $\exp[j\delta] = \cos \omega t + j \sin \omega t$, it follows from Eq. (9.5) that

$$\left(\frac{D_0}{E_0} \right) \sin \delta = \varepsilon'' \qquad (9.14)$$

and

$$\left(\frac{D_0}{E_0} \right) \cos \delta = \varepsilon' \qquad (9.14b)$$

Thus, the quotient $\sin \delta \cos \delta$ yields the value of $\tan \delta$ as defined by Eq. (9.7).

The conductivity, σ, of a dielectric is a measure of the conduction losses that take place in the presence of an external electric field. These losses result primarily from the movement of free charge carriers in the field direction; at frequencies substantially above 60 Hz, the losses begin to be increasingly or entirely determined by dipole orientation effects due to the presence of polar molecules. The charge carriers may consist of both electrons and ions; the presence of the latter may be due to electrolytic impurities or oxidation products. Improperly treated insulating papers and contaminated or oxidized insulating oils are generally responsible for the ions causing the conduction loss. Low-viscosity oils are particularly susceptible to ionic losses because ionic motion is greatly facilitated in a low-viscosity medium. In contrast, ionic mobility in solid insulations such as PE and XLPE is greatly reduced and, consequently, the $\tan \delta$ values of these insulations are very small. Indeed values of $\tan \delta < \times 10^{-4}$ suggest that the minuscule detectable losses may be caused by an electronic conduction mechanism, since the conductivity is considerably less influenced by temperature than in the case of oil-impregnated insulation. An increase in temperature enhances the ionic losses in oil-paper systems due to a decrease in the oil viscosity with temperature. The remnant or bound water content in the paper tapes also contributes to the overall losses as ions can move with ease along the polar water molecule chains. Any improper drying of papers, i.e., failure to remove all the free or unbound water, can give rise to very serious conduction losses; paper dryness can in fact be monitored by tan δ measurements.

In most insulating systems, the tan δ value exhibits an increase with the applied voltage. This is caused by the fact that a rising applied field increasingly segregates the oppositely charged ions, thus preventing their possible recombination. However, tan δ may also fall with the field due to interfacial polarization effects as ionic charge becomes trapped or piled-up at dielectric interfaces having unequal conductivities (σ) and real permittivities (ε'). The space charge process represents a complex mechanism whose behavior cannot be readily analyzed. Particular cases of the space charge or interfacial polarization processes are often referred to as the Garton or Böning effects. These mechanisms occur within the oil-filled pores of kraft paper insulation, as the ionic motions become limited by the boundaries of the pores. This manifests itself in a decrease of tan δ with the field. Cables that are characterized by a marked fall of tan δ with voltage are as suspect as those exhibiting a marked increase of tan δ with voltage. In both cases, serious ion contamination or moisture content effects are implicated.

The power dissipated, P, in the cable insulation as a result of the dielectric loss is given by

$$P = \omega C V^2 \tan \delta \qquad (9.15)$$

where V is the root-mean-square (rms) value of the applied voltage, ω is the frequency term in radians, and P the power loss in watts. Since the dielectric power loss, P, is directly proportional to the dissipation factor, it is important that the value of tan δ be small; otherwise this will not only lead to an unnecessary power loss but also to thermal instability of the cable due to dielectric heating. Furthermore, since the power loss increases with the square of the applied voltage, the seemingly sufficiently low tan δ value of 0.002 for the oil-paper system becomes too high to be tolerated at very high voltages; hence, the search for low-loss solid-liquid combinations for voltages above 500 kV and the use of low dielectric loss PE and XLPE insulations up to and including 500 kV.

9.3 BRIDGE TECHNIQUES FOR THE MEASUREMENT OF tan δ

The tan δ value of cables is often obtained by means of a Schering bridge, which was developed in the 1920s by Schering and his associates at the Berlin's Physikalische und Technische Reichsanstalt [2]. Thereafter, it quickly became the standard bridge for measuring capacitance and dielectric loss on cables at power frequencies at low and high voltages. Prior to this time these measurements were performed using the Wien bridge [3]; the Wien bridge because of its frequency sensitivity continued to serve for some time as a frequency measuring device until it ultimately was replaced even in this area of endeavor by electronic frequency counters. On the other hand the Schering bridge has become so entrenched that its use has persisted to this very day, notwithstanding the current competition presented to it by the more recently developed ratio arm bridge.

The Schering bridge is manufactured in its different variations by a number of instrument producers; because of its continued usage and historical importance, we shall attempt to present in this section a more extensive treatment of the Schering bridge. The Schering bridge regards the cable insulation as an equivalent series circuit, consisting of a capacitance in series with a resistance. The losses within the insulation as well as its capacitance are compared to those of a standard capacitor that is assumed to be loss free. The standard capacitor has either vacuum or compressed dry inert gas for its dielectric and is discharge free. In fact, the losses in such capacitors are negligible and are generally beyond the detection capability of the usual measuring instruments. With most bridges, the value of the standard capacitance is usually fixed at 100 pF, though some bridges, which are intended for measurements on low-capacitance specimens such as oils, may use values as low as 50 pF. The early standard capacitors used atmospheric air as the dielectric, and thus their losses would exhibit an increase under humid conditions; in addition, they were limited by their relatively low partial-discharge inception voltages. Present standard capacitors are constructed with compressed dry nitrogen, sulfur hexafluoride gas, or occasionally vacuum as their dielectric and are characterized by extremely low dielectric losses as well as discharge free operating voltages that for some applications may easily attain values in excess of 1000 kV.

The basic Schering bridge circuit is depicted in Fig. 9.3. Here C_s represents a standard capacitor having negligibly small losses, the series equivalent circuit of R' and C represents the specimen, and R_3, R_4, and C_4 are balancing elements of the bridge. Since the two upper bridge arms are capacitive, their impedances at the power frequency are high compared to the two lower essentially resistive arms, and, therefore, the major portion of the voltage drop occurs across the two upper bridge arms whereby the two lower arms remain close to ground potential. This constitutes an important safety feature of the Schering bridge that permits an operator to manipulate the lower bridge arms to effect bridge balance. At bridge balance the voltage drop across C_s both in phase and magnitude equals that across the specimen, as the voltage across the null detector is reduced to zero. Thus at balance, the product of the impedance terms of the opposite arms must be equal:

$$Z_3 Z_s = ZZ_4$$

or

$$\left(\frac{1}{j\omega C_s}\right)(R_3) = \left(R' + \frac{1}{j\omega C}\right)\left(\frac{1}{(1/R_4) + j\omega C_4}\right) \tag{9.16}$$

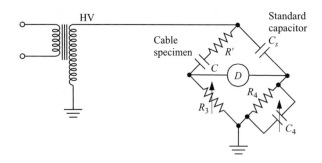

Figure 9.3 Basic Schering bridge circuit.

Equating the real and imaginary terms, the capacitance of the specimen is obtained as

$$C = \frac{R_4 C_s}{R_3} \tag{9.17}$$

and the equivalent series resistance as

$$R' = \frac{R_3 C_4}{C_s} \tag{9.18}$$

so that the dissipation factor for the series equivalent circuit of the cable insulation is thus given by

$$\tan \delta = \omega R' C$$
$$= \omega R_4 C_4 \tag{9.19}$$

From Fig. 9.3, it can be perceived that the Schering bridge views the dielectric in terms of a series equivalent circuit. Consequently, the expression for tan δ in Eq. (9.19) reflects this fact and differs from that of Eq. (9.2). In the literature, the use of the parallel equivalent circuit is prevalent, since most workers prefer the concept of a dielectric as having a large resistance shunted by a capacitance, e.g., the dc insulation resistance of a cable, as measured with a megger, is a very large quantity. The values of the capacitances in the two respective circuits are, however, equal. It is perhaps important at this juncture to delineate the equivalence between the two opposing concepts. The condition of equivalence between the two circuits requires that the admittances of the two equivalent circuits of the dielectric specimen be equal [4]. The admittance Y of the parallel circuit is obtained from Eq. (9.1) or (9.6) as

$$Y = \frac{1}{V}$$
$$= \left(\frac{1}{R} + j\omega C\right)$$
$$= (\omega \varepsilon_r'' + j\omega_r') C_0 \tag{9.20}$$

Note that here the resistance R of the parallel equivalent circuit must be distinguished from that of R' of the series equivalent circuit, since $R \gg R'$. For the series circuit, the admittance is the inverse of the series impedance Z, which is given by

$$Z = \frac{V}{I}$$
$$= R' + \frac{1}{j\omega C} \tag{9.21}$$

From Eqs. (9.20) and (9.21), the equivalence condition implies that

$$(\omega\varepsilon_r'' + j\omega\varepsilon_r')C_0 = \frac{1}{R' + 1/j\omega C} \tag{9.22}$$

Rearrangement yields

$$\varepsilon_r'' = \frac{\omega R' C^2}{C_0[1 + (\omega R'C)^2]} \tag{9.23}$$

and

$$\varepsilon_r' = \frac{C}{C_0[1 - (\omega R'C)^2]} \tag{9.24}$$

Thus,

$$\tan\delta = \frac{\varepsilon_r''}{\varepsilon_r'}$$

$$= \omega R' C \tag{9.25}$$

The result in Eq. (9.20) is identical in form to that of the bridge balance condition given in Eq. (9.19). It is well to reemphasize once more here that the small resistance value of R' in Eqs. (9.21) and (9.25) for the series circuit should not be confused with the very much larger value of R in the parallel equivalent circuit, appearing in Eqs. (9.2) and (9.20).

    The simple circuit given in Fig. 9.3 for the Schering bridge is suitable only for measurements on short lengths of cable having low-capacitance values. With longer cable lengths, such as full reel lengths of cable, large charging currents will result, and their value may exceed the current rating of the noninductive and adjustable resistance, R_3 [5, 6]. This difficulty may be obviated by inserting in parallel with R_3 a low-resistance shunt, which would be capable of carrying the anticipated charging currents. Such arrangement would necessarily require some modification of the bridge as the effective resistance across the R_3 arm would approach the value of the low-resistance shunt, rendering any balancing attempts by the variable resistor R_3 quite ineffective. A possible bridge rearrangement that remedies this difficulty is portrayed in Fig. 9.4; it has been originally proposed by Bruckmann [5].

    It can be shown that with the new arrangement, including the inserted low-resistance shunt R_{sh} [6], that

$$C = C_s R_4 \left[\frac{(R_3' + R_3'' + R_{\mathrm{sh}})}{R_3' R_{\mathrm{sh}}} \right] \tag{9.26}$$

$$\tan\delta = \omega C_4 R_4 - \omega C_s R_4 \left(\frac{R_3''}{R_3'} \right)$$

$$= \omega R_4 \left[C_4 - C_2 \left(\frac{R_3''}{R_3'} \right) \right] \tag{9.27}$$

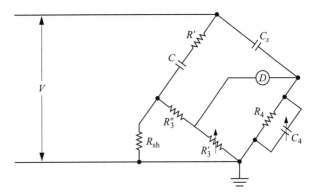

Figure 9.4 Schering bridge modification (after [5, 6]).

Since

$$C_s\left(\frac{R_3''}{R_3'}\right) \ll C_4 \tag{9.28}$$

then

$$\tan\delta = \omega R_4 C_4 \tag{9.29}$$

Precise and accurate measurements require the elimination of stray capacitances between the various bridge elements. With such particular arrangement, the standard capacitor C_s becomes a three-terminal system with guard electrodes, while the cable specimen must use terminations incorporating a guard system. The other elements of the bridge, including the detector, are enclosed within an electrostatic shield, whose potential and phase angle must be maintained equal in value to that of the detector, i.e., the potential between the guard and the detector must be zero in magnitude as well as phase. This is accomplished by means of an auxiliary bridge arm R_5C_5 connected for balancing purposes between the shield and a junction point of the main bridge. To illustrate the shielded arrangement of the Schering bridge, let us consider the basic Schering bridge circuit of Fig. 9.3. This circuit in a shielded form in fact constitutes a precision-type Schering bridge, normally employed in high-precision measurements for research purposes of short cable lengths and cable models. A shielded Schering bridge schematic for precision measurements is depicted in Fig. 9.5. Note that the low-voltage arms are enclosed in shields that are directly grounded, while the shielding enclosing the high-voltage arms and the detector eliminates stray capacitances between the former and the latter as well as ground. Furthermore, any capacitance current between the high voltage sections of the bridge and the grounded low-voltage arms simply shunts the high-voltage supply and does not influence the measurement. Likewise the capacitance current between the high-voltage portion of the bridge and the detector shield flows to ground via the auxiliary bridge arm.

It is correctly pointed out by Harris [6] that although in the shielded version of the Schering bridge the low-voltage balancing resistors R_3 and R_4 are shielded from the influence of the external high-voltage fields, their own proximity to the ground necessi-

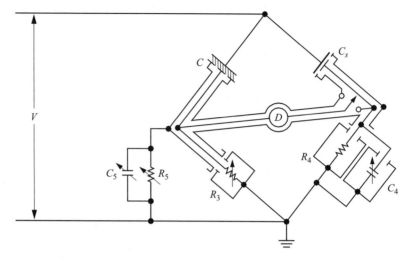

Figure 9.5 Shielded Schering bridge for precision measurements [6].

tates taking into account their self-inductive effects, L_3 and L_4, respectively. Hence, the corrected balance equation for the bridge now becomes [6]

$$\frac{-j}{\omega C_s}(R_3 + j\omega L_3) = \left(R' - \frac{j}{\omega C}\right)\frac{R_4 + j\omega L_4}{1 - \omega^2 L_4 C_4 + j\omega C_4 R_4} \tag{9.30}$$

From which equating the real and imaginary terms yields

$$C\left[1 - \left(\frac{\omega L_3}{R_3}\right)\tan\delta\right] = C_s\frac{R_4}{R_3} \tag{9.31}$$

or

$$C \approx C_s\frac{R_4}{R_3} \tag{9.32}$$

and

$$\tan\delta = \omega C_4 R_4\left[1 - \left(\frac{\omega L_4}{R_4}\right)\right] + \frac{\omega L_3}{R_3} - \frac{\omega L_4}{R_4} \tag{9.33}$$

or

$$\tan\delta \approx \omega C_4 R_4 + \omega\frac{L_3}{R_3} - \omega\frac{L_4}{R_4} \tag{9.34}$$

In practice, the residual effects are appreciably reduced or entirely eliminated by choosing a bridge arrangement such that R_3 is of the same order of magnitude as R_4 so that the last two terms cancel out in Eq. (9.34). In addition, commercial Schering bridges are direct reading instruments in which the R_3 dial is calibrated to provide a direct reading of the unknown cable capacitance C, in picofarads, while the reading on the C_4 dial is converted to provide a direct measure of the tan δ value. In the case of the C_4 dial, the resistance R_4 in parallel with C_4 is fixed at a value of $10^6/\omega$ or to 2652.6 Ω at 60 Hz to render C_4 direct reading in tan δ [6].

In some more recently manufactured bridge instruments by the Gen Rad Company, an active type of guard circuit is utilized in which an operational amplifier of unit gain is inserted into the guard circuit. In such circumstances, the guard circuit is termed an active guard circuit. The operational amplifier reduces automatically the voltage drop across the stray load capacitances to ground of the specimen, thereby eliminating measurement errors without resorting to an additional balancing operation. The active guard circuit approach utilized by the Gen Rad Company design is frequently referred to as the driven screen or guard technique [7–9]. A typical Schering bridge arrangement with a driven guard circuit is portrayed in Fig. 9.6. Since the gain of the operational amplifier is fixed at unity, the bridge potential, V_b, at point A is equal to the potential of the guard at point B. The bridge balance condition remains uninfluenced because the input impedance to the unit gain amplifier is chosen to be large in comparison to the impedance of the bridge arm comprising of R_4 and C_4. In addition, the operational amplifier is designed with a very low output impedance $\sim 1 \Omega$ [10], so that the guard circuit is effectively grounded via a low resistance.

In the more recent past, an alternative bridge has been developed to measure the capacitance and tan δ value of cables. This bridge, known as the current comparator or transformer ratio arm bridge, has the advantage when small tan δ values are to be measured. Whereas the Schering bridge can measure tan δ values down to 10^{-5}, the current comparator bridge can improve on this an order of magnitude and go down with relative ease to as low as 10^{-6}. The schematic circuit diagram of a current comparator (NRC/Guildline bridge) is shown in Fig. 9.7. Note that here the cable dielectric is represented by an equivalent parallel circuit; to distinguish between the parallel and

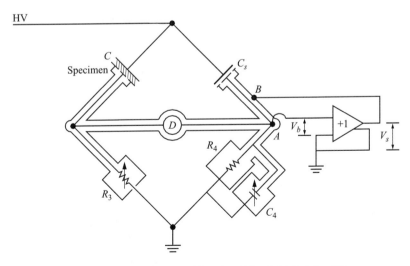

Figure 9.6 Schering bridge with active (driven) shield (after [9]).

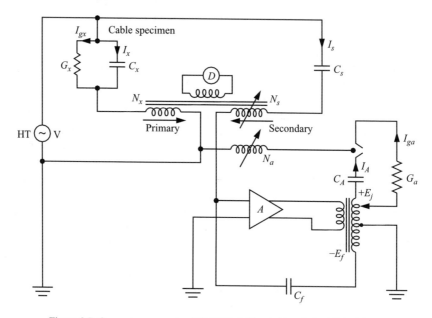

Figure 9.7 Current comparator (NRC/Guildline bridge design) [11, 12].

series representation, the subscript x is used here for the resistive and capacitive elements of the equivalent circuit for the specimen.

In the current comparator, both the primary and secondary windings carry currents of the specimen, G_x and C_x, and standard capacitor, C_s, elements, and the detector winding is used to sense the ac flux in the transformer core. When the primary and secondary ampere-turns are equal and opposite, the resultant ac flux in the core is zero. To measure the capacitance of the specimen, a switch is closed to insert the capacitor C_A into the circuit and the number of turns on the N_s side is varied until zero flux is indicated in the core, i.e., there is zero reactance in the windings (assuming negligible winding resistance). The capacitance currents are then simply [11, 12]

$$I_x = j\omega V C_x \tag{9.35}$$

and

$$I_s = j\omega V C_s \tag{9.36}$$

At balance

$$I_x N_x = I_s N_s \tag{9.37}$$

Thus, substituting in I_x and I_s gives the capacitance of the specimen C_x as [11, 12]

$$C_x = C_s \frac{N_s}{N_x} \tag{9.38}$$

Increased resolution is obtained by switching a known fraction of the C_s current into an additional comparator winding N_A. The shown amplifier is used to develop a low-voltage E_f, proportional to and in phase with the supply voltage V. At balance, the voltage drop across N_s is negligible, and

$$E_f = \frac{C_s}{C_f} V \tag{9.39}$$

and the current in the N_A winding becomes

$$I_A = j\omega E_f C_A \tag{9.40}$$

Substituting E_f into the above gives

$$I_A = j\omega V C_s \left(\frac{C_A}{C_f}\right) \tag{9.41}$$

This implies that the contribution to the ampere-turns is reduced by C_A/C_f, and at balance we have [11, 12]

$$C_x = C_s \left[\frac{N_s}{N_x} + \left(\frac{C_A}{C_f}\right)\left(\frac{N_a}{N_x}\right)\right] \tag{9.42}$$

To obtain the tan δ value of the specimen, the switch is closed to put G_a in the circuit, and now the amplifier is used to inject an in-phase resistive current component through the conductance G_a. For the in-phase current component, we have

$$I_{ga} = nE_f G_a \tag{9.43}$$

where n is a fraction of the E_f selected. Combining Eqs. (9.39) and (9.43) gives

$$I_{ga} = \left(\frac{VC_s}{C_f}\right) nG_a \tag{9.44}$$

Since

$$I_{gx} = VG_x \tag{9.45}$$

and at balance with zero flux in the core,

$$I_{gx}N_x = I_{ga}N_a \tag{9.46}$$

then from Eqs. (9.43) and (9.45) we have [11, 12]

$$G_x = nG_a \left(\frac{C_s N_a}{C_f N_x}\right) \tag{9.47}$$

Thus the dissipation factor becomes

$$\tan \delta = \frac{G_x}{\omega C_x}$$

$$= \frac{nG_a N_a}{\omega C_f N_s} \tag{9.48}$$

Seitz and Osvath [13] and Osvath and Widmer [14] described a microcomputer-controlled transformer ratio arm bridge, which is now commercially available. Its schematic circuit diagram is depicted in Fig. 9.8. The coarse and fine balances of the bridge are performed by adjusting the number of turns in the current comparator's windings N_1, N_2 and N_3, N_4. The currents in the fine balancing coils, N_3 and N_4, are proportional to the current in the standard capacitor C_s and are regulated by the multiplying digital-to-analog (D/A) converters α and β. Since N_1 is the coil of the null detector, the balance condition requires that $I_1 = 0$; thus, the ampere-turns equation at an applied voltage V may be stated as [14]

$$V(G_x + j\omega C_x)N_1 = Vj\omega C_s(N_2 + \alpha RG_1 N_4 - j\beta G_2 N_3) \tag{9.49}$$

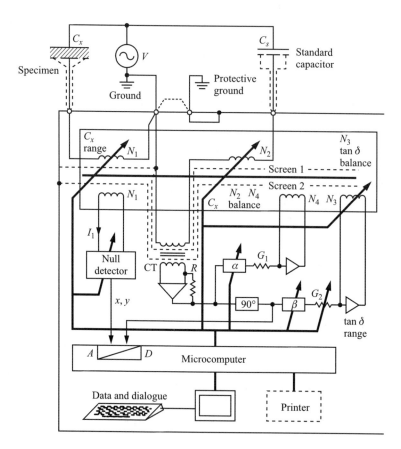

Figure 9.8 Microcomputer-controlled automated transformer ratio arm bridge (after [14]).

Equating the real terms of Eq. (9.49) yields the expression for the conductance of the cable specimen, G_x:

$$G_x = \frac{C_s(\beta R G_2 N_3)}{N_1} \tag{9.50}$$

and equating the imaginary terms results in the expression for the capacitance of the specimen, C_x:

$$C_x = C_s \frac{N_2 + \alpha R G_1 N_4}{N_1} \tag{9.51}$$

The corresponding expression for tan δ is thus

$$\tan \delta = \frac{G_x}{\omega C_x}$$

$$= \frac{N_3(\beta R G_2)}{N_2 + \alpha R G_1 N_4} \tag{9.52}$$

Since normally $R G_1 N_4 \simeq 1$ and $N_3 = N_4$, then

$$C_x \simeq \frac{C_s(N_2 + \alpha)}{N_1} \tag{9.53}$$

and

$$\tan \delta - \frac{\beta R G_2}{1 + \alpha/N_2} \tag{9.54}$$

The automated transformer ratio arm bridge is balanced by the microcomputer, and the capacitance and dissipation factor of the cable specimen are calculated in terms of the N_1, N_2, N_3, α, and β and G_2 values at balance. Protection of the computer against possible breakdown current surges at the elevated best voltages is obtained by the use of optical fiber cables. The minimum measurable tan δ value by means of the computer-controlled automated bridge is on the order of 1×10^{-5}, which is substantially higher than that of 1×10^{-7} obtainable with the utmost sensitive manually balanced ratio arm bridges.

The tan δ value obtained, using either the Schering bridge or the current compara-tor, is a measure of the dielectric loss magnitude of the oil-paper cable insulating system under test. The quality of the cable insulating system is assessed in terms of the ioniza-tion factor, IF, which is defined as the difference between the tan δ value obtained at 100 V/mil and that at 20 V/mil; i.e.,

$$\text{IF} = [\tan \delta]_{100\text{V/mil}} - [\tan \delta]_{20\text{V/mil}} \tag{9.55}$$

The permissible values of the IF are given in pertinent specifications quoted in the previous chapters on cable design and manufacture. Since under normal conditions tan δ increases with applied voltage, the IF value is essentially a measure of the tan δ tip-up voltage. This tip-up can result either from an increase in the magnitude of the losses in the solid-liquid portion of the dielectric with voltage or from partial-discharge onset. Partial discharges are unlikely to occur in high-voltage (HV) oil-pressurized cables, though their occurrence is possible in solid-type high-viscosity oil-impregnated cables. Figure 9.9 shows a typical corona partial-discharge effect in a high oil viscosity cable; the resulting $\tan \delta$ tip-up due to the partial discharges is substantially more steep and abrupt than the more gradual tip-up, which is caused by a rise in the dielectric loss with voltage occurring within the solid-liquid portion of the insulation.

Figure 9.9 demonstrates the effect of a large number of small discharging voids, whose number increases with applied voltage. If a single void were involved only, the $\tan \delta$-voltage behavior would be quite different. Figure 9.10, obtained with relatively large artificial cavities, illustrates this peculiar behavior [15]. If one assumes that the partial discharges are of the pulse form, then the latter variation of $\tan \delta$ with the applied voltage V may be described by

$$\tan \delta = \frac{C'}{C} \tan \delta' + \frac{C_T}{\omega C V^2} \sum_{j=1}^{j=i} n_j \, \Delta V_{cj} \, V_j(t) \tag{9.56}$$

where $\tan \delta'$ and C' denote the $\tan \delta$ and capacitance values due to the relatively invariant dielectric losses prior to discharge onset, C is the capacitance of the cable specimen at the applied voltage V, and C_T is the total capacitance of the partial-discharge detection circuit including the cable; n_j represent the discharge rate of the jth discharge of peak discharge voltage ΔV_{cj} occurring at the instantaneous value $V_j(t)$ of the sinusoidal voltage applied across the specimen cable. Note that the sign of the ΔV_{cj} and $V_j(t)$ terms differs when their polarities are opposite. When a single cavity is involved, the $\tan \delta$ value has its maximum at the partial-discharge inception voltage; thereafter $\tan \delta$ diminishes with applied voltage, as in Fig. 9.10, because the denominator of the second term in Eq. (9.56) increases more rapidly than the discharge power loss in the numerator. The $\tan \delta$ value may also decrease with voltage even when more than one cavity is involved, but only after all cavities have begun to discharge.

Figures 9.11 and 9.12 illustrate the influence of voltage stress with temperature as parameter upon low- and high-viscosity oil-paper cable models in the absence of partial

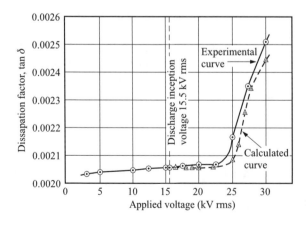

Figure 9.9 Experimental and calculated $\tan \delta$ vs. applied voltage characteristics obtained on a power cable prior to long-term voltage test [15, 16].

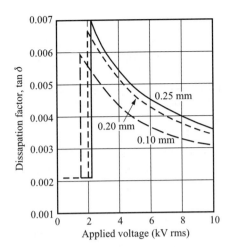

Figure 9.10 Experimental and calculated tan δ vs. applied voltage characteristics obtained on cable with artificial cavity having 0.10, 0.20, and 0.30 mm gap spacings at atmospheric pressure [15, 16].

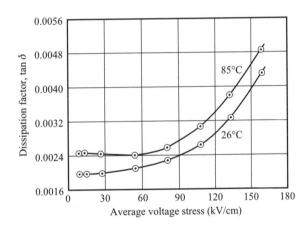

Figure 9.11 Characteristics of tan δ vs. average voltage stress at room temperature and 85°C obtained on power cable model impregnated with low-viscosity oil [17].

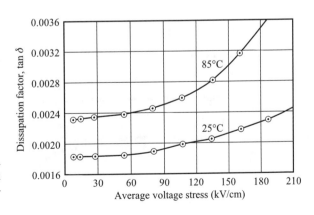

Figure 9.12 Characteristics of tan δ vs. average voltage stress at room temperature and 85°C obtained on power cable model impregnated with high-viscosity oil [17].

347

discharges [6]. Figure 9.11 exhibits a slight Garton/Böning effect [14, 18] in the 85°C characteristic over the lower field strength region. Note the small but gradual increase of tan δ with stress as opposed to its behavior in the presence of partial discharges.

When dissipation factor tests are performed upon cable specimens, considerable care must be ensured to terminate the cables properly. The cable end terminations basically serve two main purposes, namely to prevent voltage flashover at the ends and to ensure that the measured results are not influenced by spurious partial discharges occurring at stress concentration points at cable shield–insulation interfaces where the cable shield ends. The termination for bridge measurements differs from the usual high-voltage termination in that the shielding portion of the stress cone is not connected to the cable shield. The stress cone shield overlaps the cable shield slightly but is separated from it by a thin layer of insulation tape as depicted in Fig. 9.13. The guard shield of the cable termination is connected to the measuring bridge guard circuit, where through a balance procedure the potential of the guard shield is brought to the same in-phase potential as the cable shield itself. Thus the stress distribution along the entire guarded termination is uniform, thereby preventing failure at the cable ends. Any possible partial discharges or losses in the termination itself do not affect the measured values of tan δ and the specimen capacitance by the main bridge circuit, as long as the guard circuit remains balanced. Note should be made of the fact that the potential of the cable shield, and hence the guard shield, is always close to the guard potential so as to eliminate the voltage hazard to the operator. However, in some bridges, particularly those recently designed for measurements in the field, this may not be the case due to their inverse circuitry arrangement [9]. The type of null detectors normally used in the older Schering bridges were of the vibrating galvanometer type, but these have been replaced in the 1950s by tuned amplifiers or oscilloscope-type displays.

9.4 PARTIAL-DISCHARGE CHARACTERISTICS

With plastic- or rubber-extruded cables, a possibility always exists that cavities may be formed in the solid dielectric during the various extrusion steps because of faults in the extrusion process itself or because of impurities present in the material. The cavities may be either uniformly distributed along the cable length or they may occur only at certain isolated points. The radial distribution of these voids can be also varied, with

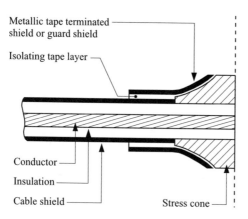

Figure 9.13 Guard shield termination schematic of an oil-paper power cable [10].

some cavities located adjacent to the conductor and insulation shields or within the insulation proper. The geometry of the cavities occluded within the insulation itself is generally of spherical or ellipsoidal shape (cf. Fig. 9.14); whereas cavities along the insulation-extruded shield interface tend to be elongated, thin, and flat in shape. Whether or not a given cavity ionizes depends on the field strength across the cavity, the gas type and pressure within the cavity, as well as the cavity thickness or diameter in the direction of the electrical field. Cavities that do not discharge at the maximum stresses to which the cable is subjected, are quite innocuous. However, cavities that undergo discharge at either the operating voltage stress or at stresses below cannot be tolerated, since the ensuing degradation of the insulation due to partial discharge will ultimately lead to breakdown of the cable insulating system. For this reason all plastic and rubber insulated cables must be tested for the presence of partial discharge to ensure that they are discharge free at the operation voltage.

Recent improvements in extrusion technology have greatly diminished the incidence or occurrence of cavity inclusions in solid-dielectric extruded power cables. The occurrence of cavities in rubber insulated cables and the early PE cables was nevertheless relatively common. However, this was largely during the days when partial-discharge pulse detection techniques were either not available or fully developed to aid in the detection of cavities and thus permit adjustment of the polymeric cable extrusion procedures to eliminate the causes of cavity formation. With present PE, XLPE, and EPR cables, the occurrence of cavities is rare and is usually confined to minute cavities along loose or poor adhering semiconducting shields or infrequently occurring cavities in the insulation wall. Occasionally, cavities are found to occur in very large numbers in the cable insulation proper, as is supported by the evidence in Fig. 9.15 obtained on a 15-kV XLPE power distribution cable. Evidently in terms of the results of Fig. 9.15, one must deduce that serious defects in the extrusion process are always possible. Such gross cavity defects, which are in many aspects reminiscent of the butyl rubber cable era, are most certainly to undergo discharge and lead ultimately to voltage breakdown of the cable insulating system. Vis-à-vis such evidence, it is clear that partial discharge standards should be rigidly enforced and maintained to prevent cavity defective cables from reaching the service field. The only possible way a cable with such gross cavity defects can operate without discharge is in a wet environment where moisture ingress may result eventually in water-filled cavities, thereby replacing the air necessary for sustaining the gaseous discharges. However, under such circumstances water tree formation would be expected to occur. Since in many instances cables in the field are subjected to varying or at times cyclic moisture levels, degradation of the cable dielectric

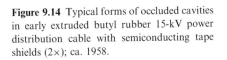

Figure 9.14 Typical forms of occluded cavities in early extruded butyl rubber 15-kV power distribution cable with semiconducting tape shields (2×); ca. 1958.

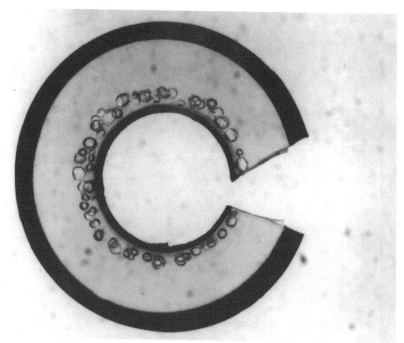

Figure 9.15 Cavities in XLPE power distribution cable with extruded semiconducting shields (3×). (Courtesy of M. Kurtz, Ontario Hydro Dobson Laboratories.)

would be anticipated to result from both discharges and water tree growth processes as the cable dielectric would fluctuate between the dry and wet states, respectively.

Before proceeding to a description of the various methods of partial-discharge detection and measurement, it is well to delve a little into the basic discharge process occurring within cavities to provide a better understanding of certain discharge quantities such as the discharge magnitude and the partial discharge inception voltage. For a flat-shaped rectangular cavity embedded within a solid insulation of dielectric constant, ε', between two circular parallel-plane electrodes subjected to an average electrical stress E, the electrical stress developed across the cavity is equal to $\varepsilon'E$. For the cavity arrangement depicted in Fig. 9.16, the voltage V_i at which the rectangular cavity, having a breakdown voltage V_b and thickness d in the field direction, will commence discharging is given by [20–22]

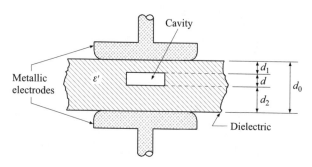

Figure 9.16 Rectangular cavity inclusion within dielectric material between two parallel-plane circular test electrodes.

$$V_i = V_b\left(1 + \frac{d_0 - d}{\varepsilon'd}\right)$$

$$= V_b\left(1 + \frac{d_1 + d_2}{\varepsilon'd}\right) \tag{9.56}$$

The partial-discharge inception voltage will also be influenced by the geometry of the cavity. Thus, if the spherical cavity shown in Fig. 9.17 is considered, then the inception voltage will be given by [23]

$$V_i = V_b\left[1 + \frac{d_0(1 + 2\varepsilon')}{3\varepsilon'd}\right] \tag{9.57}$$

where d is the diameter of the spherical cavity occluded within a dielectric material of thickness, d_0. Cavities within insulating materials may assume a variety of forms, though in the bulk of the solid dielectric for the most part they are found to be either spherical or ellipsoidal; however, adjacent to metallic or semiconducting surfaces, the cavities tend to occur as oblate spheroids or semi-elongated ellipsoids.

The value of the partial-discharge inception voltage of a cable, containing a cavity of specific size and shape, is also influenced by the coaxial geometry of the cable itself. Thus, if for simplicity, we revert back to a cavity of rectangular form as depicted in Fig. 9.16, but occluded within a dielectric wall of a cable as portrayed in Fig. 9.18, then the partial discharge inception voltage assumes the form [23]

$$V_i = E_b r_i\left(1 - \frac{1}{\varepsilon'}\right)\left[\left(1 + \frac{d}{r_i}\right)\left(1 + \frac{d_0}{r}\right)^{1/(\varepsilon'-1)}\right] \tag{9.58}$$

where E_b represents the breakdown stress of the cavity of thickness d in the field direction situated at a radial distance r_i in the bulk of the insulation having an overall thickness d_o and a dielectric constant of ε'; the radius of the cable conductor is equal to r. Evidently, the partial-discharge inception voltage will diminish the closer the cavity is moved to the surface of the cable conductor and the thinner the insulation wall thickness becomes.

The breakdown voltage or potential, V_b, of a cavity is itself a function of the type of gas and its pressure P, as well as the thickness of the cavity d, in the direction of the

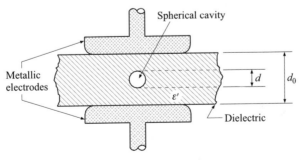

Figure 9.17 Spherical cavity inclusion within dielectric material between two parallel-plane circular test electrodes.

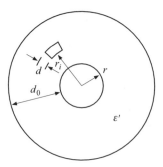

Figure 9.18 Rectangularly shaped cavity in a cable.

applied electrical field. More specifically V_b is a function of the gas pressure and gas gap separation product pd (Paschen curve) as delineated in Fig. 9.19 for a number of gases [24]. All gases exhibit a minimum in their breakdown potential at some particular pd product at which the kinetic energy of the free electrons within the gas volume of the cavity or gap is most efficient for producing the optimum number of collisions with the neutral molecules necessary to precipitate breakdown. At pd values lower than the critical pd value, i.e., at the lower gas pressures or the smaller gas volumes, the number of collisions is reduced, thereby resulting in a higher breakdown potential. Similarly, pd values exceeding the critical pd value lead to an increased breakdown potential, albeit for a different reason; now, the higher density of the gas results in too many collisions, and a substantial portion of the kinetic energy of the free electrons is squandered in various excitation processes [4]. For air, the breakdown potential at the critical pd product is approximately 320 V. In solid-dielectric extruded cables, cavities, which are free of discharges, will tend to contain air at atmospheric pressure because of the gas diffusion mechanism. However, for cavities undergoing discharge, the pressure within the cavities may fluctuate above and below the atmospheric value, depending on the degradation products produced within the cavity. Molecular chain scission due to charged particle impact upon the walls of the cavity and the reactions with the oxygen and ozone within the atmosphere of the cavity will result in the release and formation of certain gases such as hydrogen, CO, CO_2, nitrous oxides, and low-

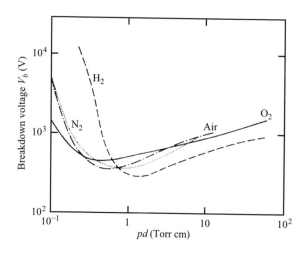

Figure 9.19 Pashen's characteristics for air, nitrogen, oxygen, and hydrogen (after Schönhuber [24]).

molecular-weight hydrocarbons as well as deposits of various solid reaction products within the cavities. As a consequence, the pressure within the cavity and the thickness of the cavity will change, causing accompanying changes in the breakdown voltage of the cavity [25].

When cavities become extremely small in size, the likelihood of their undergoing discharge is greatly reduced. First, the probability that a free electron necessary to initiate the gas breakdown process would appear in a minute gas volume due to cosmic radiation is quite low. Second, even if free electrons are injected into a microcavity as a consequence of some other process such as field emission from some unspecified asperity, the numbers of ionized particles will not be sufficient to lead to the formation of a discharge. Such is indeed the case for microcavities on the order of 5 µm [25].

In partial-discharge studies extensive use is made of an idealized equivalent circuit, which represents the cavity within a dielectric in terms of a purely capacitive network. The equivalent capacitive circuit, which is based on the work of Gemant and von Philipoff [27], Whitehead [28], and Austen and Hackett [29], is shown in Fig. 9.20. Recently, an alternative approach to evaluate the charge transfer due to partial discharge in terms of the dipole moment resulting from the charge deposited on the surface of the cavity and the associated charge induced on the terminals of the specimen has been propounded by Petersen et al. [30, 31]. Although the latter treatment of the partial-discharge process is scientifically more rigorous, its detailed approach limits its practical implementation. Accordingly, for our purposes in this section, we shall retain the capacitive equivalent circuit approach—though conceding the fact that in practice the capacitance of the occluded cavity C_v and the capacitance of the dielectric in series with the void C_s are generally unknown quantities. The capacitance C_p is a measurable quantity, since for a short cable it represents the lumped capacitance of the cable. For sufficiently long cables that act as transmission lines, the cable capacitance C_p must be replaced by the characteristic impedance of the cable, which is essentially resistive in nature and usually ranges between 20 and 40 Ω.

If a voltage is applied across a dielectric containing a cavity, then the cavity will commence discharging when the voltage across the cavity attains a value equal to its breakdown voltage V_b. Once breakdown ensues, the voltage across the cavity may collapse to zero or more often to some residual voltage V_r for the case where the cavity is not completely discharged. Considering an idealized cavity with equal breakdown and residual voltages in the two polarities of the power frequency wave, the voltage across the cavity for the condition where $V_r = \frac{1}{3} V_b$ will assume the form depicted in Fig. 9.21. It can be discerned in Fig. 9.21 that once discharge inception occurs at discharge epoch θ_1 at the apex in the first half-cycle of the power frequency wave, the subsequent discharges will repeat themselves in a regular manner leading to a fixed voltage waveform across the cavity with four discharges per cycle. This hypothe-

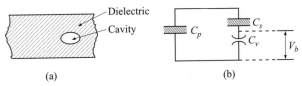

(a)   (b)

Figure 9.20 (a) Dielectric with cavity inclusion and (b) corresponding equivalent capacitive network.

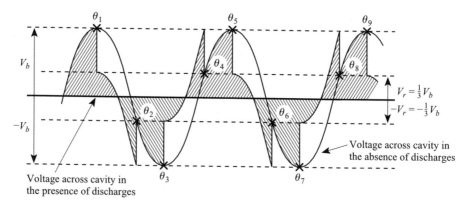

Figure 9.21 Voltage across idealized cavity with equal breakdown (V_b) and residual (V_r) voltages in two polarities and with $V_r = \frac{1}{3}V_b$.

tically idealized behavior does not exactly obtain in practice. Even with the most simple of cavities with metallic parallel-plane electrodes, it is found that the breakdown voltage and residual voltages are unequal and differ in the two polarities, resulting in discharge patterns that indicate a rapid precession of discharge epochs. Much of the observed instability reflects the influence of the statistical time lag, i.e., the time required for a free electron to appear to initiate the discharge once the applied voltage across the cavity attains its breakdown value. The longer the statistical time lag, the larger the break-down voltage developed across the cavity before actual breakdown takes place. It is generally observed that the greatest increase in the breakdown voltage takes place as the polarity of the discharge reverses, while succeeding discharges of like polarity tend to exhibit successive reductions in the breakdown voltage. Even in the absence of any variation in the breakdown and residual voltages, gradual precession of the discharge epochs would take place as a consequence of the finite discharge times, assumed to be zero in the idealized case of Fig. 9.21. When cavities with dielectric electrodes are considered, the discharge sequence becomes further compounded because of an accu-mulation of charges of like and opposite polarity on the dielectric surfaces of the cavity, as well as the occurrence of several discharge sites within individual cavities, leading to the relatively unstationary discharge patterns characterized by partial-discharge pulses of changing amplitude and migrating discharge epochs ordinarily observed with cable specimens.

The discharge rate for the idealized cavity increases in step increments of four discharges per cycle each time that the peak value of the applied power frequency sinusoidal voltage V_a increases by integer values of the breakdown voltage V_b in the two polarities; as a result, the corresponding generated step curve for the condition of $V_r = 0$ assumes the form [32]

$$n = N_0 \frac{\omega}{2\pi} \sum_{m=0}^{j} u[V_a - (m+1)V_b] \tag{9.59}$$

where $m = 0, 1, 2, , j$, n is the discharge rate in pulses per second, $\omega = 2\pi f$ is the radial frequency term, N_0 represents the initial number of partial-discharge pulse per cycle (equal to 4), and u denotes the unit step function. However, with an actual metallic parallel-plane electrode cavity, which approaches best the idealized behavior of the hypothetical cavity, the discharge rate is found to increase linearly with the voltage across the cavity. In addition, the number of discharges per cycle at discharge inception is equal to two, because the breakdown voltages in the subsequent cycles following discharge onset increase in the two polarities. Thus for a parallel-plane metallic electrode cavity, the discharge rate versus the voltage across the gap is described by the form [32]

$$n = N_0 \frac{\omega}{2\pi} \left(\frac{2V_a}{V_b} - 1 \right) u(V_a - V_b) \tag{9.60}$$

where now N_0 is equal to two discharges per cycle in lieu of the four in Eq. (9.59). When one of the parallel-plane electrodes is substituted with a dielectric surface, the dielectric surface may discharge at more than one site (particularly at voltages above the inception point), thereby augmenting the slope in the discharge rate versus applied voltage curve as may be perceived from Fig. 9.22 [33].

When partial-discharge tests are performed on cables and other electrical apparatus, the applied voltage at which discharge inception occurs is always found to be greater than that at which discharge extinction takes place. There is a simple explanation for the observed phenomenon, and the underlying reasons for this singular behavior can be readily demonstrated by considering the equivalent electrical circuit of Fig. 9.20 representing the cavity inclusion in the dielectric. The partial-discharge inception voltage, V_i, may be expressed in terms of the breakdown voltage across the cavity, V_b, the capacitance of the cavity, C_v, and the capacitance of the dielectric in series with the cavity, C_s, as [34]

$$V_i = V_b \left(1 + \frac{C_s}{C_v} \right) \tag{9.61}$$

and the partial-discharge extinction voltage is given by (cf. Fig. 9.21)

$$V_e = \frac{V_i}{2} \left(1 + \frac{V_r}{V_b} \right) \tag{9.62}$$

Since the residual voltage V_r as defined by Fig. 9.21 may in the limit be equal to 0 when the cavity is completely discharged, the discharge extinction voltage V_e can be as low as one-half of the discharge onset voltage V_i. The other extreme of the situation is

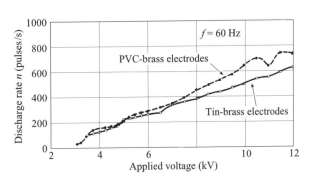

Figure 9.22 Partial-discharge rate as function of applied voltage for 0.50-mm air gap at atmospheric pressure [33].

posed when the process is characterized by a pseudoglow discharge pulse train consisting of a series of inordinately low amplitude pulses. Pseudoglow discharges arise when V_r becomes relatively large and approaches in value that of V_b. In such circumstances, it is very difficult to differentiate between the discharge inception and extinction points. However, in most cable tests the difference between the measured V_i and V_e values ranges from approximately 10 to 30%. Note that since, for the physically real cavities occluded within the cable insulation, information as regards to the values of C_s, C_v, V_r, and V_b is not available, neither V_i nor V_e can be calculated by means of Eqs. (9.61) and (9.62).

It is palpably obvious that the intensity of the discharge pulse occurring within a cable as measured by means of an external sensing circuit is proportional to the difference $V_b - V_r$ or only V_b if $V_r = 0$. That is, the detected magnitude of the discharge pulse is determined by the magnitude of the voltage collapse (or voltage step) across the discharging cavity. However, the charge transferred due to a partial-discharge pulse is contingent both upon the difference $V_b - V_r$ as well as the capacitance discharged. If we consider the equivalent circuit of the cavity represented in Fig. 9.20, then we can make the observation that for a short cable of lumped capacitance, C_p, the capacitance of the dielectric in series with the cavity $C_s \ll C_p$. Thus, essentially the capacitance discharged by the cavity is equal to C_s and, consequently, the charge transfer ΔQ during the breakdown or discharge event is given by

$$\Delta Q = C_s(V_b - V_r) \tag{9.63}$$

If for simplicity we consider a laminar cavity having an electrode surface area of A with a dielectric of dielectric constant ε', of thickness d_s in series with it such that the capacitance C_s is given by $A\varepsilon'/d_s$, then Eq. (9.63) may be expressed as

$$\Delta Q = \frac{A\varepsilon'}{d_s}(V_b - V_r) \tag{9.64}$$

Evidently, the charge transferred during the discharge event increases with the surface area discharged and decreases with the thickness of the dielectric in series with the cavity.

Had we resorted to following the approach of Petersen et al., [30, 31], the charge transferred ΔQ across the cavity during a discharge event would have assumed a more complex form given by

$$\Delta Q = -\bar{\mu} \cdot \bar{\nabla}\lambda \tag{9.65}$$

where $\bar{\mu}$ is the dipole vector due to the surface charge density q, deposited by a discharge pulse on the surface S of the cavity; hence, by definition

$$\bar{\mu} = \int_S q\bar{r}\, dS \tag{9.66}$$

where $\bar{r}$ is a radius vector that locates the position of the surface element dS. The scalar function λ in Eq. (9.65) is dimensionless and is contingent upon the position of S only. From the foregoing it is evident that the determination of the charge, ΔQ, in Eq. (9.65) in terms of the dipole moment $\bar{\mu}$ defined by Eq. (9.66), requires data on the dimensions and geometry of the discharging cavity, as well as the value of the breakdown and residual voltages, V_b and V_r, respectively.

9.5 PARTIAL-DISCHARGE MEASUREMENTS

Perhaps the greatest advantage of the simple capacitative equivalent circuit, representing the cavity occlusion within the dielectric in Fig. 9.20, is that it permits to visualize in the most simple manner the discharge sequence under alternating voltages and provides a basis for a simple calibration procedure of partial-discharge detection systems. If for simplicity we consider a sufficiently short length of cable (~ 3 m) so that it behaves as a simple lumped capacitance, then the equivalent circuit of the cable may be represented by that of Fig. 9.20, where C_p constitutes the major portion of the capacitance of the cable. Thus an abrupt voltage collapse ($V_b - V_r$) across the cavity capacitance C_v at the time of the discharge event will cause a voltage ΔV_p to develop across C_p, given by

$$\Delta V_p = (V_b - V_r)\frac{C_s}{C_s + C_p}$$

$$\approx (V_b - V_r)\frac{C_s}{C_p} \tag{9.67}$$

and the charge placed on the terminals of the cable due to a charge transfer in the cavity will approximately equal

$$\Delta Q \approx \Delta V_p \cdot C_p \tag{9.68}$$

where ΔQ is commonly referred to as the externally measured apparent charge transfer, i.e., its value is not quite the same as that of the actual charge transfer $\Delta Q'$, occurring within the cavity as defined by Eq. (9.63). Note that if we substitute the value of ΔV_p from Eq. (9.67) into Eq. (9.68), we have

$$\Delta Q \approx (V_b - V_r)C_s \tag{9.69}$$

That is, $\Delta Q \approx \Delta Q'$; hence Eqs. (9.63), (9.67), and (9.69) constitute the basis for the commonly excepted calibration procedure, whereby a charge is injected via a small calibration capacitor onto the terminals of the cable, and the resulting voltage pulse is compared to that of an actual discharge pulse amplitude to obtain the apparent charge transfer associated with the detected discharge pulse. It is important to perform partial-discharge measurement in terms of charge and not in terms of the peak voltage of the detected pulse, ΔV_p, because the latter, as Eq. (9.67) indicates, is a relative quantity, i.e., the larger the specimen's capacitance C_p, the lower the detected peak pulse voltage. In contradistinction, the charge transfer ΔQ in Eq. (9.68) is an absolute quantity dependent on the product of $\Delta V_p \cdot C_p$.

When a long length of cable is tested, which acts as a transmission line and cannot be any longer viewed as a lumped-circuit element, the capacitance C_p in Fig. 9.20 must be replaced by the characteristic impedance of the cable, Z_0. The characteristic impedance of a cable is by definition

$$Z_0 = \sqrt{\frac{R + j\omega L}{G + j\omega C}} \tag{9.70}$$

where R, L, G, and C are the respective values of the distributed resistance, inductance, conductance, and capacitance of the cable. In contrast to a low-loss high-frequency coaxial cable, the typical solid-dielectric extruded power cable exhibits appreciably more elevated losses primarily as a result of the semiconducting carbon-filled conductor and insulation shields. Thus, for both polymeric and the lossy oil-impregnated paper insulated power cables, the characteristic impedance remains a complex quantity. Nevertheless, to obtain a better insight into the signal transmission characteristics of power cables, it is helpful to consider the idealized case of a loss free line. Then the analysis becomes greatly simplified, since with R and G reduced to zero, the impedance becomes a real quantity given by

$$Z_0 = \sqrt{\frac{L}{C}} \tag{9.71}$$

where the units of Z_0 are in ohms. The absolute magnitude portion of the characteristic impedance may be approximated, using square pulse signals with a variable resistive attenuator at the far end of the cable. The attenuator setting for zero square pulse reflection from the far end is taken to correspond to the characteristic impedance of the cable. For most commercial power distribution cables, the characteristic impedance thus determined ranges from ~20 to 40 Ω. Often the lossless cable approximation defined by Eq. (9.71) is used to calculate an approximate order of the magnitude of the characteristic impedance, using the respective expressions for the inductance and capacitance per unit length of the cable,

$$L = \frac{\mu'}{2\pi} \ln \frac{r_2}{r_1} \tag{9.72}$$

and

$$C = \frac{2\pi\varepsilon'}{\ln(r_2/r_1)} \tag{9.73}$$

where r_1 is the radius of the cable conductor, r_2 the radius of the cable at the insulation shield, ε' the real value of the permittivity of the cable insulation, and μ' the real value of the permeability of the cable conductor. Thus, substituting the values of L and C in Eq. (9.71) and noting that for nonmagnetic metals $\mu' = 1$ and that $\varepsilon' = \varepsilon_0 \varepsilon_r'$, where $\varepsilon_0 = 8.85 \times 10^{-12}$ F/m yields

$$Z_0 = \frac{60}{\sqrt{\varepsilon'}} \ln \frac{r_2}{r_1} \tag{9.74}$$

It is apparent from Eq. (9.74) that the characteristic impedance of a cable increases with decreasing conductor radius and increasing insulation thickness. However, since the losses in the power cable, which principally occur at the semiconducting shields, cause the impedance of the cable to behave as a complex quantity, the cable impedance becomes a frequency-dependent quantity. Its frequency dependence characteristics may be obtained by means of a sweep frequency impedance test; such a test result, obtained on a power distribution cable, is delineated in Fig. 9.23. The characteristic impedance magnitude, $|Z_0|$, of the power cable is seen to undergo very marked, quasi-cyclic variation as a function of frequency. Peak maxima and minima of the characteristic impedance are observed at points where the rate of change of the phase angle θ with frequency is greatest. In contrast, if we consider the frequency dependence of $|Z_0|$ and of a nominal 50-Ω communications coaxial cable, the changes with frequency of both quantities are observed to be relatively small, in particular as concerns $|Z_0|$ (cf. Fig. 9.24). As the results of Fig. 9.23 attest, the characteristic impedance Z_0 of a power cable exhibits the behavior typical of a complex quantity. In this regard, it must be

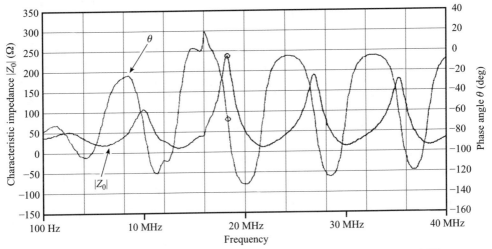

Figure 9.23 Characteristic impedance and phase angle vs. frequency of 25-kV XLPE insulated power cable.

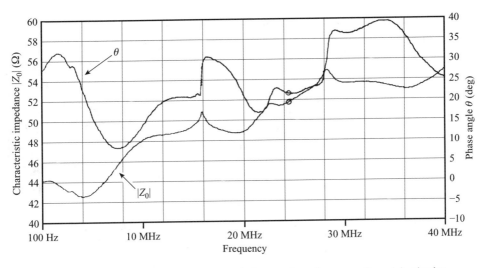

Figure 9.24 Characteristic impedance and phase angle vs. frequency of coaxial polyethylene insulated communications cable (Type RG-58).

observed that any attempt made to match the far end of a power cable by a purely resistive impedance represents at best only a rather rudimentary first approximation of Z_0.

The pronounced frequency dependence of $|Z_0|$ and θ evinced by the 25-kV XLPE power cable in Fig. 9.23 is reflected by a marked attenuation of the high-frequency components in the incident discharge pulse. Figure 9.25 depicts the frequency attenuation characteristics of an 800-ft length of the 25-kV XLPE insulated power cable terminated by its approximate nominal characteristic impedance of 37 Ω at the far end, with the unit step excitation pulse injected via a calibration capacitance of 150 pF to yield a simulated discharge pulse of 0.5 ns rise time. It is seen that signal components having frequencies beyond 40 MHz are attenuated on the average by approximately 50 dB. Thus, there is little advantage to be gained by employing high-frequency techniques for the detection of partial discharges in the relatively high-loss power cables.

When a steep front simulated discharge pulse is injected via a calibration capacitor at the near end of a power cable, its front will undergo appreciable degradation as it travels along the cable. For example, a simulated discharge pulse having a rise time of 0.5 ns injected at the near end of the 800-ft long cable characterized by the attenuation behavior delineated in Fig. 9.25 will emerge at the far end of the cable as a pulse of 31 ns rise time. Thus the detected discharge pulses in power cable specimens will be considerably slower than those that appear at the actual discharge site during the breakdown event, unless the cable is sufficiently short to behave as a lumped capacitance. Moreover, even a relatively short cable of approximated 33 ft will already behave as a distributed parameter line at higher frequencies (cf. Fig. 9.26). From the time delay between the incident and reflected pulses in Fig. 9.26, one can estimate the pulse propagation velocity in the power cable. Since the reflected pulse is behind the incident pulse, i.e., it is observed in Fig. 9.26 to arrive at the near end approximately 120 ns after the incident pulse is transmitted; the propagation velocity may be obtained from

$$v = \frac{2l}{t} \tag{9.75}$$

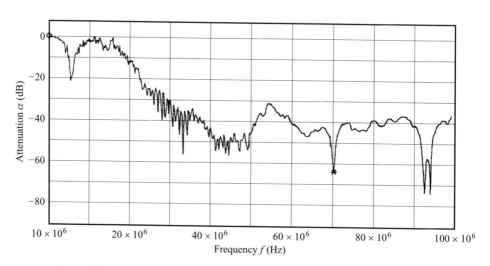

Figure 9.25 Attenuation vs. frequency of 800-ft 25-kV XLPE power cable terminated with its approximate characteristic impedance of 37 Ω.

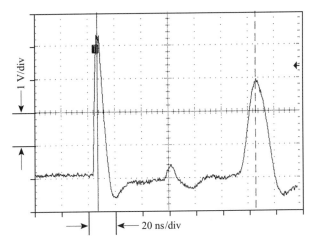

Figure 9.26 Incident and reflected pulses in open-circuited 33-ft-long XLPE power distribution cable.

20 ns/div

where l is the cable length in m and t the time delay in seconds; hence, substituting in the numerical values for l (2.012 × 10^3 cm) and t (120 ns) in Eq. (9.75) gives 1.676 × 10^{10} cm/s as the velocity of propagation, v of the discharge pulse along the power cable. This value, which is equal to 56% of the speed of light, is appreciably lower than that for a lossless polyethylene insulated coaxial communications cable of dielectric constant, ε'_r, of 2.2 for which

$$v = \frac{2.998 \times 10^{10}}{\sqrt{\mu'_r \varepsilon'_4}} \tag{9.76}$$

yields a propagation velocity value of 2.026 × 10^{10} cm/s, i.e. 68% of the value of the speed of light.

In a normal lossy power cable, the velocity of propagation will be frequency dependent because by definition

$$v = \lambda f \tag{9.77}$$

where λ is the wavelength. The propagation constant of a cable is given by

$$\gamma = \sqrt{\left(\frac{R}{l} + j\omega\frac{L}{l}\right)\left(\frac{G}{l} + j\omega\frac{L}{l}\right)}$$

$$= \alpha + j\beta \tag{9.78}$$

where l represents the unit length along the cable, and α and β are, respectively, the attenuation and phase constants, denoting the real and imaginary quantities of λ; the wavelength may also be expressed as [35]

$$\lambda = \frac{2\pi}{\beta} \tag{9.79}$$

Hence substitution of λ into Eq. (9.77) gives an alternate expression for the propagation velocity

$$v = \frac{2\pi f}{\beta} \tag{9.80}$$

If the I^2R losses were neglected in the conductor and the dielectric losses V^2G in the insulation, then for the ideal lossless cable of zero attenuation ($\alpha = 0$) with $R = 0$ and $G = 0$, Eq. (9.78) would become

$$\gamma = j\omega \frac{\sqrt{LC}}{l} \tag{9.81}$$

or

$$\beta = \frac{2\pi f \sqrt{LC}}{l} \tag{9.82}$$

and substitution of Eqs. (9.79) and (9.82) into Eq. (9.77), would lead to a frequency-independent propagation velocity for the idealized lossless cable,

$$v = \frac{l}{\sqrt{LC}} \tag{9.83}$$

Thus the velocity of propagation of discharge pulses in lossless or very low loss cables would be essentially determined by the length of the cable, its conductor size and insulation thickness, and the dielectric constant of the insulation.

Although we have seen that the characteristic impedance and attenuation factor of power cables are frequency dependent, it is often helpful to consider the cable lossless as a first approximation to gain some insight into the general signal response characteristic of the cable. If the characteristic impedance of a cable is considered to be a real quantity, then the analytical procedures may be greatly simplified. Accordingly, if the equivalent circuit of a long unterminated power cable containing a single discharging cavity is represented as in Fig. 9.27, the collapse of the voltage across the cavity, C_v, at the time of the discharge will result in an instantaneous voltage pulse, $e(t)$, appearing across the cable at the discharge site. The time integral of this voltage pulse as it propagates along the lossless cable will always remain proportional to the charge ΔQ transferred within the cavity by the discharge in accordance with the relationship [23]

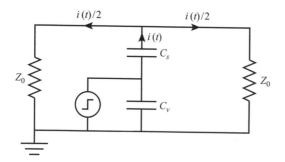

Figure 9.27 Equivalent idealized circuit of long power cable, containing single cavity of capacitance C_v.

$$\int_0^t e(t)\, dt = -\int_0^t \frac{Z_0}{2} i(t)\, dt = \frac{Z_0\, \Delta Q}{2} \tag{9.84}$$

where the factor of 2 in the denominator indicates the transmission of one half of the current, $i(t)$, toward each end of the cable. Equation (9.83) approximates very much the situation that exists in the case of SF_6 gas spacer insulated cables, which approach in their behavior a lossless transmission line. Both polymeric and oil-paper insulated power cables due to their inherent losses deviate very appreciably from the idealized lossless line behavior; their characteristic impedance is complex, and its value varies with frequency, which causes distortion in the transmitted discharge pulse. The condition for a distortionless cable requires that its velocity of propagation and attenuation factor be independent of frequency and no reflection of signals occur at any frequency [35]. A lossy transmission line behaves very much as a filter, with very substantial attenuation of the higher frequency components. A typical rapid rise time pulse depicted in Fig. 9.28 has a frequency spectrum given in Fig. 9.29; it is apparent from the latter that most of the energy of the incident discharge pulse is concentrated over the lower frequency components with the consequence that, even with the higher frequency components removed, there is more than sufficient energy remaining within the lower frequency components to excite the response of a low-frequency partial-discharge detector. It is, therefore, partly for this reason that low-frequency detectors have performed very well when applied to partial-discharge measurements on solid-dielectric extruded cables. In addition, integrating response characteristics of these detectors render them particularly suited for apparent charge measurements. With such detectors distortion of high-frequency pulses along the lossy line is of little consequence, and it is principally the attenuation of the transmitted pulse itself that is of prime concern.

In the routine partial-discharge testing of solid-dielectric insulated power cables, the two most important quantities to be determined are the partial discharge inception voltage (PDIV) and the partial-discharge extinction voltage (PDEV). The PDIV level is established by raising the applied voltage gradually across the cable and noting the value of the applied voltage at which discharges first appear; the PDEV level is then subsequently established by decreasing the voltage and observing the value at which the discharges disappear. The PDEV level is always below the PDIV level, since a lower

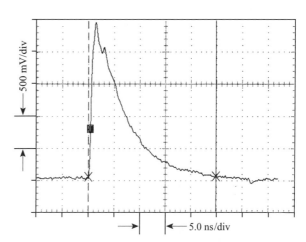

Figure 9.28 Simulated rapid rise time pulse injected via calibration capacitor across 25-kV XLPE insulated power distribution cable.

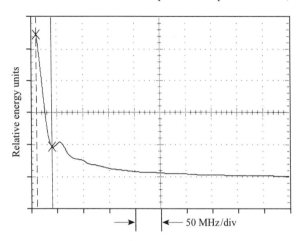

Relative energy units

← 50 MHz/div

Figure 9.29 Frequency spectrum of injected pulse displayed in Fig. 9.28.

voltage is necessary to sustain the discharge process than to initiate it [cf., Eqs. (9.61) and (9.62)]. To maintain uniformity in partial-discharge (PD) pulse detection standards, the PD detection apparatus used on cables must have a basic sensitivity of 5 pC. That is to say, all commercial PD detectors, used on cables must be capable of detecting all PD discharge pulses that are equal to or greater than 5 pC.

The PD detection circuitry used for measurements on cables is of the *RCL* type and its resonant frequency usually lies in the range from approximately 30 to 90 kHz. Relatively speaking, the detection amplifiers are of the narrow-band type, whose bandwidths rarely exceed 120 kHz. A schematic circuit diagram of a conventional PD pulse detection circuit is shown in Fig. 9.30: A typical frequency spectrum of a detected partial-discharge pulse across the detection impedance is portrayed in Fig. 9.31. Maximum sensitivity would thus be obtained by tuning an amplifier to the center frequency f_0 across the detection impedance; but this approach is not followed because of attending signal waveform distortion effects. In analyzing the PD detection circuit behavior, it is common practice to represent it with a highly simplified equivalent circuit as depicted in Fig. 9.32. Note that in this circuit, the capacitance of the coaxial cable to the oscilloscope, C_d, and the critical damping resistor, R_d, are neglected. Here R' represents the equivalent series resistance of the resonant circuit of PD detection. The other components, C_p, C_c, and L are the same as those indicated on the schematic

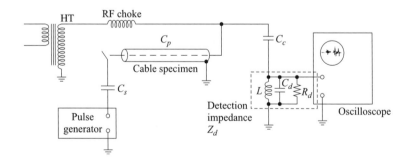

Figure 9.30 Typical RCL-type partial-discharge pulse detection circuit.

Figure 9.31 Frequency spectrum of detected
partial-discharge pulse with center frequency
f_0 at 40 kHz: vertical, 1.0 mV (rms)/div; horizontal, 10 kHz/div; frequency range, 0–90
kHz.

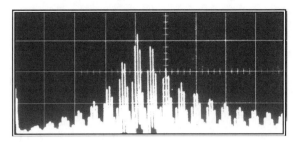

diagram in Fig. 9.30. The blocking capacitor C_c is of a sufficiently high capacitance so
that it acts as a short circuit to the high-frequency PD transients but presents a high
impedance to the 50 or 60 Hz power supply; the detection coil L on the other hand acts
as a short circuit to the power source current but presents a high impedance to the PD
currents. When a discharge takes place in the cable, C_p, it results in a charge ΔQ
suddenly appearing on its terminals. This shocks the resonant circuit into oscillation
at its natural frequency determined by the component values of C_p, C_c, and L. For
simplicity of analysis, let us first consider the case where the cable is sufficiently short
and acts as a lumped-capacitance C_p. If $i(t)$ represents the resultant loop current in the
circuit, then one can write the differential equation

$$\frac{1}{C_p}\left[\int i(t)\,dt + \Delta Q\right] + \frac{1}{C_c}\int i(t)\,dt + L\frac{di(t)}{dt} + R'i(t) = 0 \qquad (9.85)$$

where from quiescent conditions, ΔQ corresponds to the charge appearing on C_p as a
result of a partial discharge. Solving for the current $i(t)$ and then determining the
voltage drop $e(t)$ across the detection impedance, which has been assumed to be purely
inductive to maintain simplicity, we have for the discharge pulse response

$$\begin{aligned}
e(t) &= L\frac{di(t)}{dt}\\[2mm]
&= -\frac{\Delta Q}{C_p}\exp\left(-\frac{R't}{2L}\right)\cos\left(\frac{1}{LC_c}+\frac{1}{LC_p}\right)^{1/2}t
\end{aligned} \qquad (9.86)$$

where the resonant frequency is given by

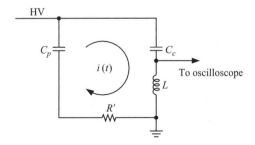

Figure 9.32 Equivalent circuit of an RCL-
type partial-discharge detector.

$$f_0 = \frac{1}{2\pi}\left(\frac{1}{LC_c} + \frac{1}{LC_p}\right)^{1/2} \tag{9.87}$$

If Eq. (9.86) is to include the small resistive drop across the detection impedance due to the negligibly small internal resistance R of the detection coil L, then the term

$$Ri(t) = \frac{R\,\Delta Q}{C_p}\exp\left(-\frac{R't}{2L}\right)\sin\left(\frac{1}{LC_c} + \frac{1}{LC_p}\right)^{1/2} t \tag{9.88}$$

must be added to Eq. (9.86).

The calibration of the detection circuit is accomplished by injecting the calibration pulses via a series capacitor C_s connected at the far end of the cable specimen. The IPCEA or NEMA specifications [16, 36] require that the injected square pulse should have a minimum pulse width of 200 µs and a charge of 20 pC; the value of C_s must be between 50 and 150 pF and the value of the square pulse voltage V must be adjusted such that the product VC_s equals 20 pC. The rise and decay times for the square calibration pulses should be equalized and must not exceed 100 ns; in addition, the calibration square pulse width must be sufficiently wide to prevent overlapping between the simulated discharge transients at the front and trailing edge of each excitation pulse [37]. According to the specifications, the 20-pC pulse injected at the far end of the cable must produce a response amplitude of at least 1 cm on the oscilloscope screen. In addition, a basic sensitivity test requires that an injected pulse of 5 pC should produce a clearly visible response with the same amplifier gain setting. Suitable noise filtering must be employed to reduce the magnitude of any extraneously caused interference signal to less than 4 pC. On an expanded time base the response pulses due to a calibration pulse would appear as shown in Fig. 9.33(b), but on the 50- or 60-Hz time base of the conventional PD detector they appear essentially as narrow single vertical line pulses. The actual wavefront of the detected signal will deviate somewhat from that of Fig. 9.33(b), depending on the length of the cable. Even at relatively low detection frequencies, cables in excess of 100 ft will act as transmission lines, and the detected signal will tend to approximate a damped sine wave transient. In fact Eager and Bahder [38] have shown that with long cable specimens, the detected signal has the form

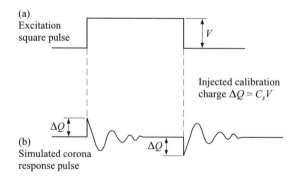

(a)
Excitation
square pulse

V

Injected calibration
charge $\Delta Q \simeq C_s V$

ΔQ

ΔQ

(b)
Simulated corona
response pulse

Figure 9.33 Simulated partial-discharge excitation and response pulses.

$$e(t) = 2V_I \exp\left(-\frac{t}{\tau_1}\right) \frac{C_c}{C_c + C_d} \frac{1 + \exp[-t(1/\tau_2 - 1/\tau_1)]}{1 - \tau_2/\tau_1}$$

$$+ 4\pi f_0 \frac{C_c}{C_c + C_d} \frac{Z_0 \, \Delta Q}{2} \exp\left[\left(-\frac{R't}{2L}\right)\right] \sin 2\pi f_0 t \tag{9.89}$$

where

$$f_0 = \frac{1}{2\pi} \left(L \frac{C_c C_p}{C_c + C_p} + LC_d\right)^{1/2} \tag{9.90}$$

where V_I is the amplitude of the propagated partial-discharge pulse of time constant τ_1, at the discharge site or cavity, τ_2 is equal to the time constant of the discharge circuit including the cable, and Z_0 is the characteristic impedance of the cable. The value of C_p represents the capacitance of the cable specimen at the resonant frequency f_0. Here R refers to the internal resistance of the detection coil L. It is now common practice to detect the partial-discharge pulse in terms of the low-frequency sinusoidal term in Eq. (9.89) as it is that portion of the detected signal that is proportional to the apparent pulse charge, ΔQ. Essentially this means that present partial-discharge detectors are of the low-frequency, narrow-band type.

Routine partial-discharge tests on power cables are generally carried out on unterminated cable specimens. Consequently, the detected partial-discharge pulse pattern, in addition to the incident detected discharge pulses at the near end of the cable also includes the reflected pulses from the far open end of the cable specimen. Hence the partial-discharge pulse train in terms of the measured transient sinusoidal portion of the signal in Eq. (9.89) may be expressed in accordance to Bahder and Eager [38] as

$$\Delta V_d'' = 4\pi f_0 \frac{C_c}{C_c + C_d} \frac{Z_0 \Delta Q}{2}$$

$$\times \left\{ \sum_{n=0}^{\infty} (1 - k) \exp\left[-\frac{R}{2L}(t - nt)\right] \sin[2\pi f_0(t - nt)] \right.$$

$$\left. + \sum_{n=0}^{\infty} (1 - k) \exp\left[-\frac{R}{2L}(t - nt)\right] \sin[2\pi f_0(t - nt)] \right\} \tag{9.91}$$

where k represents the numerical fraction of the incident-reflected apparent charge ΔQ and t is the time required for the incident pulse starting from the discharge site to reach the far end of the open circuited cable and then return to the RCL detection circuit at the near end of the cable. Depending on the length of travel by the reflected pulses, pulse superposition may occur, and for this reason the detected pulse shape should be of the form that will result in positive pulse superposition errors. That is, the measured pulse amplitude will be larger than its actual value. This is accomplished by employing amplifiers of adequate bandwidth to ensure that the first peak of oscillation of the sinusoidal exponential decaying transient will represent the highest amplitude of the signal. The term α-response is given to this type of pulse form; a representative α-response pulse is portrayed in Fig. 9.34 [39]. Extremely narrow band or tuned partial-discharge detection systems are characterized by an oscillatory

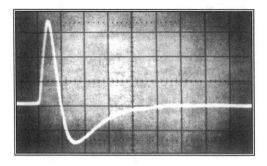

Figure 9.34 Typical α-response pulse of a partial-discharge detection system with critical damping; flat-band from 50 to 75 kHz, with lower and upper 3-dB points at 35 and 110 kHz respectively, vertical scale, 20 pC/div; horizontal scale, 10 µs/div (after Costello and Bartnikas [39]).

transient response having the first peak of oscillation lower in magnitude than the subsequent peak of oscillation. Such pulse form is commonly referred to as a β-pulse response and is depicted in Fig. 9.35. With unterminated cable specimens of sufficient length that give rise to pulse reflection, β-response pulses will in general lead to negative overlap errors should pulse integration or overlap occur involving the incident and reflected discharge pulses, i.e., the resultant measured discharge pulse amplitude may be lower than that of the incident or actual discharge pulse. Accordingly, an α-type pulse response is specified for discharge tests on cables. Since the discharge pulses are normally displayed on a 50- or 60-Hz time base, the discharge signals must be sufficiently long or "stretched" in time to be visible to the observer. Nevertheless, they should not be too long as this would again lead to undesirable pulse superposition effects.

Critical damping is employed to suppress the oscillatory nature of the pulse inherent to all RCL-type detection systems. Thus, on the basis of the critically damped α-response pulse in Fig. 9.34, the resolution limit of the detection circuit (without any signal superposition) would be approximately 40 µs as opposed to $\sim$90 µs for the underdamped β-response pulse in Fig. 9.35.

As already mentioned, the effective bandwidth of RCL-type partial discharge pulse detectors employed on extruded dielectric power cables rarely exceeds 120 kHz, though for transformer specimens it is usually extended to 300 kHz. The upper frequency limitation of these detectors essentially causes them to behave as charge integration devices; hence, they respond in a manner in which they are intentionally designed to respond, i.e., to measure the apparent charge transferred by the discharge pulses. If the input of such RCL-type detector sees a train of rapid propinquitous discharge pulses,

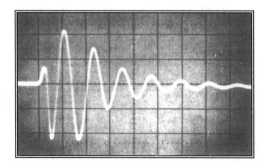

Figure 9.35 Typical β-response pulse of narrow-band or tuned partial-discharge detection system; tuned to 87 kHz with the lower and upper 3-dB points at 77 and 97 kHz, respectively: vertical scale; 20 pC/div; horizontal scale, 10 µs/div (after Costello and Bartnikas [39]).

the output of the detector will simply indicate a single response pulse, which will be an integral measure of the discrete rapid discharge pulses comprising the pulse train. This behavior of the *RCL*-type detector circuit is illustrated in Fig. 9.36, which compares its response with that of a 500-MHz bandwidth oscilloscope [40].

As discussed previously, the discharges tend to occur upon the ascending and descending portions of the power frequency, so that they cluster around the voltage zeros. When displayed on a 50- or 60-Hz time base, the discharge pulses appear essentially as compressed vertical line signals distributed around the voltage zeros, as depicted in Figs. 9.37(a) and (b). Commercially available discharge detectors often employ an elliptical trace to simulate a power frequency time base; in such a scheme the partial discharge pulse pattern appears superposed upon an elliptical trace, as in Fig. 9.37(c).

Before a cable can be tested for the presence or absence of partial discharges, great care must be exercised to ensure discharge-free cable ends. If the cable ends are poorly prepared, discharges may take place at the edges of the cut semiconducting insulation shield and the insulation proper. Since on the basis of the discharge pattern it is not possible to distinguish between the discharges actually taking place in the insulation and those arising from poorly prepared cable ends, a good but poorly terminated cable may fail a partial discharge test. In the early days of discharge testing, the cable ends were either immersed in oil cups or oil-paper tape stress relief cones were applied at the termination of the grounded shield. Whereas the latter terminations are no longer used for routine discharge tests on newly manufactured cables, the former are still extensively utilized with polymeric cables for voltages as high as 80 kV. This continuous usage centers on the ease with which the oil-cup termination can be effected and the economics associated with it. Presently extensive use is made of semiconducting paint terminations [41]. The semiconducting paint is applied as shown in Fig. 9.38. This type of termination is suitable for distribution cables rated up to 35 kV with the painted layer being approximately 1 ft in length. The nonlinear voltage-current characteristic of the stress grading paint reduces the stress concentration effectively at the cut insulation shield point as long as the conduction current along the semiconducting paint layer exceeds the displacement current in the dielectric layer. Chan et al. [42] were able to extend the voltage rating above 70 kV for this type of termination by grading in cascade the surface resistivity of the conducting paint to three equal length zones of 10, 30, and

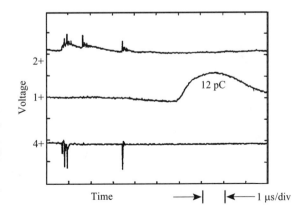

Figure 9.36 Partial-discharge pulse forms obtained with 8.0-mm needle to plane gap in oil at atmosphere pressure subjected to 12 kV rms (50 Hz). Channel 4, actual discharge pulses as detected with 500-MHz bandwidth oscilloscope (10 mV/div); channel 2, corresponding pulses at input to 300-kHz bandwidth commercial PD detector (20 mV/div); channel 1, corresponding integrated single-output pulse of commercial PD detector (100 mV/div) (after [40]).

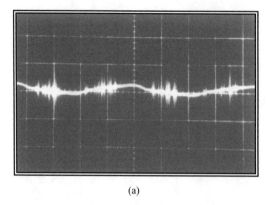

(a)

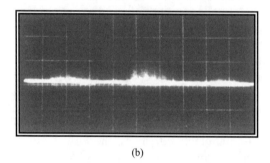

(b)

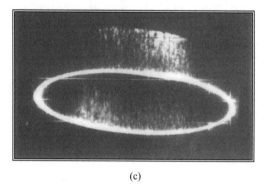

(c)

Figure 9.37 (a) Partial-discharge pulses (superimposed on 60-Hz wave. (b) Partial-discharge pulses on 60-Hz time base with power frequency component removed and pulses demodulated. (c) Intense partial-discharge pulse pattern on commercial partial-discharge detection apparatus.

100 MΩ/square extending outwards from the cut insulation shield toward the end of the bared insulation at the exposed conductor end.

Heat-shrink high-permittivity stress control tube terminations may further extend the discharge test voltage range to 120 kV. The principle of this type of termination is based on the increased alternating field, which results in the dielectric material having the lower dielectric constant of two contiguous dielectric layers. The high dielectric constant stress control tube thus assists in maintaining a more uniform electrical field at the termination point of the cut semiconducting insulation shield, as is apparent from Fig. 9.39 [42]. Since the heat shrinkable tubing must be discarded after use, the termination bears the disadvantage of increased cost.

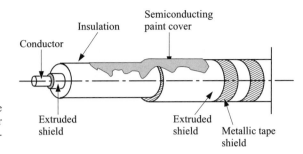

Figure 9.38   Solid-extruded-dielectric cable termination with semiconducting paint cover for partial-discharge tests at distribution voltages.

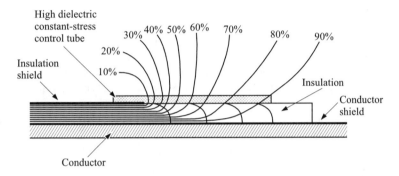

Figure 9.39 Schematic of voltage stress distribution in high dielectric constant heat-shrink tubing termination (after [42]).

For discharge tests over the higher voltage range up to 800 kV, deionized water terminations are preferred. These terminations consist of termination tubes filled with deionized water whose resistivity is maintained at a fixed elevated value; corona stress rings are located at the extremity of these tubes [41]. The water column at the cable ends extends from the insulation shield termination point to the bare conductor extremity. The voltage distribution along the cable end over which the semiconducting shield has been removed is linearly controlled by the high resistivity deionized water column, thereby decreasing the stress enhancement created at and in the vicinity of the incision point.

In North America, the partial-discharge tests on power cables are carried out in accordance with the IPCEA and Association of Edison Illuminating Companies (AEIC) specifications. In many respects the IPCEA and AEIC specifications are similar and complement and supplement each other. On newly manufactured polymeric cables, partial-discharge tests are performed prior to the alternating voltage test at the factory, and the maximum test voltage according to the IPCEA specifications is not permitted to exceed the specified voltage test value defined in the ac test. The applied voltage is gradually increased across the cable until a value is reached at which the signal to detector with the specified sensitivity indicates the presence of discharges; thereafter the applied voltage across the cable specimen is lowered at a rate not exceeding 2 kV/s to determine the partial-discharge extinction voltage (PDEV). Then as the applied voltage is subsequently raised to 20% above the specified PDEV value for the particular cable specimen under test and, provided no discharge pulses are observed, the cable is considered to have passed the IPCEA discharge-test requirements. In some cables the actual and specified values of the PDEV may differ appreciably. The IPCEA specifications require that the period of the test voltage application should not exceed 3 min.

The earlier AEIC specifications had defined a set of permissible discharge levels on polymeric cables in accordance with the empirical relation

$$\Delta Q = 5 + \left(\frac{V_T}{V_{RG}} - 1.5 \right) \times 30 \qquad (9.92a)$$

where ΔQ is the permissible discharge level in pC, V_{RG}, is the rated cable voltage phase to ground and V_T is the test voltage. The expression was valid for $(V_T/V_{RG}) \geq 1.5$ and a further constraint was again that V_T must not exceed the value of the ac factory test voltage. Equation (9.92a), though arbitrary in many respects, formed the underlying basis for partial-discharge tests on solid-dielectric extruded cables. Improved manufacturing methods and cleaner materials have resulted in polymeric cables with much higher partial-discharge inception voltages as well as correspondingly lower discharge intensities so that the permissible discharge level determined by using Eq. (9.92a) would be presently considered to be rather lenient. Present AEIC specifications [43–45] are much more demanding at the higher applied voltages as is apparent from Table 9.1. However, it is interesting to observe that for the ratio of V_T/V_{RG} equal to 1.5, Eq. (9.92a) still agrees with the value of 5 pC given in Table 9.1.

It must be borne in mind that the permissible discharge levels specified by the AEIC are arbitrary in the sense that it is not known what discharge intensities may cause eventual breakdown of the cable insulating system. By maintaining the permissible discharge charge transfers at low levels, it is hoped that a sufficient safety factor is introduced against any possible insulation degradation due to partial-discharge phenomena. It is noteworthy that the minimum specified discharge intensity value even at the lowest test voltage is not less than 5 pC. It has been established in practice that under normal ambient noise conditions in manufacturing premises, it is difficult to improve on the basic detectable sensitivity level of 5 pC. If the extraneous electrical noise were completely eliminated by means of electromagnetic shielding, the utmost sensitivity attainable would be limited by the thermally generated noise in the characteristic impedance Z_0 of the cable specimen. Assuming the cable to be a lossless trans-

TABLE 9.1 Permissible AEIC Partial-Discharge Levels for Polymeric Cables

Ratio of Test to Rated Phase to Ground Voltage (V_T/V_{RG})	Permissible PD Level for 5–46-kV Cable Class (PC)	Permissible PD Level for 69–138-kV Cable Class (PC)
1.0	5	5
1.5	5	5
2.0	5	5
2.5	5	10
3.0	5	
4.0	10	

Source: From [43]–[45].

mission line, it can be demonstrated that for low detection frequencies not exceeding 1 MHz, the charge transfer ΔQ_Υ due to thermally generated noise is given by [34]

$$\Delta Q_\Upsilon = \frac{1}{\alpha}\left(\frac{kT}{Z_0\,\Delta f}\right)^{1/2} \tag{9.92b}$$

where k is the Boltzmann constant, T the absolute temperature, α the attenuation factor, and Δf the bandwidth of the PD detector. The value of ΔQ_Υ falls within the range of 0.03–0.08 pC, depending on the bandwidth of the PD detector employed. Under laboratory conditions, measurements are often performed, using a minimum detectable sensitivity level of 0.1 pC.

In the determination of the PDIV and PDEV values, additional details on the overall testing procedure and PD detection instruments themselves are specified by the ICEA [46]. The overall detected background interference level must not exceed 4 pC. The calibration is to be carried out with a square pulse having a rise time not greater than 100 ns [in agreement with American Society for Testing and Materials (ASTM) D1868 but different from International Electrotechnical Commission (IEC) Publication 270, which specifies a rise time $\leq$ 20ns], with the charge being injected via C_s having a capacitance between 50 and 150 pF (cf. Figs. 9.30 and 9.33). Although the magnitude of the discharge pulse displayed on a cathode ray tube screen is in volts as a function of time, the recalibration of the screen in picocoulombs permits to determine the apparent charge transfer corresponding to a given voltage pulse appearing across the terminals of a cable specimen under test. Attenuation effects along the cable are taken implicitly into account by injecting the calibration signal at the far end of the cable. The attenuation factor may be determined by comparing the signal response amplitude with the calibrating pulse first injected at the far end and subsequently at the near end. It is again to be emphasized that for all discharge tests a rate of voltage rise and decrease of 2 kV/s must be employed, and the discharge extinction voltage is determined by decreasing the applied voltage from a voltage of 20% above the permissible discharge inception voltage of the cable. In addition to the measurement of the inception and extinction voltages, it may be desirable to perform an apparent charge characteristic test, which is essentially a plot of the maximum peak detectable apparent charge transfer versus applied voltage. Although this type of test has been popular on other electrical apparatus for many decades, it was only applied to cables in 1968 at which time it was referred to as the corona factor test [47]. For this test procedure, it is required that the newly manufactured cables be stored for a certain period prior to measurement in order to permit diffusion out of the insulation of a number of gaseous decomposition products associated with the crosslinking agents. For example, acetophenone in the case of XLPE insulations has been shown to have a marked effect on the partial-discharge characteristics [48]. The required rest period varies with the voltage class of the cables: 7 days for 5- to 35-kV class cables, 10 days for 46-kV class cables, and 20 days for 69- to 138-kV class cables. The peak apparent partial-discharge characteristic may be obtained with an X-Y recorder connected across commercial PD detectors, which are equipped with quasi-peak metered outputs. Figure 9.40 portrays a very pronounced hysteresis effect in the peak charge transfers versus applied voltage characteristic of a defective length of an old 15-kV in service cable. With currently manufactured EPR and XLPE cables, the hysteresis effect is much less pronounced.

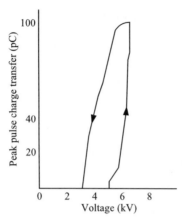

Figure 9.40 Peak apparent charge transfer characteristic obtained on a defective length of a 15-kV power distribution cable, retrieved from service and tested in dry condition (after [41, 49]).

Heretofore, the major part of discussion on partial-discharge detection and measurement was centered on low-frequency narrow-band detectors because these are principally used and specified to be used on solid extruded dielectric power distribution cables, which constitute the largest class of power cables on which routine partial-discharge tests must be performed. However, when PD measurements are made on low-loss gas insulated power transmission cables, high-frequency measurement techniques may be effectively utilized [50]. As a result of the nearly lossless character of the gas insulated cables, very high frequencies between 300 MHz and 1 GHz are employed whereby the detection impedance Z_d and the coupling capacitance C_c of the traditional circuit become incorporated as parts of a transmission line to yield the desired frequency response. With SF_6 bus ducts or cables, the detection of discharges and their location is thus carried out simultaneously by measuring the pulse arrival times at two coupling capacitors. Discharge sites may be located to within 1 m at a maximum sensitivity as high as 0.1 pC. However, the rapid detection circuitry responds essentially only to the portion of the discharge current carried by the electrons, with the result that only half of the actual apparent charge transfer associated with the discharge pulse is measured as the remainder of the charge transfer due to the slow-moving ions remains undetected. Figure 9.41 portrays some typical PD signal coupling devices suitable for use with the wide-band high-frequency discharge detection systems. A coaxial coupling device generally possesses a good frequency response but a high coupling loss, which

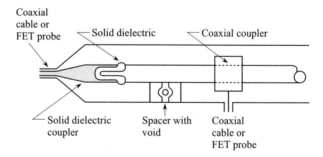

Figure 9.41 Types of couplers for partial-discharge measurements on SF_6 line specimen (after [50]).

may be eliminated by the use of a solid-dielectric coupler. A sheath coupler is also suitable when an isolated section of the cable sheath is available for this purpose; however, the latter approach requires a sheath length of 2.5 m and a Faraday cage enclosure because it does not provide adequate shielding for the measuring capacitor.

9.6 PARTIAL-DISCHARGE SITE LOCATION

Discharge site location in newly manufactured cables permits to determine the nature of the cavities giving rise to the detected discharge intensity, and often leads to the manufacturing process faults responsible for cavity formation. Whereas, location of discharge sites on cables operating in service permits to institute and enforce preventive maintenance procedures so that the faulty sections of the cables, containing the cavities, may be replaced or repaired if feasible, before serious power interruption due to failure results.

A well-proven method [51] is available for detecting cavities that are situated within the insulation bulk and at the interface between the conductor semiconducting shield and the insulation proper; however, for the method to be operative, the cable must be in an unfinished form, i.e., without an insulation shield and jacket. To locate the discharge source, the cable must be scanned using a high-voltage electrode system immersed in deionized water as shown in Fig. 9.42; the applied voltage at the center of the deionized water-filled tube is monitored, using a capacitive voltage divider formed by C_c and C_m. The electronic detection circuitry is designed to measure the partial-discharge current across Z_d; it is thus a measure of the discharge intensity, proportional both to the number of discharge pulses as well as their magnitude. The conductor of the cable is grounded and the cable is stressed from the outside of the insulation.

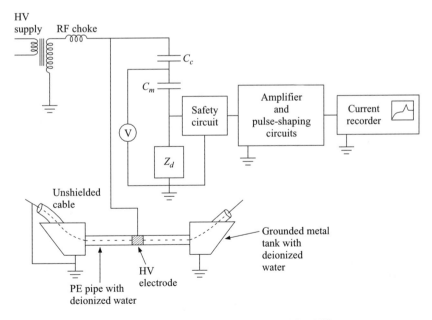

Figure 9.42 Partial-discharge fault locator (after [50]).

Distilled water is used to obtain a voltage stress distribution, which reaches maximum at the HV electrode. The complete cable length must be passed through the scanning electrode: as a cavity approaches the HV electrode, the stress increases gradually across the cavity, causing it to discharge as its inception voltage is exceeded. The discharge intensity continues to increase until the cavity is directly under the HV electrode; at this point, the discharge intensity attains a maximum. As the cavity moves past the HV electrode, the discharge intensity begins to decrease until the partial discharges become completely extinguished as the stress falls below the critical value in the vicinity of the grounded tanks. A recorder is employed to indicate on a graph the partial-discharge intensity as a function of cable length, so that the position of the discharge site may be located in the cable. With this technique, it is possible to locate cavities to less than within a foot of the cable length. Prior to the introduction of the triple extrusion process, polymeric cables were routinely scanned for discharges upon emerging from the insulation extruder. The need for this type of rigorous discharge monitoring greatly diminished with the marked reduction in the incidence of cavity defects, following the introduction of improved extrusion techniques and cleanliness of materials.

With insulation-shielded cables, discharge source location may be carried out by means of several techniques. If a well-shielded and directed X-ray source is available, then a technique based on modulated X-rays utilized by Baghurst [52] and initially proposed and tested by Eigen [53] may be employed. Baghurst's arrangement is portrayed in Fig. 9.43; the cable is scanned with a chopped or modulated X-ray beam and the partial-discharge pulses, appearing across the detection impedance of a conventional PD detector, are recovered at the synchronized modulation frequency using correlation and frequency spectral analysis techniques. The intensity of the X-ray radia-

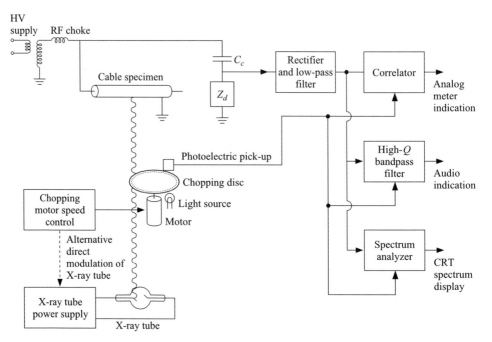

Figure 9.43 Partial-discharge site location by means of scanning X-ray radiation source (after [52]).

tion source must be suitably adjusted so that it is just sufficient to provide enough free electrons to initiate and sustain the discharge process and thus yield a pattern of regularly recurring discharges to facilitate discharge detection and site location. An unduly intense X-ray irradiation of the cavities may result in the formation of a pulse-less plasma glow within the cavities, thereby precluding the detection of discharges with a PD detector that responds only to pulse-type discharge signals.

Another electrical method for locating discharge sites in cables is based on cross-correlation techniques, whereby the discharge site within a cable is located in terms of the cross-correlation function between two detected discharge signals at the ends of the specimen cable [54, 55]. The position of the discharge site is thus determined by the time delay τ of the peak of the cross-correlation function. The value of the cross-correlation function is estimated by means of a digital system that compares the polarities of the two detected pulses as a function of time. Essentially, this constitutes in principle a partial-discharge pulse-polarity coincidence correlator; a pulse-polarity coincidence correlator, developed by Weeks and Steiner [54], is depicted in Fig. 9.44. In their arrangement, the discharge pulses are detected and amplified at the two ends of the cable specimen. Following amplification, the pulses are threshold detected whereby any voltage pulse level above a preset threshold assumes a logic level of $+1$ V, while any level below the threshold or of negative polarity becomes a logic level of 0 V. This set of newly created voltage pulses in the two detection channels is sampled at instants deter-mined by a common clock. The sampling rate is fixed at 20 MHz of a total of 2×10^4 sample points for each channel, which provides a sequential record of the discharge pulses in each detection channel at intervals of 50 μs. The information, consisting of a string of 1's and 0's at the sampled times, is stored in bit form in addressable memories; the cross correlation is thereby obtained by comparing the discharge pulse polarities in the two detection channels as a function of the time delay τ between the two channels. If $e_1(t)$ and $e_2(t - \tau)$ denote two pulse signals in the two respective channels, then the cross-correlation function is defined in terms of τ by

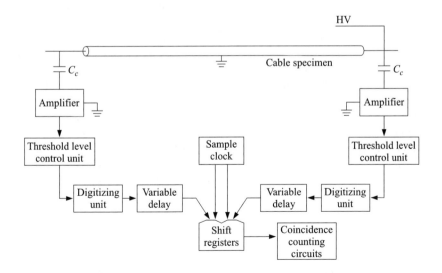

Figure 9.44 Partial-discharge pulse-polarity correlator (after [54, 55]).

$$R_{12}(\tau) = \lim_{T \to \infty} \frac{1}{T} \int_0^T e_1(t) e_2(t - \tau)\, dt$$

where T is the time of integration. By using a large number of sampling points, the authors ensure effectively a long integration time, which increases the resolution capability of their discharge site location method. To locate a discharge site between the two ends of the cable requires a time delay of

$$\tau = \frac{l}{v}$$

where l is the cable length and v is the velocity of propagation in the cable. Each discrete discharge site gives rise to a different delay time τ, which is obtained by the offset of the addresses as the strings are retrieved for comparison. A measure of the cross correlation is derived by counting the number of polarity coincidences at each delay. It is evident that for a single discharge only at the midpoint of the cable specimen, the maximum count will occur at zero delay.

Although the cross-correlation method of Weeks and Steiner [54] appears to work reasonably well with cable specimens containing a single or two discharge sites, it is questionable whether at the present stage of its development it could possibly be applicable to cables with a multiplicity of discharge sites. The cross-correlation method suffers somewhat from discharge pulse reflection effects, though the problem posed does not appear to be insurmountable. Yet it is surprising that no effort has been made to eliminate these end reflections, particularly because a resistor equal in value to the cable characteristic impedance, inserted in series with the coupling capacitance C_c at each cable end, would remedy the problem.

Time-domain reflectometry or traveling-wave methods have been used now for nearly half a century to detect faults in cables. Early attempts to locate partial-discharge sites in cables achieved only limited success because of technological limitations in the measuring analog-based circuitry [56]. If one considers an unterminated cable specimen undergoing test and containing a single discharge site as portrayed in Fig. 9.45(a), then the position of the discharge site along the cable may be determined in terms of the time of separation between the incident and the first reflected partial-discharge pulse. For this purpose the discharge pulses across the detection impedance must be critically damped and the bandwidth of the detection instruments should not exceed 10–30 MHz to minimize high-frequency distortion effects in the reflected pulses. Neither should the frequency be much below 10 MHz, since pulse resolution would suffer with attending reduction in precision as concerns the location of the discharge site.

If one considers the incident and reflected pulses in Fig. 9.45(b), then the time t_{inc} for the incident pulse to travel a distance Δl to point a is given by

$$t_{inc} = \frac{\Delta l}{v} \tag{9.93a}$$

where v is the velocity of propagation. While the time t_{refl} required for the first reflected pulse to reach point a is

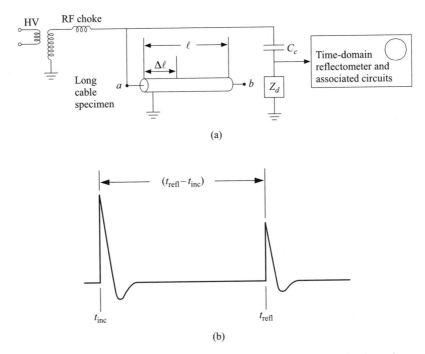

Figure 9.45 (a) Long unterminated cable specimen with single discharge site situated at distance Δl from the near end a. (b) Incident and first reflected pulse with discharge site located distance Δl from the near end a of the cable specimen.

$$t_{\text{refl}} = \frac{(l - \Delta l) + l}{\upsilon}$$

$$= \frac{2l - \Delta l}{\upsilon} \tag{9.93b}$$

Hence the separation time, or time interval between the reflected and incident pulses, is equal to

$$t_{\text{refl}} - t_{\text{inc}} = \frac{2l - 2\Delta l}{\upsilon} \tag{9.94}$$

and thus it follows that the discharge site's distance from point a is given by

$$\Delta l = \tfrac{1}{2}[\upsilon(t_{\text{refl}} - t_{\text{inc}}) - 2l] \tag{9.95}$$

where l is the overall length of the cable specimen. Had the high-voltage connection been made at the other end of the cable [point b in Fig. 9.45(a)], then a similar analysis would have shown that the distance to the discharge fault would be given by the relation

$$\Delta l = \tfrac{1}{2}\upsilon(t_{\text{ref}} - t_{\text{inc}}) \tag{9.96}$$

The traveling-wave method may be rendered ineffective due to triggering of the oscilloscope trace by multiple impulses, arising from numerous discharge sites and superposed extraneous electrical interference pulses. However, it has been shown by Beyer et al. [57] that by means of a microcomputer, using digitizing techniques, a modified version of a traveling-wave method may become most effective in locating discharge sites in cables. To locate accurately the discharging sites in terms of the incident and reflected corona transients, Beyer et al. used a controlled amplification procedure for the highly attenuated reflected pulses [58] while at the same time not overamplifying the incident pulses. Also filters were utilized to match the shapes of the incident and reflected pulses. A reflection suppressor was employed for measuring the amplitude of the discharge pulses, which was subsequently digitized by an A/D converter unit. In the computerized calculation procedure only recurring discharges are considered, and the fault position is determined in terms of an average of approximately 1000 measurements. The schematic circuit diagram is delineated in Fig. 9.46; it is indeed most remarkable that Beyer et al. [57] were able to apply their technique to a number of different cable specimens containing several of discharge sources and, perhaps for the first time, provide indisputable evidence of discharge source location by means of a modified traveling-wave method to within an accuracy of 0.2%. Unfortunately, there has been no report given as to whether this system has been applied routinely to evaluate cables in the field.

There have been a number of alternate digitized systems used in conjunction with time-domain reflectometry to locate discharge sources in cables. We shall describe here the system designed by Mashikian et al., [59–61] because it has undergone successful implementation on cable tests in the field. In their arrangement the primordial three problems of discharge source location in cables, namely extraneous noise elimination,

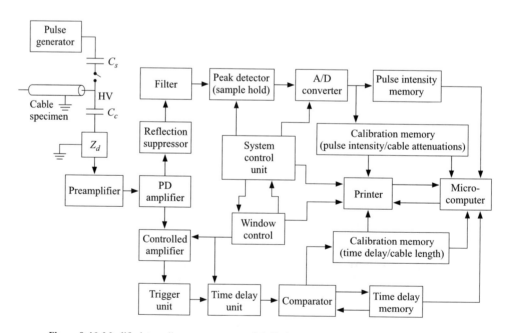

Figure 9.46 Modified traveling-wave-type partial-discharge source locator system (after [57]).

pairing of corresponding incident and reflected pulses or wave couplets, and precision of discharge site position, appear to have been effectively dealt with. The schematic circuit diagram of their PD locator, with the cable specimen in the field disconnected from the remaining power network and energized by means of a portable ac power supply, is depicted in Fig. 9.47. The high-pass filters employed are of the *RC* type, having a pass bandwidth up to 10 MHz with a lower cutoff to reject the power frequency. The detected partial-discharge pulses are transmitted via the buffer amplifier and isolation transformer to a digital storage oscilloscope, which is interfaced by means of an Institute of Electrical and Electronics Engineers (IEEE) 488 bus with a microcomputer equipped with a National Instruments type of accelerator board and Labview II software. It is here where digitally stored discharge pulse data from the oscilloscope undergo digital signal processing. It is not clear what algorithm is used by the authors [61], but there are a number of standard digital signal processing techniques [62] that are amenable to this task, such as deconvolution, linear prediction, maximum-likelihood estimation, etc. Figure 9.48 illustrates a typical PD pulse trace recorded with the time-domain reflectometry (TDR) circuitry of Fig. 9.47 obtained on a 223-m-long XLPE insulated distribution cable, with the measurement being performed at the near end (point *a*). The first pulse on the TDR trace is the incident pulse that traverses a distance $(l - \Delta l)$ from the discharge site taken to be situated a distance Δl from the far end (point *b*); its associated reflected pulse, which travels a distance $(2\Delta l + l)$, arrives later at point *a* with a time delay of t_2 with respect to the incident pulse. These two pulses are in turn reflected from the near end at point *a* and after reflection from the open circuited far end at point *b*, arrive at point *a* (the near end) t_1 seconds after the original incident pulse. This reflection process continues until the reflected pulse *couplet* is attenuated beyond the sensitivity level of the TDR detection system. Thus in reference to Figs. 9.47 and 9.48, the position of the discharge site is determined from the relation

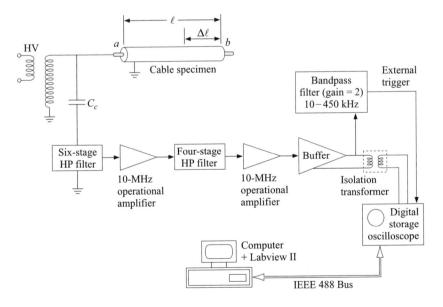

Figure 9.47 Schematic circuit diagram of PD source locator for cables (after [61]).

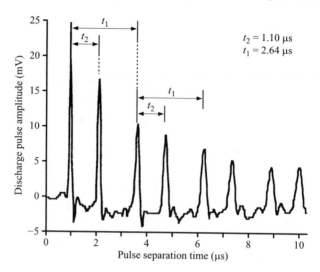

Figure 9.48 TDR incident and reflected PD signal traces obtained on XLPE insulated power distribution cable of 223 m length (after [61]).

$$\Delta l = \frac{t_2}{t_1} l \tag{9.97}$$

Since the cable specimen lies beneath the surface of the soil, Eq. (9.97) provides only the approximate location of the discharge sources from the far end of the cable in relation to the terrain surface. Because of the acclivities and falls of the cable along its meandering path, the actual discharge fault along the cable length may be appreciably displaced from the site estimation point at the ground surface. To ascertain more precisely the position of the discharge source with respect to the soil surface, Mashikian et al. [60] found that injection of a calibration pulse by means of inductive coupling to the unearthed cable specimen at the estimated discharge site location provides an effective method for pinpointing the actual discharge fault. This is accomplished by comparing the position in time of the incident and reflected current pulses of the artificially produced discharge pulse in the shorted cable specimen obtained in accordance with the circuit delineated in Fig. 9.49 with that obtained by means of

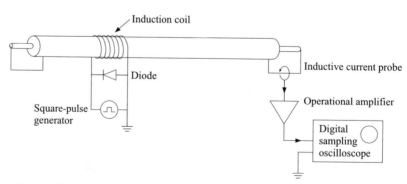

Figure 9.49 Inductively coupled discharge site simulator used to assist discharge site location (after [60]).

the TDR method under actual discharge conditions. This comparison gives an indication on the magnitude of the displacement along the cable between the actual discharge site and that simulated.

In the field testing of distribution cables, it is desirable to be able to detect discharge intensities of at least 5 pC, since the latter represents the sensitivity level to which newly manufactured cables are tested. However, the ambient electrical interference level in the field is generally in excess of 5 pC, requiring the use of suitable noise rejection filters. The most simple procedure for removing electromagnetic interference associated with the radio broadcast band is by means of band rejection or digital notch filters. In the presence of other extraneous noise such as that arising from switching, where the generated impulses bear a definite phase relationship to the applied sinusoidal voltage wave, blanking circuitry may be employed to eliminate all pulses within the applied voltage phase segments over which the interference pulses appear. Adaptive digital filtering techniques may be used in cases where direct operator intervention in the noise filtering task is not feasible. Although Mashikian et al. [61] have not used adaptive digital filtering in the discharge site locator described in Fig. 9.48, they have presented a detailed analysis of such a filter when applied to PD measurements [60]. They have demonstrated that, provided the pulse response of the full cable specimen length along with its attenuation constant α and phase constant β are known, the response due to a discharge at any point along the cable may be determined in terms of the transfer function model for the cable. Thus the pulse response due to a discharge at any point along the cable may be cross correlated with the recorded noise. Hence, the location of the PD fault corresponds to that value, which yields the maximum cross-correlation coefficient.

To circumvent the requirement of a portable high-power 60-Hz source for the field tests on long cables having large capacitance, it has been common practice to employ a portable low-frequency HV supply operating typically at 0.1 Hz [63]. In fact, a TDR system for discharge source location developed by Steennis [64] energizes the cable specimens under test in the field by means of a 0.1-Hz power source.

Although acoustical methods have proven themselves to be relatively ineffective for discharge site location in polymeric cables, in situ location of discharge sites in SF_6 gas insulated equipment can be readily accomplished using acoustical detection techniques [65] by means of the acoustical probe-type circuitry delineated in Fig. 9.50 [66]. The acoustical probe is simply moved slowly along the surface of the SF_6 line or duct to

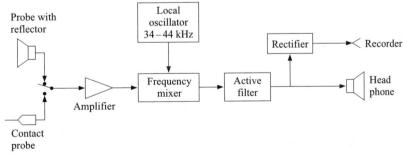

Figure 9.50 Schematic circuit diagram of typical commercial ultrasonic detector (after [66]).

establish points of maxima in the signal intensity. Compressed gas lines using SF_6 gas at 5 atm transmit ultrasonic signals much more efficiently than at 1 atm, though considerable attenuation occurs at the flanges and spacers; thus it is more expedient to scan the line in parts between flanges. The attenuation, α, between two spacers along an SF_6 line may be approximated by Kirchhoff's formula, which is valid for a tube geometry [65],

$$\alpha = (27.6 \times 10^{-5}) \frac{f^{1/2}}{r} \tag{9.98}$$

where f is the frequency in hertz, r the radius of the tube in centimeters, and α the attenuation factor in decibels per centimeter. Although the attenuation is much lower at audio frequencies ($f \leq 20$ kHz), higher frequencies must be used because of the high ambient interference over the audio frequency (af) band; hence measurements are carried out in the ultrasound region over which acoustic emission probe sensors are ideally suited. In general acoustic methods are less sensitive than electrical methods, yielding sensitivity values of $\sim$10– 25 pC [65]. However, sensitivities as high as 2 pC have been reported in conjunction with an ultrasonic detector operating at a frequency of 50 kHz [66]. The ultrasound intensity is proportional to both the discharge amplitude as well as the discharge pulse density with the consequence that an appreciable response of the ultrasonic detector to very low intensity discharges may imply a very dense discharge pattern, consisting of numerous minute pulses. In contradistinction an electrical partial-discharge detector could quite readily detect a single pulse due to a 1- or even 0.1-pC apparent charge transfer, depending on the extraneous noise level (cf. Fig. 9.41 [50]).

It must be emphasized that the presence of ultrasound in SF_6 lines is not necessarily indicative of partial discharges. Often a variation in voltage is employed to modulate the discharge intensity for ascertaining whether the acoustic response is due to a discharge source; however, such variation in the applied voltage level is not feasible under operating conditions [67]. Perhaps one of the main difficulties that afflicts acoustic detection is that it is not always possible to distinguish between the acoustical signals caused by partial-discharges and those due to the movement of metallic particles; moreover, the acoustical method is relatively insensitive to discharges emanating from cavities occluded within the epoxy spacer insulators themselves, though X-ray techniques appear to be quite effective in the latter case [52].

Radio frequency (RF) probe methods are commonly employed for locating discharge sites along power cables that are readily accessible and that have exterior metallic shields that have sufficient looseness to permit insertion of the capacitive probes underneath them. The RF probe methods function most efficiently in locating intense discharge sources and particularly on splices and joints over which metallic shields are either entirely absent or provide ineffective RF shielding. The RF probes may be either inductive or capacitive [68]. Perhaps one of the more successful portable RF capacitative probe techniques developed for use on cables is that by Morin et al. [66]. It operates at a frequency of 7 MHz, well above the low-frequency moving contact interference spectrum, the radio broadcast band, and other normal communication frequency bands; its basic circuit is depicted in Fig. 9.51. Two types of probes are utilized: One is a capacitively coupled, bidirectional contact probe consisting of a small metallic copper plate of rectangular configuration covered with a dielectric layer to avert inadvertent contact with the concentric neutral of the cable under test

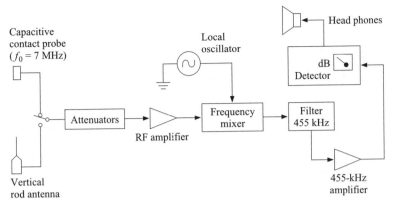

Figure 9.51 Schematic diagram of RF capacitive partial-discharge probe locator for polymeric power distribution cables (after Morin et al. [66])

as well as other grounded metallic conductors in its vicinity. The overall area and thickness of the dielectric covered-copper contact probe is adjusted to suit the sensitivity required in the measurements. The second probe is a vertical omnidirectional probe and consists of a rod with a copper conductor. The latter is essentially a rod antenna, which is employed to determine whether there is any discharge activity present in a cable specimen, i.e., it is passed through the openings of manhole covers to sense the presence of partial discharges over the entire cable length. If discharge activity is detected, the contact probe is then employed to scan the surface of the removed cable at unshielded splice joints and whenever possible by means of insertion of the capacitive problem under a loose concentric neutral. The output from the probes is applied to the input of the RF detection circuit, which comprises an attenuator in tandem with an amplifier, filter, frequency mixer, and detector with meter. As the contact probe, mounted on a 1-m-long insulating stick, is moved along the cable, maxima in the response of the meter readily indicate partial-discharge sites at unshielded splices and joints. However, at very high ambient interference levels, the discharge site location capability of the RF probe is seriously compromised.

The output of the RF partial-discharge detector in decibels is related to the apparent charge in the discharge pulse ΔQ in picocoulombs by the empirical relation [66]

$$dB = K \ln(\Delta Q) \pm b \qquad (9.99)$$

where K and b are constants and the value of ΔQ refers to a train of calibration discharge pulses at a fixed repetition rate, i.e., the meter indication in decibels is a function of the charge transfer ΔQ and its repetition rate. At a calibration pulse repetition rate of 120 per second, the RF detector sensitivity was established to be 5 pC. It is palpably evident that since, in practice, the partial-discharge pulse patterns consist of pulses of different apparent charge transfers ΔQ with their own intrinsic repetition rates, a direct calibration between the RF detector output in decibels and the average detected apparent discharge charge transfer ΔQ is not possible. Recently, there appeared commercialized versions of the RF probe method, using circuitry that responds to the peak value of the discharge pulse, thereby providing an improvement in the sensitivity of the device.

The greatest part of the discharge site location work has been centered in the polymeric distribution power cable area. As concerns oil-paper-insulated HV transmission cables, some data has been reported by Steennis [64] using a TDR technique in conjunction with a 0.1-Hz portable power source; most of the located discharge sites involve cable joints. In situ discharge site location in 275-kV XLPE transmission cables, or more specifically discharge source location in the joints of such cables, has been carried out using a method based on the balanced detection mode that has been employed on distribution cables at low frequencies [41] and on alternators at high frequencies by Stone et al. [69]. The PD location system utilized by Katsuta et al. [70] on joints on 275-kV XLPE insulated power transmission cables is depicted in Figs. 9.52(a) and (b). The metallic foil electrodes C_3 and C_4 consist of a copper sheet wrapped over the outside of the polyvinyl chloride (PVC) cable jacket, each having an equivalent capacitance of 2×10^3 pF. The capacitances C_1 and C_2 consist of the incremental capacitance of the cable (~ 1.5 pF/cm) plus the capacitance of the one-half portion of the joint, which has its metallic sheet separated at the midpoint by an insulating cylinder, as illustrated in Fig. 9.52. The detection impedance Z_d is connected across the balanced output of capacitance C_3 and C_4; Z_d comprises a balun-type circuit, whose unbalanced output is applied to a Tecktronix type P6201 high-frequency probe having an input and output impedance of $100 \, \text{k}\Omega$ and $50 \, \Omega$, respectively. The balun circuit consists of passive electrical elements and is used to couple the balanced input obtained across C_3 and C_4 to the unbalanced input of the high-frequency probe. The input impedance of the balun circuit is between 420 and 830 Ω in the frequency range between 1 and 10 MHz. The authors quote a measurement sensitivity of approximately 1 pC at the optimum detection frequency of 10 MHz; however, there appears to be some uncertainty in the calibration procedure, which involves the injection of the calibration pulse across C_1 and C_2. The calibration problems arise from the complexity of the measurement circuit configuration, which essentially consists of two separate transmission lines: one along the cable itself and the other one along the line formed

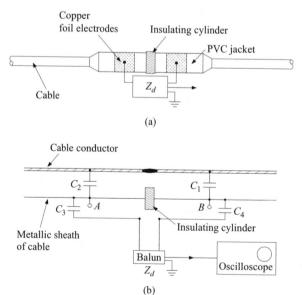

(a)

(b)

Figure 9.52 Partial-discharge source location on joints of 275-kV XLPE power transmission cable: (a) schematic connection diagram and (b) equivalent circuit (after [70]).

between the cable shield and the ground itself. Their equivalent circuit acts in some respects as a normal PD detection network, i.e., if discharges emanate from the B side of the cable or joint, the C_1 represents the coupling capacitor C_c in the conventional type of detection circuit. Conversely, if the discharge sites occur on the A side of the cable, then C_2 assumes the role of C_c [cf. Fig. 9.52(b)]. The partial-discharge source location technique of Katsuta et al. [70] would be expected to perform well in determining whether discharges occur at or in the vicinity of the cable joints; however, this would entail installation of metallic foil electrodes and a joint-dielectric separation ring at each joint.

Since the resistivity of a semiconducting insulation shield is still relatively high, it may be viewed as a lossy capacitance. This attribute of the semiconducting shield permits the implementation of portable PD capacitive probes for the scanning of any unshielded sections (e.g., splices, joints, etc.) of the polymeric cables. But over the shielded sections, which comprise the major portion of the cables, inductive-type probes must be utilized. The performance of the inductive probes compares favorably with that of the capacitive probes, though they tend to be more susceptible to extraneous interference [70a]. Their performance may be enhanced through the use of ferrite cores, which renders them frequency selective and results in sensitivity levels of about 2 pC.

9.7 DISCHARGE PULSE PATTERN STUDIES

Heretofore we have confined our discussion on partial-discharge measurements specifically as concerns methods of detection and the location of discharge sources whose intensity exceeds the permissible level of 5 pC as dictated by the specifications on cables. We have intentionally omitted other active areas of partial-discharge measurement because at present the quality of cables is essentially assessed in terms of go–no go type of partial-discharge standards, i.e., cables that contain discharging cavities are rejected and those that are free of discharges are deemed acceptable. However, for many decades research in the area of partial discharges has been conducted with the purpose of determining the discharge energy or discharge pulse amplitude threshold or level below which the associated degradation of the cable insulation could be considered to be negligible in regard to cable life. Thus far this objective has eluded all our efforts, principally because the discharge intensity is an indeterminate function of time, i.e., usually the discharge intensity measured at the commencement of a partial-discharge aging test bears no relationship to that just preceding the final failure event. Considerable interest centers on the interpretation of the form of the partial-discharge pulse patterns themselves insofar as they may bear a definite relationship to cavities of certain size and geometry or even electrical trees. We shall not delve deeply into these areas of endeavor in this chapter, since such information may be found elsewhere [15]. However, it behooves us to do some justice to these long sustained and often arduous efforts by alluding at least in a cursory manner to some of the more fundamental studies and their possible implication as concerns the practical testing of cables.

Much effort has been expended on partial-discharge pulse pattern studies, particularly as concerns partial-discharge pulse height distribution measurements and discharge pulse pattern recognition using intelligent machines. Although neither one of these approaches is employed in the routine partial-discharge testing of cables, there are some practical possibilities for these experimental procedures. Partial-discharge pulse spectrometry began first with simple discharge pulse counting circuits [71, 72] and single-channel variable-threshold pulse height analyzers [73], before progressing to the early more sophisticated multichannel analyzers [74], and finally to storage and treatment of the obtained data by microcomputers [75]. A typical pulse height distribution measurement system suitable for application to cables is depicted in Fig. 9.53. In order that each partial discharge pulse be registered as a discrete event, the cable specimen must be suitably terminated at the far end by its characteristic impedance to eliminate pulse signal reflections, which would be otherwise recorded by the pulse-height analyzer as separate pulses of lower magnitude. The termination of the cable at the far end consists of a discharge free capacitor in series with resistor R_0, whose value is equal to the approximate characteristic impedance of the cable specimen. The value of capacitor C is usually made equal to that of blocking capacitor C_c of the partial-discharge detector circuit. The purpose of C is to assume the entire power frequency voltage drop across its terminals and thereby prevent a short circuit across the low resistance of R_0. The relatively high capacitance of C acts as a short circuit to the arriving partial-discharge pulses at the far end of the cable, which, as a consequence of their high frequency content, only see the resistance R_0. The output of the multichannel pulse height analyzer is calibrated to yield a plot of the discharge pulse rate, n as a function of the pulse charge transfer ΔQ. The integral of such discharge pulse height distribution curve or more correctly pulse probability density function is dimensionally equal to the total discharge current I given by

$$I = \sum_{j=1}^{m} n_j \, \Delta Q_j \tag{9.100}$$

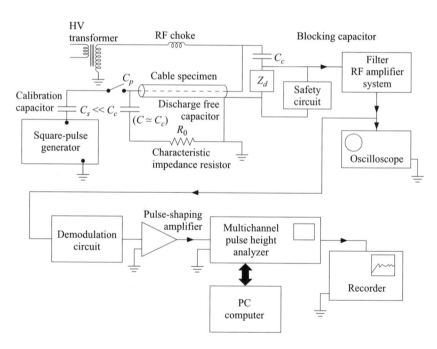

Figure 9.53 Terminated cable with partial-discharge pulse height distribution measuring apparatus (after [76]).

where j refers to the jth discharge pulse. It should be emphasized that the above result may be equally well obtained without the hardware of a multichannel analyzer by using in lieu of it an analog-to-digital converter in conjunction with the appropriate software to carry out the processing of the digitized discharge pulse data with a PC computer. The development some time ago of circuitry to obtain the discharge pulse time interval distribution [77] together with the discharge pulse phase distribution with respect to the applied voltage wave [78, 79] has permitted to determine also the total energy dissipated per unit time by the discharge pulses in a given distribution pattern, because the discrete charge transfers ΔQ associated with each discharge pulse could be related to the instantaneous value of the applied voltage V_j at which they occur. Hence the power P, dissipated by the partial-discharges, may be expressed as

$$P = \sum_{j=1}^{m} n_j \, \Delta Q_j \, V_j \tag{9.101}$$

Since the charge transfer ΔQ_j is defined by

$$\Delta Q_j = C_T \, \Delta V_j \tag{9.102}$$

then

$$P = C_T \sum_{j=1}^{m} n_j \, \Delta V_j \, V_j \tag{9.103}$$

where ΔV_j represents the voltage change or discharge pulse magnitude across the cable specimen at the time of the discharge within a cavity and C_T is the total capacitance of the detection circuit, including that of the cable specimen. Because the polarity of ΔV_j may be positive or negative, the calculation of the discharge power loss in terms of Eq. (9.103) must take into account the relative polarities of ΔV_j and V_j.

From Eq. (9.103) it is inferred that the discharge power increases with the discharge pulse magnitude and its repetition rate. Thus the total discharge power loss produced by numerous minute discharge pulses may have the same magnitude as that produced by a much smaller number of large-magnitude pulse discharges. However, the pitting of the dielectric's surface within the cavity, subjected to very high magnitude discharges, will be more pronounced in depth; it may be argued that if the intense discharges remain fixed over a given surface without any spatial migration, the high-intensity discharges may precipitate dielectric breakdown more rapidly. But, if the individual discharges migrate freely and randomly between the different sites over the dielectric's surface within the cavity, then the time to insulation failure should depend more on the total discharge energy expended in the surface irrespective of the discharge magnitudes. There have been numerous investigations carried out to study the degradation rate and its character of cable insulating materials [25, 48], and although, it is well known that insulating materials deteriorate more rapidly when subjected to higher intensity discharges, it has not been possible to establish a discharge energy or discharge pulse magnitude threshold below which the discharges may be tolerated without any concern for eventual catastrophic breakdown of the cable insulating system. The inability to arrive at such permissible discharge energy threshold level has been impeded by a number of factors, which are inextricably associated with the complexity of the discharge process itself. For example, the discharge energy even in the most well-defined cavities changes with time as the cavity walls become coated with degradation products, some of which (e.g., oxalic acid) are electrically conducting. Thus, the discharges may disappear with time and then reappear subsequently or continue increasing in intensity until breakdown ensues. Consequently, the initial discharge power loss observed on a given insulating system need not bear any definite relationship to the time to failure. Moreover, the nature of the discharge may undergo change itself: the pulse-type discharges may become replaced by a pseudoglow discharge (a discharge consisting of minute pulses of slow rise time) or even a pulseless glow discharge to which a conventional partial-discharge pulse detector may not respond [16, 32, 33]. In such circumstances, degradation due to discharges may proceed within the cavity without any external indication. In view of the sporadic nature of the partial-discharge process, it is justifiable to insist on an absolute absence of even the lowest detectable intensity of discharges to ensure long service cable performance. It is thus not surprising that cable specifications continue to reflect this stance and, consequently, specifications based on arbitrary discharge energy levels are unlikely to be proposed in the near future.

Different cavity sizes and geometries in cable insulating systems will lead to particular discharge pulse distribution patterns. The term pulse distribution pattern may refer to a number of different displays such as that depicting the discharge pulses over a power frequency time base on the screen of a conventional partial-discharge detector [cf. Fig. 9.33(a)], i.e., the pattern has a certain pulse density, pulse amplitude distribution, and pulse phase relationship to the power frequency wave. This particular pulse density–amplitude–phase display will also have a given pulse height distribution when viewed on the screen of a multichannel pulse height analyzer, and if viewed on a three-dimensional display in a computerized system it will indicate clearly the separation between the adjacent pulses, their magnitude, and phase relationship. Following the advances made in the area of intelligent machines, a considerable effort has been directed to determine whether machine intelligence can be used to determine cavity size, geometry, and number as well as their location within the cable. This is an area into which experienced partial-discharge test personnel had delved into for many years on a much simpler level, e.g., practical testing of cables has always involved making the distinction between what signals are due to actual discharges occurring within cavities and what signals are attributable to extraneous noise. However, the problem posed by extraneous interference in cable manufacturing plants has been more or less adequately solved because it has a simpler solution. For example, such interference does not vary with applied voltage unless it is the result of discharges from sharp points in the high-voltage connection leads or from a defective testing transformer. Some interference pulses tend to occur at random without any phase relationship to the applied voltage wave or at some fixed phase relationship points that do not vary with the amplitude of the applied voltage. The modulated broadcast band signals can be readily distinguished on the screen of a conventional partial-discharge detector and eliminated, if necessary, by band rejection filters. In contrast, the distinction between the discharge patterns of two different cavity sizes becomes much more difficult even for a very experienced observer, particularly if the difference in size of the two cavities is small, e.g., 1.0 and 1.5 mm diameter.

Artificial neural networks have a great potential in the area of pattern recognition [80–82]. However, it should be emphasized that the success of PD pattern recognition, using the neural networks, is highly contingent upon the information provided to train and develop the recognition capability of the neural networks. Such information consists of a number of features that are extracted from the PD pattern so as to optimize the likelihood of the correct interpretation of the available PD data. The most common features that have been generally provided to the neural networks to recognize the PD patterns are the apparent charge, ΔQ, the pulse discharge rate, n, and the pulse phase relationship, ϕ. However, it is found that there may arise certain disparities in the measurement of these parameters especially when the statistical time lag becomes unduly long, causing a discharge to occur at a voltage considerably in excess of the normal inception voltage. Thus, the PD pattern is altered, in a statistical sense [83], due to the variations in the value of the breakdown voltage at which the discrete discharges take place, thereby affecting not only the pulse amplitude but also the pulse discharge sequence or phase relationship [32]. Consequently, there are some valid arguments vis-à-vis the suitability in the use of these parameters as the only guiding basis in the PD recognition task. In addition, when discharges between dielectric surfaces are involved, the occurrence of multiple discharge sites and their statistical distribution over the dielectric surface due to fluctuations in the deposited surface charge density will lead to further randomness in the discharge sequence. Alternatively, one may use features that describe the shape of the PD pulses. Yet the approach based on the discharge pulse features does not circumvent entirely the statistical effects. Since the pulse rise time and shape are influenced also by the overvoltage across the cavity due to the statistical time lag [26], this approach only represents a slight improvement.

For simplicity, to illustrate the PD pattern recognition capabilities of neural networks, we shall consider the multilayer perceptron (MLP) neural network and employ the pulse shape features. The MLP neural network structure topology is a feedforward network composed of an organized topology of processing elements or neurons; its general structure is depicted in Fig. 9.54. Here the circles denote the neurons or processing elements; W_i, W_e, and W_0 represent the weights of the input, hidden, and output layers, respectively. The signals relating to the partial-discharge pulses entering each neuron are summed and then processed through a sigmoid function to produce the output. The sigmoid function is given by

$$Z_j = \left[1 + \exp(\bar{\mathbf{A}} \cdot \bar{\mathbf{W}}_j - w_{mj}\theta_j)\right]^{-1} \tag{9.104}$$

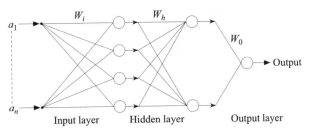

Figure 9.54 General topology of multilayer perceptron neural network.

The left hand side of Eq. (9.104) denotes the sigmoid function that defines the output of each neuron in terms of the input and the weights $\overline{W}_j$; here w_{mj} refers to the weight associated with the internal threshold θ_j. The vector $\overline{A}$ is the input to the network, and it consists of the set of attributes or features $a_1, a_2, \ldots, a_n$ (cf. Fig. 9.54) of the discharge pattern to be classified. If we decide to select the features of the partial-discharge pattern in terms of the discharge pulse shape, then a_1 is the rise time of the pulse, a_2 its magnitude, a_3 its decay time, a_4 its area, etc. Had we chosen to work with partial-discharge pulse height distribution data, the a_1 would be the discharge pulse amplitude or charge transfer, a_2 its repetition or discharge rate, and a_3 its phase relation to the applied voltage. Irrespective of our choice of the features describing the discharge process, the input of the information to the MLP propagates through the neurons or processing elements in the forward direction. It is to be noted that a learning step constitutes the means by which the neural network adapts to the desired output. The adoption is performed through a change in the weights, which gradually converge to values that ensure that each input vector produces the correct output. Essentially, the learning procedure dictates to the neuron how to modify its weight in response to a certain stimulus. The MLP neural network's learning scenario is called back propagation, and it consists of being provided with both the input patterns and the desired response in a supervised fashion. It is then put through a series of iterations where in each iteration it compares its own output with the desired responses. If there is a match, no change is imposed on the network; in the absence of a match, the weights are altered using the gradient search technique to minimize the mean-square difference between the desired response and the actual output. Once the learning process is completed, the MLP neural network when presented with a new partial-discharge input pattern should assign it the correct classification.

The success rate of the MLP neural network in Fig. 9.54 in distinguishing between the discharge pulse forms characterizing 1-, 2-, and 3-mm size idealized cylindrical cavities has been found to be better than 80%; the performance was somewhat poorer when it was required to differentially be between the partial-discharge pulse shapes associated with 1- and 1.5-mm size cavities [80]. The MLP neural network was demonstrated to be equally capable of distinguishing between the discharge pulse forms arising from discharges occurring in electrical trees as opposed to those in cavities [84]. Work carried out on discharges within cavities, comparing different neural network paradigms (nearest neighbor classifier, learning vector quantization, and the multilayer perceptron), indicated approximately equivalent performance, with the exception that the learning vector quantization paradigm was somewhat better in recognizing the differences between the discharge patterns of smaller void sizes [85]. Notwithstanding these encouraging results, it must be conceded that at the present the indiscriminate usage of neural networks for discharge pattern recognition on actual cable insulating systems would be rather premature. In such circumstances the neural networks would have to deal with complex discharge patterns involving numerous different cavity sizes occurring at various parts along the cables. Serious erroneous predictions would be expected, since the neural networks cannot possibly interpret a complex discharge pattern on the basis of the limited learning given to them using a number of idealized cavities of fixed geometry. Much more work is required in this area before neural networks can be applied routinely on cables. It is also to be noted that some recent work is being carried out in the area of Fuzzy logic, as concerns partial-discharge pattern recognition; thus far the obtained performance appears to be comparable in some limited respects to that of neural networks.

9.8 ELECTRICAL AGING MECHANISMS

Aging tests are carried out on power cables to predict their life expectancy under in-service operating conditions. Ideally an aging test should subject the cable insulating system to typical voltage, mechanical and thermal stresses that are encountered in service; however, this would result in unacceptably long testing times. Hence aging tests must be accelerated and be of as short duration as possible as long as the induced aging mechanisms remain, as much as possible, identical to those occurring under actual service conditions. The most effective means to carry out accelerated aging is by subjecting the cable to enhanced voltage stresses. The importance of accelerated electrical aging test methods has been already recognized for more than half a century [86, 87]. It has now become common practice to carry out accelerated electrical aging tests on cables either by subjecting the cable specimens to an elevated voltage stress for a fixed time period or for a time period sufficiently long to cause failure of the cable specimens. In the former case the insulating system of the cable specimen is rapidly degraded to some level, which can then be related to the properties of the cable insulating system that correspond to those attained over much longer periods under actual operating conditions. For example, it is normal to compare the ac breakdown voltage of the aged cables with that of their initial values in their original virgin state. Thus the degree of depression of the ac breakdown voltage from its initial value constitutes a measure of the degree of degradation [88] that has occurred in the cable during the fixed time test. If one proceeds with this method of test for a series of increased fixed time periods, the results may be utilized to obtain an estimate of the cable lifetime by extrapolation, i.e., the point in time at which the ac breakdown strength remaining falls to the operating voltage level of the cable. Such a testing sequence assumes that the ultimate failure will be precipitated by the conditions accelerated in the experimental procedure. The alternative approach to accelerated electrical aging involves subjecting the cable insulating system to progressively lower electrical stress values and continuing the aging until electrical failure ensues at each different value of the electrical stress. In this section as well as the next we shall principally concern ourselves with aging mechanisms and tests as applied to solid polymeric cables. The oil-paper insulating system has proven its reliability and longevity over many decades of service and does not require much additional aging evaluation. Most current efforts in the area of aging studies are thus directed toward the prediction of the life expectancy of polymeric cables.

In the accelerated electrical aging test, the controlled parameter is the voltage stress E and the determined variable is the time to failure t. The expression relating E and t is the well-known inverse power law [89]

$$t = KE^{-n} \qquad (9.105a)$$

where K is a constant and the exponent n is indicative of the quality of the cable in the presence of the degradation process involved [90]. For partial-discharge induced degradation n is ~ 4, for water treeing mechanisms $2 \leq n \leq 4$ [91], for contaminant effects in polymers under dry conditions $8 \leq n \leq 10$ [92], and for deterioration mechanisms in high-purity polymer cables with smooth extruded semiconducting shields $n \geq 15$ [93]. Aging test data are very frequently characterized by an appreciable dispersion in the individual time to failure points of the cable specimens evaluated. As early as 1971 Occhini [94] took a more rigorous statistical approach when he applied a two-parameter Weibull distribution for voltage breakdown analysis on cables. Using the controlled parameter of voltage stress E and the determined variable of the time to failure t, he expressed the cumulative failure probability, $P(E, t)$, as

$$P(E, t) = 1 - \exp\left[\left(\frac{E}{E_s}\right)^b \left(\frac{t}{t_s}\right)^a\right] \qquad (9.105b)$$

where E_s is an electrical field scale parameter and t_s is the time scale parameter; a and b are, respectively, the time and electrical field shape parameters. Equation (9.105b) is related to the inverse power law because at constant probability, the exponent n in Eq. 9.105a, equals the ratio b/a.

The concept of a threshold electrical stress, below which no deterioration of the solid insulation can take place, was introduced by Montsinger [95] in 1935 when he devised the general form of the inverse power law equation that included the threshold stress term as well. Based on the antecedent work of Dakin and Studniarz [96], Bahder et al. [97] advanced further the concept of a threshold electrical field below which degradation is absent; the value of this threshold field was taken to be that of the partial-discharge inception stress. It was argued that enhanced fields at protrusions are sufficiently high to ionize cavities as small as $\sim 10\,\mu m$ in diameter at normal operating voltage, though recent evidence would suggest that cavity sizes substantially in excess of $10\,\mu m$ would be required to sustain a partial-discharge process [26]. The cyclic charge efflux from the cavities was conjectured to migrate along preferential channels within the insulation in the process degrading the polymer and leading to electrical tree channel enlargement culminating in breakdown. The time to breakdown, t, at an operating stress E of frequency f was stated as

$$t = \frac{1}{fb_1[\exp\ b_2(E - E_r) - 1][\exp(b_3 E) + b_4]} \qquad (9.106)$$

where E_r is the threshold field for discharge onset and b_1, b_2, b_3, and b_4 are constants that depend on the polymer material, temperature, and geometry. Equation (9.106) yields a straight line at high stresses on an $\ln E$ versus $\ln t$ plot, similar to the inverse power law model. However, at intermediate stresses, Eq. (9.106) leads to a curve that is concave downwards and, therefore, infers shorter failure times than would be implied by the inverse power law model; but with increasing lower stresses the failure time approaches an infinite value.

It was postulated by Zeller et al. [98] that at the asperities above a certain threshold field the injected charge gives rise to regions of elevated charge mobility wherein degradation of the polymer proceeds either due to chemical changes, dielectric heating, or bond scission, thereby leading to electrochemical and electromechanical fracture and, eventually, to low-intensity discharge and thence an electrical tree. The investigations of Auckland and Arbab [99] on amorphous polymers below their glass transition temperature (T_g) lend some support to the hypothesis of Zeller et al. [98]. However, the suggested mechanism still requires confirmation on semicrystalline polymers above T_g of the type used in cables, where the polymer is in the elastomeric state and over which range aging has been demonstrably shown to depend upon the size, number and shape of contaminants, protrusions and other defects.

The inverse power law defined by Eq. (9.105a) requires the electrical aging to be carried out at different values of the controlled parameter of voltage stress at fixed values of thermal and mechanical stress. There has been extensive early work done on the thermal degradation characteristics of electrical insulating papers [100]. Montsinger [101] derived a relationship between the tensile strength of thermally degraded paper and the remaining life, t, but it was Dakin [102] who provided a more rigorous basis to the thermal degradation model of Montsinger by the application of the chemical reaction rate theory showing that

$$t = A \exp\left(\frac{\Delta H}{kT}\right) \qquad (9.107)$$

where A represents a constant, ΔH the activation energy, T the absolute temperature and k the Boltzmann constant. The equation predicts a straight-line relationship when $\log t$ is plotted against $1/T$, so that extrapolation to the operating temperature yields the time t equivalent to the expected life of the insulating system. It must be emphasized that Eq. (9.107) is valid only for first-order chemical reactions and must, therefore, be modified when aging factors other than thermal become important. For example, non-linearities in the $\log t$ versus $1/T$ characteristic may arise with polyethylene extruded insulation when accelerated aging is carried out above its melting point. Thus, intrinsic aging factors such as recrystallization can change the deterioration mechanisms, which can produce non-linearity. Publication 216 of the IEC [103] outlines the techniques for deriving the temperature indices or thermal endurance profiles so that performance of these tests with applied voltage as a parameter allows electrical aging to be taken into account.

The question of mechanical aging has been dealt with both using the inverse power law approach by Sletback et al. [104] as well as from the point of view of chemical rate theory principles by Zhurkov [105], who demonstrated the time to failure t due to mechanical stress to have the form

$$t = t_0' \exp\left(\frac{\Delta H_m - \gamma S_m}{kT}\right) \qquad (9.108)$$

where t_0' is a preexponential constant, ΔH_m the activation energy of the fracture process, γ the activation volume for sub-microcrack formation, and S_m the mechanical stress. The model visualizes aging as a stress-induced elongation of molecular chains that results eventually in localized bond scission at the various defect or stress enhancement sites, leading to microcracks or fissures that coalesce, culminating in mechanical failure. The foregoing considerations reveal a remarkable similarity between the electrical and mechanical aging models, notwithstanding that the nature of the stresses causing the aging differs.

Perhaps the oldest and the most general multiple-stress aging model is that proposed by Eyring [106] in 1941; it is based on the rate theory, and all subsequent models based on the rate theory may essentially be regarded as modification of Eyring's approach. The multiple-stress model combines the thermal stress either with mechanical or electrical stress and gives the aging rate t' as [106, 107]

$$t' = \frac{kT}{h} t_0' \exp\left(-\frac{\Delta G}{kT}\right) \exp\left(\frac{cE}{kT}\right) \tag{9.109a}$$

where h is the Planck constant, ΔG the Gibbs free energy, E the electrical stress, and t_0' and c are constants. Endicott et al. [108], in applying Eq. (9.109a) in a modified form to analyze aging data obtained on capacitors, included the electrical stress, in addition to the temperature term; their derived relation was dimensionally correct and had the form

$$t' = t_0' \exp\left(-\frac{-b}{kT}\right) \exp\left[S\left(c + \frac{d}{kT}\right)\right] \tag{9.109}$$

where t_0' is the degradation rate in the absence of applied stress, S is a stress function, and b, c, and d are constants independent of time, temperature, and stress. Endicott et al. [33] were successful in demonstrating a direct relationship between Eq. (9.109b) and Eq. (9.105) for the inverse power law, by substituting the stress function S in Eq. (9.109b) with a function of voltage V of the form $\ln V$, which led directly to the inverse power law model of Eq. (9.105a). The modified equation of Endicott et al. was combined with that of the thermal aging model by Simoni [109] to deduce an equation of the form

$$t = K_b t_0 \frac{\exp[-B \,\Delta T - cE + bE \,\Delta T]}{E/E_\gamma + \Delta T/\Delta T_\gamma - 1} \tag{9.110}$$

wherein he introduced a threshold temperature T_γ (for the applied field $E = 0$) in addition to a threshold electrical field gradient E_γ below both of which no aging is assumed to occur; here t denotes the time to failure, $\Delta T = 1/T - 1/T_\gamma$ and K_b, t_0, B, b, and c are constants. In the application of Eq. (9.110), test data are used to generate a surface, which relates the failure time t as a function of E and T. It is to be emphasized that Eq. (9.110) cannot predict thermomechanical aging and assumes only first-order chemical reactions.

There have been numerous other aging models proposed that will not be dealt with here; an excellent assessment of various aging models as concerns their applicability has been presented elsewhere by Montanari [110]. We shall necessarily be obliged to confine our discussion to the inverse power model, since it is by far the most common approach followed in practice as concerns accelerated aging of power cables. There exists a vast amount of accumulated experimental aging data for comparative purposes obtained using the inverse power law and, moreover, the performance characteristics of cables as well as the aging mechanisms are classified in terms of the exponent n values. In fact, the value of n constitutes an important design parameter for present power transmission cables.

The most important stresses and variables that influence the aging rate of the cable insulating system are voltage and mechanical stresses, temperature, and the environment. The rate of thermal aging increases with temperature because of enhanced chemical reaction rates, diffusion and solubility of organic compounds, and wetting of contaminants. The oxidation rates in oil-impregnated-paper cables become significant above 100°C as the oxygen within the cellulose molecules is liberated and reacts with carbon to form CO gas; embrittlement occurs and the paper loses mechanical strength. In solid polymers such as polyethylene and ethylene-propylene-rubber, oxidation constitutes the primary chemical reaction [111]; it is estimated [90] that for a 500-kV XLPE cable [112], the volume of free gas (conceivably mostly air) adjacent to a corrugated metallic jacket is ~20% of the total volume, which (omitting the amount of dissolved oxygen) represents still a copious amount of oxygen to initiate oxidation of the cable polymer. However, the oxidation rate of the latter is somewhat retarded by the barrier of the semiconducting sheath and the limited diffusion rate in the bulk of the polymer [113]. The crystalline regions in both low-density polyethylene (LDPE) and XLPE undergo melting over the temperature range extending from 60 to 110°C as demonstrated by the data obtained using differential scanning calorimetry [114]. It is thus apparent that at 90°C some of the melted smaller and less ordered crystallites undergo oxidation; also the originally amorphous regions become subjected to oxidation once the residual antioxidants are depleted. At temperatures in excess of 90°C when all the crystalline regions are in a molten state, oxidation will progress in regions that, at the lower temperatures, were not susceptible to oxidation [90]. Oxidation within the insulation may initially be somewhat offset by migration of antioxidants from the semiconducting shields into the insulation bulk [115, 116]. Oxidation in distribution cables exposed to a moisture environment has been found to lead to reduction in the ac breakdown strength [45]; however, polymeric transmission cables are expected to be immune to this effect, since the amount of residual moisture formed during the crosslinking process is <200 ppm [117]. In transmission cables subjected to very large fields, oxidation may affect adversely the electrical characteristics because of the higher charge carrier and trap concentration within the more oxidized regions; furthermore, increased temperatures will enhance the detrapping rate and thus provide additional charge carriers for conduction [118]. High temperatures will also augment the solubility of the residual crosslinking agent by-products, such as acetophenone, cumyl alcohol, and α-methyl styrene, although their exact influence upon aging has not been ascertained.

Both the 60-Hz and impulse breakdown strengths diminish with temperature; this phenomenon is commonly attributed to changes in the structural properties of the polymers (cf. Chapter 2). At 130°C, which represents the emergency temperature limits of XLPE cables, the depressed value of the mechanical modulus of XLPE would imply the possibility of thermomechanical damage as concerns the thicker insulation walls of the higher voltage class cables.

Many polymers, particularly rubbers, tend to degrade chemically more rapidly when stressed both thermally and mechanically in the presence of oxygen [90]. For example, this may cause embrittlement of the semiconducting shields and an increase in their resistivity, ultimately leading to cracking, onset of partial discharge, and a reduction in the impulse strength. Elevated temperatures and thermal cycling can also induce physical changes such as thermal expansion, relaxation of mechanical stresses, and polymer melting with subsequent recrystallization. The expanding polymer exerts appreciable pressure on the concentric neutral of the cable [119], which after attaining a maximum with temperature commences to diminish above the polymer melting range (~105–110°C); the foregoing effect has been found to result in concentric neutral wire indentation and penetration into the insulation of XLPE distribution cables [120].

Recrystallization of polymers during thermal cycling has been observed to lead to the formation of cavities, that subject the cable insulation to degradation by partial discharges [121]. This data is in part supported by the work of Barlow et al. [122], who have found an increase in the number of cavities with time in LDPE, which they ascribed to secondary recrystallization, i.e., a protracted crystallization that occurs in the amorphous phase below the primary crystallization temperature. Since at constant temperature crystallinity of polymers increases very slowly with time, recrystallization may be more pronounced if the insulation is subjected to a constant temperature in the range from 60 to 90°C [90]. On the beneficial side, it must be remarked that, with thermal expansion of the insulation, compressive forces hinder the formation of cavities. However, since cables in service are exposed to different load cycles and aging tests are normally designed to simulate load cycling conditions, constant-temperature behavior in the accelerated aging of cables has not received the attention that it deserves.

Thermal treatment is normally applied to XLPE transmission cables to remove most of the volatiles such as acetophenone, cumyl alcohol, and α-methyl styrene, whose combined concentrations are usually on the order of 2000 ppm [90]. The residual amounts of volatiles that remain following the thermal treatment become trapped under the applied hermetic metallic sheath [123]. Aging tests carried out up to 1.7 times the rated voltage with temperature cycling from ambient to 105°C for a period of 6–12 months, including examination of in-service aged cables, have not revealed any definite trends in their influence on the aging characteristics of cables [119, 124]. Exposure of XLPE above 150°C causes water formation from cumyl alcohol decomposition, which is a crosslinking by-product of dicumyl peroxide [125]. In situ molding of splice joints gives rise to water condensation and the attending formation of microcavities during the cooling stage from the high temperatures; both water content and the number of cavities are known to increase with the amount of antioxidant and oxygen in the polymer. The water-filled cavities act as nucleation points for water and electrical trees and thus reduce the life expectancy of cables [90].

In examining the effect of electrical stress on the aging rate of cables, one must consider the manner in which electrical aging progresses when cavities, contaminants, and electrical and water trees are involved. Gas-filled cavities when subjected to electrical stress may undergo discharge, which results in the chemical and physical degradation of the cavity walls. Whether or not a given cavity will discharge is contingent upon its discharge inception stress, which is dependent both on the diameter of the cavity and the statistical time lag, i.e., the availability of a free electron to initiate the discharge at the instant the ac applied voltage across the cavity attains its inception voltage. Table 9.2 gives values of the peak 60-Hz alternating breakdown voltage strength determined under uniform field conditions in terms of the Paschen curve data for air at atmospheric pressure [126, 127] and the mean time of the occurrence of a free electron due to a cosmic radiation rate of 3 electrons/s/cm^3 for spherical cavities of varying diameter. It is apparent that as the cavity size decreases, the inception stress increases markedly so that very small cavities are unlikely to be sites for partial discharges. In addition, the smaller the cavity volume, the lower is the probability of an available electron to initiate the discharge. Moreover, it has recently been demonstrated by Novak and Bartnikas [26] that cavity sizes much below 50 μm are unlikely to support a partial-discharge process. For this reason the calculated ac breakdown values for the gaseous volumes of the 5-, 10-, and 20-μm size cavities are inapplicable and, hence, have been intentionally omitted from Table 9.2. Different forms of

TABLE 9.2 Calculated Discharge Inception Stress at 60 Hz and Free-electron Availability in Air at Atmospheric Pressure as Function of Cavity Size

Cavity Diameter (μm)	Calculated Peak ac Breakdown Strength (kV/mm)	Mean Time Between Free Electrons in Cavity
5	—	361.69 years
10	—	20 years
20	—	2.5 years
50	12.6	59 days
100	9.5	7.4 days
200	7.2	22 h
500	5.2	1.4 h
1000	0.5	10.6 min

Source: From [126] and [127].

discharge (pulse, pseudoglow, and pulseless) [33] may occur in cavities, with their character being primarily determined by the resistivity of the cavity surface, its diameter, gas pressure, photoemission intensity and overvoltage across the cavity [128, 129]. The degradation rate of the cavity walls will depend to some extent on the type of discharge occurring within the cavity. Very rapid rise time pulse discharges in the nanosecond and subnanosecond range have been observed by Holboll et al. [130], and it has been demonstrated in terms of a mathematical model by Bartnikas and Novak [131] that their rise time decreases with photoemission at the cathode and the overvoltage. True glow-type discharges may occur in cavities in which the electrical field is too low to cause discharge channel constriction necessary for the detectable spark or pulse-type discharges to occur.

Contaminants (e.g., metallic or fibrous particles or degraded particles of polymer) are deleterious as they cause local electrical field enhancement. The magnitude of the enhancement stress is contingent upon the relative permittivity, the radius of curvature of the contaminant particles, and their position in the insulation. The stress enhancements, from either contaminants or protrusions at interfaces, can lead to the initiation of electrical trees and eventual breakdown. As pointed out in Chapter 2 the values of stress enhancement, considering a number of different geometrically shaped asperities, may range from ∼50 to 500 times the uniform electrical field magnitude. Formulas for calculating these stress enhancements, whose magnitude is determined by the size and geometry of the protrusions or contaminants, have been summarized by Eichhorn [132]. Should a microcavity be situated adjacent to an electrical stress concentration point where electrons are being injected due to field emission, then there would be a copious supply of free electrons available to initiate the partial-discharge process in the microcavity provided the gas volume within could sustain the discharge.

It is apparent from the foregoing considerations that a polymeric cable, which contains an appreciable number of cavities and contaminants, will have a predisposition to age more rapidly. If we confine our discussion here to a polymeric transmission cable dielectric that is free from water ingress, the effects of water trees may be omitted. It is true that residual moisture in the insulation (≤200 ppm) may cause the formation of bow tie water trees. However, it has been observed that the size of bow tie water trees in the case of a dry-cured hermetically sealed 138-kV XLPE transmission cable is limited to lengths of only between 30 and 70 μm (cf. Chapter 2). It is generally found that bow tie water tree growth is inhibited by the restricted amount of moisture available in hermetically sealed insulations and, consequently, its occurrence tends to be regarded as innocuous. In contrast, electrical tree growth and propagation, which is intrinsically associated with cavities and contaminants, constitutes a mechanism that can readily precipitate breakdown once an electrical tree is initiated under high electrical field conditions. Hence to reduce the aging rate of polymeric transmission cables, rigorously controlled limits are imposed on the maximum size of cavities, contaminants, and protrusions.

Stress enhancements due to cavities [133], contaminants, and protrusions, which cause charge injection ultimately leading to the formation of electrical trees, have been proposed as a possible cause of electrical aging [96, 97]. Injection of charge carriers into polyethylene gives rise to changes in the electron density as well as in the positions of the atoms on the molecular sites occupied by the moving or trapped charges as well as on the molecules in their proximity [118]; consequently, the injected charges have both intramolecular and intermolecular influences within the polymer structure [134, 135]. The injected charge carriers undergo hopping from site to site along the individual molecular chains and if the latter were periodic in nature—i.e., there would be a constant spacing between the trapping or hopping sites—it would be possible to determine the probability of finding charge carriers at any particular site. However, the disorder even along the relatively well defined polyethylene molecular chains precludes this sort of calculation [134, 136]. Even though the charge trapping site distribution over the entire electrical tree cannot be predicted, the base of the tree stem is always formed at the charge injection point on the asperity, contaminant, or cavity producing the stress enhancement. Since the potential barriers at the asperity-insulation interface are high, thermally controlled injection of electrons or holes is low at moderate fields ($\leq 10^6$ V/cm). But as the fields at the asperities may readily attain values of 10^7 V/cm or greater, electron injection via tunneling through potential barriers can take place, while hole injection will occur by the mechanism of electron tunneling from the valence band. The large charge carrier density augments the conductivity at the injection point; the electrical field within the injected space charge region is small but increases in the direction of the opposite electrode (semiconducting shield) with falling space charge density. It is within the high space charge density region that electrical tree channel propagation is initiated. The electrical stress at which charge injection begins is considered to be the threshold stress for aging. Although charge injection appears to be a sine qua non condition for electrical tree initiation, the exact mechanism by which an electrical tree commences to propagate from a semiconducting surface is not known [132]. Auckland and Arbad [28] believe tree initiation at asperities to be caused by mechanical fatigue failure resulting from the vibrating electrostatic forces produced by the alternating voltage. A rather plausible explanation, substantiated with some experimental evidence, is that of photodegradation caused by ultraviolet radiation produced by electron-hole recombination or electron collisions with electroluminescence centers in the material's lattice [137]. Energetic or hot electrons, which denote electrons that acquire additional energy from the external field and thus are no longer in thermal equilibrium with the polymer matrix, have also been implicated in the electrical tree growth mechanism [98].

It has been established that certain gases, notably oxygen, when dissolved in a polymer stabilized with antioxidants may significantly reduce the electrical tree inception voltage [138, 139]. However, the role of oxygen on long-term electrical tree initiation for alternating voltages and during transient overvoltages must still be ascertained with respect to the effects of different types of additives, such as antioxidants and ultraviolet stabilizers.

9.9 ACCELERATED ELECTRICAL AGING TESTS

A multistress accelerated aging test must take into account the mechanical strain in the cable that may originate from manufacturing processes, bending and flexing of the cable during installation, and thermal expansion and contraction arising from load variations while in service [90]. In a solid-dielectric extruded polymeric cable the inner and outer layers of the insulation cool first as the cable exits the extruder, thus leaving the central portion of the insulation in a state of strain. As a result partially crystalline polymeric insulations exhibit anisotropic behavior, whereby expansion above the melting point of the crystallites in the longitudinal, radial, and transverse directions becomes unequal [119]. This manifests itself in a shrinkback at the cable ends in the longitudinal direction. If the insulation is free to retract, the length of the shrinkback may attain several millimeters; if this is not the case, then separation of the polymeric cable insulation may occur at joints or terminations. While in service, the cable is subjected to a thermomechanically induced strain that arises from a difference in the thermal coefficient of expansion between the insulation and the conductor and metallic sheath. For example, the thermal expansion coefficient of XLPE is approximately 20 times that of the conductor. The expansion of the insulation exerts pressure upon the insulation shield, metallic neutral, and sheath whose magnitude may exceed the rupture strength of the copper neutral tapes and cause penetration of the concentric neutral wires into insulation and result in severe kinking of distribution cables [119]. With the cable subjected to either bends or long vertical drops, a strain develops due to the lateral forces that a conductor exerts upon the insulation. Since the compression modulus of the polymer insulation decreases with temperature, the conductor will penetrate into the insulation should the lateral force exceed the modulus. Table 9.3 illustrates the lateral force effects at elevated temperature obtained with a 275-kV cable [140].

Accelerated electrical aging test data on polymeric high-voltage transmission cables is analyzed using the half power law defined by Eq. (9.105a). However, since

TABLE 9.3 Conductor Displacement in XLPE 275-kV Cable Resulting from Lateral Stresses

Temperature (°C)	Lateral Force (bars)	Conductor Displacement (mm)	Displacement as Percentage of Insulation Thickness (%)
105	8	4.2	17
120	8	6.6	26
105	27	14.2	56

Source: From [140].

the desired or expected service life or design life time t_0 must appear in explicit form, it is common to express Eq. (9.105a) in normalized form [89, 141]:

$$\frac{t_t}{t_0} = \left(\frac{V_0}{V_t}\right)^n \tag{9.111}$$

where t_t is the testing time or the time to failure, V_0 the service or operating voltage of the cable, and V_t the test or the accelerated aging voltage. Hence a plot of the accelerating test voltage, V_t, as a function of the time to failure t_t, on a log-log graph, yields a curve from which the service life or remaining life t_0 may be estimated by extrapolation. The slope of the resulting curve, which often approximates a straight line, is equal to $-1/n$. The value of the exponent n for solid-dielectric extruded power transmission cables using LDPE and XLPE insulation ranges from ~8 to 20 [90]. Multistress accelerated electrical aging of power transmission cables requires thermal or load current cycling, so that the values of n thus determined using Eq. (9.111) represent an average value over a range of temperatures (e.g. from ambient temperature to 90°C, also frequently including a number of thermal cycles at 105°C corresponding to emergency operation). The accelerated aging or test voltage V_t used is usually between 1.5 and 2.0 times the rated operating voltage V_0; a frequently used value of V_t is that of ~$1.7V_0$ or $\sqrt{3}V_0$. The overall testing periods may extend up to a year or longer. In any accelerated electrical aging test, the longer the test duration and the lower the accelerating voltage, the greater will be the accuracy of the predicted life in service under normal operating conditions. The aim should be to avert introducing aging mechanisms into the accelerated aging test that do not occur in service and obliterating any aging mechanisms that do occur in service. In an attempt to circumvent the very high test voltages required for the accelerated aging tests on transmission cables, some accelerated aging tests are performed on cables with reduced wall thickness [124, 142]. However, there is a tendency to employ full-sized cables in the accelerated tests wherever it is economically feasible to do so. It is interesting to note that frequently previous aging test data is employed to devise an accelerated electrical aging protocol, i.e., if the exponent n values are known approximately for certain cable designs, it is possible to determine the most appropriate accelerated aging test voltage and duration to yield the desired life time t_o (e.g. 30 years at the operating voltage V_o). For example, for $n = 9$, a test period of 80 days at $\sqrt{3}V_0$, results in a life time of 30 years at the rated voltage V_0; whereas the required test period is reduced to 15 days for $n = 15$ in terms of Eq. (9.111) [90].

In the area of polymeric power transmission cables, there are no standardized accelerated electrical aging procedures, with the consequence that cable manufacturers tend to devise their own particular aging tests. Figure 9.55 depicts an accelerated electrical aging assembly, arranged for test on two 500-kV phase-to-phase rated XLPE insulated power transmission cables, having an overall length of 500 m each [142]. One cable was manufactured with an aluminum sheath and the other cable specimen has a stainless steel sheath. Both cable lengths contained four premolded joints and six SF_6 end terminations and a snaking layout was used within the tunnels. The cables were subjected to an alternating voltage of 450 kV phase-to-ground for a test period of 246 days (6000 h), which was equivalent to an in-service life time of 30 years using an exponent n value of 15. Thermomechanical stresses were induced by means of load

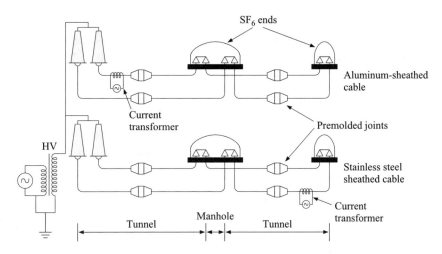

Figure 9.55 Layout of two 500-kV XLPE power transmission cable systems undergoing accelerated electrical aging tests in the field (after [142]).

current cycling, with rated current being circulated in the 2500 mm^2 conductor for a period of 8 h followed by a zero load current period for 16 h. The conductor was allowed to attain a temperature of 90°C for 205 cycles and 105°C for 41 cycles. There is considerable variation in the length of the testing period employed by different manufacturers and testing laboratories (c.f. Table 9.4) [90]. Much of this difference arises from the different values of the exponent n in the inverse power law selected in designing the accelerated aging test protocol.

In contradistinction to the accelerated electrical aging tests on polymeric power transmission cables, the tests on polymeric power distribution cables are always carried out in a wet environment. Thus aging phenomena in distribution cables are inextricably associated with the inception and propagation mechanisms of water trees and the attending failure events. Degradation due to water trees essentially progresses in four stages: permeation of moisture through the insulation, the subsequent initiation of water trees in the presence of moisture ingress, their propagation and ultimately

TABLE 9.4 Summary of Some Typical Accelerated Electrical Aging Test Conditions for Polymeric Power Transmission Cables

Test Voltage (kV)	Stress (kV/mm)	Testing Time	Temperature (°C)	Insulation Type	Phase-to-Phase Cable Rating (kV)	Operating Stress (kV/mm)	Reference
	24.0	1900 h	50	XLPE	420	13.0	142
374	25.0	1 yr	95	XLPE	420	15.0	124
550	27.0	0.5 yr	90/105	XLPE	500	15.5	112
225	17.5	6000 h	75	LDPE	225	10.0	143
440	26.0	6000 h	80/85	LDPE	400	13.5	144
176	10.6	0.5 yr		XLPE	154	5.3	145

Source: From [90].

electrical tree inception and its rapid propagation and electrical breakdown of the water-treed insulation. In North America and some other countries, distribution cables are installed without hermetic sheaths and are thus directly exposed to moist environments such as when they are installed or buried directly in moist soil or placed within water containing pipes or conduits. To simulate the water environment during the accelerated electrical aging tests, the distribution cables are either inserted in pipes containing water [43, 44, 46] or directly immersed in water tanks [147–149].

References [43] and [44] describe the AEIC accelerated water treeing test for power distribution cables. The AEIC accelerated electrical aging test procedure entails the testing of four cable specimens for a period of 120 days; following this period the degree of deterioration or aging in two of the aged specimens is evaluated by subjecting them to an impulse test; the degradation of the remaining two specimens is assessed in terms of an ac breakdown test. Following these tests an additional two specimens are aged: one specimen is aged for 6 months and the other specimen for 12 months; the degree of aging in both of these distribution cable specimens is assessed in terms of an ac break-down test. Originally, the AEIC aging test procedure gave the option of performing the accelerated electrical aging using either a 60-Hz or 1-kHz power source. However, the present specifications dispense with the 1-kHz test frequency and require the aging tests to be carried out at the power frequency of 60 Hz. It is apparent that from the statistical point of view, the AEIC accelerated electrical aging test procedure does not involve a sufficient number of test specimens to provide a representative statistical sample to permit the determination of confidence limits concerning the test results.

Since temperature constitutes an important parameter in any accelerated electrical aging test, the temperature at the conductor must be known; irrespective of whether the aging tests are carried out in conduits or baths, the maximum temperature usually occurs in that part of the cable immediately external to the water interface. The temperature of the conductor within the section of the cable, which is immersed in the water medium is dependent on the size of the conduit or pipe. It is for this very reason that the AEIC test, which is described in References [43] and [44], explicitly specifies the conduit size to ensure that aging tests at different laboratories are performed under similar temperature conditions to provide a basis for comparison of the experimental results obtained. The axial variation of temperature along the length of the cable specimen may be reduced by testing shorter lengths in the conduits, using the arrangement depicted in Fig. 9.56. To obtain a representative statistical sample of the test results, at least 10 specimens should be tested simultaneously, with one additional energized specimen serving as a tempera-

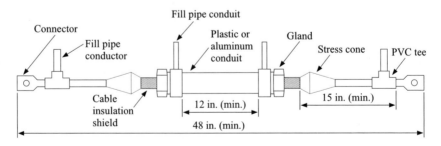

Figure 9.56 Polymeric power distribution cable model specimen in conduit with stress cones, water inlets and connectors for accelerated electrical aging (after [146]).

ture monitor with a thermocouple directly attached to the conductor. Alternatively axial variations in temperature may be adequately dealt with by immersing longer lengths of cable specimens in water-filled tanks. Thus shorter lengths of cable for any diagnostic tests could be obtained by cutting the longer test specimens, thereby ensuring that all shorter lengths have been subjected to the same temperature profiles during the course of their aging. Figure 9.57 portrays one possible arrangement of cable test specimens in a water tank [149]. A total number of 10 specimens are employed with each specimen having an overall length of ≥ 5 m (16 ft) with the length submerged in water being ≥ 3.7m (12 ft). As individual specimens fail with time, they must be replaced by other identical specimens of the same length, if cable heating is necessary to maintain constant water temperature. Thermocouples are employed to monitor the water specimen's surface and conductor temperatures. The latter measurement is carried out on the specimen with no voltage applied to its conductor.

There is now a large volume of accumulated experimental evidence that indicates that both the actual location of the water and its molar content exert a very pronounced effect on the aging rate of polymeric cables in wet environments. Cables, which do not contain strand-blocking materials that obstruct water ingress into the cable along the conductor, should be aged with water in the conductor as well as outside as indicated in Fig. 9.56. It has been demonstrated that the presence of water within the stranded conductor may reduce the time to failure by a factor of 10 [147]. The water admitted into the conductor strands should be deionized to avert corrosive effects of various salts upon the conductor. In contradistinction the water on the outside of the distribution power cable should be of some fixed specified ionic content, which represents more realistically the condition of the water to which the cables may be exposed in their particular environments. It has been established that the ionic impurities either dissolved in the water or originally dispersed in the insulation and semiconducting shield

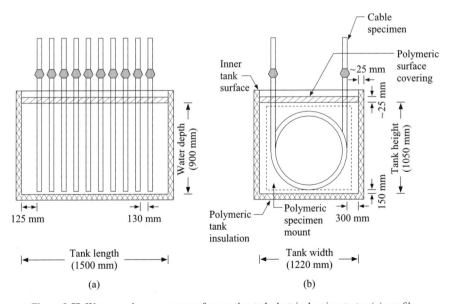

Figure 9.57 Water tank arrangement for accelerated electrical aging tests: (a) profile view; (b) end view. (After an IEEE guide proposal [149]).

augment the water tree initiation and propagation rates and that particularly the latter
is a function of the ion type, concentration, radius, and size [91, 150–152]. It is apparent
that the electrolytic or ionic impurities and their concentration in the groundwater vary
appreciably with the seasons of the year and the geographical locality [153]; moreover,
since synergistic effects may arise due to possible interactions between different ions
affecting water tree growth, a standard solution must be employed. The most simple
and expedient approach is to make use of a weak salt solution—either NaCl or
$CuSO_4$—having a standard concentration between 0.05 and 0.10 M. During the course
of the aging test evaporation losses must be replenished by the addition of deionized
water in order to avert an increase in the ion concentration with time.

   For accelerated electrical aging to occur over reasonably short periods of time, e.g.,
one year, it has been common practice to subject the distribution cables to applied
voltages as high as four times their rated operation voltage. However, if the source
frequency is increased above 50 or 60 Hz, then the accelerating voltage level may be
reduced as a result of the compensation effect of an increased aging rate at the elevated
test frequency. If the degradation rate is caused by partial-discharges, then it is palpably
evident that an increase in the test frequency leads directly to an augmented degradation rate because the number of discharge pulses per unit time is proportional to the
number of cycles per unit time (frequency), as long as the duration of each cycle is very
long to the electron transit times across the cavities undergoing discharge. Although the
relationship between tree growth and the source frequency is appreciably more complex, there is much experimental evidence to indicate that both electrical [154] and water
tree [155, 156] propagation rates increase with frequency.

   Accelerated frequency, as accelerated electrical aging at elevated voltages, requires
some precaution. The dissipation factor, tan δ, should be sufficiently low at the test
frequency to ensure against undue dielectric heating. Figure 9.58 shows that for typical

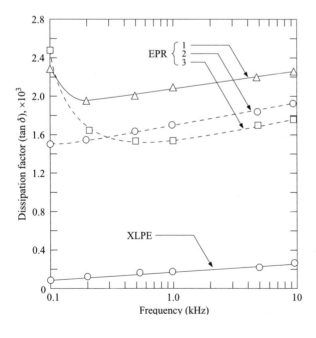

Figure 9.58 Dielectric loss characteristics of
XLPE and three typical EPR formulations
[146].

EPR and XLPE formulations, the $\tan \delta$ values at frequencies up to 10 kHz differ only slightly from those at the power frequency. In accordance with Eq. (9.15), the power loss in a dielectric material is directly proportional to the $\tan \delta$ value and varies with the square of the rms phase to ground applied voltage V. If we introduce the thermal resistivity term $\rho (°C\,m/W)$ into Eq. (9.15), then the resultant temperature rise, ΔT in degrees Celsius due to the dielectric losses may be expressed as [146]

$$\Delta T = \pi f \rho \varepsilon_0 \varepsilon_r' V^2 \tan \delta \tag{9.112}$$

where f is the test frequency, ε_0 the permittivity in vacuo (8.854×10^{-12} F/m) and ε_r' the real value of the relative permittivity or dielectric constant. For example, if in reference to Fig. 9.54 a value of 2×10^{-3} is used for $\tan \delta$ at 10 kHz corresponding to the EPR 1 formulation, then for a 15-kV EPR cable tested at 15 kV phase to ground yields a ΔT value of 1°C for $\varepsilon_r' = 3.0$ and $\rho = 3.0°C\,m/W$. Had the cable been tested at three times the rated phase-to-ground voltage, i.e., 45 kV, then the corresponding temperature rise would have been 9°C; and, finally were the test frequency reduced from 10 to 1 kHz, but with the applied voltage still at 45 kV, then the corresponding value of ΔT would be reduced to 0.9°C. Evidently, a test frequency of 1 kHz and a test voltage of three times the rated value appears to be a reasonable choice.

Figure 9.59 depicts the water tree growth initiation time as a function of frequency obtained on cable models of XLPE, PE, and an EPR formulation at room temperature. The experimental results indicate that in the case of the XLPE and PE specimens, the water tree growth initiation time is not much reduced with frequency when the test frequency exceeds 600 Hz. It is noteworthy that the tree growth initiation time diminishes at a lower rate with frequency for the EPR compound than for PE and XLPE.

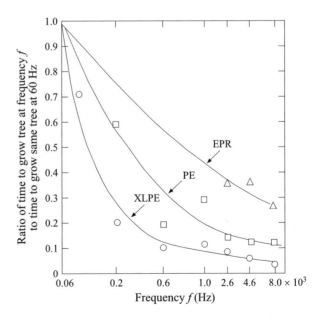

Figure 9.59 Water tree propagation rate as function of frequency for PE, XLPE, and EPR [157].

Little appears to be gained by extending the accelerated aging test frequency beyond 1 kHz. Accumulated experimental data suggest that for frequencies up to 1 kHz, the aging mechanisms do not differ appreciably from those occurring at 60 Hz [146]; consequently, it is common practice to utilize a test frequency of 1 kHz, when carrying out accelerated electrical aging tests at frequencies above 60 Hz. The frequency acceleration factor may be expressed empirically as

$$\text{Acceleration factor} = \left(\frac{f}{f_0}\right)^k \tag{9.113}$$

where f is the test frequency, f_0 is the power frequency (50 or 60 Hz), and the exponent k is a constant that ranges from 0.45 to 0.70 [155, 156], with its value depending on the insulating material, the mechanism of failure, voltage gradient, temperature, and mechanical strain.

Either power amplifiers or motor-generator sets may be employed as high-frequency test sources for accelerated electrical aging tests. In the power amplifier arrangement, depicted in Fig. 9.60, a variable audio oscillator provides the desired test frequency to an audio frequency (af) power amplifier. The latter, which has its own dc power supply, amplifies the ac signal from the oscillator and a high-frequency transformer in tandem with the regulator of the af power amplifier provides the required high voltage. A series resonance circuit, having a Q value of $\geqslant 15$, is adjusted by means of a variable inductance coil to resonate with the cable capacitance at the desired test frequency. A second transformer enables the variable inductance to operate at a low voltage. The power amplifier units are usually designed with a rating of 25 kVA within a frequency range of 5 to 10^4 Hz. The frequency may be controlled to an accuracy within $\pm5\%$. A fault current limiter is normally employed to limit the breakdown current to permit analysis and examination of the region at and in the vicinity of the fault.

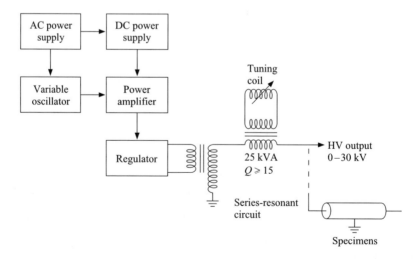

Figure 9.60 Schematic diagram of typical power amplifier test set [146].

Motor-driven alternator sets have the advantage of a higher power output and greater reliability without the undesirable susceptibility of the solid-state power amplifiers to transient pulses. However, since these units operate at constant speed, they have the disadvantage of little frequency control; often the frequency of operation is fixed at 400 Hz, this being the common power frequency on ships, where such generators are commonly utilized. Figure 9.61 shows a schematic circuit diagram of a typical motor-generator set.

It must be conceded that if there occurs a certain amount of damage or deterioration in the insulation per given cycle at the power frequency, then one obvious method to accelerate electrical aging is to employ more elevated test frequencies. Accelerated aging by means of increased frequency represents an effective means of acceleration in which the prime aging parameters of applied voltage and temperature may be maintained realistically nearer to those existing under normal solid polymeric distribution cable operating conditions. The sparse data obtained with the AEIC test procedure using the earlier 1982 higher frequency option should not be interpreted as a failure of the 1-kHz test frequency to accelerate the aging of full sized cables. The latter test was principally unacceptable from the statistical point of view in that only two test specimens were employed and thus were insufficient to statistically distinguish between sets of data having any scatter. It was demonstrably shown by Silver and Lukac [157] that aging of polymeric distribution cables in a wet environment at a stress of 3.4 kV/mm and 1 kHz yield similar results as the aging at an elevated electrical stress of 6.0 kV/mm and a power frequency of 60 Hz. As test specimens, the authors used 15-kV rated distribution cables, having a 1/0 American Wire Gage (AWG) (53.5 mm^2) aluminum, class B compressed strand conductors. All cable specimens had an insulation thickness of 4.45 mm, with extruded conductor and insulation shields of 0.38 and 0.75 mm thickness, respectively. In their accelerated electrical aging test, they adhered to the early AEIC test protocol [43, 44]. At 60 Hz, their corresponding test voltage was three times the rated to ground voltage ($3V_0$), while at 1 kHz it was reduced to approximately $2V_0$. The load cycle consisted of allowing rated current to circulate for 8 h, with the current reduced to zero for the remaining 16 h at 5 days per week. The specimens were preconditioned and exposed to water both from within and outside the cables. The conductor temperature in air for the thermoplastic cable insulations was fixed at 75°C

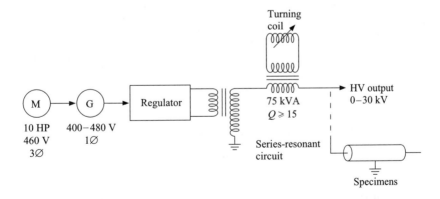

Figure 9.61 Schematic diagram of typical motor-generator test set [146].

and for the thermosetting cable insulations at 90°C. The aging rate of the specimens was assessed in terms of their ac breakdown strength as well as the ac breakdown strength retention at the end of each testing time period compared to that of a virgin cable. Figures 9.62 and 9.63 illustrate the former approach and Figs. 9.64 and 9.65 the latter. As regards the nomenclature in the figures HMWPE refers to high-molecular-weight polyethylene, TRHMWPE to tree-retardant HMWPE, TRXLPE-1 and TRXLPE-2 to two different tree-retardant XLPE formulations, EPR-1 to an ethylene-propylene-rubber compound based on an amorphous polymer with a mineral filler content, and EPR-2 to an EPR based on a crystalline polymer with a low mineral filler content without any oil additive. It can be readily perceived from the experimental results that the trend in the accelerated aging behavior is essentially similar irrespective as to whether aging is assessed in terms of the reduction in the ac breakdown strength or the ac breakdown strength retention as a function of time. As to be anticipated, the water tree retardant compounds are characterized by an intrinsically lower aging rate. The highest aging rate is exhibited by the XLPE insulated distribution cable specimens. Their poor performance vis-à-vis the HMWPE is in part attributed by Silver and Lukac [157] to the extruded semiconducting shields in the currently manufactured HMWPE insulated cables and to the adverse effects of preconditioning on the XLPE insulated cables whereby the crosslinking by-products, which tend to retard the propagation of water treeing, are expelled in a relatively short time. The results obtained by Silver and Lukac [157] clearly demonstrate that the use of a higher test frequency permits a substantial reduction in the test voltage. A test voltage that approaches closer to the voltage of operation in the field allows a more valid simulation of the degradation mechanisms that normally would occur in service. It is interesting to note that very pronounced reductions in aging time with frequency have been also found with miniature XLPE cable models tested under wet conditions, as illustrated in Fig. 9.66 [158].

It is often common practice to precondition solid-dielectric insulated distribution cables prior to subjecting them to an accelerated aging test under wet conditions. This

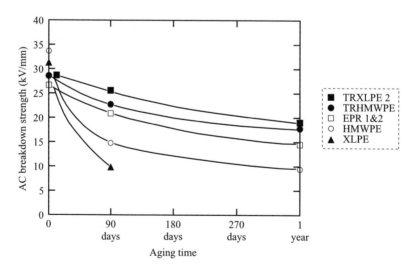

Figure 9.62 AC breakdown strength vs. aging time of different polymeric distribution cables aged at constant stress of 3.4-kV/mm at 1 kHz (after [157]).

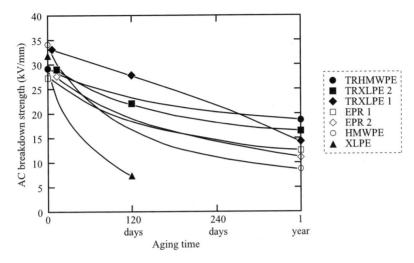

Figure 9.63 AC breakdown strength vs. aging time of different polymeric distribution cables aged at constant stress of 6.0 kV/mm at 60 Hz (after [157]).

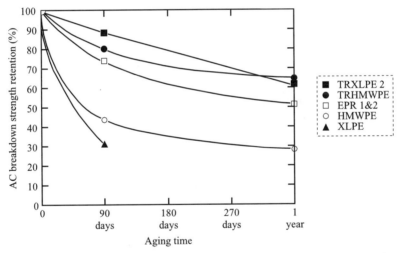

Figure 9.64 AC breakdown strength retention vs. aging time of different polymeric distribution cables aged at constant stress of 3.4 kV/mm at 1 kHz (after [157]).

tendency is in cognizance with the fact that the remnant crosslinking by-products (principally in the of crosslinked polymeric cables) and the residual mechanical strain are known to influence water tree inception and propagation behavior. The effects of mechanical strain upon water tree growth have been studied by Densley and co-workers at the National Research Council, and Fig. 9.67 depicts water tree patterns in heat annealed and nonannealed polyethylene specimens [159]. The mechanical strain in polymeric cables may be readily eliminated through thermal treatment given by means of load current cycling; great care must be exercised to obviate high temperature

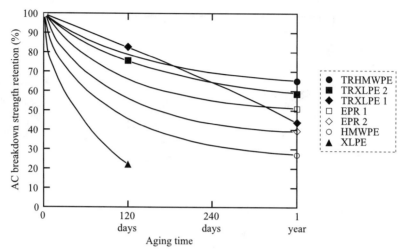

Figure 9.65 AC breakdown strength retention vs. aging time of different polymeric distribution cables aged at constant stress of 6.0 kV/mm at 60 kHz (after [157]).

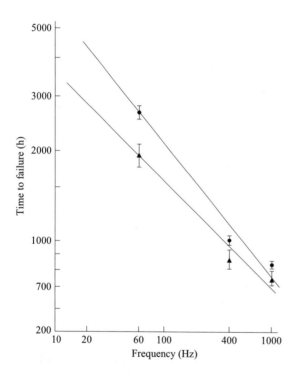

Figure 9.66 Time to failure vs. frequency of XLPE miniature cables for 50% (●) and 5% (▲) probabilities of failure (after [158]).

rises that may result in morphological changes that can affect adversely the long-term dielectric behavior of the polymer. The thermal treatment given to relieve mechanical strain will also remove the crosslinking by-products, whose remnant level in the polymer depends on the amount of cross-linking agent used as well as upon the extrusion and curing conditions.

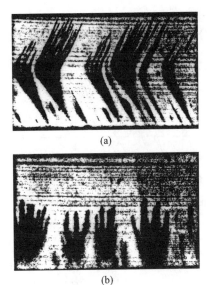

Figure 9.67 Water tree configurations in (a) nonannealed and (b) annealed polyethylene specimens [159].

(a)

(b)

The subjection of polymeric cables to normal load cycling while in service relieves quickly the residual mechanical strains and removes very gradually the remnant crosslinking by-products, as the latter migrate or diffuse out of the insulation. Preconditioning of a crosslinked polymeric cable insulation prior to its accelerated electrical aging test under wet conditions is intended to render the cable into its worst possible condition before the tests actually commence, i.e., to produce a large reduction in the level of the crosslinking products present. Since the latter steps concern primarily XLPE cables, preconditioning is most effectively carried out by conductor heating at 90°C for 30 days [160]. It has been found that already after 600 h the crosslinking by-product levels of acetophenone, methyl styrene, and cumyl alcohol fall, respectively, to < 1, < 1, and < 100 ppm [161]. Whereas by comparison cables following two years of operation in service indicate a remnant level of 40 ppm crosslinking by-products—representing a substantial decrease from an initial value of 3000 ppm; finally after 8–10 years of operation, very little of these by-products remains [162]. With respect to the preconditioning procedure, it is noteworthy that some European practices on water-treeing-related aging tests also call for preconditioning of distribution cables at 80°C for a period of 30 days [163].

The choice of temperature for aging tests on polymeric power distribution cables under wet conditions does not present itself as a trivial decision. Figure 9.68 demonstrates conclusively that temperature cycling reduces the time to failure as compared to that obtained when crosslinked cable models are aged at room temperature. Since many distribution cables experience very little load cycling while in service, a constant test temperature may be more representative. The omission of thermal cycling would reduce the incidence of bow tie water trees, which form as a result of water condensation caused by temperature cycling. The selection of a constant test temperature has the intrinsic advantage of maintaining the thermal stress parameter fixed, thereby leading to more consistent results between accelerated aging tests carried out at different institutions. A constant test temperature of 40°C has been already proposed by CIGRÉ

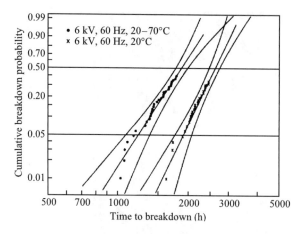

Figure 9.68 Cumulative Weibull breakdown voltage probability vs. time to failure of XLPE cable models at constant test voltage of 6 kV at 60 Hz; ● temperature cycling between 20 and 70°C; ×, at constant temperature of 20°C (after [158]).

Working Group 21–11 [164]; moreover, since it has been demonstrated that the number of vented water trees and their propagation is quasi-independent of temperature up to 50°C [165], a fixed water bath temperature of any value within 20–50°C would appear to be justified [160]. The upper temperature limit of 50°C is a prudent choice in view that unduly high test temperatures may induce failures in XLPE cables due to thermal degradation of the polymer in the presence of water. Very few cables in service would be expected to undergo this form of degradation, since most distribution cables are operated well below their maximum temperature limit. If it is decided to perform the accelerated electrical aging test at constant temperature, then the preconditioning procedure must include a brief load cycling period to remove the residual mechanical strain present in the cable specimen prior to commencing the aging test.

Long-term accelerated electrical aging tests performed on distribution cables under wet conditions fall essentially into two categories, namely fixed-time duration tests and time-to-failure tests. In the former tests the cable specimen is electrically aged for a fixed time period at the end of which the degree of aging in the specimen is assessed in terms of voltage breakdown tests, while the time-to-failure tests are tests in which the specimen is voltage stressed until failure ensues.

In preset or fixed-time accelerated aging tests under wet conditions, the selected electrical stress must not be too high so as to lead to failure mechanisms that may not be representative of those occurring under normal operating stresses in service; also the testing time should be sufficiently long to allow the degradation mechanisms to develop to the same degree as they do in service. To produce measurable and relatively rapid degradation, the cable specimen must be stressed at a voltage approximately three times the rated value [146] for a period ca. one year [166]. Notwithstanding the efforts made to replicate the water tree growth behavior observed in the field, salient differences remain between that observed in the field and the water tree growth behavior obtainable with accelerated electrical aging tests. Under normal aging in service, vented water trees, once initiated, propagate rapidly at the beginning and then at a lower rate; however, the growth of bow tie trees is eventually stemmed, even though some bow tie water trees, originating from large water-soluble contaminants, attain appreciable lengths [159, 160]. While in an accelerated aging test at elevated temperature and electrical stress, many more bow tie water trees are observed, and their inordinately large size and number are greater than those of the vented water trees. Yet when the duration of accelerated electrical aging test is extended and the temperature decreased, the water tree patterns become more similar to those characterizing field-aged cables.

Since a preset time accelerated electrical aging test subjects a specimen for a restricted time to an electrical stress, a diagnostic test procedure must be devised to determine whether any degradation has occurred and to obtain an estimate of the life remaining. For this purpose a simple 60-Hz ramp or step-voltage test is utilized because the ac breakdown strength constitutes a most effective indicator of the degree of deterioration in the dielectric insulating system of the cable [88]. Polymeric distribution cables, which are operated in service or undergo accelerated aging at higher electrical field gradients, exhibit subsequently a lower ac breakdown strength (cf. Fig. 9.69). The alternating voltage may be raised linearly or in equal steps of a few kilovolts every 5–30 min until dielectric breakdown ensues. The AEIC specifications [43, 44] recommend time steps of 5 min duration with voltage increments of 1.6 kV/mm (40 V/mil). The degree of aging or deterioration in the cable dielectric is assessed either in terms of amount by which the ac breakdown strength falls below that of a virgin cable specimen or the percentage retention of the nominal dielectric breakdown strength of the virgin cable. A further constraint imposed by the AEIC specifications [43, 44] is that of an *ad arbitrium* minimum ac dielectric strength requirement of 10 kV/mm; aged cables whose ac breakdown strength is below this value are considered to have failed the accelerated aging test.

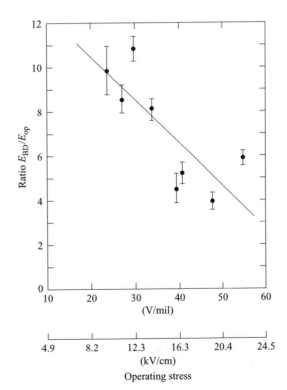

Figure 9.69 Ratio of mean ac breakdown stress E_{BD} to operating stress E_{op} vs. cable operating stress in service (after [88]).

There is some evidence of cable failures in the field following lightning activity; for this reason it is argued that an impulse superimposed on the alternating voltage waveform may constitute a more effective diagnostic test to assess the degree of aging [160, 167]. Hartlein et al. [167] found that positive 70-kV impulses applied at the negative peak of the sinusoidal alternating voltage wave are most effective in precipitating dielectric failure of the cable. An effective diagnostic procedure in the case of the time to breakdown test would thus be to apply 10 such positive impulses at the negative peak of the sinusoidal voltage wave once every 30 days of the accelerated aging test [160]. It should be emphasized that the ac applied voltage should be reduced to its rate value when applying the impulses, so as to simulate the conditions in the field.

Certain analytical tests constitute an important addendum to the impulse and ac voltage tests. By limiting the value of the breakdown current, it is possible to obtain some pertinent details on the nature of the failure mechanisms involved. Cross sections of the cable in the vicinity of the breakdown site can be removed and analyzed for defects relating to water trees, halos [88], and cavities in particular reference to their geometrical configuration, size, density, and distribution within the dielectric bulk or volume. Chemical and physical analysis as concerns impurity content, volume density, water content, and degree of oxidation of the polymer should also be performed.

From the foregoing considerations on the aging of power distribution cables under wet conditions, one can define a meaningful test protocol for two types of accelerated electrical aging tests, namely the preset or fixed-time test procedure and the time-to-breakdown test procedure [160]. For reasonable statistical certainty, the statistical sample should consist of 10 specimen cables, each 10 m long with a 5-m effective test section and 2.5 m for HV termination at each end. The voltage regulation of the electrical power source should be at least 1%. Since the specimens are to be highly overstressed electrically, some partial-discharge may be present. Thus partial-discharge measurements should be made on each cable specimen prior to the commencement and following the completion of the aging test (in the case of the fixed-time test procedure). All cables failing the partial-discharge test at the outset of the aging test should be removed and replaced with discharge-free specimens. It would be interesting to monitor any discharge activity that may arise in the course of the aging test, but this may not be possible due to the presence of discharges form the various interconnections comprising the experimental arrangement. Table 9.5 specifies the pre-set of fixed-time test protocol, using water tanks as the immersion medium for the power distribution cable specimens. The experimental data should be analyzed using Weibull statistics to ensure that the differences in experimental values are significant [89]. The determined mean values should be complemented by calculation of the 95% confidence intervals.

To reduce the time required for breakdown in the time-to-breakdown accelerated electrical aging tests, there is a tendency to use unduly high electrical stresses. At very high stresses the extrapolated inverse power law characteristics predict erroneously short times to failure at the rated operating stresses: this error is caused by aging mechanisms active at the high stresses that differ significantly from those aging mechanisms that prevail under the normal electrical stresses characterizing the cable operating conditions in the field. For this reason it is important not to exceed test voltage levels by more than four times the rated value. Moreover, if an acceleration frequency of 1 kHz is employed, then the test voltage level may be reduced to three times the rated voltage and still yield acceptably short times to failure. The three to four times the rated voltage levels correspond to voltage gradients from approximately 6–8 kV/mm for the usual

TABLE 9.5 Preset (Fixed-Time) Accelerated Electrical Aging Test Procedure for Polymeric Power Distribution Cables Immersed in Water Tanks

Test voltage	$2.5V_0$ or $3.0V_0$
Frequency	60 Hz (for $3.0V_0$) or 1 kHz (for $2.5V_0$)
Water location	Inside conductor and outside insulation shield
Water impurity content in tank	$0.05M$ NaCl concentration
Water in conductor	Deionized water
Fixed water tank temperature	20–50°C
Mechanical stress	Minimum bending radius
Partial-discharge monitoring before/after aging (optional during aging)	Detection of all discharges ≥ 5 pC
Aging time	1 year, followed by diagnostic tests
Cable specimens	10 specimens, with an overall length of 10 m and effective testing length of 5 m
Diagnostic test	AC breakdown (limited breakdown current) or impulse test as per AEIC [43, 44]; physical and chemical analysis of test specimens

Note: Prior to the commencement of the preset time accelerated electrical aging test, the cable specimens are to be preconditioned at rated temperature for 30 days.
Source: From [160].

range of insulation thickness employed for distribution cables. Table 9.6 delineates the time-to-failure test protocol, using water tanks as the immersion medium for the polymeric power distribution cable specimens. Over the test duration of 2–3 years indicated in the time-to-breakdown test procedure, it is more than likely that some of the specimens within the statistical sample of 10 may not fail within the prescribed time limit. In such circumstances some truncation of the test data with additional statistical analysis becomes necessary [168]. Evidently if this situation transpires, the time-to-breakdown

TABLE 9.6 Time-to-Breakdown Accelerated Electrical Aging Test Procedure for Polymeric Power Distribution Cables Immersed in Water Tanks

Test voltage	$3V_0$ or $4.0V_0$ with superimposed impulses (10 of 70 kV per month)
Frequency	60 Hz (for $4.0V_0$) or 1 kHz (for $3V_0$)
Water location	Inside conductor and outside insulation shield
Water impurity content in tank	$0.05M$ NaCl concentration
Water in conductor	Deionized water
Fixed water tank temperature	20–50°C
Mechanical stress	Minimum bending radius
Partial-discharge monitoring before aging (optional during aging)	Detection of all discharges ≥ 5 pC
Aging time	2–3 years
Cable specimens	10 specimens, with an overall length of 10 m and an effective testing length of 5 m
Diagnostic test	Failed specimens are to be examined by means of various analytical techniques for trees, halos, water content, oxidation, and impurity content

Note: Prior to the commencement of the time-to-breakdown accelerated electrical aging test, the cable specimens are to be preconditioned at rated temperature for 30 days.
Source: From [160].

test becomes a combination of the preset time and time-to-breakdown tests. To extricate oneself from this dilemma, it may be expedient to terminate the time-to-breakdown test after 80% of the specimens have failed and then proceed with the usual statistical analysis method. In the long-term time-to-breakdown aging test not only should great care be exercised to replenish the volume of the water lost through evaporation with attending adjustments of the molarity of the solution but duplicate cables must be inserted in the water tank with each removal of a failed specimen. This is a sine qua non condition for maintaining constant thermal conditions within the water bath when temperature control is achieved via conductor heating.

It is apparent from the description of the two accelerated electrical aging tests on polymeric power distribution cables under wet conditions that from the practical point of view there are certain advantages and disadvantages inherent with each test procedure. The time-to-breakdown test provides data on the actual time-to-failure values but requires much longer periods of time to complete—with the attending long time requirements of continuous use of high-voltage apparatus. In the latter respect, the preset time test has the advantage of providing aging data more quickly, but then such aging data is incomplete since in a strict sense the time-to-failure data cannot be extracted from the experimental results. However, both types of test serve certain needs and will continue to be utilized, with their use being contingent upon the time available for performing the test, the urgency for the aging data, and the equipment available for performing the tests. It is thus important that the test procedures in each of the two aging methods be carefully specified and followed to permit comparison of accelerated electrical aging data between different laboratories.

REFERENCES

[1] R. Bartnikas, "Dielectrics and Insulators," in *The Electrical Engineering Handbook*, R.C. Dorf, Ed., CRC Press, Boca Raton, FL, 1993, Chapter 52.

[2] H. Schering, *Die Isolierstoffe der Electrotechnik*, Julius Springer, Berlin, 1924.

[3] M. Wien, "Messung der Inductionsconstanten," *Annalen der Physik Chemie*, Vol. XLIV, 1891, pp. 688–712.

[4] A. von Hippel, *Dielectrics and Waves*, Wiley, New York, 1954.

[5] Bruckmann, *Revue générale de l'électricité*, Vol. 17, 1925, p. 881.

[6] F. K. Harris, *Electrical Measurements*, Wiley, New York, 1952.

[7] E. F. Hasler, "An amplifier-detector for Schering bridge measurements at power frequency," *Electrical Energy*, Vol. 2, 1958, pp. 372–377.

[8] R. Bartnikas, "High Voltage Measurements," in *Engineering Dielectrics, Vol. IIB, Measurement Techniques*, R. Bartnikas, Ed., STP926, ASTM, Philadelphia, 1987, Chapter 3.

[9] H. Schwab, *Hochspannungsmeßtechnik*, Springer, Berlin, 1969.

[10] Tettex Aktiengesellschaft, "Hilfszweige der Typenserie 2000 für Wechselstromme Brükke," Technical Report BG 2900'D-5-8.66, Zürich, 1966.

[11] N. L. Kusters and O. Petersons, "A transformer-ratio-arm bridge for high voltage capacitance measurements," *IEEE Trans. Commun. Electronics*, Vol. 82, 1963, pp. 606–611.

[12] Issue 2: Operating Instructions for Model 9910A HV Capacitance Bridge, Guildline Instruments Ltd., Smith Falls, Ontario, Oct. 1975.

[13] P. Seitz and P. Osvath, "Microcomputer-controlled transformer ratio arm bridge," Third International Symposium on High Voltage Engineering, Milan, 1979.

[14] P. Osvath and S. Widner, "A high voltage-precision self balancing capacitance and dissipation factor measuring bridge," *IEEE Trans. Instrumentation Measurement*, Vol. IM-35, 1986, pp. 19–23.

[15] R. Bartnikas, "Pulsed corona loss measurements in artificial voids and cables," CIGRÉ Proc., Part II, Report 202, Paris, 1966, pp. 1–37.

[16] R. Bartnikas and E. J. McMahon, Eds., *Engineering Dielectrics, Vol. I, Corona Measurement and Interpretation*, STP 669, ASTM, Philadelphia, 1979.

[17] R. Bartnikas, "Dielectic losses in solid-liquid insulating systems—Part I," *IEEE Trans. Electrical Insulation*, Vol. EI-5, 1970, pp. 113–121.

[18] P. Böning, *Elektrische Isolierstoffe: Ihr verhalten auf Grund der Ionenadsorption an inneren Grenzflachen*, Vieweg, Braunschweig, 1938.

[19] C. G. Garton, "Dielectric loss in thin films of insulating liquids," *J. IEE*, Vol. 88, Part III, 1941, pp. 23–40.

[20] S. I. Reynolds, "On the behavior of natural and artificial voids in insulation under internal discharge," *IEEE Trans. Power Apparatus Syst.*, Vol. 77, 1959, pp. 1604–1608.

[21] T. W. Dakin and D. Berg, in "Theory of gas breakdown," *Progress in Dielectrics*, Vol. 4, J. B. Birks and J. Hart, Eds., Haywood, London, 1962, pp. 151–198.

[22] H. C. Hall and R. M. Russek, "Discharge inception and extinction in dielectric voids," *Proc. IEE*, Vol. 101, 1954, pp. 47–55.

[23] J. H. Mason, "Discharge detection and measurement," *Proc. IEE*, Vol. 112, 1965, pp. 1407–1423.

[24] M. J. Schönhuber, "Breakdown of gases below Paschen minimum: Basic design data of high voltage equipment," *IEEE Trans. Power Apparatus Syst.*, Vol. PAS-88, 1969, pp. 100–107.

[25] M. Garcia-Gamez, R. Bartnikas, and M. Wertheimer, "Synthesis reactions involving XLPE subjected to partial discharges," *IEEE Trans. Electrical Insulation*, Vol. 22, 1987, pp. 199–204.

[26] J. Novak and R. Bartnikas, "Ionization and excitation behavior in a microcavity," *IEEE Trans. Dielectrics Electrical Insulation*, Vol. 2, 1995, pp. 724–728.

[27] A. Gemant and W. von Philipoff, *Zeitschrift technische Physik*, Vol. 13, 1932, pp. 425–430.

[28] S. Whitehead, *Dielectric Breakdown of Solids*, Oxford University Press, Oxford, 1953, pp. 163–233.

[29] A. E. W. Austen and W. Hackett, "Internal discharges in dielectrics—their observation and analysis," *J. IEE*, Vol. 91, Part I, 1944, pp. 298–322.

[30] G. C. Crichton, P. W. Karlsson, and A. Pedersen, "Partial discharges in ellipsoidal and spheroidal voids," *IEEE Trans. Electrical Insulation*, Vol. 24, 1989, pp. 335–342.

[31] A. Pedersen, G. C. Crichton, and I. W. McAllister, "The functional relation between partial-discharges and induced charge," *IEEE Trans. Dielectrics Electrical Insulation*, Vol. 2, 1995, pp. 535–543.

[32] R. Bartnikas, "Discharge rate and energy loss in helium at low frequencies," *Archiv Electrotechnik*, Vol. 52, 1969, pp. 348–359.

[33] R. Bartnikas, "Some observations on the character of corona discharges in short gap spaces," *IEEE Trans. Electrical Insulation*, Vol. EI-6, 1971, pp. 63–75.

[34] G. Mole, "Basic characteristics of internal discharges in cables," IEEE Underground Distribution Conference, Conf. Record Supplement 69CNPWR, Anaheim, May 12–16, 1969, pp. 198–207.

[35] E. W. Kimbark, *Electrical Transmission of Power and Signals*, Wiley, New York, 1955.

[36] "Cross-linked-thermosetting polyethylene-insulated wire and cable for the transmission and distribution of electrical energy," IPCEA Pub. No. 5066-524 or NEMA Pub. No. WC 7-1971.

[37] R. Bartnikas, "Effect of pulse rise time on the response of corona detectors," *IEEE Trans. Electrical Insulation*, Vol. EI-7, 1972, pp. 3–8.

[38] G. S. Eager and G. Bahder, "Discharge detection in extruded polyethylene insulated power cables," *IEEE Trans. Power Apparatus Syst.*, Vol. PAS-86, 1967, pp. 10–34.

[39] D. A. Costello and R. Bartnikas, "Partial discharges in primary distribution cables," IEEE Underground Distribution Conference, Conf. Record Supplement, 69 CNPWR Anaheim, May 12–16, 1969, pp. 179–195.

[40] M. Pompili, C. Mazzetti, and R. Bartnikas, "Simultaneous measurements of PD in oil using conventional narrow band and ultra-wide band detection systems," Proc. XIIth International Conf. on Conduction and Breakdown in Dielectric Liquids, IEEE Conf. Record 96CH35981, Roma, July 15–19, 1996, pp. 185–188.

[41] N. B. Timpe, in *Engineering Dielectrics, Vol. I, Corona Measurement and Interpretation*, R. Bartnikas and E. J. McMahon, Eds., STP 669, ASTM, Philadelphia, 1979.

[42] J. C. Chan, P. Duffy, L. J. Hiivala, and J. Wasik, "Partial discharge—Part VIII: PD testing of solid dielectric cable," *IEEE Electrical Insulation Mag.*, Vol. 7, 1991, pp. 9–20.

[43] AEIC CS5-95 Specifications for Crosslinked Polyethylene Insulated Power Cables Rated 5 through 46 kV, AEIC, New York, 1994.

[44] AEIC CS6-95 Specifications for Ethylene Propylene Rubber Insulated Power Cables Rated 5 through 69 kV, AEIC, New York, 1996.

[45] AEIC CS7-95 Specifications for Crosslinked Polyethylene Insulated Power Cables Rated 69 through 138 kV, AEIC, New York, 1993.

[46] *Guide for Partial Discharge Test Procedure*, ICEA Publication T-24-380, 1980.

[47] G. S. Eager, G. Bahder, and D. A. Silver, "Corona detection experience in commercial production of power cables with extruded insulation," *IEEE Trans. Power Apparatus Syst.*, Vol. PAS-88, 1969, pp. 342–364.

[48] M. Garcia-Gamez, R. Bartnikas, and M. R. Wertheimer, "Modification of XLPE exposed to discharges at elevated temperature," *IEEE Trans. Electrical Insulation*, Vol. 25, 1990, pp. 153–155.

[49] N. B. Timpe and S. V. Heyer, "Laboratory and field partial discharge studies by a utility," *IEEE Trans. Electrical Insulation*, Vol. EI-12, 1977, pp. 159–164.

[50] S. Boggs and G. Stone, "Fundamental limitations in the measurement of corona and partial discharges," *IEEE Trans. Electrical Insulation*, Vol. EI-17, 1982, pp. 143–150.

[51] F. H. Gooding and H. B. Slade, "Corona level scanning of high voltage power cables," *AIEE Trans. on Power Apparatus and Systems*, Vol. 16, Dec. 1957, pp. 999–1009.

[52] A. H. Baghurst, "A new method for the location of partial-discharge sites using modulated X-rays," 1985 Annual Report, Conf. on Electrical Insulation and Dielectric Phenomena, IEEE Conf. Record No. 85 CH2165, Buffalo, Oct. 20–24, 1985, pp. 471–476.

[53] D. Eigen, "Cable scanning method for short bursts of highly penetrating radiation," U.S. Pat. 3,466,537, 1966.

[54] W. L. Weeks and J. P. Steiner, "Instrumentation for the detection and location of incipient faults in power cables," *IEEE Trans. Power Apparatus Syst.*, Vol. PAS-101, 1982, pp. 2328–2335.

[55] W. L. Weeks and J. P. Steiner, "Improvement for the instrumentation for partial discharge location," 1984 IEEE Transmission and Distribution Conference, Kansas City, April 29–May 4, 1984.

[56] L. Galand, "Localisation des décharges partielles dans les câbles par une méthode d'ondes stationnaires," *Rev. général d'électricité*, Vol. 80, 1971, pp. 399–405.

[57] M. Beyer, W. Kann, H. Borsi, and K. Feser, "A new method for detection and location of distributed partial discharges (cable faults) in high voltage cables under external interference," *IEEE Trans. Power Apparatus Syst.*, Vol. 101, 1982, pp. 3431–3438.

[58] M. Beyer and H. Borsi, "Teilentladungsmessung on Hochspannungskabeln—Ursachen für Messfehler und Möglichkeiten zu ihrer Vermeidung," *Elektrizitäts Wirtschaft*, Vol. 76, 1977, pp. 931–936.

[59] C. H. Knapp, R. Bansal, M. S. Mashikian, and R. B. Northrop, "Signal processing techniques for partial discharge site location in shielded cables," *IEEE Trans. Power Delivery*, Vol. 5, 1990, pp. 859–865.

[60] M. S. Mashikian, F. Palmieri, R. Bansal, and R. B. Northrop, "Location of partial discharges in shielded cables in the presence of high noise," *IEEE Trans. Electrical Insulation*, Vol. 27, 1992, pp. 37–43.

[61] M. S. Mashikian, R. Luther, J. C. McIver, J. Jurcisin, and P. W. Spencer, "Evaluation of field aged crosslinked polyethylene cables by partial-discharge location," *IEEE Trans. Power Delivery*, Vol. 9, 1994, pp. 620–628.

[62] J. R. Steiner, P. H. Reynolds, and W. L. Weeks, "Estimating the location of partial discharges in cables," *IEEE Trans. Electrical Insulation*, Vol. 27, 1992, pp. 44–59.

[63] G. Bahder, G. S. Eager, R. Suarez, S. M. Chalmers, W. H. Jones, and W. H. Mangrum, "In service evaluation of polyethylene and crosslinked polyethylene insulated cables rated 15 to 35 kV," *IEEE Trans. Power Apparatus Syst.*, Vol. PAS-96, 1977, pp. 1754–1766.

[64] E. F. Steennis, "Diagnostic testing of distribution cable systems," Appendix 12-A-3, Minutes of the Insulated Conductors Committee of the IEEE Power Engineering Society, April 14–17, 1996, Houston.

[65] L. E. Lundgaerd, G. Tangen, B. Skyberg, and K. Faugstad, "Acoustic diagnosis of gas insulated substations: a theoretical and experimental basis," *IEEE Trans. Power Delivery*, Vol. 5, 1990, pp. 1751–1758.

[66] R. Morin, A. Gonzales, P. Christophe, and B. Poirier, "Partial discharge detector for in service cables and joints under normal network operating conditions," Proc. IEEE Transmission and Distribution Conference, Dallas, Sept. 23–28, 1991, p. 23.

[67] R. Bartnikas, "Partial discharge diagnostics: a utility perspective," in *Gaseous Dielectrics VII*, L. G. Christophorous and D. R. James, Eds., Plenum, New York, 1994, pp. 209–220.

[68] T. W. Dakin, C. N. Works, and J. S. Johnson, "An electromagnetic probe for detecting and locating discharges in large rotating machine stators," *IEEE Trans. Power Apparatus Syst.*, Vol. PAS-88, 1969, pp. 251–257.

[69] G. C. Stone, H. G. Sedding, N. Fujimoto, and J.M. Braun, "Practical implementation of ultrawideband partial discharge detectors," *IEEE Trans. Electrical Insulation*, Vol. 27, 1992, pp. 70–81.

[70] G. Katsuta, A. Toya, K. Muraoka, T. Endoh, Y. Sekii, and C. Ikeda, "Development of a method of partial-discharge detection in extra high voltage crosslinked polyethylene insulated cable lines," *IEEE Trans. Power Delivery*, Vol. 7, 1992, pp. 1068–1079.

[70a] R. Morin, R. Bartnikas, and G. Lessard, "In-service location of partial discharge sites in polymeric distribution cables using capacitive and inductive probes," Proc. IEEE Transmission and Distribution Conference, April 11–16, 1999, New Orleans.

[71] R. Bartnikas and G. C. d'Ombrain, "A Study of corona discharge rate and energy loss in spark gaps," *IEEE Trans. Power Apparatus Syst.*, Vol. 84 Suppl., 1963, pp. 366–375.

[72] R. Bartnikas and J. Levi, "Improved pulse discharge rate measuring apparatus for ionization discharge studies at low frequencies," *Rev. Scientific Instruments*, Vol. 37, 1966, pp. 1245–1251.

[73] R. Bartnikas and J. Levi, "A simple pulse-height analyzer for partial-discharge rate measurements," *IEEE Trans. Instrumentation Measurement*, Vol. IM-18, 1969, pp. 341–345.

[74] R. Bartnikas, "Use of multichannel analyzer for corona pulse height distribution measurements in cables and other electrical apparatus," *IEEE Trans. Instrumentation Measurement*, Vol. IM-22, 1973, pp. 403–407.

[75] R. Bartnikas and R. Morin, "Corona pulse charge transfer at elevated frequencies," *IEEE Trans. Electrical Insulation*, Vol. EI-18, 1983, pp. 458–461.

[76] R. Bartnikas, "Corona pulse probability density function measurements on primary distribution cables," *IEEE Trans. Power Apparatus Syst.*, Vol. PAS-94, 1975, pp. 716–723.

[77] J. C. Bapt, Bui-Ai, and C. Mayoux, "Corona frequency analysis in artificial cavities in epoxy resins," *1973 Annual Report, Conference on Electrical Insulation and Dielectric Phenomena*, NAS INRC, Washington, DC, 1974, pp. 282–288.

[78] S. Kärkkäinen, "Internal partial discharge pulse distributions. Physical mechanisms and effects on insulations," D. Tech. Thesis, Helsinki Institute of Technology, Finland, 1976.

[79] A. Kelen, "The functional testing of HV generator stator insulation," Proc. CIGRÉ, Paper 15-03, Paris, 1976, pp. 1–16.

[80] A. A. Mazroua, M. M. A. Salama, and R. Bartnikas, "PD pattern recognition with neural networks using the multilayer perceptron technique," *IEEE Trans. Electrical Insulation*, Vol. 28, 1993, pp. 1082–1089.

[81] E. Gulski and A. Krivda, "Neural networks as a tool for recognition of partial discharges," *IEEE Trans. Electrical Insulation*, Vol. 28, 1993, pp. 984–1001.

[82] A. Schnettler and V. Tryba, "Artificial self-organizing neural network for partial-discharge source recognition," *Archiv Elektrotechnik*, Vol. 76, 1993, pp. 149–154.

[83] R. J. Van Brunt, "Stochastic properties of partial discharge phenomena," *IEEE Trans. Electrical Insulation*, Vol. 26, 1991, pp. 902–948.

[84] A. A. Mazroua, R. Bartnikas, and M. M. A. Salama, "Neural network system using the multilayer perceptron technique for recognition of PD pulse shapes due to cavities and electrical trees," *IEEE Trans. Power Delivery*, Vol. 10, 1995, pp. 92–96.

[85] A. A. Mazroua, R. Bartnikas, and M. M. A. Salama, "Discrimination between PD pulse shapes using different neural network paradigms," *IEEE Trans. Dielectrics Electrical Insulation*, Vol. 1, 1994, pp. 1119–1131.

[86] E. H. Rayner, "High-voltage test and energy losses in insulating materials," *J. IEE*, Vol. 49, 1912, pp. 3.

[87] D. M. Robinson, *Dielectric Phenomena in High Voltage Cables*, Chapman and Hall, London, 1936.

[88] R. Bartnikas, S. Pélissou, and H. St-Onge, "A-C breakdown characteristics of in-service aged XLPE distribution cables," *IEEE Trans. Power Delivery*, Vol. 3, 1988, pp. 454–462.

[89] W. D. Wilkens, "Statistical methods for the evaluation of electrical insulating systems," in *Engineering Dielectrics, Vol. IIB, Electrical Properties of Solid Insulating Materials: Measurement Techniques*, R. Bartnikas, Ed., STP 926, ASTM, Philadelphia, 1987, Chapter 7.

[90] R. J. Densley, R. Bartnikas, and B. Bernstein, "Multiple stress aging of solid dielectric extruded dry-cured insulating systems for power transmission cables," *IEEE Trans. Power Delivery*, Vol. 9, 1994, pp. 559–571.

[91] E. Favrie and H. Auclair, "Effect of water on electrical properties of extruded synthetic insulations," *IEEE Trans. Power Apparatus Syst.*, Vol. PAS-90, 1980, pp. 1225–1234.

[92] S. Nagasaki, N. Yoshida, M. Aihara, S. Fujiki, N. Kato, and M. Nakagawa, "Philosophy of design and experience on high voltage XLPE cables and accessories in Japan," Proc. CIGRÉ, Paper No. 21, 1988.

[93] L. Deschamps, "The aging of extruded dielectric cables," JICABLE / EPRI / CEA Workshop Cables '89, EPRI Report EL-7090, 1990, pp. 5c.1–5c.19.

[94] E. Occhini, "A statistical approach to the discussion of the dielectric strength of electric cables," *IEEE Trans. Power Apparatus Syst.*, Vol. PAS-90, 1971, pp. 2671–2682.

[95] V. M. Montsinger, "Breakdown curve for solid insulations," *AIEE Trans. Power Apparatus Syst.*, Vol. 54, 1935, pp. 1300–1301.

[96] T. W. Dakin and S. A. Studniarz, "The voltage endurance of cast and molded resins," 13th IEEE/NEMA Electrical/Electronics Insulation Conf., Boston, 1977, pp. 318–321.

[97] G. Bahder, T. Garrity, C. Sosnowski, R. Eaton, and C. Katz, "Physical model of electric aging and breakdown of extruded polymeric insulated power cables," *IEEE Trans. Power Apparatus Syst.*, Vol. PAS-101, 1982, pp. 1378–1388.

[98] H. R. Zeller, T. H. Baumann, and F. Stucki, "Microscopic models of aging in solid dielectrics," IEEE International Conf. on the Properties and Applications of Dielectric Materials, IEEE Conf. Record No. 88CH2587-4, Vol. 1, Xi'ian, June 1985, pp. 13–15.

[99] D. W. Auckland and M. N. Arbab, "Growth of electrical trees in solid insulation," *Proc. IEE*, Vol. 136A, 1989, pp. 73–78.

[100] E. Brancato, "Insulation aging: a historical and critical review," *IEEE Trans. Electrical Insulation*, Vol. EI-13, 1978, pp. 308–317.

[101] V. M. Montsinger, "Loading transformers by temperature," *AIEE Trans.*, Vol. 49, 1930, pp. 776–792.

[102] T. W. Dakin, "Electrical insulation deterioration treated as a chemical rate phenomena," *AIEE Trans., Part I (Communications and Electronics)*, Vol. 67, 1948, pp. 113–122.

[103] IEC Publication 216, "Guide for the determination of thermal endurance properties of electrical insulation materials," Part I, II, III and IV, IEC, Geneva, 1980.

[104] J. Sletback, C. W. Reed, and S. Hirabayashi, "Multiple stress behavior of insulating materials and insulation structures," CIGRÉ, Paper 15-03, 1983.

[105] S. N. Zhurkov, "Kinetic concept of the strength of solids," *International J. Fracture Mechanics*, Vol. 1, 1965, pp. 311–323.

[106] S. Glasstone, K. J. Laidler, and H. E. Eyring, *The Theory of Rate Processes*, McGraw-Hill, New York, 1941, p. 196.

[107] A. A. Frost and R. G. Pearson, *Kinetics and Mechanisms*, Wiley, New York, 1961.

[108] H. S. Endicott, B. D. Hatch, and R. G. Sohmer, "Application of the Eyring model to capacitor aging data," *IEEE Trans. Component Parts*, Vol. CP-12, 1965, pp. 34–41.

[109] L. Simoni, "A general approach to the endurance of electrical insulation under temperature and voltage," *IEEE Trans. Electrical Insulation*, Vol. EI-16, 1981, pp. 277–289.

[110] G. C. Montanari, "Electrical life threshold models for solid insulating materials subjected to electrical and multiple stresses," *IEEE Trans. Electrical Insulation*, Vol. 27, 1992, pp. 974–999.

[111] K. D. Wolter, J. F. Johnson, and J. Tanaka, "Polymer degradation and its measurement," in *Engineering Dielectrics, Vol. II B, Measurement Techniques*, R. Bartnikas, Ed., STP 926, ASTM, Philadelphia, 1987, Chapter 5.

[112] K. Ogawa, T. Kosugi, N. Kato, and Y. Kawawata, "The world's first use of 500 kV XLPE insulated aluminum sheathed power cables at the Shimogo and Imaichi Power Stations," *IEEE Trans. Power Delivery*, Vol. 5, 1990, pp 26–32.

[113] P. Rohl and B. Andress, "Oxidation degradation of polyolefin insulations due to surface influences," *Siemens Forschung und Entwicklungsberichte*, Vol. 16, 1987, pp. 117–121.

[114] K. Iida, S. Nakamura, M. Ieda, K. Ito, and G. Sawa, "A dielectric study of oxidation in the amorphous and crystalline regions of low-density polyethylene with antioxidant," *Polymer J. (Tokyo)*, Vol. 19(8), 1987, pp. 905–913.

[115] B. Andress, P. Fisher, H. Repp and P. Rohl, "Diffusion losses of additives in polymeric cable insulations," 1984 IEEE International Symposium on Electrical Insulation, Conf. Record 84-CH1964-6-EI, Montréal, 1984, pp. 65–67.

[116] S. Haridoss, "Migration of vinyl acetate from semiconductive shields to insulation of power cables," IEEE International Symposium on Electrical Insulation, Conf. Record 90-CH2727-6, Toronto, 1990. pp. 281–285. See also EPRI Report EL-6207, "Evaluation of Diagnostic Techniques for Cable Characterization," EPRI, Palo Alto Feb. 1989.

[117] A. Bulinski, S. Bamji, J. Densley, J. P. Crine, S. Haridoss, and B. Bernstein, "Water-treeing in a heavily oxidized cross-linked polyethylene insulation," Proc. 6th International Symposium on High Voltage Engineering, Vol. 1, Paper 13-06, 1989, p. 4.

[118] R. Bartnikas, "Influence of space charges on the short and long term performance of solid dielectrics," Proc. IInd International Conf. on Space Charge in Solid Dielectrics, IEEE/ASPROM/Soc. Française du Vide, Antibes, April 2–7, 1995, pp. 9–33.

[119] H. E. Orton, R. G. Fletcher, D. M. Castlidge, J. F. Bradley, M. J. Colwell, and J. Y. Wong, "Elevated temperature operation of XLPE distribution cable systems," *JICABLE*, Paper A.1.3, 1987, pp. 14–22.

[120] E. Dorison, A. Royere, H. Auclair, Y. Lecoq, and J. Midoz, "Changes in the specifications of HV and EHV cables associated with installations and operating conditions," CIGRÉ, Paper 21-03, Paris, 1984.

[121] C. W. Melton, D. Mangaraj, and M. Epstein, "Morphology of thermoplastic and cross-linked polyethylene cable insulation," 1981 Annual Report, Conf. on Electrical Insulation and Dielectric Phenomena, IEEE Conf. Record 81-CH1668-3, Whitehaven, PA., 1981, pp. 299–305.

[122] A. Barlow, L. A. Hill, and M. F. Maringer, "Possible mechanism of microvoid formation in polyethylene insulated high voltage cables," *IEEE Trans. Power Apparatus Syst.*, Vol. PAS-102, 1983, pp. 1921–26.

[123] K. Kaminaga, T. Harada, M. Ono, T. Kojima, Y. Sekii, and M. Marumo, "Research and development of 500 kV XLPE cables," 1986 IEEE International Symposium on Electrical Insulation, IEEE Conf. Record No. 86 CH2196-4-DEI, 1986, pp. 29–36.

[124] W. Boone, E. F. Steennis, P. A. C. Bentvelsen, and A. M. F. J. Van Der Laar, "Development and trial of EHV XLPE cables in the Netherlands," CIGRÉ, Paper 20-20, 1984, p. 7.

[125] Y. Miyashita, H. Itoh, and T. Shimomura, "Void Formation in XLPE at High Temperatures," 1985 International Conf. on Prop. and Applic. of Diel. Matls, IEEE Conf. Rec. 85-CH2115-4, (Xi'ian), 1985, pp. 670–674.

[126] T. W. Dakin, G. Luxa, G. Opperman, J. Vigreux, G. Wind, and H. Winkelnkemper, "Breakdown of gases in uniform fields: Paschen curves for nitrogen, air and SF_6," *Electra*, No. 32, 1974, pp. 61–82.

[127] J. D. Cobine, *Gaseous Conductors*, McGraw-Hill, New York, 1958.

[128] R. Bartnikas and J. P. Novak, "On the spark to pseudoglow and glow transition mechanism and discharge detectability," *IEEE Trans. Electrical Insulation*, Vol. 27, 1992, pp. 3–14.

[129] R. Bartnikas and J. P. Novak, "Effect of overvoltage on the risetime and amplitude of PD pulses," *IEEE Trans. Dielectrics Electrical Insulation*, Vol. 2, 1995, pp. 557–566.

[130] J. T. Holboll, J.-M. Braun, N. Fujimoto, and G. C. Stone, "Temporal and spacial development of partial discharges in spherical voids in epoxy related to the detected electrical signals," 1991 Annual Report, Conf. on Electrical Insulation and Dielectric Phenomena, Conf. Record no. 91 CH3055-1, Knoxville, 1991, pp. 581–588.

[131] R. Bartnikas and J. P. Novak, "Different forms of partial discharge and their terminology," *IEEE Trans. Electrical Insulation*, Vol. 28, 1993, pp. 956–968.

[132] R. M. Eichhorn, "Treeing in solid organic dielectric materials," in *Engineering Dielectrics, Vol. IIA, Electrical Properties of Solid Insulation Materials: Molecular Structure and Electrical Behavior*, R. Bartnikas and R. M. Eichhorn, Eds., STP 783, ASTM, Philadelphia, 1983, Chapter 4.

[133] R. Bartnikas, "Performance characteristics of dielectrics in the presence of space charge," *IEEE Trans. Dielectrics Electrical Insulation*, Vol. 4, 1997, pp. 544–557.

[134] H. J. Wintle, "Conduction processes in polymers," in *Engineering Dielectrics, Vol. IIA, Electrical Properties of Solid Insulating Materials: Molecular Structure and Electrical Behavior*, R. Bartnikas and R. M. Eichhorn, Eds., STP 783, ASTM, Philadelphia, 1983, Chapter 3.

[135] R. Bartnikas, "Direct current conductivity measurements," in *Engineering Dielectrics Vol. IIB, Electrical Properties of Solid Insulating: Measurement Techniques*, R. Bartnikas, Ed., STP 783, ASTM, Philadelphia, 1987, Chapter I.

[136] N. F. Mott and E. A. Davis, *Electronic Processes in Non-Crystalline Materials*, Clarendon Press, Oxford, 1971.

[137] S. S. Bamji, A. T. Bulinski, and R. J. Densley, "Degradation of polymeric insulation due to photoemission caused by high electric fields," *IEEE Trans. Electrical Insulation*, Vol. 24, 1990, pp. 91–98.

[138] C. Laurent, C. Mayoux, and S. Noël, "Dielectric breakdown of polyethylene in divergent fields: role of dissolved gases and electroluminescence," *J. Appl. Phys.*, Vol. 54, pp. 1532–1539.

[139] S. S. Bamji, A. T. Bulinski, and R. J. Densley, "The role of polymer interface during tree initiation in LDPE," *IEEE Trans. Electrical Insulation*, Vol. EI-24, 1986, pp. 639–644.

[140] E. H. Ball, H. W. Holdup, D. J. Skipper, and B. Vecellio, "Development of cross-linked polyethylene insulation for high voltage cables," CIGRÉ Paper 21-01, 1984.

[141] R. Morin, J. P. Novak, R. Bartnikas, and R. Ross, "Analysis of in-service aged stator bars," *IEEE Trans. on Energy Conversion*, Vol. 10, 1995, pp. 645–654.

[142] M. D. Andersen, J. Jorgensen, O. K. Nielsen, and S. H. Poulsen, "Design, manufacturing and installation of XLPE Cables in Denmark," CIGRÉ, Paper 21-8, 1986.

[143] E. Favrie and H. Auclair, "225 kV low density polyethylene insulated cables," IEE Conf. Progress in Cables and Overhead Lines for 220 kV and Above, Publ. No. 176, 1979, pp. 155–159.

[144] R. Jocteur, E. Favrie, H. Auclair, and B. Bhuicq, "Development of 400 kV links with low density polyethylene insulation," CIGRÉ, Paper 21-09, 1986.

[145] S. H. Lee, K. M. Han, and W. K. Park, "Development and installation of XLPE insulated high voltage power cables in Korea," CIGRÉ, Paper 21-04, 1986, p. 6.

[146] R. Bartnikas, H. C. Doepken, R. M. Eichhorn, G. W. Rittmans, and W. D. Wilkens, "Accelerated life testing of wet cables specimens at frequencies above 60 Hz," *IEEE Trans. Power Apparatus Syst.*, Vol. PAS-99, 1980, pp. 1575–1585.

[147] R. Lyle, "Effect of testing parameters on the outcome of accelerated cable life test," *IEEE Trans. Power Delivery*, Vol. 2, 1988, pp. 434–439.

[148] M. D. Walton, J. T. Smith, and W. A. Thue, "Accelerated aging of extruded dielectric power cables, Part I: Control and monitoring methodology," *IEEE Trans. Power Delivery*, Vol. 7, 1992, pp. 596–602.

[149] Guide for accelerated aging tests for medium voltage extruded electric power cables using water-filled tanks, WG 12-35 Accelerated Electrical Aging, IEEE Insulated Conductors Committee, 1996, (draft).

[150] J. Sletback and E. Ilstad, "The effect of service and test conditions on water tree growth," *IEEE Trans. Power Apparatus Syst.*, Vol. PAS-102, 1983, pp. 2069–2076.

[151] H. Kato and A. Veda, "The influence of ionic substances on water treeing," *Dainichi Nippon Cable Rev.*, No. 64, 1979, pp. 23–31.

[152] M. J. Given and B. H. Crichton, "The effects of anions on the growth rate of water trees," Proc. of the Second International Conf. on Conduction and Breakdown in Solid Dielectrics, Erlangen, 1986, pp. 257–261.

[153] G. Tchobanoglous and E. D. Schroeder, *Water Quality*, Addison-Wesley, Reading, MA, 1987.

[154] L. J. Frisco, "Frequency dependence of electric strength," *Electro-Technology*, August 1961.

[155] G. Bahder, T. W. Dakin, and J. E. Lawson, "Analysis of treeing type breakdown," Proc. CIGRÉ, Paper 15-05, Paris, 1974.

[156] J. Sletbak and A. Botne, "A study of inception and growth of water trees and electrochemical trees in polyethylene and cross-linked polyethylene insulations," *IEEE Trans. Electrical Insulation*, Vol. EI-12, 1977, pp. 383–389.

[157] D. A. Silver and R. G. Lukac, "Factors affecting dielectric strength of extruded dielectric cables in a wet environment," JICABLE (International Conf. on Polymer Insulated Power Cables, Paris, 1984, pp. 99–104.

[158] A. Bulinski, S. Bamji, and J. Densley, "The effects of frequency and temperature on water tree degradation of miniature XLPE cables," *IEEE Trans. Electrical Insulation*, Vol. EI-24, 1986, pp. 645–650.

[159] R. Bartnikas, R. J. Densley, and R. M. Eichhorn, "Accelerated aging tests for polymer cables under wet conditions," *IEEE Trans. Power Delivery*, Vol. 6, 1991, pp. 929–937.

[160] R. Bartnikas, R. J. Densley, and R. M. Eichhorn, "Long term accelerated aging tests on distribution cables under wet conditions," *IEEE Trans. Power Delivery*, Vol. 11, 1996, pp. 1695–1701.

[161] K. Abdolall, "Design of an accelerated aging test for polymer insulated distribution cables," 1990 IEEE International Symposium of Electrical Insulation, Conf. Record No. 90-CH2727-6, Toronto, 1990, pp. 343–347.

[162] M. A. Martin and R. A. Hartlein, "Correlation of electrochemical treeing in power cables removed from service and cables tested in the laboratory," *IEEE Trans. Power Apparatus Syst.*, Vol. PAS-99, 1980, pp. 1547–1605.

[163] W. Boone, "A Comparison of two accelerated water treeing tests: UNIPEDE and AEIC," Presented to the IEEE-ICC Working Group 12-35, Accelerated Electrical Aging, Minutes of the 98th Meeting of the Insulated Conductors Committee (ICC) of the IEEE Power Engineering Society, St. Petersburg Beach, Florida, Nov. 6–9, 1995.

[164] R. G. Schroth, W. Kalkner, and D. Fredrich, "Test methods for evaluating the water tree aging behavior of extruded cable insulations," Proc. CIGRÉ, Paper 15/21-01, Paris, 1990.

[165] A. Bulinski, C. Au, S. Bamji, and R. J. Densley, "Effects of temperature on the growth of vented water trees in laboratory molded polyethylene specimens," 1987 Annual Report, Conference on Electrical Insulation and Dielectric Phenomena, IEEE Conf. Record 87CH2462-O, Gaithersburg, Oct. 18–22, 1987, pp. 440–447.

[166] E. F. Steennis, C. C. Van den Heuvel, and W. Boone, "Accelerated aging to predict water tree behavior in extruded cables," JICABLE (International Conf. on Polymer Insulated Cables), Paris, 1987, pp. 161–167.

[167] R. Hartlein, V.S . Harper, and H. W. Ng, "Effect of voltage impulses on extruded dielectric cable life," *IEEE Trans. Power Delivery*, Vol. 4, 1989, pp. 829–841.

[168] G. C. Montanari and M. Cacciari, "Progressively censored aging tests on XLPE insulated cable models," *IEEE Trans. on Electrical Insulation*, Vol. EI-23, 1988, pp. 365–372.

Chapter 10

FIELD TESTS AND ACCESSORIES FOR POLYMERIC POWER DISTRIBUTION CABLES

H. D. Campbell
W. T. Starr

10.1 INTRODUCTION

Decisions about whether to test cable systems in the field generally depend on economics as much as technical issues. An acceptable test immediately after installation is a common procedure because it provides an overview of the workmanship *quality* during installation (a technical consideration), and it can be performed without interrupting production (an economic consideration). Field tests also create an excellent communication medium between the contractor, the cable suppliers, and the testers, which is desirable both from the economics and technical points of view.

The direct-current (dc) overvoltage test with charging current measurement has been popular for both testing of cable systems at installation and for periodic maintenance tests. This is partly because the test is simple, and the equipment is available and inexpensive. This reasoning does not hold completely for the megavolt range, but dc testing is popular for extra-high-voltage (EHV) cables too. Periodic maintenance tests during scheduled power shutdown times are less widely used. When they are used, it is because of fear of a random failure in the middle of a production run; again the decisions for the need of testing in the latter case involve both economical and technical considerations.

Recently the popularity of dc testing has been waning. It has been discovered that crosslinked polyethylene (XLPE) cable installed in wet locations may be severely damaged by the application of prescribed levels of dc voltage, a practice considered essentially harmless in the past [1]. This change in attitude toward dc tests has been influenced by two factors. First, there have been techniques developed for salvaging cables, degraded by water trees; these techniques do not work on cables that have failed when subjected to a dc test. Second, there have been developed viable alternative test methods to dc testing. Alternating-current (ac) testing at very low frequency (VLF), such as 0.1 Hz, is flourishing because inexpensive and lightweight computer-controlled power supplies with sufficient kilo-volt-ampere capacity to excite miles of cable up to voltages even higher than service voltage levels are becoming available (cf. Chapter 1). Very low frequency testing is being applied to measure also the dissipation factor of the dielectric. The result can be used to diagnose a wet dielectric and to monitor field drying processes. Also it may be employed to measure the discharge inception and extinction voltages of long lengths of installed cable. Lastly, VLF may be used as a power source in the location of partial discharges to pinpoint discharge sites where problems may eventually lead to failure. This may require superimposing impulses to initiate the discharge.

The current test data, accumulated in 1997, has resulted in a ban on direct voltage field testing of XLPE cable that has been subjected to wet conditions. However, thus far insufficient information has been gathered to permit extension of the ban to ethylene-propylene-rubber (EPR) and ethylene-propylene-diene monomer (EPDM) insulated cables [1]. It should be emphasized that the scope of this chapter is limited to dry cable and to cable that is kept dry in service. Information on the effects of moisture on extruded insulants, for cable and on rapidly changing technologies on how to test wet cable, or cable that may possibly be wet, is given in [2–16] (cf. Chapters 2 and 9).

Justification for testing of a cable system either immediately after installation or during service depends largely on four factors: (i) the particular characteristics of the load, (ii) the judgment of the cable system owner on the overall probability that the test will reduce costs, (iii) the availability of reliable information on the effectiveness of the testing itself to discriminate between good and bad cable, and (iv) any potential reduction in cable life imposed by the testing. It is important to remember that the *reliability* of the information depends upon the history of the test. This includes the number of tests performed, especially the number of times that a given test has been used on the specific system. Two rather different situations may be illustrative. Case (a) may involve an EHV cable several kilometers long for which little of the test history is available. Repair of a failure that may occur during a test would be very expensive, but power may be maintained by alternate circuits. In this case, the testing may be judged to pose too high a risk. Case (b) may represent a relatively low cost medium-voltage system supplying a large industrial plant in which the manufacturing process is such that a short service interruption could be very costly. In this case, the cable and its accessories cost little and are easy to repair, and a convenient store of spare cable and components is available. Much historical experience is available to guide the plant manager. it is almost certain that periodic power shutdown times will be allowed for testing. The installation of redundant power distribution could be recommended if the frequency of failures becomes excessive. Many situations do not support such simple decisions and tend to be decided on the basis of engineering judgment, or what the cynic might term ignorance, for the documentation on the efficacy of such testing modes is more convincing to some than to others.

10.2 ALTERNATING-CURRENT OVERVOLTAGE TEST

The ac overvoltage test is specified in industry standards as 80% of the acceptance test voltage; one would expect that the applied alternating test voltage should be reduced from the factory test values to account for deterioration of the insulation during the installation and subsequent to service in the field. The ac overvoltage test is a logical proof voltage test to apply to installed ac cable, because the voltage stress distribution is the same as occurs in service. This renders the partial-discharge inception and extinction voltages and the damaging characteristics of partial discharge similar to those that may occur in service. It is also logical because ac tests of many kinds are used in the factory, e.g., partial discharge, load cycling, high potential withstand (proof voltage), dielectric strength, and time-to-failure versus voltage tests. The manufacturer, therefore, knows much more about the effects of applying ac potentials than he does about the effects of dc. One would expect that the test method applied in the field would be similar to the factory qualification test, which consists of a 5-min application of a 200-V/mil ac average stress. Assuming that 80% of this stress level were used on a 15-kV XLPE cable, the test would be 5 min at 28 kV. A large amount of information is available, which shows that the effect of voltage on the time to failure of XLPE cable follows an inverse power law with an exponent often exceeding 8 (though it may be as low as 4.8 for moist XLPE). That is, the life decreases approximately with the 8th power of the ratio of the test voltage to the operating voltage. Using this power law, a 5-min test of a 15-kV cable at 28 kV is equivalent to about 19 days at 110% of the line to ground voltage, V_g (9.54 kV). This is only about 0.24% of a 20-year expected life span and does, therefore, not constitute a serious risk. But including the possibility of moisture effects, the life corresponding to 5 min at 28 kV is reduced from 19 days to 14.6 h at 9.54 kV (a $19 \times 24/14.6$ or 31:1 reduction in life due to moisture). This implication is often enough to frighten the cable engineer away from ac tests.

The damaging potential of ac resides in the presence of partial discharge and the time of voltage application. The inception and extinction voltages of partial discharge are theoretically linked to the zero-to-peak and the peak-to-peak voltages. A 5-min test of a 15-kV cable at 28 kV entails subjecting the cable specimen to $60 \times 60 \times 5 = 18,000$ cycles at $\sqrt{2} \times 28 = 39.6$ kV peak. This is not a severe test for present-day XLPE power distribution cable, because there is generally no presence of partial discharges. Load cycle tests per stand are generally run at 35 kV. With dc, on the other hand, the fact that the voltage applied is high (say, 60 kV for 15 min for a 15-kV rated cable) means that the occurrence of partial discharges is virtually certain, and a significant level of polarization will occur in the dielectric. Under these circumstances, if the cable is suddenly shorted, either inadvertently or by a breakdown, a ringing or oscillating voltage condition will occur, and the released trapped charge in its path to ground could cause a weakening of the dielectric. Nevertheless, the ac overvoltage test is almost never used. The difficulty with the test, except for low-voltage systems of modest lengths, centers on the large reactive load that the cable poses even with modest voltages: the large kilovolt-ampere demand requires expensive equipment.

10.3 DISSIPATION FACTOR (POWER FACTOR) TEST

The measurement of the dissipation factor or the tangent of the loss angle, δ, of the insulation (the power factor is equal to $\sin \delta$) is an effective way to determine deterioration that could result from contamination of aging effects. If the test stress is made equal to the operating stress, any losses due to partial discharges at operating voltage would be included. There are two difficulties with this: The elevated reactance of the cable would create a large load on the test equipment as in the case of the ac overvoltage test, and it may not be practical to isolate the cable from earth. To a limited extent these two difficulties can be resolved by employing the Doble test [17]. This test utilizes a Voltmeter-Wattmeter-Ammeter method of measuring the dissipation factor at a low applied voltage. The test may be useful in certain special conditions, but it must be remembered that cable insulations currently employed have a very low dissipation factor such that wattmeter methods are not adequate unless the insulation has been very badly aged or contaminated. If a means can be found to isolate the cable ground from earth, a Schering bridge or a current comparator bridge (cf. Chapter 9) can be used to provide reliable readings. Such instruments have the advantage in that they are very sensitive, and they allow the test to be performed at operating voltage where losses due to partial discharges would be included in the measurement. Here reference should be made to the studies carried out at the Wartz Mill HV test facility where techniques have been developed for interrupting the shield of EHV cables so that power factor tests can be made on joints using a Schering bridge [18]. In this context the authors themselves have observed an incipient failure in a high-voltage oil-filled cable test specimen, which was monitored using a Schering bridge. When the dissipation factor began to rise rapidly and had reached 1%, the test was terminated. Extensive tracking in one of the high-voltage terminations was determined to be the cause. If the test specimen were a long length of cable instead of a short length, the loss would hardly have been detected even with a Schering bridge because the localized loss at the termination would have represented a much smaller loss relative to the entire cable length (cf. Chapter 9). Even when using sensitive bridge instruments, it is obvious that incipient faults resulting from localized sites of intensive discharge cannot be positively identified. The dissipation factor value is a measure of the total dielectric loss. It does not differentiate between a dielectric that has a small, uniform increase in loss and one that has a catastrophically high localized loss.

It is interesting to note that included in the extensive investigations on the service life of solid dielectric cables [usually polethylene (PE) or XLPE constructions] since the early 1970s have been field tests of dissipation factor, partial discharge, and insulation resistance [19]. The dissipation factor measurements were made with a specially constructed inverted transformer ratio arm (current comparator) bridge at the cable rated voltage levels. When specimens of these cables were later examined in the laboratory, some correlation was found between the high dissipation factor levels and severe water treeing. Subsequent work [20] has led to a proposal for field dissipation factor testing to identify cables having unacceptable deterioration. It is not yet know how practical and definitive such a test procedure would be.

10.4 INSULATION RESISTANCE TEST

The insulation resistance test is, of course, very simple to perform but yields very little useful information when applied to high-voltage cables, unless the measured value is very low relative to permissible values given in industry standards or manufacturers' handbooks. Insulation resistance values can vary enormously between different types of high-quality insulation. Even a drop of insulation resistance by a factor of 100 compared to a reading of a few months previously does not necessarily mean the cable is unserviceable. On the other hand, a cable on the verge of failure can easily exhibit a very good reading. The measurement can be of practical value when trouble is suspected, but usually only to confirm previous indications of a defective circuit. Precautions should be taken to ensure that there is no surface leakage at terminations, otherwise a false reading may be obtained. Low insulation resistance of PE or XLPE cables may indicate the presence of water treeing in the insulation [19].

10.5 PARTIAL-DISCHARGE TEST

The great value of the partial-discharge test rests in its sensitivity to very low level discharges occurring at discrete sites in the cable, unlike the dissipation factor test that merely indicates the sum of such discharges as a small added loss at a much lower effective sensitivity. It will be recalled that the partial-discharge test was developed in the 1950s when polymeric insulated cables, namely high-voltage rubber, e.g., butyl, and high-voltage thermoplastic, e.g., polyethylene were beginning to become an acceptable alternative to medium-voltage oil-paper insulated lead-covered (PILC) cables. Whereas oil-paper cables were proved by many years of operating experience to be highly reliable products without partial-discharge testing at the factory (except by the increase in dissipation factor with electrical stress; i.e., the ionization factor test), the polymeric dielectrics were known to be far less resistant to this form of aging mechanism. There is little history of partial-discharge testing, as we know it today, on oil-paper dielectrics. It follows that the partial-discharge test in the field can, in practice, primarily be considered for polymeric insulated cables (cf. Chapter 9).

Experience with partial-discharge testing in the field indicates that the normal factory test methods are not practical. Testing at 60 Hz requires large kilo-volt-ampere power supplies not practical for mobile test equipment. High-frequency noise in the field environment (power distribution circuits) usually requires the cables to be tested in a balanced mode (two cables of identical construction and length tested simultaneously). A satisfactory technique using a specially constructed bridge-type discharge detector designed to operate at 0.1 Hz is described in [19]. The slow rate with which partial-discharge field testing has been gaining acceptance over the past two decades must be attributed to its relative complexity, cost, and lack of definitive evidence of what discharge level constitutes an unacceptable condition. However, the situation has been changing rapidly over the last few years due to the availability of more precise discharge site locating techniques and procedures (cf. Chapter 9).

10.6 DIRECT-CURRENT OVERVOLTAGE TEST

It is the dc high-voltage field test that continues to be by far the most commonly used for acceptance and maintenance tests in the field. With the development of reliable solid-state high-voltage rectifiers, the dc high-voltage test set has become relatively simple, low cost, and low weight compared to high-voltage ac test sets. Since only the leakage current must be supplied, the power requirements are very low. Current rating is typically 10 mA. The physical size and weight of test sets are largely dependent on the voltage rating. Test sets rated over 100 kV may be mounted on wheels so that they are readily mobile in the field and can be easily moved about by the operator. Units rated at 300 kV and above can be easily transported in small trucks. The authors are familiar with a modular test set rated up to 600 kV that can be assembled in the field in about an hour. The total shipping weight is about 100 kg. The output power of a 300-kV set at 10 mA is 3 kW. The load current is normally only 1 or 2 mA so that the power supply to the set is easily accommodated by the common 115/230 V circuits. The low power, low cost, and ease of transportation are great inherent advantages of the dc equipment.

10.7 DIRECT-CURRENT TEST PROCEDURES

There are two fundamentally different approaches. The first approach is simply to apply the test voltage for the required time as specified in the industry standards. This is a proof test to demonstrate that the cable will withstand the test voltage without failure. It is often used as an acceptance test. No particular significance is given to the load current, monitored from the output milliammeter, except to note that it is about normal and remains approximately constant. The other testing mode is to record carefully the load current as a function of time and applied voltage. From the data of the apparent resistance of the specimen during the test, certain conclusions may sometimes be drawn on the condition of the insulation. This is the more usual mode in maintenance tests. Although the second procedure demands experienced interpretation of the results, both test procedures are quite straightforward, thus dc field testing combines the advantages of relatively portable, low-cost equipment with low-cost testing procedures.

10.8 INTERPRETATION OF TEST RESULTS

The dc overvoltage test is very simple in concept. It consists simply of the application of a high dc stress and the subsequent observation of current response. The current response is illustrated in Fig. 10.1. It is assumed in Fig. 10.1 that a potential has been applied from a constant direct voltage source in series with a resistance to illustrate better the current components. In practice, of course, the dc test set operator raises the voltage slowly. The current components shown in Fig. 10.1 are defined as follows: i_g is the geometric capacitance current that depends on the time constant RC, where R is the series output resistance of the test set and C is the capacitance of the cable under test. In a practical case, where the voltage is high and the cable capacitance large, it may take several minutes for the current to approach zero; i_a is the absorption or polarization current that decays asymptotically to zero. The rate of change is normally very small after about 10 min, but it may require many hours to reach zero; i_c is the conduction current that should remain constant at any given temperature and electrical stress; i_t is the total current equal to $i_g + i_a + I_c$, which for testing purposes is nearly constant in a satisfactory cable when the specified test voltage is reached and may be considered for the purposes of the test to be the true conduction current.

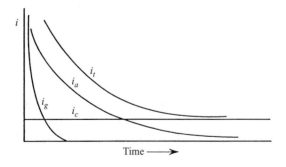

Figure 10.1 Current response components following the application of dc voltage step.

It is apparent from the brief outline of the current response to the applied voltage that it is not, in general, very practical to compare the apparent resistivity of the cable under test to the resistivity values given in handbooks of industry standards. The values are measured at a few hundred volts after an electrification of 1 min and are only nominal resistivities obtained under a definite prescribed procedure. The apparent resistivity is not only a function of the insulation under test but of the temperature and electrical stress. It decreases both with increasing temperature and increasing stress. In practical tests, the conduction current increases fairly linearly with stress, but an increase in temperature of 10°C may double the conduction current.

To measure reasonably accurately the apparent conduction current within the cable insulation, joints, and terminations, it is necessary to ensure that the external surfaces of the terminals are clean and dry, and for voltages above 50 kV or so also to ensure that there is no partial discharge at the terminals or high-voltage test lead. It has been shown that leakage resulting from partial discharges at high-voltage terminals can be effectively eliminated by the snug fitting of an insulating container such as a polyethylene bag. The test lead should be a tube of sufficient diameter to be discharge free. Sharp points can be eliminated by the application of conducting putty. Of course, most of these precautions are unnecessary if one is only interested in a simple high-voltage proof test. It is evident that a high degree of reproducibility of the apparent resistivity is not practical. If, however, the same procedure is chosen, it should not be changed. The duration of the full test voltage should be at least 15 min. The cable should be at ambient temperature with no appreciable temperature gradient through the insulation wall.

Experienced test engineers have had some success [21] in interpreting the leakage current response in terms of defects in oil-paper, rubber, and varnished cambric insulations. Experience with polyethylene insulations indicates that breakdown is likely to occur without any prior change in leakage current. If tests are repeated every year or so, then the previous records are helpful in determining if the leakage current magnitude is normal. High leakage currents that stabilize after the test voltage is reached may be due to surface leakage within joints or terminations. Leakage currents that increase with time indicate a localized defect where the leakage current is of sufficient magnitude to cause an increase in temperature. Since an increase in temperature of 10°C is roughly sufficient to double the localized current, an unstable condition may be reached where the current increases very rapidly to failure. It is probable that insulations having a very high volume resistivity, e.g., polyethylene, are not subject to this failure mode, because the electrical field must be very near its breakdown value before any appreciable heating occurs.

10.9 QUESTION OF TEST LEVELS

It has been shown in the previous section that the leakage current response from a high direct voltage can possibly indicate defects before actual breakdown occurs. Since there is very little partial discharge associated with high dc stresses, it is not surprising that dc withstand levels are approximately equal to impulse withstand levels. Energy losses in the dielectric, associated with an alternating field; i.e., the polarization losses, are not significant. Thus, it is to be expected that destructive time effects that occur with very high ac stress should be much reduced. Experience, as is well known, verifies this hypothesis. Direct voltage breakdown is much less time dependent than alternating voltage breakdown. These facts tend to indicate that it is practical to employ very high stress levels near the impulse withstand voltage or at least at the basic impulse level. This technique would perhaps be possible if the insulation were of a uniformly high quality, in which case the test would not be necessary. The sections that follow indicate how a variety of factors may influence the test level.

10.10 DIRECT STRESS VERSUS ALTERNATING STRESS CONSIDERATIONS

Although, as we have seen, certain indications of the insulation condition can be found by careful observation of the dc leakage current, it must be remembered that, in fact, the insulation is being overstressed compared to the operating stress. Furthermore, it is a different stress mode, a mode that creates a different stress distribution than the service alternating voltage. The dc stress distribution is a function of resistance not capacitance. The consequences of this fact are illustrated in a highly oversimplified manner in Fig. 10.2. If the resistivities of the insulation layers ρ_1 and ρ_2 are equal, then the stress distribution is the same as the ac stress distribution; but if either by design or as a result of contamination the outer region has a higher resistivity than the inner, i.e., $\rho_2 \gg \rho_1$ (as the resistivities can vary by orders of magnitude), the stress in region ρ_2 could rise above the withstand level. The increase in stress, fortunately, increases the conductivity and tends to be self-correcting; nevertheless, it can be much higher than the test voltage apparently indicates. Under normal service, the capacitance of the outer section could also cause an unfavorable stress distribution, but permittivities of materials do not, in general, vary so much as resistivities. Thus failure could result from a dc test, even though the insulation may be quite satisfactory for ac conditions.

Conversely, one can postulate a condition where a high dc stress may be ineffective in detecting a serious defect. If there is a defect, say a small cavity or a hole partially through the insulation, its equivalent circuit may be represented as in Fig. 10.3. Here C_1 represents the capacitance of a gas-filled cavity or hole; R_1 the resistance of the insulation adjacent to the cavity and R_2 the resistance of insulation separating the cavity from the conductor. A steady-state nonalternating voltage, e, is applied so that the cavity C_1 breaks down. When the voltage across C_1 collapses, the full voltage e will be applied to

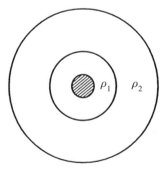

Figure 10.2 Resistivity variation in cable insulation layers.

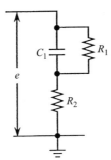

Figure 10.3 Equivalent dc circuit of cable dielectric containing gas-filled cavity.

R_2. Complete breakdown will not occur unless the applied voltage e is near the withstand voltage of the undamaged insulation. Note that the defect dimensions are small compared to the total wall thickness. If R_2 does not fail, the discharge ceases and C_1 becomes recharged and the cycle is repeated. The recurrent frequency of the partial discharges is a function of the time constant $R_1 C_1$, but it will be very much less than the recurrent frequency in an alternating 60-Hz field [21a]. Here the gap will discharge at least 120 times per second at or slightly above the corona inception voltage. A failure may not result from a dc test, but the insulation may be quite unsatisfactory for ac conditions. The foregoing also explains that there are partial discharges associated with direct stress, but they are much less severe than those occurring with ac stress.

It becomes apparent that the application of a very high dc stress may overstress some part of the insulation. In practical cable testing the voltage distribution in joints and terminations may be unfavorable, because the combination of materials and design concepts are based on ac stresses. The concept of dc test stresses near to the impulse withstand level cannot, in general, be applied. On the other hand, as shown in Fig. 10.3 quite serious defects will not be detected at moderate stresses. These conditions have led to quite wide differences of opinion on how test levels should be calculated from the rated voltage and, in particular, on the actual magnitude of the test levels themselves.

10.11 PRACTICAL TEST LEVELS

The dc test voltages normally used in North America and Europe for factory test and field acceptance tests have been determined from an equivalent ac test voltage: a value based on some factor of the ac test voltage, usually 2.4 or 3.0. These factors probably were derived from the concept of equivalent safety factors, since the dc breakdown values are roughly two or three times the 5-min ac breakdown values. The proponents of higher dc test values believe that the present factors are too small, that the stress is not sufficient to detect defects, and that much higher values could be quite safely used. The opponents to higher dc values argue that unfavorable stress distribution and possibly other mechanisms could cause breakdown when, in fact, the cable is quite adequate for the intended service. Experience over many years proves that the current practice provides reliable cables. Field data accumulated from maintenance testing indicates that many of the failures occur in joints and terminations, often from unsatisfactory workmanship. On the other hand, dc test levels about 80% of the basic impulse level (BIL), much higher than conventional values, employed by some engineers have yielded apparently good results. The difficulty, as in so much of electrical testing of insulations, is that the variables are not precisely known. The theory of electrical breakdown indicates that there is not much logic in applying a dc stress of less than half the withstand value to an insulation system of uniform properties, for it can be stressed to much higher values without distress. Variations of properties within a practical system will always be present. It is not possible for one stimulus, the dc stress, to characterize the response of these variations to another stimulus, the ac service stress. This is not to say that the voltage levels currently employed are optimum. It is quite possible that higher values should be used, but more data must be accumulated before substantial revisions could be justified.

As the result of work, extending over several years, of the Insulated Conductors Committee (ICC) of the Institute of Electrical and Electronics Engineers (IEEE) on cable dc test methodology, a document [22] has been prepared to assist the user, giving details of test procedures and voltage test levels.

10.12 JOINTS AND TERMINATIONS

When tests are carried out on cables in the field, the entire cable systems are being tested. The terminations as well as any cable joints, which are an integral part of the cable system, must withstand the same voltage as the cables themselves. A failure in a termination or joint may have the same disastrous effects as that in the cable itself as far as the load being supplied is concerned. For this reason, it will be of considerable value here to present some data on cable accessories that form part of the overall cable system. Accessories for solid-dielectric cables have advanced from the early techniques of laminated tape construction to the present premolded modular construction. There are fundamental differences in the two constructions, and it is logical that the solid-dielectric cable should have a solid-dielectric accessory just as the laminated paper cable has a laminated dielectric in the accessory. The design and manufacture of accessories for solid-dielectric cables has been highly developed by the accessory manufacturers. The discussion here is essentially an introduction to the field of accessory design. Some basic principles will be examined and followed by a general review of current practices.

The electric field in any shielded cable, since it is a coaxial system, is purely radial. The axial stress is zero. When a joint or termination is introduced, the stress distribution is no longer completely radial. Joints cannot, in general, be made without introducing an increase in diameter of both the conductor and the insulation shield. At these transitions in diameter, an axial or longitudinal component of stress is introduced. Again, when a termination is applied, there is obviously a longitudinal component of stress between the high-voltage terminal and the grounded cable shield.

The traditional method of controlling this longitudinal stress is by the accessory geometry, i.e., capacitive stress control. A common example is the familiar stress cone shown in an earlier chapter (Fig. 8.5). It may be seen that the longitudinal stress along the profile is a function of its diameter and flow. A rigorous analytical solution of this three-dimensional electric field is difficult. An approximate expression given by Brazier, with the assumption that the field is everywhere radial, was derived as follows [23]: A section of the stress control profile of a cable termination is shown in Fig. 10.4. If the insulation shield is simply terminated at R_s, very high longitudinal stresses will occur at the end of the shield because of the abrupt change in geometry. The shielded profile R_s to R is utilized to relieve the stress concentration. In Fig. 10.4, V is the potential at the

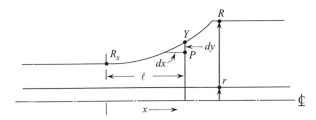

Figure 10.4 Stress control profile of cable termination.

conductor of radius r, v is the potential at any point P at radius y, g is the longitudinal stress at point P at radius y, g_r is the radial stress at radius y, R is the maximum radius of profile, and R_s is the radius of the cable insulation shield.

In a coaxial geometry g_r varies inversely as the radius, so that in general

$$g_r = \frac{k}{y} \tag{10.1}$$

where k is an arbitrary constant. Integration between the limits y and Y gives

$$v = k \int_y^Y \frac{dy}{y}$$

$$= k \ln\left(\frac{Y}{y}\right) \tag{10.1a}$$

and

$$V = k \ln\left(\frac{Y}{r}\right) \tag{10.1b}$$

Therefore in Fig. 10.4, we have the relationship

$$\frac{v}{V} = \frac{\ln(Y/y)}{\ln(Y/r)} \tag{10.2}$$

The longitudinal stress at point P is given by

$$g = \frac{dv}{dx}$$

$$= \frac{dv}{dy}\frac{dy}{dx} \tag{10.3}$$

Hence, in accordance with Eqs. (10.2) and (10.3),

$$g = \frac{V}{Y}\frac{\ln(y/r)}{\ln^2(Y/r)}\frac{dy}{dx} \tag{10.4}$$

From Eq. (10.4) it can be seen that the longitudinal stress within the profile is always less than that on the profile itself. If P lies on the profile, $y = Y$ and Eq. (10.4) becomes

$$g = \frac{V}{Y} \frac{1}{\ln(Y/r)} \frac{dY}{dx} \tag{10.5}$$

To maintain the profile as short as possible, it is necessary to keep g constant (i.e., $g = G$); therefore,

$$\frac{1}{V} \int_0^\ell G \, dx = \int_{R_s}^Y \frac{1}{Y \ln(Y/r)} \, dy$$

or

$$G\left(\frac{\ell}{V}\right) = \ln\left[\frac{\ln(Y/r)}{\ln(R_s/r)}\right] \tag{10.6}$$

and

$$\exp\left(\frac{G\ell}{V}\right) = \frac{\ln(Y/r)}{\ln(R_s/r)} \tag{10.7}$$

Transposition of terms yields

$$Y = r \exp\left[\ln\left(\frac{R_s}{r}\right) \exp\left(\frac{G\ell}{V}\right)\right] \tag{10.8}$$

Also, from Eq. (10.6), we obtain

$$L = \frac{V}{G} \ln\left[\frac{\ln(R/r)}{\ln(R_s/r)}\right] \tag{10.9}$$

where L is the overall axial length of the profile.

The difficulty with this derivation is the assumption that the field is everywhere radial, then showing that quantitatively it is not. The quick answer is that since the slope of the profile is small, the stress is nearly radial and the assumptions are valid. More importantly, the expression (10.8) has been confirmed by experimental results [23] and many years of use in practical applications [24, 25]. In practice, the use of several dielectrics, having different permittivities, in the same accessory modifies the electrical field and the contour of the stress cone, but these conditions can be readily included in the calculations [24–26].

Expression (10.8) represents the equation of the profile curve when the longitudinal stress is constant on the profile $R_s R$ and when the contour $R_s R$ is shielded. It is obvious that there is still stress enhancement at point R where the shield is terminated; however, both the longitudinal and radial stresses have been reduced because of the increase in diameter. If the diameter is large enough, the stress enhancement will be within safe limits. In high-voltage cables, it may be necessary to extend effectively the profile shielding by employing a stress control ring as shown in Fig. 10.5, which tends to change the edge geometry of the shield termination to a cylindrical geometry. In medium-voltage cables a simple stress profile is normally sufficient. In principle, at least, stress control in joints is simpler than in terminations because it is not necessary to terminate the shield. The shield is continuous over the complete joint. It should be noted that in practice it is not necessary to construct a stress control profile as given in Eq. (10.8). The profile can approximate a straight line, provided the slope of the profile is small. The double logarithmic curve is simply the most efficient curve in terms of space.

Although approximate analytical solutions of the stress profile have proven to be adequate, the control of stress at the end of the built-up termination profile can only be solved by empirical methods, analog techniques, or computer-aided designs. The very large number of calculations required by the segregation of the electric field into a grid of many small elements, the finite-element technique, can readily be handled by modern computers. These computer-aided numerical methods have virtually superseded analog methods, i.e., electrolytic tanks and resistive networks, in solving practical electrical field problems in general [27–29] and are utilized in many modern cable applications including the relatively simple designs required for distribution cable joints and terminations. Computer-aided designs (CAD) have been employed to introduce new, more cost-efficient designs employing new materials and new solutions to an old problem [30–32].

10.13 SOME CURRENT PRACTICES

The discussion will be limited to solid-dielectric extruded primary distribution cables. Accessories in the polymeric transmission cable category, 69 kV and above, are a rather specialized field and reference should be made to Chapter 4. Three methods of joint construction are employed: taped, modular, and molded. The terminals for transmission cables are in some respects simpler to design than joints because the cable insulation remains undisturbed; those used on solid-dielectric cables are normally modular and follow the high-voltage paper cable approach. Solid-dielectric distribution cable accessories are normally either of taped or modular construction.

10.13.1 Taped Designs

Before circa 1960, when underground primary distribution cables were usually oil-impregnated PILC cables and very little rubber or plastic insulated cable was designed for voltages over 10 kV, joints and terminations for solid-dielectric extruded cables were constructed by building up the required insulation reinforcements with tapes. This technique was a natural extension to the methods employed for paper insulated cables. The electrical stresses were low compared to PILC cables and the distribution cables of today. Self-bonding tapes of some kind were generally used to reduce partial discharge

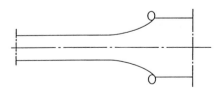

Figure 10.5 Stress control ring.

within the voids that could occur in the overlapped laminations. A common tape was self-amalgamating polyethylene (SAP). Some designs employed an open-weave tape that was later filled with resin. Because of the modest electrical stress in the cables, joints and terminations were easily made by hand-wrapped tapes in the field. These designs have been largely superseded by modular designs that require very little field labor and benefit from factory-controlled quality. Some taped designs still find application in industry (as opposed to utility) because of their versatility. A representative taped joint is shown in Fig. 10.6 Note the pencilled cable insulation and the semiconducting (SC) tape over the compression-type ferrule. The hand-wrapped rubber tape, temperature rated in compatibility with the cable insulation, forms a cohesive mass of thickness and contour desired to maintain low radial and longitudinal stresses. The semiconducting layer and copper braid insulation shield are protected by an overall vinyl tape.

10.13.2 Shrink Back

Polymeric cables have a tendency to exhibit a phenomenon called *shrink back*. This shrink back is similar to the action of heat-shrinkable polymers. When these cables are heated above their crystalline melting point (about 90°C), the insulation tends to contract at a discontinuity such as a joint due to the relaxation of locked-in forces (cf. Section 10.13.3 for further discussion). The phenomenon is particularly significant at 130°C emergency temperatures when a joint may become disturbed due to the relative movement of the cable insulation. The problem has been addressed by improvements in cable design and manufacture. Modern designs with extruded semiconducting conductor shields are much less prone to the phenomenon and joint design is able to accommodate safely any tendency for insulation movements.

10.13.3 Modular Designs

Modular accessory design for distribution voltages (usually 25 kV or less) may control the electric field either by field shields (stress cones) or by high permittivity stress control tubes [33, 34]. A field shield design is shown in Fig. 10.7. The main body of the joint is factory molded in one piece, including the inner SC ferrule shield, the insulating wall, and the outer SC shield. The EPR is formulated to have a low modulus so that a reliable interface fit is obtained at the cable-joint insulation interface. Silicone grease is applied to ensure further that interface voids are eliminated. Note that the

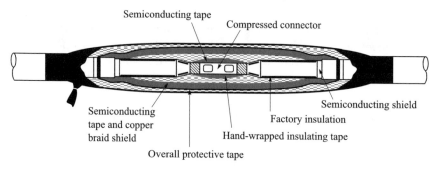

Figure 10.6 Typical taped joint for a polymeric power distribution cable. (Courtesy of 3M Canada, Inc.)

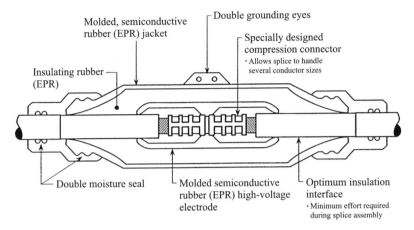

Figure 10.7 Modular joint for a polymeric power distribution cable. (Courtesy of 3M Canada, Inc.)

ferrule shield, a kind of Faraday cage, eliminates any need for insulation pencilling. Longitudinal stresses are controlled by the SC profiles. It is readily apparent that field labor for such a joint is much less than for a taped joint.

Another perhaps more recent development in modular designed terminations and joints involves the use of a stress control tube. This design is an example of capacitive grading as described in Chapter 8. Instead of discrete capacitors formed by concentric layers of insulated electrodes, a high-permitivity material ($\epsilon_r' \geq 25$) is employed that forms a capacitive grading at the connector or ferrule and cable shield terminations, as illustrated in Fig. 10.8. Table 10.1 gives the jacket, tape shield, and semiconducting shield layer cutback lengths as a function of the cable conductor size for the joint depicted in Fig. 10.8. The stress control layer that extends from the cut insulation

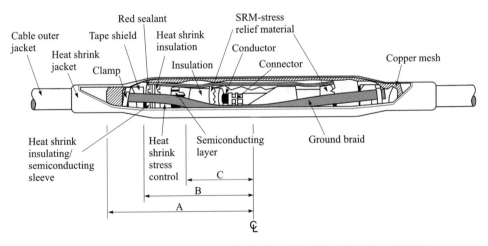

Figure 10.8 A 15-kV polymeric power distribution cable splice with high-permittivity stress control tube, where A, B, and C refer to the jacket, tape, and semiconducting layer lengths as defined in Table 10.1. (Courtesy of Raychem Corp.)

TABLE 10.1   Cutback Lengths (in.) of High-Permittivity Stress Control Tube Joint Shown in Fig. 10.8 as Function of Cable Conductor Size of 15-kV Tape or Wire Shielded Jacketed Power Cable

Conductor Size	Jacket Cutback (A)	Tape Shield Cutback (B)	Semiconducting Layer Cutback (C)
No. 2-4/0 AWG	9.5	8.0	4.5
250–350 kCM[a]	10.5	9.0	5.0
500–750 kCM	11.5	10.0	6.5
750–1000 kCM	12.0	10.5	6.5
1250–2000 kCM	14.0	12.0	9.0

kCM = thousands of circular mils (1 kCM = 1.217 mm^2)
(Courtesy of Raychem Corp).

shields to the ferrule has a sufficiently low impedance such that the axial field at the shield and ferrule discontinuities is controlled by a capacitive network, as illustrated in Fig. 10.9. Computer-aided designs such as these and high-technology polymers now enable more cost-effective and reliable products to be developed.

One of the difficulties with modular designs is the absolute necessity of eliminating voids in the cable-accessory interface. This problem can be eliminated by utilizing materials of low modulus, normally EPR, to accommodate tolerances in cable diameters and possibly to include a range of conductor sizes in one accessory kit. This low modulus technique has been extended by a method in which a core, which is inserted in the accessory tube during manufacture, keeps the unit extended until after it is slipped over the cable insulation. A Cold Shrink stress control tube termination design of this technique is shown in Fig. 10.10. The silicone insulator and stress control tube form an integral unit that is permitted to shrink on the prepared cable end by unwinding the special core [35]. Table 10.2 gives the BIL levels achievable with the stress control tube terminations depicted in Fig. 10.10 as a function of the cable conductor size, service voltage and outside insulation diameter.

A second method of providing a void-free fit is to use a heat-shrinkable material that will fit tightly over the cable. After each heat-shrinkable component is positioned over the prepared cable, it is shrunk into place by the application of heat using typically a portable glass flame torch. This technique is used to construct the joint (Fig. 10.8) and the terminal (Fig. 10.11). In Fig. 10.11 the stress control tube and high-voltage outer tube are applied separately one over the other. The sheds, where required for outdoor applications, are heat shrunk onto the outer tube to complete the assembly. The length

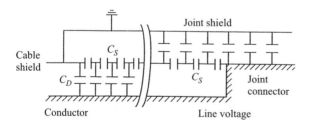

Figure 10.9 Heat-shrinkable joint equivalent circuit (after [31]).

TABLE 10.2 BIL Levels of Stress Control Tube Terminations of Type Shown in Fig. 10.10 as Applied on Polymeric Power Distribution Cables with Different Voltage Ratings and Dimensions

BIL (kV)	Cable insulation OD range		Conductor Size Range AWG and kCM (mm²)				
	in.	mm	5 kV	8 kV	15 kV	25 kV	55 kV
95	0.32–0.59	8.2–15.0	8–4	8–6			
95	0.44–0.89	11.2–22.7	2–3/0	4–2/0			
110	0.64–1.08	16.3–27.4	4/0–400	3/0–300	2–4/0 (35–120)	2–1/0 (35–50)	
150	0.64–1.08	16.3–27.4	4/0–400	3/0–300	2–4/0 (35–120)	2–4/0 (35–120)	
150	0.72–1.29	18.3–32.8	300–500	250–500	2/0–300 (70–150)	2/0–250 (70–150)	
150	0.83–1.53	21.1–38.9	500–750	350–700	4/0–500 (120–240)	250–800 (125–400)	
150	1.05–1.80	26.7–45.7	700–1500	600–1250	500–1000 (240–500)	900–1750 (500–800)	
150	1.53–2.32	38.9–58.9	1750–2000	1500–2000	1250–2000 (625–1000)		
200	0.72–1.29	18.3–32.8	300–500	250–500	2/0–300 (70–150)	2–4/0 (35–120)	2–2/0 (35–70)
200	0.83–1.53	21.1–38.9	500–750	350–700	4/0–500 (120–240)	2/0–250 (70–150)	2–4/0 (35–120)
200	1.05–1.80	26.7–45.7	700–1500	600–1250	500–1000 (240–500)	250–800 (125–400)	3/0–600 (95–325)
200	1.53–2.32	38.9–58.9	1750–2000	1500–2000	1250–2000 (625–1000)	900–1750 (500–800)	700–1500 (400–725)

(Courtesy of 3M Corp.)

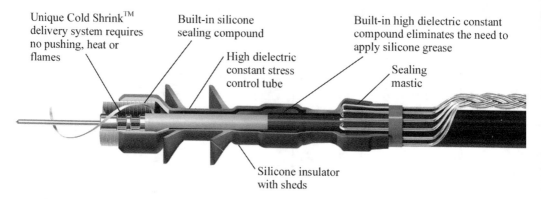

Figure 10.10 Cold Shrink stress control tube termination for a polymeric power distribution cable. (Courtesy of 3M Corp.)

of the termination delineated in Fig. 10.11 is contingent upon the voltage rating of the polymeric power distribution cable and is specified in Table 10.3.

Heat-shrinkable materials have a practical preshrink to postshrink diameter ratio of up to 4:1. The phenomenon results from crystalline or partially crystalline materials that have been subjected to a crosslinking process. Crosslinked polyethylene is a well-known example. When such materials are heated above the crystalline melting point, they become rubbery or elastomeric because of the crosslinked molecules. If the material is permitted to cool in an extended or stretched form, the crystals will reform in a lattice of the extended geometry, i.e., the materials will be frozen in the extended form. When the material is heated, it will shrink to its natural relaxed state. This phenomenon is commonly known as *elastic memory* and is utilized in the accessories of Figs. 10.8 and 10.11 to form tightly fitting units. The larger diameter reduction when the tubes are heated permits a range of cable sizes to be accommodated in one kit. Silicone grease or stress relief mastic may be employed at critical points such as the discontinuity of the semiconducting shielding where the heat-shrunk material may not conform perfectly to the contour of the substrate.

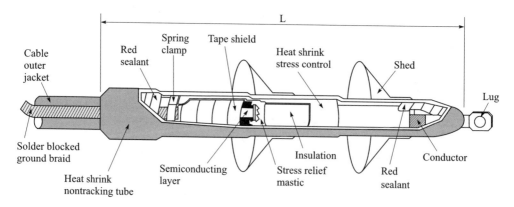

Figure 10.11 Heat-shrinkable 5/15-kV shielded polymeric power distribution cable stress control tube termination of an alternative design. (Courtesy of Raychem Corp.)

TABLE 10.3 Length of Cable Termination, Presented in Fig. 10.11, as a Function of Voltage Rating of Polymeric Power Distribution Cable

Voltage Class Cable (kV)	Length Termination (in.)	Number of Sheds
5 and 8	12	1
15	18	2

(Courtesy of Raychem Corp.)

Termination designs employed for paper insulated cables (PILC) traditionally had porcelain insulators or potheads as environmental protection. Solid-dielectric cables, not hygroscopic like PILC, do not require sealed porcelain terminals but still require the nontracking feature of the porcelain designs especially in contaminated atmospheres. This fact means that the termination and shed assembly must be carefully selected for the ability to resist tracking. Methyl silicone rubbers, polyalkylmethacrylates, and polytetrafluoroethylene (PTFE) are polymers, which are basically track resistant because they do not produce carbon in the presence of surface discharges that occur when an energized cable termination is subjected to wet conditions. Unfilled carbon chain polymers generally tend to form carbon, which yields a conductive path (a track) that short-circuits the insulation. Materials suitable for this application must also possess a high resistance to erosion in such conditions. Crosslinked polyethylene, butyl rubber, EPDM, and crosslinked ethylene vinyl acetate (EVA) materials, when compounded with the proper fillers, may be made sufficiently resistant to surface tracking and erosion for use in contaminated atmospheres.

It must be emphasized that the foregoing described modular accessories require careful installation; the installation instructions are always supplied. However, workmen often compete on minimizing installation time. There are some regions of an accessory that require special care with different kinds of products, and it would be well to foster some competition in the quality of work in these areas. One of these is the region of the cutback where the polymeric insulation shield is terminated. A ring cut is made, and the insulation shield is removed between the cut and the end of the cable. This cut is ideally a perfect step which fits into a matching step in the accessory. Also, the cut should be in a plane that is perpendicular to the axis of the cable. In addition, it is extremely important that the knife not damage the surface of the cable dielectric. Any cut in the surface creates a weakness in the dielectric at which a tree may be initiated during a lightning stroke, and later progress to a failure by discharge at operating voltage. Manufacturers measure the quality of the product themselves by observing the partial-discharge extinction voltages (PDEVs) of the assemblies made by new employees and by the PDEVs of the accessories assembled by installers in classes held at the customer's facility. Manufacturers have also arranged for tests to be applied to in-service terminations by applying a sensitive partial-discharge (PD) detection device to the accessories at the cutback region. One of the authors has used a sensing antenna mounted on a long insulating rod and connected to a battery-operated pocket radio used as an audio signal indicator of partial discharge. Other and similar methods are also described in Chapter 9.

10.13.4 Tests

Tests on joints including those used for development, quality control, and routine procedures are similar to cable tests, since a joint is essentially designed to permit a cable discontinuity to perform equally to the cable itself. On the other hand, terminations have additional requirements and imposed constraints. First, as indicated previously, the termination external surface is subjected to axial electrical fields and must be designed to contain these stresses in environments often hostile, i.e., the surfaces become contaminated and tend to track. Terminations are subjected to overvoltage tests both in the dry and wet conditions to ensure adequate external flash-over withstand capabilities. Long-term tests are performed to improve the ability to withstand successfully polluted atmospheres. Generally, in a correctly designed termination, the internal flash-over will be significantly higher than the external value. Of course, terminations should also meet cable requirements for partial discharge. Detailed test requirements may be found in standards IEEE 48, IEEE 404, and ANSI C119.2 [36–38].

The load cycle tests of the above standards are the most searching tests for compatibility of an accessory to a cable. Various diagnostic tests can be applied to determine the stability of the cable/accessory interface during the test. Measurement of the PDEVs are strongly recommended for solid-dielectric extruded power distribution cables and their accessories. With solid-dielectric cable/accessories, it is common for the PDEV to be measured and any subsequent significant drop in its value to be considered as a serious fault.

10.13.5 Separable Connectors

The developments of premolded stress relief devices based on EPR elastomers have been of great importance in the growth of Underground Residential Distribution (URD) systems. It has greatly facilitated the multitude of connections and switching facilities necessary in these systems. These connectors are well documented [26, 39, 40] and usually rated for 200-A service on 15- and 25-kV systems, the current rating being limited by the pin and socket type of connector. The principles of stress control are similar to conventional joints. These devices must be designed to act as circuit breakers in the sense that they must be capable of being opened and closed under load.

REFERENCES

[1] W. A. Thue, Private communication.
[2] M. S. Mashikian, R. Luther, J. C. McIver, J. Jurcisin, and P. W. Spencer, "Evaluation of field aged crosslinked polyethylene cables by partial discharge location," *IEEE Trans. Power Delivery*, Vol. 9, 1994, pp. 620–638.
[3] H. Emmanuel, M. Kuschel, C. Steineke, D. Pepper, R. Plath, and W. Kalkner, "A new voltage dielectric test system for insulation diagnosis and partial discharge measurement," Nordic Insulation Conference, Bergen, 1996.
[4] B. Holmgren, P. Werelius, R. Eriksson, and U. Gaevert, "Dielectric measurements for diagnosis of XLPE insulation," International Conference and Exhibition on Electricity Distribution (CIRED), IEE Conference Publication, No. 438, Birmingham, 1997.

[5] M. Kruger, "Isolierstoffprüfung verlegter Hochspannungs-Polyathylenkabel für Nennspannungen für 10 bis 30 kV mit 0.1 Hz," Dissertation Technisches Universität, Graz, Austria.

[6] U. Gaevert, R. Eriksson, and P. Werelius, "XLPE cable diagnosis by measurement of dielectric losses as a function of frequency and voltage," International Conference and Exhibition on Electricity Distribution (CIRED), IEE Conference Publication No. 3.06.1, Brussels, 8–11 May, 1995.

[7] M. Kruschel, R. Plath, I. Stepputat, and W. Kalkner, "Diagnostic techniques for service-aged XLPE-insulated medium voltage cables," REE, Special Cable, Paris, 1996.

[8] B. Bernstein, N. N. Srinivas, E. K. Duffy, and W. Starret, "Effect of dc testing on extruded cross-linked polyethylene insulated cables," TR-101245 EPRI Research Project Report 2436-01 Interim Report, Jan. 1993.

[9] R. A. Hartlein, V. S. Harper, and H. W. Ng, "Effects of voltage impulses on extruded dielectric cable life," *IEEE Power Delivery*, Vol. 4, 1989, pp. 828–841.

[10] R. A. Hartlein, V. S. Harper, and H. Ng, "Effects of voltage surges on extruded dielectric cable life: Project update," *IEEE Trans Power Delivery*, Vol. 9, 1994, pp. 611–619.

[11] R. J. Densley, R. Bartnikas, and B. Bernstein, "Multiple stress aging of solid-dielectric extruded dry-cured insulation systems for power transmission cables," *IEEE Trans. Power Delivery*, Vol. 9, 1994, pp. 559–571.

[12] R. Bartnikas, R. J. Densley, and R. M. Eichhorn, "Long term accelerated aging tests on distribution cables under wet conditions," *IEEE Trans Power Delivery*, Vol. 11, 1995, pp. 1695–1701.

[13] C. Katz, G. W. Seman, and B. S. Bernstein, "Low temperature aging on XLPE and EPR insulated cables with voltage transients," *IEEE Trans. Power Delivery*, Vol. 11, 1995, pp. 34–42.

[14] C. Katz and M. Walker, "An assessment of field aged 15 and 35 kV ethylene propylene rubber insulated cables, *IEEE Trans. Power Delivery*, Vol. 11, 1995, pp. 25–33.

[15] M. M. Epstein, L. H. Stember, B. S. Bernstein, R. Bartnikas, and S. Pélissou, "Service aged XLPE cables: insulation characteristics and breakdown strength," 1986 IEEE International Symposium on Electrical Insulation, IEEE Conference Record 86 CH2 196-4-DEI, Washington, D.C., pp. 23–27, 1986.

[16] R. Bartnikas, S. Pélissou, and H. St-Onge, "AC breakdown characteristics of in-service aged XLPE distribution cables," *IEEE Trans Power Delivery*, Vol. 3, 1988, pp. 454–462.

[17] F. C. Doble, "The ac dielectric loss and power factor method for field investigation of electrical insulation," AIEE Conference Paper CP 41-132, 1941.

[18] ERC Manufacturers, "EHV taped cable operating committee, ERC manufacturers 500/550 kV research project—Preliminary test results from Waltz Mill," *IEEE Trans. Power Apparatus and Syst.*, Vol. PAS 92, 1973, pp. 978–998.

[19] B. Bahder, G. S. Eager, R. Suarez, S. M. Chalmers, W. H. Jones, and W. H. Mangrum, "In service evaluation of polyethylene and cross linked polyethylene insulated power cables rated 15 to 35 kV," *IEEE Trans Power Apparatus Syst.* Vol. PAS-96, 1977, pp. 1754–1766.

[20] G. Bahder, C. Katz, G. S. Eager, E. Leber, S. M. Chalmers, W. H. Jones, and W. H. Mangrum, "Life expectancy of cross linked polyethylene insulated cables rated 15 to 35 kV," *IEEE Trans. Power Apparatus Systems*, Vol. PAS-100, 1981, pp. 1581–1590.

[21] R. A. Nelson and P. H. Ware, "Practical aspects of direct current testing of electric cables in the field," AIEE Conference Paper CP 62-532, 1962.

[21a] R. J. Densley, in *Engineering Dielectrics, Vol. I, Corona Measurement and Interpretation*, R. Bartnikas and E. J. McMahon, Eds, STP669, ASTM, Philadelphia, 1979, Chapters 11 and 12.

[22] ANSI/IEEE Standard 400-1995, IEEE "Guide for making direct voltage tests on power cable systems in the field."

[23] L. G. Brazier, "Joints, sealing ends and accessories for pressure cable," *Proc. IEE*, Vol. 93, Part 2, 1946, pp. 415–434.

[24] H. D. Short, "A theoretical and practical approach to the design of high voltage cable joints," *AIEE Trans. Power Apparatus Systems*, Vol. 68, 1949, pp. 1275–1283.

[25] H. D. Short, "The HST termination for UD cables in theory and practice," IEEE Conference Record No. 69C1-PWR and No. 69C-PWR (Sup) IEEE Conference on Underground Distribution, Anaheim, CA, May 1969, pp. 12–16.

[26] N. M. Sacks, "Development of separable connector system for underground power distribution," *IEEE Trans. Power Apparatus Syst.*, Vol. PAS-91, 1972, pp. 986–991.

[27] O. W. Anderson, "Laplacian electrostatic field calculations by finite elements with automatic grid generation," *IEEE Trans. Power Apparatus Syst.*, Vol. PAS-92, 1973, pp. 1485–1492.

[28] M. F. Scott, J. M. Mattingley, and H. M. Ryan, "Computation of electric fields: recent developments and practical applications," *IEEE Trans. Electrical Insulation*, Vol. E1-9, 1974, pp. 18–25.

[29] O. W. Anderson, "Finite element solution of complex potential electric fields," *IEEE Trans. Power Apparatus Syst.*, Vol. PAS-96, 1977, pp. 1156–1161.

[30] A. M. S. Katahoire, M. R. Raghuveer, and E. Kuffel, "Determination of stress cone profiles for termination of high voltage XLPE cables," *IEEE Trans. Power Apparatus Syst.*, Vol. PAS-101, 1982, pp. 3804–3809.

[31] G. J. Clarke, and L. H. Katz, "Advances in multiconductor jointing systems for medium voltage solid dielectric cables using heat shrinkable components," IEEE Conference Paper No. PC1-81-34, 1981.

[32] K. Sanjo, K. Shiraoka, N. Yasuda, K. Yatsuka, and K. Kawano, "New type cable accessories for power distribution," *IEEE Trans. Power Apparatus Syst.*, Vol. PAS-101, 1982, pp. 4484–4489.

[33] R. J. Pennek and D. D. Nyberg, "Improvements in non-tracking materials for outdoor cable terminations," 7th IEEE/PES Conference on Transmission and Distribution, Atlanta, April 1979, pp. 341–346.

[34] A. E. Blake, G. J. Clarke, and W. T. Starr, "Improvements in stress control materials," 7th IEEE/PES Conference on Transmission and Distribution, Atlanta, April 1979, pp. 264–270.

[35] R. J. T. Clabburn, *et al.* "Advances in cable jointing techniques using heat shrinkable components," Electric Energy Conference, Sydney NSW, Australia, 13–17 Oct. 1980, pp. 44–51.

[36] IEEE 48, IEEE standard test procedures and requirements for alternating-current cable terminations: 2.5–765 kV.

[37] IEEE 404, IEEE standard for cable joints for use with extruded dielectric cable rated 5,000–138,000 V, and cable joints for use with laminated dielectric cable rated 2,500–500,000 V.

[38] ANSI C119.2, IEEE standard for separable insulated connectors for power distribution systems above 600 V.

[39] R. H. Arndt and H. N. Tachick, "Underground primary connector reliability," 1971 IEEE Conference on Underground Distribution, Detroit, Sept. 27–Oct. 1, 1971, pp. 393–396.

[40] J. W. Simpson and S. R. Gilligan, "Connector modules—an aid to economical and flexible URD system design," 1971 IEEE Conference on Underground Distribution, Detroit, Sept. 27–Oct. 1, 1971, pp. 411–416.

POWER CABLE SYSTEMS

G. Ludasi

11.1 INTRODUCTION

In most power systems, cables are only an appendage to overhead lines. The high costs of underground cables restrict their use to areas where overhead lines are not practicable or aesthetically acceptable. The most important reasons for choosing underground cables versus overhead lines are as follows: (a) heavily built-up urban or industrial regions, leaving little or no room for overhead lines; (b) the costs of rights-of-way for an overhead line exceed the extra cost of undergrounding; (c) congestion of overhead lines near substations or generating stations; (d) crossing of wide waterways, large railway yards and superhighway interchanges; (e) crossing of an overhead line corridor by another electric line; (f) aesthetic considerations, public laws and regulations; and (g) safety of aviation. Consequently, cable system design can rarely be independent and has to be often subordinated to overall system design. The only exceptions where cables are the predominant factor of system design, are the networks of large, very densely built-up cities and large, heavy industrial complexes.

11.2 COMPARISON OF OVERHEAD LINES AND CABLES

Parallel operation of overhead and underground lines is restricted on account of their different electrical characteristics. We shall compare here briefly the characteristics of overhead and underground cables (cf Table 11.1).

11.2.1 Resistance

The permissible load of an overhead wire is limited by thermal expansion (sag) and the annealing temperature of the conductor. Losses are generated only in the conductor and are directly transmitted to ambient air. Whereas cable loading is mainly limited by the aging of the insulating material. Other limiting factors, such as fatigue induced by thermal expansion and contraction and prevention of the drying out of the soil, will be dealt with later. To ensure a cable life of at least 30–50 years, cables must be operated at much lower temperatures than are bare overhead conductors [1].

Losses in cables are not only generated in the conductor but also in the insulation (dielectric loss); this loss is constant and independent of the load carried. Further losses are generated in the cable shield and sheath, the enclosing metallic (magnetic) duct (pipe), and the metallic cable supports (trays). The heat generated by these losses is

449

TABLE 11.1 Principal Parameters of 120-kV, 200-MVA Overhead and Underground Lines

Parameter	Overhead	Underground
Conductor material	ACSR	Copper
Cross section, kcmil	795	1250
Cross section, mm^2	252	633
	(equivalent copper)	
Resistance, Ω/km	0.08	0.0357
Inductance		
(positive and negative sequence), Ω/km	0.455	0.199
Capacitive charge, MVAR/km	0.052	2
Conductor loss		
(at 1000 A), W/m	80	38.6
Dielectric loss		
(oil-paper insulation), W/m	—	4.46
Sheath loss (eddy), W/m	—	0.367

transmitted to the ambient through the following media: (a) cable insulation, (b) cable jacket, (c) duct space, (d) duct material, (e) encasing material, and (f) soil and backfill.

For the transmission of a given load, the cable conductor is comparatively larger than the equivalent overhead conductor. Thus the corresponding cable resistance is much lower than the resistance of the overhead line designed for the same load.

11.2.2 Inductance

Inductance is mainly defined by the quantity $\ln(s/r)$, where s is the spacing between phases and r is the radius of the conductor. It is evident that overhead lines have higher inductances on account of the wider spacing and smaller conductors utilized. Also pipe-type and three-conductor cables have lower inductances than lines consisting of individual single-conductor cables because of the closer spacing of the former [2–4].

11.2.3 Capacitance

Capacitance decreases with spacing between the phases and clearance to ground. Thus cables draw much higher charging currents than overhead lines on account of the closer spacing between phases and the proximity of the conductor and the grounded cable insulation shielding or sheath. Also the permittivity of cable insulating materials is generally 2.3–4 times that of air, thereby further augmenting the value of capacitance.

11.2.4 Overall Parameter Effects

It can be seen from the above considerations that the transmission of an equivalent amount of power will require different sizes of an overhead as opposed to an underground cable. In Table 11.1 a typical comparison is provided for the case of a 200-MVA load transmission at 120 kV. It will be noted that although the capacitive charge for an equivalent underground cable is much higher, the losses per meter in the cable are about a half of those in the overhead line (primarily because of lower I^2R losses in the larger conductor). When operated in parallel with overhead lines, cables will take over most of the load unless series reactors are installed. Caution has to be exercised as the cable capacitance together with the series reactance may form a resonant circuit. Also the charging current of a long high-voltage cable line may reduce its load-carrying ability or in an extreme case it may even exceed its current carrying-capacity; costly inductive compensation may be necessary at the ends and possibly on the route [5].

11.3 RADIAL POWER SYSTEMS

Radial systems are of the lowest cost as well as the simplest and easiest to design but offer no alternate supply route in case of failure. Because of long times required to repair underground cable and joint failures, alternate supply routes have to be provided.

The simplest design consists of distribution circuits laid out in a tree pattern, mostly without sectionalization. Branches of the circuit trees are intertwined, so that each customer or distribution transformer is supplied by two taps from different cables. Switches (manual or automatic, one normally closed, the other normally open) are provided at each primary customer to permit transfers from the normal supply line to stand by, and vice versa. At transformer vaults and pedestals separable connectors or transfer switches are usually installed to facilitate transfers. The principle of this type of radial distribution system is depicted in Fig. 11.1; for the sake of clarity, only two circuits without branches are shown.

The main advantage of this design is the ease of transferring loads to the standby circuit as the system has to be designed for each feeder to have sufficient capacity to accept these additional loads in contingency situations. Its disadvantage is the relative difficulty of fault locating in the tree pattern. The installation of fuses in the branches, and sectionalizing the trunk reduce the extent of outages and facilitate fault location.

11.3.1 Radial Branches for Underground Residential Distribution

Concentric neutral cables are usually installed for such districts. Figure 11.2 shows a typical example where each transformer vault or pedestal is fed by taps from different cables, one for normal supply, the other for standby. The two cables may be connected to different overhead or underground circuits. Another solution is connecting the cables to different phases of the same overhead distribution circuit, in case no second supply line is conveniently available near by.

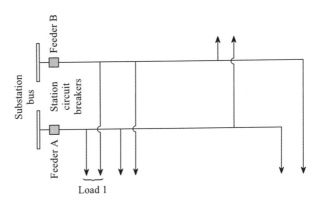

Figure 11.1 Radial feeders providing duplicate service [1].

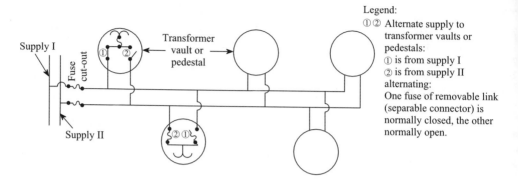

Figure 11.2 Dual radial supply for underground district.

Another solution, shown in Fig. 11.3, is connecting the transformers to a cable in a chainlike arrangement. Sectionalizing devices are installed at each transformer location, all of them normally closed, except one normally open at midpoint. In case of a fault, the defective cable has to be first identified, then disconnected at both ends. Only then can service be restored by closing the normally open disconnect and by replacing the fuse that cleared the fault. The installation of fault indicators for each cable section accelerates the process identifying the faulted cable.

Comparing the solutions shown in Figs. 11.2 and 11.3, it is evident that the first scheme (Fig. 11.2) requires more cable whereas the second necessitates additional sectionalizing devices. With the first scheme (Fig. 11.2) each transformer has to be individually transferred to its standby circuit, but it is not necessary to locate the fault first. With the second scheme (Fig. 11.3) only three switching operations (disconnecting the

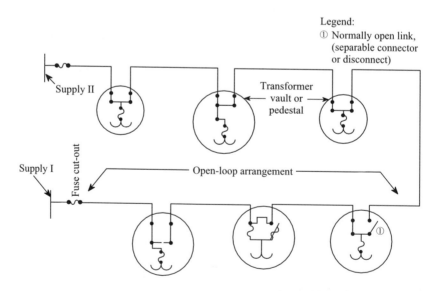

Figure 11.3 Open-loop radial supply for underground district.

faulted cable on both ends and replacing the fuse that cleared the fault) are necessary to restore service, but the faulted cable has to be identified first.

11.3.2 Secondaries

Secondary distribution systems are usually radial, much like overhead secondaries. Interconnections (normally open or closed) between transformers are rare. The only exceptions are the secondary networks (Fig. 11.4) of the very high density city centers in large metropolitan areas [1].

11.4 LOOPED SYSTEMS

11.4.1 Open Loop

Apart from the radial system described in Section 11.3 (tree pattern, Fig. 11.1), this is the most common form of underground distribution. An elemental primary distribution loop is shown in Fig. 11.5. The loop is normally fed radially from both ends with one switch near midpoint normally open. In case of trouble the faulted section has to be identified and isolated. Service is then restored by closing the normally open switch and the circuit breaker that cleared the fault. The installation of fault indicators at the ends of each cable section accelerates the process identifying the faulty cable.

This basic distribution loop is usually extended with the addition of dual radial supply branches and/or branches forming a radial open loop with a branch of an adjacent circuit. On extensive circuits some branches may be equipped with fuses. Sensitive customers receive dual feeds by taps from different circuits (Fig. 11.6).

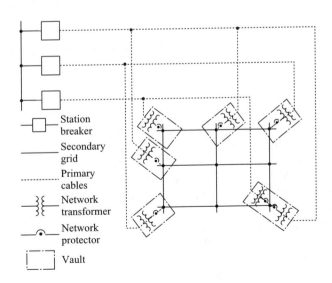

Station breaker

Secondary grid

Primary cables

Network transformer

Network protector

Vault

Figure 11.4 Typical secondary network [1].

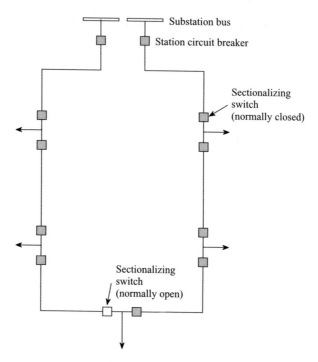

Figure 11.5 Open-loop distribution [1].

11.4.2 Closed Loop

Underground transmission and subtransmission systems are normally built to operate as closed loops to ensure continuity of service. The initial loop consists of two circuits feeding a single load. The loop may be extended to feed additional loads. A simple loop with three loads is shown in Fig. 11.7.

The addition of two feeders (*a*) and (*b*), and extensions (*e*) to supply two additional loads (*a*) and (*b*), are shown in Fig. 11.8. Note that the line between loads 2 and 3 in Fig. 11.7 has been looped to the new load (*b*), and extended (*e*) to new load (*a*) in Fig. 11.8.

Selective protective relaying will isolate the fault by opening the circuit breakers of the defective element, and the loads will be shared by the sound circuits remaining in service. The calculation of current distribution is the system planner's task and thus is not covered by this book; basics are covered in [2] and [3].

When incorporating transformers in the loop system, load distribution will be determined by the choice of the transformers, since the transformer impedance exceeds the cable impedance by an order of magnitude. A typical layout is shown in Fig. 11.9. This makes it possible to operate nonidentical cables in parallel [2]. Even the parallel operation of overhead and underground transmission lines is feasible by the insertion of transformers and by completing the connection at distribution voltage. Figure 11.9 actually depicts such a layout.

In this type of loop system the transformers are sometimes installed in different substations and parallel operation is accomplished through a number of distribution

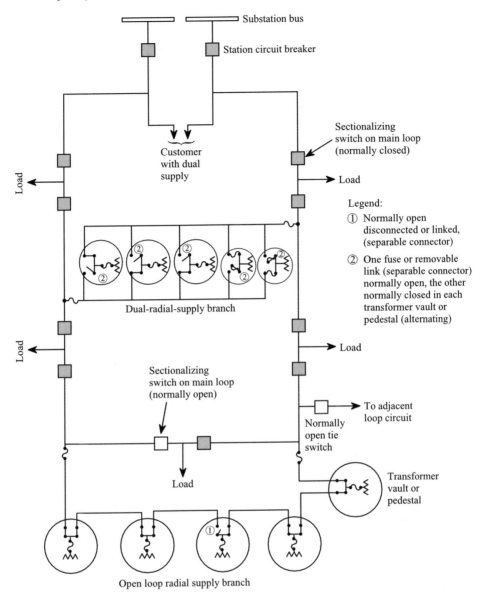

Figure 11.6 Extended open-loop distribution.

loops like the one shown in Fig. 11.5 with the following difference: *All sectionalizing switches are normally closed*, thus the open loop is turned into a closed one. Both station circuit breakers have to open in case of a distribution cable fault. In case of a loss of power supply to one of the substation buses, protective relays have to open the station circuit breakers on the affected bus, while power distribution is maintained from the other end of the distribution loops.

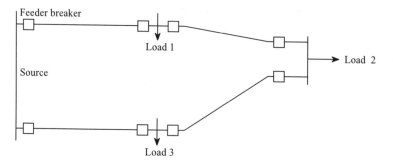

Figure 11.7 Simple loop system [1].

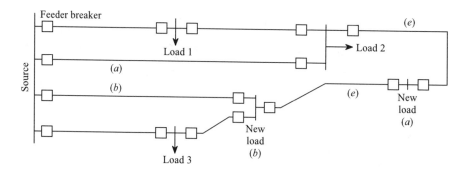

Figure 11.8 Reinforced loop system [1].

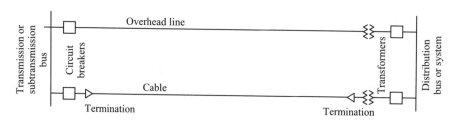

Figure 11.9 Loop system with transformers.

11.5 CURRENT-CARRYING CAPACITY: RATING EQUATIONS

Thermal considerations are at the center of cable system design. Primary sources of heat in a cable are the ohmic losses I^2R in the conductor and the dielectric losses in the insulation; the latter are usually negligible up to about the 60-kV level (cf. Table 11.5) but assume an increasing importance with the rise of the operating voltage and become a critical factor of design at 300 kV. High losses in the cable insulation present the main problem in cable development at 500 kV and above. However, low-loss insulating materials, such as polypropylene–paper laminates (PPL) and high-purity extruded polyethylene, have made it possible to install 500-kV cables, even without forced cooling, in Japan.

Also, there are induced losses in the cable sheath and shield or armor, as well as losses in the pipe or the metallic duct. The dielectric loss is voltage dependent and relatively constant regardless of load; thus it contributes heat as long as the cable is energized. The sheath and pipe associated losses are mainly caused by induction in alternating current (ac) cables and are a function of conductor current. Special measures, described in Section 6.7 and in [6], [6a], [6b], and [6c], must be taken for high-power single-conductor cables to eliminate or substantially reduce sheath losses. Pipe, armor and duct losses may be reduced by the use of nonmetallic or nonmagnetic materials. Return currents may also generate heat in cable sheaths, armor, pipe, or duct.

The created heat must be transmitted to the ambient earth or air; the heat flow is generated by the temperature gradient $\Delta\theta$. The basic mathematical description of heat flow is given by Fourier's law of heat conduction:

$$\Delta\theta = -P\rho_T\left(\frac{\Delta d}{S}\right) \tag{11.1}$$

where $\Delta\theta$ (in kelvin) is the temperature gradient, P is the heat loss in watts, ρ_T [in kelvin-meter per watt (K·m/W)] is the temperature drop with one watt being transmitted over a unit of surface over a unit length, better known as the thermal resistivity of the material, Δd is the thickness of the material in meters, and S is the surface area (in meters squared) over which the heat flows. The preceding formula is also known as Ohm's law of heat conduction, $\Delta\theta$ being analogous to voltage drop, P being analogous to current, and ρ_T pertaining to the resistivity. Fourier's law is the basis of current-carrying capacity calculations.

In most countries loading tables have been issued for different types of cables, for different voltages and conductor sizes, and many commonly encountered installation conditions. For the United States and Canada "IEEE Standard Power Cable Ampacity Tables" [7] covers most common cable types and installation conditions; they are based on the Neher–McGrath calculating method [8] traditionally used in North America; thus English units are used in [7] and [8] except for temperature, which is given in degrees Kelvin or Celsius; cross sections are American Wire Gauge (AWG) and thousands of circular mils (kcmil). These tables, however, cannot cover every type of problem, such as groups of unevenly loaded cables, and special designs. We shall therefore describe the most important calculating methods and give references to enable the reader to solve more complex problems.

The International Electrotechnical Commission (IEC) standard [9] was initially derived from the Neher–McGrath method [8] but metric units were used. This international standard has been updated and expanded to cover practically all design aspects; it has been divided into three parts. The parts are further divided into sections, each section being published separately [9]. This and the immediately following sections refer for guidance to this wide-ranging group of standards continuously updated and expanded by an international group of experts; accordingly metric [Système International (SI)] units are used here. Also, for the sake of convenience, the symbols used in the IEC-287 group of standards [9] are used here with two exceptions: (i) for losses, P is used here like in the other chapters of this book, instead of W which is the symbol for loss in IEC 287 [9]; (ii) for the voltage, V is used here like in the other chapters of this book, instead of U which is the symbol for voltage in IEC 287 [9]. Current-carrying capacity calculations are further treated in detail in the compendium *Rating of Electric Power Cables* [10]. The IEC-287 group of standards [9] only treats steady-state loads (100% load factor), accordingly Sections 11.5–11.7.4.10 cover steady-state loading only.

Cyclic loading and short-term overloading (emergency loading) are treated by the IEC-853 group of standards [11]. The Neher–McGrath method [8] also covers cyclic loading (and emergency loading to a certain extent); its treatment of cyclic loading is much simpler than that of IEC. A comparison of the two methods is given in Section 11.8.4. In this chapter cyclic loads are treated in Sections 11.8–11.8.3 on the basis of Neher and McGrath, whereas emergency (short-term or transient) ratings are dealt with in Sections 11.9–11.9.9 on the basis of IEC-853-2 [11b].

In current-carrying capacity (the term *ampacity* is often used instead) calculations for extra high-voltage (EHV) cables 230 kV and above the *capacitive charging current* has to be *included* in the *permissible current I*, which is in this case the *vector sum of the load and charging currents* [2, 3].

The temperature rise above ambient of an ac cable can be expressed as [9a]

$$\Delta\theta = (I^2 R + \tfrac{1}{2} P_d) T_1 + [I^2 R(1 + \lambda_1) + P_d] n T_2 + [I^2 R(1 + \lambda_1 + \lambda_2) + P_d] n (T_3 + T_4)$$

$$(11.2)$$

where $\Delta\theta$ is the conductor temperature rise above ambient due to heat generated by the losses in the cables (in kelvin).

Note that here the ambient temperature is the temperature of the surrounding medium under normal conditions, at the location where the cables are installed or are to be installed, including the effect of any local source of heat, but not the increase of temperature in the immediate neighborhood of the cables due to the heat arising from the cables proper.

Also here I is the current flowing in one conductor in amperes, R is the ac resistance of the conductor at maximum operating temperature (for a unit length) in ohms per meter, P_d is the dielectric loss of the insulation surrounding the conductor (for a unit length) in watts per meter, T_1 is the thermal resistance of one conductor to sheath (for a unit length) in kelvin-meter per watt, T_2 is the thermal resistance of bedding between sheath and armor (for a unit length) in kelvin-meter per watt, n is the number of conductors in the cable (conductors of equal size and carrying the same load), λ_1 is the ratio of sheath losses to the total of the conductor losses in that cable, λ_2 is the ratio of armor losses to the total of the conductor losses in that cable, T_3 is the thermal resistance of the outer serving of the cable (for a unit length) in kelvin-meter per watt, and T_4 is the thermal resistance of the surrounding medium (ratio of cable surface temperature rise above ambient to the losses per unit length) in kelvin-meter per watt.

From Eq. (11.2) the permissible current rating is obtained by expliciting I:

$$I = \left[\frac{\Delta\theta - P_d[0.5 T_1 + n(T_2 + T_3 + T_4)]}{R T_1 + n R(1 + \lambda_1) T_2 + n R(1 + \lambda_1 + \lambda_2)(T_3 + T_4)} \right]^{0.5} \qquad (11.3)$$

For cables in direct contact with the soil or a backfill material that is losing its thermal conductivity when drying out, the permissible load may be limited by θ_x (in degrees Celsius) the critical temperature of the soil, or of another surrounding medium, where moisture migration is initiated.

$$I = \left[\frac{\Delta\theta_x - nP_d T_4}{nRT_4(1 + \lambda_1 + \lambda_2)} \right]^{0.5} \tag{11.4}$$

where $\Delta\theta_x = \theta_x - \theta_a$ is the critical temperature rise of the soil above ambient, in this case both θ_x and $\Delta\theta_x$ are taken at the cable surface, and θ_a is the ambient temperature (in degrees Celsius) as previously defined in the note to (11.2). The permissible current rating shall be the lower of (11.3) or (11.4).

Special problems, such as cables or ducts directly exposed to solar radiation and buried cables where partial drying out of the soil occurs, are treated in detail by [9] and [10].

11.5.1 Direct-Current Cables

In direct-current (dc) cables skin and proximity effects, dielectric losses, induced losses in cable screens, shields and sheaths, and magnetization losses are zero. Therefore Eq. (11.3) is simplified:

$$I = \left[\frac{\Delta\theta}{R'T_1 + nR'T_2 + nR'(T_3 + T_4)} \right]^{0.5} \tag{11.3a}$$

where R' is the dc resistance per unit length of the conductor (in ohms per meter) at the maximum operating temperature.

Similarly Eq. (11.4) is simplified:

$$I = \left[\frac{\Delta\theta_x}{nR'T_4} \right]^{0.5} \tag{11.4a}$$

11.6 CALCULATION OF LOSSES

11.6.1 Resistance of the Conductor

The ac resistance of the conductor at its maximum operating temperature θ is calculated for a unit length by the following formula [Eq. (11.5)], which has to be modified [Eq. (11.6)] for pipe-type cables; see Section 11.6.1.3:

$$R = R'(1 + y_s + y_p) \tag{11.5}$$

and R' is calculated as

$$R' = R_0[1 + \alpha_{20}(\theta - 20)] \tag{11.5a}$$

where R is the ac resistance of the conductor at maximum operating temperature for a unit length (in ohms per meter), R' is the dc resistance of the conductor at maximum operating temperature for a unit length (in ohms per meter), y_s is the skin effect factor, y_p is the proximity effect factor, R_0 is the dc resistance of the conductor at 20°C for a unit length (in ohms per meter), α_{20} is the temperature coefficient of resistance at 20°C, θ is the maximum operating temperature (in degrees Celsius).

The value of R_0 is calculated as

$$R_0 = \frac{10^6 \rho}{A_c} \tag{11.5b}$$

where A_c is the nominal cross-sectional area of the conductor (in millimeters squared) taking into account the effect of stranding (to convert thousands of circular mils into millimeters squared, multiply by 0.5067075), and ρ is the resistivity of the conductor material in ohm-meters at 20°C. Values for ρ and α_{20} are given in Table 11.2.

To calculate the resistance of multilayer stranded conductors, Eqs. (11.15), (11.16), and (11.16a) may be applied here for a layer-by-layer calculation, rather than Eq. (11.5b). For this application read R_0 instead of R_s, and read ρ (at 20°C) instead of ρ_{sn} (at operating temperature), accordingly the resistance R_{sn} of the individual layers is also taken at 20°C for the calculation of R_0.

11.6.1.1 Skin Effect

The most used computation method is given in IEC 287-1-1 [9a] for the calculation of the skin effect factor:

$$y_s = \frac{x_s^4}{192 + 0.8 x_s^4} \tag{11.5c}$$

where

$$x_s^2 = \frac{8\pi f}{R'}(10^{-7})k_s \tag{11.5d}$$

and f is the supply frequency in hertz. Values for k_s are given in Table 11.3. For hollow conductors IEC Standard 287-1-1 [9a] provides (Eq. (11.5e)]

TABLE 11.2 Electrical Resistivities and Temperature Coefficients of Metals at 20°C

Material	Resistivity, ρ ($\Omega \cdot$m)	Temperature Coefficient (α_{20}) K^{-1}
Conductors		
Copper	1.7241×10^{-8}	3.93×10^{-3}
Aluminum	2.8264×10^{-8}	4.03×10^{-3}
Sheaths and armor		
Lead or lead alloy	21.4×10^{-8}	4.0×10^{-3}
Steel	13.8×10^{-8}	4.5×10^{-3}
Bronze	3.5×10^{-8}	3.0×10^{-3}
Stainless steel	70×10^{-8}	Negligible
Aluminum	2.84×10^{-8}	4.03×10^{-3}

Source: From [9a].

TABLE 11.3 Skin and Proximity Effects (Experimental Values)

Conductor Type	Whether Dried and Impregnated or Not	k_s	k_p
Copper			
Round, stranded	Yes	1	0.8
Round, stranded	No	1	1
Round, compact	Yes	1	0.8
Round, compact	No	1	1
Round, segmental (Milliken)	—	0.435[a]	0.37[a]
Hollow, helical stranded	Yes	Eq. (11.5e)	0.8
Sector shaped	Yes	1	0.8
Sector shaped	No	1	1
Aluminum[b]			[c]
Round, stranded	either	1	
Round, 4 segment (Milliken)	either	0.28	
Round, 5 segment (Milliken)	either	0.19	
Round, 6 segment (Milliken)	either	0.12	

[a] Values under consideration by IEC applicable to conductors of a cross-sectional area up to 1500 mm^2 (with or without central duct), with four segments, where all layers have the same direction of lay.
[b] Values under consideration by IEC.
[c] For k_p there are no accepted experimental results with aluminum conductors, it is recommended by IEC to use values given for similar copper conductors.
Source: From [9a].

$$k_s = \frac{d'_c - d_i}{d'_c + d_i}\left(\frac{d'_c + 2d_i}{d'_c + d_i}\right)^2 \tag{11.5e}$$

where d_i (in millimeters) is the inside diameter (diameter of the central duct) of the conductor and d'_c (in millimeters) is the outside diameter of the equivalent solid conductor having the same inside diameter.

Equations (11.5c) and (11.5d) apply to most conductor types and sizes, provided $x_s \leq 2.8$. Reference [10] provides for a number of special cases.

Skin effect is very important in large conductors, it can be reduced by using hollow-core or segmental (Milliken) construction or the combination of hollow-core and segmental. This combination is often called a *hollow-core Milliken* conductor.

11.6.1.2 Proximity Effect

For most three-phase applications IEC 287-1-1 [9a] gives the following for the calculation of the proximity effect factors:

$$y_p = \frac{x_p^4}{192 + 0.8x_p^4}\left(\frac{d_c}{s}\right)^2\left[0.312\left(\frac{d_c}{s}\right)^2 + \frac{1.18}{[x_p^4/(192 + 0.8x_p^4)] + 0.27}\right] \tag{11.5f}$$

where

$$x_p^2 = \frac{8\pi f}{R'}(10^{-7})k_p \tag{11.5g}$$

and d_c is the diameter of the conductor (in millimeters), s is the distance between conductor axes (in millimeters). Values for k_p are given in Table 11.3. Equations (11.5f) and (11.5g) apply to most conductor types and sizes, provided $x_p \leq 2.8$. Reference [10] provides for a number of special cases.

Proximity effect is important in closely spaced large single-conductor cables and in three-phase cables with large conductors, its effect may be reduced the same way as skin effect. Proximity effect is further reduced by increasing the spacing between cables. On the other hand circulating currents in the sheaths and screens or shields increase with spacing (cf. Section 11.6.3.1) leading to increased losses there, unless special measures (cf. Section 11.6.3.2) are taken to eliminate or reduce sheath (shield) currents.

11.6.1.3 Skin and Proximity Effects in Pipe-Type and SL Cables

As indicated in Section 11.6.1 in conjunction with Eq. (11.5), this formula has to be modified when applied to pipe-type cables. The modification consists of the introduction of a multiplier applicable to the skin and proximity effect factors. Neher and McGrath [8] and IEC 287-1-1 [9a] both use a 1.7 multiplier (empirical). Following research results in the United States this multiplier is in the process of being reduced to 1.5 for pipe-type cables by IEC; accordingly an amendment to this effect is expected from IEC; also the *Underground Transmission Systems Reference Book* [2] already uses the revised value of 1.5 for pipe-type cables. Pending final approval by IEC, this chapter uses the conservative value 1.7. The modified equation is

$$R = R'[1 + 1.7(y_s + y_p)] \tag{11.6}$$

11.6.2 Dielectric Loss

Dielectric loss is voltage dependent but variations are small, taking into account the changes of voltage normally occurring in power systems. It is therefore taken as being constant when calculating current-carrying capacities.

The magnitude of dielectric loss P_d for a unit length of cable (watts per meter) is given by

$$P_d = 2\pi f C V_0^2 \tan \delta \tag{11.7}$$

where C is the capacitance of one phase per unit length in farads/meter (F/m), V_0 is the maximum system voltage to ground in volts. Dielectric loss is negligible at voltages lower than those in Table 11.5.

The dielectric loss factor $\tan \delta$ has been discussed in several previous chapters; typical standard values are given in Table 11.4.

The capacitance of the cable for Eq. (11.7) is calculated by

$$C = \frac{\varepsilon_r}{18 \ln(D_i/d_c)} 10^{-9} \tag{11.7a}$$

TABLE 11.4 Values of Relative Permittivity (ε_r) and Dielectric Loss Factor tan δ for Power Cables at Power Frequency

Cable Type	ε_r	tan δ
Cables insulated with impregnated paper		
Solid-type, fully impregnated, preimpregnated, or mass-impregnated nondraining	4	0.01
Oil-filled, self-contained		
Up to $U_0 = 36$ kV	3.6	0.0035
Up to $U_0 = 87$ kV	3.6	0.0033
Up to $U_0 = 160$ kV	3.5	0.0030
Up to $U_0 = 220$ kV	3.5	0.0028
Oil pressure, pipe-type	3.7	0.0045
External gas pressure	3.6	0.0040
Internal gas pressure	3.4	0.0045
Cables with other kinds of insulation		
Butyl rubber	4	0.050
EPR		
Up to and including 18/30 (36) kV cables	3	0.020
Greater than 18/30 (36) kV cables	3	0.005
PVC	8	0.1
PE (HD and LD)	2.3	0.001
XLPE		
Up to and including 18/30 (36) kV cables (unfilled)	2.5	0.004
Greater than 18/30 (36) kV cables (unfilled)	2.5	0.001
Greater than 18/30 (36) kV cables (filled)	3.0	0.005
Polypropylene-paper laminate (PPL)[a]	2.8	0.0014[b]

[a] Under consideration by IEC.
[b] 0.0021 for oil-pressure, pipe-type cable [2]; 0.001 for oil pressure, pipe-type cable [7].
Source: From [9a].

TABLE 11.5 Lowest Phase to Ground Voltage Where Dielectric Losses Have to Be Taken into Account

Cable Type	V_0 (kV)
Cables insulated with impregnated paper	
Solid type	38
Oil-filled and gas pressure	63.5
Cables with other types of insulation	
Butyl rubber	18
EPR	63.5
PVC	6
PE (HD and LD)	127
XLPE (unfilled)	127
XLPE (filled)	63.5
PPLP[a]	63

[a] Under consideration by IEC.

where ε_r is the relative permittivity of the insulation (typical standard values are given in Table 11.4), D_i is the external diameter in millimeters of the insulation inside the screen or shield, d_c is the outside diameter of the conductor, measured over the conductor screen if any, in millimeters.

Equation (11.7a) applies to circular conductors within a concentric cylinder but may be used for oval conductors if the appropriate geometric mean diameter is substituted for D_i and d_c.

11.6.3 Losses in Cable Screens, Shields, and Sheaths

The ratio of sheath (screen or shield) losses to the total of conductor losses (λ_1) consists of *circulating current losses* (λ_1') and *eddy current losses* (λ_1''). Accordingly,

$$\lambda_1 = \lambda_1' + \lambda_1'' \tag{11.8}$$

11.6.3.1 Circulating Current Losses

For three solidly bonded single-conductor cables in equilateral (trefoil) formation, and also for three equidistant regularly transposed single-conductor cables in flat formation with solid bonds at every third transposition:

$$\lambda_1' = \frac{R_s}{R} \frac{1}{1 + (R_s/X)^2} \tag{11.9}$$

where R_s is the resistance per unit length of sheath (shielding) at its maximum operating temperature (in ohms per meter), and for tubular sheaths it is calculated as

$$R_s = \frac{\rho_s 10^6}{d\pi(\Delta d_s)}[1 + \alpha_{20}(\theta_s - 20)] \tag{11.10}$$

Also in Eq. (11.9) X (in ohms per meter) is the reactance per unit length of sheath or insulation shielding to be calculated as

$$X = 4\pi f 10^{-7} \ln\frac{2s}{d} \quad \text{for equilateral (trefoil) formation} \tag{11.11}$$

$$X = 4\pi f 10^{-7} \ln\left[2(2)^{1/3}\,\frac{s}{d}\right] \quad \text{for flat formation} \tag{11.11a}$$

In Eqs. (11.10), (11.11), and (11.11a) d is the mean diameter of the sheath (in millimeters), s is the distance between cable axes (in millimeters), Δd_s is the thickness of the sheath (in millimeters), ρ_s is the resistivity of the sheath material given in Table 11.2, and α_{20} is its temperature coefficient of resistance given in Table 11.2.

For calculating the resistance of concentric neutrals Eq. (11.10) has to be replaced with Eq. (11.16).

For pipe-type and armored SL cables, Eq. (11.9) has to be modified as described in Section 11.6.3.7.

Cables with magnetic armor, other configurations of laying, and parallel circuits are dealt with in [9a], [9b], and [10]. Further corrections have to be applied to λ_1' to account for variations in spacing between the cables [9a].

The following example will illustrate the method for calculating the sheath loss in a lead-sheathed, oil-filled cable, having a copper conductor of 1500 kcmil or 760.06 mm^2 to be operated at $\theta = 75°C$; we are also given that

$$\theta_s = 50°C \qquad \text{(approximate sheath temperature)}$$

$$d_i = 0.6\,\text{in} = 15.24\,\text{mm} \qquad \text{(conductor inside diameter)}$$

$$d_c = 1.543\,\text{in} = 39.19\,\text{mm} \qquad \text{(conductor outside diameter)}$$

$$d = 2.243\,\text{in} = 56.97\,\text{mm} \qquad \text{(mean diameter of sheath)}$$

$$\Delta d_s = 0.13\,\text{in} = 3.3\,\text{mm} \qquad \text{(thickness of sheath)}$$

$$s = 9\,\text{in} = 228.6\,\text{mm} \qquad \text{(axial spacing between cables)}$$

Substituting the numerical values in the appropriate expressions gives

$$R_0 = \frac{1.7241}{760.06} \times 10^{-2} = 2.2684 \times 10^{-5}\,\Omega/\text{m}$$

$$R' = 2.2684\left[1 + \frac{3.93}{1000}(75 - 20)\right] \times 10^{-5} = 2.7587 \times 10^{-5}\,\Omega/\text{m}$$

$$R_s = \left(\frac{21.4}{56.97\pi 3.3} \times 10^{-2}\right)\left[1 + \frac{4}{1000}(50 - 20)\right] = 4.0581 \times 10^{-4}\,\Omega/\text{m}$$

$$k_s = 0.72 \quad \text{and} \quad k_p = 0.8$$

Also,

$$x_s^2 = \frac{8\pi 60}{2.7587} \times 0.72(10^{-2}) = 5.466(0.72) = 3.9357$$

$$x_p^2 = \frac{8\pi 60}{2.7587} \times 0.8(10^{-2}) = 5.466(0.8) = 4.373$$

$$y_s = \frac{3.9357^2}{192 + 0.8(3.9357)^2} = 7.57845 \times 10^{-2}$$

$$y_p = \frac{4.373^2}{192 + 0.8(4.373)^2}\left(\frac{1.543}{9}\right)^2\left[0.312\left(\frac{1.543}{9}\right)^2 + \frac{1.18}{\{4.373^2/[192 + 0.8(4.373)^2]\} + 0.27}\right]$$

$$= 8.8574 \times 10^{-3}$$

$$R = 2.7587(1 + 0.0757845 + 0.0088574) \times 10^{-5} = 2.9922 \times 10^{-5}\,\Omega/\text{m}$$

$$X = 4\pi 60(10^{-7})\ln\left(2\frac{9}{2.243}\right) = 1.57 \times 10^{-4}\,\Omega/\text{m}$$

Finally,

$$\lambda_1' = \frac{4.0581 \times 10^{-4}}{2.9922 \times 10^{-5}} \left[\frac{1}{1 + (4.0581/1.57)^2} \right] = 1.7657$$

It is seen that in this particular example the sheath losses exceed the conductor loss by 76.57%, which for an aluminum sheath of similar dimensions is found to be slightly less at 61%.

The example illustrates that high-power single-conductor cables have excessively high sheath losses when constructed with solidly bonded sheaths, and special measures have to be taken to eliminate or limit such losses (cf. Section 11.6.3.2).

It should be pointed out that sheath losses remain at acceptable levels in solidly bonded cables where the conductor current is maintained at less than 500 A.

For loads up to about 750 A sheath losses may be kept at an acceptable level by laying the single-conductor cables of a circuit in a triangular configuration with the three cables touching. For the three cables in the preceding example, the spacing (s) is thus reduced to 2.613 in (66.37 mm), assuming a jacket thickness of 120 mils (approximately 3 mm). As a result losses in a lead sheath are reduced to about 30% of the losses in the conductor, or about one sixth of the losses with a 9-in. spacing. In an aluminum sheath the losses are reduced to less than 10% of conductor losses, or about $\frac{1}{16}$ of the losses with a 9-in. spacing. On the other hand the ac conductor resistance is increased 9% due to proximity effect. Also the thermal resistance (T_4) is increased (see Section 11.7) on account of mutual heating.

Sheath eddy currents (λ_1'') are usually negligible in solidly bonded single-conductor cables where $\lambda_1' \gg \lambda_1''$, except for special cases such as heavily loaded ac cables with large Milliken conductors. These special cases are extensively treated in [9a], and [10].

11.6.3.2 Losses in the Sheaths or Shields of Specially Bonded Systems

Special sheath-bonding systems have to be used to eliminate or reduce sheath circulating currents in high-power cables; these systems have been described in Section 6.7, and [6], [6a], [6b], and [6c]. There are no sheath circulating currents flowing in single-point bonded systems and in electrically balanced cross-bonded systems, therefore $\lambda_1' = 0$, but λ_1'' has to be calculated according to Section 11.6.3.3. However, when the three minor sections forming a major section are unequal, λ_1' is determined by multiplying its value calculated according to Section 11.6.3.1 by the correction factor, c_F,

$$c_F = \left(\frac{p + q - 2}{p + q + 1} \right)^2 \tag{11.12}$$

where $p = l_2/l_1$ and $q = l_3/l_1$; here l_1 is the shortest minor section, whereas l_2 and l_3 are the longer minor sections. Circulating currents being greatly reduced, eddy current losses have to be calculated according Section 11.6.3.3.

11.6.3.3 Eddy Current Losses in Sheaths or Shields

In Single-Conductor Cables. Eddy current losses are significant when circulating currents are eliminated or greatly reduced by single-point bonding or cross bonding. The eddy current loss factor λ_1'' in the sheaths of single-conductor cables is calculated as

$$\lambda_1'' = \frac{R_s}{R}\left[g_s\lambda_0(1 + \Delta_1 + \Delta_2) + \frac{(\beta_1 t_s)^4}{12 \times 10^{12}}\right] \qquad (11.13)$$

where

$$g_s = 1 + \left(\frac{t_s}{D_s}\right)^{1.74}(\beta_1 D_s \times 10^{-3} - 1.6) \qquad (11.13a)$$

and

$$\beta_1 = \left[\frac{8\pi^2 f}{10^7 \rho_s}\right]^{0.5} \qquad (11.13b)$$

Here ρ_s is the resistivity of the sheath material at operating temperature taken from Table 11.2 (in ohm-meters), t_s is the thickness of the sheath (in millimeters), D_s is the outside diameter of the sheath (in millimeters); for corrugated sheaths the mean value of the outside diameter should be taken as D_s.

For lead sheaths g_s may be taken as unity and the term $(\beta_1 t_s)^4/12 \times 10^{12}$ in Eq. (11.13) becomes negligible. For aluminum sheaths with $D_s > 70\,\text{mm}$, and also for aluminum sheaths of greater than usual thickness, both terms are significant in Eq. (11.13) and must be considered.

For Three Single-Core Cables in Trefoil Formation

$$\lambda_o = 3\left(\frac{m^2}{1 + m^2}\right)\left(\frac{d}{2s}\right)^2 \qquad (11.13c)$$

where

$$m = \frac{2\pi f}{R_s} \times 10^{-7} \qquad (11.13d)$$

whenever $m \le 0.1$, Δ_1 and Δ_2 may be neglected.

Also for Trefoil Formation

$$\Delta_1 = (1.14m^{2.45} + 0.33)\left(\frac{d}{2s}\right)^{(0.92m+1.66)} \qquad (11.13e)$$

and $\Delta_2 = 0$.

For single-conductor cables in other formations calculating methods are to be found in [9a], whereas [9b] treats double circuits.

In Cables with Large Segmental (Milliken) Type Conductors. With a reduced proximity effect, eddy current loss λ_1'' cannot be ignored, and it even has to be increased by applying a correction factor to the result of Eq. (11.13). References [9a] and [10] treat these cases in detail, as mentioned at the end of Section 11.6.3.1.

In Three-Conductor Cables with a Common Sheath and No Magnetic Armor. λ_1' is negligible and the eddy current loss factor λ_1'' *for round conductors* is given by

$$\lambda_1'' = \frac{3R_s}{R}\left\{\left(\frac{2c}{d}\right)^2\left[1+\left(\frac{R_s}{2\pi f}\times 10^7\right)^2\right]^{-1} + \left(\frac{2c}{d}\right)^4\left[1+4\left(\frac{R_s}{2\pi f}\times 10^7\right)^2\right]^{-1}\right\} \quad (11.14)$$

where c is the distance between the axis of one conductor and the axis of the cable (in millimeters), and d is the mean diameter of the sheath (in millimeters).

Eddy current losses in the sheaths of other types of three-conductor cables (such as cables having oval- or sector-shaped conductors and armored cables) are treated in [9a] and [10].

11.6.3.4 Calculation of Losses in Nonmagnetic Armor or Sheath Reinforcement

It is customary to treat the sheath and nonmagnetic reinforcement as a single entity if they are solidly bonded together at both ends and on the route of long lengths. For concentric cylinders (round or oval) the value of d is calculated as the root-mean-square (rms) value of the diameters of the bonded components: shield, sheath, reinforcement, and armor. For the calculation of the combined resistance R_s, the bonded components are considered to be connected in parallel; therefore,

$$\frac{1}{R_s} = \frac{1}{R_{s1}} + \frac{1}{R_{s2}} + \frac{1}{R_{s3}} + \cdots \quad (11.15)$$

where R_{s1}, R_{s2}, etc. are the resistances of the components in (ohms per meter):

The following are to be considered for the calculation of the resistances (R_{s1}, etc.) of different types of taped or wired components:

  (i) For a long lay construction with bare tapes in contact with each other, the resistance is based on an equivalent cylinder having the same mass of material per unit length and the same inside diameter as the layer of tapes (or wires);

 (ii) when two or more layers of tapes are in contact with each other within a short lay helicoidal construction, twice the calculated value according to (i) must be employed;

(iii) for a single layer, short lay construction (circumferential tapes), R_s is considered to be infinite and the losses can be neglected;

 (iv) for a single layer of tapes wound at approximately 54° to the cable axis, the same calculation procedure as outlined in (i) should be employed and the calculated value shall be multiplied by 2;

(v) in a long lay construction with insulated wires, or with a gap between the individual bare tapes (wires), there are no eddy currents, therefore $\lambda_1'' = 0$, where as for circulating current losses (λ_1') the resistance R_{sn} is to be calculated as

$$R_{sn} = L_{sn}\rho_{sn}10^6/(n_{sn}A_{sn}) \tag{11.16}$$

where

$$L_{sn} = \left[1 + \left(\frac{\pi d_{sn}}{l_{sn}}\right)^2\right]^{0.5} \tag{11.16a}$$

and is known as the lay factor or lay-length factor, further on ρ_{sn} is the resistivity (in ohm-meters) of the material of the layer n at its operating temperature, n_{sn} is the number of wires or tapes in the layer n, A_{sn} is the cross-sectional area (in millimeters squared) of an individual wire or tape in the layer n, d_{sn} is the mean diameter (in millimeters) of the layer n, l_{sn} is the length of lay (in millimeters) of the layer n. The ratio l_{sn}/d_{sn} is known as the lay ratio of layer n.

11.6.3.5 Losses in Magnetic Armor

When magnetic armor materials are employed, both the resistances and reactances are influenced by the magnetic properties of the metals and a series of formulas must be used as detailed in [9a] and [10].

11.6.3.6 Concentric Neutral Cable

The material for concentric neutral conductors is specified to be annealed copper (wires or tapes) by U.S. and Canadian standards [12 and 13]. They are generally constructed according to item (v) of Section 11.6.3.4, and their resistance is to be calculated accordingly [Eqs. (11.16) and (11.16a)]. American [12] and Canadian standards [13] specify a lay ratio l_{sn}/d_{sn} of 6–10. Cable manufacturers strive for the most efficient use of material, therefore, in calculations it is customary to assume a lay ratio of 10 and a corresponding lay factor $L_{sn} = 1.05$.

Single Phase Systems. Both U.S. [12] and Canadian [13] standards specify that for single-phase applications the concentric neutral conductor's cross-sectional area shall not be less than 98% of the nominal cross section of the copper central conductor (60% of the nominal cross section of a central conductor made of aluminum); this is known as a *full neutral*. Unless a substantial portion of the return current takes other parallel paths (such as adjacent neutral conductors solidly bonded to the concentric neutral for its full length or adjacent metallic pipes solidly bonded to it), the current in the concentric neutral may be taken to be equal to the current in the central conductor. By definition

$$\lambda_1' = \frac{I_s^2 R_s}{I_c^2 R_c} \tag{11.16b}$$

where I_s and I_c are, respectively, the currents in the concentric neutral and the central conductor, where as R_s and R_c are, respectively, the resistances (in ohms per meter) of the concentric neutral and the central conductor, each at its own maximum operating temperature. If $I_s \approx I_c$, λ_1' is reduced to

$$\lambda_1' \approx \frac{R_s}{R_c} \tag{11.16c}$$

and with both currents and resistances being equal, λ_1' may be taken as unity. As mentioned in item (v) Section 11.6.3.4, $\lambda_1'' = 0$.

Three phase systems. For three single-phase concentric neutral cables forming a three-phase circuit, both U.S. [12] and Canadian [13] standards specify that each concentric neutral conductor's cross-sectional area shall not be less than 33% of the nominal cross section of the copper central conductor (20% of the nominal cross section of a central conductor made of aluminum); this is known as a *one-third neutral*. Cross sections other than *full* or *one-third* may be chosen for concentric neutral conductors, if required, to withstand fault currents (cf. Section 11.10) or to carry circulating and return currents. The three cables may be laid separately (pulled into individual ducts) or cabled together to form a triplex assembly.

Skin and proximity effects are calculated according to Sections 11.6.1.1 and 11.6.1.2. Equation (11.9) is used to calculate circulating current losses in the concentric neutrals (λ_1'). As mentioned earlier in item (v) of Section 11.6.3.4, eddy current losses in the concentric neutral conductor (λ_1'') are equal to *zero*.

11.6.3.7 Pipe-Type Cable Losses and Losses in the Sheaths of SL Cables

As mentioned in Section 11.6.1.3, a 1.7 multiplier (which is in the process of being reduced to 1.5 for pipe-type cables and for sheath losses in SL cables) has to be applied to the y_s and y_p values for the calculation of conductor losses. Similarly, losses in the insulation shielding (sheath) and skid wires are also increased by applying the same multiplier to Eq. (11.9). Therefore, the expression for λ_1' must be modified to read

$$\lambda_1' = \frac{R_s}{R} \frac{1.7}{1 + (R_s/X)^2} \tag{11.17}$$

where, for pipe-type cable, R_s is calculated using the same procedure as described in Section 11.6.3.4 for nonmagnetic armors; for SL cable R_s is calculated by Eq. (11.10). For oval conductors and cores use is made of equivalent diameters. These equivalents are calculated as the rms values of the major and minor axes. For the reactance X, Eq. (11.11) shall be used. For armor losses in SL cables IEC-287-1-1 [9a] provides details.

11.6.3.8 Losses in Steel Pipes

For the determination of pipe losses (λ_2) empirical formulas have been developed in the United States for pipe sizes and steel types used there. For a closely bound triangular configuration, the value of λ_2 is given by

$$\lambda_2 = \frac{0.0115s - 0.001485d_d}{R} \times 10^{-5} \tag{11.18a}$$

while for a cradled formation

$$\lambda_2 = \frac{0.00438s + 0.00226d_d}{R} \times 10^{-5} \tag{11.18b}$$

where d_d is the inside diameter of the pipe (in millimeters), s is the axial spacing of the conductors in the pipe (in millimeters), and R is the ac resistance per unit length of one conductor at the maximum operating temperature (in ohms per meter).

When the three phases of a pipe-type cable are not closely bound together in a triangular formation for pulling into the pipe, they will assume random positions. In this case, therefore, the mean value of the results calculated with Eqs. (11.18a) and (11.18b) shall be used for further calculations.

Formulas (11.18a) and (11.18b) apply only if the power frequency is 60 Hz; for 50 Hz a multiplication factor of 0.76 has to be applied to λ_2.

11.7 THERMAL RESISTANCE OF CABLES

For heat transmission between concentric cylinders the integration of Eq. (11.1), the Ohm's law of heat conduction, leads to

$$\theta_1 - \theta_2 = P \frac{\rho_T}{2\pi} \ln\left(\frac{r_2}{r_1}\right) \tag{11.19}$$

where P is the heat loss in a unit length (watts per meter), θ_1 (in degrees Celsius) is the temperature of a cylinder of radius r_1, and θ_2 (in degrees Celsius) is the temperature of a cylinder of radius r_2, whereas ρ_T (in kelvin-meter per watt) is the thermal resistivity of the material between the concentric cylinders. Accordingly, the thermal resistance (T_n) per unit length of a cylindric wall (in kelvin-meter per watt) is

$$T_n = \frac{\rho_T}{2\pi} \ln\left(1 + \frac{2\Delta d}{d_1}\right) \tag{11.20}$$

where d_1 is the inside diameter of the cylindric wall (in millimeters), and Δd is the thickness of the cylindric wall (in millimeters). The quantity $\ln(1 + 2\,\Delta d/d_1)$ is known as the geometric factor and is designated by the letter G. Metallic shielding layers are considered to be part of the adjacent metallic component (conductor or sheath), whereas semiconducting screens (shields) are added to the insulation (semiconducting screens are assumed to have the same thermal properties as the adjacent dielectric insulating layers); hence, in thermal calculations the appropriate dimensions must be adjusted accordingly. Formula (11.20) applies to jackets, pipe coatings, bedding layers, nonmetallic ducts, and cylindric insulation walls when the appropriate inside diameter, thickness, and resistivity values are substituted. Table 11.6 is giving the thermal resistivities (ρ_T) of materials used in cable construction.

TABLE 11.6 Thermal Resistivities of Materials

Material	Thermal Resistivity, ρ_T (K·m/W)
Insulating Materials	
Paper insulation in solid-type cables	6.0
Paper insulation in oil-filled cables	5.0
Paper insulation in cables with external gas pressure	5.5
Paper insulation in cables with internal gas pressure:	
Preimpregnated	5.5 ~ 6.5
Mass impregnated	6.0
PPL[a]	4.5 ~ 6.5
PE	3.5
XLPE	3.5
Polyvinyl chloride:	
Up to and including 3-kV cables	5.0
Greater than 3-kV cables	6.0
EPR:	
Up to and including 3-kV cables	3.5
Greater than 3-kV cables	5.0
Butyl rubber	5.0
Rubber	5.0
Protective Coverings	
Compounded jute and fibrous materials	6.0
Rubber sandwich protection	6.0
Polychloroprene	5.5
PVC:	
Up to and including 35-kV cables	5.0
Greater than 35-kV cables	6.0
PVC/bitumen on corrugated aluminum sheaths	6.0
PE	3.5
Materials for Duct Installations	
Concrete	1.0
Fiber	4.8
Asbestos	2.0
Earthenware	1.2
PVC	6.0
PE	3.5

[a] Thermal resistivity of PPL has not been standardized yet.
Source: From [9c].

11.7.1 Thermal Resistance of Insulation

11.7.1.1 Single-Core Cables

For the calculation of the thermal resistance of the insulation of single-core cables, Eq. (11.20) is written as

$$T_1 = \frac{\rho_T}{2\pi} \ln\left(1 + \frac{2\,t_1}{d_c}\right) \tag{11.21}$$

where d_c is the diameter of the conductor (in millimeters) including its metallic screen if any, and t_1 is the thickness (in millimeters) of the insulation including the semiconducting screens over the conductor and the insulation. For corrugated sheaths t_1 is based on the mean internal diameter of the sheath.

11.7.1.2 Three-Conductor Cables

For *belted cables* IEC 287-2-1 [9c] gives

$$T_1 = \frac{\rho_T}{2\pi} G \tag{11.22}$$

where T_1 is the thermal resistance between one of the conductors and the common sheath (in kelvin-meter per watt), whereas G is the geometric factor given in Fig. 11.10.

IEC-287-2-1 [9c] also provides a digital method for calculating the geometric factor G, instead of using the graphical method given in Fig. 11.10.

For *three-conductor cables with a metallic insulation screen (shield)* over the individual cores IEC-287-2-1 [9c] provides a correction factor K called the *screening factor* given in Fig. 11.11. Accordingly Eq. (11.22) is modified as

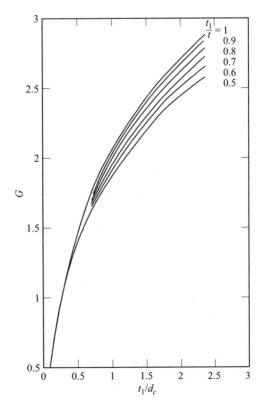

Figure 11.10 Geometric factor G for three-conductor belted cables with circular conductors [9c].
t is thickness of insulation between conductors,
t_1 is thickness of insulation between conductor and sheath,
d_c is diameter of conductor (circular).
For oval conductors d_c shall be taken as the geometric mean of the major and minor diameters of the oval conductor.

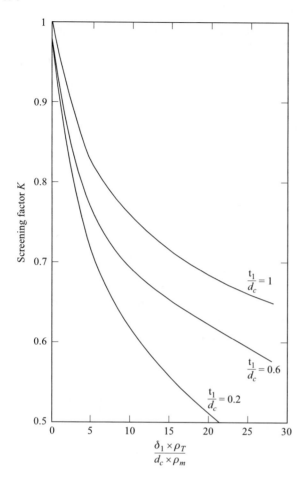

Figure 11.11 Correction factor K (screening factor) to be applied to the thermal resistance of an unscreened three-conductor cable with circular conductors for the calculation of the thermal resistance of a corresponding screened (shielded) cable: [9c].

δ_1 is thickness of metallic screen on core,
ρ_T is thermal resistivity of insulation,
d_c is diameter of conductor (circular),
t_1 is thickness of insulation between conductor and screen,
ρ_m is thermal resistivity of screening material: 27×10^{-4} K·m/W for copper, 48×10^{-4} K·m/W for aluminum.
For oval conductors d_c shall be taken as the geometric mean of the major and minor diameters of the oval conductor.

$$T_1 = K \frac{\rho_T}{2\pi} G \qquad (11.23)$$

In this type of cable the insulation screens are touching each other and the common sheath (shield) of the three insulated cores, therefore $t_1/t = 0.5$.

IEC-287-2-1 [9c] also provides a digital method for calculating the screening factor K required to establish the thermal resistance of three-conductor screened cables, instead of using Fig. 11.11.

Equations (11.22) and (11.23) and Figs. 11.10 and 11.11 were developed for *solid-type paper insulations* where the thermal resistivities of the different elements of the insulations (conductor and belt insulations), and fillers are *equal*. This is not the case with extruded insulations.

Today's *three-conductor cables with extruded insulation* have extruded semiconducting screens over each conductor (conductor screen) and over each individually insulated core (insulation screen). The thermal resistivity of the filler between the cores is usually *different* from that of the insulation extruded over the individual conductors. A recent study [14], involving finite-element computations on 228 different cable constructions, led to the following formula:

$$T_1^{\text{fitted}} = T_1^{\text{IEC}} + 0.031(\rho_f - \rho_i)e^{0.67t_1/d_c} \tag{11.24}$$

where

$$T_1^{\text{IEC}} = \frac{\rho_i}{2\pi} G \tag{11.24a}$$

and ρ_i in kelvin-meter per watt is the thermal resistivity of the insulating material extruded over the individual conductors, ρ_f in kelvin-meter per watt is the thermal resistivity of the filler material, t_1 (in millimeters) is the thickness of the insulation extruded over the individual conductors including the semiconducting extruded screens of the conductor and the insulation, $t_1/t = 0.5$ in this type of cable as the insulated cores are touching each other and the common sheath (shield) of the three insulated cores, d_c (in millimeters) is the diameter of the conductor (this shall be taken as the geometric mean of the major and minor diameters of an oval conductor), the geometric factor G may be found with the help of Fig. 11.10 or calculated with the corresponding digital method. The result, T_1^{fitted}, is the thermal resistance between one conductor and the common shield (sheath).

Results obtained with Eq. (11.24) were found to coincide with the results of the finite-element computations. This calculating method [14] is under consideration by IEC.

For *three-conductor cables with extruded insulation and metallic screens over each insulated core* a current study, involving finite-element computations, is aiming at the development of a formula similar to Eq. (11.24). For the time being, it is recommended [14] for this type of cable to begin the computation of the thermal resistance using Eq. (11.24) and then multiply the result by the screening factor K.

The thermal resistance T_1 of the insulation of *pipe-type* and *SL cables* is calculated the same way as for single-conductor cables [Eq. (11.21)].

For *other conductor types and other cable constructions* [9c] and [10] provide formulas and calculating methods.

11.7.2 Thermal Resistance of Coverings over Sheaths, Shields, Armor, and Pipe (Oversheaths, Jackets, Bedding, Outer Serving)

To establish values of thermal resistances (T_2 and T_3) for different layers of cable coverings, Eqs. (11.20) and (11.21) may be applied to all concentric cylindrical elements by substituting the appropriate diameters (d_x), thickness (t_x), and thermal resistivities (ρ_T).

For jackets over corrugated sheaths, the value of the thermal resistance (in kelvin-meter per watt) is calculated as

$$T_3 = \frac{\rho_T}{2\pi} \ln\left[\frac{D_{oc} + 2t_3}{[(D_{oc} + D_{it})/2] + t_s}\right] \tag{11.25}$$

where t_3 (in millimeters) is the thickness of the jacket, t_s (in millimeters) is the thickness of the sheath, D_{oc} (in millimeters) is the diameter of the imaginary coaxial cylinder that just touches the crests of the corrugated sheath, and D_{it} (in millimeters) is the diameter of the imaginary cylinder which just touches the inside surface of the troughs of the corrugated sheath.

For the calculation of the thermal resistance of the *fillers and bedding under the armor of SL cables*, graphical and digital methods are provided in references [9c] and [10]

11.7.3 Pipe-Type Cables, Cables in Metallic Ducts

This calculating method is used when losses in the pipe or duct (designated here as λ_2) cannot be neglected.

(i) *For pipe-type and single-conductor cables* the thermal resistance T_1 of the insulation of each core between the conductor and the metallic screen, shield, or sheath is to be calculated using Eq. (11.21) for single-core cables.

For other types of three-conductor cables T_1 is calculated using Eq. (11.22), (11.23), or (11.24).

(ii) The thermal resistance T_2 is made up of two parts:

1. *For pipe-type and single-core cables* the thermal resistance of the covering, if any, over the screen, shield, or sheath of each core: The value to be substituted for part of T_2 in the rating Eq. (11.3) is *the value per cable; that is, the value for three cores is one third of the value of a single core.* The value per core is calculated according to Section 11.7.2. For oval cores of single-conductor cables the geometric mean radius of the major and minor diameters $(d_M d_m)^{1/2}$ shall be used in place of the diameter for a circular core assembly.

For three-conductor cables the thermal resistance of the covering, if any, over the common screen, shield, or sheath of the *three-cores*: The value to be substituted for part of T_2 in the rating equation (11.3) is *the value per cable (three cores)* and is derived according to Section 11.7.2.

2. The thermal resistance of the gas or liquid between the surface of the cores [cable(s) and the pipe or duct]: This resistance is calculated in the same way as that part of T_4 that is between a cable and the internal surface of a duct, as given in Section 11.7.4.1. The value derived will be *per cable* and should be added to the quantity calculated in part 1 above before substituting in the rating Eq. (11.3).

(iii) The thermal resistance T_3 of any external covering on the pipe or duct is dealt with in Section 11.7.2. The thermal resistance of the metallic pipe is negligible.

11.7.4 External Thermal Resistance

The external thermal resistance T_4 of a cable in a duct consists of three parts:

$$T_4 = T_4' + T_4'' + T_4'''$$

(11.26)

where T_4' (in kelvin-meter per watt) is the thermal resistance of the gas (air) or liquid between the cable surface and the internal surface of the duct, T_4'' (in kelvin-meter per watt) is the thermal resistance of the duct itself (the thermal resistance of a metal pipe is negligible), T_4''' (in kelvin-meter per watt) is the external thermal resistance of the duct.

11.7.4.1 Thermal Resistance between Cable and Duct or Pipe (T_4')

For cables in duct, and cable diameters in the range 25–100 mm, IEC-287-2-1 [9c] provides the following formula [Eq. (11.27)], originally a part of the Neher and McGrath study [8]. The same formula also applies to pipe-type cables for the calculation of the thermal resistance of the space between the cores and the pipe (cf. Section 11.7.3), provided the equivalent diameter D_e of the three cores is within the range 75–125 mm:

$$T_4' = \frac{U}{1 + 0.1(V + Y\theta_m)D_e} \tag{11.27}$$

where U, V, and Y are constants that are given in Table 11.7 for different installation conditions, D_e is the external diameter of the cable (in millimeters); when more than one cable is installed in the same duct and for pipe-type cables (cf. Section 11.7.3), D_e becomes the *equivalent diameter* of the group, and its value is calculated as follows:

- Two cores or cables: $D_e = 1.65 \times$ outside diameter of cable or core.
- Three cores or cables: $D_e = 2.15 \times$ outside diameter of cable or core (this is the case of a pipe-type cable, and also of a group of three concentric neutral cables pulled into the same duct).
- Four cores or cables: $D_e = 2.50 \times$ outside diameter of cable or core.

Further on θ_m is the mean temperature (in degrees Celsius) of the medium in the duct between the cable and the duct or pipe. This value must be estimated at first and subsequently revised if necessary.

The thermal resistance of ducts or pipes, designated usually as T_4'', is evidently zero if metallic ducts or pipes are used. The thermal resistance of nonmetallic ducts is calculated according to Section 11.7.2.

For exceptional installations where, on account of the above-mentioned restrictions Eq. (11.27) cannot be applied, Ref. [10] reviews and provides more elaborate

TABLE 11.7 Values of U, V, and Y Constants for Use in Eq. (11.27)

Installation Condition	U	V	Y
In metallic conduit	5.2	1.4	0.011
In fiber duct in air	5.2	0.83	0.006
In fiber duct in concrete	5.2	0.91	0.010
In asbestos cement			
Duct in air	5.2	1.2	0.006
Duct in concrete	5.2	1.1	0.011
Gas pressure cable in pipe	0.95	0.46	0.0021
Oil pressure pipe-type cable	0.26	0.0	0.0026
Plastic ducts		Under consideration	
Earthenware ducts	1.87	0.28	0.0036

Source: From [9c].

methods of calculation, some of them involving finite element techniques, as the surfaces of the cables (cores) and the duct wall are not necessarily isotherms.

11.7.4.2 Cables, Ducts, or Pipes Laid in Free Air

For a cable, duct, or pipe laid in free air, and *protected from solar radiation*, the thermal resistance T_4 between the cable (duct, pipe) and the ambient air is given by

$$T_4 = \frac{1}{\pi D_e^* h(\Delta\theta_s)^{1/4}} \tag{11.28}$$

where

$$h = \frac{Z}{(D_e^*)^g} + E \tag{11.28a}$$

and D_e^* (in meters) (D_e^* is in meters throughout this section) is the external diameter of a cable (duct, pipe), h is the heat dissipation coefficient, obtained either from the above formula (11.28a) using the appropriate values of constants Z, E, and g given in Table 11.8 or from curves provided in [9c].

Cables (ducts, pipes) having a nonmetallic surface or covering should be considered to have a black surface. For cables without a covering, such as plain lead or armored, the value of h is to be multiplied by a factor of 0.88.

Further on $\Delta\theta_s$ (in kelvin) is the excess cable (duct, pipe) temperature above ambient. This temperature may be obtained by a graphical method given in [9c] or calculated by the iterative process described hereafter:

Calculate the coefficient K_A (the parameters are those of the rating Eqs. (11.2) and (11.3):

$$K_A = \frac{\pi D_e^* h}{(1 + \lambda_1 + \lambda_2)}\left[\frac{T_1}{n} + T_2(1 + \lambda_1) + T_3(1 + \lambda_1 + \lambda_2)\right] \tag{11.28b}$$

and $\Delta\theta_d$ (in kelvin), which is the excess temperature above ambient to account for the dielectric losses:

$$\Delta\theta_d = P_d\left[\left(\frac{1}{1 + \lambda_1 + \lambda_2} - \frac{1}{2}\right)T_1 - \frac{n\lambda_2 T_2}{1 + \lambda_1 + \lambda_2}\right] \tag{11.28c}$$

Then substitute the results of Eqs. (11.28b) and (11.28c) into the following equation:

$$(\Delta\theta_s)_{n+1}^{1/4} = \left[\frac{\Delta\theta + \Delta\theta_d}{1 + K_A(\Delta\theta_s)_n^{1/4}}\right]^{1/4} \tag{11.29}$$

where $\Delta\theta$ in kelvin is the permissible conductor temperature rise above ambient.

Start the iteration by substituting the value of $(\Delta\theta_s)^{1/4} = 2$ into Eq. (11.29) and continue the iteration until $(\Delta\theta_s)_{n+1}^{1/4} - (\Delta\theta_s)_n^{1/4} \leq 0.001$.

TABLE 11.8 Values for Constants Z, E and g for Black Surfaces of Cables (Pipes) in Free Air

No.	Installation	Z	E	g	Mode
1	Single cable	0.21	3.94	0.60	$> 0.3 D_e^*$
2	Two cables touching, horizontal	0.29	2.35	0.50	$> 0.5 D_e^*$
3	Three cables in trefoil	0.96	1.25	0.20	$> 0.5 D_e^*$
4	Three cables touching, horizontal	0.62	1.95	0.25	$> 0.5 D_e^*$
5	Two cables touching, vertical	1.42	0.86	0.25	$> 0.5 D_e^*$
6	Two cables spaced D_e^*, vertical	0.75	2.80	0.30	$> 0.5 D_e^*$
7	Three cables touching, vertical	1.61	0.42	0.20	$> 1.0 D_e^*$
8	Three cables spaced D_e^* vertical	1.31	2.00	0.20	$> 0.5 D_e^*$
9	Single cable	1.69	0.63	0.25	
10	Three cables in trefoil	0.94	0.79	0.20	

Notes:

Item 1. Values for a "single cable" also apply to each cable of a group when they are spaced horizontally with a clearance between cables of at least 0.75 times the cable overall diameter.

Items 1–8. Installation on noncontinuous brackets, ladder supports or cleats, D_e^* not greater than 0.15 m.

Items 9 and 10. Clipped directly to a vertical wall (D_e^* not greater than 0.08 m).

Source: From [9c].

The external thermal resistance T_4 of cables exposed to solar radiation is treated in detail in [9c]. For other configurations not shown on Table 11.8, IEC-287-2-2 [9d] provides methods of calculation. The effect of wind is covered in [10].

For *cables in unfilled troughs* see Section 11.7.4.10, part B.

For *cables in nonmetallic ducts* the thermal resistances between cable and duct, as well as that of the duct itself, established according to Section 11.7.4.1, are to be added to T_3.

Cables in metallic ducts like pipe-type cables are first to be treated according to Section 11.7.3, before calculating T_4 as per this section.

11.7.4.3 Thermal Resistance of a Single Buried Cable or Pipe

All thermal resistance formulas of cables installed under ground are derived from the Kennelly formula, which states that

$$T_4 = \frac{\rho_T}{2\pi} \ln(2u) \tag{11.30}$$

where ρ_T is the thermal resistivity of the soil (in kelvin-meter per watt) and

$$u = \frac{2L}{D_e} \tag{11.30a}$$

where L in millimeters is the distance from the ground surface to the cable or pipe axis, D_e (in millimeters) is the external diameter of cable or pipe including its outer covering. It is to be noted that Eq. (11.30) only applies when $U > 10$, otherwise the following expression is used for the calculation of the external thermal resistance of a *single buried cable or pipe*:

$$T_4 = \frac{\rho_T}{2\pi} \ln[u + (u^2 - 1)^{1/2}] \tag{11.31}$$

11.7.4.4 Influence of Soil Conditions on the Design of Underground Lines

The thermal resistivity, ρ_T may be as low as 0.3 K·m/W in a good-quality crushed limestone mixture contained within a moist environment. Its value may increase to about 0.8 K·m/W for the same mixture but in dry state. On the other hand, its value may approach roughly 3.4–3.7 K·m/W when dry sand or clay surroundings are involved. In the case of dry concrete, a value of 0.85 K·m/W is generally employed [8]; this value may vary according to composition: IEC [9c] puts a value of 1.0, another source [2] uses a value of 0.8, whereas in the Institute of Electrical and Electronics Engineers (IEEE) ampacity Tables 0.6 K·m/W was assumed to be the thermal resistivity of concrete in duct banks [7]. The wide range or spread in the ρ_T values underlines the importance for the control of the surrounding soil of a cable. Using good-quality backfill material, if available, may greatly increase the loading capability of the cable by ensuring a medium of good thermal conductivity [15–29]. In [9e] are described the prevailing soil conditions and backfills used in different countries with the associated ρ_T values used for calculation purposes.

The presence of moisture in the soil is also of very great importance [15–28]. Temperatures exceeding a certain limit (this is generally taken as 50–55°C) and high heat flux concentration for prolonged periods at the soil interface may cause moisture migration away from the cable and progressive drying out of the soil, leading to a rapid increase of ρ_T resulting in overheated cables. This phenomenon is given the name of thermal runaway. The modified rating formula [Eq. (11.4)] serves to calculate the maximum allowable current limiting the temperature of the cable–soil interface to the critical temperature where moisture migration is initiated.

11.7.4.5 *Thermal Resistance of Groups of Buried Cables (Not Touching)*

The calculation is performed by superposition, provided there is sufficient separation (generally $s \geq 2D_e$) between cables, so that they do not distort each other's heat field. Groups of cables touching or very closely spaced are dealt with in Section 11.7.4.6.

(A) Unequally Loaded Cables. The method suggested by IEC-287-2-1 [9c] is to calculate the temperature rise at the surface of the cable under consideration caused by the other cables of the group; then subtract this rise from the value of $\Delta\theta$ used in the rating Eq. (11.3) when calculating the maximum permissible current of the cable under consideration. The total loss per unit length in each cable (P_k) has to be estimated beforehand, these estimates can be subsequently adjusted if deemed necessary as a result of the calculation.

The temperature rise ($\Delta\theta_{kp}$) at the surface of the cable under consideration caused by the loss P_k in cable k:

$$\Delta\theta_{kp} = \frac{\rho_T}{2\pi} P_k \ln\frac{d'_{pk}}{d_{pk}} \tag{11.32}$$

where the distances d_{pk} and d'_{pk} are shown in Fig. 11.12 ; in Fig. 11.12 cable k' is the mirror image projection of cable k with respect to the ground surface.

The temperature rise $\Delta\theta_p$ above ambient at the surface of the pth cable, whose rating is being determined, caused by the losses generated in the other $q - 1$ cables in the group, is given by the sum

$$\Delta\theta_p = \Delta\theta_{1p} + \Delta\theta_{2p} + \cdots + \Delta\theta_{kp} + \cdots + \Delta\theta_{qp} \tag{11.33}$$

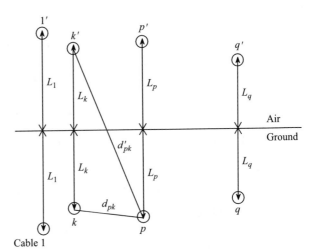

Figure 11.12 Diagram showing group of q cables and their reflection in ground-air surface [9c].

where the term $\Delta\theta_{pp}$ is excluded from the summation in Eq. (11.33).

The value of $\Delta\theta$ in the rating Eq. (11.3) is then reduced by the amount $\Delta\theta_p$ and the rating of the pth cable is determined using a value T_4 corresponding to the thermal resistance of the single isolated cable at the position p. This calculation is to be performed on all cables in the group and repeated, if necessary, to avoid overloading any of the cables.

(B) Equally Loaded Identical Cables. The maximum permissible load of the hottest cable determines the rating of the group. It is usually possible to decide from the configuration which cable will be the hottest and to calculate the rating of this one. In case of doubt a further calculation for another cable is advisable. The following formula is derived from Eq. (11.33) and establishes the modified value T_4 of the pth cable deemed to be the hottest, taking into account the mutual heating of the group. This modified value of T_4 is substituted directly into the rating equation [(Eq. (11.3)] as well as the established maximum conductor temperature rise $\Delta\theta$ which is also entered there unaltered:

$$T_4 = \frac{\rho_T}{2\pi} \ln\left\{ [u + (u^2 - 1)^{1/2}]\left[\left(\frac{d'_{p1}}{d_{p1}}\right) \left(\frac{d'_{p2}}{d_{p2}}\right) \cdots \left(\frac{d'_{pk}}{d_{pk}}\right) \cdots \left(\frac{d'_{pq}}{d_{pq}}\right) \right] \right\} \tag{11.34}$$

where $u = 2L/D_e$, here D_e (in millimeters) is the overall diameter of the pth cable and L (in millimeters) is the distance of its axis from the ground surface; there are $q - 1$ terms in the long expression in square brackets, the term d'_{pp}/d_{pp} being excluded. The distances d_{pk} and d'_{pk} are shown in Fig. 11.12. The long expression within the square brackets, the "mutual heating effect factor," has been assigned the letter F by Neher and McGrath [8], as well as in Section 11.8.2.

(B.1) For three identical cables having approximately equal losses, laid in a horizontal plane, equally spaced apart, Eq. (11.34) can be simplified as

$$T_4 = \frac{\rho_T}{2\pi} \left\{ \ln[u + (u^2 - 1)^{1/2}] + \ln\left[1 + \left(\frac{2L}{s_1}\right)^2 \right] \right\} \tag{11.35}$$

where s_1 (in millimeters) is the axial separation of two adjacent cables in a horizontal group of three, not touching; $u = 2L/D_e$ where L (in millimeters) is the distance of the cable axis from the ground surface; and D_e (in millimeters) is the external diameter of one cable.

The value of T_4 is for the center cable (the hottest of the group) and is substituted directly in the rating equation (11.3).

For single-conductor cables forming a three-phase circuit, Eq. (11.35) is valid only when the sheath (shield) losses (losses in concentric neutrals) are negligible (sheaths, shields, or concentric neutrals are single-point grounded or cross bonded) or almost equal (the cables in a solidly bonded system are transposed).

(B.2) For other cable combinations of equally loaded identical cables, and also for the case where the shield (sheath) currents are unequal, IEC-287-2-1 [9c] provides simplified forms of Eq. (11.34).

11.7.4.6 Groups of Buried Cables (Identical) Equally Loaded and Touching

Superposition should not be applied to cable configurations involving axial spacing less than twice the overall cable diameter. Accordingly the formulas in this section, given in IEC-287-2-1 [9c], were developed empirically, analytically (modeling heat transfer with an electric field), or by finite-element studies.

(A) Three Single-Core Cables, Flat Formation. For this configuration IEC-287-2-1 [9c] provides

$$T_4 = \rho_T [0.475 \ln (2u) - 0.346] \quad \text{for } u \geq 5 \tag{11.36}$$

where $u = 2L/D_e$, here L (in millimeters) is the distance of the cable from the ground surface, and D_e (in millimeters) is the external diameter of one cable.

Equation (11.36), given in IEC-287-2-1 [9c], was originally derived empirically; a recent finite-element study, referred to in [10], found that the formula gives good results for the outer cables but underestimates the temperature of the central cable, which is the hottest. The same study found that Eq. (11.34), or its simplified form, Eq. (11.35), yields close results for the hottest cable but slightly overestimates its temperature. The same study proposes a new formula applicable when the cable surfaces are nonisothermal [10].

(B) Three Single-Core Cables, Trefoil Formation. For this configuration L is measured to the center of the trefoil group and D_e is the external diameter of one cable. The result T_4 is the external thermal resistance of any one of the cables.

(B.1) For cables with metallic sheaths or shields, IEC-287-2-1 [9c] proposes

$$T_4 = \frac{1.5}{\pi} \rho_T [\ln (2u) - 0.63] \quad \text{for } u \geq 4 \tag{11.37}$$

In this case IEC-287-2-1 [9c] also recommends to multiply the thermal resistance of the outer covering T_3, as calculated according to Section 11.7.2, by a factor of 1.6. This increase of the thermal resistance of the outer covering is necessary on account of the heat transfer from the trefoil being restricted to its outer surface. The finite element study mentioned in part A of this section confirmed the validity of Eq. (11.37) but found the value 1.6 of the factor too high. This, however, has a negligible effect on the cable rating.

(B.2) For cables that are partially metal covered (helically laid shield or armor wires covering 20–50 % of the cable circumference), according to IEC-287-2-1 [9c], Eq. (11.37) applies to this type of long lay cable shield: the lay 15 times the diameter under the wire shield, individual wires having a diameter of 0.7 mm giving a total cross-sectional area 15–25 mm^2. In this case not only should T_3 be increased as for cables in part B.1 but also the insulation's thermal resistance T_1, as calculated according to Eq. (11.21), should be multiplied by the following factors:

T_1 to be multiplied by 1.07 for cables up to 35 kV and by 1.16 for cables from 35 kV to 110 kV

T_3 to be multiplied by 1.6

The clause in [9c] referred to here is under consideration by IEC.

(B.3) For cables with nonmetallic sheaths, IEC-287-2-1 [9c] proposes

$$T_4 = \frac{\rho_T}{2\pi}[\ln(2u) + 2\ln(u)] \qquad (11.38)$$

The finite-element study mentioned in part A confirmed the validity of Eq. (11.38). The clause in [9c] referred to here is under consideration by IEC.

(C) Concentric Neutral Cables, Touching, Directly Buried in Trefoil Formation. Let us calculate the percentage of coverage for some typical cable constructions:

Wired full concentric neutrals usually provide better than 50% coverage. The external thermal resistance of this type of cable may therefore be calculated according to the above part B.1.

Some *wired concentric half neutrals* provide less than 50% coverage, while others may provide better. The external thermal resistance of this type of cable should therefore be calculated only after verification of the coverage; the calculating procedure (part B.1 or B.2) should be chosen accordingly.

Wired concentric one third neutrals usually provide less than 50% coverage, in most of such cases therefore the value of T_4 may be calculated according to part B.2.

Flat straps mostly provide better than 50% coverage, accordingly the value of T_4 is mostly calculated according to part B.1.

Wired, less than one-third neutrals will mostly provide better than 20% coverage, in such cases T_4 may be calculated according to part B.2. However, *less than 20% coverage* may be provided in a few cases by concentric neutrals. This type of construction is to be treated according to part B.3.

Note: American [12] and Canadian [13] standards prescribe the minimum size of wires used in the concentric neutral: 16 AWG in a jacketed cable, 14 AWG in an unjacketed cable. The Canadian standard also limits the perpendicular distance between strands to 7 mm. Thus a concentric neutral cable, made according to the Canadian standard [13], has a minimum metallic coverage of 15.5%. Local regulations may also limit the spacing between strands, thus indirectly prescribing minimum coverage.

11.7.4.7 Thermal Resistance of Duct or Pipe (T_4'')

As mentioned earlier, the thermal resistance of the duct wall is calculated according to the principles outlined in Section 11.7 for coaxial cylinders:

$$T_4'' = \frac{\rho_T}{2\pi}\ln\left(\frac{D_o}{D_d}\right) \qquad (11.39)$$

where D_o (in millimeters) is the outside diameter of the duct, D_d (in millimeters) is the inside diameter of the duct, and ρ_T in (in kelvin-meter per watt) is the thermal resistivity of the duct material (cf. Table 11.6). For metallic ducts ρ_T can be taken as zero.

11.7.4.8 External Thermal Resistance of Ducts or Pipes (T'''$_4$)

For ducts directly buried, the thermal resistance is calculated in the same way as for a cable, using the appropriate formulas from Sections 11.7.4.3, 11.7.4.5, and 11.7.4.6. The external diameter D_e of the cable is replaced in this case with the external diameter of the duct (pipe) including the external covering, if any, of the latter. The depth of burial L is measured between the center of the duct (pipe) and the ground surface.

11.7.4.9 Cables or Ducts (Pipes) Embedded in Special Backfill, Duct Banks

In this section T_4 implies the total external thermal resistance of a cable embedded in a special backfill, whereas T_4''' implies the part of the external thermal resistance between the duct (pipe) and ambient.

For a group of cables or ducts embedded in a special backfill material of low thermal resistivity, or duct banks encased in concrete, IEC-287-2-1 [9c] has adopted the Neher–McGrath method [8]. This is a two-stage procedure:

1. The thermal resistance (T_4 for embedded cables, T_4''' for encased ducts) is calculated according to Sections 11.7.4.3, 11.7.4.5, 11.7.4.6, and 11.7.4.8 assuming an overall surrounding medium having the uniform thermal resistivity ρ_c of the special backfill or concrete.

2. A correction T_4^c is added to T_4 or T_4''' to take into account the difference between the thermal resistivities of the concrete or special backfill ρ_c and the surrounding soil ρ_e:

$$T_4^c = \frac{N}{2\pi}(\rho_e - \rho_c)\ln[u + (u^2 - 1)^{1/2}] \tag{11.40}$$

Here N is the number of loaded cables in the duct bank or encased in the special backfill, ρ_c (in kelvin-meter per watt) is the thermal resistivity of concrete or backfill material in contact with the cables or ducts, ρ_e (in kelvin-meter per watt) is the thermal resistivity of the surrounding soil (or "final backfill"), and

$$u = \frac{L_g}{r_b} \tag{11.40a}$$

where L_g (in millimeters) is the depth of the center of the duct bank (or the envelope of special backfill material), r_b (in millimeters) is the equivalent radius of the duct bank (or the envelope of special backfill material). The value of r_b is determined from

$$\ln r_b = \frac{1}{2} \frac{x}{y} \left(\frac{4}{\pi} - \frac{x}{y} \right) \ln \left(1 + \frac{y^2}{x^2} \right) + \ln \frac{x}{2} \qquad (11.40b)$$

where x and y (in millimeters) are, respectively, the shorter and longer sides of the duct bank or the envelope of the special backfill material. *Equation (11.40b) is only valid when $y/x < 3$.* To summarize:

$$T_4^{tot} = T_4 + T_4^c \qquad (11.40c)$$

or

$$T_4^{\prime\prime tot} = T_4^{\prime\prime\prime} + T_4^c \qquad (11.40d)$$

Here, for directly buried cables, T_4^{tot} is the value to be entered in the rating equation (11.3) in the place of T_4; for cables in ducts or pipes $T_4^{\prime\prime tot}$ is entered in Eq. (11.26) in the place of $T_4^{\prime\prime}$, and then this result of Eq. (11.26) is entered as T_4 in the rating Eq. (11.3).

A finite-element study [30] demonstrates that Eq. (11.40a) has further limitations. The discrepancies between the finite-element and Neher–McGrath results tend to increase when the height of the duct bank (envelope of special backfill material) is greater than its width. To overcome these restrictions, and also to cover the case of cables close to the ground surface, reference [30] provides a table and a diagram giving geometric factors G_b to be substituted directly in Eqs. (11.40) and (11.40b); the table in [30] is included in [10].

The above methods assume that the periphery of the duct bank (envelope) is isothermal; this may not be suitable for certain cable configurations. Special cases, such as nonuniform environments, require solutions by finite-element calculations or other numerical methods.

Municipal authorities in most large cities do not permit the use of native soil when refilling excavations in city streets; they specify graded stone screenings ("final backfill"). Graded stone screenings have excellent thermal qualities. In case *an excavation wider than the duct bank (envelope)* becomes necessary, a thermal resistivity value close to that of the final backfill may be substituted for ρ_e. However, it may be prudent to perform finite-element calculations of T_4 or $T_4^{\prime\prime\prime}$ for some typical situations: such as standard duct banks covered with approved stone screenings (final backfill) surrounded with typical native soils of known thermal resistivity, other utility installations or foundations nearby interfering with heat distribution.

A *simple approximation* is given in [1] to determine the minimum trench dimensions required to achieve a given value of *effective thermal resistivity*: "If the soil from the trench is replaced by a backfill having a given thermal conductivity superior to that of the original soil, the thermal resistance of the backfill within the trench, in series with that of the surrounding soil, is equal to the total thermal resistance which would be encountered if the *pipe (cable or duct bank)* were buried in a homogeneous soil having the maximum acceptable *(effective)* thermal resistivity (ρ_{eff})" (pp. 10–8 and 10–15). The following formula provides a close approximation of the required amount of special backfill:

$$\rho_{\text{eff}} = \rho_f + (\rho_o - \rho_f)\,e^{-(w/h)} \tag{11.40e}$$

where w is the width of the trench, h is the depth to the bottom of the *imbedded* structure, (width and depth to be in the same units), e is the base for natural logarithm, ρ_f and ρ_o (in kelvin-meter per watt) are, respectively, the thermal resistivities of the fill material and the original surrounding soil. The result ρ_{eff} is the thermal resistivity to use as ρ_e when calculating T_4^c for concrete duct banks in trenches filled with a special (final) backfill; also ρ_{eff} is taken as the uniform thermal resistivity ρ_T when calculating T_4 or T_4''' for cables, pipes, or ducts directly embedded in a special backfill, which is also used for the final filling of the trench.

11.7.4.10 Cables in Buried Troughs

(A) Cables in Troughs Filled with Sand. For cables in troughs filled with sand IEC-287-2-1 [9c], in a simple approach, advises to use a value of 2.5 K·m/W for the thermal resistivity of dry sand filling, whereas for troughs filled with special backfill it recommends to use the appropriate dry-state thermal resistivity value of the special filling material.

Instead of this approximation other analytic methods may also be used to meet particular installation conditions:

(A.1) Filled buried troughs may be treated in calculations as "cables imbedded in an envelope of special backfill" (cf. Section 11.7.4.9), with the thermal resistivity of the sand or other filling material being substituted for ρ_c.

(A.2) To *shallow burial* and *surface troughs* (where the cover of the trough is flush with the ground surface and shallow burial implies cables less than 0.6 m from the ground surface) the approach in part A.1 cannot be fully applied on account of the effect of solar radiation. There are several approaches possible that could easily yield widely different results; three of those are outlined and compared in [10].

(i) The first method is a two-stage iterative one: The thermal resistivity T_4^{tot} is calculated first according to Section 11.7.4.9. On account of the cables' closeness to the ground surface, the use of the "extended values of geometric factor" [30] is indicated for the calculation of T_4^c. In the second stage a further adjustment [31] to account for solar radiation (and wind effect) is added to T_4^{tot}. This latter adjustment is dependent of the total losses in the cables, losses which have to be estimated at the beginning, then adjusted progressively during the iterations according to the value I obtained from the rating Eq. (11.3). The pertinent formulas from reference [31] are also given in [10].

In moderate climates this method might yield overly pessimistic results as a few hours of extreme solar radiation may not lead to total drying of the surrounding media.

(ii) This is the approximation in IEC-287-2-1 [9c]: With this method the thermal resistivity values of the surrounding media in the dry state are used for calculations carried out according to Section 11.7.4.9, but the effect of solar radiation is not considered.

(iii) Another approximation used in Great Britain [10] is to apply a considerable seasonally adjusted increase (up to 25 K) to the ambient soil temperature normally taken for the calculation of cable rating; on the other hand the effect of solar radiation is not considered.

(B) Cables, Ducts, or Pipes in Unfilled Troughs. For cables, ducts, or pipes in unfilled troughs, covered, with the top flush with the soil surface and exposed to free air IEC-287-2-1 [9c] provides an empirical formula to calculate the temperature rise (θ_{tr} in degrees kelvin) of the air in the trough above the ambient air:

$$\Delta\theta_{tr} = \frac{P_{tot}}{3p} \tag{11.41}$$

where P_{tot} (in watts per meter) is the total power dissipated in the trough per unit length, and p (in meters) is that part of the trough perimeter that is effective for heat dissipation. Any portion of the perimeter that is exposed to sunlight is therefore not included in the value of p.

The rating of a particular cable in the trough is then calculated as for a cable in free air (cf. Section 11.7.4.2), but the ambient temperature shall be increased by $\Delta\theta_{tr}$.

During the iterative calculating process described in Section 11.7.4.2, the result of Eq. (11.41) is entered in Eq. (11.29) for the calculation of $(\Delta\theta_s)^{1/4}$ and also in the rating Eq. (11.3). The value P_{tot} entered in Eq. (11.41) being estimated, its value has to be verified and revised using the value I obtained with the rating Eq. (11.3). On account of P_{tot} being estimated, and not only $(\Delta\theta_s)^{1/4}$, the iterating process becomes more involved.

11.8 CYCLIC LOADING

The IEC group of standards 287 [9], as spelled out in the title, is only dealing with current rating at 100% load factor, accordingly Sections 11.5–11.7.4.10 of this chapter are also restricted to the calculation of steady-state loads. For cyclic loads the long-standing American practice has been the introduction of "loss factor" into the calculation of the thermal circuit. This method is further developed and summarized in the Neher–McGrath study [8], and it is also the basis of the current issue of the IEEE ampacity tables [7]. This calculating method is based on the following principle: A buried cable's environment has a high thermal capacitance; therefore, its temperature remains relatively steady and is a function of the average of the losses during repeated load cycles. On the other hand, the conductor's temperature immediately follows the fluctuations of the current; its maximum temperature therefore is a function of the peak load; the adjacent cable parts also follow load fluctuations rather quickly up to a fictitious diameter D_x. The external thermal resistances T_4 or T_4''' and T_4^c (for T_4^c refer to Sections 11.7.4.9 and 11.8.3) are therefore modified to include the loss factor to reduce all thermal resistances outside the fictitious diameter D_x. The temperature of cables in free air follows load fluctuations, therefore their cyclic rating is the same as their steady-state rating.

The daily load factor is defined as

$$\text{Load factor} = \frac{\sum_{i=0}^{23}(I_i)}{24 I_{max}} \qquad (11.42)$$

where I_i in amperes are the hourly load readings for a period of $i = 24$ hours.

The daily *loss factor* μ ("loss-load factor" according to IEC [11], and CIGRE [32] terminology) is defined as

$$\mu = \frac{\sum_{i=0}^{23}(I_i^2)}{24 I_{max}^2} \qquad (11.42a)$$

For a typical daily load curve, if the load factor is known, [1], [2], and [8] provide the following relation for the calculation of the loss factor:

$$\mu = 0.3\,(\text{load factor}) + 0.7\,(\text{load factor})^2 \qquad (11.42b)$$

The calculation of the fictitious diameter D_x in the Neher–McGrath work [8] is derived from AIEE Committee Report [33], with the conclusion that a value of $D_x = 8.3$ in. is applicable to most cases. The same report [33] also states that theoretically D_x should vary as the square root of the product of the diffusivity and the length in time of the load cycle. Therefore in case the diffusivity δ of earth is known, [8] provides the following formula using all *English units*:

$$D_x = 1.02[\delta\,(\text{length of cycle in hours})]^{1/2} \qquad (11.42c)$$

In [33] the diffusivity was taken as $\delta = 2.75\,\text{in}^2/\text{h}$, if this is substituted into Eq. (11.42c), the result obtained is $D_x = 8.3$ ins. IEC-853-1 [11a] states that for cyclic load calculations it is not necessary to know the thermal diffusivity accurately, and therefore a value of $0.5\,(10)^{-6}\,\text{m}^2/\text{s}$ was used for the tabulated functions given in the standard; this value roughly corresponds to a soil thermal resistivity of 1 K-m/W and a moisture content of about 7%. It also equals the thermal diffusivity value of $2.75\,\text{in}^2/\text{h}$ given in the Neher–McGrath work [8]. Equation (11.42c) is given here in *SI (metric)* units:

$$D_x = 6.12 \times 10^4[\delta\,(\text{length of cycle in hours})]^{1/2} \qquad (11.42d)$$

Here D_x is in millimeters, while δ is in square meters per second. For most cyclic loading calculations the value of δ can be taken as $0.5\,(10)^{-6}\,\text{m}^2/\text{s}$ according to IEC-853-1 [11a] and 853-2 [11b]. These standards also recommend, for very dry soil conditions, to calculate the thermal diffusivity as

$$\delta = \frac{(10)^{-3}}{\rho_T d(0.82 + 0.042\eta)} \qquad (11.42e)$$

where δ (in square meters per second) is the thermal diffusivity of the soil or other surrounding medium, ρ_T (in kelvin-meter per watt) is its thermal resistivity, d (in kilogram per cubic meter) is its dry density, and η is its moisture content in percent of dry mass.

Thermal diffusivity is defined as the inverse of the product of volumetric specific heat, and thermal resistivity: $\delta = 1/(c\rho_T)$.

For a 24-h load cycle, when substituting $0.5 (10)^{-6}\,\mathrm{m^2/s}$ for δ in Eq. (11.42d), we obtain $D_x = 212\,\mathrm{mm}$, which corresponds to the value given in inches in the Neher–McGrath study [8]. A refinement to the Neher–McGrath method [8] is proposed by a European source [34] referred to in [10]: Formulas for the value of D_x are proposed for different types of load cycles, and thermal resistivities (ρ_T) of the medium surrounding the cable, pipe, or duct.

In practice, according to [8], T_4 or T_4''', as the case may be, is partitioned by D_x, and the part representing the thermal resistance outside the diameter D_x is multiplied by the loss factor (μ). As far as the dielectric loss is concerned, no reduction due to the loss factor applies because the dielectric loss magnitude is considered to be constant. Therefore the modified values of T_4 or T_4''' and T_4^c (for T_4^c refer to Section 11.8.3) shall be used only in the denominator of the rating Eq. (11.3) and not in the numerator, as the thermal resistance values in the numerator apply to the dielectric loss, which is constant. Often in large cables $D_x < D_e$; in such circumstances $\ln(D_x/D_e)$ is negative.

The IEC group of standards 853 [11], based on CIGRE studies [32], provides calculating procedures (involving exponential integrals) for load ratings with different load cycles, including the required tables, diagrams, nomograms, and a computer method. Finite-element calculations are indicated in special situations, such as complex environments, outside heat sources, or when a number of adjacent cables are subjected to different load cycles.

As demonstrated by an example [10], the permissible cyclic load calculated with the Neher–McGrath method [8] is slightly higher (about 2%) than the result obtained by the more precise but laborious IEC procedure [11]; this difference is reduced by half through the refinement (calculating the value D_x instead of using 212 mm or 8.3 in.) mentioned above [10, 34].

For basic configurations, Sections 11.8.1–11.8.3 provide modified formulas of T_4, T_4''', and T_4^c according to the Neher–McGrath method [8], adapted to include the parameters D_x and μ for the calculation of cyclic loads; as mentioned above, these modified values are to be used only in the denominator of the rating equation (11.3). The value I obtained is the permissible peak value of current during the load cycle, such that the conductor attains, but does not exceed, the permissible maximum temperature allowed for this cycle.

11.8.1 External Thermal Resistance of a Single Buried Cable, Duct, or Pipe

For the calculation of cyclic loads Eq. (11.31) is adapted as

$$(T_4 \text{ or } T_4''') = \frac{\rho_T}{2\pi}\left\{\ln\frac{D_x}{D_e} + \mu \ln[u + (u^2 - 1)^{1/2}]\right\} \qquad (11.43)$$

where $u = 2L/D_x$.

11.8.2 External Thermal Resistance of Groups of Equally Loaded Identical Cables

For the calculation of cyclic loads Eq. (11.34) is adapted as

$$(T_4 \text{ or } T_4''') = \frac{\rho_T}{2\pi} \left\{ \ln\frac{D_x}{D_e} + \mu \ln\left[u + (u^2 - 1)^{1/2}\right] F \right\} \tag{11.44}$$

where $u = 2L/D_x$, and F is the "mutual heating effect factor" as defined in conjunction with Eq. (11.34).

For three identical cables having approximately equal losses, laid in a horizontal plane, equally spaced apart, when calculating the rating under cyclic loading, Eq. (11.35) is adapted as

$$(T_4 \text{ or } T_4''') = \frac{\rho_T}{2\pi} \left\{ \ln\frac{D_x}{D_e} + \mu \ln[u + (u^2 - 1)^{1/2}] + \mu \ln\left[1 + \left(\frac{2L}{s_1}\right)^2\right] \right\} \tag{11.45}$$

where $u = 2L/D_x$.

11.8.3 Cables, Ducts, or Pipes Embedded in Special Backfill, Duct Banks

The result T_4^c of Eq. (11.40) has to be multiplied by the loss-factor μ for the calculation of cyclic loads, and T_4 or T_4''' shall be calculated according to Eq. (11.43), (11.44), or (11.45) as the case may be.

11.8.4 Comparison of the Neher–McGrath and IEC Methods

The basis of the two methods is the same and they follow the same principles. The IEC-287 group of standards [9] evolved from the Neher–McGrath work [8] but includes developments since 1957. IEC-287 also treats in much more detail a much wider range of cable types and laying configurations than the Neher–McGrath study.

The main practical difference between the two is the use of English measure by Neher and McGrath and SI (metric) units by IEC. This difference is further accentuated by the use of 10-base logarithms by Neher and McGrath, while IEC uses natural logarithms; as a result *geometric factors have to be converted* using the conversion factor between 10-base and natural logarithms, whenever the value given is in the terms of the other method. In the Neher–McGrath method the conversion factors, required on account of these differences, have been incorporated in the constants of every equation. In the formulas in the Neher–McGrath work the factor (π) or the common symbol (f) for frequency hardly ever appear, as the constant (π) and the frequency value 60 have also been incorporated in the constants of the individual equations. On account of this, the use of the Neher–McGrath work is limited to 60 Hz (unless frequency-dependent parameters are properly modified). For all these reasons identical equations take a completely different form in the two versions. Parameters in the tables are in SI units in the IEC documents and in English units in the Neher–McGrath study. For the number of conductors in a cable the symbol is n in both methods, but in IEC-287 it is included in the rating equations, whereas in the Neher–McGrath work it is incorporated in the thermal resistance formulas. For these reasons, *diagrams, formulas, and data can only be exchanged between the two methods after conversion and proper adaptation (modification);* Section 11.7.4.9 illustrates such an adaptation.

As mentioned in Sections 11.5 and 11.8 the IEC-287 group of standards only treats steady-state conditions (100% load factor). Another pair of standards, IEC publications 853-1 [11a] and 853-2 [11b], cover cyclic loads and emergency ratings (short-term overloading); on the other hand the Neher–McGrath study covers steady-state conditions (100% load factor) but also includes the simple method described in Sections 11.8–11.8.3 for cyclic loads. The Neher–McGrath work provides a formula for short-term emergency ratings but refers the user to four earlier studies for the calculation of transient thermal resistance. The Neher–McGrath work does not cover the problem of drying of the soil due to heating by loaded power cables.

In the present chapter the IEC-287 and 853 approaches are followed. Neher–McGrath procedures (already adapted to IEC-287 in its pertinent clause) have been adopted in Section 11.7.4.9 for encased cables, ducts, or pipes. For cyclic loading in Sections 11.8–11.8.3 the Neher–McGrath procedure has been adopted with the required adaptations. Sections 11.9–11.9.9 illustrate the approach of IEC-853-2 for the treatment of short-term overloading, however, IEC-853 does not cover encasements with thermal resistivities different from the surrounding soil; a solution to this is included in the notes to Eq. (11.51) according to [35] and [36] mentioned in [10].

11.9 SHORT-TERM OVERLOADING

As mentioned in Section 11.8, the IEC group of standards 853 [11] provides procedures for the calculation of variable (transient and cyclic) loading. In all calculations according to IEC 853 [11], and accordingly throughout Sections 11.9–11.9.10, all linear dimensions (diameters, depths, etc.) are expressed in meters. When dealing with transient loads, parallel capacitances have to be added to the chain of resistances representing the thermal circuit in the calculation of steady-state loading. These capacitances represent the thermal capacitances per unit length Q [joule per meter-kelvin, abbreviated J/(m·K)] of different elements of the thermal circuit: components of the cable (conductor, insulation, sheath or shield, jacket, armor etc.), the space between cable and duct or pipe, the duct or pipe itself, and the pipe coating. The *thermal capacitance per unit length* Q (in joule per meter-kelvin) is defined as the product of the cross-sectional area of the different elements of the thermal circuit, as defined above, (in square meter), and the volumetric specific heat (values are given in Tables 11.9 and 11.10 in joule per cubic meter-kelvin) of that same element of the thermal circuit.

TABLE 11.9 Properties of Conducting Materials

Material	Reciprocal of temperature Coefficient of Resistance at 0°C (K)	Volumetric Specific Heat (J/m³·K)	Resistivity at 20°C (Ω·m)
Conductors			
Copper	234.5	3.45×10^6	1.7241×10^{-8}
Aluminum	228	2.5×10^6	2.8264×10^{-8}
Sheaths, screens, and armour			
Lead or lead alloy	230	1.45×10^6	21.4×10^{-8}
Steel	202	3.8×10^6	13.8×10^{-8}
Bronze	313	3.4×10^6	3.5×10^{-8}
Stainless Steel	[a]	3.8×10^6	70×10^{-8}
Aluminum	228	2.5×10^6	2.84×10^{-8}

[a] The temperature coefficient of resistance is negligible; thus no correction is necessary in Section 11.9.6.3.
Source: From [11b].

Thermal resistances T (in kelvin-meter per watt) are calculated according to Sections 11.7.1–11.7.4.3.

IEC 853-1 and 2 provide guidance (text and tables) as to which clauses to use to solve problems under different conditions; for this choice the product TQ, the *time constant* has to be known. Here T is the *total* thermal resistance (excluding the "earth

TABLE 11.10 Thermal Properties of Insulating Materials

Material	Thermal Resistivity (K·m/W)	Volumetric Specific Heat [J/(m^3·K)]
Insulating materials		
Paper insulation in solid-type cables	6.0	2.0×10^6
Paper insulation in oil-filled cables	5.0	2.0×10^6
Paper insulation in cables with external gas pressure	5.5	2.0×10^6
Paper insulation in cables with internal gas pressure		
(a) Preimpregnated	$5.5 \sim 6.5$	2.0×10^6
(b) Mass impregnated	6.0	2.0×10^6
Oil	7.0[a]	1.7×10^6
Nitrogen at 15 atm	[a]	0.017×10^6
PE	3.5	2.4×10^6
XLPE	3.5	2.4×10^6
Polyvinyl chloride		
Up to and including 3-kV cables	5.0	1.7×10^6
Greater than 3-kV cables	6.0	1.7×10^6
EPR		
Up to and including 3-kV cables	3.5	2.0×10^6
Greater than 3-kV cables	5.0	2.0×10^6
Butyl rubber	5.0	2.0×10^6
Rubber	5.0	2.0×10^6
PPL	$4.5 \sim 6.5$	2.0×10^6
Protective coverings:		
Compounded jute and fibrous materials	6.0	2.0×10^6
Rubber sandwich protection	6.0	2.0×10^6
Polychloroprene	5.5	2.0×10^6
PVC		
Up to and including 35-kV cables	5.0	1.7×10^6
Greater than 35-kV cables	6.0	1.7×10^6
PVC/bitumen on corrugated aluminum sheaths	6.0	1.7×10^6
PE	3.5	2.4×10^6
Other components		
Semiconducting XLPE and PE	2.5	2.4×10^6
Semiconducting EPR	3.5	2.1×10^6
Materials for duct installations		
Concrete	1.0	1.9×10^6
Fiber	4.8	2.0×10^6
Asbestos	2.0	2.0×10^6
Earthenware	1.2	1.7×10^6
PVC	6.0	1.7×10^6
PE	3.5	2.4×10^6

[a] The thermal resistance of the oil or gas filling in pipe-type cables is calculated according to Section 11.7.4.1.
Source: From [11b, 37, 52].

portion") of a cable or a separate single core, and Q is the *total* thermal capacitance (excluding the "earth portion") of a cable or a separate single core.

Tables 11.9 and 11.10 give properties of materials for the calculation of thermal transients.

Caution: Throughout IEC-853 [11], and accordingly in Sections 11.9–11.9.10, time in the formulas is expressed in seconds, whereas in the text time is expressed in either hours, minutes, or seconds.

11.9.1 Representation of the Dielectric: Long-Duration Transients ($>\frac{1}{3}$ of *TQ*, Also for cyclic rating)

11.9.1.1 Single-Core Cables

Figure 11.13 depicts the equivalent circuit, representing the dielectric, used in calculations for transients of long duration: greater than $\frac{1}{3}$ ($\frac{1}{2}$ for three-core cables) of the product *TQ*, where T and Q, respectively, are the totals of the thermal resistances and capacitances of cable, space between cable and duct or pipe, duct or pipe, and pipe coating; IEC 853-2 [11b] also states that the equivalent circuit in Fig. 11.13 and the ensuing formulas are usually suitable for transient durations greater than 1 h.

The value p in Fig. 11.13 is given by

$$p = \frac{1}{2 \ln(D_i/d_c)} - \frac{1}{(D_i/d_c)^2 - 1} \tag{11.46}$$

known as the Van Wormer coefficient [38] for long-duration transients. In Eq. (11.46) D_i and d_c, respectively, are the external diameters of the dielectric and the conductor of a single-conductor cable. For thermal calculations metallic screening layers are to be considered to be part of the adjacent metallic component (conductor or sheath), while semiconducting layers (including metallized carbon paper tapes) are considered as part of the insulation.

In Figure 11.13 for a single-conductor cable T_1 is the thermal resistance of the dielectric as calculated by Eq. (11.21), while Q_i and Q_c, respectively, are the total thermal capacitances of the dielectric and the conductor also of a single-conductor cable.

11.9.1.2 Three-Core Cables

The three-core cable is replaced by an equivalent single-core construction dissipating the same conductor losses; the equivalent parameters are to be established as follows:

The equivalent single-core conductor has a diameter:

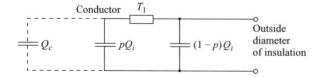

Figure 11.13 Representation of dielectric for transients of duration greater than $\frac{1}{3}$ times *TQ* [11b].

$$d_c = D_i e^{-(2\pi T_1/\rho_i)} \tag{11.46a}$$

where D_i, the equivalent diameter, is the real diameter of the three-core cable over the insulation (under the sheath) and T_1 is the equivalent thermal resistance of the dielectric taken as one-third of the value calculated for the appropriate three-conductor construction according to Section 11.7.1.2, and ρ_i is the thermal resistivity of the dielectric.

The thermal capacitances are calculated according to the following assumptions:

(a) The actual conductors are considered to be all inside the diameter of the equivalent single-core conductor; the remainder of the area within the equivalent conductor diameter is considered to be occupied by insulation.

(b) The space between the equivalent single-core conductor and the sheath is considered to be completely occupied by insulation (in three-conductor oil-filled cables this space is filled by the oil in the ducts and the remainder by oil-impregnated paper).

The factor p is then calculated using the dimensions of the equivalent single-conductor cable and then the thermal capacitance of the insulation based on assumption (b) is allocated as shown in Fig. 11.13.

11.9.2 Representation of the Cable: Long-Duration Transients (Also for cyclic rating)

The first part of the thermal circuit (cable portion) simulating the cable is a two-section network shown in Fig. 11.14. Generally, the first section includes the thermal capacitance of the conductor lumped with the inner portion of the dielectric, along with the thermal resistance of the dielectric, while the second section includes the thermal capacitance and resistance of the rest of the cable plus the space between the cable and the duct (pipe) and the duct (pipe) itself.

In Sections 11.9.2.1–11.9.2.4 symbols are defined only where they are first used.

11.9.2.1 Self-Contained Cables with Impregnated, Laminated (Taped), or Extruded Insulation, Unarmored, and Thermally Similar Constructions

Formulas for the constants in Fig. 11.14 are

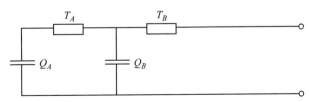

Figure 11.14 Equivalent cable network for transient response calculations [11b].

$$T_A = T_1 \tag{11.47a}$$

$$T_B = q_s T_3 \tag{11.47b}$$

$$Q_A = Q_c + pQ_i \tag{11.47c}$$

$$Q_B = (1 - p)Q_i + \frac{Q_s + p'Q_j}{q_s} \tag{11.47d}$$

where: T_1, Q_c, Q_i, and p are defined in Sections 11.9.1.1 and 11.9.1.2; Q_s is the thermal capacitance of the sheath and reinforcement; and

$$q_s = \frac{\text{losses in (conductor + sheath)}}{\text{losses in conductor}} \tag{11.47e}$$

which is used to take account of the extra losses occurring in the sheath. Also, T_3 is the thermal resistance of jacket (oversheath, outer covering) calculated according to Section 11.7.2 and Q_j is the thermal capacitance of jacket (oversheath, outer covering).

Jackets (oversheaths, outer coverings) are to be treated like the insulation of a single-core cable. The thermal capacitance is allocated according to the principle of Eq. (11.46) and Fig. 11.13; it has also been found convenient to omit its outer part as its role is insignificant in determining the thermal transient of the conductor. Accordingly,

$$p' = \frac{1}{2 \ln(D_e/D_s)} - \frac{1}{(D_e/D_s)^2 - 1} \tag{11.47f}$$

where D_e and D_s are the outer and inner diameters of the jacket.

11.9.2.2 Oil-Filled (Liquid-Filled) Pipe-Type Cables (High Pressure)

Formulas for the constants in Fig. 11.14 are

$$T_A = T_1 + \frac{q_s T_o}{2} \tag{11.47g}$$

$$T_B = \frac{q_s T_o}{2} + q_e T_3 \tag{11.47h}$$

$$Q_A = Q_c + pQ_i \tag{11.47i}$$

$$Q_B = (1 - p)Q_i + \frac{Q_o}{q_s} \tag{11.47j}$$

where T_1, Q_i and p are defined in Section 11.9.1.1;

T_o = thermal resistance of oil in pipe calculated according to Sections 11.7.3 and 11.7.4.1

T_3 = thermal resistance of covering on pipe calculated according to Sections 11.7.2 and 11.7.3

Q_o = thermal capacitance of oil in pipe

$$q_s = \frac{\text{losses in conductors and screens}}{\text{losses in conductors}}$$

$$q_e = \frac{\text{losses in (conductors + screens + pipe)}}{\text{losses in conductors}}$$

11.9.2.3 Gas Pressure Pipe-Type Cables (No Filling Material or Armor) Also Three Single-Conductor Cables in Metallic Duct

Formulas for the constants in Fig. 11.14 are

$$T_A = T_1 \tag{11.47k}$$

$$T_B = q_s T_2 + q_e T_3 \tag{11.47l}$$

$$Q_A = Q_c + pQ_i \tag{11.47m}$$

$$Q_B = (1 - p)Q_i + \frac{Q_s}{q_s} + \frac{0.5Q_p}{q_e} \tag{11.47n}$$

where, T_2 = thermal resistance of gas (air) in pipe (duct) calculated according to Sections 11.7.3 and 11.7.4.1

Q_p = thermal capacitance of pipe or metallic duct

$$q_e = \frac{\text{losses in (conductors + sheaths + pipe)}}{\text{losses in conductors}}$$

11.9.2.4 Cables in Ducts (Nonmetallic)

Formulas for the constants in Fig. 11.14 are

$$T_A = T_1 \tag{11.47o}$$

$$T_B = q_s(T_3 + T_4' + T_4'') \tag{11.47p}$$

$$Q_A = Q_c + pQ_i \tag{11.47q}$$

$$Q_B = (1 - p)Q_i + \frac{Q_s + Q_j + 0.5Q_d}{q_s} \tag{11.47r}$$

where T_4' = thermal resistance of air space in duct calculated according to Section 11.7.4.1

$T_4'' = $ thermal resistance of duct calculated according to Section 11.7.4.7

$Q_d = $ thermal capacitance of the duct.

11.9.2.5 Other Types of Cables and Installations

In Sections 11.9.2.1–11.9.2.4 the most common types of cable construction were described. For the treatment of particular problems not covered, the reader is referred to [11b] and [32].

11.9.3 Long-Duration Partial Transient of the Cable (Also cyclic loading)

The following procedure is to be followed for the calculation of the transient response of a cable's thermal network (cable portion, Fig. 11.14) to the application of a load current; here the cable is considered to be in isolation that is with the right-hand pair of terminals in the thermal network short-circuited.

$$M_o = \tfrac{1}{2}[Q_A(T_A + T_B) + Q_B T_B] \tag{11.48a}$$

$$N_o = Q_A T_A Q_B T_B \tag{11.48b}$$

$$a = \frac{M_o + \left(M_o^2 - N_o\right)^{1/2}}{N_o} \tag{11.48c}$$

$$b = \frac{M_o - \left(M_o^2 - N_o\right)^{1/2}}{N_o} \tag{11.48d}$$

$$T_a = \frac{1}{a-b}\left[\frac{1}{Q_A} - b(T_A + T_B)\right] \tag{11.48e}$$

$$T_b = T_A + T_B - T_a \tag{11.48f}$$

where Q_A, Q_B, T_A, and T_B are taken from Sections 11.9.2.1–11.9.2.4.

Substituting the results of the group of Eqs. (11.48) into Eq. (11.49) we obtain the transient temperature rise $\theta_c(t)$ of the conductor above the outer surface of the cable (duct or pipe) at the time t after the application of the transient load:

$$\theta_c(t) = P_c[T_a(1 - e^{-at}) + T_b(1 - e^{-bt})] \tag{11.49}$$

where P_c is the power loss per unit length in a conductor or an equivalent conductor based on the maximum conductor temperature attained. The power loss is assumed to be constant during the transient.

The attainment factor conductor to cable surface $\alpha(t)$ is found as

$$\alpha(t) = \frac{\theta_c(t)}{P_c(T_A + T_B)]} \tag{11.50}$$

and is defined as

$$\alpha(t) = \frac{\text{temperature rise across cable at time } t}{\text{steady-state temperature rise across cable}}$$

11.9.4 Long-Duration Partial Transient of the Cable Environment (Also cyclic loading)

The cable environment is the second part of the thermal circuit. The methods for calculating this partial transient are given in Sections 11.9.4.1 and 11.9.4.2.

11.9.4.1 Buried Cables (Directly or in Ducts)

The transient response of the cable environment (earth portion) is calculated by an exponential integral formula. Equation (11.51) gives the temperature rise above ambient of the earth interface (outer surface of a buried cable, outer surface of a buried duct or pipe) of the hottest cable (duct or pipe) in a group of cables with similar loading.

$$\theta_e(t) = \frac{\rho_T}{4\pi} P_I \left\{ -Ei\left(\frac{-D_e^2}{16 t\delta}\right) - \left[-Ei\left(-\frac{L^2}{t\delta}\right) \right] \right\}$$

$$+ \frac{\rho_T}{4\pi} \sum_{k=1}^{k=N-1} P_I \left\{ -Ei\left(-\frac{d_{pk}^2}{4 t\delta}\right) - \left[-Ei\left(-\frac{d'^2_{pk}}{4 t\delta}\right) \right] \right\} \qquad (11.51)$$

where ρ_T = soil thermal resistivity
  D_e = external surface diameter of cable or duct (diameter of soil interface)
  δ = soil thermal diffusivity: Eq. (11.42e) (usually taken as 0.5 (10^{-6}) m^2/s, cf., Section 11.8)
  t = time from moment of application of heating transient
  L = axial depth of burial of hottest cable
  d_{pk} = distance from center of cable k to center of hottest cable p (Fig.11.15)
  d'_{pk} = distance from the center of image of cable k to center of the hottest cable p (Fig. 11.15)
  N = number of cables (ducts)
  P_I = total I^2R power (conductor, shield, sheath, and armor) loss of each cable in group (per cable, per unit length), in case of ducts or pipes: total I^2R power loss within one duct or pipe
  $-Ei(-x)$ = exponential integral function: values are given by nomograms on Figs. 11.16 and 11.17, also by approximate equations in notes to these nomograms in [11], and in references [39–41].

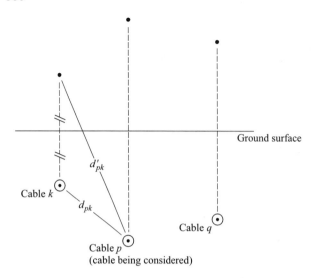

Ground surface

d'_{pk}

Cable k

d_{pk}

Cable q

Cable p
(cable being considered)

Figure 11.15 Cables (ducts) or circuits of touching cables (groups of ducts touching) and their images [11].

Notes to Eq. (11.51)

1. For values $x \geq 8$ it is sufficient [11] to put $-Ei(-x) = 0$.

2. The summation always includes $N - 1$ cables, excluding the hottest. The summation is not needed if only a single cable (duct or pipe) is considered.

3. Usually, the second term in Eq. (11.51) and also the second term in the summation become negligible at normal depths of laying and for transients shorter than 24 hours [11b].

4. Equation (11.51) reflects the principles of the effect of mutual heating adopted for steady-state calculation in Section 11.7.4.5. Accordingly, the summation in (11.51) lends itself to be adapted to the treatment of unequal transient loads [10, 36]. In the first two terms in (11.51) P_I is to represent the losses in the hottest cable (duct or pipe) p, whereas in the summation P_I is to be replaced by the loss P_k of each individual cable (duct or pipe) in the group of $N - 1$, which excludes the hottest.

5. Equation (11.51) was written for a uniform surrounding medium; therefore IEC-853 cannot cover duct banks or cables (ducts or pipes) imbedded in special backfill or concrete. Reference [10] mentions recent studies [35, 36], which propose to solve the problem of calculating the transient loading of encased cables, ducts, or pipes by adopting the well-known principles of the Neher–McGrath method to steady-state loading described in Section 11.7.4.9. For steady-state calculations a correction T_4^c has been introduced [Eq. (11.40)]; following the same principle, it is proposed for transient calculations that the correction $\theta_e^c(t)$ be added to Eq. (11.51). Also in this case ρ_T in Eq. (11.51) has to be replaced by ρ_c. For a single cable or duct

$$\theta_e^c(t) = \frac{\rho_e - \rho_c}{4\pi} P_I \left\{ -Ei\left(-\frac{r_b^2}{4t\delta}\right) + \left[-Ei\left(-\frac{L_g^2}{t\delta}\right)\right] \right\} \qquad (11.51a)$$

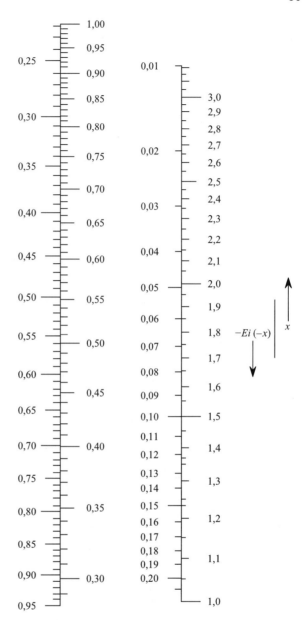

Figure 11.16 Nomogram for exponential integral $0.30 \le x \le 3.0$ [11].

where all parameters are defined in Section 11.7.4.9 and in conjunction with Eq. (11.51), but all linear dimensions are to be in meters.

6. The sum of the results of Eqs. (11.51) and (11.51a) has to be entered in Eq. (11.52) for $\theta_e(t)$.

7. This procedure, as proposed in [10], [35] and [36], may further be extended to groups of embedded cables or duct banks according to the principles mentioned earlier in these notes.

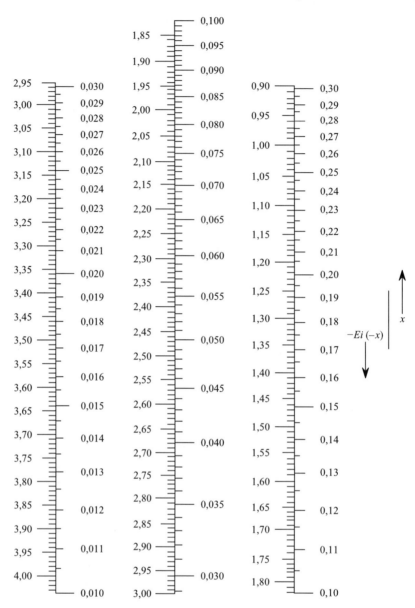

Figure 11.17 Nomogram for exponential integral $0.01 \leq x \leq 0.30$ [11].

11.9.4.2 Cables (Ducts) in Air

For cables in air the cable environment partial transient (earth portion) is irrelevant. The complete transient $\theta(t)$ is obtained by replacing T_B by $T_B + T_C$ in the constituent terms [(Eqs. (11.48a–11.48f)] of the formula for $\theta_e(t)$ [Eq. (11.49)]. The calculation of T_C is given in Section 11.9.6.2.

11.9.5 Short-Duration Transients
(Duration $< \frac{1}{3}$ of TQ)

Sections 11.9.1–11.9.4 have dealt with long-duration transients only. The proce-dure for short-duration transients is similar in principle but more elaborate, as the equivalent thermal circuit representing the conductor and the dielectric is replaced by a two-part one according to Van Wormer [38]. This yields a network of more than two sections which is then reduced to two sections with a method given in an appendix of reference [11b]. The reader is referred to [11b], [32b] and [32c] for calculating procedures and formulas.

Short-duration overloads, calculated on the basis of permissible short-term emer-gency temperatures given for the materials of the dielectrics, may lead to thermo-mechanical problems in the cable and accessories due to the rapid rate of expansion of the conductor compared to the much slower rates of the dielectric and the sheath. In liquid-filled cables rapid heating may lead to damaging transient pressures.

11.9.6 Calculation of the Complete Temperature Transient

11.9.6.1 Buried Cables (Directly or in Ducts)

The total transient rise $\theta(t)$ of the conductor above ambient is obtained by simply adding to the cable partial transient $\theta_c(t)$ the product of the environment (earth portion) partial transient $\theta_e(t)$ and the conductor to cable surface attainment factor $\alpha(t)$:

$$\theta(t) = \theta_c(t) + \alpha(t) \cdot \theta_e(t) \qquad (11.52)$$

where: $\theta(t)$ = transient temperature rise of conductor above ambient

$\theta_c(t)$ = transient temperature rise of conductor above the cable (duct) surface, assuming losses (P_c) from conductor surface: Eq. (11.49)

$\theta_e(t)$ = transient temperature rise of cable (duct) surface above ambient for period $t = 0$ to (t), assuming total losses (P_I) from the cable (duct) surface: Eqs. (11.51) and (11.51a)

$\alpha(t)$ = attainment factor for the transient temperature rise between the conduc-tor and the outside surface of the cable (duct): Eq. (11.50)

11.9.6.2 Cables (Ducts) in Air

In this case the total transient rise $\theta(t)$ of the conductor above ambient is obtained by Eq. (11.53), where a modified value $\theta_c(t)$ has to be used:

$$\theta(t) = \theta_c(t) \qquad (11.53)$$

The modified value $\theta_c(t)$ is obtained by replacing T_B by $T_B + T_C$ in Eqs. (11.48a–11.48f). This group of equations provides the modified constituent terms used in Eq. (11.49) for the calculation of the modified value of $\theta_c(t)$.

The value of T_C is obtained as follows:

$$T_C = q_s \cdot T_4 \text{ for cables described in Sections 11.9.2.1 and 11.9.2.4}$$

$$T_C = q_e \cdot T_4 \text{ for cables described in Sections 11.9.2.2 and 11.9.2.3}$$

where T_4 is the external resistance of a cable (duct) in free air as calculated in Section 11.7.4.2, Eq. (11.28), q_s and q_e are given in Sections 11.9.2.1–11.9.2.4.
    For other cable types and configurations the reader is referred to [11b].

11.9.6.3 Correction to Transient Temperature Response for Variation in Conductor Losses with Temperature

During transients the temperature of the conductor changes with time, and its resistance varies accordingly. Therefore conductor loss also changes even though the load current is assumed constant for the duration of the transient. Allowance for this variation is given by the following formula, which provides the corrected conductor temperature $\theta_\alpha(t)$ at time t after the initiation of the transient.

$$\theta_\alpha(t) = \frac{\theta(t)}{1 + \alpha[\theta(\infty) - \theta(t)]} \tag{11.54}$$

where

$$\alpha = \frac{1}{\beta + \theta_i} \tag{11.54a}$$

and

$\theta(t) =$ conductor transient temperature rise above ambient without correction for variation in conductor loss, and is based on conductor resistance at end of the transient: Eq. (11.52) or (11.53)
$\theta(\infty) =$ conductor steady-state temperature rise above ambient
    $\alpha =$ temperature coefficient of electric resistivity of the conductor material at the start of the transient
    $\beta =$ reciprocal of temperature coefficient of resistance of conductor at 0°C (Table 11.9)
    $\theta_i =$ conductor temperature at the start of transient

11.9.7 Dielectric Loss

In Sections 11.9.1–11.9.6 and Section 11.9.8, the assumption is that the cable has been energized long enough, before the initiation of the transient, for the temperature rise of the conductor due to dielectric loss to have reached steady state. Therefore a constant temperature rise due to dielectric loss is simply being added to the rise generated by the transient load current.
    In case a cable is switched on with a load, both transients (load and dielectric loss) have to be calculated and added. Particular attention has to be given to this possibility

when calculating pressure transients in liquid-filled cables. A fundamental description of pressure transients (during heating and cooling) and pressurizing systems is given in Chapters 6 and 7, the detailed study of the hydrodynamics in liquid-filled cables is not within the scope of this volume. The reader is referred to [42–45] on this subject.

For oil-flow and oil-pressure calculations [11b] in an appendix provides information regarding transient oil temperatures within the dielectrics of self-contained liquid-filled cables. For pipe-type cables similar information is given for the oil space between cores and pipe.

11.9.7.1 Dielectric Losses in Cables at Voltages Up to and Including 275 kV

Long durations, exceeding $\frac{1}{3}$ of *TQ*: It is assumed that one half of the dielectric loss is produced at the conductor with the other one half at the screen over the insulation. The cable's thermal circuit is constructed as outlined in Section 11.9.2.1 with the coefficient p calculated according to Section 11.9.1. Half of the dielectric loss $(\frac{1}{2}P_d)$ originates at the conductor; to account for the other half the coefficients q have to be taken equal to 2.

Short durations $(\leq \frac{1}{3}$ of *TQ*): The procedure is basically the same as for long durations, but the coefficient p has to be calculated and the thermal circuit constructed according to the appropriate clauses of [11b].

11.9.7.2 Dielectric Losses in EHV Cables, Higher than 275 kV

Because of the importance of the dielectric loss, and the thickness of the insulation at these voltage levels, Morello [44] proposed a more elaborate formula for the factor p for both long- and short-duration transients.

$$p_d = \frac{[(D_i/d_c)^2 \ln(D_i/D_c)] - [\ln(D_i/d_c)]^2 - (1/2)[(D_i/d_c)^2 - 1]}{[(D_i/d_c)^2 - 1][\ln(D_i/d_c)]^2} \tag{11.55}$$

where D_i = external diameter of insulation
D_c = external diameter of conductor

The rest of the transient thermal circuit, outside the sheath, is the same as for transient conductor losses.

Half of the dielectric loss $(\frac{1}{2}P_d)$ originates at the conductor. Therefore, as in Section 11.9.7.1, the coefficients q have to be taken equal to 2.

11.9.8 Emergency Ratings

Emergency ratings are calculated according to the following procedure, principles, and assumptions. It is assumed that prior to the emergency a constant current I_1 has been applied so that a steady-state temperature has been reached. Subsequently, at a time defined by $t = 0$, an emergency load current I_2 is applied for a given time t. The question is, how large I_2 may be so that the conductor temperature does not exceed a specified value, taking into account the variation of the electrical resistivity of the conductor with temperature.

The emergency current I_2 is calculated as

$$I_2 = I_R \left\{ \frac{h_1^2 R_1}{R_{max}} + (R_R/R_{max}) \left[\frac{r - h_1^2(R_1/R_R)}{\theta_R(t)/\theta_R(\infty)} \right] \right\}^{1/2} \tag{11.56}$$

The method is only valid if $I_2 \leq 2.5 I_R$.

The parameters in Eq. (11.56) are to be calculated and defined as follows:

$h_1 = I_1/I_R$ Here I_R is the steady-state (100% load factor) rated current at the maximum permissible normal operating temperature [Eq. (11.3)].

$r = \theta_{max}/\theta_R(\infty)$ Here θ_{max} is the maximum permissible temperature rise above ambient at the end of the emergency loading period, and $\theta_R(\infty)$ is the conductor temperature rise above ambient in the steady-state, namely the maximum permissible normal operating temperature rise.

$\theta_R(t)$ Conductor temperature rise above ambient after the application of I_R, neglecting variation of conductor resistance, the value of $\theta_R(t)$ is given by Eq. (11.52) or (11.53), as the case may be.

Further on R_1 is the ac resistance of the conductor before the application of emergency current, namely at the conductor temperature corresponding to the steady-state current I_1 [Eqs. (11.2), (11.3), (11.5)]. Also R_{max} is the ac resistance of the conductor at the end of the emergency loading period [Eq. (11.5)], while R_R is the ac resistance of the conductor at the maximum permissible steady-state operating temperature [Eq. (11.5)].

Any change of dielectric loss with temperature is neglected, and the steady-state temperature rise due to dielectric loss is calculated as

$$\theta_d = P_d[\tfrac{1}{2} T_1 + n(T_2 + T_3 + T_4)] \tag{11.56a}$$

where P_d is the dielectric loss using Eq. (11.7), and the thermal resistances T_1, T_2, T_3, and T_4 have been calculated according to Section 11.7.

Before substituting into Eq. (11.56), θ_d has to be subtracted from the values of θ_{max}, $\theta_R(\infty)$ and $\theta_R(t)$; these adjusted values are then used for the calculation of I_2.

11.9.9 General Remarks about Overloads and Emergency Ratings

Cables rarely operate at their maximum admissible temperature. This leaves room for temporary overloading without exceeding permissible operating temperatures. An example in reference [11b] illustrates the possibilities. In this particular example the maximum steady-state operating temperature permitted under normal conditions cannot be exceeded. This conservative approach is understandable in view of the high (EHV) operating voltage of 400 kV, and the large 2000 mm^2 conductor.

Cable data in the example: 400 kV single circuit, consisting of three single-core self-contained oil-filled cables, cross-bonded, with a 2000 mm² copper conductor, directly buried at a depth of 1000 mm with a spacing of 300 mm between the cable axes, 10°C ambient ground temperature, *85°C maximum operating temperature.*

The thermal resistivity of the surrounding soil is assumed to be 1.0 K-m/W, with a diffusivity of 0.5 (10^{-6}) m²/s.

Calculated values: 30.3 W/m conductor (copper) loss, 13.35 W/m dielectric loss with the cable having been energized long enough to reach steady-state, 2.1 W/m loss in the lead sheath, *1580 A continuous* rated current.

Results of the calculations in IEC-853-2 [11b]

Cyclic load: 50.4% loss-load factor, (with a given daily load curve) 2022 A *maximum peak* load, 28% higher than the continuous rated 1580 A (100% load factor).

Emergency current rating: 6 h duration

After *continuous* operation at 60°C with a load of 1195 A or 75.6% of 1580 A (the continuous rating). The *permissible overload* is *2247 A*, 42% higher than continuous *but only 11% higher than cyclic*, with the conductor at a temperature of *85°C* at the end of the emergency.

By *allowing a maximum conductor temperature of 95°C for emergencies of short duration*, the *permissible short-term term (6 hr) overload is 2523 A*, 60% higher than continuous, *25% higher than cyclic.*

(The last entry for a 95°C emergency temperature has been added by the author for further information.)

This illustrates the benefits of increasing the temperature by just 10 K at the end of the emergency. As far as the allowable emergency temperature of an HV or EHV dielectric is concerned, it is 105°C for today's liquid-filled paper dielectrics (80 to 90°C, depending upon the duration, for cables manufactured before the 1960s or 1970s), and up to 130°C for today's extruded synthetic dielectrics. Most utilities however, impose 95°C for their liquid-filled cables, and 100–110°C for extruded synthetic insulated dielectrics. As mentioned earlier, the main reasons for this restriction are mainly problems of expansion loops, excessive transient pressure changes in liquid-filled dielectrics, and different rates of expansion of the conductor compared to the other components (dielectric and sheath) of the cable during short duration transients calculated on the basis of the final allowable temperature of the dielectric. Further questions concerning cable loading are treated in Section 11.11.2.

Further Emergency Load Calculations (based on the above examples in reference [11b]): Emergency current rating: 6-h duration

After *continuous* operation at 51.5°C with a load of 1054 A or $\frac{2}{3}$ of 1580 A (the continuous rating). The *permissible overload is 2420 A*, 53% higher than continuous, *20% higher than cyclic*, and still with a maximum conductor temperature of 85°C *only at the end of the emergency.*

By *allowing a maximum conductor temperature of 95°C for emergencies of short duration*, the *permissible short-term (6 hour) overload* is *2666 A, 69% higher than* continuous, *32% higher than cyclic*.

11.9.10 Other Methods for the Calculation of Short-Term Overloading

Conformal transformation, finite-element, and finite-difference methods are used to calculate cable ratings for complex environments and to find solutions where load fluctuations among different cables do not coincide in time, or where other heat sources (steam pipes, etc.) are present. Such mostly numerical methods have been of great use in the development of equations to be incorporated in analytical calculating methods, examples can be found in Sections 11.7.4.6, 11.7.4.9 [30], and 11.9.4.1 [35, 36]; also [46], using conformal transformation, provides refinements to the Neher–McGrath method by modeling the heat distribution with an electric field where isotherms are replaced in the model by the equipotential lines of an electric field and the lines of heat flow are replaced by the lines of electric flux. These methods are treated in more detail in [10].

A *simplified rapid approximation for short-term loading* is given in [34] (clause 29.7, 1970 edition): A single-loop thermal network is used, the temperature dependency of the conductor resistance is neglected (it is taken at its maximum value at the end of the transient), T is taken as the total thermal resistance of the cable (and duct) in kelvin-meter per watt, Q in joule per meter-kelvin is the thermal capacitance of the conductor, $\Delta\theta_\infty$ in kelvin is the steady-state temperature rise of the conductor due to the joule losses generated by a given transient current I_t, and t in seconds is the length of the application of the transient load I_t. The temperature rise $\Delta\theta_t$ (in kelvin) due to the transient I_t at the end of the period t is given by

$$\Delta\theta_t = \Delta\theta_\infty\left[1 - \exp\left(-\frac{t}{TQ}\right)\right] \tag{11.57}$$

The same source also provides the following for short-time rating factor:

$$n_s = \left\{\frac{1}{1 - \exp[-t_s/(TQ)]}\right\}^{1/2} \tag{11.57a}$$

where n_s is defined as the ratio of the allowable transient current I_s for the duration t_s in seconds, and the I_m steady-state current generating the temperature rise $\Delta\theta_s$ permissible at the end of the application of the transient load I_s. Accordingly,

$$I_s = n_s I_m \tag{11.57b}$$

For equations (11.57), (11.57a) and (11.57b) it is assumed that the cable is at ambient temperature when the transient load is applied.

11.10 FAULT CURRENTS

The *magnitude of short-circuit current* is a function of overall system parameters, it may be calculated with methods given in [3].

For the *duration of short-circuits* the following contingency situation is usually considered: The primary breaker fails to clear the fault; therefore, the clearing time of the backup breaker is taken as the duration of the short circuit. On systems with reclosing breakers the most onerous situation usually occurs under the reclosing sequence.

Mechanical aspects, such as electromagnetic forces and expansion (thrust) induced by fault currents, have to be considered when choosing the type of cable and the cable support system, such as structures, ties, and clamps. These forces and movements are in addition to those induced by weight, loading (normal and emergency), and variation of ambient temperature.

The guiding principles regarding the mechanical aspects of cable system design related to short circuits are outlined in the relevant ICEA [47, 48] and IEC [49–51] standards, even though the principal subject of these documents is the calculation of the allowable short-circuit current, and its duration based on established temperature limits.

Electromagnetic forces are perpendicular to the axis of the cable. They also tend to induce "bursting" (radial separation of the strands) of the cable conductor. The *maximum current (peak value)*, generating the force (bursting) a particular cable can resist, depends upon cable construction and design and thus has to be provided by the cable manufacturer and supported by experimental results.

A *perpendicular force* between adjacent single-phase cables of the same circuit is generated by a phase-to-phase short circuit. This force is uniformly distributed along the line and its value in newtons per meter (N/m) is given by

$$F = \frac{\mu_0 I^2}{2\pi s} \tag{11.58}$$

where $\mu_0 = 1.256 \times 10^{-6}$ H/m is the magnetic permeability of air
  I = rms value of the short-circuit current in amperes
  s = cable spacing in metres

A factor of 2 should be allowed to account for the instantaneous value of current.

This distributed force results in a force at the supporting cleats or restraints, and it is a direct linear function of the spacing between these supports. This distributed bending force generates a bending moment; M (in newton-meter) in the cable (sheath):

$$M = \frac{FL^2}{12} \tag{11.58a}$$

where L is the spacing (in meters) between cleats of a rigid installation; in a flexible installation L is the distance between restraints or between a restraint and the adjacent cleat. The resulting strain in the sheath is

$$e = \frac{Md}{2(fm)} \qquad (11.58b)$$

where d is the outside diameter of the sheath or shield (in meters), and fm (in newton-meter squared) is the flexural rigidity of the cable based on its short-term properties. The latter may be calculated theoretically as the sum of the flexural rigidities of all the components: conductor, insulation, sheath (shield), jacket, and armor. The flexural rigidity of the individual components may be calculated theoretically as the product of Young's modulus, and the moment of inertia based on the geometry of the particular component. Young's modulus, however, is not only a function of the elastic properties of the materials but greatly depends upon cable construction and manufacturing procedures. Accordingly, this information has to be provided by the cable manufacturer and supported by experimental data.

In *flexible-type installations* the spacing between cleats is much wider than in rigidly restrained cables, as cleats in the latter have to be closely spaced to prevent deflection under axial thrust. Therefore restraints (spacers between the spaced single-conductor cables of the same circuit or straps around cables installed in close trefoil formation) have to be provided to hold the cables together against the perpendicular force generated by short circuits.

On *directly buried cables* firmly contained in a well-compacted backfill, the effect of this perpendicular force is of little or no consequence.

Thrust in the direction of the cable axis is a problem that becomes critical under short-circuit conditions. Because of the very short duration of short circuits, practically no heat is transmitted from the conducting parts of the cable to the adjacent insulating material. As a result the conductor tends to expand rapidly while no expansion is induced in the adjacent insulating portion.

When a cable is rigidly contained in its environment (directly buried or firmly held in position by closely spaced cleats), longitudinal expansion or contraction is restricted and a thrust develops upon heating. Cleats and the supporting structure have to be designed to support this force, the weight of the cable, plus the perpendicular forces mentioned above. The cleats have to be spaced to prevent deflection of the cable under all the compressive forces.

The thrust in a cable component is the product of the relaxation coefficient, the coefficient of expansion of the metallic component, the temperature rise of the metallic component, Young's modulus for the metallic component, and the cross-sectional area of the component. As mentioned earlier, Young's modulus and also the relaxation coefficient are to be established experimentally and also depend upon the rate of temperature rise and decline. The total thrust in a rigid installation is the sum of the thrusts in the components due to temperature changes caused by variations of the load, and ambient temperature changes; to this, one has to add for liquid-filled self-contained cables the thrust due to oil-pressure which is the product of oil pressure and the cross-sectional area under the sheath. Finally, the momentary thrust due to the temperature rise under short circuits has to be added to the above.

A *distributed radial linear pressure* is generated throughout the curves by the axial cable thrust. The value per unit length of this radial linear pressure p_{sc} in newtons per meter is

$$p_{sc} = \frac{C}{r} \tag{11.59}$$

where C (in newton) is the total axial thrust in the cable, including the thrust due to short circuit, and r is the *inner* radius of the bend in meters.

In *rigid installations* the cleats and the supporting structure have to support this force, plus the forces mentioned earlier.

This radial pressure in the curves is critical during short-circuit conditions in any type of installation, as it may force the central conductor against the insulation or the concentric conductor against the jacket, inflicting damage particularly to thermoplastics; the value p_n in pascals (newtons per meter squared) of this pressure is given by

$$p_n = \frac{C_n}{rd} \tag{11.59a}$$

where C_n is the total momentary thrust in the central conductor in newtons, r is the *inner* radius of the bend in meters, and d in meters is the diameter of the particular conductor; for a concentric conductor's effect on the jacket, C_n is the sum of the thrust in both conductors, central, and concentric.

The radial pressure in the curves increases with the decrease of the bending radius [Eq. (11.59a)]. Therefore IEC standards [49–51] emphasize the importance of respecting the regulations or standards prescribing minimum bending radii; in case these prescriptions cannot be followed, it is recommended to reduce the allowable temperature (see Sections 11.10.1–11.10.1.3) by 10°C. Also the EPRI report on cables under fault conditions [52] finds that short-circuit withstand temperatures may decrease by as much as 15% when cables are tested bent to a 36-in. radius in comparison with straight samples.

In *flexible systems* cables are arranged on cleats in a wave pattern and thrust is alleviated by letting the cable deflect laterally to accommodate expansion. Cables pulled into ducts or pipes are a limited form of a flexible system (see Section 11.15.1).

The *interface* between rigid and flexible systems requires special arrangements (buried or clamped curves, special joints, expansion loops of special design) to prevent the transfer of axial thrust from the rigid to the flexible portion of the route.

Joints and terminations have to sustain all the above mechanical forces. Accessory design has to take account of cable insulation shrink-back within the accessories under sudden heating during short-circuits and tensile forces in the cable conductor due to cooling following short-circuits. Certain types of joints and terminations, such as those with soldered connectors, impose lower temperature limits than the maximum values given for the cable. Terminations also have to sustain wind forces, the weight of accumulated ice, snow, and sleet, plus the forces transmitted by the connected busbars or overhead lines; the latter forces may be reduced if flexible leads are inserted between the termination and the bus or line.

Sudden expansion of impregnation compounds may lead to damaging internal pressure.

Induced voltages in the sheaths or shields of specially bonded cable systems (Section 11.6.3.2) are treated in detail in [2, 3, 6, 6a, 6b, 6c].

11.10.1 Calculation of the Thermally Permissible Short-Circuit Current

Temperatures (heat) generated by short-circuit currents in the cable conductor, sheath, screen, shield, and armor shall not exceed the allowable short-circuit temperature rating of the adjacent insulation or jacket. The allowable temperature of lead sheath may be lower than that of the adjacent insulating components. The short-circuit rating of connectors may also be a governing factor.

11.10.1.1 Adiabatic Method

On account of the short duration of short-circuits it is assumed, in the large majority of cases, that all the heat developed by the fault current is completely contained within the conductor (central or concentric) and other current-carrying components, such as metallic sheath, shield or screen, and armor. The basic formula given in IEC (International) standards [37, 49, 50] is

$$I_{AD}^2 t = K^2 S^2 \ln\left(\frac{\theta_f + \beta}{\theta_i + \beta}\right) \tag{11.60}$$

where I_{AD} in amperes is the short-circuit current (rms over the short-circuit's duration) calculated on an adiabatic basis, t in seconds is the duration of the short circuit, S in millimeters squared is the cross-sectional area of the current-carrying component, θ_f in degrees Celsius is the final temperature of the particular current-carrying component at the end of the short circuit, θ_i in degrees Celsius is the initial temperature of the particular current-carrying component at the beginning of the short circuit, β in kelvin is the reciprocal of the temperature coefficient of resistance of the current-carrying component at 0°C, K is a constant depending of the material of the current-carrying component and it is given in tables of [37, 49, 50] or it may be calculated:

$$K = \left(\frac{\sigma_c(\beta + 20)10^{-12}}{\rho_{20}}\right)^{1/2} \tag{11.60a}$$

where σ_c (in joule per cubic meter-kelvin) is the volumetric specific heat of the current-carrying component, and ρ_{20} (in ohm-meter) is the electrical resistivity of the current-carrying component at 20°C. Values of β, σ_c, and ρ_{20} are given in Table 11.9. If the fault current is known, the final temperature (θ_f) can be determined:

$$\theta_f = (\theta_i + \beta)\exp\left(\frac{I_{AD}^2 t}{K^2 S^2}\right) - \beta \tag{11.60b}$$

The basic formulas [Eqs (11.60) and (11.60a)] are given in English units (cross-sectional area in circular mils) in the ICEA (U.S.) standards [47, 48], which also use 10-base logarithms, whereas IEC (International) standards [37, 49–51] use natural logarithms and metric (SI) measure.

For the insulation over central conductors the ICEA standard [47] provides equations derived from the basic equations for copper and aluminum conductors that already include in the constants the physical characteristics of the metals, leaving only the temperatures (initial and final) to be substituted, with the cross-sectional area, the current, and its duration as unknown. The user's task is further simplified, as the standard [47] also provides a series of curves for short-circuit durations of 1–100 cycles, where recommended temperatures (maximum normal operating for "initial," and maximum short-circuit temperature for "final") for the most common types of insulation have already been substituted in the equations, with cross sections of 10 AWG to 1000 kcmil on the abscissa and short-circuit currents up to 100 kA on the ordinate. Maximum short-circuit temperature limits recommended by ICEA for materials of the insulation are given in Table 11.11.

IEC standards [49–51] in most cases provide more conservative values than ICEA [47] for the short-circuit withstand temperature of insulation over the central conductor of cables; some of these withstand values in the IEC standards also depend of voltage and some adjustments are under consideration. The same IEC standards also state that the allowable temperatures given apply to short-circuit durations of 5 s or less.

According to ICEA Standard P-45-482 [48] the values in Table 11.12 apply to materials both over and under the current-carrying screen, shield, sheath, or concentric conductor and the lower temperature value shall prevail; whereas according to IEC standards [49–51] distinct tables apply to insulations and jackets, and the lower temperature value shall prevail. However, only the jacket is to be considered if a thermally insulating cushion separates the insulation from the current-carrying shield or sheath over it.

Some allowable values given in IEC standards [49–51] for jacketing materials are more conservative than those of ICEA [48] and they are applicable to durations of 5 s or less according to IEC. Permissible values given by IEC are under review, and it is proposed that notes in the forthcoming revisions allow higher short-circuit temperatures than those in the present standards, provided they are supported by experimental data. The revisions under review of IEC standards also include a warning concerning

TABLE 11.11 Maximum Allowable Temperature for Materials of Cable Insulation[a]

Material	Temperature (°C)
Paper, rubber, varnished cloth	200
Paper	
In self-contained liquid-filled cables	250[b]
In liquid-filled pipe-type cables	200[c]
Thermoplastic	150
XLPE or EPR	250

[a] For PPL insulation standard values have not yet been established. Reference [2] quotes 150°C for PPL.
[b] This value is taken from IEC-986 [50], as ICEA Standard P-32-382 [47] does not cover liquid-filled cables.
[c] This value is taken from IEC Standard Project 61443 [51], as ICEA Standard P-32-382 [47] does not cover liquid-filled cables.
Source: From [47].

TABLE 11.12 Maximum Allowable Temperature of Materials in Contact with the Current-Carrying Screen, Shield, Sheath, or Concentric Conductor

Material	Temperature ($°C$)[a]
Crosslinked (thermoset)	350
Paper, thermoplastic, varnished cloth	200
Paper	
in self-contained liquid-filled cables	250[b]
in liquid-filled pipe-type cables	200[b]

[a] For lead sheaths this temperature is limited to 200°C (IEC limits pure lead sheaths to 170°C, alloy lead to 200°C).
[b] Values are from IEC standards (see notes to Table 11.11), as ICEA Standard P-45-482 [48] does not cover liquid-filled cables.
Source: From [48].

single cores in trefoil: Recommended temperatures should be applied with care due to the possibility of high temperatures in the center of the trefoil.

For the calculation of the effective cross-sectional area of different types of metallic screens, sheaths (shields), and concentric conductors, guiding formulas and tables are provided in both ICEA [48], and IEC [37] standards. For composite shields the total effective area is the sum of the effective areas of the components. The contact resistance between overlapped turns of shields, or between layers of shields, may decrease due to aging and also on account of forces or movement during short circuits; this has to be accounted for when calculating the effective cross section.

In case special connectors are present in a current-carrying component, IEC recommends the following allowable temperatures under short-circuit in that particular component:

Exothermic welded joint: 250°C

Soldered joint or connector: 160°C

Compressed connector (involving mechanical deformation): 250°C

Welded joint or connector: Same as the conducting component

For mechanical (bolted) connectors IEC suggests to consult the manufacturer.

The temperature of any current-carrying cable component made of copper, aluminum, or steel is limited by the material with which it is in contact; where there is a thermally insulating cushion between the current-carrying part and the adjacent insulation or jacket, IEC standards [49–51] recommend that the lower of 350°C, or the allowable jacket temperature, not to be exceeded.

11.10.1.2 Comparison of the Adiabatic and Nonadiabatic Methods

The adiabatic method always yields results that are on the safe side. During the cooling periods between faults interrupted and reestablished with automatic reclosing circuit breakers, significant amounts of heat may be dissipated from the current-carrying cable components. In such cases calculating with the nonadiabatic method may lead to increases of the permissible short-circuit current, or to smaller cross sections for the same current and duration. Nonadiabatic calculations may also be justified for longer-duration single, uninterrupted short-circuits, in particular for small cross sections. ICEA standards [47, 48] suggest nonadiabatic calculations for short-circuits in excess of 2 s. IEC-949 [37] suggests that using the nonadiabatic method may be advantageous for screens, shields, sheaths, and possibly for small conductors with cross-sections less than 10 mm^2 (approx. 7 AWG) especially when used as screen wires (which is also the case of concentric neutrals); it is also suggested by IEC that the nonadiabatic method yields negligible improvements for a short-circuit duration to conductor (copper) cross-sectional area ratio of less than 0.1 s/mm^2 ($t/S < 0.1$), and that the adiabatic method may be used in this case.

11.10.1.3 Nonadiabatic Method

Section 11.10.1.2 summarized the limitations of the adiabatic method and indicated under what conditions nonadiabatic procedures may yield savings.

When cables with extruded insulations and thin metal tape shields replaced lead-covered impregnated paper insulated cables, it became important to introduce more rigorous calculating methods for these shields under short-circuits with reclosing. A procedure was developed and confirmed by tests on cable samples [53]. This method is one of the foundations for more advanced procedures [37, 52] published 15 years later.

Tests were conducted and reported [54] with other types of shields, and test results were compared with the calculated values of the study mentioned above.

Further tests were conducted and reported on cables with round wire shields [55]. Temperature, like in previous work, was plotted as a function of time from the moment of fault initiation throughout the heating and cooling periods of the reclosing cycle, until final clearing of the fault by the reclosing circuit breaker. Measured values were compared with results obtained by a computer program based on [53].

The results of continued research were published in [52]:

EPRI Report EL-3014, Research Project 1286-2 (April 1983)

1. Numerous further tests [52] with different reclosing sequences were conducted on different types of extruded dielectric cables using several combinations of insulating and jacket materials. Cable samples were aged, unaged, installed straight and bent, and subjected to normal and emergency operating conditions. The conclusions of the report apply to round wire and flat strap shielded XLPE insulated cables with various commonly used jacket materials in extruded or sleeved jacket constructions.

2. Two computer programs were also developed: one for round wire shields, the other for flat strap shields. All these represent typical cable constructions and recloser operating sequences encountered on utility power distribution systems. The programs provide shield temperature as a function of time throughout the duration of the short circuit (reclosing sequence). The computer programs take into account the change of contact surface area (in percent) during short circuits (first, second, third reclosing sequence) between shield wires or straps and the adjacent media (extruded insulation screen on the inside, jacket material on the outside); this because it has been found during this and previous research that the difference

between measured and calculated values was due to variation of the contact surface area between the metallic shield and the surrounding media.

3. A simplified laboratory test procedure was developed to help evaluate cable designs and operating conditions (different reclosing sequences and short-circuit durations) other than those investigated under this project. As a result of these tests other (new) products and different operating conditions may be adapted to the above computer program.

4. The tests revealed that, for the products tested, the allowable short-circuit temperatures in the ICEA standards were too conservative, the report recommends increased maximum withstand temperatures for specific cable materials and constructions.

5. The report also provides information on different parameters influencing withstand temperatures under short circuits, such as jacket thickness and materials, difference between extruded and sleeved jackets, behavior of round wire and flat strap shields, installation in bends compared with straight cables, aging, etc.

IEC Standard 949: Calculation of Thermally Permissible Short-Circuit Currents, Taking into Account Nonadiabatic Heating Effects [37]. The calculating procedure in this standard is the following:

(1) Calculate the adiabatic short-circuit current I_{AD} [Eq (11.60)].
(2) Calculate the modifying factor (ε) to take into account the effect of nonadiabatic heating.
(3) The product of the results in (1) and (2) is the permissible short-circuit current (I):

$$I = \varepsilon I_{AD} \tag{11.61}$$

also if the short-circuit current (I_{SC}) is known:

$$I_{AD} = \frac{I_{SC}}{\varepsilon} \tag{11.61a}$$

the result I_{AD} may be substituted into Eq. (11.60b) to obtain the final temperature θ_f at the end of the short-circuit.

The standard [37] provides procedures, series of formulas, curves, and tables for the calculation of the modifying factor ε for cable conductors and several types of shields; calculated values were matched with results obtained with other methods ([32c] among others), also for screens and sheaths reference [53] was a principal source. As is the case with the EPRI method (see earlier, clause 2 under EPRI Report, EL-3014), adjustments are included in the IEC formulas [37] to allow for variations in the contact surface area. Tables 11.9 and 11.10 give the physical properties of materials. Allowable temperatures according to IEC are given in [49–51]. No provision is given in IEC Standard 949 [37] regarding the temperature drop during the "off periods" of the reclosing sequence.

11.11 CABLE SYSTEM ECONOMICS

The problem of finding the economical cross section of the conductor of an electric circuit was first approached by William Thomson (Lord Kelvin) in 1881. His assumption that the cost of electric wiring was a linear function of conductor cross section, was substantially correct for the low-voltage installations of those times. Thus the capital cost may be expressed as

$$CI(S) = A_t S \qquad (11.62a)$$

where $CI(S)$ in dollars (or any other currency unit) is the installed capital cost of a circuit as a function of the cross-sectional area of its conductor, A_t (in dollars per square millimeter) is the cost component factor related to conductor size (the slope of the linear function), and S (in square millimeters) is the cross-sectional area of the conductor.

The yearly carrying charges in dollars (or any other currency unit) of the capital investment, as a function of the cross-sectional area, can be expressed as

$$CI_y(S) = \frac{p}{100} CI(S) = \frac{p}{100} A_t S \qquad (11.62b)$$

where p is the yearly percentage value of the carrying charges.

The yearly cost of losses, also as a function of cross section is

$$CJ_y(S) = F_y \frac{I_{eff}^2}{S} \qquad (11.62c)$$

where F_y is a factor that includes the unit cost of energy, plus material and circuit parameters defining the effective resistance; I_{eff} is the equivalent constant current generating the same total yearly joule losses as those generated by the varying current I which is a function of time (t); I_{eff}^2 is defined as

$$I_{eff}^2 = \frac{1}{8760} \int_0^{8760} I^2(t)\, dt \qquad (11.62d)$$

The total yearly cost $CT_{Ty}(S)$, as a function of cross section, is

$$CT_{Ty}(S) = CI_y(S) + CJ_y(S) \qquad (11.62e)$$

The optimum conductor size (S_{eco}) is obtained by equating to zero the derivative of Eq. (11.62e) with respect to S:

$$\frac{dCT_{Ty}(S)}{dS} = \frac{dCI_y(S)}{dS} + \frac{dCJ_y(S)}{dS} = \frac{p}{100} A_t - F_y \frac{I_{eff}^2}{S^2} = 0 \qquad (11.62f)$$

By expliciting S from Eq. (11.62f), the optimum conductor size (S_{eco}) is obtained:

$$S_{\text{eco}} = I_{\text{eff}} \left(\frac{F_y}{A_t(p/100)} \right)^{1/2} \tag{11.62g}$$

By rearranging Eq. (11.62g):

$$\frac{p}{100} A_t = F_y \frac{I_{\text{eff}}^2}{S_{\text{eco}}^2} \tag{11.63a}$$

and multiplying both sides of Eq. (11.63a) by S_{eco}, it becomes evident that

$$CI_y(S_{\text{eco}}) = CJ_y(S_{\text{eco}}) \tag{11.63b}$$

Accordingly, *Thomson's (Kelvin's) rule* states: The optimum conductor size for a given load is the one where the yearly cost of losses equals the yearly carrying charges of the installed capital cost of the circuit.

    This rule is illustrated in Fig. 11.18(a), with cross section (square millimeters) on the abscissa, and yearly cost (dollars) on the ordinate: the total yearly cost $CT_{\text{Ty}}(S)$, which is the sum of linear function $CI_Y(S)$ (yearly carrying charges of installed capital cost) and the hyperbolic function $CJ_Y(S)$ (yearly cost of losses), reaches its minimum where the lines representing the two component functions (linear and hyperbolic) bisect.

    It is evident that Thomson's approach is approximate only in most cases, as several cost components are independent of cross section. Accordingly, Kapp proposed to add the constant C_K to Eq. (11.62a):

$$CI_K(S) = A_t S + C_K = CI(S) + C_K \tag{11.64a}$$

where $CI_K(S)$ is the total installed capital cost of a circuit as a function of the cross-sectional area of its conductor, A_t is the cost component factor related to the portion of capital cost which is the function of conductor size (the slope of the linear function), C_K is the portion of the installed capital cost, independent of conductor size, and $CI(S)$ is the portion of the installed capital cost that is a linear function of conductor size (S):

$$CI(S) = A_t S \tag{11.64b}$$

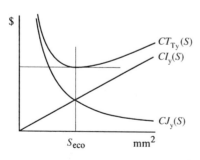
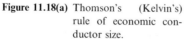
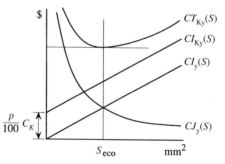

Figure 11.18(a) Thomson's (Kelvin's) rule of economic conductor size.

Figure 11.18(b) Kapp's rule of economic conductor size.

As can be seen, the mathematical expression of $CI(S)$ has not changed here, but it has assumed a different meaning under Kapp's treatment of the problem.

The total yearly carrying charges of the capital investment:

$$CI_{Ky}(S) = \frac{p}{100} CI_K(S) = \frac{p}{100} C_K + \frac{p}{100} A_t S \qquad (11.64c)$$

The total yearly cost $CT_{Ky}(S)$ as a function of cross section is

$$CT_{Ky}(S) = CI_{Ky}(S) + CJ_y(S) = \frac{p}{100}(C_k + A_t S) + F_y \frac{I_{eff}^2}{S} \qquad (11.64d)$$

where the yearly cost of losses $CJ_y(S)$ is taken from Eq. (11.62c).

Upon the differentiation of Eq. (11.64d) the constant C_K is eliminated; the result therefore is the same as in Eq. (11.62f); this also leaves Eq. (11.62g) and (11.63b) unchanged under Kapp's approach.

Accordingly, *Kapp's rule* states: The optimum conductor size for a given load is the one where the yearly cost of losses equals the yearly carrying charges of that portion of the installed capital cost, which is a direct linear function of conductor size.

This rule is illustrated in Fig. 11.18(b). The graphic illustrations of the Thomson (Kelvin) and Kapp rules are shown side by side [Figs. 11.18(a) and (b)]. The total yearly cost $CT_{Ky}(S)$ reaches its minimum, where the curve representing the yearly cost of losses $CJ_y(S)$ bisects the line representing the portion of installed capital cost (yearly carrying charges) $CI_y(S)$ dependent of conductor size (S); the line representing the yearly carrying charges of total installed capital cost $CI_{Ky}(S)$ is obtained by shifting the line $CI_y(S)$, representing its portion that is a function of cable size, by the addition of its constant portion $(p/100)C_K$, as a result the curve representing the total yearly cost $CT_{Ky}(S)$ is shifted by the same amount when compared to the Thomson method.

11.11.1 Calculating Procedure

IEC Standard 287-3-2 [9f] provides a procedure for calculating the economic current range for a given conductor and also the economic conductor size for a given load. The IEC standard [9f] follows Kapp's principle for the determination of economic conductor size, however:

- For the cost of loss due to load: Instead of the yearly cost, IEC uses the present value of the cost of joule losses during N years of the projected "economic life" of the cable: CJ in *dollars* or any other currency unit.
- For capital cost: Instead of the yearly carrying charges of the installed capital cost, IEC uses the installed cost of the length of cable; CI in dollars or any other currency.
- The total cost of the system (installing and operating the cable during its projected economic life, expressed in present values) CT in dollars or any other currency is

$$CT = CI + CJ \qquad (11.65)$$

11.11.1.1 Calculating Cost of Joule Losses

The present value of the cost of joule losses, as defined in Section 11.11.1, is given by IEC [9f]:

$$CJ = I_{max}^2 RlF \tag{11.66}$$

where I_{max} in amperes is the maximum load current in the first year, l in meters is the length of the cable, and R in ohms per meter is the *apparent* ac resistance of the cable per unit length including not only skin and proximity effects but also the shield (sheath) and armor (pipe) losses:

$$R\{or\ R(S)\} = \frac{1}{S} \rho_{20} B[1 + \alpha_{20}(\theta_m - 20)]10^6 \tag{11.66a}$$

Here S (in millimeters squared) is the effective cross-sectional area of the cable conductor, ρ_{20} (in ohms per meter) is the conductor resistivity at 20°C (Tables 11.2, 11.9), α_{20} is the temperature coefficient of conductor resistance at 20°C (Table 11.2), and B is the factor allowing for skin and proximity effect plus shield (sheath) and armor (pipe) losses:

$$B = (1 + y_p + y_s)(1 + \lambda_1 + \lambda_2) \tag{11.66b}$$

where y_p, y_s, λ_1, and λ_2, are defined in Sections 11.6.1 and 11.6.3.

Also θ_m (in degrees Celsius) is the mean operating temperature of the conductor during N years of the projected economic life of the cable; IEC [9f] suggests to assume for most calculations that

$$\theta_m = \left[\frac{\theta - \theta_a}{3}\right] + \theta_a \tag{11.66c}$$

where θ (in degrees Celsius) is the maximum rated conductor temperature, and θ_a (in degrees Celsius) is the average ambient temperature. For more precise estimates of mean temperature IEC Standard 287–3-2 [9f] provides a calculating procedure in an appendix.

Further on F is a coefficient (in dollars per watt), which includes the parameters for the calculation of CJ with the exception of current and resistance:

$$F = N_P N_c (TP + D) \frac{Q}{1 + (i/100)} \tag{11.66d}$$

where N_P is the number of phase conductors per circuit, N_c is the number of circuits (each of them carrying a load of the same type and magnitude), P in dollars per watt-hour) is the cost of energy at the relevant voltage level, D is the yearly demand charge in dollars per watt over a year, i is the discount rate (percent) excluding the effect of inflation (IEC Standard 287-3-2 [9f] does not consider the effect of inflation as it expects inflation to have the same effect on both the cost borrowing and the cost of energy), and T in hours/year is the operating time at maximum joule loss, or it is the number of hours per year necessary for the maximum current I_{max} to produce the same yearly energy loss (watt-hour per year) as the actual load current I varying as a function of time (t). T is defined as

$$T = \int_0^{8760} \frac{I^2(t)\, dt}{I_{max}^2} \tag{11.66e}$$

or

$$T = 8760\mu_y \tag{11.66f}$$

and here μ_y is the yearly loss-factor ("loss-load factor" according to IEC [9f] and [11], also according to CIGRE [32]), and it is defined as

$$\mu_y = \frac{\int_0^{8760} I^2(t)\, dt}{8760 I_{max}^2} \tag{11.66g}$$

A study of losses in underground transmission and distribution cables by a working group of the Insulated Conductors Committee (ICC) of IEEE [56] proposes empirically developed formulas for a close approximation of μ_y for systems where the yearly load-factor is known:

- For transmission systems:

$$\mu_y = 0.3 \text{ (yearly load factor)} + 0.7 \text{ (yearly load factor)}^2 \tag{11.66h}$$

- For distribution systems:

$$\mu_y = 0.2 \text{ (yearly load factor)} + 0.8 \text{ (yearly load factor)}^2 \tag{11.66i}$$

For loss factors varying during the life of the cable, formulas are provided by [10].

In Eq. (11.66d) Q is a coefficient to account for the discount rate, plus for increases in energy cost and load over the life (N years) of the cable.

$$Q = \sum_{n=1}^{N} \left(r^{(n-1)} \right) = \frac{1 - r^N}{1 - r} \tag{11.66j}$$

where

$$r = \frac{[1 + (a/100)]^2 [1 + (b/100)]}{1 + (i/100)} \tag{11.66k}$$

and a in percent is the yearly rate of increase in load, b in percent is the yearly rate of increase in the cost of energy. The effect of inflation is not included for the reason mentioned in the comments to Equation (11.66d).

For varying rates of increase in energy cost and load during the life of a cable, formulas are provided by [10].

11.11.1.2 Total Cost

Total cost is calculated by combining Eqs. (11.65) and (11.66):

$$CT = CI + I_{max}^2 R l F \tag{11.67}$$

Utilities usually have to borrow for the purchase, construction, and installation of plant. According to the usual accounting practice of utilities, the interest accounted for during construction is included in the installed cost CI, which is accounted for as fully paid capital assets upon the in-service date. After this date interest and other costs become parts of the rate structure. The cost of joule losses is projected to the same date for the calculation of their present value. The IEC standard [9f] suggests it is convenient to use this date (the date of purchase) as a point of reference.

11.11.1.3 Determination of Economic Current Range for Given Conductor Size

The upper and the lower limits of the economic range for a given conductor size are given by the following equations:

$$I_{max\,L} = \left[\frac{CI - CI_1}{Fl(R_1 - R)} \right]^{1/2} \quad \text{lower limit} \tag{11.68}$$

$$I_{max\,U} = \left[\frac{CI_2 - CI}{Fl(R - R_2)} \right]^{1/2} \quad \text{upper limit} \tag{11.69}$$

where CI (in dollars) is the installed cost of the length of cable having the size of conductor being considered, R (in ohms per meter) is the *apparent* ac resistance [Eq. (11.66a)] per unit length of the conductor size being considered; subscripts 1 and 2, respectively, designate the same parameters (installed cost and apparent ac resistance) of the next smaller (1) and the next larger (2) standard conductor.

Examples given in an appendix of IEC Standard 287-3-2 [9f] demonstrate that cost variations for a given load are not critically dependent of conductor size within a fairly wide range near the economic value.

11.11.1.4 Economic Conductor Size for Given Load

For the purpose of calculating economic conductor size S_{ec}, Eq. (11.67) is transformed to express CT, CI, and R as functions of the conductor cross-section S:

$$CT(S) = CI(S) + I_{max}^2 R(S) l F \tag{11.70a}$$

IEC Standard 287-3-2 [9f] recommends to plot calculated values CI over the range of conductor sizes considered. It is usually possible to find a linear relationship over a limited range by fitting a line to the calculated values. This yields the linear function CI(S) expressed as follows according to Kapp's principle:

$$CI(S) = l(AS + C) \tag{11.70b}$$

where l (in meters) is the length of cable, A (in dollars per meter per millimeter squared) is the cost component factor related to conductor size (the slope of the fitted linear function), and C (in dollars per meter) is the constant component of the cost independent of cross section. The value of C is at the point where the line representing the fitted linear function $CI(S)$ bisects the ordinate.

Combining Eqs. (11.70a) and (11.70b) we obtain

$$CT(S) = l[(AS + C) + I_{max}^2 R(S)F] \tag{11.70c}$$

After substituting $R(S)$ [Eq. (11.66a)] into Eq. (11.70c), the optimum conductor size (S_{ec}) (in millimeters squared) is obtained by equating to zero the derivative of Eq. (11.70c) with respect to S, and solving to obtain S_{ec}:

$$S_{ec} = I_{max} 1000 \left\{ \frac{F\rho_{20} B[1 + \alpha_{20}(\theta_m - 20)]}{A} \right\}^{1/2} \tag{11.70d}$$

where I_{max} (in amperes) is the maximum load in the first year, F is given in Eq. (11.66d), a_{20} is given in Table 11.2, B is given in Eq. (11.66b), θ_m may be estimated with the help of Eq. (11.66c) or Appendix B of [9f], and A is described in conjunction with Eqs. (11.70b) and (11.70c).

For ρ_{20} in Eq. (11.70) IEC [9f] recommends the following: The economic conductor size is unlikely to be identical to a standard size, and so it is necessary to provide a continuous relationship between resistance and size. This is done by assuming a value of resistivity for each conductor material. The values recommended in IEC 287-3-2 [9f] for ρ_{20} are 18.35×10^{-9} for copper and 30.3×10^{-9} for aluminum. These values are not the actual values for the materials but are compromise values chosen so that conductor resistances can be calculated directly from nominal conductor sizes, rather than from the actual effective cross-sectional area.

The economic size being unknown, y_p, y_s, λ_1, and λ_2 have to be established for an assumed cable construction. Recalculation may be warranted if the calculated economic size is too different from the assumed one.

The calculated S_{ec} is unlikely to be equal to a standard size, therefore the total cost of the next larger and smaller standard sizes shall be calculated and the most economical chosen.

Upon the differentiation of Eq. (11.70c) the constant component C is eliminated. Thus it does not affect the value of the optimum conductor size S_{ec} in Eq. (11.70). In the evaluation of the installed cost $CI(S_{ec})$ and total cost $CT(S_{ec})$, required for establishing the definitive value of S_{ec}, however, the constant component C cannot be ignored and real costs (which automatically include the variable and constant portions) have to be calculated. Thus CI, CI_1, and C_2 for Eqs. (11.68) and (11.69) also have to be based on real cost figures that automatically include both variable and constant components. At this stage of the calculation, conductor and cable construction are known; therefore, real values have to be used: conductor resistivity is to be taken from Table 11.2 or Table 11.9. Also skin and proximity effects as well as shield (sheath) losses have to be calculated according to actual cable dimensions.

11.11.1.5 *Dielectric Loss and Losses Due to Charging Current*

Dielectric loss and charging current are voltage dependent and therefore permanently present while the cable is energized. They have to be accounted for at 100% load factor for the economic life of high-voltage cables with rated voltages exceeding the values given in Table 4.5; the calculation of dielectric loss is given in Section 11.6.2. Charging currents are treated in [2, 3, 10, 56].

A proposed amendment to IEC Standard 287-3-2 concerning dielectric loss suggests to add the present value of these losses to the values of CI, CI_1, and CI_2 in Eqs. (11.68) and (11.69) when calculating the economic current range for a given cross section; the same treatment could be applied to charging current losses.

For the calculation of the optimum cross-section S_{ec} for a given load, the same amendment suggests first to calculate S_{ec} using Eq. (11.70d) but ignoring the voltage-dependent losses. Then the total cost CT of this size and the next two smaller standard sizes is calculated, including the cost of voltage dependent losses and the most economical chosen. Formulas for the cost of both types of voltage-dependent losses are given in [10].

11.11.2 Design Considerations

Progress in the domain of dielectric materials has been leading to increased operating temperatures; one of the consequences is increased energy losses in cables. At the same time the cost of energy has also been rising. It has therefore been recognized that system design should not only be based on lowest initial investment cost, but energy losses in the cable over its economic life should also be considered; the result is larger conductor size than thermally required, lower operating temperature, and lower loss of energy.

Examples in the appendix of IEC Standard [9f] demonstrate that savings are very important when load factor is high (0.9 in the series of examples), and initial planning is based on the cable reaching its permissible normal operating temperature under normal operating conditions. The consideration of economic factors is of prime importance for "dedicated-use" cables where the cable is attached to another component of the system and the load to be carried by the cable has to be matched to the loading capability of that other component. An incomplete list of such uses is station getaways to overhead distribution lines, underground or underwater sections inserted into distribution feeders or overhead transmission lines, high-voltage and extra-high voltage underground connections of overhead transmission lines to large substations or power plants, cable connections to generators or transformers, among others. In such cases important savings are possible by increasing the "matched" conductor size to the economic value.

The reliability of cables is very high, but repair times are much longer than those of overhead lines. At distribution voltages one has to consider a day or days for a repair, at transmission voltages it is a week or weeks; the higher the voltage, the longer the repair duration. Planning for contingencies is therefore one of the first concerns in underground system design. Allowable temperatures for short-term overloads are also discussed briefly in Section 11.9.9. Because of the duration of contingency conditions involving cables, emergency temperatures allowed are mostly the same as the normal operating ones or slightly higher, particularly in transmission systems. Emergency temperatures as high as 130°C (for insulations rated 90°C continuous) or

140°C (for insulations rated 105°C continuous) are allowed according to American Standards [12, 57] for XLPE and EPR insulations for limited periods (1500 h cumulative during the lifetime of the cable). The pertinent Canadian standard [13] specifies 90°C for normal operation and quotes "normal industry practice" accepting 100 h maximum at 130°C in any 12 consecutive months (500 h cumulative during the lifetime of the cable). Such high temperatures are useful for real short-duration overloads, such as transferring load to a substation getaway cable on account of a fault on the overhead portion of an adjacent circuit, or a load transfer due to an easily repairable defect in a transformer vault or pedestal, or a load transfer because a pressurized (liquid-filled or gas pressure) cable has to be taken out of service on account of loss of pressure due to a leak in the piping.

During normal operating conditions, on underground systems designed for contingencies involving cables, each circuit carries about 50–75% of its allowable emergency current: 50% in a simple loop, 75% in a system with four cables designed to function with three during emergencies. These percentages assume an even distribution of the load. In most of such cases the economic cross section and the one designed for emergency loading will be close; the larger one is to be chosen. The emergency periods, because of their short duration compared to the lifetime of the cables, do not have to be considered in economic loading calculations.

Other considerations (mechanical and thermal design, fault currents, voltage drop, rationalization of conductor sizes to reduce inventories, transportation clearances and weight limits, bending radii, and possibly other reasons) will often prevail over economic choice of conductor size.

11.12 CHOICE OF SYSTEM VOLTAGE

Lowest investment cost is the prevailing criterion in the choice of the voltage for a transmission or distribution system, apart from fundamental technical criteria. These system planning studies involve generating stations, lines and substations of the bulk transmission system, choice of transformers and switchgear, the fundamental layout of the subtransmission and distribution system, among others.

Generally, the cost of losses in conductors is taken into consideration during the final planning stage when the definitive choice of conductor sizes is made for the voltage levels already chosen. For cable conductors calculations have to be made as discussed in Section 11.11.1.

For choosing the supply voltage for a particular load or group of loads within a given system, the optimum cross sections S_{ec} have to be determined for that particular load and for the voltage levels considered; then total costs CT have to be calculated for each voltage level using the appropriate S_{ec} values. Transformer costs and losses have to be considered as well.

Cost studies for pipe-type cable systems in the United States indicate that with natural cooling the economical performance increases with voltage for high-power lines (400–500 MVA range) up to about the 345-kV level; increasing further the voltage to 500 kV is counterproductive when only the economic aspects are taken into consideration. The main reasons for this are the higher temperature drop in the thicker insulation requiring the use of very large conductor cross sections to lower the joule losses, plus the presence of high dielectric losses and charging currents [5, 58]. The feasibility of naturally cooled high-power EHV lines has improved recently through the introduction of low-loss insulations: synthetic (PE and XLPE) and liquid-filled taped (PPL). Jointing techniques are also progressing with the introduction of premolded components for higher insulation levels. The progress in technology so far has not been able to offset the economic disadvantage of EHV cables 400 kV and above.

Effective cost studies cannot easily be standardized as cost components vary according to local conditions. For example, a study by a Canadian utility shows that for transmitting power into a city center: (a) a single 315-kV line carrying 500 MVA installed in a duct bank, providing spare ducts for the installation of another line in the future, costs approximately the same as (b) a double circuit 120-kV line carrying a total load of 400 MVA. For the solution possibility (a) was chosen. It provides a higher initial capacity than (b); also another line may be added in the future without excavation and the elimination of the 315- to 120-kV transformation more than compensates for the added cost of 315 kV (versus 120 kV) switchgear. To conclude, in most cases, cable system design is only a part of the overall system planning, especially in the case of transmission and subtransmission.

11.13 CABLE SELECTION AND INSTALLATION METHODS

To facilitate the choice of cable type and installation method, the advantages and disadvantages of each have been listed. It has been attempted to number the features to provide a tool for cross referencing; in the subsequent sections, Section 11.13.1 deals with directly buried cables, Section 11.13.2 with cables installed in ducts, and Section 11.13.3 with pipe-type cables. Thus, for example, in the selection process one should compare the advantage (i) of a directly buried cable to the disadvantage (i) of a cable in a duct or a pipe-type cable; when comparing the two latter types of cables, similarities may be identified noting advantage (i) of the two types of cables.

11.13.1 Directly Buried Cables

Burying the cables directly is the oldest and still the most common cable laying method. Its main advantages are as follows:

(i) Direct heat transfer from the cable surface to the ground, thereby yielding the optimum loading capability. Caution has to be exercised, however, as moisture migration may be initiated if the cable–soil interface temperature exceeds a "critical limit" for extended periods (Eqs. (11.4) and (11.4a), plus [9, 10]). The use of a suitable thermal backfill material around the cable will alleviate this problem [21, 22, 24–26, 28].

(ii) Cables and joints are firmly held by the surrounding earth or backfill material so that the cables cannot move due to heating and cooling induced by the cyclic variation of the load.

(iii) Pulling tension is practically nonexistent on level ground as the cable is moving forward on rollers as it is being laid into the trench; synchronized motorized rollers or haulers (on slopes) are often used for long lengths and heavy cables. Another installation method for heavy cables is attaching them throughout their entire length to the winch rope to remove all pulling forces from the cables.

(iv) Ease of guiding the cable around corners by means of rollers and guides.

(v) Cable length between joints is not limited by the pulling tension due to friction.

(vi) Smaller jointing bays than for the duct system, as no expansion loops are necessary.

(vii) Once a fault has been located, repairs are limited to the immediate area of the defect.

The main disadvantages of direct burying are as follows:

(i) During construction, work in the trench has to be carried out simultaneously in at least three adjacent cable sections: (a) while the trench is being dug and its bottom is being prepared for cable laying in one section, (b) in the next section the cable laying is carried out (this operation has several stages: positioning of rollers and guides, cable pulling, positioning of the cable in the trench, and removal of rollers and guides), (c) in the following section separators between cables are placed, the cables are covered with fine material or thermal backfill, mechanical cable protection (slabs, warning strips, etc.) are installed, and finally the trench is filled and the surface is reinstated. This widely extending and lengthy working procedure, hampering traffic and interfering extensively with residential and business properties along the route, has precluded its use in most North American cities. Therefore, direct burial of cables in North America is limited to suburban areas, very wide roads, or private rights of way.

(ii) New trenches must be dug for every addition.

(iii) Wide space occupancy: Access to each cable has to be secured for repairs. Therefore cables cannot be positioned one above the other (except for trefoil groups of cables belonging to the same circuit).

(iv) Lack or weakness of mechanical protection.

(v) Difficulty in fault locating.

(vi) Extensive digging for repairs or replacement.

(vii) Unforeseen obstacles encountered on the route are bound to cause costly delays, and possible waste of cable, as cable lengths are ordered in advance according to plan and cable laying immediately follows excavation, leaving no time for adjustments.

(viii) For high-voltage (HV) or extra high voltage (EHV) directly buried installations when liquid-filled cables are used, the cables are of the self-contained type. In this type of cable and its accessories, the oil pressure must be carefully maintained during installation work, any oil to be used must pass through a high degree degasification process, and precautionary measures have to be taken to avoid oil spills. This remark only applies to the prevalent cable type used for direct burial in HV and particularly in EHV applications. It is meant to help the comparison with pipe-type cables.

It is evident from the economical point of view that, notwithstanding the disadvantages of directly buried cables, the latter represent the lowest cost method of installation for single circuits on routes where no other circuits are contemplated for the future.

11.13.2 Cables Installed in Ducts

The prevailing cable installation method practiced in North American city centers is that of duct laying. Again we shall proceed to enumerate the various advantages and disadvantages of the method for comparison purposes.

The duct laying technique is characterized by the following advantages:

(i) The trench work may be limited easily to 300- to 500-m sections.

(ii) For a relatively low increase in cost, ducts may be added along the first line to ensure the easy addition of future circuits.

(iii) The ducts may be laid in several layers as normally no access to the individual cable is necessary between manholes.

(iv) Concrete encasing of the ducts provides good mechanical protection.

(v) Defects in a joint may be detected by visual inspection. It is evident that leaks of liquid-filled cables installed in ducts are easy to locate as oil usually appears at the duct mouth (to have this advantage, the duct-line profile should facilitate drainage of the ducts toward the manholes). Fault location with instruments does not require pinpointing of the exact location of the defect but only needs to identify the defective length.

(vi) Repairs do not require digging as the damaged cable is pulled out and replaced. To accelerate the restoration of service, it is customary to pull the replacement cable into a spare duct before the removal of the defective one.

(vii) The exact cable cutting lengths may be given to the cable manufacturer after the completion of each individual duct section as trenching and cable pulling are not simultaneous. This makes it easier to adjust plans when unforeseen underground obstacles are encountered.

There are disadvantages to the installation of cables in duct systems:

(i) The air space within the duct between cable and duct wall limits the heat transfer from the cable to the environment. Thus larger conductor cross sections are necessary in ducts than for directly buried cables. On the other hand the concrete encasing ensures thermal stability in the area immediately surrounding the cables. Thus the risk of "thermal runaway" is reduced unless the duct bank is filled with heavily loaded cables. The use of a suitable thermal backfill material around the duct-bank will alleviate the risk of thermal runaway [21–26, 28].

(ii) Cable expansion due to load cycling has to be planned carefully to prevent sheath and insulation damage (Section 11.15.1).

(iii) The cable route (curves, offsets) and the distances between manholes need to be planned carefully to keep pulling tensions and sidewall pressures within allowable limits (Section 11.14).

(iv) To avoid 90° turns in the duct passages, manholes for distribution cables are usually placed at street corners. On the other hand, city planning authorities generally attempt to keep large manholes for transmission (subtransmission) cables away from intersections to conserve underground space for the crossing of different utilities and also for the sake of traffic flow. Another alternative for

90° turns is the construction of large-radius curves in the duct runs at the street corners. However, curves greatly increase pulling tension and they are costly to plan and to build.

(v) With frequent turns and offsets inevitable in city center cable routes, manholes must be spaced closely.

(vi) The size of manholes has to allow for cable training for jointing, for expansion loops or cable clamping; or both (Section 11.15.1).

(vii) An entire cable length must be replaced in case of a fault.

(viii) For HV or EHV installations in ducts when liquid-filled cables are used, the cables are of the self-contained type. In this type of cable and its accessories the oil pressure must be carefully maintained during installation work, any oil to be used must pass through a high degree degasification process, and precautionary measures have to be taken to avoid oil spills. This remark only applies to the prevalent cable type used for installation in ducts in HV and particularly in EHV applications. It is meant to help the comparison with pipe-type cables.

In conclusion, it must be emphasized that the laying of cables in ducts represents definitely the most economical method for multiple circuits, especially when numerous future additions are planned. Even though direct burial is positively more economical for single circuits, its practice in North America is limited because of its often unacceptable length of trench work; duct work is also preferred on account of its sturdiness and better resistance against dig-ins (cable damage due to accidents resulting from various excavation activities).

11.13.3 Pipe-Type Cables

Pipe-type cables constitute the prevailing form of underground transmission in the United States. They have gained wide acceptance at a time when bare lead-sheathed cables were plagued by sheath fatigue and corrosion problems. Even though the application of extruded jackets has later solved the problem of sheath corrosion, and now sheath fatigue can be prevented by providing for proper cable expansion in the design stage, pipe cables have maintained their preeminence for underground power transmission in the United States.

The outstanding advantage of pipe-type cable lies in the ruggedness of the pipe as a supporting structure. Steel pipes are rarely severed completely when accidentally hit by heavy earth-moving machinery. Concrete duct banks also have a high mechanical resistance but are bound to be treated by mistake as abandoned foundations to be demolished when encountered by the operators of heavy construction machinery.

Section 7.9 describes a particular problem of EHV pipe-type cables with heavy insulation walls and large conductors: cyclic cable movement induced by load cycles leads to the dislocation of the insulating tapes, causing electrical breakdown of the cable. This problem has not been encountered in self-contained cables where the sheath applied over the heavy EHV insulation firmly contains the tapes preventing uncontrolled dislocation. The addition of spacers, called spiders, can solve the problem in the vicinity of joints. The same difficulty has also been encountered away from joints; it is believed therefore, that tight control of paper lapping, and design changes of the insulation and its shielding, will prevent tape dislocation in future EHV pipe cable installations. The problem has been recognized early in France where pipe-type cables are widely used in congested cities but have been limited to 225 kV and 330 MVA. Due to progress in the manufacture of heavy extruded insulation walls for HV and EHV cables, directly buried cables with extruded insulation have been installed for new lines in France for a number of years, rather than pipe-type cables.

The following brief summary lists the various advantages of pipe-type cables:

(i) Similar to duct systems, trenching for pipe laying may be limited to 300- to 500-m sections.

(ii) Additional cable-carrying pipes and oil-return pipes for cooling may be laid initially in the same trench if required for projected load increases. Since pipe work is more expensive than duct laying, the initial capital outlay for anticipated load growth is higher for pipes than for ducts.

(iii) The oil-return pipes for future cooling may be positioned above their proper cable-carrying pipes, though an adequate horizontal clearance needs to be maintained between adjacent cable-carrying pipes.

(iv) The pipe system provides unsurpassed mechanical protection.

(v) Longer pulling lengths are possible than in ducts on account of the low coefficient of friction between the pipe and the skid wires over the cable core.

(vi) Repairs can be done normally without excavation when cables only have been damaged and no damage occurred to the pipe, but freezing is necessary to retain the oil in the unaffected parts of the cable.

(vii) Exact cable cutting lengths may be given to the cable manufacturer after the completion of each individual pipe section, as trenching and cable pulling are not simultaneous. This makes it easier to adjust plans when unforeseen obstacles are encountered.

As with the two previous cable systems, pipe-type systems also possess a number of inherent disadvantages, which are given in the following brief list:

(i) The thermal resistance of a liquid-filled pipe cable is much lower than that of a cable surrounded with air in a duct; however, the close spacing of the three cores within a pipe increases the thermal resistance. Magnetization and eddy current losses in the pipe also contribute heat and thus for equal current-carrying capacities the pipe-type requires the largest cross section, and the directly buried cable system the smallest. Heat transfer to the environment is direct from the pipe; thus moisture migration may be initiated if the pipe–soil interface temperature exceeds a "critical limit" for extended periods [Eqs. (11.4) and (11.4a), and [9, 10]]. The use of a suitable thermal backfill material around the pipe will alleviate this problem [21, 22, 24–26, 28].

(ii) In EHV cables with thick insulation walls, tape dislocation induced by load cycling may lead to the breakdown of the insulation.

(iii) The cable route (curves, offsets) and the distances between manholes have to be planned carefully, even though the lower coefficient of friction in pipes than that in ducts reduces pulling tensions. On the other hand, the absence of a sheath accentuates the importance of close control of side-wall pressures.

(iv) Corner manholes cannot be used for pipe-type cables. Large-radius curves must therefore be constructed in the pipe line at street corners. Steel pipes are required to be bent in advance; therefore all obstacles at such turns of the cable run must be located with precision well before pipe laying; this increases planning and construction preparation costs.

(v) Difficulty of fault locating, in particular leakage from the pipe into the ground with the cables remaining intact.

(vi) The joint sleeves must be sufficiently long to allow the spreading of the cable cores to leave adequate clearance for splicing. In the manholes, space needs to be provided on both sides of the joint for eventual freezing of the cable and for the positioning of the joint sleeves.

(vii) Following a cable fault, the three phases of an entire cable length must be replaced.

(viii) Splicing is very lengthy as cable cores cannot be straightened within the splice, precluding the use of wide paper rolls. Thus, narrow paper tapes must be hand applied onto the splice. The sleeve can only be closed after the completion of taping over the three phases; therefore a 345-kV cable remains exposed at each splice location for about 10 days. To prevent contamination, the relative humidity in 345-kV system manholes is maintained below *30%* during splicing, whereas a *50%* humidity level is acceptable for self-contained cables as their jointing is a less than 24-h operation. The risk of oil spills during installation work of pipe-type cables is smaller than that of liquid-filled self-contained cables. On the other hand, much larger quantities of oil are lost as a result of a pipe break than due to a similar defect of a self-contained liquid-filled cable.

(ix) Throughout installation work, the entire pipe system has to be kept filled with dry nitrogen to prevent contamination. Manholes have to be thoroughly ventilated to prevent asphyxiation by nitrogen emanating from the pipes. This is particularly critical during the lengthy jointing operation.

(x) Manholes must be located so that three reels can be positioned for pulling under the cover of a tent and a nitrogen blanket as the three phases are simultaneously pulled in together. The cables also require protection from the atmosphere during manufacture, transportation, and storage, as they do not possess a protective sheath.

(xi) Very large quantities of oil and large-capacity reservoirs with pumping plants are required for pipe-type cables; the oil in self-contained cables is only a fraction of that of pipe-type cables, thus requiring smaller reservoirs and simpler oil-feeding systems. On the other hand, oil degasification for pipe-type cable installation work is of a lower degree than for self-contained cables installed in ducts or directly buried.

The outstanding economic consideration in favor of a liquid-filled pipe cable system is that it represents the lowest cost installation technique for HV and EHV single circuits in congested city centers. Because the oil is in direct contact with the cable and contained in a sealed pipe system of large cross-sectional area, the later addition of forced cooling is relatively easy.

11.14 CABLE PULLING

Rifenburg's study [59] is a summary of fundamental mechanics governing the conveying of a flexible cylindric object (or a group of such objects) in the direction of its axis over flat, curved, or sloped surfaces against the forces of gravity and friction. Even though the title specifies "Pipe-line design for pipe-type feeders", Rifenburg's formulas have been applied with equal success to cable-pulling into ducts and trenches. Reference books [1, 2] extensively quote Rifenburg's formulas.

The basic relation for the pulling force applied to a single cable or a group of cables pulled into a straight horizontal pipe or duct is

$$F = \mu w L \tag{11.71}$$

where F in pound-force (kilogram-force) is the pulling force, μ is the effective coefficient of friction of a cable or a group of cables pulled into a duct or pipe, w [in pounds per foot (kilograms per meter)] is the total weight (mass) per unit length of the cable or group of cables being pulled, and L [in feet (meters)] is the length of the cable section being considered. Equation (11.71) in SI units is

$$F = \mu w L 9.81 \tag{11.71a}$$

where F is in newtons, w is in kilograms per meter, and L is in meters.

According to experience with lead-sheathed impregnated paper insulated cables pulled into fiber ducts, the *basic* coefficient of friction may be taken for calculations as 0.4, if the duct is not prelubricated and the lubricant is only applied to the cable upon its entry into the duct. If the lubricant is spread entirely over the duct run prior to pulling and the cable is also lubricated upon its entry to the duct, then the basic coefficient of friction may be reduced to 0.3 or possibly 0.25. In plastic ducts 50–65% of the above-cited values may be used for calculation purposes. Basic and effective coefficients of friction are defined in Section 11.14.3.

When cables with extruded insulation replaced lead-covered impregnated paper-insulated cables, it became necessary to establish experimentally confirmed coefficients of friction, allowable pulling forces, and side-wall pressures, as well as minimum bending radii. A testing program was undertaken involving 23 types of cables rated 600 V to 138 kV with copper and aluminum conductors from 1/0 (54 mm^2) to 2500 kcmil (1266 mm^2). The conclusions of the testing program are included in a two-volume report [60]. The second volume of the report is a *Cable User's Guide*, which includes a calculating method based on [59] and a corresponding computer program CABLEPUL. Some of the most important findings of the report are permissible pulling forces and side-wall pressures greatly exceed previously accepted values but compression-type pulling eyes on aluminum conductors significantly limit the pulling force. Therefore the report proposes epoxy-filled pulling eyes for aluminum conductors. Dynamic coefficients of friction are a function of normal force between cable and duct wall (they are greater on straight runs and low-tension bends, lower on high-tension bends). Therefore, the report recommends two levels of dynamic coefficients of friction. The type of lubricant greatly influences the dynamic coefficient of friction on straight runs and low-tension bends but has a much lesser influence on high-tension bends. Pulling speed has little influence on dynamic coefficient of friction. In three-cable pulls (when the cables are not cabled together, *triplexed*), it shall be assumed that the pulling force is distributed on two conductors only. Coefficients of friction are higher on groups of three cables in the same duct than on a single cable per duct. Pulling grips are much less effective on three-conductor cables with a common shield or jacket than on single conductor cables or groups of three single-conductor cables with individual grips.

As mentioned above, formulas in [60] are based on those in Rifenburg [59] but are written in a form different from those in [1] and [2]. The specific distinctions will be pointed out in sections 11.14.3 and 11.14.5. Here basic formulas are quoted in the form used in [60] and other recent material. The symbols used here are consistent with those of other chapters in this book and therefore different from [1], [2], [59] and [60]: F for force (instead of T), μ for coefficient of friction (instead of K), w for total weight (mass) per unit length of the cable or group of cables being pulled (instead of W), α for angle (instead of θ); diameters d are defined in Sections 11.14.1 and 11.14.1.1.

Friction and side-wall pressure between the pulling rope or winch cable and the duct wall also have to be considered [60]. In bends gouging of the duct by the pulling rope or winch cable may lead to wedging of a small-diameter conductor in the groove cut by the rope or winch cable. For pulling pipe-type cable into an internally coated pipe, a jacketed winch cable has to be used.

11.14.1 Clearance between Cable and Duct or Pipe

Regulations (electric codes) usually regulate the number and size of cables allowed to be installed within a duct. In nonregulated applications:

(a) For a single cable a minimum $\frac{1}{2}$ in. (12.7 mm) clearance is deemed desirable between cable and duct. This may be reduced to $\frac{1}{4}$ in. (6.35 mm) for straight runs. Clearance also has to be verified to ensure the safe passage of the pulling eye within the duct, particularly through small-radius curves.

(b) For three cables Rifenburg [59] recommends a minimum clearance of $\frac{1}{4}$ in. (6.35 mm) between cable and pipe. Most utilities, however, maintain a $\frac{1}{2}$ in. (12.7 mm) minimum; reference [60] suggests the same minimum clearances between cable and duct. Clearance is calculated with the following equation:

$$c_L = \frac{d_d}{2} - 1.366d + 0.5(d_d - d)\left[1 - \left(\frac{d}{d_d - d}\right)^2\right]^{0.5} \tag{11.72}$$

where c_L is the clearance between cable and duct or pipe, d_d is the inside diameter of the duct or pipe, and d is the overall diameter of one single-conductor cable (in case of a pipe-type cable d is the nominal outside diameter (OD) of the insulation shield plus 1.5 times the thickness of the skid wire whereas [60] suggests $d = 1.05$ times nominal cable OD) all in inches (millimeters).

11.14.1.1 Jam Ratio and Configuration in Duct or Pipe

When the ratio d_d/d_n is equal or close to 3, the risk of jamming is high especially in curves, therefore [60] suggests the following:

$$\frac{1.05d_d}{d_n} < 2.9 \tag{11.73a}$$

or

$$\frac{d_d}{1.03d_n} > 3.1 \tag{11.73b}$$

where d_n in inches (millimeters) is the nominal overall diameter of one single-conductor cable.

For pipe-type cable [59] suggests to calculate the jam ratio:

$$\text{Jam Ratio} = 1.05(d_d/d) \tag{11.73c}$$

The diameters have been defined in Section 11.14.1 for pipe-type cable. Most designers avoid ratios between 2.9 and 3.1; others refrain from values of 2.95–3.05 [2], reference [1] suggests values <2.8.

With a ratio <2.4 the cables tend to remain in a trefoil configuration whereas with a ratio >3 they will assume a cradled formation. Over the in-between range a trefoil is the most probable formation in the straight sections and a cradled configuration in the curves.

For four-cable groups [60] provides formulas.

11.14.2 Choice of Lubricant

Lubricants should not induce any harmful effects on the duct, cable sheath, or jacket materials. For the choice of the most appropriate lubricant, it is recommended to consult references [61–63].

11.14.3 Pulling Forces in Pipe-Type Cables and Ducts

A correction has to be applied to groups of cables pulled into a common duct or pipe on account of the interacting normal forces between the individual cables and the duct or pipe wall; these forces produce a greater pressure against the pipe or duct than the linear force generated during the pulling of a single cable. This multiplier w_c is called the *weight correction factor* and it is a function of the configuration the cables assume in the pipe or ducts.

For three cables in a cradled configuration:

$$w_c = 1 + \frac{4}{3}\left(\frac{d}{d_d - d}\right)^2 \tag{11.74a}$$

For three cables in trefoil formation:

$$w_c = \frac{1}{\left[1 - \left(\dfrac{d}{d_d - d}\right)^2\right]^{1/2}} \tag{11.74b}$$

For three cables in a duct: $d = d_n$, as defined in Section 11.14.1.1. For pipe-type cables the diameters have been defined in Section 11.14.1.

Reference [60] also covers the case of a group of four cables pulled into the same duct.

In practice the weight correction factor is a multiplier applied to the friction coefficient; therefore the effective coefficient of friction μ in Eq. (11.71) is given:

$$\mu = w_c \mu_0 \qquad (11.75)$$

where μ_0 is the basic coefficient of friction, and at the same time the effective coefficient of friction of a single cable; thus $w_c = 1$ for a single cable.

Accordingly it is convenient to rewrite Eq. (11.71):

$$F = w_c \mu_0 wL \qquad (11.76)$$

or for SI units:

$$F = w_c \mu_0 wL9.81 \qquad (11.76a)$$

Note: Equation (11.76) differs from the formula for T_3 on page 3-17 of reference [1] because in the format used here w stands for total weight (mass) per unit length of the cable or group of cables being pulled (instead of W which is the weight per unit length of a single cable).

A series of tests, carried out earlier, have yielded an average (basic) coefficient of friction of $\mu_0 = 0.23$ in uncoated pipes slightly preoiled prior to the pulling operation. A thin coating with the actual pipe filling oil is the only admissible form of lubrication in pipe-type cables. In coated pipes, a series of tests have yielded an average (basic) coefficient of friction of $\mu_0 = 0.173$.

For lead-sheathed cables with taped insulation traditional values of basic coefficient of friction are given in Section 11.14, reference [60] provides basic coefficients of friction for cables with extruded insulation.

11.14.4 Allowable Pulling Force

For cable types with extruded insulation covered by the EPRI report [60], it recommends the maximum allowable pulling force in lb_f when pulling grips are used, and the maximum allowable pulling stress (pound-force per square inch) in the conductor in case pulling eyes are fitted to the cable.

For pipe-type cable [2] quotes an allowable stress of $10\,lb_f/kcmil$ ($12,735\,lb_f/in^2$ or $8.9\,kg_f/mm^2$ or $87.7\,MPa$) in a copper conductor and $6\,lb_f/kcmil$ in an aluminum conductor; it is also assumed that only two of the three conductors carry the pulling force. Rifenburg [59] suggests a $10,000\,lb_f/in^2$ maximum stress in copper conductors of pipe-type cables with the assumption that the three cables carry the pulling force; a 15% reduction of the calculated total pulling force is suggested to allow for uneven distribution of the force.

For impregnated paper insulated lead-covered cables the same $10,000\,lb_f/in^2$ ($7\,kg_f/mm^2$ or $69\,MPa$) stress is generally allowed for copper conductors but with the pulling force limited to $6000\,lb_f$ ($2720\,kg_f$ or $26.6\,kN$) for hollow-core liquid or gas-filled cables. When pulling with a grip over the lead sheath, the stress in the lead sheath is generally limited to 1500 lb_f/in^2 for pure lead, 2000 $lb/in.^2$ for arsenical lead.

11.14.5 Pulling Force in Bends and on Slopes

For the calculation of pulling force the route has to be subdivided into sections according to geometry: straight, uphill or downhill slope, curve (bend) in a horizontal or vertical plane. Pulling force has to be calculated section by section progressively starting at the cable reel: The point where the cable exits each particular section is automatically the point of entry to the following section; accordingly the pulling force at the exit of each section (F_2) is equal to the pulling force (F_1) at the entry of the next section. The total pulling force is the one (F_2) at the exit of the last section at the winch. To make an allowance for the force needed to unwind the cable from the reel, it is customary to add a 50-ft (15-m) straight section at the reel. To select the direction of pull, the calculation should be performed for both directions, and the one yielding the lowest force chosen.

Volume 2 of reference [60] provides formulas for the calculation of practically all geometric forms possible. The following group of equations (11.77) is given for the most common geometries.

Straight horizontal pull:

$$F_2 = F_1 + \mu w L \tag{11.77a}$$

where μ the effective coefficient of friction is defined by Eq. (11.75), L in feet (meters) is the length of the section, and w in pounds per foot (kilograms per meter) is the *total* weight (mass) per unit length of the cable or group of cables being pulled and the forces F are in pound-force (kilogram-force). Equation (11.77a) in SI units is

$$F_2 = F_1 + \mu w L 9.81 \tag{11.77b}$$

where the forces F are in newtons, w is in kilograms per meter, and L is in meters.

As has been done in Eq. (11.77b), when using SI units Eqs. (11.77c)–(11.77f) have to be modified: In the second part of the equation the value of mass per unit length (w given in kilograms per meter) has to be multiplied by 9.81 in order to obtain the value of the forces in newtons.

Horizontal bend (curve):

$$F_2 = F_1 \cosh \mu\alpha + [F_1^2 + (wr)^2]^{1/2} \sinh \mu\alpha \tag{11.77c}$$

where α in radians is the angle of the curve, and r is its inner radius in feet (meters).

Slope (straight), pulling up:

$$F_2 = F_1 + Lw(\sin\alpha + \mu\cos\alpha) \tag{11.77d}$$

where α is the angle of the slope.

Slope (straight), pulling down:

$$F_2 = F_1 - Lw(\sin\alpha - \mu\cos\alpha) \tag{11.77e}$$

Vertical dip (small offset, small angle): In case $F_1 > rw$, the cable is lifted off the bottom of the duct or pipe and the following formula applies:

$$F_2 = F_1 \exp(4\mu\alpha) + rw[\exp(4\mu\alpha) - 2 \exp(3\mu\alpha) + 2 \exp(\mu\alpha) - 1] \qquad (11.77f)$$

Here α in radians is the angle of each half curve of the dip: straight to point of inflection, inflection to bottom, bottom to inflection, inflection to straight; the four angles are deemed to be equal.

In case $\mu\alpha < 0.1$, the second part of Eq. (11.77f) may be neglected [1]. The effect of the dip may be neglected, and the section considered to be straight, if $F_1 \leq rw$ or $r < \mu L$ where L is the total length of the dip [2]. For angular couplings [64] provides a calculating method.

A refinement to the cited calculations of pulling forces in bends (curves) is proposed [64] to account for the stiffness of cables with heavy insulation walls and large conductors. Cables are bent upon entering a curve and straightened when exiting curves. This cyclic deformation of the cable involves loss of energy, and the force used has to be added to those generated by friction and gravity.

As mentioned near the end of Section 11.14, the *Cable User's Guide* by EPRI [60] transformed the classic Rifenburg formulas [1, 2, 59]. In the classic literature equivalent lengths are calculated section by section (the equivalent length of a straight horizontal section is its real length) with the final L_2 value being used in Eq. (11.76) to calculate the pulling force. This method is convenient as long as the coefficient of friction may be considered uniform for the whole pulling section, which is generally true for the conditions studied by Rifenburg: practically no low radius bends, except at the terminations. EPRI [60] found that the effective coefficient of friction μ changes throughout the pull due to variations in the value of the basic coefficient of friction μ_0, and the weight correction factor w_c. As mentioned in Section 11.14, according to the findings of EPRI, the dynamic coefficient of friction is a function of the normal force between cable and duct wall. This explains the variable nature of μ_0. Also the formation (trefoil or cradle) the cables assume within the duct or pipe may change throughout the route (Sections 11.14.1.1 and 11.14.3) leading to changes of the weight correction factor w_c. The transformed formulas [60] let us calculate the forces F directly section by section. Also the calculation of side-wall pressure is more convenient with the transformed formulas. In this chapter the transformed Rifenburg formulas are quoted.

11.14.6 Composite Curves and Angular Offsets

It is customary to build a large radius curve or an offset up to 5°, using angular couplings with straight lengths of conduit, especially in fiber and cement duct lines. In this type of construction each angular coupling is to be treated as an individual curve; the assumed radius (r) of the curve depends upon the rigidity of the cable, and its assumed value for the calculations has to be established experimentally. For some time, a Canadian utility has been using successfully an assumed radius of 1 ft for calculating the pulling forces on their system of 120-kV single-phase self-contained cables installed in fiber ducts. Reference [64] proposes a method for calculating the forces over angular couplings.

11.14.7 Sidewall Pressure

The sidewall pressure per unit length of cable, p_s, is calculated using the relation [1, 59]

$$p_s = \left[\left(\frac{F_2}{r} \right)^2 + w^2 \right]^{1/2}$$

(11.78a)

In most cases gravity is negligible; therefore, the simplified version may be used for a single-conductor cable:

$$p_s = \frac{F_2}{r}$$

(11.78b)

For three conductors in cradled configuration, the center conductor produces the greatest pressure against the pipe or duct wall; the sidewall pressure against it is obtained:

$$p_s = \frac{(3w_c - 2)F_2}{(3r)}$$

(11.79a)

Similarly, for a triangular configuration, as the sidewall pressure is divided between two conductors:

$$p_s = \frac{F_2 w_c}{2r}$$

(11.79b)

For four cables pulled into the same duct, a formula is provided in [60].

Generally, the allowable sidewall pressure is established and experimentally proven by the cable manufacturer for each cable type. For most cable types in general use today, EPRI conducted a series of tests and the allowable sidewall pressures are given in Volume 2 of reference [60].

For pipe-type cables a sidewall pressure of 400 lb/ft is considered a safe value [1]. For transmission (HV and EHV) cables a more recent source [2] quotes much higher values:

Taped insulations (paper and PPL)	1000 lb$_f$/ft	1490 kg$_f$/m
XLPE, EPR with tightly applied jacket	2000 lb$_f$/ft	2990 kg$_f$/m
XLPE, EPR with jacket (sleeve type)		
not applied tightly	1500 lb$_f$/ft	2235 kg$_f$/m

11.14.8 Bending Radii

Bending radii during the installation are primarily governed by the allowable pulling forces and sidewall pressures. The final bending radius, which is the smallest allowable radius for a cable in its permanent position after installation, is defined by the cable structure. Therefore it has to be established by the manufacturer and confirmed with tests. For taped insulations it is mainly defined by the gap between turns, called the butt gap, and the flexural behavior of the sheath and covering. For cables with extruded insulations it depends upon the flexural qualities of the insulation, plus construction of the shielding and the covering.

For the most common cable types with extruded insulation currently being installed on 5– 46-kV systems, American [12, 57] standards specify final bending radii (inside radii) as given in Table 11.13.

The Insulated Cable Engineers Association (ICEA), formerly the Insulated Power Cable Engineers Association (IPCEA), also issues standards not only for bending radii but also for cable reel diameters. Some of their recommendations are quoted in [1] and [60].

National and international standards and specifications governing cable manufacture and testing define drum diameters for bending performance tests, but most of them do not cover bending radii during installation and final training into position. Electrical codes and some local regulations establish requirements for the bending of certain cable types used in buildings and plants.

Some international practices (Canadian and Japanese [65], also German [34]) are quoted in Tables 11.14 to 11.17.

It is evident from the above that overseas recommendations are much more conservative than the general North American practice; this difference is even more pronounced when bending test diameters in cable manufacturing and testing standards are compared.

As most duct runs and manholes were originally designed for paper insulated lead-sheathed cables, it is reasonable to expect that modern extruded dielectrics can perform just as well or better under the conditions designed for their forerunners, provided proper clearance between cable and duct is respected (Section 11.14.1). This may impose limitations as to conductor size in older ducts on account of the insulation walls being slightly heavier in extruded dielectrics than in taped ones; also for three-phase installations triplex constructions prevail. The circumscribing circle of a triplex

TABLE 11.13 ICEA Specified Bending Radii for Solid Dielectric Extruded Cables Rated 5 Through 46 kV

Concentric Neutral Single-Conductor [12]	Single-conductor	8 times the overall diameter
	multiplexed	8 times the diameter of the individual cable or 5 times the overall diameter, whichever is greater
Shielded (utility) cables [57]		
Tape shield	Single conductor	12 times the overall diameter
	multiplexed	12 times the diameter of the individual cable or 7 times the overall diameter, whichever is greater
Wire shield	single-conductor	8 times the overall diameter
	multiplexed	8 times the diameter of the individual cable or 5 times the overall diameter whichever is greater
Lead sheath		12 times the overall diameter of the individual cable

Source: From [12, 57].

TABLE 11.14 Canadian Practice on Bending Radii for Smooth Aluminum-Sheathed Solid-Type Impregnated-Paper Cables

Cable Outside Diameter (mm)	Minimum Bending Radius As Multiple of Cable Outside Diameter
≤ 50	15
> 50	18

Source: From [65].

TABLE 11.15 Japanese Practice on Ratio of Bending Radius/Outside Cable Diameter for Smooth Aluminum-Sheathed Solid-Type Impregnated-Paper Cables

Outside Diameter over Aluminum Sheath (mm)	Minimum Bending Radius as Multiple of Diameter Over Aluminum Sheath		
	≤30	30–50	≥50
Single-conductor cable	40	50	60
Three-conductor cable	23	23	23

Source: From [65].

TABLE 11.16 Canadian Practice on Bending Radii for Corrugated Aluminum-Sheathed Solid-Type Impregnated-Paper Cables

Cable Outside Diameter (mm)	Conductor Cross Section (mm²)	Minimum Bending Radius as Multiple of Cable Outside Diameter	
		20 kV and less	Above 20 kV
≤25		8	
>25	≤760	8	10
	>760	10	12

Source: From [65].

TABLE 11.17 German Practice on Bending Radii for Impregnated-Paper Cables

Cable Type	Ratio of Minimum Bending Radius to Cable Outside Diameter	
	Lead Sheath	Smooth Aluminum Sheath with Diameter ≤ 50 mm
Multi-core cables	15	25
Single-core cables	25	30

Source: From [34].

group is larger than the outside diameter of the common sheath and jacket of an equivalent three-phase paper insulated metal-sheathed cable.

11.14.9 Cable Training in Manholes and at Terminations

A cable training radius in the range of 70–80% of the cable drum diameter is generally recommended in Canada for high-voltage cables. If the cable is firmly anchored or embedded in its final position after training, and is thus not subjected to movements induced by the load cycles, sharper bends, approaching that of the cable drum, may be acceptable. In this case mandrels and guides must be used to avoid excessive local bending during cable training operations. For oil-filled cables rated 69–120 kV, the design practices followed in Canada are given in Table 11.18. At higher voltages, the prevailing ratio is approximately 18 for reinforced lead and corrugated aluminum sheath.

11.14.10 Curves on the Cable Route

Minimum radii, twice the radius specified for the cable fixed in its final position, are generally considered safe, provided the pulling tension and sidewall pressure remain within allowable limits. Slightly sharper curves may be acceptable if there is little or no pulling tension at the curve, and guides are used to prevent excessive local bending.

11.15 CHOICE OF CABLE ROUTE AND MANHOLE LOCATION

The ideal cable route is naturally the straightest and shortest one. It is evident when examining Equation (11.77c) and easily conceivable that the influence of curves on the pulling force increases with distance from the cable reel. Manholes, where reels are to be placed for pulling, should therefore be located near curves, and duct runs should be as straight as possible when approaching the winch. Downhill pulls are obviously easier than uphill ones. With the limited slopes of streets, curves usually have a much greater influence on the pulling force than a normal uphill pull. Sharp reverse bends are to be avoided as much as possible. Manholes should be located where cable installation equipment, such as reels, winches, and splicing shelters, will cause the least possible inconvenience to traffic and to the surrounding area (residents, business, industry). In general, mechanical problems and accessibility are of prime importance to cable system designers in the preparation of cable route plans.

TABLE 11.18   Canadian Practice for Cable Training of 69–120 kV Oil-Filled Cables

Type of sheath	Ratio of Bending Radius to Outside Diameter of Sheath
Lead	≥ 12
Smooth Al	16–18
Corrugated Al	≥ 12

11.15.1 Manhole Design for Duct Installations

Cable in ducts will expand and contract under load cycles. The cyclic heating and cooling will have two mechanical effects on the cable: First, a lateral movement within the duct, commonly known as snaking, and second, a longitudinal movement toward the manhole. The friction between the cable and the duct will tend to restrain the longitudinal movement and induce snaking; some stresses within the cable will also develop due to these restraints. However, in spite of these restraints, considerable longitudinal movement will take place, inducing cyclic flexing of the cable in the manhole; the longitudinal movement of the cable will be further limited to a certain degree by the spring action of the cable within the manhole.

Figure 11.19 is a schematic presentation of a typical cable layout in a manhole, depicting the cable and the joint in the contracted (full line) and expanded (dashed line) positions; the sketch mostly uses the symbols and nomenclature of the main references [1, 66–70].

In Fig. 11.19, M is the amount of cable movement at the duct mouth, L is the length of the cable offset, which is the axial distance between wipe (or cable clamp or joint cradle) and duct mouth, H is the width of the cable offset, G is the lateral movement of the cable, $kl/2$ is the straight portion of cable near the joint (clamp or cradle) and duct mouth, r_t is the radius of cable training, and ϕ is the angle of the arc of cable training. Figure 11.19 shows an installation allowing free lateral movement of the cable and joint within the manhole. Such an arrangement, if practicable, yields the lowest strain in the cable and, consequently, longer cable life than with firmly clamped joints. References [1, 67–69] include slightly different formulas for the calculation of the strain e, based on the geometric parameters in Figure 11.19. According to [1], e in percent is given by

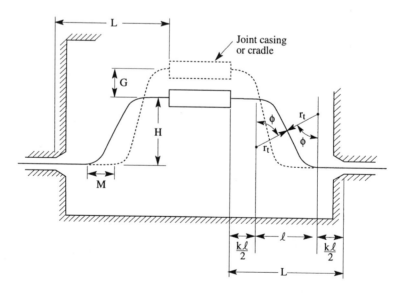

Figure 11.19 Cable training dimensions in a typical manhole.

$$e = 600 \frac{rM}{L^2 + H^2} \left[1 + \left(\frac{L}{H}\right)^2 \left(1 - \frac{M}{L}\right) \right]^{1/2} \tag{11.80a}$$

from [68]

$$e = \frac{400r}{L^2 + H^2} [(H^2 + 2LM - M^2)^{1/2} - H] \tag{11.80b}$$

from [69]

$$e = \frac{r600(2)^{1/2}}{L^2 + H^2} [H^2 + 2LM - H(H^2 + 2LM - M^2)^{1/2}]^{1/2} \tag{11.80c}$$

where r represents the outer radius of the sheath.

According to [66], for a particular case examined, there is a discrepancy of 10–15% between the values obtained using Eqs. (11.80a) and (11.80c). Also there is a discrepancy of approximately 40% between those of Eqs. (11.80a) and (11.80b). Equation (11.80a) yields the best safety factor, while Eq. (11.80b) gives the lowest stress, still on the conservative side when compared to the experimental values measured by the author of [66]. In spite of their inherent discrepancies, all the preceding formulas uniformly reveal that increasing the value of L only is almost ineffectual and that increasing H contributes more to the reduction of cyclic strain than does any increase in the value of L. Since the minimum bending radius r_t is determined by the cable construction (Section 11.14.8), a series of acceptable H and L values may be established geometrically. It is advisable to use the criterion that $kl/2$ be at least 6 in (152 mm). However, this straight portion of the cable near the joint also depends upon joint design.

The cyclic cable movement M still remains to be established. The cyclic cable movement at the mouth of the duct is mainly a function of the coefficient of expansion of the cable, and the length of the duct section between manholes; it is also influenced by the stiffness of the cable, the friction in the duct, the space occupancy of the duct, the springiness of the cable in the offset within the manhole (which also restrains the cable movement at the duct mouth). Further on all the deviations, such as offsets and curves, also restrain cable movement within the duct run. A method to calculate M and e, using a series of formulas, is provided in [66].

Reference [70] gives a combination of experimental values and a calculating method that makes use of the experimentally established parameters of certain cable types for determining the value of the cable movement M. Based on this work, diagrams and a table are provided in [1], giving experimental and computed data for certain types of lead-sheathed paper insulated cables to provide a practical tool for estimating cable movement (M).

Experimental work on very stiff high-current high-voltage cables quoted in [68] showed that the movement M at the duct mouth approximated the free bar expansion for up to 350 ft (107 m) of the cable; on this basis it was assumed that the cable movement at the duct mouth (M) only was increasing with the manhole-to-manhole distance up to a length of 700 ft (213 m). Hence the assumption: The construction of manhole-to-manhole duct sections longer than 700–800 ft (214–244 m) does not appear to aggravate the problem of cable expansion into the manhole. It should be noted that the cable in this project had an impregnated "sandwich-type" jacket over the lead sheath, thus the coefficient of friction between cable and duct was high.

More recently tests were conducted on a 1310-ft (400-m) practically straight installation with PVC ducts; the cables had a PE jacket and they were lubricated. Thus the coefficient of friction was low. Two HV cables (one self contained liquid-filled, the other with XLPE insulation, both single-conductor with a lead sheath) were subjected to load cycling to 65°C, and also to 90°C. The movement M at the duct mouth was considerably higher than estimated according to the previous paragraph. The data obtained are used in the development of computer-based modeling of the behavior of cables in ducts [71]. It should be noted that these cables had considerably larger cross sections than those in the previous paragraph. This adds stiffness to the cable. Also the springy and resilient character of the extruded insulation increases the stiffness of the cable, in particular at temperatures near ambient. The extruded structure becomes more malleable at temperatures approaching the rated level. Higher stiffness and lower coefficient of friction should explain the higher movement M at the duct mouth.

Reference [70] suggests an optimum manhole offset small enough to provide a restraining force, limiting the displacement M. This type of training may, however, give rise to excessive stresses at the wipe attaching the joint sleeve to the cable sheath; joints, joint supports (cradles), wipes (welds), and possibly the cable sheath need to be reinforced to cope with these forces.

Another solution, used under special circumstances, consists of restricting all cable expansion to the duct by eliminating offsets in the manhole. Joints are precisely aligned with the ducts, while joints and cables in the manhole are firmly clamped; joint, clamping, and cable sheath reinforcement must all be designed to cope with the resulting compressive forces. This system without offsets was originally developed for integrally cooled cables, where the cooling water circulates in the duct and immediately surrounds the cables [72]. These cables usually have corrugated aluminum sheaths that perform well under axial compression. This solution has also been applied to an installation in a very congested downtown area [73], where no space existed for manholes sufficiently large to permit conventional cable training with expansion loops. This system utilized naturally cooled lead-sheathed cables with special reinforcing over the sheath to prevent buckling. The system's advantage rested largely in the greatly reduced manhole sizes. Nevertheless, it also possessed two important drawbacks, in that exceedingly high duct banks were required in at least the vicinity of manholes, as the ducts needed to be spaced vertically and aligned (spaced) with the centerlines of the joints. Further on with no offsets, and within the confines of short manholes, the cable ends could not be moved back and forth to permit their insertion into the connector, hence special connectors, onerous to install, had to be developed.

For the lead alloys, normally used for cable sheathing in North America, Fig. 11.20 gives the sheath fatigue life, that is, the number of cycles to mechanical breakdown as a function of maximum strain [1]. The tests were carried out using a cyclic load of either 360 or 1230 cpd (cycles per day). For other types of sheaths, the fatigue values must also be determined experimentally. For corrugated aluminum sheaths the profile of the corrugations, and the type of work hardening in the manufacturing process, greatly influence the fatigue characteristics.

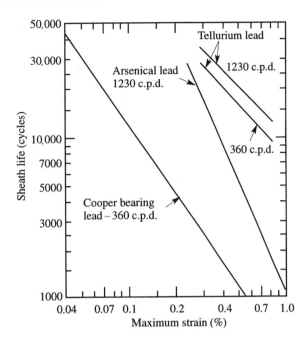

Figure 11.20 Lead sheath life vs. strain. (after [1]).

For a particular type of cable, tests are necessary to establish the fatigue performance of the cable: (a) The cable's characteristics under thermal expansion, flexing and compression have to be established to predict its behavior (possibly by computer-based modeling) in the duct and, as a result, the movement M at the duct mouth. (b) Sheath fatigue testing must be carried out on a manhole flexing machine inducing the anticipated movement M in order to record the values of strain e at different locations throughout the bend between the duct mouth and the joint (Fig. 11.21). This test is usually continued until mechanical breakdown occurs. The number of cycles has to be counted throughout the test; the number of cycles sustained before the failure defines the anticipated life span of the cable.

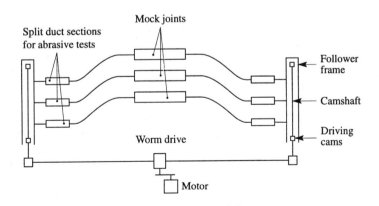

Figure 11.21 Schematic of manhole testing machine.

REFERENCES

[1] Edison Electric Institute, *Underground Systems Reference Book*, Edison Electric Institute, New York, 1957 (first edition 1931 by the National Electric Light Association).

[2] EPRI, *Underground Transmission Systems Reference Book*, Electric Power Research Institute, Palo Alto, CA, 1992.

[3] Westinghouse, *Transmission and Distribution Reference Book*, Westinghouse Electric Corporation, Pittsburgh, 1956.

[4] E. R. Thomas and R. H. Kershaw, "Impedance of pipe-type cables," *IEEE Trans. Power Apparatus Syst.*, Vol. PAS-84, Oct. 1965 pp. 953–965.

[5] Federal Power Commission, "Underground power transmission" Report by the Advisory Committee on Underground Transmission, Federal Power Commission, Washington, DC, April 1966.

[6] IEEE Standard 575, "IEEE guide for the application of sheath-bonding methods for single-conductor cables and the calculation of induced voltages and currents in cable sheaths," IEEE, Piscataway, NJ, 1988.

[6a] CIGRE Study Committee No. 21, Working Group 07, "The design of specially bonded cable systems, Part I, *Electra*, No. 28, 1972, pp. 55–81.

[6b] CIGRE Study Committee No. 21, Working Group 07, "The design of specially bonded cable systems, Part II, *Electra*, No. 47, 1976, pp. 61–86; also see Erratum in No. 48, p. 73.

[6c] CIGRE Study Committee No. 21, Working Group 07, "Guide to the protection of specially bonded cable systems against sheath overvoltages," *Electra*, No. 128, Jan. 1990, pp. 46–62.

[7] IEEE Standard 835, "IEEE standard power cable ampacity tables."

[8] J. H. Neher and M. H. McGrath, "The calculation of the temperature rise and load capability of cable systems," *AIEE Trans. Power Apparatus Syst.*, Vol. 76, Oct. 1957, pp. 752–772.

[9] IEC, "Electric cables—calculation of the current rating," IEC 287 standard series (new expanded number: 60287), International Electrotechnical Commission, Geneva, Switzerland.

[9a] IEC, "Part I: Current rating equations (100% load factor) and calculation of losses—Section 1: General," IEC Standard 287-1-1 (new expanded number 60287-1-1), 1994.

[9b] IEC, "Part 1: Current rating equations (100% load factor) and calculation of losses—Section 2: Sheath eddy current loss factors for two circuits in flat formation," IEC Standard 287-1-2 (new expanded number 60287-1-2), 1993.

[9c] IEC, "Part 2: Thermal resistance—Section 1: Calculation of thermal resistance," IEC Standard 287-2-1 (new expanded number 60287-2-1), 1994.

[9d] IEC, "Part 2: Thermal resistance—Section 2: A method for calculating reduction factors for groups of cables in free air, protected from solar radiation," IEC Standard 2872-2 (former number 1042), (new expanded number 60287-2-2), 1995.

[9e] IEC, "Part 3: Sections on operating conditions—Section 1: Reference operating conditions and selection of cable type," IEC Standard 287-3-1 (new expanded number 60287-3-1), 1995.

[9f] IEC, "Part 3: Sections on operating conditions—Section 2: economic optimization of power cable size," IEC Standard 287-3-2 (new expanded number 60287-3-2), (former number 1059), 1995.

[10] G. J. Anders, *Rating of Electric Power Cables*, McGraw-Hill, New York, 1998, IEEE Press, Piscataway, NJ, 1997.

[11] IEC, "Calculation of the cyclic and emergency current rating of cables," IEC-853 standard series (new expanded number 60853), International Electrotechnical Commission, Geneva, Switzerland.

[11a] IEC, "Part 1: Cyclic rating factor for cables up to and including 18/30 (36) kV," IEC Standard 853-1 (new expanded number 60853-1), 1985.

[11b] IEC, "Part 2: Cyclic rating of cables greater than 18/30 (36) kV and emergency ratings for cables of all voltages," IEC Standard 853-2 (new expanded number 60853-2), 1989.

[12] Insulated Cable Engineers Association, "Standard for concentric neutral cables rated 5 through 46 kV," ICEA Publication 649, South Yarmouth, MA, 1996.

[13] Canadian Standards Association, "Concentric neutral power cable," CSA Standard C68.2, Toronto, ON. 1988.

[14] G. J. Anders, A. K. T. Napieralski, and W. Zamoiski, "Calculation of the internal thermal resistance and ampacity of 3-core unscreened cables with fillers," *IEEE Trans. Power Delivery*, Vol. 13, July 1998, pp. 699–705.

[15] W. Z. Black, J. G. Hartley, R. A. Bush, and M. A. Martin. "Thermal stability of soils adjacent to underground transmission power cables," Final report of EPRI Research Project 7883, Sept. 1982.

[16] H. Brakelmann, *Physical Principles and Calculation Methods of Moisture and Heat Transfer in Cable Trenches*," ETZ Report 19, VDE Verlag, Berlin, 1984.

[17] F. Donazzi, E. Occhini, and A. Seppi, "Soil thermal and hydrological characteristics in designing underground cables," *Proc. IEE*, Vol. 126, June 1979, pp. 506–516. (Discussion: *Proc. IEE*, Vol. 127, Pt. C, 1980, pp. 333–338.)

[18] AIEE Insulated Conductors Committee, "Soil thermal characteristics in relation to underground power cables," Parts I–IV, *AIEE Trans. Power Apparatus Syst.*, Vol. 79, 1960, pp. 792–856.

[19] A. N. Arman, D. M. Cherry, L. Gosland, and P. M. Hollingsworth, "Influence of soil-moisture migration on the power rating of cables in H.V. transmission systems," *Proc. IEE*, Vol. 111, May 1964, pp. 1000–1016. (Discussion: *Proc. IEE*, Vol. 112, pp. 1000–1016.)

[20] A. G. Milne and K. Mochlinski, "Characteristics of soil affecting cable ratings," *Proc. IEE*, Vol. 111, May 1964, pp. 1017–1039. (Discussion: *Proc. IEE*, Vol. 112, pp. 1000–1016.)

[21] J. I. Adams and A. F. Baljet, "Thermal properties of cable backfill materials—laboratory analysis," *Ontario Hydro Res. Report*, Vol. 17, 1965, pp. 6–13.

[22] N. A. Halfter and S. Y. King, "Experimental investigation of moisture migration in power cable backfills," *Elect. Eng. Trans. Institution Engineers (Australia)*, 1975, pp. 50–57.

[23] S. Y. King, P. Y. Foo, and N. A. Halfter, "External thermal resistances of shallow buried cables," *Elect. Engr. Trans. Institution Engineers (Australia)*; 1975, pp. 44–49.

[24] S. Y. King and N. A. Halfter, *Underground Power Cables. Some Aspects of Their Thermal Environment*, Hong Kong University Press, 1977.

[25] S. Y. King and N. A. Halfter, *Underground Power Cables*, Longman, London, 1982.

[26] H. N. Cox, H. W. Holdup, and D. H. Skipper, "Developments in UK cable-installation techniques to take account of environmental thermal resistivities," *Proc. IEE*, Vol. 122, Nov. 1975, pp. 1253–1259. (Discussion: *Proc. IEE*, Vol. 123, pp. 1245–1248.)

[27] K. Mochlinski, "Assessment of the influence of soil thermal resistivity on the ratings of distribution cables," *Proc. IEE*, Vol. 123, Jan. 1976, pp. 60–72. (Discussion: *Proc. IEE*, pp. 1245–1248.)

[28] J. V. Schmill, "Lowering the thermal resistivity of the soil for improved performance of underground cables," IEEE-PES Paper No. A77538-2, Summer Power Meeting, Mexico City, July 1977.

[29] IEEE Standard 442, "Guide for soil thermal resistivity measurements," IEEE, Piscataway, NJ, 1981.

[30] M. A. El-Kady and D. J. Horrocks, "Extended values for geometric factor of external thermal resistance of cables in duct Banks," *IEEE Trans. PAS*, Vol. 104, No. 8, Aug. 1985, pp. 1958–1962.

[31] J. D. Endacott, H. W. Flack, A. M. Morgan, H. W. Holdup, F. J. Miranda, D. J. Skipper, and M. J. Thelwell, "Thermal design parameters used for high capacity EHV cable circuits," CIGRE, Report 21-03, CIGRE, Paris, France, 1970.

[32] CIGRE Working Group 02 (Cable rating Factors) of Study Committee No. 21 (H.V. Cables), "Current rating of cables for cyclic and emergency loads," CIGRE, Paris, France.

[32a] CIGRE, "Part 1: Cyclic ratings (load factor less than 100%) and response to a step function," *Electra*, No. 24, Oct. 1972, pp. 63–87.

[32b] CIGRE, "Part 2: Emergency ratings and short duration response to a step function." *Electra*, No. 44, Jan. 1976, pp. 3–16.

[32c] CIGRE, "Computer method for the calculation of the response of single-core cables to a step function thermal transient," CIGRE Report, *Electra*, No. 87, March 1983, pp. 41–64.

[33] AIEE Committee Report, "Symposium on temperature rise of cables," *AIEE Trans.*, Vol. 72, Part III, June 1953, pp. 530–562.

[34] L. Heinhold, *Power Cables and Their Application, Part I*, Siemens, Berlin, Germany, 1970, 3rd ed., 1990.

[35] J. F. Affolter, "An improved methodology for transient analysis of underground power cables using electrical network analogy," M.E. Thesis, 1987, McMaster University, Hamilton, Ontario, Canada.

[36] G. J. Anders and M. A. El-Kady, "Transient ratings for buried power cables, Part 1: Historical perspective and mathematical model," *IEEE Trans. Power Delivery*, Vol. 7, No. 4, Oct. 1992, pp. 1724–1734.

[37] IEC, "Calculation of thermally permissible short-circuit currents, taking into account non-adiabatic heating effects," IEC Standard 949 (new expanded number 60949), International Electrotechnical Commission, Geneva, Switzerland, 1988, confirmed 1997.

[38] F. C. van Wormer, "An improved approximate technique for calculating cable temperature transients," *AIEE Trans. Power Apparatus Syst.*, Vol. 74, 1955, part 3, pp. 277–281.

[39] NBS, *Tables of Exponential Integrals, Vol. 1*, National Bureau of Standards, Washington, DC, 1940.

[40] NBS, *Tables of the Exponential Integral for Complex Arguments*, National Bureau of Standards Applied Mathematics Series, 51, Washington, DC, 1958.

[41] M. Abramowitz and I. Stegun, *Handbook of Mathematical Functions*, Dover, New York, pp. 228–231.

[42] F. H. Buller and J. H. Neher, "The short time transient temperature rise of self-contained oil-filled cable systems with particular reference to oil demands," *IEEE Trans. PAS*, Vol. PAS-84, Sept. 1965, pp. 761–770.

[43] F. H. Buller, J. H. Neher, and F. O. Wollaston, "Oil flow and pressure calculations for self-contained oil-filled cable systems," *AIEE Trans. Power Apparatus Syst*, Vol. 75, April 1956, pp. 180–194.

[44] A. Morello, "Transient pressure variations in oil-filled cables," *l'Electtrotecnica*, Vol. XLV, 1958, pp. 569–582 (in Italian, an English translation is available).

[45] AIEE Committee Report, "Oil-flow and pressure calculations for pipe-type cable systems," *AIEE Trans. Power Apparatus Syst.*, Vol. 74, April 1955, pp. 251–261.

[46] S. M. Sellers and W. Z. Black, "Refinements to the Neher–McGrath model for calculating the ampacity of underground cables," *IEEE Trans. Power Delivery*, Vol. 11, No. 1, Jan. 1996, pp. 12–30.

[47] ICEA, "Short-circuit characteristics of insulated cable," Publication P-32-382, Insulated Cable Engineers Association, Inc., South Yarmouth, MA, 1994.

[48] ICEA, "Short-circuit performance of metallic shields and sheaths on insulated cable," Publication P-45-482, Insulated Cable Engineers Association, Inc., South Yarmouth, MA, 1994.

[49] IEC, "Guide to the short-circuit temperature limits of electric cables with a rated voltage not exceeding 0.6/1.0 kV," IEC Standard 724 (new expanded number 60724), International Electrotechnical Commission, Geneva, Switzerland, 1984.

[50] IEC, "Guide to the short-circuit temperature limits of electric cables with a rated voltage from 1.8/3 (3.6) kV to 18/30 (36) kV," IEC Standard 986 (new expanded number 60986), International Electrotechnical Commission, Geneva, Switzerland, 1988.

[51] IEC, "Short-circuit temperature limits of electric cables with rated voltages above 30 kV ($U_m = 36$ kV)," IEC 61443 Project, International Electrotechnical Commission, Geneva, Switzerland.

[52] EPRI, "Optimization of the design of metallic shield-concentric conductors of extruded dielectric cables under fault conditions," EPRI EL-3014, Research Project 1286-2, Final Report April 1983, Electric Power Research Institute, Palo Alto, CA.

[53] R. C. Mildner, C. B. Arends, P. C. Woodland, "The short-circuit rating of thin metal tape cable shields," *IEEE Trans. PAS*, Vol. PAS-87, Mar. 1968, pp. 749–759.

[54] M. A. Martin, D. A. Silver, R. G. Lukac, and R. Suarez, "Normal and short circuit operating characteristics of metallic shielded solid dielectric power cable," *IEEE Trans. PAS*, Vol. PAS-93, Mar./Apr. 1974, pp. 601–613.

[55] M. A. Martin and A. W. Reczek, "The transient temperature rise of round wire shields of extruded dielectric cables under short circuit condition," Insulated Conductors Committee of the IEEE Power Eng. Soc., Minutes of the 57th Meeting, Nov. 17–19, 1975, Appendix F-1.

[56] Insulated Conductors Committee (of IEEE), Task Group 7-39, Cost of Losses, "Loss evaluation for underground transmission and distribution cable systems," *IEEE Trans. Power Delivery*, Vol. 5, No. 4, Nov. 1990, pp. 1652–1650.

[57] ICEA, "Standard for utility shielded power cables rated 5 through 46 kV," Publication S-97-682, Insulated Cable Engineers Association, Inc. South Yarmouth, MA.

[58] EPRI Report, *Underground Power Transmission*, Arthur D. Little, Boston.

[59] R. C. Rifenburg, "Pipe-line design for pipe-type feeders," *AIEE Trans. PAS*, Vol. 72, Dec. 1953, pp. 1275–1288.

[60] EPRI, *Maximum Safe Pulling Lengths for Solid Dielectric Insulated Cables, Vol. 1, Research Data and Cable-Pulling Parameters, Vol. 2, Cable User's Guide*, EPRI EL-3333; Research project 1519-1, February 1984, Electric Power Research Institute, Palo Alto, CA.

[61] IEEE Standard 532, "Guide for selecting and testing jackets for underground power cables," IEEE, Piscataway, NJ, 1993.

[62] IEEE Standard 1026, "IEEE recommended practice for test methods for determination of compatibility of materials with conductive polymeric insulation shields and jackets," 1995.

[63] IEEE Standard 1210, "IEEE standard test for determining compatibility of cable-pulling lubricants with wire and cable," IEEE, Piscataway, NJ, 1996.

[64] I. Iordanescu and J. Tarnowski, "PULLFLEX—New software for duct-cable pulling forces," *IEEE Trans. Power Delivery*, Vol. 11, No. 2, April 1996, pp. 676–682.

[65] IEEE Standard 635, "IEEE guide for selection and design of aluminum sheaths for power cables," IEEE, Piscataway, NJ, 1989.

[66] H. Hata, "On the design of cable offsets in manholes," *Sumitomo Electric Tech. Rev.*, No. 10, Oct. 1967, pp. 32–40.

[67] E. S. Halfmann, "Critical inside dimensions for power cable manholes," *AIEE Trans.*, Vol. 69, part II, 1950, pp. 1576–1581.

[68] J. Banks and F. O. Wollaston, "Mechanical behaviour of high voltage cables in underground ducts," CIGRE Proceedings, report 201, Paris 1960.

[69] C. S. Schifreen, "Cyclic movement of cables—Its causes and effects on cable sheath life," *AIEE Trans.*, Vol. 63, 1944, pp. 1121–1139.

[70] C. S. Schifreen, "Thermal expansion effects in power cables," *AIEE Trans.*, Vol. 70, part I, 1951, pp. 160–170.

[71] J. Tarnowski, M. Iordanescu, C. Hardy, and J. Nourry, "Optimization of expansion loops of 120 kV XLPE cables," JICABLE Conference, Paper D.2.13, 1995, Paris, France.

[72] U. Müller, E. F. Peschke and W. Hahn, "The first 380 kV cable bulk power transmission in Germany," CIGRE Proceedings, Paper 21-08, Paris 1976.

[73] S. Kozak, J. T. Corbett and F. J. Bender, "Features of the new 138 kV self-contained oil-filled cable system for Detroit Edison," *IEEE Trans. PAS*, Vol. PAS-94, 1975, pp. 949–958.

[74] W. Z. Black, "Short circuit temperature ratings of buried extruded dielectric cables," *IEEE Trans. PAS*, Vol. PAS-97, Mar./Apr. 1978, pp. 362–369.

[75] AEIC, "Underground extruded power cable pulling guide," AEIC Task Group 28, Guide: AEIC G5, Association of Edison Illuminating Companies, Birmingham, AL, 1990.

Chapter 12

CRYOGENIC AND COMPRESSED GAS INSULATED POWER CABLES

K. D. Srivastava

12.1 INTRODUCTION

The electric supply industry has come to occupy a dominant position among industries in all industrialized nations. The average compound rate of growth in the earlier part of this century has been 7.5% per year in North America. It is clear that with the growing consciousness of energy conservation and environmental damage, such rates of growth could not continue indefinitely. However, the factors influencing the growth are so complex that forecasts made for more than 25–30 years are likely to be in considerable error. The growth rate in the last two decades, however, in the industrialized countries of Western Europe and North America, has been considerably lower and in some instances negative. On the other hand, the developing countries of Asia, Africa, and South America are experiencing significant growth in the use of electric power. These worldwide rates of growth represent capital investments in the electric utility systems of several billion dollars per year.

Probably the greatest uncertainty in forecasting arises from four factors: (i) the role of oil and natural gas in electric power generation; (ii) the impact of environmental control legislation—public pressure on location of power stations and transmission lines is mounting and will continue in the foreseeable future; (iii) the impact of utility deregulation and cost of capital financing; and (iv) unconventional energy sources for electricity generation are not likely to make more than a local impact, although these unconventional energy sources will tend to slow down the growth of conventional power stations.

Despite the many difficulties, high-capacity transmission systems seem likely to grow to ensure reliable operation of the highly interconnected power system in North America and Europe. The role of peaking power, generated by local power stations of a "nonpolluting" nature, will also continue to grow. Virtually all high-capacity transmission systems built to date are overhead. Presently, in North America alternating current (ac) transmission systems at 345, 500, and 765 kV are well established. We have seen in the last few years significant research and development (R&D) effort on 1100 kV ac and high-voltage direct-current (HVDC) systems. High-voltage direct-current is being exploited increasingly for long-distance power transfers and for submarine interconnections. Planning studies show that transmission system circuit loading will exceed 2500 MVA, and transmission voltages will move to higher levels in the developing countries [1, 2].

The transmission costs, including both the capital and operating costs, are much less for overhead lines than for any underground system proposed or devised so far. The relative cost factors may be as high as 20–30, depending upon terrain, urbanization, and the technology proposed. In addition to the principal approaches to practical underground transmission discussed in other chapters, several new technologies have emerged, for example, compressed gas transmission lines and cryogenic power cables [1]. Reference [2] lists some highly speculative ideas, such as microwaves and lasers. For microwaves, energy conversion equipment will be very expensive and there will be severe mechanical problems of tolerances and alignment. The technical feasibility, however, of very high power lasers for energy transmission is, at this stage, very uncertain. High-power laser systems have been developed for nuclear fusion research and are utilized for artificial triggering of lightning.

Cryogenic systems include both resistive and superconductive designs. The lapped tape insulation appears to be preferred by most designers. However, at least in one design, vacuum was proposed both as thermal as well as electrical insulation; except for a flurry of activity in the early 1970s, this design has not attracted much commercial interest.

It is clear that there are no instant solutions to the problems created by rapid growth in energy consumption, utility deregulation, and the adverse environmental impact. Major technical factors that could reduce the cost of underground transmission are new techniques for trenching and backfill, cable splicing, and improved quality and quality control of insulating materials. Joint rights-of-way and utilization of waste heat in urban areas also reduce the cost differential between the overhead and underground transmission. Moreover, environmental concerns may help to narrow the differential in other ways. The new "aesthetically designed" overhead structures, as well as the landscaping of line routes, substantially increases the cost of overhead lines by as much as a factor of 2. Perhaps the short-range goals of the electric supply and cable industry are to extend the voltage range of systems, improve the power transfer capability, improve the life of high-voltage extruded dielectric cables, reduce the time and cost of cable splicing, and develop a taped synthetic insulation for extra high voltage (EHV) cables. Considerable experience has been accumulated in France with 225-kV cables based on voltage gradients of 8–9.5 kV/mm, but test results suggest that a working gradient of 15 kV/mm for 400-kV plastic cables is feasible. The long-range goals must concern new cable systems capable of carrying 4–10 times the load of existing cables.

Finally, one should not underestimate the impact of large underground transmission capacity on dynamics of an existing power system. Integration of cable systems into power transmission networks is very important. Cables in a transmission system introduce new sets of parameters: lower resistance and reactance, higher capacitive mega-volt-ampere (MVA), and higher cost factors. In a mixed transmission system cables may thus get overloaded when run in parallel with higher impedance aerial lines. This could require series capacitor and/or shunt reactor compensation and may further increase the costs. Moreover, the changes in system reactance will require higher interrupting capabilities of circuit breakers; ratings between 80,000 and 100,000 A are already being envisaged. Reference [3] discusses optimal use of hybrid overhead line underground cable transmission systems to solve aesthetic and environmental problems at critical locations.

12.2 COMPRESSED GAS INSULATED TRANSMISSION LINE SYSTEM

The basic technology of gas-insulated transmission lines (GITL) is well established, and gas-insulated substations (GIS) have been in use for 30 years for system voltages up to 765 kV. For long bus ducts there is also the technical expertise of gas/oil pipelines. Along with the development of GIS technology in the 1970s, short runs of GITL (up to several hundred feet) have been in use in several locations around the world, for voltages up to 500 kV [4]. Usually such links were employed for underground hydro power stations, major highway crossings, and similar applications in congested urban areas. In the last two decades there has been a steady growth with 7 or 8 new GITL installations, each over a few hundred feet in length. Up to 1993 approximately 130 installations, totaling more than 30 km, were in use worldwide (Fig. 12.1). Power utilities in North America and Europe are facing increasing opposition to plan and install new high-voltage overhead lines. In addition to the aesthetic and the general environmental considerations, there is growing concern about possible health hazards of electromagnetic fields. These considerations favor a gradual move toward underground transmission networks, particularly if the relative cost of GITL, as compared to overhead lines, could be brought to 10 or below. There has been a very significant increase in R&D effort in this area in the 1990s as evidenced by technical publications [5–9].

Basically a single-phase GITL system consists of two coaxial cylinders, the inner being the load current conductor and the outer the grounded pressure vessel to contain the gas. The cylinders are maintained coaxial by means of solid-dielectric spacers at intervals along the cable. Because of thermal expansion and construction problems, this simple structure is modified in practice to include expansion joints and various facilities for field assembly, monitoring and maintenance of the system. A typical section of a GITL is shown schematically in Fig. 12.2. This simple construction results in the following advantages for a GITL system: (1) Low losses—Dielectric losses in a compressed gas are very small and the spacers, occupying only a small percentage of the cable volume, can be of very low loss material; therefore, the dielectric losses in a GITL are almost negligible. This should be contrasted with a composite dielectric cable at 345 kV on a typical duty cycle where dielectric losses can be as high as 26 W per circuit

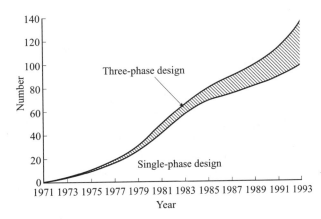

Figure 12.1 Growth in GITL installations.

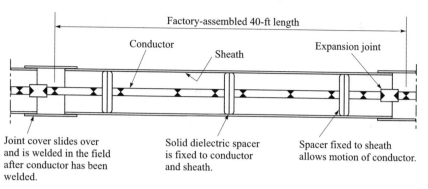

Figure 12.2 Typical GITL section.

meter. Because the total loss on such a line is limited to 72 W per meter by heat dissipation only 46 W per meter is left for I^2R losses. In addition, the construction enables the use of large conductor and sheath cross sections, which would be impossible in a conventional cable; and this significantly reduces the I^2R losses. It is perhaps worth mentioning that a flexible GITL would probably require smaller conductors [10]. The term *cable* for the system illustrated in Fig. 12.2 is somewhat of a misnomer, but the problems of manufacturing a flexible compressed gas cable are presently far from being solved and, therefore, the rigid ducted system will be referred to as a cable. (2) Good thermal performance—the compressed sulfur hexafluoride (SF_6) gas has a high heat transfer capability and the relatively large size of the sheath facilitates dissipation of the heat. In addition, the materials used for solid spacers, such as silica- or alumina-filled epoxy, can tolerate much higher temperatures than oil-impregnated paper and, therefore, higher conductor temperatures are possible. (3) Low charging current—the dielectric constant of the main insulant is nearly unity and the geometric capacitance is small, therefore long critical lengths are possible. (4) Simplicity and ease of terminations—the central conductor continues through a porcelain shell that acts as a bushing and may be filled with compressed SF_6 gas. Other pothead designs can be adapted for use with GITL systems.

12.2.1 Conductor and Sheath

Typical installations such as that reported by Cookson [4] consist of three parallel isolated single-phase lines, similar to that illustrated in Fig. 12.2, filled with SF_6 at a pressure of 50 psig. The conductor and sheath are usually rigid high-conductivity aluminum alloy tubes. The outer enclosure may also be steel, especially for longer runs. The conductor dimensions are chosen so that the radial thickness is slightly greater than the skin depth at power frequencies, and the diameter is chosen to keep the electric field at the conductor surface within the desired range. The insulation and the electric field requirements are discussed later. The conductor expansion joint movement is 0.5 in. for a temperature differential of 90°C. The sheath must combine good electrical conductivity with high mechanical strength. When the cable is to be buried, it is assumed that backfill will prevent movement, and hence expansion joints are unnecessary, but the thermal stresses in the conductor become large and the welding must be designed to handle these. Typical conductor and sheath dimensions, for one manufacturer, with the associated electrical characteristics are given in Table 12.1. It should be noted that there is considerable variation in the sheath diameter among manufacturers (see Table 12.2). Conservative designs use larger outer diameter that helps with the overall insulation integrity [11]. Three-phase designs at lower voltages are used in congested urban areas in several countries, and three-phase designs for higher voltages have been proposed and tested [12, 13].

12.2.2 Insulating Gas

As stated above, SF_6 is the most commonly used gas in any GITL system due to its high dielectric strength. However, before a choice is made of operating pressure for the gas and the design stress levels, some problems particular to GITL insulation must be considered. Sulfur hexafluoride has a low boiling point compared with other highly electronegative gases, but still it liquefies at approximately 200 psi at room temperature. It is essential that the gas does not condense in the system if insulation is to be maintained, and therefore, although the dielectric strength of SF_6 increases almost linearly with pressure, it is necessary to limit the pressure in the cable to ensure that the gas does not liquefy under cold start conditions. For SF_6 this limits the pressure approximately to 50 psig at 20°C. If the gas is dried so that the water content is not more than 30 ppm, then no moisture condensation occurs down to −40°C. Another factor in the choice of gas pressure is the requirement that the cable must be able to support normal line voltages if the gas pressure falls to atmospheric. This is a practical constraint on any cable as the possibility of a gas leak cannot be ignored. It is obviously no use operating a cable at a very high pressure to improve the insulation if the resultant high stress would cause breakdown at atmospheric pressure.

The maximum stress that can be tolerated in the system is, as in the case of solid and liquid insulation, determined by secondary processes rather than by the idealized

TABLE 12.1 Typical Dimensions for GITL Systems

System voltage, kV	138	230	345	500
Conductor OD, in.	3	4	5	6.75
Sheath OD, in.	8.5	12	15	20
Sheath wall, in.	0.25	0.30	0.375	0.375
Power rating underground, MVA	300	620	1200	2200
Current rating underground, A	1230	1550	2000	2530
Losses, W/circuit-foot	36	38	46	55
Charging power, MVA/circuit-mile	0.76	2.05	4.8	10.9
Critical length, miles	390	300	250	200
Effective resistance, Ω/phase foot	7.93	5.26	3.84	2.80
Capacitance, μF/phase mile	0.106	0.103	0.107	0.115
Inductance, H/phase mile	316	340	337	337
Surge impedance, Ω	54.5	57.2	56.1	54.1
Withstand voltage, 60 Hz, 1 min-kV rms	240	395	460	690
BIL, kV-crest	550	900	1050	1550
Switching surge insulation level, kV-crest	460	745	870	1290
Momentary current rating (1 s), kA	80	100	100	100
Open air current rating, 30°C rise	750	1800	3400	6500
Peak operating gradient, V/mil	76	90	108	115
Phase separation, centers-inches	14.5	18	22	28

TABLE 12.2 GITL Dimensions for Different Manufacturers

Parameter	Manufacturer 1		Manufacturer 2		Manufacturer 3		Manufacturer 4	
	230 kV	550 kV	230 kV	550 kV	230 kV	550 kV	230 kV	550 kV
Inner diameter or enclosure, mm	410	550	335	508	315	465	400	540
Field at inner conductor, kV/mm	1.82	3.14	2.23	3.40	2.38	3.72	1.93	3.35
Ratio of fields for the two voltages	1.73		1.52		1.56		1.74	
Outer diameter of inner conductors, mm	130	190	86	178	109	160	102	149

behavior of the gas. The solid-dielectric spacers are subject to the same degradation processes as any other solid dielectric in cable use; these crucial components will be discussed separately. The gas itself is inert and not prone to degradation under normal conditions provided that water does not contaminate the system. But discharges in SF_6 in the presence of moisture can produce highly reactive compounds that can cause deterioration of solid insulation. The most important factor in the reduction of the insulation strength of the gas is the contamination introduced during the assembly of the line in the field. A rigid line must be transported in sections and joined in the field. Most designs are assembled by welding. It is not feasible to assemble such a coaxial system in a trench and to maintain "clean room" levels of contamination. A certain amount of contamination by conducting and nonconducting particles is inevitable. Any water introduced into the system during assembly could be removed by gas circulation and drying. The most important type of contaminant is free conducting particles introduced into the system during welding or abrasion of metallic surfaces due to thermal cycling or during transportation. Several laboratory investigations of the influence of different types of conducting particles, both free and fixed to the electrodes, have been made [14–16]. It has been found that at a certain voltage a free particle in a horizontally mounted system is lifted from the inner surface of the outer coaxial electrode, and under electric stress it can oscillate between the electrodes. Small spark discharges can occur between a particle and an electrode; these may be detected as bursts of corona. Breakdown of the gas gap usually occurs when the particle is close to the high-voltage electrode. The result of particle contamination is severe reduction of the electric strength. This is shown in Figs. 12.3, 12.4, and 12.5 [15], which show the effect of filamentary and idealized spherical particles. There is a maximum breakdown voltage at 0.4 MPa (30 psig); this maximum value may be as much as a factor of 4 below the clean breakdown value. With spherical particles there is no maximum in breakdown voltage in the pressure range reported. Measurement of breakdown voltage for different electrode sizes has shown that, unlike a clean system, the gas with particles breaks down at a voltage not solely dependent upon the field at the electrode. Particle motion and corona play a crucial role in ac breakdown. It has also been shown that a particle can remain in the midgap region for many cycles. Microdischarges between the particles and the electrodes, and field enhancement produced by the particles on an electrode surface, appear to play an important role in the breakdown. For a satisfactory performance the particle contamination should be eliminated [16]. Most troublesome particles

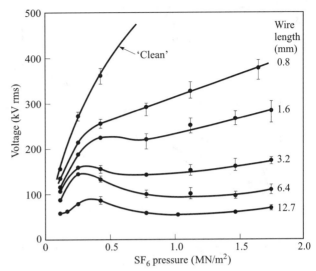

Figure 12.3 Effect of conducting filaments on breakdown; filament diameter 0.4 mm.

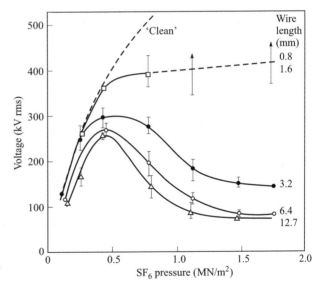

Figure 12.4 Effect of conducting filaments on breakdown; filament diameter 0.1 mm.

can be eliminated by providing particle traps. In one design it takes the form of a perforated electrode close to the outer sheath that allows mobile particles to travel into a field-free region where they are then trapped. Use of dielectric coatings on the inner conductor and the inside surface of the outer sheath are sometimes used, especially for direct current (dc) bus duct/GITL systems, to reduce the deleterious effects of metallic particle contamination. In any case, some protective coatings may be necessary to avoid mechanical damage during shipment and installation [17, 18].

Almost all GIS/GITL currently use SF_6, which is a highly efficient dielectric gas. Although SF_6 is nontoxic, it has been linked to global warming [19]. Moreover, SF_6 is an expensive gas, and for long bus duct runs some cheaper alternatives must be found.

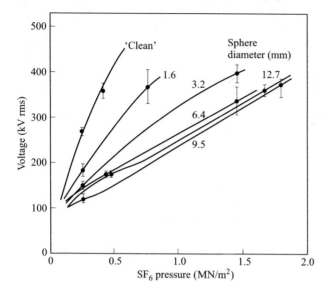

Figure 12.5 Effect of aluminum spheres on breakdown.

Commercial quality nitrogen or nitrogen/SF_6 mixtures have been proposed and tested [6, 20].

Considerable literature exists on the breakdown characteristics of SF_6 mixtures with N_2 and other gases [20]. Under ideal uniform field conditions the breakdown strength of SF_6 is $\approx (p)8.9\,\text{kV/mm}$, where p is the pressure in atmospheres. The area and surface roughness of electrode, and other factors, affect the achievable breakdown strength of SF_6 in GITL. The actual breakdown occurs at "weak points" and is defined by extreme value distributions, such as [4]

$$p(s) = 1 - \exp\left[-\lambda s \left(\frac{E - E_0}{E_d}\right)^m\right] \tag{12.1}$$

where, S is the electrode area (in cm^2), E is the breakdown field, E_d is the ideal breakdown field of $\approx (p)8.9\,\text{kV/mm}$, and empirical expressions for λ, E_0, and m are given as

$$E_0 = \frac{E_d}{1 + 0.355 p^{0.7}} \quad \text{kV/mm}$$

$$m = 7.4 \qquad \lambda = \lambda_0 \exp(1.15p)$$

Here λ_0 is dependent upon the material and the finish of the electrodes and may be taken as 100 for aluminum. For SF_6/N_2 mixtures the ideal uniform field breakdown voltage reduces with increasing fraction of N_2 in the following manner:

SF_6(%)	Breakdown [(p)kV/mm]
100	8.86
50	7.90
20	6.52
10	5.70
5	5.00

A mixture containing 20–40% SF_6 may offer a reasonable compromise for longer GITL.

12.2.3 Solid Spacers

The insulating spacers that hold the inner conductor coaxial must be spaced sufficiently close to prevent mechanical distortion of the coaxial system under short-circuit conditions. The solid insulator must be able to retain its strength at the operating temperature of the cable. Because of the excellent dielectric properties combined with ease of fabrication, the most common material used in spacer construction is epoxy resin, usually filled with refractory material such as silica or alumina, to allow an increase in the temperature of operation. Operational experience shows that spacers form a weak link in the electrical insulation of the cable. The most obvious cause of this is the existence of voids between the insulator and the electrodes. Unlike the case of voids in a solid cable, a discharge in a void at the spacer-conductor junction usually results in a complete breakdown by surface discharge across the insulator surface. The field in a void in a solid insulator in a uniform field gap is given by εE_0, where E_0 is the field in the absence of the void, provided that the void is small compared with the thickness of the insulator of dielectric constant ε. The effect of such voids upon the insulation of a coaxial gap bridged by a disk spacer is shown in Fig. 12.6 [21]. Bad cohesion is defined as the existence of a visible gap between the disk insulator and the central conductor. It can be seen that the poor contact reduces the breakdown strength by as much as 65%.

In practice it is difficult to maintain a void-free junction between a solid insulator and the central conductor. Differential thermal expansion between the insulator and the metal may easily create voids. A satisfactory way of eliminating this problem is to use a shielding electrode [21]. This is an extension of the normal electrode surface designed to reduce the electric stress at the critical region of the solid-electrode-gap triple interface, where a small discharge may readily trigger a complete breakdown. An illustration of

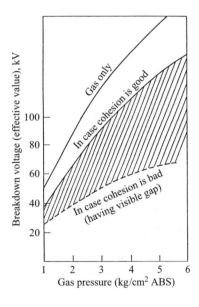

Figure 12.6 Influence of a spacer on the breakdown of a coaxial electrode system in SF_6 [21].

one type of shielding electrode and its effectiveness is shown in Figs. 12.7(a) and (b). It can be seen that the additional electrode increases the dielectric strength of the spacer practically to that of the unbridged gap. In some designs the spacer supporting the central conductor is a disk or a cone, or some modification of it, that completely closes the pipe at that point. Several workers have explored the effect of spacer profile on the breakdown strength and various degrees of success have been obtained with disks thicker at the central conductor than at the outer edge. Figure 12.8 shows a section through a typical disk insulator with equipotential lines included to illustrate how the field at the triple junction is reduced.

There are some advantages in using a spacer design that does not close off the pipe. It is useful to be able to circulate the insulating gas for purification. This is possible if the disk insulator has "holes" or is replaced by three post-type insulators. Once again

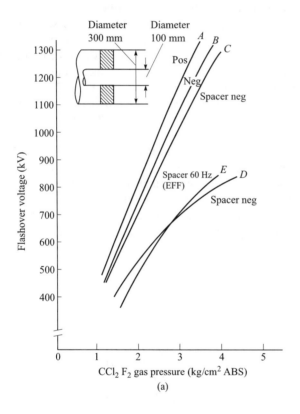

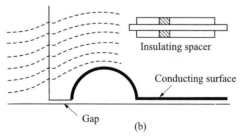

Figure 12.7 Influence of a spacer on flashover voltage: A, gas only for one 40-μs impulse; B, gas only for impulse; C, spacer with shielding electrode; D, spacer without shielding; E, spacer for 60 Hz. POS, positive polarity; NEG, negative polarity; EPF, effective value.

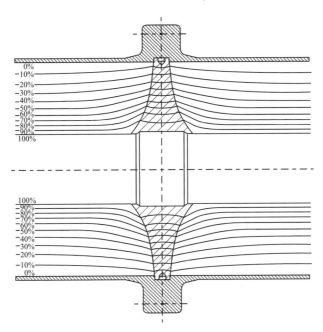

Figure 12.8 Potential distribution for a contoured spacer.

there is need to control the stress at the junctions. This has been successfully done using extensions of the central conductor as internal shielding electrodes [22]. The electrodes are profiled to produce a low field at the junction and concentrate the stress in the volume of the solid dielectric where the strength is high (see Fig. 12.9). This system can be designed to make the solid support electrically as strong at the remainder of the cable. However, the field experience with post insulators has not been good. Moreover, insulator surface is known to acquire a static charge, which may attract metallic debris to its surface. Although shielding electrodes prevent flashover, the actual mechanism of flashover is not fully understood at present. If a metallic particle attaches to an insulator or to the central conductor near an insulator, the insulation of the system may be severely compromised [22, 23]. Moreover, charges on the spacers are known to alter the V-t characteristics of a GITL [24, 25]. If a spacer does fail, it will be destroyed by the power arc that follows, leading to system outage. The gaseous insulation to some extent is self-healing but not the spacers; therefore, a robust spacer design is absolutely essential.

Reference [4] gives a good review of the state-of-the-art spacer design to the early 1980s. A large variety of spacer shapes have been tried, but the industry has now adopted a few simple shapes, viz. cones, disks, single posts, or tri-posts. Three-dimensional electric field modeling is essential. The insulator material must be able to withstand internal and surface field of 4.1 kV/mm (rms) for continuous operation and up to 17 kV/mm under lightning surge voltages. In recent years the tendency has been to limit bulk fields to below 4 kV/mm (rms). There is also the concern of long-term failure of spacers due to the presence of internal voids too small (<0.5 mm diameter) to detect by conventional partial-discharge (PD) testing. Improved PD detection down to 0.01 pC are possible [26]. X-ray illumination of spacers during life testing to detect small

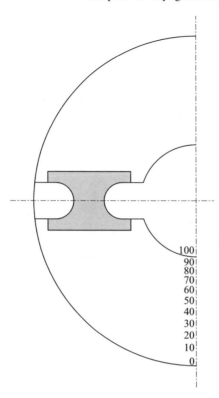

Figure 12.9 Shielded post insulator [22].

voids has also been proposed [27]. Under certain switching operations a section of GITL may have trapped dc charge. This will distort the field at spacers and may deposit charge on it from nearby high-field protrusions on the central conductor. Weakly conducting surface coatings on spacers have been proposed [28].

12.2.4 Power Rating

A typical GITL of rigid construction is manufactured in lengths of about 40 ft containing three solid spacers and one expansion joint, tested at 60 Hz and impulse voltages and then shipped filled with dry nitrogen. The sections are welded in situ. At the moment the civil engineering cost of a GITL amounts to more than half the total cost. Any change in design that would simplify the installation, such as a flexible line, could greatly reduce the cost per mile. If the cable is to be used underground, the critical factor is the rate at which the power losses can be dissipated to the surroundings. The underground rating of a three-phase GITL may be determined from Eqs. (12.2) and (12.3) [29],

$$\text{MVA} = \frac{(33.9)V(\text{TRS})}{\rho(R + R_S)(\text{LF})K} \tag{12.2}$$

$$K = \ln\left[1 + \left(\frac{2L}{\text{RO}}\right)^2\right] + \ln\left[1 + \left(\frac{2L}{B + \text{RO}}\right)^2\right] + \ln\left[1 + \left(\frac{2L}{B - \text{RO}}\right)^2\right] \tag{12.3}$$

where V is the system voltage (phase to phase) in kilovolts, TRS is the temperature rise of the sheath above ambient in degrees Celsius, ρ is the thermal resistivity of earth in degrees Celcius-centimeters per watt, R is the conductor resistance, in ohms/phase foot, R_S is the sheath resistance in ohms/phase foot, L is the burial depth to center, in inches, B is the phase spacing between centers, in inches, RO is the sheath outer radius, in inches, and LF is the load factor. The constant K is given by Eq. (12.3). Conductor and sheath resistance depend on the resistivity of the material and the diameter and thickness of the cylinders. At a given voltage level the insulation requirements dictate a certain minimum flashover gap. While the optimum use of the dielectric strength of the gas is obtained with the ratio of the outer to inner cylinder diameters equal to 2.73, other ratios may be useful when a certain MVA rating is desired. Thus for a 230-kV GITL the MVA rating increases almost linearly as RO is increased (keeping the insulating gap between conductor and sheath constant), while the cost per MVA-mile is nearly constant in a certain range. This is shown in Fig. 12.10. In this way the GITL is able to adapt readily and economically to different power requirements [29].

The factors of temperature rise of the sheath, the loss factor, and earth thermal resistivity exert a square root influence on the MVA rating. The rating is sensitive to burial depth but is not a strong function of the phase spacing, except for very close phase spacings, where the above formula is slightly pessimistic [30]. Other publications for ampacity assessment are [31] and [32].

The present use of such a cable or ducted bus is confined to short distances, but an expansion of application is expected by the year 2000. A flexible gas insulated cable that

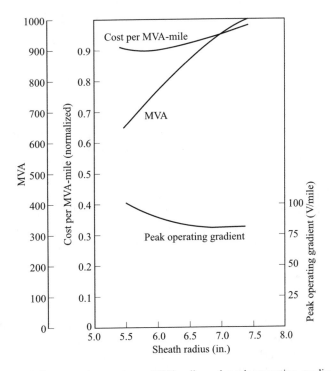

Figure 12.10 Power rating, cost per MVA-mile and peak operating gradient for typical GITL systems [29].

could be laid by existing cable laying techniques would be a significant advance. R&D work on ultra high voltage (UHV) designs for single- and three-phase GITL systems is progressing in the United States and elsewhere [5–9].

12.2.5 Field Experience

Worldwide there is almost 30 years of accumulated experience with both GIS and GITL technology. The longest GITL link was installed in Germany, it is 700 m in length and operates at 400 kV, 2000 A. The highest voltage of GIS installation is 765 kV in the Republic of South Africa. Over the past decade industry organizations have conducted surveys of field experience with GIS installations. From these surveys the following aspects emerge as being relevant to the GITL applications as well: Arc/discharge breakdown by-products of SF_6; control of metallic particle contamination; quality control of support spacers; impact on insulation integrity of very high frequency overvoltages generated by disconnect switch operations; advanced monitoring and diagnostic equipment; insulation testing in the field. Considerable laboratory data is now available on the electrical discharge by-products of SF_6 [20]. The presence of water vapor greatly exacerbates the production rates of some highly corrosive by-products, such as HF. The by-products encountered are SF_4, SF_2, S_2F_{10}, SO_2, SOF_2, SOF_4, SO_2F_2, S_2OF_{10}, $S_2O_2F_{10}$, and HF. All these compounds are considerably more reactive and toxic than pure SF_6. Compounds such as S_2F_{10} and $S_2O_2F_{10}$ are considered extremely toxic.

The presence of oxygen and water vapor greatly contribute to the toxicity and corrosiveness of the discharge by-products. Industry standards have now been established for the safe handling of SF_6 installations exposed to electric arcs and discharges [33].

It is now recognized that, in spite of the high-level quality control of cleanliness during manufacture and installation, a low level of metallic particle contamination is internally generated within GIS/GITL equipment. Perhaps moving parts, such as expansion joints, and strongly adhering machining debris are the sources. The presence of metallic contamination can be detected through partial discharge and acoustic sensors. Some manufacturers oversize the outer enclosure, thus reducing the electric field responsible for metallic particle levitation and movement. The use of particle traps and dielectric coatings has also been practiced and explored [16–18].

In a 1992 CIGRE (Conférence Internationales des Grands Réseaux Electriques) survey [34] support spacers were identified as a continuing area of concern for reliability by the user groups. Usually the factory quality control requires a partial-discharge level of 1 pC. Recent laboratory studies show that the factory test procedures are inadequate to detect very small voids (<1 mm) due to the long statistical delays for initiating a discharge in such a small cavity. With repeated exposure to very fast transients over its lifetime even such small voids could lead to failure [26].

In GIS, the frequency of disconnect switch operations depends upon the power system complexity and the operational requirements. It is known that SF_6 insulated GIS can generate a very large number of very fast transient (VFT) overvoltages [35, 36]. The presence of trapped charge on the phase conductor after a disconnect operation may further exacerbate the combined dc/VFT insulation stress. There have been reported failures of equipment connected to GIS that may have been due to these fast transients [37], and laboratory investigations with oil-paper insulation show that the breakdown stress is lower under VFT than under lightning or switching surge voltages [38].

In the 1990s there has been considerable progress with on-line monitoring and diagnostic equipment for GIS, and some of it is relevant to GITL installations; for example, UHF detection of PD, monitoring quality of gas, pressure, humidity, and presence of metallic particle contamination [39–41].

Field testing of GIS/GITL installations is another area in which very significant progress has been made in the past decade. Readily transportable high-voltage test equipment, of course, limits the range of tests that could be performed on-site. Resonant ac test transformers, lower voltage impulse generators, and dc test sets fall into this category. Past experience indicates that dc testing is not to be recommended, specially if the presence of metallic particles is suspected [42, 43].

12.3 CRYORESISTIVE CABLES

For many years cryogenic cables have been regarded as potentially economic alternatives for high-capacity transmission systems. It is well known that the resistivities of aluminum, copper and berylium can be reduced by factors up to 1000 below their room temperature values at liquid nitrogen or liquid hydrogen temperatures. Table 12.3 lists typical physical parameters of interest. Figure 12.11 shows the above relationships graphically. It should be noted that impurities in copper can reduce the advantage due to cooling. To obtain a resistivity ratio of 500 for copper, 99.999% purity is necessary. The cryoresistive cables fall into two main categories: (i) liquid nitrogen cooled, vacuum insulated cable and (ii) taped insulation, liquid nitrogen/hydrogen cooled cable. Below the second type of cable design is discussed first.

12.3.1 Taped Insulation Cables

General Electric Company in the United States was awarded a contract several years ago to recommend a cryoresistive cable design based on either liquid nitrogen or liquid hydrogen. The findings have been reported in several studies [44–46]. Initially the company favored a hydrogen-cooled cable. Paper and other synthetic tape materials were evaluated. Typical values of voltage stress at breakdown for impregnated paper are shown in Table 12.4. Typical $\tan \delta$ values are shown in Fig. 12.12. Conceptual designs of cryoresistive cables proposed by GE are shown in Fig. 12.13. Reference [46] gives a comparative cost breakdown for the three designs, and the following

TABLE 12.3 Metal Resistivity and Refrigeration Needs for Different Insulants

Insulant	Operating Temp. (K)	Relative Metal Resistivity	Refrigeration Ratio (watts input/watts load)	
			Practical	Theoretical
Liquid				
Oil	293 (room)	1	1	1
Nitrogen	77	1/8	6–10	3
Hydrogen	20	1/5000	50–100	14
Helium	4			
Superconductors	—	—	500–1000	75

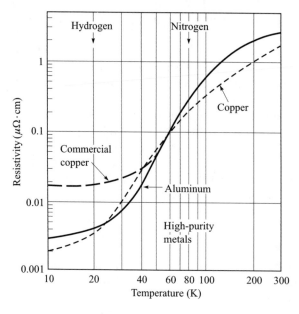

Figure 12.11 Variation of resistivity with temperature at cryogenic temperatures.

TABLE 12.4 Breakdown Voltage Stresses for Different Impregnants

Impregnant	Voltage Stress at Breakdown (V/mil)
Oil at 23°C	725
Boiling liquid H_2 at 20 K	950
Boiling liquid N_2 at 77 K	1000
Boiling liquid H_2 at 14 K	1025

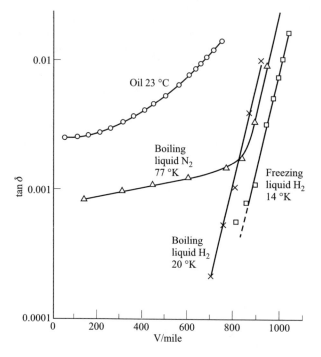

Figure 12.12 Dissipation factor, tan δ, versus voltage stress. Paper at 50% RH when vacuum impregnated with liquid nitrogen and hydrogen. Paper vacuum dried before vacuum impregnation with oil.

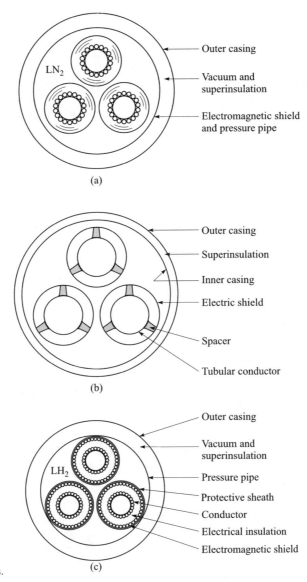

Figure 12.13 GE cable proposals.

conclusions were reached by the company. A flexible cable, much like conventional oil-paper cables shows a cost advantage over a rigid conductor concept. A free fluid dielectric appears to be less predictable than conventional impregnated taped paper system. Liquid-hydrogen-cooled flexible systems show a slight cost advantage over liquid nitrogen systems. In addition, liquid hydrogen systems should result in smaller overall dimensions and may result in lower manufacturing and installation costs.

However, hydrogen-cooled systems would appear to be technologically somewhat more difficult to develop than comparable liquid nitrogen systems. Moreover, there is the additional consideration of safety. Liquid nitrogen systems should therefore be preferred for the near future, at least. Finally, cable capacities above 3000 MVA at voltages close to 500 kV would appear to be economically feasible.

The above company described tests on a 345/500-kV prototype liquid nitrogen cable using a synthetic paper taped insulation [45]. The tests showed that the dissipation factor was far less than the assumed design figure of 0.0003. At 290 kV the measured dissipation factor was 0.000005. Voltage stress tests results were as follows: 425 kV/mil at the conductor surface was maintained for 5 days; failure occurred when the stress was raised to 640 V/mil. The cable had been designed to withstand a maximum stress of 400 V/mil at the conductor surface. It is interesting to note that the experience gained with the above lapped paper cryoresistive cable designs is proving to be very useful with the new high-temperature superconductive cable concepts.

Developments in cryoresistive cables have continued in Japan, Korea, and other countries [47, 48]. In Japan, a prototype 10-GVA at 500 kV with segmented aluminum conductor and synthetic paper insulation was evaluated at liquid nitrogen temperature. More recently in Korea development has continued for a 154-kV prototype using aluminum conductor and synthetic polyethylene tape insulation. The first prototype has been evaluated for 800 MVA. The basic design is similar to a pipe-type cable with hollow conductors for liquid nitrogen.

12.3.2 Vacuum-Insulated Cryocable

Only one manufacturer had proposed a design that combines the thermal and electrical insulation functions by using vacuum insulation. The design concept is very simple [49–51]. The main conductor is cooled by liquid nitrogen, while the outer shield is vacuum insulated. The U.S. Electric Research Council and the National Science Foundation had supported work on this design at the M.I.T. National Magnet Laboratory in Cambridge, Massachusetts.

A 138-kV specimen of the cable was tested at the EPRI Waltz Mill test facility near Pittsburgh, Pennsylvania. The main area of doubt in this design is the reliability of vacuum insulation on such a large scale. Due to insufficient commercial interest in this design, it is highly unlikely that a practical system of this type will emerge in the foreseeable future. In any case new high-temperature superconductors, operating at liquid nitrogen temperature and using conventional cable manufacturing technology, offer a more acceptable and promising alternative.

12.4 SUPERCONDUCTIVE CABLES

The idea of using superconductors for power cables was initially mentioned in 1961. However, the technology of using superconductors at power frequencies has been developing over the past decades and it is only now that such applications are even technically feasible.

The phenomena of superconductivity was discovered earlier in this century by the Dutch scientist Kamerlingh Onnes. The basic property of superconductivity is graphically shown in Fig. 12.14. Superconductivity can only exist between certain limits of current, magnetic field, and temperature (Fig. 12.15). These values are called critical

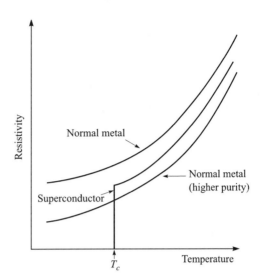

Figure 12.14 Change in resistance near absolute zero.

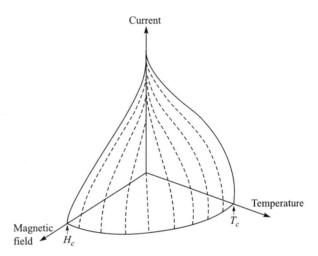

Figure 12.15 Boundary of superconductivity in three-dimensional space of current, field, and temperature.

current I_C, magnetic field H_C, and temperature T_C, respectively. The phenomena is common for metals, but for pure metals the highest known critical field is only 20 A/m for niobium. It was only in the early 1960s that alloys were formed, that had critical fields in the region of 1000 A/m.

Superconducting alloys are normally classified as type I or type II, according to their behavior in the presence of an external magnetic field. Figures 12.16(a)–(c) show the idealized magnetization curves for the two types. For type I [Fig. 12.16(a)] currents flow on the surface of the superconductor in such a way as to totally exclude the penetration of magnetic field inside the metal; the metal appears to be diamagnetic with zero effective permeability. Throughout the range up to H_C, the magnetization curve is reversible. Type II materials exhibit similar behavior up to a characteristic field H_C (Fig. 12.16). But at a field H_{C1}, instead of going normal, they permit penetration of field up to another characteristic field H_{C2}; beyond H_{C2}, type II materials also go

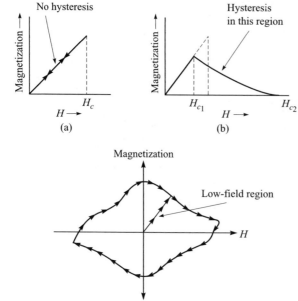

Figure 12.16 Characteristics of (a) type 1 and (b) type II superconductors. (c) Hysteresis loop for high fields in type II superconductor.

normal. Between H_{C1} and H_{C2}, the magnetization process is not reversible and energy is dissipated in the superconductor. The losses are quite large for ac operation, but stable operation is possible by using twisted filamentary (or ribbon type) conductors. Table 12.5 shows the properties of selected low-temperature superconductors.

During the 1960s significant development work in low temperature superconducting cables was conducted in England, France, and the United States. A general review of the prospects of cryogenic cables, as seen in the late 1960s, is given in [52]. In this discussion we will initially examine the two U.S. projects: Union Carbide Corporation in Yonkers, New York, and the Brookhaven National Laboratory on Long Island, New York [53–56]. These projects are typical of the design concepts for cables using low-temperature superconducting alloys. The technological experience gained with

TABLE 12.5 Parameters for Some Low-Temperature Superconductors

Material	Critical Temperature, T (K)	Critical Field, H_C (kOe)	Critical Current Density, J_c at 5 kOe (A/cm^2)
Pb	7.2	0.8	0
Nb	9.3	2 to 20 (depends on degree of cold working)	10^6
V	5.3	1	
NbTi	10.0	120	10^6
NbZr	10.8	90	10^6
Nb$_3$Sn	18.3	230	10^7
VaGa	15 to 16	220	10^7
Nb$_3$Al	19.3	360	
VaSi	17.2		

these projects is proving very useful for the new developments based on high-tempera-ture superconductors.

12.4.1 Union Carbide Design

The Union Carbide group recognized the advantages of working at higher vol-tages, and they proposed ac designs up to 345 kV. Moreover, they abandoned flexible conductors and made vigorous efforts to simplify the thermal insulation by eliminating the intermediate liquid nitrogen coolant and used helium vapor instead. Their manu-facturing costs were reasonable when compared with oil-paper insulated pipe-type cables. Table 12.6 gives their data. In the 1970s their view was that it will take at least 15–20 years of R&D work before a superconducting cable design can be fully developed! How true! The areas of research on this project included: (i) systems analysis to explore potential applications of such a cable up to the year 2000, (ii) conductor system development, (iii) dielectric system, (iv) cryogenic system, and (v) terminations. Information on emergency capacity, switching surges, fault currents, line lengths, and power requirements had to be gathered. This included the choice of substrate material (copper, aluminum, etc.) and choice of superconductor (Nb, Nb_3Sn, filaments or tapes, etc.). The program started in November, 1971, and the first phase led to a laboratory prototype design. The test results for the prototypes are described in [53]. The project was terminated.

12.4.2 Brookhaven Design

The Brookhaven work was also started in 1971 and the design chosen was a relatively low voltage flexible cable system. The operating temperature in this design was higher than that in the Union Carbide design. Figure 12.17 and Table 12.7 sum-marize this particular design [54–56]. In 1982 two superconducting cables were put under test. The test facility was designed to simulate a three-phase 138-kV, 1000-MVA cable system. The two lengths tested were each 115 m long and the operating

TABLE 12.6 Typical Dimensions for Union Carbide Cables

Line voltage, kV	69	138	230	345
Conductor diameter, in.	0.79	1.57	2.63	3.94
Shield diameter, in.	2.48	3.94	6.05	8.68
Outer cable diameter, in.	9.55	13.24	18.41	24.72
Insulation thickness, in.	2	2	2	2
Magnetic field strength, G	1000	1000	1000	1000
Dielectric strength, kV/cm	200	200	200	200
Power, MVA	423	1690	4710	10,590
Relative total expense	2.86	1	0.51	0.334
Cost Breakdown (%)				
Stabilized conductor	9.6	14.6	20.1	26.3
Cryogenic enclosure	24.3	25.6	25.6	23.7
Terminals (5-mile spacing)	4.1	3.9	3.5	2.8
Installation	32.2	26.1	21.9	19.8
Refrigeration (5-mile spacing)	16.6	14.7	11.6	8.3
Helium	1.9	4.2	7.4	10.7
Annual capitalized expenses	11.3	10.9	9.9	8.4

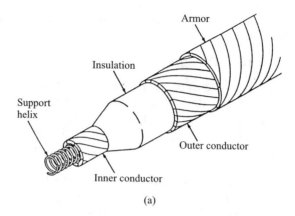

(a)

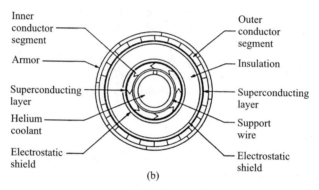

(b)

Figure 12.17 Flexible superconducting coaxial cable: (a) isometric view and (b) cross section (not to scale).

temperature was 9 K. An emergency three-phase rating of 1430 MVA was maintained for 1 hour. Although a 5-year development plan was formulated in mid-1980s to transfer technology to manufacturers, the new developments in high-temperature ceramic superconductors forced the abandonment of this project in 1987 [56].

TABLE 12.7 Modified Brookhaven Superconducting Cable Design

Length installed (each)	115 m
Outside diameter (over armor)	5.84 cm
Inner conductor diameter	2.95 cm
Operating temperature	Average 7 K, 8 K maximum
Operating H_e density	100 kg/m^3, minimum
Operating pressure	1.55 MPa, typical
Rated 60 Hz voltage	80 kV line-neutral
Rate 60 Hz current	410 A continuous, 6000 A for 30 min
Operating voltage stress	10 MV/m
Operating surface current density	422 A/cm
Maximum continuous power	1000 MVA; three phase
Cable conductor loss	0.2 W/m
Cable dielectric loss at 80 kV	< 0.06 W/m
Cable impedance	24 Ω (calc.)
	25 Ω (meas.)

12.4.3 General Comments on Low-Temperature Superconducting Cables

These cables have certain operational advantages. Chief among them are: (i) critical lengths may be as long as several hundred miles; (ii) short-duration fault currents can be carried by the system without the need for tripping healthy load currents, (iii) extended overload "cycles" are possible to design into such systems; (iv) cable rating is independent of soil conditions. The chief disadvantages are: (i) the economic capacities of these systems are very large, and overall security of the transmission system should be considered; (ii) the technology is very new and has to be "tested" and proven, only then realistic cost estimates can be arrived at; (iii) the world supply of helium is very limited.

Since the early 1970s, when the R&D effort in designing commercial superconducting cables was initiated in earnest, there has been considerable progress in understanding the requirements of electrically insulating such cables. The technical problems facing an insulation engineer are not insignificant. The industry expects a life of some 40 or more years. The insulation has to withstand not only the operating and transient electric stress, have low dielectric losses, but also be mechanically robust. The low-temperature properties of various insulations (gases, liquids, and solids) have been studied and are summarized in an excellent review [57]. In the early phases of development it was assumed that the superconducting power cables would operate at higher system voltages, thus taking advantage of the large current-carrying capacity of a superconductor. Some developments in the former Soviet Union and the United Kingdom were, however, in the voltage range of 15–35 kV. The present thinking is that the above voltage range is too low, and several R&D projects are aiming at voltages in the subtransmission range (138–345 kV three phase). Typical insulating materials are cellulose paper, polypropylene, or polyethylene paper/tape and even extruded crosslinked polyethylene. The research so far clearly indicates that liquid nitrogen impregnated insulation could operate at 20–50% higher electric stress than liquid helium impregnated tapes. At temperatures higher than liquid nitrogen it may even be feasible to utilize existing cable manufacturing equipment. Notwithstanding the above manufacturing difficulties for power applications, several R&D teams around the world are actively engaged in developing superconducting power cables [58].

The insulation design for any cable for a given operating voltage is not a simple matter. An overriding consideration is insulation integrity, that is, low failure probability of the dielectric. For ac operation a cable is considered to be "matched" with respect to its reactive load when equal amounts of energy are stored in electric and magnetic fields; that is,

$$\frac{E}{J_S} = \frac{377}{(\varepsilon_r)^{1/2}} \tag{12.4}$$

where E is the electric stress in the insulation, J_S the surface magnetic field at the conductor, and ε_r the relative permittivity of the insulation. Note that any increase in J_S implies an increase in E. In conventional cables the above matching is not economical, and reactive compensation elsewhere in the power system is needed. Typically, conventional cables operate at $J_S = 13$ A/mm and E lies between 10 and 20 kV/mm. So, they are very poorly matched. For superconducting cables J_S has to satisfy the other critical parameters of loss magnitude, critical magnetic field, and temperature. Typical surface fields for superconductors are between 35 and 45 A/mm, which implies E to be between 8.9 and 11.4 kV/mm. Also, for superconducting cables the dissipation factor must be low to minimize refrigeration needs. For low-temperature operation at liquid He this should be below 10^{-4}. References [57] and [58] give detailed data of the dielectrics commonly used in such cables.

12.4.4 High-Temperature Superconducting Power Cables

Early in the 1980s several research groups around the world began exploring the fascinating, but quirky, area of ceramic superconductors. In the early stages, the researchers could not get a ceramic such as barium–lead–bismuth to superconduct at a temperature higher than 13 K. Then in 1986 a team of researchers from IBM in Zurich published a paper reporting that the transition temperature of 30 K was possible for a ceramic (lanthanum–barium–copper oxide). The progress since 1986 in designing ceramic superconductors for higher temperatures has been astounding and very rapid. For example, in March, 1988, researchers from IBM-Alamaden (United States) reported a transition temperature of 125 K for a thallium–barium–calcium–copper oxide ceramic. In other words, it was possible to get a higher operating temperature than liquid nitrogen. However, before ceramics could be used for power applications, they must be carefully processed and manufactured into tapes or wires. This has not proven to be easy! Much less progress has occurred in bulk materials, which may be more ductile. The other important practical matters are the critical magnetic field and current density (Fig 12.16). An excellent review of these early days of development in high-temperature superconducting (HTS) ceramics appears in Ref. 59.

Very early in the development of power applications of HTS (for cables, transformers, and rotating machines) three considerations were crucial: (i) the mechanical brittleness of ceramics, which made it difficult to produce workable conductors, (ii) poor coupling for current flow in a polycrystalline material; and (iii) the phenomenon of "fluxcreep," which results in a strong magnetic field penetrating the superconductor and increasing power losses. Basically the new HTS materials are oxide ceramics, and whole families of new ceramics may be custom designed. For power applications, initially the new HTS ceramics were grouped as from barium, copper, oxygen, plus rare earths, or lanthanum, copper, oxygen, plus small quantities of barium, strontium, and calcium. As the technology has progressed, there are essentially two emerging leading materials, namely, BSCCO and YBCO. BSCCO contains bismuth, strontium, calcium, copper and oxygen; currently two formulations are available BSCCO-2212 and BSCCO-2223. The other group contains titanium, barium, calcium, copper, oxygen (TBCCO-1223) and YBaCuO (YBCO-123) with yttrium, barium, copper, and oxygen. At liquid nitrogen temperature the effective critical current densities of these ceramic compounds are $> 10^3 \, A/cm^2$ [60]. Two different manufacturing techniques are being used to produce conductors: (i) powder-in-tube, and (ii) coated tape. In the first case, the HTS ceramic powder is packed inside a metallic tube (silver or silver alloy). By a repeated process of swaging, extruding and restacking, and finally subjecting it to a thermomechanical treatment that aligns and textures the HTS grains. For producing coated tape conductors, ion beam deposition of yttria-stabilized zirconia on a pliable nickel base is the first step, followed by depositing YBCO material.

Regarding target current densities, in the United States, the R&D sponsors have established a minimum of 10 kA/cm^2. For the BSCCO material a critical current of 18 kA/cm^2 at 77 K and zero magnetic field has been achieved. For the YBCO tape, however, critical currents, at zero magnetic field, of 1 MA/cm^2 have been reported. Researchers report critical current densities of over 100 kA/cm^2 in a field of 4 T.

Most of the R&D activities, utilizing HTS materials, is concentrated in Europe, Japan, and the United States. In Europe, both ends of the power capability are being explored (up to 3000 MVA and down to 500 MVA). In the United States, the Electric Power Research Institute (EPRI) and the Department of Energy (DOE) are taking the lead in funding R&D. Retrofitting existing pipe-type cables, in the 115-kV/400-MVA range, is being given a high priority and a 6-year cooperative project (HTS material and cable manufacturers and the funding agencies) was initiated in 1993.

In the United States two different design concepts are proposed: room temperature dielectric (RTD) and cryogenic dielectric (CD). Each single phase design is conceptually similar to the low-temperature design shown in Fig 12.17. The essential difference is that in the room temperature design only the central conductor, in each phase, is at the liquid nitrogen (LN2) temperature. The cryogenic dielectric design has the complete coaxial conductor system in a cryostat. The room temperature design is obviously cheaper in capital cost but has higher losses. Both RTD and CD designs are suitable for pipe-type cables, but RTD is easier for retrofitting in urban areas. In urban areas, the retrofit option for pipe-type cables may be very attractive. In one study, the capacity increases from a number of upgrade options for an existing 115-kV, 222-MVA cable system are shown in Table 12.8.

In a U.S. study, for long-distance (100-mile) transmission of 900 MW, comparison was made between a conventional 345 kV polypropylene paper cable with appropriate reactive shunt compensations and a 230-kV room temperature dielectric HTS cable, with a midpoint shunt compensation. The capital costs of the HTS cable are estimated to be 80% of the conventional cable. It was considered premature to estimate the differences in operating costs [61].

In Europe, a consortium of several cable manufacturers, with support from the European Commission carried out a similar technical/economic study for transmission of power in the range 500–3000 MVA [62]. Again, the cost of HTS material and the critical current density were identified as the crucial comparison parameters. Capacity upgrades of up to five times, with loss reductions in the range of 25–50%, are

TABLE 12.8 Improvement in Power Capacity Using HTS Cable for 115-kV, 222-MVA Pipe-Type Cables

Upgrade Method	New Capacity (MVA)	Over Base Improvement (%)
Reconductor	260	17
Force cooled	300	35
Reconductor + force cooled	400	80
HTS, room temperature dielectric	500	125
HTS, cryogenic dielectric	1000	350

Source: From [61].

considered feasible, if the critical current densities are in the 100 kA/cm^2 range, at zero magnetic field.

R&D projects are also underway in Japan and in Korea [47, 48, 63]. References 60 and 64–66 give an excellent review of the current experience with superconducting cables with suggestions for continuing R&D. The first 115-kV prototype suitable for retrofitting up to 400 MVA capacity has been recently factory tested [65].

12.5 ECONOMIC CONSIDERATIONS

Bulk power transmission over long distances is a fact of life in most industrialized nations. It is recognized that societal and environmental pressures will continue to mount and alternatives to overhead lines must be found to transport bulk electric power. Conventional transmission cables not only have technical limitations due to inherent capacitance but are also very expensive to install and maintain. In any evaluation of compressed gas insulated or cryogenic cable comparison with overhead lines may be appropriate. Factors to be considered are approval/authorization protocols, technology, reactive load, availability and overload capacity, tolerance to faults, losses, direct and indirect costs, extent of land occupation, terrain, audible noise and electromagnetic interference, electric and magnetic fields, and life cycle expectancy [1]. In the subtransmission voltage range (70–150 kV), the relative direct and indirect costs of conventional cables are expected to be 8–10 times those for aerial lines. For the long-distance transmission voltages (> 400 kV) the cost differential may be 20–30 times! A recent survey of user and public concerns and relative costs of overhead versus underground transmission provides an interesting overview [67]. The most frequently cited reasons for going underground are public pressure, aesthetic/environmental impact, topography, and technical considerations. From the data provided by 58 utilities from 19 different countries, the following ranges of capital cost penalty were noted: 3.6–16 times for 110–219 kV, 5.1–21.1 times for 220–362 kV, and 13.6–33.3 times for 363–764 kV.

From the survey, as expected, it is clear that the environmental impact of underground power transmission is less severe than for overhead lines. The life cycle assessment would, however, favor overhead lines.

There is always uncertainty in evaluating costs of new systems of technology. Several R&D groups have worked out the relative costs of the various systems of underground cables [4]. It is clear from these studies that, even for very high capacity systems, the economic advantage of high-temperature superconducting cables is uncertain. A comparative cost study of new systems was commissioned by the U.S. Electric Research Council in the early 1970s. The study was conducted by A.D. Little Inc. and the results for various assumed locations and conditions are given in Ref. 68. The study made a set of comprehensive recommendations, which have stood the test of time and are summarized below. Classical cables must bear the brunt of transmitting increasing power loads into congested cities. To obtain greater power handling capacity and cost reduction, the following principal areas of investigation are suggested: (i) the development of a thorough understanding of the thermal characteristics of underground (particularly pipe-type) cables and the development of cable systems having improved thermal design; (ii) the analytical and experimental development of economical and efficient forced cooling techniques for all cable systems, including the development of cooling equipment and associated hardware, e.g., potheads and terminations; (iii) the development of tape insulating materials with lower losses and thermal resistance; and (iv) the development of semiautomatic and less expensive splicing techniques.

The development of advanced naturally cooled, extruded-dielectric cables offers the most significant reductions in transmission costs for circuit capacities below 1000 MW.

Such cables have been developed with higher operating voltages (345 kV and above) and larger conductor sizes. If sodium is used as the conductor, transmission costs can be reduced to approximately one third of present costs in circumstances where they can be used. To achieve these objectives, R&D programs will be required to advance the basic understanding of voltage breakdown in dielectric solids and development of economical manufacturing techniques with excellent quality control.

Future demands for very low loss systems at high powers (5000 MW and up) and at higher power densities will probably require cryogenic systems. The intervening years can be usefully employed by: (i) developing fundamental information upon which to base final system designs offering lower costs and high reliability, and (ii) demonstrating, through a series of installations of gradually increasing size and complexity, that power can be transmitted reliably by cryogenic techniques. It is interesting to note that high-temperature superconductors offer hope that existing pipe-type cables in urban areas could be retrofitted to increase the power transmission capabilities.

At powers around 1000 MVA, gas-spacer cables are more economical than the currently available alternative of conventional cables. Their capacities are well matched to those of overhead circuits at the same voltage and forced-cooled gas-spacer cables at high voltages may be the most economical alternative at power levels of several thousand MVA until HTS cryogenic cables are available. If the cost differential between aerial lines and gas-spacer cables can be brought down to 10 or less, there would be a rapid growth of such systems around urban areas.

High-voltage, high-power gas-spacer cables have the drawback of requiring a very large trench, hence a system with three phases in one sheath and of a flexible cable is preferable. An integrated system for the installation of underground power cables would offer additional advantages. This would involve designs compatible with a modern continuous laying technique, the development of a better backfill that can be laid along with the cable.

In the late 1970s, the Philadelphia Electric Company of the United States conducted a theoretical study to compare various underground transmission systems [69]. The study was sponsored by the U.S. Department of Energy, and its goal was to evaluate various systems for transmitting bulk power in the range of 10,000 MVA over a distance of approximately 100 km. The study considers problems associated with the selection of right-of-way, losses, enclosures, terminations, system planning, substation design and compensation, cable design and ampacity, short-circuit ratings, maintenance, and overall economics. This study did not use the criterion of "dollars per MVA-mile" since it was felt that such a comparison often provides fictitiously low values for high-ampacity systems. After conducting load flow and stability analysis, the usable ampacity of some systems is found to be well below the theoretical value. Consequently, total cost of an application is the only accurate method for comparison. A 500-kV aerial/underground system offers the lowest cost followed by 230-kV superconducting, 500-kV SF_6 insulated, and 500-kV self-contained oil-filled systems. Although over the intervening period the economic factors have changed, the costs of the various alternatives are likely to be in the same relative order. Also, it is highly unlikely that bulk power at the level of 10,000 MVA would be transmitted over underground cables. In the early part of the next century, we expect to see commercial applications of HTS cables in urban areas and compressed SF_6/N_2 cables in power corridors approaching major load centers. A philosophical approach to economic evaluation of various underground transmission alternatives is discussed by Moran [70].

ACKNOWLEDGMENTS

Much of the material in this chapter is derived from [71]. Assistance of Professor J.D. Cross in preparing [71] and the permission of the Sandford Educational Press to use the material is gratefully acknowledged.

REFERENCES

[1] P. Couneson, J. Lamsoul, X. Delree, and X. van Merris, "Bulk power transmission by overhead lines and cables: comparative assessment and principles adopted in Belgium for future development of the HV network," CIGRÉ Paper 21/22-09, 1996.

[2] G. D. Friedlander, "Is power to people going underground," *IEEE Spectrum*, Vol. 9, 1972, pp. 62–71.

[3] E. C. Bascom, D. A. Douglass, G. C. Thomann, and T. Aabo, "Hybrid transmission: aggressive use of underground cable sections with overhead lines," CIGRÉ Paper 21/22-10, 1996.

[4] A. H. Cookson, "Gas insulated Cables," *IEEE Trans. Electrical Insulation*, Vol. EI-20, 1985, pp. 859–890.

[5] H. Koch and V. Cousin, "From gas insulated switchgear to cross-country cables," Proc. JICABLE, Paper A.5.1, 1995, pp. 118–121.

[6] E. Thuries, V. D. Pham, P. Roussel, and M. Guillen, "420 kV three phase compressed nitrogen insulated cable," Proc. JICABLE, Paper 1.5.2, 1995, pp. 122–127.

[7] C. Aucourt, C. Boisseau, and D. Feldmann, "Gas insulated cables: from the state of the art to feasibility for 400 kV transmission," Proc. JICABLE, Paper 1.5.4, 1995, pp. 133–138.

[8] E. Thuries, A. Girodet, P. Roussel, and M. Guillen, "Underground gas insulated transmission line," CIGRE Paper 21/22-05, 1996.

[9] T. Nojima, M. Shimizu, T. Araki, H. Hata, and T. Yamauchi, "Installation of 275 kV-3.3 km gas-insulated transmission line for underground large capacity transmission in Japan," CIGRE Paper 21/23/33-01, 1998.

[10] B. F. Hampton, D. N. Browing, and R. M. Myers, "Outline of a flexible SF_6 insulated EHV cable," *Proc. IEE*, Vol. 123, Part 2, 1976, pp. 159-165.

[11] N. Cuk, K. K. Nishikawara, G. G. McCrae, and P. T. B. Adams, "Specification and application of SF_6 compressed gas insulated switchgear—a utility's point of view," *IEEE Trans. On Power Apparatus Syst.*, Vol. PAS-99, 1980, pp. 2241–2250.

[12] A. H. Cookson, T. F. Garrity, and R. W. Samm, "Research and development in the United States on three-conductor and UHV compressed gas insulated transmission lines for heavy load transmission," CIGRE Paper 21-09, 1978.

[13] T. Takagi, H. Hayashi, T. Higashino, S. Nishihara, and K. Itaka, "Dielectric strength of SF_6 gas and 3-core type CGI cables under inter-phase switching impulse voltages," *IEEE Trans. Power Apparatus Syst.*, Vol. PAS-93, 1974, pp. 354-361.

[14] A. H. Cookson, P. C. Bolin, H. C. Doepken, Jr., E. R. Wootton, C. M. Cooker, and J. G. Trump, "Recent research in the United States on the effects of particle contamination reducing the breakdown voltage in compressed gas insulated systems," CIGRE Paper 15-09, 1976.

[15] Electric Power Research Institute, "Investigation of high voltage particle initiated breakdown in gas insulated systems," Report EL-1007, March 1979.

[16] U.S. Department of Energy, "Elimination of particle effects in SF_6 insulated transmission systems," Report DOE/ET/29336-1, August 1983.

[17] K. D. Srivastava and R. G. van Heeswijk, "Dielectric coatings—effect on breakdown and particle movement," *IEEE Trans. Power Apparatus Syst.*, Vol. PAS-104, 1985, pp. 22–31.

[18] M. M. Morcos and K. D. Srivastava, "Control of metallic particle contamination in compressed SF_6 switchgear through conductor coatings," *Proc. Nordic Insulation Symp*, NORD-IS, 1995, 1996, pp. 371–378.

[19] CIGRE Working Group 23.10, "SF_6 and the global atmosphere," *Electra*, No. 164, 1996, pp. 121–131.

[20] L. G. Christophorou and R. J. van Brunt, "SF_6/N_2 mixtures—basic and hv insulation properties," *IEEE Trans. Dielectrics Electric Insulation*, Vol. 2, No 5, 1995, pp. 952–1003.

[21] K. Itaka and G. Ikeda, "Dielectric characteristics of compressed gas insulated cables," *IEEE Trans. Power Apparatus Syst*, Vol. PAS-89, 1970, pp. 1986–1994.

[22] C. M. Cooke and J. G. Trump, "Post type support spacers for compressed gas insulated cables," *IEEE Trans. Power Apparatus Syst.*, Vol. PAS-92, 1978, pp. 1441–1447.

[23] A. K. Chakrabarti, R. G. van Heeswijk, and K. D. Srivastava, "Spacer involvement in free particle initiated breakdown on compressed gas insulated systems," *IEEE Trans. Electrical Insulation*, Vol. EI-22, 1987, pp. 431–438.

[24] V. N. Maller and K. D. Srivastava, "Insulator surface discharge studies in SF_6 under switching transients" CIGRE Paper 15-02, 1986.

[25] N. G. Trinh, G. Mitchell, and C. Vincent, "Influence of an insulating spacer on the $V - t$ characteristics of a coaxial gas-insulated cable. Part I: Study of a reduced scale coaxial conductor. Part II: Tests on EHV buses," *IEEE Trans. Power Delivery*, Vol. PWRD-3, 1988, pp. 16–34.

[26] J. M. Braun, G. L. Ford, N. Fujimoto, S. Rizzetto, and G. C. Stone, "Reliability of GIS EHV epoxy insulators: the need and prospect for more stringent acceptance criteria," *IEEE Trans. Power Delivery*, Vol. PWRD-8, 1993, pp. 121–131.

[27] N. Achatz, A. Diessner, and E. Kynast, "X-ray induced partial discharges in small hollow glass spheres in GIS insulators," Proceedings of the 9th International Symposium on HV Engineering, Graz, Austria, Paper 5599, 1995.

[28] S. A. Boggs and Y. Wang, "Trapped charge induced field distortion on GIS spacers," *IEEE Trans. Power Delivery*, Vol. PWRD-10, 1995, pp. 1270-1275.

[29] B. O. Pederson, H. C. Doepken, and P. C. Bolin, "Development of a compressed gas insulated transmission line," *IEEE Trans. Power Apparatus Syst.*, Vol. PAS-90, 1971, pp. 2631–2638.

[30] J. C. Cronin and A. Conangla, "Heat flux-temperature relationships for closely spaced underground power transmission conductors," *IEEE Trans. Power Apparatus Syst.*, Vol. PAS-90, 1971, pp. 1246–1252.

[31] CIGRE Working Group 21.12, "Calculation of the continuous rating of single core rigid type compressed gas insulated cables in still air with solar radiation," *Electra*, No. 106, 1986, pp. 23–31.

[32] CIGRE Working Group 21.12, "Calculation of the continuous rating of three core rigid type compressed gas insulated cables in still air and buried," *Electra*, No. 125, 1989, pp. 103–111.

[33] International Electrotechnical Commission, "High voltage switchgear and controlgear—use and handling at SF_6 in high voltage switchgear and controlgear," Report IEC-1634, 1995.

[34] D. Kopejtkova, T. Molony, S. Kobayashi, and I. M. Welch, "A twenty five year review of experience with SF6 gas insulated substations," CIGRÉ Paper 2101, 1994.

[35] CIGRE Working Group 33/13.09, "Very fast transient phenomena associated with GIS," Paper 33-13, 1988.

[36] CIGRE Working Group 15.03, "GIS insulation properties in case of VFT and DC stress," Paper 15-201, 1996.

[37] A. M. Sahni, C. Baliga, D. Raina, and J. K. Tiku, "Site investigations and study of very fast transient overvoltages in 245 kV gas insulated switchgear of Tata Electric Companies," CIGRE Paper 15/21/31-01, 1996.

[38] A.J. Vandermaar, M. Wang, J. B. Neilson and K. D. Srivastava, "The electrical breakdown characteristics of oil-paper insulation under fast-front impulse voltages," *IEEE Trans. Power Delivery*, Vol. PWRD-9, 1994, pp. 1926–1935.

[39] Working Group 15.03, "Effects of particles on GIS insulation and the evaluation of relevant diagnostic tools," CIGRE Paper 15-103, 1994.

[40] J. S. Pearson, O. Farish, B. F. Hampton, M. D. Judd, D. Templeton, B. M. Pryor, and I. M. Welch, "Partial discharge diagnostics for gas insulated substations," *IEEE Trans. Dielectrics Electrical Insulation*, Vol. DEIS-2, 1995, pp. 893–905.

[41] C. J. Jones, O. Beierl, E. Colombo, G. Ebersohl, B. Henderson, S. Kobayashi, K. Pettersson, and P. Wesler, "Guidelines for monitoring, control and supervision of GIS incorporating advanced technologies," CIGRE Paper 23-203, 1996

[42] K. Feser, R. Sun, A. Eriksson, and K. Fröhlich, "On-site dielectric testing of GIS: theoretical and practical considerations," *IEEE Trans. Power Delivery*, Vol. PWRD-6, 1991, pp. 615–625.

[43] Working Group 21.12, "Review of long term field tests of compressed gas insulated cables," *Electra*, No. 127, 1987, pp. 86–89.

[44] M. J. Jeffries and K. N. Mathes, "Insulation systems for cryogenic cables," *IEEE Trans. Power Apparatus Syst.*, Vol. PAS-89, 1970, pp. 2006–2014.

[45] M. J. Jeffries, S. H. Minnich, and B. L. Belanger, "High voltage testing of a high capacity, liquid nitrogen cooled cable," *IEEE Trans. Power Apparatus Syst.*, Vol. PAS-92, 1973, pp. 514–519.

[46] Electric Power Research Institute, "Resistive cryogenic cable," Report EL-503, October, 1977.

[47] T. Mizukami, M. Fukasawa, K. Sugiyama, S. Kuma, and H. Nagana, "Prototype tests on cryoresistive cable," *IEEE Trans. Power Apparatus Syst.*, Vol. PAS-99, 1980, pp. 528-535.

[48] K. S. Ryu, S. S. Oh, J. Y. Koo, M. K. Choi, S. C. Hwang, Y. Kim, and K. Chang, "Development and research activities on the cryogenic power transmission cable in Korea," CIGRE Paper 21/22-08, 1996.

[49] P. Graneau, "Economics of underground transmission with cryogenic cables," *IEEE Trans. Power Apparatus Syst.*, Vol. PAS-89, 1970, pp. 1–7.

[50] S. B. Afshartous, P. Graneau, and J. Jeanmonod, "Economic assessment of a liquid nitrogen cooled cable," *IEEE Trans. Power Apparatus Syst.*, Vol. PAS-89, 1970, pp. 8–16.

[51] Electric Power Research Institute, "Development and field trial of VI-LN2 cryocables—Stage 1," Report 7813-1, February, 1976.

[52] S. Neal, "Cryogenic transmission in the power industry of the future," *Proc. Am. Power Conf.*, Vol. 30, 1968, pp. 1210–1217.

[53] L. K. Eigenbrod, H. M. Long, and J. Notaro, "Conceptual design and economic analysis of the superconducting ac power cable system," *IEEE Trans. Power Apparatus Syst.*, Vol. PAS-89, 1970, pp. 1995–2005.

[54] US Atomic Energy Commission, "*Underground transmission of power by superconducting cables,*" E. B. Forsythe, Ed., Report BNL 5032, March, 1972.

[55] Electric Power Research Institute "Superconducting cable systems," Report EL-402, Vols. I and II, May, 1977.

[56] E. B. Forsyth and R. A. Thomas, "Operational test results of a prototype superconducting power transmission system and their extrapolation to the performance of a large system," *IEEE Trans. Power Delivery*, Vol. PWRD-1, 1986, pp. 10–18.

[57] F. Krähenbühl, B. Bernstein,, M. Danikas, J. Densley, K. Kadotani, M. Kahle, M. Kosaki, H. Mitsui, M. Nagao, J. Smit, and T. Tanaka, "Properties of electrical insulating materials at cryogenic temperatures: a literature review," *IEEE Electrical Insulation Mag.*, Vol. 10, No. 4, 1994, pp. 10–22.

[58] E. B. Forsyth, "The dielectric insulation of superconducting power cables," *Proc. IEEE*, Vol. 79, No. 1, 1991, pp. 31–40.

[59] K. Fitzgerald, "Superconductivity: facts vs. fancy," *IEEE Spectrum*, Vol. 25, 1988, pp. 31–41, May.

[60] U. Balachandran, Guest Editor, "Superconductivity in electric power: a special report," *IEEE Spectrum*, Vol. 34, 1997, pp. 18–49.

[61] D. W. Von Dollen, N. G. Hingorani, and R. W. Samm, "High temperature superconducting cable technology," Proc. of the IEE Conf. on Power Cables and Accessories: 10kV–500kV, Pub. No. 382, , November, 1993, pp. 253–257.

[62] P. Metra, S. Ashworth, R. J. Slaughter, and E. M. Hughes, "Preliminary analysis of performance and cost of high temperature superconducting power transmission cables," Proc. of the IEE Conf. on Power Cables and Accessories: 10kV–500kV, Pub. No. 382, November, 1993, pp. 248–252.

[63] M. Kosaki, "Research and development of electrical insulation of superconducting cables by extruded polymers," *IEEE Electrical Insulation Mag.*, Vol. 12, No. 5, 1996, pp. 17–24.

[64] Electric Power Research Institute, "Design concepts for a superconducting cable," Report TR-103631, September 1994.

[65] M. M. Rahman, N. Nassi, L. Gherardi, and D. V. Dollen, "Design development and testing of the first factory made high temperature superconducting cable for 115 kV–400 MVA," CIGRE Paper 21-202, 1998.

[66] R. D. Blaugher, in *Advances in Cryogenic Engineering*, L. T. Summers, Ed., Plenum, New York, pp. 883–898, 1996.

[67] Working Group 21/22.01, "Comparison of overhead lines and underground cables for electricity transmission," CIGRE Paper 21/22-01, 1996.

[68] A. Nicol, "Planning for more economic underground power transmission," *Proc. IEEE Underground Transmission Conf.*, 72-CHO-608-O-PWR (Supp.), 1972, pp. 339–356.

[69] Department of Energy, "Evaluation of the economical and technological viability of various underground transmission systems for long feeds to urban load areas," Report HCP/T-2055/1, December, 1977.

[70] J. A. Moran, Jr., "Comparative cost analyses for bulk power transmission systems," *Proc. IEEE Underground Transmission Distribution Conference*, Pub. no. 76 CH 1119-7-PWR, pp. 442–445, 1976.

[71] K. D. Srivastava and J. D. Cross, "Recent developments in power cable systems," in *Power Cable Engineering*, R. Bartnikas and K. D. Srivastava, Eds., Sandford Educational Press, Waterloo, Ontario, Canada, 1987, Chapter XII.

Chapter 13

UNDERWATER POWER CABLES

Richard T. Traut

13.1 INTRODUCTION

Use of high-voltage power cables to transmit electrical energy across bodies of water is a well-established practice. In cases where load centers, such as populated islands, are isolated by water from the power transmission source, cables are often a necessity. Independent generating systems as a primary source are often not practical or economical for such isolated load centers. Most commonly, underwater alternating-current (ac) systems are used to transmit power 40–50 km or less, and direct-current (dc) underwater systems are used for transmission and/or major system interchange over 25 km or more. As system length increases, dc transmission becomes economically and technically advantageous. For very long lengths (over 50 km) only dc transmission is feasible.

Rather than being a separate technology on its own, underwater power cable technology is really an offshoot of underground power cable technology, Underwater cables that are well designed to withstand the bottom conditions on ocean, river, and lake beds include components to provide strength and protection. The primary difference between most underground and underwater cables is the addition of armor. The design of this armor will vary considerably depending upon the design objectives. Unlike underground cables, which are inherently well protected from external hazards, underwater (or submarine) cables must successfully withstand the hazards of being located on the seabed where they can be damaged by anchors, storms, dredges, or other marine hazards. They must also be designed to withstand safely the rigors of handling, installation, and recovery. In addition to the added strength and protection built into underwater cables, these cables also generally differ from land cables in the manner in which they are handled. Underwater cables are larger and heavier than land cable counterparts and may also be supplied in very long continuous factory lengths. Unlike land cables, it is seldom if ever practical to perform field splices on underwater cables during installation except at the shore transition. This is because it is most desirable to install such cables by a continuous and uninterrupted method. To transport such long and heavy cables it is possible to take the finished cable up in large circular or oval cribs, rotating drums, or turntables mounted on the installation vessel. The cable may be loaded directly from the factory to vessel or may be transported across country in railroad gondola cars, extra large reels, or special containers. Then the cable may be loaded on the vessel.

This chapter is intended to provide the reader with a description of the design considerations for underwater power cables. It would not be possible to go into detail in each area due to practical limits on the chapter length. It is hoped that the text will sufficiently describe the design considerations to give the reader a good general understanding. Further description and details are available to the interested reader by consulting cited references.

13.2 UNDERWATER POWER CABLE DESIGN

In general, power cables for underwater transmission of power may be classified in one of two categories. The first are *conventional* underwater cables. The design and fabrication of these cables are mainly based upon industry standards and specifications [1–9]. Usually, no single industry specification will cover all aspects of the design and manufacturing, and thus it is common for more than one of these documents to be referenced. The second category is for those cables that are *specially designed* for unusual service [10–19]. These cables include design features that are different from conventional cables. In the following sections we will explore the design and application of underwater power cables. Some examples of underwater high-voltage ac and dc power cables are shown in Table 13.1.

13.2.1 Configuration

The configuration of nearly all underwater cable systems ever installed has been either single conductor or three conductor. Examples of single-conductor and three-conductor underwater cables are shown in Figs. 13.1 and 13.2. Three-conductor cables are used for ac power transmission whereas cables for both ac and dc application can be single-conductor cables.

13.2.1.1 Single-Conductor Cables

In general, when the diameter of the individual single phases of a cable system exceeds about 5 cm, it becomes economically attractive to provide single-conductor cables rather than three-conductor configurations. This is very general, however, and the decision is highly influenced by several important factors. These include, first, the system length and the consideration of cable joints. The maximum length of single-conductor cable that a given cable manufacturer can supply is, of course, inversely related to the cable diameter. The limitations are mainly imposed by handling, processing, and testing limits. For example, the cable producer has finite limits on the amount of continuous cable of a given diameter that can be taken up on, or in, process tanks, reels, or platforms. If such volume and/or weight capacity is exceeded, it is certain that multiple lengths must be made and joined together by a reliable technique.

Second, handling is an important factor. Many cables can be coiled and, therefore, it is possible to handle extremely long lengths by storage in huge stationary tanks or by coiling on suitable platforms. If the cable is noncoilable, then the length that can be handled as one piece is limited by the volume and weight capacity of the rotating take-up drum or turntable. A means of cable transport that allows such long-lengths storage

Figure 13.1 Single-conductor ac submarine power cable. (Courtesy of Simplex Technologies, Inc.)

TABLE 13.1 Characteristics of Some High-Voltage Submarine Power Cables

Cable	Voltage	Power (MW)	Conductor (mm²)	Insulation type	Hermetic Sheath	Armor Configuration	Approx. Wt. in Seawater (kg/m)	Approx. Maximum Depth (m)	Approx. Installation Tension (kg)
Sardinia-Corsica-Italy (1965)	200 kV dc	300	Copper, oval (420)	Solid/paper	Lead	Single, round galv. steel		451	8000
British Columbia–Vancouver									
1956	138 kV ac	120	Copper (228)	Gas filled	Lead	Single, round galv. steel	18.6	183	6800
1969	300 kV dc	360	Copper, oval (405)	Solid/paper	Lead	Single, round galv. steel		189	4100
1975	300 kV dc	400	Copper (405)	SCOF	Lead	Single, round galv. steel			
1982, 1981	525 kV ac	1100	Copper, Keystone (1620)	SCOF	Lead	Double, flat copper	50.6	396	26,300
Cook Strait (1965)	250 kV dc	600	Copper (507)	Gas filled	Lead	Single, round galv. steel	30.2	256	9500
Hokkaido-Honshu (1978)	250 kV dc	600	Copper (608)	SCOF	Lead	Single, round galv. steel		290	10,900
Skagerrak (1976, 1977)	250 kV dc	500	Copper, Keystone w/center wire (810)	Solid/paper	Lead	Double, round galv. steel	29.8	549	21,800
OTEC Riser (Development)									
Oil/paper	138 kV ac	100	Copper (279)	SCOF	Lead	Double, round A16X stainless	19.5	1830	44,000
XLPE	138 kV ac	100	Copper (177)	XLPE	Lead	Double, round A16X stainless	20.4	1830	45,800

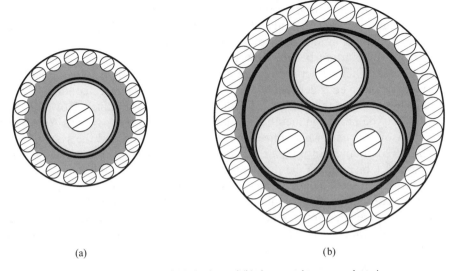

(a)   (b)

Figure 13.2 Cross sections of (a) single- and (b) three-conductor ac submarine power
cable. (Courtesy of Simplex Technologies, Inc.)

and handling must be selected early in the design process. In general, very long continuous lengths are shipped via a cable vessel or barge or by rail in gondola cars or on
special flat-car-mounted submarine cable reels.

Third, an important limitation on the continuous lengths of single-conductor ac
cables that can be made is the ability to perform appropriate 50- or 60-Hz tests. If
overvoltage ac withstand, ionization factor, or partial-discharge tests are required on
the full lengths, then the practical limits of the available equipment will set the maximum lengths that can be made. Where only dc testing is necessary, factory lengths of
more than 200 km can be produced.

Fourth, installation tensions constitute an important parameter. In some cases,
due to cable size and/or weight, and limitations on the ability to handle extremely high
installation tensions, single-conductor cables (as opposed to three-conductor cables) are
mandatory. This will occur where the weight of the three-conductor cable in seawater is
so great as to prohibit handling with available equipment.

Lastly, an important consideration in the choice of three single conductors versus
one three-conductor cable is redundancy and repairability in the event of either damage
to or failure of the cable. In one sense, the three singles may possibly be considered to
have three times the probability of damage from anchors, dredges, etc. From another
sense, the damage is simpler to repair for single-conductor cables, although the cost of
the repair may be close to that for a comparable three-conductor cable. Most of the cost
of repair is usually associated with damage location, mobilization, recovery, and rein-
stallation rather than with the actual reconstitution of the damaged cable. In this
context, the marginal cost for a three conductor (3C) versus a one-conductor (1C)
cable repair is only higher by a small percentage of the total repair operation costs.

Where three single conductors are judged to be vulnerable to damage, it is common
practice to install four single conductors, including one spare. This is obviously an
economic decision involving the additional costs of the spare length versus the value

of lost revenue and requirements for uninterrupted electrical transmission. Important elements in an economic study of redundancy also include probability of damage or failure, mobilization and repair costs (estimated) at all times and weather conditions, and the availability of backup power sources.

In cases where either a three-conductor or three single-conductor cable is feasible, it is important to consider redundancy. It would take two three-conductor cables (one a spare) to equal the redundancy of four single-conductor cables. However, the two 3C cables may be half as vulnerable to damage as the four 1C cables. Also consider, for comparison, the case where one cable has been damaged and the system must operate without redundancy until repair is accomplished. During this interval (which could be days, weeks, or months) the three intact single-conductor cables may be three times as vulnerable to a second damage incident as the single 3C cable. In either case, during this interval, a second incident would interrupt power transmission for a significant time. In any case where the choice is between ac single-conductor cables and three-conductor cables, it is important to weigh carefully initial installed cost against the costs of interrupted power and/or cable repair.

13.2.1.2 Three-Conductor Cables

As pointed out, the application for three-conductor underwater cables is the transmission of three-phase 60 Hz (or 50 Hz) power. A three-conductor power cable, as contrasted to three single-conductor cables of comparable size performing the same electrical function, has the following advantages. First, it is characterized by lower losses and improved voltage regulation; while capacitive charging currents and dielectric losses will be comparable to three 1C cables, the electrical losses due to induced currents in the sheath and armor will be significantly lower with 3C cables. In fact, in a balanced three-phase load condition, the 3C sheath and armor losses are negligible. Voltage regulation will also be better with the 3C option, because these losses are low and because phase-to-phase inductance will be lowest with a 3C cable. Second, a three-conductor cable is associated with lower initial installed cable cost. Installed cost includes both the cost of the cable and the cost of installation. While generalizations can be misleading, it can be stated that in many cases the installed cost of one 3C cable will be lower than the installed cost of three 1C cables. Up to some practical limit in weight and size this is true. In some cases, 3C cables rated as high as 138 kV have been installed. The practical limit where the 3C initial installed cost is lower will generally cover all underwater cables 46 kV and below. Moreover, the lower losses just described result in lower operating cost, and thus the operating economics will favor the 3C option. This is particularly true for very long cable systems.

Three-conductor cables have the disadvantage of shorter manufacturing lengths and increased jointing requirements. The ability to supply long unspliced 3C underwater cables is often limited by cabling machine volume and weight capacity. As compared to the three 1C option, there will usually be significantly more joints required in a given long 3C circuit length. This disadvantage is overcome by existing jointing technology for most 3C cable types. It is also offset by the ability to perform a series of ac tests on the relatively short lengths that can be cabled.

In the last decade, significant manufacturing progress has been made in the capability to produce large 3C underwater cables using the oscillating lay approach to cabling the phase conductors followed by unidirectional lay armoring. This has resulted in virtually unlimited length continuous three-conductor underwater power cables that have coiling and handling properties comparable to conventionally cabled designs. In addition to the greater manufacturing efficiency of this technique, it also dramatically reduces the required number of 1C splices.

Also, greater installation tensions are required for 3C underwater cables. This is only a real disadvantage in the case where the cable weight and water depth result in tensions that exceed existing available equipment. In general, a 3C armored cable will have a total weight of 2.5–3 times that of a single-conductor cable electrical counterpart.

13.2.2 Mode as Related to Configuration

The mode of power transmission, ac or dc, is fundamental to the cable design. If the system is to be ac [20–23], then the possible configurations are generally both 1C and 3C cables with options for spares. If the system will be dc, then it is virtually certain that only 1C designs will be considered. Unlike ac cables, there is no electrical advantage to very close proximity of the bipolar dc conductors and, in fact, mutual heating would result in reduced cable ampacity. In a bipolar system (i.e., ± 250 kV dc) two cables are required, and a third (spare) may be specified. Underwater cables for dc operation are usually sized to carry substantial power and thus tend to be large and heavy. Cabling of two dc cables together would be generally unfeasible for handling and installation and would require frequent joints. If such design was used, and if damage or failure occurred to one cable, the other cable may also be made inoperable. Thus, a significant advantage of dc transmission, the ability to operate at half power with only one cable functional, would be lost.

It would be useful here to define briefly the types of cable systems that have been applied for underwater use. These are defined in terms of the type of insulation, i.e., ac or dc.

13.2.2.1 Alternating-Current Systems

Solid paper insulated lead covered (PILC) cables in 1C and 3C configuration can be used up to 69 kV, 60 Hz, rated phase-to-phase voltage. Generally, usage has been at 35 kV or lower and they have provided reliable service. Self-contained oil-filled (SCOF) and self-contained gas-filled (SCGF) cables can be designed in 1C or 3C configuration. As 3C cables, SCOF types have been used up to 138 kV. The SCGF types have been used up to 138 kV in 1C configuration [24, 25]. The SCOF cables are easily the dominant cable type for underwater 60 Hz power transmission above 46 kV, and 1C SCOF cables are used up to 525 kV [26–32]. SCGF cables were used in some systems in the past, but are not commonly supplied now. Below 46 kV the use of self-contained-type cables is not common.

Extruded solid-dielectric cable systems are commonly used at 46 kV and below in 3C and 1C configuration, and 1C extruded dielectric cables have been installed for underwater systems up to 150 kV [33–37]. The most common types of extruded insulations for underwater power cables have been ethylene-propylene-rubber (EPR), crosslinked polyethylene (XLPE), and polyethylene (PE). Most applications have been at 35 kV and lower.

Pipe-type cable or high-pressure oil-filled (HPOF) systems have been used for relatively short (less than 3 km) underwater systems up to 230 kV ac. The use of this type of cable for underwater systems can be expected to provide the same high level of electrical reliability that has been experienced with land-based pipe-type cables. However, in underwater use, these cables are also subject to the same external hazards (anchors, dredges, etc.) that must be considered in selecting any underwater cable system. In such cases, pipe-type cables can be the most difficult and costly to repair in the marine environment. Pipe-type cables are also subject to pulling length limitations and also to limitations regarding possible vertical suspension.

13.2.2.2 Direct-Current Systems

Solid PILC cables have been successfully used in underwater systems rated up to ±250 kV dc, and in the future this may possibly be extended to ±400 kV dc or more [38–45]. With rare exception, these cables have always been supplied in a 1C configuration. This is the dominant type of cable used for underwater dc transmission today.

In 1C configuration SCOF cables are also well suited to dc application, with voltage levels of ±100 to ±600 kV being practical levels for application. The SCGF cables have been used up to ±250 kV dc [46]. For very long cable systems, SCOF cables may not be practical due to hydraulic limitations. Pipe-type cable systems for ±600 kV dc use have been developed [47, 48], but no underwater application of such cable for dc transmission has yet been made. Extruded dielectrics have not been developed and proven for high-voltage dc transmission, and except for a very short section on land for the original England–France cross-channel cable, no high-voltage dc application of significance is known.

Steady technical advances in ac/dc conversion over the last several decades has greatly increased the incidence of long high-voltage dc cable systems [49–85].

13.2.3 Electrical Core Design

In the design and selection of a long underwater power cable system, there are three fundamental economic considerations the designer, and purchaser, should regard. These are reliability, initial cable system cost, and operating costs.

Both supplier and purchaser are very interested in the basic reliability of the system. Particularly in the case of underwater cables, where cost of repair tends to be substantially greater than for land systems, and service outages are also usually longer, the reliability is key to the overall economics of the system. For this reason, for cable design, both suppliers (in designing) and purchasers (in specifying) will be conservative in the sense of using only well-proven technology. Both suppliers and purchasers require well-documented and thorough proof of technical suitability before changes in the technology can be accepted. Such proof is most readily accepted in the form of long years of successful operating experience. For new innovations or extensions of the technology, thorough research and development (R&D) programs are necessary. These R&D programs are essentially designed to prove over a "short" period of months, or (more likely) years, the technical suitability over anticipated decades of use. There are also many other factors, beyond the basic design suitability, that have great effect on cable reliability. These factors include manufacturing techniques, quality control, installation, and propensity of the cable system (due to route location) to external damage.

The second fundamental economic consideration is the initial installed cable system cost. This item is dominated, for long underwater cables, by the cost of the cable. In general the cost of the cable installation will be smaller, and a smaller percentage of the total cost, as the system considered becomes longer and longer. The economy of scale nevertheless leaves the cable itself (except for very short systems) as the most important cost factor. Therefore, in designing and specifying the cable, the objective is to make the cable as economical as possible without sacrificing technical suitability.

The operating costs are defined as the daily costs of operating the cable, and one part of these costs is dominated by the cable design chosen. This part is the cable electrical loss. While no cable system can be 100% efficient and operate with no electrical losses, the designer and specifier of the cable do exercise some degree of control of the level of these losses. Within the range of practical possibilities the selection of system voltage and rated load, as well as the design of the conductor, insulation, shields, sheath, and armor, can be made to improve cable efficiency and lower the losses. In this way some control can be exercised at the time the cable is designed, over the cost of operating losses during the system lifetime. All three of the economic considerations just described have traditionally been important factors. In more recent years, with the great increase in energy costs, more and more attention has been paid to the cost of operating electrical losses. Particularly in the case of long underwater power cable systems, the designer and purchaser must include in the decision process an analysis of the cost of losses over the system lifetime versus the initial cable cost. When this is done, it is sometimes seen that the lowest initial cost cable design option is not the one that will result in lowest overall cost. With these economic considerations in mind, the technical design requirements will now be discussed.

13.2.3.1 Conductor

In previous chapters the options that are available for power cable conductor design were presented and discussed. For underwater cables the same considerations apply and will not be discussed again here. There are a few points that should be considered further as regards conductors for underwater power cables. For very long cables the conductor may need to be oversized in order to bring voltage regulation within an acceptable value. For example, if a 10% voltage regulation is needed, but not quite achieved, and system voltage and load are fixed, one primary means of improving the voltage regulation to the anticipated level would be by increasing the conductor size. However, because both reactance and resistance enter into the voltage regulation equation, voltage regulation change is not linearly inversely proportional to conductor size change. Increasing the conductor size to improve voltage regulation for long ac cables reaches a point of diminishing returns, because the increase in inductance is nonlinear. Also as compared to a nonarmored 1C power cable (operating under the same conditions), the conductor of an armored 1C cable will be somewhat larger. This is due to the induced I^2R losses in the armor, which have the effect of causing the conductor to reach a somewhat higher operating temperature. Therefore, to achieve the proper conductor temperature, the underwater cable conductor size is increased. If nonmagnetic armor materials are used, this effect is reduced. Another important point to consider is that in the case of 1C SCOF cable, which requires a central oil passage within the conductor, the size and hydraulic flow characteristics of this channel are very important. In all cases a sufficient oil pressure must be maintained, and the condition of *oil starvation*, where outgassing is encouraged, must be prevented. In some cases the cross-sectional area of the SCOF conductor must be greater than required electrically in order to meet mechanical requirements. For a given configuration of center-channel conductor, there are both electrical and mechanical lower limits in size.

For high-voltage dc 1C PILC-type cables, it is important that the conductor be designed to reduce the percentage of conductor cross-sectional space. These strand interstice spaces will be filled with the impregnating compound. Regarding the cable lead sheath expansion and fatigue, it is important to keep the impregnating compound volume as low as possible. This is because the compound, when heated by the conductor, is the primary cause of the lead sheath expansion and contraction during operation. It is also desirable to reduce these spaces to a minimum to limit axial flow of compound. To achieve this, in the extreme, a solid metal conductor (100% metal) cross section would be most desirable but impractical to produce or handle. Therefore, to make a conductor practically with minimized interstrand spaces, either a compaction process is used on concentric stranded constructions or the strands are preshaped to a noncircular geometry that is designed to result in the minimization of these spaces. However, practically, even the most effective of these processes still results in such spaces occupying about 3–5% of the conductor cross-sectional area.

Whenever the cable size and/or water depth will result in installation or retrieval tensions that are more than about 10–15% of the cable armor strength, it is good practice to analyze thoroughly the tension and strain that will occur in the cable conductor and sheath. For special dynamic or deep water applications, this is absolutely mandatory. The degree of elongation (strain) occurring in these cable core components must be determined by analysis. With this information, the fatigue of the metallic core components can be determined in order to ensure that the design is mechanically suitable to the application.

For power cables that include optical fibers, the long-term strain imposed on the fibers is critical to determining the long-term reliability of the fibers. The permanent strain on the fibers is directly influenced by the cable design, how it is handled, and what residual strain is added when the cable is subjected to its maximum installation and/or recovery tension. In addition, the fibers may be subject to additional axial strain due to the normal daily current loading of the conductors as well as the unusual additional strain caused by emergency loading conditions. To ensure high long-term optical fiber reliability, the fibers must be proof-tested to stress levels, during their manufacture, that ensure that critical flaws are removed. The designer and specifier of the underwater power cable, that incorporates optical fibers needs to understand the fracture mechanics theory of optical glass fibers and design accordingly. For a discussion on the mechanical properties of optical fibers, the reader is referred to Chapters 15 and 17.

13.2.3.2 Insulation and Shields

The earlier material on insulations and shields is referenced here to avoid redundancy. As specially regards underwater power cables, a number of points are made regarding these important cable components. For extruded power cables for underwater use it is not uncommon for a somewhat thicker insulation wall to be specified, often corresponding to the 133% voltage insulation level. It is not possible to quantify accurately the degree of increase in expected life or reliability that results from such a conservative specification. However, it is only possible to say that the addition of this extra assurance will lower electrical stresses and hence result in improvement of the lifetime or reliability to an unknown degree. In addition, it is common practice to specify that extruded insulations and shields for underwater power cables be applied by modern triple extrusion techniques. Such cables can be manufactured to exacting levels of adhesion and smoothness between the insulation and extruded shields. In underwater extruded power cables this good adhesion is especially important in order to preclude the formation of interfacial voids. This is particularly true for underwater power cables because they typically experience more bends under high mechanical tensions than a typical land-type cable.

The insulation of SCOF cables for dynamic and/or deep-water power cables has some unique requirements. First, the mechanical properties and requirements for the paper tapes in combination with the impregnating oil must be known and well understood. The compressibility of the impregnated paper and oil under the various combinations of hydrostatic, oil, and inward radial pressures due to armor and reinforcement tensions must be carefully analyzed. The transferral of axial loads from the armor to the sheath and conductor is an important area of engineering analysis. In addition, SCOF-type submarine cables for dynamic applications must be designed to withstand repeated bending, under tension, at known bend curvatures. This implies that the long-term characteristics of the paper at the butt spaces must be maintained over the cable life to allow sliding of the paper interfaces without detriment.

In the case of dc applications it must be noted that electrical stresses distribute essentially resistively while in ac cables the stresses distribute capacitively. Capacitive properties change only slightly with temperature, whereas resistance properties change greatly with temperature. This results in the electrical stress (in a dc cable with insulation all at the same temperature) being distributed very much like that in an ac cable. In this state, the maximum electrical stress occurs at the innermost insulation layer. However, when conductor heating occurs, a temperature gradient appears radially across the insulation wall, and the maximum per unit stress may occur elsewhere. Therefore, it is necessary, for dc cables, to analyze the dc stress for every conceivable temperature gradient pattern that can occur to ensure that design levels are never exceeded [86–93]. Greater detail for use in the analysis of dc stress in both steady-state and transient conditions is given in the cited references; an in depth discussion on dc stress distribution phenomena is presented in Chapter 14.

The effective pressure experienced by the cable insulation for PILC cables (which are normally nonpressurized) and for extruded dielectric cables will tend to improve the dielectric strength of the cables. It is well known that, under compression, cable insulations tend to exhibit improved electrical properties. For extruded dielectrics the effect does not appear to be as dramatic as for laminated paper dielectrics. The 60-Hz electrical strength in paper/oil dielectrics is most dramatically influenced by pressure increase, while the increase that can be expected in dc and impulse strength is relatively small. Therefore, in 60-Hz PILC cables a reasonable increase in the dielectric strength of underwater sections is expected to occur as a function of depth, but impulse strength will be practically unaffected.

For SCOF-type cables of the low pressure (nominal 3.5 kg/cm^2) design, the enhancement of 60-Hz dielectric strength in the submarine portion can be substantial. It is known that when the pressure on such a cable is raised from 3.5 to 14 kg/cm^2, the 60-Hz dielectric strength increases by 30–40% (as in a pipe-type cable) due largely to void suppression. In a low-pressure SCOF system pressured on shore at 3.5 kg/cm^2 with oil specific gravity of 0.93, seawater specific gravity of 1.025, and water depth of 137 m, the pressure differential would be about a positive 2.46 kg/cm^2 on the oil with respect to the seawater pressure at 137 m. The operating pressure on the paper/oil insulation at this point would be effectively between about 14 and 16.5 kg/cm^2. Thus, the cable insulation in this SCOF cable at a 137-m depth section would have dielectric properties similar to a pipe-type cable.

In SCOF cables it is standard practice always to maintain a positive pressure with respect to seawater pressure so that if damage to the sheath should ever occur, oil flow outward from the core would keep water out. By oil consumption at the pressurized ends of the cable, the leak would be detected, and generally the cable could remain operational while mobilization to find and repair the leak takes place [94, 95]. In the case of PILC cable the insulation will be at the seawater hydrostatic pressure plus the radial pressure imposed by the armor (underwater power cables are installed with some residual tension constantly on the cable). In the case of PILC cable, a crack in the lead sheath can be expected, in time, to result in an electrical failure due to insulation contamination by seawater. In the case of SCGF cables, due to the absence of an internal pressure supplied by the oil static head, it is necessary to apply very high gas pressures. For the deepest power cable systems under consideration (1830–2140 m), gas pressures exceeding 210 kg/cm^2 would be needed and the sheath reinforcement would be very substantial and expensive. Thus there appear to be practical limits to the SCGF concept regarding depth. On the other hand SCGF systems could, at least theoretically, overcome long-length hydraulic limitations of SCOF cables. The SCGF insulation possibilities at such distances would be limited to dc applications, however, due to the prohibitive ac charging currents that would occur.

13.2.3.3 Sheath Design

Every paper/oil insulated cable must be completely and positively contained within a hermetic sheath. In virtually all long paper/oil-type cable underwater systems that have ever been installed (except pipe-type), the sheath has been made of lead. Although straight aluminum, corrugated copper, and corrugated aluminum sheaths have been studied regarding such application, there appear to be fundamental limitations to the approach. Perhaps the major drawback is the inability to *grip* tightly the cable core with the copper or aluminum sheath and to transfer loads radially from the armor strength member to the conductor. Thus during installation in the vertical plane in even a few hundred feet of water, the cable conductor would be required to support the electrical insulation and itself. The armor would be substantially nonfunctional, supporting itself and the sheath. For deep-water cables this could result in damage to the cable core. Also, for deep-water application, the corrugated sheath would be subject to crushing due to the hydrostatic pressure unless special design features were incorporated. Overall, in the future there may be specific underwater applications where copper or aluminum sheaths may be used, but some fundamental innovations must be devised and developed before this can become a practical reality. Therefore, in this discussion the use of lead sheaths will be the focus of attention.

As described previously (cf. Chapters 2, 5, and 6), lead sheaths are applied by well-proven techniques to all SCOF and PILC cables. For extruded dielectric cables rated above 46 kV, it is not uncommon to specify lead sheathing also in order to exclude moisture from the extruded insulation and thereby reduce the possibility of treeing. At this time opinion is divided in the power cable engineering community regarding the need for hermetic sheaths on extruded power cables. There is no question about the need for such sheaths on all paper/oil cables.

Lead sheaths for underwater power cables have been proven to be suitable for long trouble-free service. All major underwater crossings have used lead sheaths. There are many lead alloys and the alloy selection depends on the required performance. In North America copper-bearing lead sheaths are commonly used for PILC cables. Arsenical lead alloys are sometimes specified where cables will be subjected to certain types of fatigue causing applications. Lead sheaths must also have fatigue properties that allow the cable to be manufactured and installed without acquiring cumulative fatigue beyond a low value [96–99]. Figure 13.3 shows the fatigue properties of various lead alloys as a function of bending strain and frequency. In the case of lead-sheathed cables that must withstand many low-magnitude strains during the cable lifetime, a special lead alloy will probably also be specified. Many considerations enter into the choice of lead alloy type, including manufacturing capability, quality, handling requirements, and performance requirements. The performance of the lead will also be influenced by the jacket and reinforcement that is used.

For long dc underwater cables that are essentially PILC cables, it is important that the cable sheath be designed to withstand circumferential strain cycling. This mechanical requirement involves expansion and contraction of the electrical core within the lead sheath during thermal excursions of the conductor as loads vary. By far the most important influence on this behavior is the impregnating compound, which has a relatively high coefficient of expansion and is not very compressible. When heating occurs, the core expands. Conductor temperature for this type of cable is limited to a maximum of 55°C. The core expansion causes circumferential strain in the lead sheath. When the cable cools after the load current is changed, the core contracts and some or all of this strain is relieved. Although this sheath behavior is of a low magnitude, it is possible that

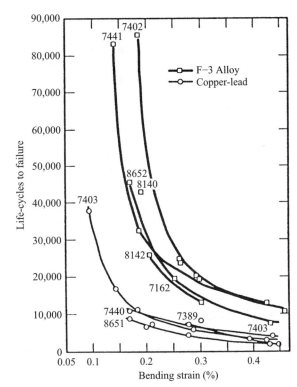

Figure 13.3 Lead sheath fatigue characteristics (after [96]).

over years of conductor thermal excursions circumferential fatigue can occur and result in axial cracking of the sheath. To limit the sheath strains to levels that will not result in such destruction, the thermal behavior of the cable must be well understood by the designer, and sound design principles must be used to specify the lead sheath. The designer must also be sure to specify a lead alloy that can be produced uniformly for practical long-length manufacture.

13.2.3.4 Reinforcement and Jacket Design

Although lead sheaths have an unusually good resistance to corrosion, it is necessary to cover the sheath with a jacketing material to provide both electrical isolation from the armor and sea and to provide mechanical protection. In all SCOF cables and in some PILC cables, a sheath reinforcement is also added. This reinforcement may be in the form of metal tapes or synthetic reinforced tapes with neoprene or other material. If metal tapes are used, they may usually be applied at angles of 55° to 65° in such a way that they do not overlap but have butt spaces designed to accommodate cable bending. A second metal tape will be applied to center over the butt space of the first tape. If the tapes are applied directly over the lead sheath, it is always necessary to apply suitable bedding tapes over the lead to prevent the metal tapes from damaging the lead sheath during bending.

The design must provide a practical means of continuous electrical interconnection between the lead sheath and the metal tapes. In effect the metal reinforcing tape in this location must be grounded continually to the sheath to prevent arcing due to a potential difference in sheath and metal tape. The use of a conductive bedding tape may also be advisable between the metal tape layers to prevent or reduce abrasion. Over the metal tapes a taped and cured jacket of neoprene or an extruded jacket may be used. This jacket must have sufficiently high electrical strength to withstand both the standing voltage that occurs on long insulated cable sheaths and higher magnitude transients that may occur during switching operations.

On very long underwater 1C cables it is necessary to provide sealed shorting connections, periodically along the cable length, between the lead sheath and the sea/armor wires [100]. These connections must provide a means of shorting out the sheath to the effective seawater/armor ground. At the same time the connection must not allow seawater to enter beneath the jacket and cause corrosion. For a given cable construction the spacing of these sheath shorting connections can be derived from the following formula:

$$V_{j\max} = V_d \left(\frac{C_d}{C_d + C_j} \right)^{(1-\exp[-Bx])} \tag{13.1}$$

where

$$B = \tfrac{1}{2} v (C_d + C_j) R_s \tag{13.2}$$

Here x is the distance from feeding-in point, $V_{j\max}$ is the maximum voltage across the jacket, V_d is the voltage between conductor and sheath, C_d is the capacitance (in farads) between the conductor and sheath, C_j is the capacitance (in farads) between the sheath and armor or sea, v is the propagation velocity of the insulating material, and R_s is the resistance of the metal sheath in ohms. Such connections are required on both ac and dc

1C cables of very long length. On 3C cables operating with various degrees of unbalanced loading, it is also possible to develop substantial sheath voltages on very long lengths. However, these voltages are not usually high enough to endanger the jacket and to require sheath-to-sea shorting connections.

Sheath-to-sea connections must be both adequately designed and properly installed if they are to perform their function effectively. Inadequate connections can result in a potential danger to the cable integrity, which could be more hazardous than the high sheath-to-ground voltage that they are intended to prevent. Basically, such connections are meant to prevent a dielectric breakdown of the sheath protective jacket. When such a puncture occurs, then a path will be created for sheath currents (I_s) to flow to ground through some resistance (R). Thus the possibility of a *hot spot* is created. If the $I_s^2 R$ values should fall within the critical value range, this could result in serious damage to the lead sheath.

Another means of applying reinforcement entails the use of fiberglass, Kevlar (a duPont trademark for aramid fiber synthetics), or other nonmetallic material, in the form of a fabric tape. Generally such materials are applied as embedded components of a neoprene (or other material) taped and vulcanized jacket. These types of reinforcements have been used successfully on some significantly long low-pressure SCOF underwater cable systems [101]. Because such materials are nonmetallic, the problem of shorting them to the sheath is eliminated. However, such jackets still require sheath to sea to armor shorting connections in long cable systems.

Still another way of applying metal sheath reinforcing tapes is by first applying the extruded or taped jacket followed by the metallic reinforcement tapes. This procedure uses the jacket itself as bedding for the metal tapes and will tend to eliminate any chance of the metal tapes causing mechanical damage to the lead sheath during bending and handling. While it would be possible to apply another jacket over these reinforcing tapes, to protect them, it is not usually done. The seawater is usually free-flooding up to the sheath jacket. The reinforcing metal tapes must be designed to resist corrosion in this case. Of course, in this configuration for reinforcement, it is also necessary to provide the special sheath to reinforcing tape/sea/armor connections across the jacket. On nonpressurized cables (PILC and solid-dielectric extruded types) jackets are also applied over the lead sheaths. These jackets are often specified according to the Insulated Cable Engineers Association (ICEA), the Association of Edison Illuminating Companies (AEIC), or the International Electrotechnical Commission (IEC) standards. The specified thicknesses are compared in Table 13.2. Also, in Table

TABLE 13.2 Average Thickness of Sheath Coverings in Accordance with AEIC Specifications

				Reinforced Neoprene			
Calculated Approximate Diameter over Sheath		Polyethylene Thickness		Neoprene Compound Thickness		Total Covering	
mm	in.	mm	in	mm	mils	mm	mils
0–38.10	0–1.500	2.03	80	1.02	40	2.03	80
38.11–57.15	1.501–2.250	2.29	90	1.52	60	2.79	140
57.16–76.20	2.251–3.000	2.79	110	1.52	60	2.79	140
Over 76.20	Over 3.000	3.18	125	1.52	60	3.56	140

Source: From [4, 5].

13.2 the polyethylene jacket thicknesses for jackets over non-lead-sheathed cables are shown. In general, extruded cables operating at 60 Hz voltages 46 kV and below do not require a hermetic sheath. It should be understood that there is no universal agreement on this subject. In any case the provision of an extruded jacket over extruded cables will benefit the cable by reducing the corrosive effects of seawater on the cable metallic shield. It has been seen on communication cables recovered after decades of deep ocean service that the extruded polyethylene outer jacket has preserved the copper shield in virtually original condition.

The same can be expected for extruded power cables. It should be understood, however, that only a metallic material will provide a hermetic sheath. Any polymer will transmit water vapor. The debate over the need for hermetic sheaths on extruded insulation cables is likely to continue for many years. Whether or not such sheaths are used, it is generally desirable to include an extruded cable jacket. For convenience, Table 13.3 provides relationships for some common metric measures.

13.3 POWER TRANSMISSION REQUIREMENTS

This section will generally describe power transmission requirements and design characteristics as they uniquely apply to underwater cables. Chapter 11 has already provided an excellent discussion on the subject of cable transmission of electrical power.

13.3.1 System Power Transfer Requirements

The load that underwater cables must carry is, of course, the fundamental consideration for cable design. The main difference that may occur with underwater power cables, as compared to land-based systems, is that of system length. Most land-based cable systems do not exceed a few kilometers in length, whereas it is not unusual for underwater systems to exceed 20 km and systems of 250 km and longer are being considered. Mainly, this long length impacts upon the cable design in regard to voltage regulation and losses. In very long ac systems, the capacitive charging current becomes so great as to limit the circuit length.

13.3.2 Selection of Voltage and Mode

The voltage for long underwater crossings is based upon the load, voltage regulation needs, and the particular system requirements. For both ac and dc systems, the most effective power transmission will be obtained by use of the highest voltage prac-

TABLE 13.3 Some Common Equivalents

Metric Unit	English Unit
1 mm^2	1.55×10^{-3} in.^{2} = 1.9735 kcmil (thousand circular mils)
1 mm	3.937×10^{-2} in.
1 m	39.37008 in. = 3.2808 ft
1 km	3280.84 ft = 5.3899×10^{-1} nautical miles
	(1 nautical mile = 6087 ft)
	= 6.21371×10^{-1} statute miles (1 statute mile = 5280 ft)
1 kg/m	0.6720 lb/ft
1 W/m	3.084×10^{-1} W/ft
1 kg/cm^2	14.2233 psi

tically possible for the selected type of cable [56, 61, 67, 72, 77, 102–119]. The practically possible voltage will be based upon the proven reliability of the cable type at that level and upon the power transmission system that exists, or will be designed, at both ends of the cable system. In some cases it is desirable to transform the underwater cable voltage to a particular level to provide cost-effective underwater cable operation.

In general, the transmission efficiency will vary nearly directly with the square of the ratio of voltage change. For example, the operating efficiency of a 69 kV cable system designed to carry the same mega-volt-ampere (MVA) load as a 34.5-kV cable system will be approximately equal to the ratio $(69^2/34.5^2)$ or four times improved.

The system designer must consider not only efficiency but also reliability and initial installed cable system cost. In addition the cable system designer must consider reactive compensation, as needed, to provide for capacitive charging current in long high-voltage ac cables. Considering long underwater crossings, studies show that for each ac voltage class there are limits to the length over which efficient power transmission can be obtained. Therefore, the system designer must choose the cable type and voltage level to provide the most cost-effective underwater transmission system within practical limits.

For dc transmission, there are no apparent limitations regarding length. However, as with ac systems, it is always more efficient to transmit more power by using a higher-voltage lower-ampacity system rather than a lower-voltage higher-ampacity system. Therefore, for underwater cable systems the highest voltage for the selected cable type, consistent with reliability, should be selected. Transmission at direct voltages is not generally economical for short systems, because the cost of ac/dc and dc/ac conversion plants at each end of the underwater crossing is relatively high. For longer systems dc becomes very economically attractive. For very long systems, where the limits of possible ac systems are exceeded, dc transmission becomes the only viable choice.

13.3.3 Ampacity and Electrical Losses

Ampacity is the load current that a given cable can carry under a given set of conditions without exceeding a specified conductor temperature. Nearly all electrical losses in both ac and dc systems are fundamentally I^2R losses (in watts per unit length) that result from current flowing in the metallic components of the cable. In underwater cables of the 3C ac type, the losses occurring under balanced loading of the three phases are reasonably comparable to those in an unarmored land cable. In effect the loss currents in the shields and armor of such a cable are nullified by the counteracting fields of the load currents of each of three phases.

In 1C ac cables, however, the losses are characteristically much higher and in some cases can be prohibitively high as a percentage of total power transmitted. In such cases the loss currents in the armor become significant. The effect may be reduced, however, by raising the design voltage and lowering the design current. The armor and sheath loss currents will essentially vary directly with the magnitude of the load current. Therefore, this approach will result in lower percent losses. Another approach is to use nonmagnetic armoring materials, such as copper, bronze, aluminum, or stainless steel to achieve significantly lower armor losses. However, as will be discussed later, such an approach is often not both technically and economically feasible. A third approach is to vary the cross-sectional areas of the conductor, shields, and armor (within practical limits) to reduce the losses. This approach can often be used to find ways of providing significant reductions in operating losses.

Figure 13.4 illustrates the radial thermal resistances that influence the ampacity of a typical underwater power cable. All thermal resistances may be derived from the thermal resistivity expressed in degrees Celsius-centimeters per watt. As with any power cable, the resistance to radial heat flow, outward from the conductor (as well as radially outward from any metallic component that is producing heat from its own loss currents) of each layer internal to the cable and all external material(s) around the cable, determines the temperature that the conductor and cable components will reach with a given load current. The symbol identification included in Fig. 13.4 lists the various thermal resistances that are typically taken into consideration in performing ampacity calculations for underwater power cables. In general, metallic component layers have a relatively insignificant contribution to reduction of ampacity due to their very low thermal resistance (high thermal conductivity). Cable insulations and jackets always have relatively high thermal resistivity as do any other components that are nonmetallic. Armor bedding, interserving and outer serving layers will have lower thermal resistivity if they are allowed to be free-flooding. To obtain cable designs with the greatest ability to dissipate heat radially, air or gas entrapment between any component layers should be eliminated or minimized. Still air has a thermal resistivity that is about one order of magnitude higher than that of cable dielectrics. Another important precaution for underwater power cables regards the actual thermal resistivity of the material composition surrounding the cable. There may be a tendency for the engineer to presume a certain maximum composition thermal resistivity based upon the premise that it must be relatively low because the material is completely saturated with water. This may not in fact be true because of two distinct possibilities. First, certain suspensions, organic suspensions, and muds have very high thermal resistivity in spite of the fact that they are saturated with water [120]. Second, the "thermal bottleneck" of the cable system may exist at the shore landing due to material composition, less than full water saturation, and temperature. It is advised that if any uncertainty exists about the true thermal resistivity along the underwater power cable route the true values should

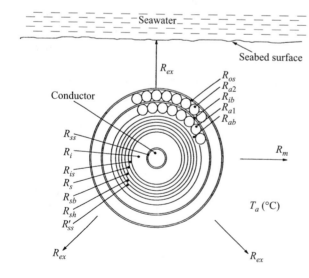

Figure 13.4 Illustration of radial thermal influences in a typical underwater power cable. Thermal resistances: R_{ss}, semiconducting shield; R_i, insulation; R_{is}, insulation semiconducting shield; R_s, shield; R_{sb}, shield binder; R_{sh} sheath (i.e., lead); R'_{ss}, sheath supportive jacket; R_m, effective added thermal resistance due to proximity of another cable; R_{ab}, armor bedding; R_{a1}, armor layer 1; R_{ib}, interbedding (or interserving); R_{a2}, armor layer 2; R_{os}, outer serving; R_{ex}, external composition (e.g., sand, mud). Ambient temperature of external composiiton (°C), T_a.

be determined. In the long run it is less expensive to perform actual samplings and measurements to determine the true values along the route as compared to the cost of replacing a cable damaged by overheating.

Methods to calculate ampacity [121–124] have been modified to include the effects of armor on cable ampacity. Power transmission cables used under direct voltages are inherently more efficient due to the absence of sheath and armor losses. All of the significant electrical loss in dc cables is due to I^2R heating of the conductor. The only other loss is the dielectric loss current that flows from conductor to shield, and this is an insignificant loss. In addition, the charging current is very small, as compared with ac cables. Dielectric loss in ac systems usually is not among the most important losses. In extruded cables, this loss is generally insignificant. In paper/oil cables the dielectric loss generally becomes significant only in the higher stress designs. Dielectric loss in these cables is related to the operating temperature, since the dielectric loss current increases substantially at the higher operating temperatures.

13.3.4 Additional Shield and Sheath Design Considerations

The sizing of the shield and sheath, which act together as the electrostatic shield and ground (or neutral) of the cable, must be carefully considered. There are no particular areas where the fundamental electrical design of these components is usually different from land cables. Early in the design work, for practical manufacturing and handling reasons, a lead sheath thickness is selected, usually in line with industry standards such as AEIC 1 and 4 [4, 5]. This thickness is selected based primarily on the past experience where it was found that buckling of the lead sheath would occur if the lead sheath thickness was not in proper ratio to the cable diameter. From the electrical point, such sheaths generally possess adequate electrical conductance to (in parallel with the insulation shielding tapes) perform the functions of a grounding component. As with any grounding component, the lead sheath is limited by the duration of time that it can sustain a fault current without fusing or exceeding a specified maximum sheath operating temperature.

In normal operation the sheath and shield in both ac and dc systems will operate at a voltage that is above ground potential. The magnitude of this voltage will be a function of the impedance and the current involved in the shield/sheath. For very long underwater systems this voltage is kept to an acceptable level by suitable grounding of the sheath to the sea/armor at periodic intervals along the length.

For the shore landings of underwater cable systems, the land and air sections will often have lower ampacity than the underwater portions. To increase this ampacity, the cable armor is sometimes removed. Also the armor and shield may be grounded at or below the water line, and the land and air section of the cable shields may be operated with open-circuit shields. This is accomplished by isolating the shield at the termination end from ground. This eliminates the circulating shield current and greatly improves the ampacity of these sections. However, the engineer must ensure that there is no possibility of personnel contact with this non-directly-grounded shield end section because standing and transient voltages on the shield would pose an additional safety hazard. If calculations indicate that the ampacity cannot be raised sufficiently by these measures it is then necessary to size the cable differently for the land/air sections and provide a transition joint to the underwater portion.

13.3.5 Additional Factors Regarding Cable Sizing and Losses

Spacing of 1C cables is particularly important for submarine cables in that substantial spacing is the rule. These cables will usually fan out from the shore landings and be spread, typically, between 15 and 150 m apart. The spacing is influenced by the water depths since this also affects the amount of spacing that is possible. For 1C ac cables, mutual heating reduces, and cable losses tend to increase rapidly as spacing increases from 0.3 to 3 m. Then, from 3 m to 30 m these effects change more gradually. Beyond 30 m there are very slight increases in losses with increased spacing.

Shore end sections are often trenched-in both on land and to some point below mean low water. This is to provide additional protection to the cable and inaccessibility to the public. In order to calculate the ampacity of these sections, the thermal resistivity and temperature can usually be readily determined. The cable designers must also consider the untrenched cable lying on the bottom regarding possible burial. They must determine first what depth of burial is likely or possible due to shifting bottom sediments and must also determine the thermal resistivity and temperature of this material. The ampacity of such sections can then be checked. For most ocean bottom sediments thermal resistivities of 50–90°C-cm/W have been used. In some underwater environments such as lake bottoms, canals, rivers, and some ocean locations, thermal resistivity can be very high, even exceeding 200°C-cm/W. When there is any serious uncertainty about the properties of the sediments along the proposed cable route, it is prudent to perform a thorough bottom route survey including core samplings and thermal resistivity measurements.

The temperature of the underwater surrounding is also a great influence on the cable ampacity. This ambient temperature for the underwater environment will be as low as 3–5°C at depths of 1200 m and lower and generally 10°C or lower at depths exceeding 365 m. Temperatures of surface ocean waters rarely exceed 30°C, even in the Tropics. In inlets or very gradually sloping shore landings, the shallow water is usually the warmest ever encountered. It can be seen from this that the deepest portions of an underwater cable crossing are generally most thermally favorable to the cable. Therefore, as a result, the ampacity limitation is usually in the air, land, or shore burial sections of an underwater cable. At these locations the higher ambient temperature, higher soil thermal resistivity, closer spacing, and burial condition all work in the direction of limiting cable ampacity.

The load factor is, of course, an important influence on the cable sizing. Cables that are designed for 50%, 75%, or some other load factor below 100% can be smaller and less costly. If load patterns are well known, and will remain as established, it may be possible and desirable to size the cable accordingly.

System length can impose some important limitations on the electrical performance of underwater ac cables. It should first be understood that the calculation of ampacity for various sections of the cable only determines the current that can be carried at a specified conductor temperature. Although the ampacity of all sections may satisfy the load requirements at the stated voltage, it should be realized that the cable design may still not be desirable. This is possible because voltage regulation may not be within desirable limits. Voltage regulation is defined as the change in voltage, as a percentage of system zero-load voltage, which occurs at the load end of the system during full load. This is generally not a problem in typical fairly short length land systems. In long underwater cable crossings, it can become significant because both cable reactance and resistance are directly proportional to system length. For a given load current and system power factor, the voltage drop that occurs will be independent of the rated voltage level. Thus, if the voltage level is not high enough to begin with, the voltage regulation can be an unacceptably high percentage. The solution is to specify an appropriate high-voltage level. Not only will the higher voltage allow the voltage regulation to decrease, but the load current can be reduced and thereby losses can be reduced and (possibly) a smaller conductor size can be used. The increased voltage cable selected will likely have higher capacitive charging current dependent upon the cable geometry, insulation dielectric constant, and voltage selection. The capacitive charging current per phase in amperes per kilometer is given by

$$I_s = \frac{0.34768 \, Vf \varepsilon'_r}{1000 \ln(D/d)} \tag{13.3}$$

Here, V is the phase-to-neutral voltage, f is the frequency in hertz, ε'_r the dielectric constant of cable insulation, D is the diameter over the cable insulation, and d is the diameter under the cable insulation. The charging current in very long ac systems can be of sufficiently high value to limit the system length. The charging current adds vectorially to the load current. No difficulties are to be anticipated, as long as the resultant current is at an acceptable level. In general, the capacitive charging current problem is not commonly encountered and can be modified by design. It is also possible to compensate for some, or all, of the charging current by the use of inductive reactance at the sending end.

13.4 ARMOR AND EXTERNAL PROTECTION DESIGN

13.4.1 Design Considerations

The primary functions of the armor are to provide strength to the cable and to provide protection against external damage. The degree of strength provided in the armor of conventional cables that are installed in conventional depths is generally anywhere from 3 to 20 or more times the strength actually required for installation. Often, due to the dimensional requirements for obtaining adequate armor coverage with round armor wires, cable designs have far more steel armor than is required to provide adequate strength. However, the protection of the cable core, provided by the armor, is certainly as important as strength in most cases. Therefore, even though a cable may have far more strength than required, using a reasonable factor of safety, the given amount of armor may be desirable from the protection standpoint. When conventional cables are designed using industry standards (ICEA) and are for relatively shallow depths (150 m or less) the above generally applies. For all underwater cable design, the armor design needs to be coordinated with the overall system protection against external hazards to the cable [17, 125–137].

When the cable design extends to specialty applications such as dynamic and/or deep-water power cables, it is extremely important that the mechanical behavior of the cable be understood. This requires a complete analysis of the cable mechanical properties during handling, installation, operation, and recovery. Examples of two dynamic deep-water power cables are shown in Figs. 13.5 and 13.6. Power cables that include

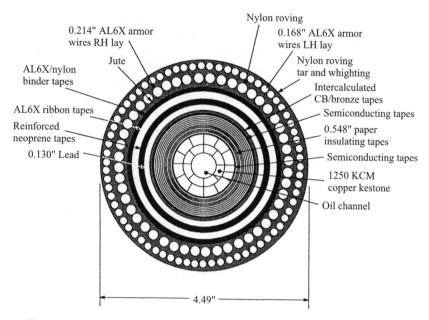

Figure 13.5 Dynamic application 138 kV [ocean thermal energy conversion (OTEC)] self-contained oil-filled prototype cable (RH, right hand; LH, left hand; CB, carbon black).

optical fibers should also be treated as specialty applications from the design and reliability viewpoint. A brief discussion of understanding and controlling the fiber strain is given as the last paragraph of Section 13.2.3.1.

The modulus of elasticity (Young's modulus) of the armoring material, as well as the cross-sectional area of the armor material, are two of the most important factors in determining the mechanical behavior. Young's modulus is defined by the relation

$$\text{Young's modulus} = \frac{\text{stress}}{\text{strain}} = \text{Modulus of elasticity} \tag{13.4}$$

as an example, we note that for steel

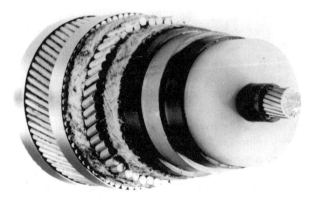

Figure 13.6 Dynamic application 138 kV (OTEC) crosslinked polyethylene prototype cable. (Courtesy of Simplex Technologies, Inc.)

$$\text{Young's modulus} = \frac{7030 \text{ kg/cm}^2}{0.00344861} = 2.0385 \times 10^6 \text{ kg/cm}^2$$

Thus steel under 7030 kg/cm^2 stress will exhibit a 0.00344861 unit strain (0.344861%). It should be understood that specifying a higher cable breaking strength does not necessarily mean that the cable will experience lower strain (elongation). For example, there are many steels available with nominal tensile strengths in the range of 50,000 psi (3515 kg/cm^2) to over 300,000 psi (21,090 kg/cm^2). Two cable designs, with steel armor strength as the only variable, would exhibit very nearly the same strain characteristics at a given elastic range tensile loading. This is because the Young's modulus of both is very nearly the same. Therefore, for a given type of armor material, the design is constrained to increasing armor cross-sectional area, changing the angle of application, changing the Poisson's ratio of the cable, or any combination of these, in order to reduce the strain (elongation) of the cable design. Typical values for Young's modulus are given in Table 13.4.

The Poisson's ratio of the cable (and of each layer) is an expression of the change in cable diameter as a function of mechanical axial tension. It is applied in the formula: $d' = (1 - V\varepsilon)d$, where d' is the diameter under the new tension, V is Poisson's ratio, ε is the unit axial strain, and d is the original diameter.

Example: Using the strain or elongation of 0.00344861 from the preceding example for a cable of 10 cm original diameter and assuming the applicable Poisson ratio is 4.0, gives $d' = [1 - 4(0.00344861)]10 = 9.8621$ cm. Thus a reduction in core diameter of about 1.4% is seen to take place.

Based upon experience, the designer estimates the Poisson's ratio for a given cable construction. Where the value is critical to the verification of the analysis of behavior, the Poisson's ratio must be determined or verified experimentally.

The lay angle (α) is given as

$$\arctan \alpha = \frac{D_p \pi}{\text{lay length}} \tag{13.5}$$

where D_p is the armor pitch diameter (the diameter of the center point of the armor layer). For conventional armored cables the lay length of the armor wires is specified by ICEA to be 7–12 times the pitch diameter. Therefore, the armor lay angle α may be expressed independently of the actual pitch diameter as

$$\arctan \alpha = \frac{\pi}{M} \tag{13.6}$$

TABLE 13.4 Some Typical Values for Young's Modulus

Material	Young's Modulus
Steel	29×10^6 psi (2.0389×10^6 kg/cm^2)
Copper	17×10^6 psi (1.1952×10^6 kg/cm^2)
Bronze	13×10^6 psi (9.1399×10^5 kg/cm^2)
Aluminum	10×10^6 psi (7.0307×10^5 kg/cm^2)
Kevlar	$11–19 \times 10^6$ psi ($7.7338 \times 10^5 - 1.3358 \times 10^6$ kg/cm^2)
Lead	$0.7–1.0 \times 10^6$ psi ($4.9215 - 7.0307 \times 10^4$ kg/cm^2)

where M is the lay length multiplier $7, 8, \ldots, 12$. From this, the typical range of lay angles in conformance with ICEA are given in Table 13.5.

Armor lay angle is, of course, an important factor in the strain that will occur in the cable. By considering the armor as a helical spring, it is possible to envision more clearly the effect of lay angle on the cable armor and cable core strain. It should be stated at the beginning that in every case when an armored cable is placed under mechanical tension, the load is shared by all components to some extent. To look at extremes, the total cable strain for both armor and cable core would be identical if the armor lay angle were $0°$. In this case the percentage of total cable tension taken by the armor would be maximum, but also, a part of the total would be taken by the sheath and conductor. At the other extreme, if the armor lay angle were $89°$, the armor would be a spring shape; and with tension applied, the armor would offer extremely little resistance to elongation; the core would stretch according to its modulus of elasticity and would take the greatest proportion of the tension. While this example is simple, it can be seen that the lay angle has a great effect on the elongation that will be imposed on the cable core. The standard lay angles based on 7–12 times the pitch diameter prove to be satisfactory for conventional cables that are not under high tension or installed in very deep water. As long as the levels of strain that occur on the metal components of underwater cables during installation and retrieval result in mechanical stresses well within the known yield points, this conventional design approach is satisfactory.

Strain in cable components is an important consideration. If the stress on the armor exceeds the yield point, this means that the mechanical behavior of the cable has been permanently altered and that the armor could fail. In all cases, by design, this must be avoided. Strain in other cable components is generally designed to be well within the yield point for static conventional underwater cables. If the cable is designed for a special use, it may be necessary even to design the strain on certain cable (not armor) components to nearly reach the yield point, but this must be carefully analyzed and well understood in each particular design case.

Underwater power cables that include optical fibers must be designed to ensure the long-term reliability of the fibers (see last paragraph of Section 13.2.3.1). Unlike metals, glass fibers fail by a phenomenon known as static fatigue (cf. Chapter 15). This has to do with a lifetime summation of fiber strain versus time, which produces failure at a critical glass surface flaw (submicron level in size) if the lifetime summation of strain exceeds a certain value. These critical flaws are eliminated at fiber manufacture by screening 100% of the fiber through a predetermined stress level mechanical proof test.

TABLE 13.5 Armor Lay Angle
as a Function of M

M	α
7	24.17
8	21.44
9	19.24
10	17.44
11	15.94
12	14.67

Poisson's ratio is also an important factor in the assessing of the load distribution and strain in cable components. If the cable is totally noncompressible as tension is applied (a nonreal example), then Poisson's ratio would not apply. In real cases, however, radial dimensional changes do occur as any cable is placed under increasing tension. This results in a decrease in the pitch diameter and an increase in the cable core strain. The Poisson's ratio is primarily influenced by the compression of the cable components, the level of tension, armor cross-sectional area, and armor lay angle. The armor lay angle is the primary factor in the level of radial (inward) force that the armor wires impart to the core. For design estimation, this radial armor wire force is approximately proportional to $(\sin \alpha_1^2 - \sin \alpha_2^2)$ when considering two design lay angles.

13.4.2 Armor Size and Specification

The sizing of armor wires is usually specified by reference to the Birmingham wire gage (BWG). For example, the diameters of some common BWG sizes are given in Table 13.6.

This gage system is the same as the Old English Gauge System but is distinctly different from the American wire gage (AWG) and the (British) Standard Wire Gauge (SWG) system. The numbering method in this system (as in the different gage systems) is in inverse order to the wire size, thus a larger number identifies smaller wires. The reason for this is that the gage sizes are in a logical sequence of draw-down operations (from wire rod).

It is not uncommon to specify armor sizes that are not standard BWG. Special sizes are generally readily obtainable and add little to the cost. For most conventional power cables ICEA guidelines are used, as are laid out in Table 13.7.

In addition, whether a standard steel wire is used or not, standard practice is to specify galvanized steel with zinc coating in conformance to ICEA as shown in Table 13.8. The breaking strength of wire is usually specified as a minimum acceptable value. It is common for actual supplied wire to somewhat exceed this value. The designer specifies the wire strength according to the needs of the particular underwater cable. In general, the highest strength wires (300,000 psi or 21,090 kg/cm^2) can be obtained in the smaller sizes, but the larger size wires can also be obtained in strengths of 200,000 psi (14,060 kg/cm^2) or more. The cost of high-strength wires is somewhat higher.

TABLE 13.6 Sizing of Round Steel Armor Wires

BWG Size No.	Diameter	
	mm	in.
00 (2/0)	9.652	0.380
0	8.636	0.340
1	7.620	0.300
2	7.214	0.284
3	6.579	0.259
4	6.045	0.238
5	5.588	0.220
6	5.156	0.203
7	4.572	0.180
8	4.191	0.165
10	3.404	0.134
12	2.769	0.109

TABLE 13.7   NEMA/ICEA Guidelines for Submarine Power Cable Armor[a]

(a) Size of Galvanized Steel Armor Wire for Submarine Cable

Calculated Diameter of Cable under Jute Bedding		Nominal Size of Armor Wire		
in.	mm	BWG	mils	mm
0–0.750	0–19.05	12	109	2.77
0.751–1.000	19.08–25.40	10	134	3.40
1.001–1.700	25.43–43.18	8	165	4.19
1.701–2.500	43.21–63.50	6	203	5.16
2.501 and larger	63.53 and larger	4	238	6.05

(b) Thickness of Jute Bedding for Armored Cable

Calculated Diameter of Cable under Jute Bedding		Minimum Thickness of Jute Bedding (Round Wire Armored Cable)					
		Metal Taped		Sheathed or Jacketed		Nonsheathed or Nonjacketed	
in.	mm	mils	mm	mils	mm	mils	mm
0–0.450	0–11.43	30	0.76	45	1.14	80	2.03
0.451–0.750	11.46–19.05	45	1.14	45	1.14	80	2.03
0.751–1.000	19.08–25.40	45	1.14	65	1.65	95	2.41
1.001–2.500	25.43–63.50	65	1.65	80	2.03	110	2.79
2.501 and larger	63.53 and larger	65	1.65	95	2.41	125	3.18

[a] NEMA: National Electrical Manufacturers Association.

TABLE 13.8   NEMA/ICEA Guidelines for Zinc Coating and Wire Size Tolerance

Nominal Diameter of Coated Wire			Minimum Weight of Zinc Coating	
			Exposed Wire Surface	
mils	mm	Size, BWG	oz/ft^3	g/m^2
238	6.05	4	1.00	305
220	5.59	5	1.00	305
203	5.16	6	1.00	305
165	4.19	8	0.90	275
134	3.40	10	0.80	244
109	2.77	12	0.80	244
83	2.11	14	0.60	183

Mandrel Diameter for Adherence of Coating Tests			Tolerances in Diameter, Nominal Diameter of Coated Wire		
Wire Diameter		Mandrel Diameter			in. (tolerance)
mils	mm		mils	mm	
238–134	6.05–3.40	3× wire dia.	238–166	6.05–4.22	±0.005
133 and smaller	3.38 and smaller	2× wire dia.	165–109	4.19–2.77	±0.004
			108–65	2.74–1.65	±0.003

13.4.3 Armor Material

Galvanized steel is by far the most common underwater power cable armoring material. However, there have been important crossings where aluminum [101, 138, 139] and copper [28, 29] have been used as the armoring material. In these cases the cables have been 1C ac types where reduction in the cable core size (as compared to the use of ordinary magnetic steel wires) and in the cost of cable electrical losses could be realized by the use of such nonmagnetic armor materials. Special steel alloys have been evaluated for use in dynamic deep-water applications [11]. The use of ordinary steel for such applications would be technically unfeasible due to corrosion in the dynamic environment.

The use of synthetic fibers, such as Kevlar, for underwater power cables has not yet become practical. While Kevlar has a strength-to-weight ratio 1.25–7.5 times that of common steels, the cost per unit weight presently is in the range of 18–30 times that of steel. While Kevlar can provide the strength required, it has little protective value such as against impact. It is likely that Kevlar will continue to be extremely attractive for lightweight electromechanical and optical-electro-mechanical power cables.

The choice of armor material is constrained by economic considerations, corrosion properties, available strength, and Young's modulus of the material. Steel will likely continue to be the material of choice for the vast majority of underwater power cables due to relatively low cost and the high strength-to-weight ratio.

13.4.4 Armor as Protection

The armor also provides protection to the cable core. The armor directly protects the jacket. On lead-sheathed cables, for example, the jacket is both an electrical insulation and an anticorrosive protection for the sheath. The jacket must itself be protected against severe abrasion or puncture such as could occur on sharp sea-bottom rocks. The armor gives this protection. It also protects the cable core against anchors, dredges, and other hazards. The armoring affords a substantial degree of impact protection, but no practical submarine cable can be totally protected against, for example, large ship anchors. It is possible, of course, to supply two or more layers of armor, and this is indeed done for especially hazardous routes. It is usually possible to make some estimate of the probability of damage. Past history may prove useful.

The usual construction of submarine cables involves the use of bedding of a treated jute or synthetic material. This bedding primarily acts as a cushion and prevents the steel wire from damaging the underlying jacket or sheath. The steel wires, applied under tension, sink down into this compressible material. In some cases, an asphalt compound is also applied at this stage. It is a well-proven procedure to apply an asphalt compound to the cable at the closing die, then to apply one or more layers of synthetic overserving such as nylon. These layers are then typically saturated with asphalt compound so that the final surface of the cable is cohesive and black. In addition the tar/synthetic surface provides a coefficient of friction that is much higher than the bare serving. The surface then has good properties for gripping the installation cable drum or linear engine track. This tough, resilient layer also provides important protection to the galvanized armor wire. Without this protection, the armor wires would become abraded during handling and installation. A substantial amount of zinc would be removed and the steel corrosion protection would be reduced.

Armor protection is important for cables that require long-term mechanical protection and the ability to be recovered. The use of galvanized steel wire, which constitutes standard practice, is almost always one means of corrosion protection. Table 13.8 shows some industry guidelines regarding galvanizing.

Although the subject of corrosion is a complex one, a brief review of some of the fundamental mechanisms of corrosion may be useful here. The electrochemical theory perhaps best explains steel corrosion in underwater cables. Recall that when any metal forms the electronegative element (anode) of an electrolytic cell it is attacked or corroded and passes into solution. The presence of both oxygen and water is necessary for corrosion of steel to occur at usual temperatures. For deep-ocean cables, the low oxygen content of the water is favorable to the steel, and low corrosion rates are the rule. However, cable in shallower water and on land is more subject to corrosion. Table 13.9 shows the galvanic series of alloys in seawater.

Any metal in the table is electropositive to any element following it and electronegative to any element preceding it. When bare steel is placed in seawater, it will usually be anodic. When the steel is galvanized, the zinc coating becomes the anode and the steel is the cathode. Thus the zinc, through slow dissolution, prevents or delays the steel corrosion. When the armor is also coated with tar and protected from the effects of water-borne sediments by synthetic outer coverings, the corrosion is stopped or greatly reduced. In some cases, where it is possible to isolate the armor wire sufficiently from the seawater ground, a slight negative dc potential can be applied to the armor and thereby force the steel to be a cathode rather than an anode, because it is at a slightly lower potential than the surrounding sea or soil. In practice it has been found that all usual forms of corrosion are prevented when the cathodic protection makes the cable armor 0.25–0.3 V negative to the surrounding. For underwater power cables this is not usually a practical solution, because the armor is not sufficiently isolated from the sea by the usual coatings. Another means of reduction of corrosion is the placement of zinc anodes along the cable, thus causing the steel to be the cathode. The corrosion of the zinc, however, requires replacement of the anodes periodically, which could be impractical.

The concept of coating steel to prevent corrosion has been a quite successful approach. In the case of armor wire the most attractive method has generally been to apply an extruded jacket. Several polymers, including high-density polyethylene, have been used with a high degree of success to provide permanent encapsulation of the individual wires. This approach has been used extensively on 35-kV and lower voltage cables. Use on dc cables appears feasible, and use on ac cables rated above 35-kV is also feasible. It is necessary to determine that the jacketed wires will not be broken down electrically due to normal operating standing voltages on the wires or due to transients. Special measures to avoid this can be applied. Another method to protect the armor wires from corrosion is the use of water blocking to encapsulate the galvanized wires followed by extrusion of a tough polymeric jacket. This also requires that armor-wire-induced voltage be kept to acceptable limits. In the case of double armor nonjacketed wires, a thin synthetic interlayer may be used to protect the two layers of metal from abrasion when the cable bends.

TABLE 13.9 Galvanic Series of Alloys in Seawater

Noble or cathodic	Platinum Gold Graphite Titanium Silver
	{ Chlorimet 3 (62 Ni, 18 Cr, 18 Mo) { Mastelloy C (62 Ni, 17 Cr, 15 Mo)
	{ 18-8 Mo stainless steel (passive) { 18-8 stainless steel (passive) { Chromium stainless steel, 11–30% Cr (passive)
	{ Inconel (passive) (80 Ni, 13 Cr, 7 Fe) { Nickel (passive)
	Silver solder
	{ Monel (70 Ni, 30 Cu) { Cupronickels (60-90 Cu, 40-10 Ni) { Bronzes (Cu–Sn) { Copper { Brasses (Cu–Zn)
	{ Chlorimet 2 (66 Ni, 31 Mo, 1 Fe) { Hastelloy B (60 Ni, 30 Mo, 6 Fe, 1 Mn)
	{ Inconel (active) { Nickel (active)
	Tin Lead Lead-tin solders
	{ 18-8 Mo stainless steel (active) { 18-8 stainless steel (active)
Active or anodic ↓	Ni-Resist (high Ni cat. iron) Chromium stainless steel, 13% Cr (active)
	{ Cast iron { Steel or iron
	2024 aluminum (4.5 Cu, 1.5 Mg, 0.6 Mn) Cadmium Commercially pure aluminum (1100) Zinc Magnesium and Magnesium alloys

Note: Alloys will corrode in contact with those higher in the series. Curly brackets enclose alloys so similar that they can be used together safely.
Source: Fontana and Green, *Corrosion Engineering*, New York, McGraw-Hill.

13.4.5 Special Application Designs

For deep-ocean power cables there are great hazards to the use of single-layer armor designs. The high tension causes the armor to develop a large torque. Should the tension be lost or released momentarily, the slack section or cable near the bottom could instantly "throw a loop." Then, resumption of tension would seriously damage the cable. Two fundamental approaches have been used to reduce or eliminate this great risk. The first is the use of antitwist tapes where a relatively thin metal tape is applied at a fairly high angle counter to the direction of the armor lay. This metal tape is applied before the armor. Both theoretically and practically, this approach is the more limited and difficult to provide uniformly over the cable length, although it has been used with adequate success on a few major installations. The second approach is the more desirable and positive from the technical standpoint. This is the use of contra-helical torque-balanced armor. Two layers of armor are applied with opposite lay directions. When both armor layers are of the same material, the fundamental relationship for the maximum torque unbalance is given by

$$\left(1 - \frac{\sin \alpha_1 A_1 D_{p1}}{\sin \alpha_2 A_2 D_{p2}}\right) \times 100 = K \tag{13.7}$$

where α_1 is the lay angle of armor layer one, A_1 is the cross-sectional area of armor layer one, D_{p1} is the pitch diameter of layer one, α_2 is the lay angle of armor layer two, A_2 is the cross-sectional area of armor layer two, and D_{p2} is the pitch diameter of layer two; K is the approximate maximum torque unbalance as a percentage. This formula is suitable as an approximation of torque balance.

More precise sets of formulas also include the effect of cable core torque, Poisson's ratio, cable stiffness, load sharing, and constructional stretch. In addition, the Poisson's ratio for each layer must be known, and the changing pitch diameter and armor angle must be taken into account using an iterative computer program to evaluate torque balance at all anticipated tensions including retrieval. The same program also evaluates cable component stress and strain characteristics as a function of load tension.

In the case of dynamic underwater power cables, the motions of the moving support structure must be defined, and a spectrum of tension loading can then be defined for the torque-balanced cable. In such duty, only a well-torque-balanced design would be desirable. To date, the deepest high-voltage power transmission cables have been installed to about 585 m. It is anticipated that similar cables may be installed to 2135-m depths. This application will require torque-balanced design, and the weight in water will be critical to the design of installation vessel and equipment. For other special applications, where weight is the most critical factor, the use of Kevlar appears feasible, and some experimental cables have been made and tested. Early problems with inter-strand abrasion have mostly been overcome. The use of this very strong synthetic to replace the steel armor in other specialized underwater power cables will likely receive greater industry attention in the future.

13.5 UNDERWATER POWER CABLE MANUFACTURE

The manufacture of extruded dielectric and of the several types of paper power cables has been covered in Chapters 3–7 and will not be repeated here. In addition to the manufacture of the sheathed electric core, there are four primary manufacturing functions for submarine power cable: splicing, jacketing, armoring, and handling.

13.5.1 Splicing

The splicing of lengths of 3C cable in the submarine cable factory is carried out either before the jacketing operation or before the armoring operation. The jointing of these cables by proven techniques is a well-established procedure, and modern techniques for evaluating the finished splice quality assure a high level of reliability. The advantages of the factory environment (as opposed to field splicing) offer higher quality splices. Properly designed and manufactured splices will have no effect on the reliability of the installed system. Such splices are designed to withstand the numerous bends of manufacture and installation without detriment; 1C splices and 3C splices are readily made in the factory.

13.5.2 Jacketing

In cases where splicing is done before jacketing, very long continuously jacketed lengths are then supplied. Such jacketed cables in extremely long lengths are then taken up on turntables or coiled in tanks or on platforms. In tanks, the water-tight integrity of the jacket can be tested by immersing the cable and performing an insulation resistance test of the jacket (providing that a nonjacketed shield or ground underlies the jacket).

13.5.3 Armoring

Large heavy power cables are routinely armored in virtually any continuous length. The armoring machine must be stopped periodically to replenish bobbins. The manufacturing operation is planned so that welds in armor wires are spaced far apart and only a small percentage of the wires in the complete layer are welded at any one time. Continuous qualification of welders, quality control, and replacement of the galvanized coating over the weld are standard operating procedures. In cases where jacketed armor wires are welded, the jacket is also reconstituted over the weld.

Depending upon machine capacity and configuration, it is possible to apply bedding, two armor layers (either unilay or counter lay), several tar immersions, and several serving/interserving layers in a one-pass operation. Round armor wires are applied using either planetary or rigid bobbins while flat armor wires are applied using rigid bobbin machines. Preforming of flat armor wires is usually necessary. Preforming of round armor wires is done, where needed, to provide greater armor package stability and/ or to enhance armor uniformity.

13.5.4 Handling

The way in which a cable can be handled is part of the design. For example, the smaller the lay angle of the armor (the longer the lay length), the larger the handling sheaves and capstans must be to avoid excessive strain in the wires and to prevent circumferential movement of the wires. Single-layer and double-layer (unidirectional lay) cables can be coiled or can be taken up on a reel or turntable. Large-diameter double-armored contra-helical designs generally cannot be coiled and must be taken up on a turntable or reel. When an armored cable is coiled, for every full turn the cable must rotate 360°. As a result, the rotation in degrees per foot must be established by the designer to set the minimum coiling diameter. When coiling is done, it should properly be done in the direction to unlay the armor. When this is done, compressive forces are imposed on the armor, and these forces must either be balanced or relieved by slight expansion of the armor wire pitch diameter, by tension on the cable core, or by a combination of these. If the armor is constrained from expansion outwards, as by metal bands or a serving with poor elongation properties, very substantial core tension can occur. The cable design must provide for this expansion. If it is not considered, the phenomenon of "bird caging" may occur. Figure 13.7 shows a large power cable being coiled in a 12-m diameter tank. The cable is being coiled clockwise, because it has a left-hand-lay armor.

13.5.5 Testing

Power cables for underwater use are made to the same standards previously discussed. The limitation on cable length regarding ac testing is of course a real constraint, and this has previously been discussed in Chapter 8. However, dc testing on any length and ac and impulse testing by end sampling procedures are frequently done. In addition, factory splices are subjected to special tests and X-raying to assure their proper construction. In the case of 3C cables, lengths are generally short enough to allow full ac factory tests which are done before splicing. Testing after installation consists usually only of dc tests and the measurement of metal component resistances and jacket resistance.

Special development testing programs for new concepts and new applications of submarine power cables are being pursued. One such test program during the 1980s first performed full-scale testing of 138-kV extruded and self-contained cables to simultaneously evaluate very high mechanical cyclic tensions, cyclic bending, high voltage, and load currents while immersed in seawater. These tests and others are leading the way to increased understanding of the characteristics of underwater power cables.

13.6 CABLE TRANSPORT

Cables that exceed normal reel lengths can be supplied in virtually any continuous length necessary. In the case of shipment via rail in gondola cars it is readily possible to ship about 50,000-ft. lengths of 4-in. diameter cable. Such shipment involves the

Figure 13.7 Coiling large-diameter underwater power cable in 12-m-crib. (Courtesy of Simplex Technologies, Inc.)

chaining together of cars for trans-continental shipment. Such cables must be coilable. Figure 13.8 illustrates a long power cable loaded into gondola cars. Cables somewhat more than 2 in. in diameter can be shipped in greater than 20,000-ft lengths by way of giant railcar-mounted reels. The very longest and heaviest cables are readily shipped by water on barges or special ships [140]. The cables may either be coiled or taken up on a special turntable.

13.7 UNDERWATER POWER CABLE INSTALLATION

Before installing a submarine cable [141–146], and often before cable design can be completed, information regarding the characteristics of the cable route must be established. In many cases a recorded side-scan sonar survey will reveal sufficient information. In cases where there is uncertainty about such items as sediment type or thermal resistivity, obstructions, location of other cables, etc., divers or unmanned submersibles may be used to obtain visual evidence and/or samples. The route must be well planned and the most favorable weather, tidal, and ocean current conditions determined. Maintaining steerage and position control during the entire marine installation is important. Before the installation begins, two or more electronic navigation range points are established and manned on shore. Using these, modern navigation methods allow precise placement of the cable.

The cable end, suitably sealed and protected, is fed from the crib or turntable to a winch line from shore, and the cable is fed into the water near shore with flotation attached at usual intervals of 1.5–3 m as it is slowly let overboard. Once the cable is ashore and pulled up the shore trench to position, the flotation is removed by divers working from the shore out to the installation vessel and the cable is settled to the bottom or into the underwater trench section. When all flotation is removed, the cable vessel begins to make way on course. Tension is maintained by a fleeting cable drum or by a linear cable machine. In the case of the cable drum, some back tension is applied between the crib or turntable and the drum. This is to maintain frictional contact of the cable on the drum.

Figure 13.8 Long underwater power cable loaded in railroad gondola cars with center crib designed to maintain minimum coiling radius. (Courtesy of Simplex Technologies, Inc.)

The cable laying requires close coordination between the installation manager, navigator, helmsman, and the cable machine operator. The installation manager calls the orders and is in absolute control. The machine operator, according to a previously engineered plan, pays out cable at a rate that maintains the cable at a previously determined acceptable payout speed range and a prescribed laying angle, usually between 50° and 80° from the horizontal. The tension is related to this angle by the following expression:

$$T = wd \frac{1}{1 - \cos \theta} \tag{13.8}$$

where T is the installation tension, w is the weight per unit length, d is the water depth, and θ is the laying angle, from the horizontal. Normally, once underway, power cable will be laid at a rate of 2–4 knots per hour. For cables only a few miles in length, by far the greatest amount of time is required for the shore and/or platform landings, with the second end being usually the more time consuming. Figures 13.9 and 13.10 show power cables being installed.

The second shore (or platform) landing is more difficult since the free end is not available to payout readily to shore. Therefore, to procure that end, there are essentially three methods that can be used. One is to pay the cable out in a bight with flotation attached, then to fix the length and cut on the vessel, cap the end, and then proceed as before to feed the end to shore. The second way is essentially the same concept except that a floating sheave is used to feed out the bight and is winched in from shore or controlled by a tugboat. This second method is more predictable and controllable and preferable if the seas are anything but calm. The third method consists of forming a figure 8 stack of cable on a free deck by feeding out cable from the crib or turntable, through the engine, and having a crew of experienced cable handlers lay the cable on deck in this way so as to allow the cut end to be pulled off by winch from shore (with flotation), as shown in Fig. 13.11.

A set of formulas to predict the normal maximum installation tension T_m and retrieval tension T_{mr} are derived from CIGRE [147, 148] documents and have been

Figure 13.9 A 1983 installation of 3C ac 37-kV submarine cable weighing 87 kg/m, diameter 12 cm, and 7.9 km in length. (Courtesy of Simplex Technologies, Inc.)

Figure 13.10 A 1983 installation of 12-cm 3C 15-kV double-armored submarine cable from large reel, cable weight = 22 kg/m. (Courtesy of Simplex Technologies, Inc.)

Figure 13.11 Installation of shore end by means of flotation. (Courtesy of Simplex Technologies, Inc.)

found to be quite suitable. The maximum installation tension T_m in kilograms can be expressed as

$$T_m = 1.15wd + 2155 \tag{13.9}$$

where w is the weight of cable in water in kilograms per meter and d is the water depth in meters. The maximum retrieval tension is given by

$$T_{mr} = 1.3wd + 4080 \tag{13.10}$$

The last factor in each equation is a safety factor that adequately provides for such things as ship heave. For most cable designs, and for determination of cable vessel equipment requirements, these formulas are fully adequate. The difference in the multiplier constants 1.15 and 1.3 is mainly due to the fact that during laying the water drag, related to the speed of the vessel, is working to reduce the cable tension by effectively adding an upward vertical component to cancel out some of the tension. In the case of retrieval just the opposite is true. Also when cables are recovered an additional component of tension (included in the CIGRE formula) may be added due to so-called adhesion or the necessity to pull the cable up from the sediment in which it has become submerged.

Unlike communication cables, power cables are normally laid with residual tension. This is both to maintain cable position and to prevent the possibility of hockling. In uneven bottom conditions where the cable could form in a catenary between two projections, it is important that the cable not be laid with a catenary that can be vibrated or swung by ocean currents. Over sufficient time, such a condition will likely result in fatigue of the cable. To assure that the cable was indeed installed properly and according to plan, it is common practice to provide a second side-scan sonar route survey and/or visual inspection by remote vehicle or divers. It is also generally desirable to videotape, with recorded navigational coordinates, each unburied cable along its entire route. This documentation can prove to be of considerable value.

REFERENCES

[1] ICEA/NEMA Standards Publication, Steel Armor and Associated Coverings for Impregnated-Paper-Insulated Cables, ICEA Pub. S-67-401; NEMA Pub. No. WC-2-1980, Washington, DC.

[2] ICEA/NEMA Standards Publication, Crosslinked-Thermosetting-Polyethylene-Insulated Wire and Cable, ICEA Pub. No. S-66-524, NEMA Pub. No. WC-7, 1996, Washington, DC.

[3] ICEA/NEMA Standards Publication, Ethylene-Propylene-Rubber-Insulated Wire and Cable, ICEA Pub. No. S-68-516, NEMA Pub. No. WC-8-1996, Washington, DC.

[4] AEIC Specifications for Solid-Type, Impregnated-Paper-Insulated Lead Covered Cable, AEIC CS1-90, New York, 1990.

[5] AEIC Specifications for Impregnated-Paper-Insulated Low and Medium Pressure Self-Contained Liquid Filled Cable, AEIC CS 4-93, New York, 1993.

[6] AEIC Specifications for Cross-Linked Polyethylene Insulated Shielded Power Cables Rated 5 through 46 kV, AEIC CS 5-94, New York, 1994.

[7] AEIC Specifications for Ethylene Propylene Rubber Insulated Shielded Power Cables Rates 5 Through 69 kV, AEIC CS6-96, New York, 1996.

[8] C. C. Barnes, Submarine Telecommunication and Power Cables, IEEE Monogram Series 20, London, 1977.

[9] "IEEE guide to the factors to be considered in the planning, design, and installation of submarine power and communication cables," IEEE Std 1120–1990, New York.

[10] C. Bazzi, P. Gazzana-Priaroggia, and M. A. Durso, "Proposed design of a 345 kV cable connection for an offshore generating station," IEEE Underground Transmission/Distribution Conference (76 CH 1119-7 PWR), Sept. 28–Oct. 1, 1976, pp. 480–488.

[11] Simplex Wire and Cable Company, "Riser segment design of underwater electric power transmission cable system," U.S. Dept. of Energy, Report OR0/5359-1, Oct. 1978.

[12] Pirelli Cable Systems, "Bottom segment design-underwater cable power transmission system," U.S. Dept. of Energy, Report DOE/ET/20324-1, Nov. 1978.

[13] "Phase I—Preliminary Prototype Cable Design Criteria—Hawaii Deep Water Electrical Transmission Cable Demonstration Program" State of Hawaii Dept. of Planning and Economic Development, April 1982.

[14] "Riser cable development for ocean thermal energy conversion plants," U.S. Dept. of Energy, Report DOE/ET/29180-2 Dist. Category UC97a, July 1983.

[15] J. E. Soden, R. T. Traut, H. St.-Onge, and D. Train, "Testing of a high voltage XLPE cable for dynamic submarine application," JICABLE, International Conference Polymer Insulated Power Cables, Paris, March 5–10, 1984, pp. 314–320.

[16] S. P. Walldorf, G. A. Chapman, and G. N. Okura, "Cable selection methodology for deep water power transmission system applications," *IEEE Trans. Power Delivery*, Jan. 1986, pp. 34–40.

[17] J. Larsen-Basse, "Preliminary evaluation of the Hawaii deep water power cable for potential of failure due to erosion–corrosion," Oceans 87–Proceedings, IEEE, Sept.–Oct. 1987, pp. 445–448.

[18] R. Eaton and W. A. Bonnet, "Hawaii Deep Water Cable program: Submarine power cable installation in an adverse environment," CMAE 1988, Proceedings of the Seventh International Conference on Offshore Mechanics and Arctic Engineering, ASME, 1988, Vol. 1, pp. 525–533.

[19] U. Arnaud, G. Bazzi, and D. Valenza, "Proposal for a commercial interconnection among the Hawaiian Islands based on results of the Hawaii Deep Water Cable Program," *IEEE Trans. Power Delivery*, Oct. 1992, pp. 1661–1666.

[20] D. M. Farnham, G. B. Shanklin, S. H. Cunha, and D. H. Short, "The St. Lawrence River high voltage submarine cable crossing" *AIEE Trans. Power Apparatus Syst.*, Vol. 78, 1959, pp. 98–185.

[21] J. G. Thornton, "132 kV submarine cables crossing Botany Bay, New South Wales, Australia," CIGRE Proc., Report No. 202, Paris, 1964.

[22] G. Bazzi, G. Monti, A. Halesani, S. Balli, and G. Porta, "132 kV ac power cables and the optical fiber cable for the submarine intertie Italy–Elba Island," *IEEE Trans. Power Delivery*, Vol. 4, No. 1. 1989, pp. 58–67.

[23] J. H. Cooper and M. J. Polasek, "Planning and installation of the 138 kV South Padre Island submarine cable," *IEEE Trans. Power Delivery*, Oct. 1993.

[24] T. Ingledow, R. M. Fairfield, E. L. Davey, K. S. Brazier, and J. N. Gibson, "British Columbia–Vancouver Island 138 kV submarine power cables," *Proc. IEE*, Vol. 104A, 1957, pp. 485–504.

[25] I. M. Crabtree and E. T. O'Brien, "Performance of the Cook Strait 250 kV dc submarine power cable 1964–1985," CIGRE, Report 21.01, 1986, pp. 1–6.

[26] L. Elgh, C. T. Jacobsen, B. Bjurstrom, G. Hjalmarsson, and S. O. Olsson, "The 420 kVac submarine cable connection between Denmark and Sweden," *CIGRE Proc.*, Report 21-02, Paris, 1974.

[27] M. Takaoka, et al. "Development of 500 kVdc oil-filled cable," *Fujikura Tech. Rev.*, December 1975, pp. 3–9.

[28] E. Crowey, J. E. Hardy, L. R. Horne, and B. G. Prior, "Development programme for the design, testing and sea trials of the British Columbia mainland to Vancouver Island 525 kV ac submarine cable link," *CIGRE Proc*, Paper No. 21-10, Paris, 1982.

[29] R. G. Foxall, K. Bjorlow-Larsen, and G. Bazzi, "Design, manufacture and installation of a 525 kV alternating current submarine cable link from Mainland Canada to Vancouver Island," CIGRE Report 21-14, 1984.

[30] C. A. Arkell and B. Gregory, "Design of self-contained oil-filled cables for UHV DC transmission," CIGRE Report 21-07, 1984.

[31] J. Grzan, E. I. Hahn, R. V. CasaLaina, and J. O. C. Kansog, "The 345 kV Underground/Underwater Long Island Sound Cable Project," *IEEE Trans. Power Delivery*, July 1993, pp. 750–759.

[32] U. Arnaud, A. Bolza, F. Magnani, and E. Occhini, "Long Island Sound submarine cable crossing 345 kV, 750 MVA," CIGRE Proc. 34th Session, Vol. 1, p. 21-306.1–7.

[33] H. B. Slade, "High voltage rubber-insulated submarine power cables," *Proc. IEE*, Vol. 100, Part 11a, 1953, p. 263.

[34] N. D. Kenney and M. J. Koulopoulos, "115 kV and 138 kV polyethylene insulated cable installations and cable evaluation data," CIGRE Proc., Paper 21-07, Paris, 1968.

[35] The Okonite Company, "The development of a high voltage dc cable," EPRI Report EL-606, Palo Alto, CA, January, 1978.

[36] K. Zbindon, E. Barragan, and R. Pederson, '150-kV and 110-kV XLPE submarine cable installations between Morcotte and Brusino, Switzerland," *IEEE Electrical Insulation Mag.*, Vol. 4, No. 2, 1988, pp. 11–14.

[37] Y. Maekawa, A. Yamaguchi, Y. Sekii, M. Hara, and M. Marumo, "Development of DC XLPE cable for extra-high voltage use," *Transactions IEE Japan*, Part B, Vol. 114-B, Issue 6, June 1994, pp. 633–641.

[38] H. J. Miller, C. T. W. Sutton, and A. M. Morgan, "Flat pressure cables for submarine installations," CIGRE Proc, Paper No. 216, Paris, 1958.

[39] C. C. Barnes, J. C. E. Coomber, J. Rollin, and L. Clavreul, "The British-French direct current submarine cable link," CIGRE Proc., Paper No. 210, Paris, 1962.

[40] G. von Geijer, S. Smedsfelt, and L. Ahlgren, "The Conti-Skan HVDC Project," IEEE Conference Paper 63-1096, Summer General Meeting and Nuclear Radiation Effects Conference, Toronto, Ontario, Canada, June 16–21, 1963.

[41] S. Smedsfelt, L. Ahlgren, and V. Mets, "Operational performance and service experience with the Konti-Skan and Gotland HVDC Projects," *IEE Conf. Publ. 22*, 1966 Part I, pp. 11–16.

[42] P. Gazzana-Priaroggia and G. L. Palandri, "200 kVdc submarine cable interconnection between Sardinia and Corsica and between Corsica and Italy," CIGRE Proc., Paper 21-05, Paris, 1968.

[43] I. Eyraud, L. R. Horne, and J. M. Oudin, "The 300 kV direct current submarine cables transmission between British Columbia Mainland and Vancouver Island," CIGRE Proc., Paper 21-07, Paris, 1970.

[44] O. Hauge, A. Berg, J. N. Johnsen, G. Wettre, and K. Bjorlow-Larsen, "The Skagerrak HVDC Cables," CIGRE Proc., Paper 21-05, 30 August–7 September, 1978.

[45] O. Hauge, J. N. Johnsen, T. A. Holte, and K. Bjorlow-Larsen, "Performance of the +/− 250 kV HVDC Skagerrak submarine cables; further development of the HVDC paper-insulated mass-impregnated (solid type) submarine cable," *IEEE Trans. Power Delivery*, Vol. 3, Jan. 1988, pp. 1–5.

[46] A. L. Williams, E. L. Davey, and J. N. Gibson, "The 250 kVdc submarine cable interconnection between the North and South Islands of New Zealand," *Proc. IEE*, Vol. 113, 1966, pp. 121–133.

[47] G. Bahder, G. S. Eager, G. W. Seman, F. E. Fisher, and H. Chu, "Development of ±400/600 kV high and medium-pressure oil-filled paper-insulated dc power cable system," *IEEE Trans. Power Apparatus Syst.*, Vol. PAS-97, Nov./Dec. 1978, pp. 2045–2056.

[48] A. L. McKean and E. M. Allam, "Progress in development of a ±600 kVdc cable system," Paper No. F-79-613-1, Summer IEEE Power Meeting, Power Engineering Society, July, 1979.

[49] S. S. Gorodetsky, "200–400 kV direct current cables," CIGRE Proc., Report 206, Paris, 1958.

[50] J. Malval, J. Clade, L. Csuros, and G. S. H. Jarrett, "Operational performance of the direct-current cross-channel link," CIGRE Proc., Report No. 417, Paris, June, 1964.

[51] P. Fourcade and C. C. Barnes, "Service experience with the Anglo-French dc cross-channel cable," *IEE Conf. Publ. 22*, 1966, Part 1, pp. 26–28.

[52] O. Hauge, "HVDC cable for crossing the Skagerrak Sea between Denmark and Norway," CIGRE Proc., Paper No. 21-07, Paris, 1974.

[53] CIGRE, HV Insulated Cables Study Committee, "Recommendations for tests on dc cables for a rated voltage up to 550 kV," *Electra*, No. 32, Jan., 1974, pp. 83–88.

[54] N. G. Hingorani, "The reemergence of dc in modern power systems," *EPRI J.*, June 1978.

[55] R. L. Hauth and G. D. Bruer, "Interconnecting ac/dc systems," Conference on Renewable Energy Technology, Honolulu, December, 1980.

[56] S. Minemure, T. Imai, T. Kihara, T. Otonari, and H. Hashimoto, "±250 kV direct current submarine cable for Hokkaido-Honshu link," CIGRE Proc., paper 21-03, Paris, 1980.

[57] Bonneville Power Authority, *Transmission Line Reference Book, HVDC to ±600 kV*, EPRI, Palo Alto, CA.

[58] J. A. Moran, Jr. and J. A. Williams, "HVDC cables—an overview of design practices and experiences," Proc. of the Symposium on Urban Applications of HVDC Power Transmission Conference, 24–26 Oct. 1983, pp. 131–147.

[59] Y. Thomas, "Main design features of the French part (2000 MW France–UK HVDC link)," *Rev. Gen. l'Electricite*, No. 2, 1985, pp. 104–113.

[60] C. A. Arkell, E. H. Ball, K. J. H. Hacke, N. W. Waterhouse, and J. B. Yates, "Design and installation of the UK part of the 270 kVDC cable connecting between England and France including reliability aspects," 1986 CIGRE International Conference on Large High Voltage Electric Systems, Report 21.02, pp. 1–12.

[61] L. Rebuffat, G. Bazzi, and G. Pavini, "400 kV, 1000 MVA submarine cable connection between Sicily and mainland Italy," *Energia Elettrica*, Vol. 64, No. 7–8, July–Aug. 1987, pp. 301–313.

[62] Q. Bui-Van, G. Beaulieu, H. Huynh, and R. Rosenqvist, "Overvoltage studies for the St. Lawrence River 500 kV DC cable crossing," *IEEE Trans. Power Delivery*, Vol. 6, July 1991, pp. 1205–1215.

[63] L. Carlsson, A. Nyman, L. Willborg, and G. Hjalmarsson, "The Fenno–Skan HVDC submarine cable transmission. System and design aspects, commissioning and initial operating experience," AC and DC Power Transmission Conference, 17–20 Sept. 1991, Publ. No. 345, pp. 344–349.

[64] M. T. O'Brien, J. E. Larsen, and G. Hjalmarsson, "Installation of 350 kV HVDC submarine power cables for uprating the HVDC system in New Zealand," *CIGRE Proc. 34th Session*, Vol. 1, Sept. 1992, Report 21-308, pp. 1–5.

[65] J. C. Gleadow and W. G. McElhinney, "High voltage DC transmission: New Zealand HVDC link upgrade," CIGRE 1993 Regional meeting: South-East Asia and Western Pacific Conference Papers, Oct. 1993, pp. 443–454.

[66] G. Luoni and D. Povh, "HVDC submarine cables," CIGRE 1993 Regional Meeting: South–East Asia and Western Pacific Conference Papers, Oct. 1993, pp. 402–413.

[67] B. Ekenstierna and A. Nyman, "Baltic HVDC interconnection," CIGRE International Colloquium on HVDC and flexible AC Power Transmission Systems, 1993, pp. 6.3/1-11.

[68] M. T. O'Brien, J. E. Larsen, and G. H. Hjalmarsson, "New Zealand HVDC transmission: 350 kV power cable specification and overload capability," CIGRE International Colloquium on HVDC and Flexible AC Power Transmission Systems, 1993, pp. 1.4/1-10.

[69] J. Kure, "HVDC submarine link for transmission of electrical energy" *Trans. Distribution Int.*, Dec. 1993, Vol. 4, Issue 4, pp. 10–14.

[70] N. Bell, Q. Bui-van, D. Couderc, G. Ludasi, P. Meyere, and C. Picard, "+/− 450 kVDC underwater crossing of the St. Lawrence river of a 1500 km overhead line with five terminals" *CIGRE Proc. 34th Session*, Vol. 1, 1993, pp. 21-301/1-11.

[71] G. Hjalmarsson, J. Thoren, U. Grape, G. Malmquist, R. Eriksson, and M. Kvarngren, "After Service Analysis of the 32-Year Old HVDC Cable Gotland 1," CIGRE Proc. 34th Session, Vol. 1, 1993, pp. 21-302/17.

[72] Se il Kim, Suk Jin Lee, J. L. Haddock, and M. H. Baker, "System design characteristics of the 300 MW submarine link to Cheju," CIGRE Proc., 1993, pp. 29–44.

[73] J. L. Haddock, F. G. Goodrich, and Se il Kim, "Design aspects of Korean mainland to Cheju island HVDC transmission," *Power Tech. Int.*, 1993, pp. 125, 126, 128–130.

[74] A. Nyman, K. Jaaskelainen, M. Vaitomaa, B. Jansson, and K-G. Danielsson, "The Fenno-Skan HVDC link commissioning," *IEEE Trans. Power Delivery*, Vol. 9, 1994, pp. 1–9.

[75] E. Illstad, "World record HVDC submarine cable," *IEEE Electrical Insulation Mag.*, Vol. 10, Issue 4, July–Aug. 1994, pp. 64–67.

[76] "HVDC link extends the German network north," *Mod. Power Syst.–Suppl. Issue*, Feb. 1994, pp. 51–54.

[77] T. G. Nielsen, A. Canelhas, and B. S. Hansen, "Kontek link has the word's longest underground HVDC cable," *Mod. Power Syst.*, Vol. 14, Issue 3, March 1994, pp. 39–42.

[78] T. J. Hammons, K. H. Chew, and T. C. Chua, "Competitiveness of renewable energy from Iceland via the proposed Iceland/UK HVDC submarine cable link," Universities Power Engineering Conference 1995—Conference Proceedings, Vol. 1, pp. 69–73.

[79] T. Karlsson and G. Liss, "HVDC transmissions with extremely long DC cables—control strategies," Stockholm Power Tech International Symposium on Electric Power Engineering, June 1995, vol. 2, pp. 24–29.

[80] D. Valenza and G. Cipollini, "HVDC submarine power cable systems—state of the art and future developments," Proceedings EMPD '95—1995 International Conference on Energy Management & Power Delivery (Cat. No. 95TH8130 IEEE), Vol. 1, pp. 283–287.

[81] E. Guonason, J. Henje, P. Shepherd, and D. Valenza, "A 550 MW HVDC submarine cable link: Iceland–UK–continental Europe," IEE Third International Conference on Power Cables and Accessories 10 KV-500 KV, Conf. Publ. No. 282, pp. 220–224.

[82] A. Fujimora, T. Tanaka, H. Takashima, T. Imajo, R. Hata, T. Tanabe, S. Yoshida, and T. Kakihana, "Development of 500 kV DC PPLP-insulated oil-filled submarine cable," *IEEE Trans. Power Delivery*, Vol. 11, Jan. 1996, pp. 43–50.

[83] "MPS (more power submarine) cables could be mind blowing for 1200 MW links," *Mod. Power Syst.*, July 1996, Vol. 16, Issue 7, pp. 43–46.

[84] T. H. Carlsen, D. Lysheim, T. R. Time, J. Rittiger, W. Schultz, D. Troger, and R. Witzmann, "Feasibility study for increased power exchange between Norway and continental Europe by new HVDC link," Sixth International Conference on AC and DC Power Transmission, Publ. No. 423, 1996, pp. 100–105.

[85] P. M. Smith, I. W. Whitlock, and J. Balkwill, "The effect of HVDC submarine cables on shipping with magnetic autopilots," Sixth International Conference on AC and DC Power Transmission, Publ. 423, 1996, pp. 76–80.

[86] P. Gazzana-Priaroggia, P. Maschio, and N. Palmiera, "Surge performance of impregnated paper insulation for HVDC cables," IEE Conf. Publ. 22, 1996, Part I, pp. 325–333.

[87] J. M. Oudin and H. Theveion, "Theory of dc cables, calculation of gradient and its correlations with breakdown gradient," CIGRE Proc., Paper 208, Paris, 1966.

[88] G. Bahder, F. G. Garcia, and A. S. Brookes, "Insulation coordination in high voltage dc cables," CIGRE Proc., Paper 21-03, Paris, 1972.

[89] B. Nyberg, K. Herstad, and K. B. Larsen, "Numerical methods for calculation of electrical stresses in HVDC cables with special application to the Skaggarak cable," *IEEE Trans. Apparatus Syst.*, Vol. PAS-94, Mar./April 1975, pp. 491–497.

[90] W. G. Lawson, F. K. Padghem, and P. Metra, "The effect of polarity-reversals on the dielectric strength of oil-impregnated paper insulation for HVDC cables," *IEEE Trans. Power Apparatus Syst.*, Vol. PAS-97, May/June, 1978, pp. 884–892.

[91] G. Luoni, P. Metra, and E. Occhini, "Dc and ac thermal stability of oil impregnated paper," *IEEE Trans. Power Apparatus Syst.*, Vol. PAS-98, Jan./Feb. 1979, pp. 149–158.

[92] B. M. Weedy and S. A. M. Shehata, "2 GW Britain–France H.V.D.C. link transition joint–steady state stresses," *IEEE Trans. Power Delivery*, Vol. 1, July 1986, pp. 1–6.

[93] B. M. Weedy and S. A. M. Shehata, "2 GW Britain–France H.V.D.C. link transition joint–transient stresses," *IEEE Trans. Power Delivery*, Vol. 1, July 1986, pp. 7–12.

[94] F. H. Buller, J. H. Neher, and F. O. Wollaston, "Oil flow and pressure calculations for self-contained oil-filled cable systems," *AIEE Trans. Power Apparatus System.*, Vol. 75, April, 1956, pp. 180–192.

[95] N. Klein and D. Schieber, "Radial oil flow and pressure differences in cable insulations," *AIEE Trans. Power Apparatus Syst.*, Vol. 81, April 1962, pp. 72–81.

[96] L. F. Hickernall, A. A. Jones, and C. V. Snyder, "F-3 lead alloy—an improved cable sheathing," *AIEE Trans. Power Apparatus Syst.*, Vol. 70, 1951, pp. 1273–1285.

[97] C. F. Gohn and W. C. Ellis, "The fatigue test as applied to lead cable sheath," *Proc. ASTM*, Vol. 51, 1951.

[98] D. G. Havard, "Selection of cable sheath lead alloys for fatigue resistance," *IEEE Trans. Power Apparatus Syst.*, PAS, Vol. PAS-96 Jan./Feb. 1977, pp. 80–87.

[99] P. Anelli, F. Donazzi, and W. G. Lawson, "The fatigue life of lead alloy E as sheathing material for submarine power cables," *IEEE Trans. Power Delivery*, Vol. 3, Jan 1988, pp. 69–75.

[10] G. Maxhio and E. Occhini, "Overvoltages on anti-corrosion sheaths of high voltage cables with particular reference to long submarine cables," CIGRE Proc., Paper 224, Paris, 1964.

[100] R. C. Waldron, "115 kV submarine cable crossing of Puget Sound," *IEEE Trans. Power Apparatus Syst.*, Vol PAS-84, 1965, pp. 746–755.

[101] J. Sallard, R. Tellier, D. M. Cherry, and C. C. Barnes, "Problems arising from the design and construction of high voltage dc submarine cable systems," CIGRE Proc., Paper 415, Paris, 1960.

[102] P. Gazzana-Priaroggia and G. Maschio, "Continuous long length ac and dc submarine HV power cables—The present state of the art," *IEEE Trans. Power Apparatus Syst.*, PAS, Vol. 92, 1973, pp. 1744–1749.

[103] A. Morello, C. Tsralli, and J. McConnell, "Power transmission by way of submarine cables," Offshore Technology Conference, Paper No. OTC 2257, 1975.

[104] H. M. Brinser, "Submarine power cables," Oceans '76, Second Annual Combined Conference, Sept. 13–15, 1976, pp. 3B1–3B4.

[105] J. N. Johnsen, T. A. Holte, K. B. Larsen, and P. B. Larsen, "Submarine power cable development in Norway," CIGRE Proc., Paper 21-12, Paris, 1980.

[107] K. Bjorlow-Larsen, "Submarine power cables," *Electrical Commun.*, June 1983, Vol. 58, No. 2, pp. 150–154.

[108] K. Bjorlow-Larsen, "Submarine power cables in Scandinavian waters," *Mod. Power Syst.*, Norway suppl., June 1988, pp. 19, 21, 23.

[109] G. Feld, R. L. Reuben, D. G. Owen, and A. E. Crockett, "Power cables and umbilicals. Conductor strain under pure bending," Proc. First International Offshore and Polar Engineering Conference, Edinburgh, Scotland, Aug. 1991.

[110] S. Aleo, G. Bazzi, F. Corbellini, F. Magnani, and D. Valenza, "Design, manufacturing, installation, protection of the 345 kV 750 MVA submarine power cables in New York," *Energia Elettrica*, 1992, Vol. 69, No. 4, pp. 169–173.

[111] G. Feld, D. G. Owen, R. L. Reuben, and A. E. Crockett, "Mechanical behaviour of the metallic elements of submarine cables as a function of cable loading," *Eng. Structures*, May 1995, Vol. 17, No. 4, pp. 240–253.

[112] L. J. W. Bartholomew, "Interconnection of the Isle of Wight with the main transmission system of the British Electricity Authority by 33 kV submarine cables," CIGRE Proc., Paper 218, Paris, 1952.

[113] G. Barclay and A. L. Verheil, "Operation and maintenance of British Columbia Hydro and Power Authority mainland—Vancouver Island 132 kV submarine cable interconnect," *IEEE Trans. Power Apparatus Syst.*, 1963, PAS-82, pp. 876–884.

[114] G. Bazzi, G. Monti, A. Malesani, S. Balli, and G. Porta, "The 132 KV A.C. power cables and the optical fiber cable for the submarine intertie Italy–Elba Island," *IEEE Trans. Power Delivery*, Vol. 4, Jan. 1989, pp. 58–67.

[115] C. A. Arkell, S. J. Galloway, E. B. Parsons, B. G. Woodcock, and D. E. Woolmer, "Design, manufacture and installation of 150 kV submarine cable system for the Java-Madura interconnection," *IEE Proc.*, 1989, Vol. 136, No. 3, pp. 121–129.

[116] L. Pottonen, "Joint operation of the Finnish power system with neighbouring countries," *Power Tech. International, 1993* UK, pp. 121–123.

[117] B. Ekenstierna and A. Nyman, "Power link in the deep: The Swedish–German link," *Trans. Distribution*, Vol. 46, Issue 3, March 1994, pp. 110–121.

[118] R. Fermo, U. Guida, G. Poulet, F. Magnani, and S. Aleo, "150 KV system for feeding Ischia Island," IEE Third International Conference on Power Cables and Accessories 10 KV–500 KV. Conf. Publ. No. 382, pp. 213–219.

[119] M. Abderrazzaq and B. A. T. Al Zahawi, "The Jordan–Egypt Red Sea cable interconnection project," Sixth International Conference on AC and DC Power Transmission, IEE Publ. No. 423, London, pp. 124–127.

[120] "Underwater cable fault due to thermal runaway," *Electrical World*, 1971, Vol. 175, No. 9, pp. 52–53.

[121] J. H. Neher and M. H. McGrath, "The calculation of the temperature rise and load capability of cable systems," *AIEE Trans. Power Apparatus Syst.*, Vol. 76, 1957, pp. 752–772.

[122] "Calculation for the continuous current rating of cables (100% load factor), IEC Pub. No. 287.

[123] S. L. Cress, Y. Yakov, and D. Motlis, "Temperature rise of submarine cable on riser poles," *IEEE Trans. Power Delivery*, Vol. 6, Jan. 1991, pp. 25–33.

[124 L. J. Bohmann, D. O., Wiitanen, J. M. Wilson, and J. Zipp, "Impedance of a double submarine cable circuit using different types of cables within a single circuit," *IEEE Trans. Power Delivery*, Vol. 8, Oct. 1993, pp. 1668–1674.

[125] R. D. Taylor, "Increasing the life of undersea cables," *Undersea Technology*, June 1972, Vol. 13, No. 6, pp. 33–35.

[126] V. Ciallella, F. Farneti, E. Gabriel, M. Salieri, and E. Sesto, "External mechanical protections of the SA.CO.I submarine cables: Characteristics and operational experience," IEE 2nd International Conference on Progress in Cables and Overhead Lines for 220 kV and Above, London, Sept. 4–6, 1979.

[127] G. Bazzi, "Submarine cables: A very good reliability judgement," *TE Int. J.* March–April 1983, Vol. 7, No. 2, pp. 41–49.

[128] W. C. Engelmann, A. J. von Alt, and W. F. Searle, Jr., "A different approach to the repair and protection of submarine cable systems," *IEEE Trans. Power Apparatus Syst.*, Jan. 1984, pp. 168–173.

[129] Y. Kato, A. Watanabe, H. Konishi, T. Kawai, Y. Inoue, and M. Sanpei, "Cable section fault detection for HVDC line protection," *IEEE Trans. Power Delivery*, Vol. 1, July 1986, pp. 332–336.

[130] "Methods to prevent external mechanical damage to submarine cables," CIGRE Paper 21-12 by WG 21.06, 1986.

[131] Y. Z. Mesenzhnik, A. G. Korchagin, L. Y. Prut, and I. S. Tyan, "Reliability of submerged electrical systems and methods of forecasting this reliability," *Elektrichestvo*, July 1986, No. 7, pp. 22–26.

[132] T. H. Birkeland, "Corrosion of undersea cables. Correct design reduces the problem," *Elektro*, 1987, No. 5, pp. 18–20.

[133] U. Arnaud, G. Bazzi, and D. Valenza, "Advantages and disadvantages of embedment to prevent external mechanical damage to submarine cables," *IEEE Trans. Power Delivery*, Vol. 5, Jan. 1990, pp. 54–57.

[134] F. Farneti, B. Riot, G. Bazzi, and C. Morris, "Reliability of underground and submarine high voltage cables," CIGRE Symposium, Electric Power Systems Reliability, Montreal, Quebec, Canada, 16–18 Sept. 1991, pp. 2-07/1-6.

[135] M. Nakamura, N. Nanayakkara, H. Hatazaki, and K. Tsuji, "Reliability analysis of submarine power cables and determination of external mechanical protections," *IEEE Trans. Power Delivery*, Vol. 7, April 1992, pp. 895–902.

[136] T. Nishimoto, T. Miyahara, H. Takehana, and F. Tateno, "Development of 66kV XLPE submarine cable using optical fibers as a mechanical-damage-detection-sensor," *IEEE Trans. Power Delivery*, Vol. 10, Oct. 1995, pp. 1711–1717.

[137] J. Larsen-Basse, "Abrasion of 0.85% C-steel under natural Hawaiian marine conditions," *Tribology Trans.*, Vol. 38, No. 3, July 1995, pp. 672–678.

[138] P. Gazzana-Priaroggia, J. Piscioneri, and S. Margolin, "The Long Island Sound submarine cable interconnection," *IEEE Trans. Power Apparatus Syst.*, Vol. PAS-90, 1971, pp. 1863–1873.

[139] D. M. Chamberlain and S. W. Margolin, "The Long Island Sound submarine cable interconnection-operating experience," IEEE Transmission and Distribution Conference, (79 Ch 1399-5-PWR), April 1–6, 1979, pp. 290–298.

[140] J. E. Soden, D. O. Libby, and J. R. Spillers, "A study of ocean cableships applied to submarine power cable installation," 8th Ocean Energy Conference, Washington, DC, June, 1981.

[141] T. K. Ingledow, D. T. Hollingsworth, and A. L. Williams, "The 138 kV Vancouver submarine power cable, with special reference to the theory and control of the laying of heavy submarine power cables," CIGRE Proc., Paper 216, Paris, 1958.

[142] W. A. Gallotte, "115 kV cable crossing of Puget Sound," *IEEE Trans. Power Apparatus Syst.*, Vol. PAS-84, 1965, pp. 737–745.

[143] L. Rebuffat, G. M. LanFranconia, F. Magnani, U. Arnaud, and G. Monti, "Installation of submarine power cables in difficult environmental conditions—The experience with 400 kV Messina cable," CIGRE Report 21–10, 1984, pp. 1–8.

[144] F. B. Jaafer, M. Saito, N. Yamada, and M. Matsuda, "Installation of 132 kV oil-filled submarine cable at Pulau Langkawi, Malaysia," IEEE/CSEE Joint Conference on High Voltage Transmission Systems in China, Oct. 1987, pp. 831–837.

[145] A. Durai, S. Saibir, and H. Soerotaroeno, "Special technical features associated with the installation of submarine power cables in Indonesian waters," Seventh Conference on Electric Power Supply Industry, 15–22 Oct. 1988, Vol. 3A, pp. 3.57/1-15.

[146] Y. Nakamura, T. Kuroshima, M. Takeuchi, T. Sanpei, S. Suzuki, S. Ishikura, H. Inoue, and T. Uematsu, "Installation of 66 kV XLPE power-optical fiber composite submarine cable and water pipe for the Trans-Tokyo Bay highway," *IEEE Trans. Power Delivery*, Vol. 10, July 1995, pp. 1156–1167.

[147] J. M. Oudin, I. Eyraud, and L. Constantin, "Some mechanical problems of submarine cables," CIGRE Proc., Report 21-08, Paris, 1972.

[148] CIGRE, HV Insulated Cables Study Committee, "Recommendations for mechanical tests on submarine cables," *Electra*, No. 68, Jan. 1980, pp. 31–36.

Chapter	**HIGH-VOLTAGE DIRECT-CURRENT**
14	**CABLES**

C. Doench, K. D. Srivastava

14.1 INTRODUCTION

Although low-voltage direct-current (dc) cables for distribution networks were in use at the very early stages of electric power utilization, the major advancements in dc cable technology are closely related to the growth of high-voltage power transmission. More specifically to the growth of high-voltage direct-current (HVDC) transmission since the early 1950s. The economics of HVDC transmission often becomes more attractive when substantial lengths of cable are required for river or marine crossings. The first HVDC cable link was commissioned in 1954 between the Swedish mainland and the Island of Gotland in the Baltic Sea. This was a 100-km link utilizing a 100-kV, mass impregnated oil-paper submarine cable. Table 14.1 shows the present status of other HVDC cable projects and proposed new installations. It should be noted that, except for a few cases, almost all installations are for marine applications. The technology and installation design practices developed for alternating-current (ac) submarine cables are also applicable to HVDC cables.

Any HVDC cable installation is an integral part of a larger electric power transmission network. The performance requirements, such as the insulation level and power transfer capability, and the installation specifications are determined by the overall power system planning for each project. A technical/economic evaluation is always carried out at this stage. Although the ex-factory cost of a HVDC cable is usually much less than the equivalent high-voltage alternating-current (HVAC) cable, the dc cable has other costs of terminal equipment associated with it. Reference [1] reviews the economic/technical comparison between HVAC and HVDC overhead transmission links. The methodology of economic/technical comparison is also applicable to cable links. In addition to the terminal equipment needed for dc interconnections, the reliability and security requirements call for dedicated telecommunication links for HVDC installations.

High-voltage direct-current links are often more economical for bulk power transfer applications, especially for interconnections between major power utilities. As Table 14.1 shows, in many instances, such bulk power transfer projects, cross-national jurisdictions, and therefore in economic/technical evaluation factors such as national energy policies, currency exchange rate contracts, and the necessity of long-term power purchase agreements have to be considered.

Practical designs for HVDC cables are dominated by oil-paper technology. The other cable design alternatives, such as extruded polyethylene cables, compressed gas

TABLE 14.1 HVDC Cable Links[a]

Year	Location	Type of Cable[b]	Voltage (kV)	Length (km)	Comments
1954	Swedish mainland to Gotland	Submarine solid type	100	100	1 pole
1961	England/France, Cross channel	Submarine solid type	100	65	2 poles
1965	New Zealand, Cook strait	Submarine high pressure gas filled	250	42	2 poles
1966	Italy to Sardinia	Submarine solid type	200	120	1 pole
1969–1971	UK (Kingsnorth–Beddington–Willesden)	Land based oil filled	266	40	2 poles
1969	Canadian mainland to Vancouver Island	Submarine solid type	266	27	2 poles
1970	Swedish mainland to Gotland to supplement 1954 cable	Submarine oil filled	150	96	2 poles
1973	Majorca/Menorca	Submarine oil filled	200		Initially a 3-phase 132 kV connection
1976	Canadian mainland to Vancouver Island	Submarine oil filled	266	27	2 poles
1977	Japan (Hokkaido–Honshu)	Submarine oil filled	250	65	2 poles
1977	Norway/Denmark (Skagerrak)	Submarine solid type	250	65	2 poles
1983	Swedish mainland to Gotland to supplement 1970 cables	Submarine oil filled	150	100	1 pole
1984–1986	England/France, Cross channel	Submarine solid type	270	32	2 poles
1986	Italy to Corsica		200	415	1 pole
1987	Swedish mainland to Gotland to supplement 1983 cable	Submarine oil filled	150	103	2 poles
1988	Sweden/Finland, Konti-Skan		285	150	1 pole
1989	Sweden/Finland, Fenno-Skan		400	200	1 pole
1991	New Zealand, Cook Strait to supplement 1965 cable	Submarine solid type			
1992	Canada St. Lawrence River Crossing, Hydro Quebec	Oil filled. Installed in a tunnel	500		2 poles

[a]There is insufficient information available on several projects, for example, installations in the former USSR, the Baltic Cable project, the Scotland/Ireland Cable project, the Greece/Italy cable project, and the Iceland/Scotland cable project.

[b]All the cables installed, so far, use impregnated-paper insulation.

insulated cables, and high-temperature superconducting cables, are still in the development stage. Of those alternatives, only extruded polymeric cables are viable for submarine applications. There are three designs of oil-paper cables in present-day use, namely the impregnated paper insulated "solid"-type, the pressurized gas-filled preimpregnated paper insulated-type and the oil-filled paper insulated-type. The pipe-type cable has also been proposed, but so far there are no HVDC installations using this technology.

The paper insulated solid-type cables have the major share of present installations. These were first used in the Swedish Gotland link (1954) followed by the England/France Cross Channel Link (1961), Konti-Skan Scandinavian link (1965), Italian Sardinia link (1965), and the Canadian Vancouver Island links (1969). The first oil-filled HVDC cable was used in England for a land-based project. The nonpressurized solid-type cable may be competitive with oil-filled cables for long marine links since additional cooling is not often needed. The only application of the pressurized gas-filled HVDC cable is for the Cook Strait interconnection in New Zealand. The oil-filled cable technology has been successfully used for the second Canadian Vancouver Island link (1976) and the Majorca–Menorca link (1973), which was initially operated at 132 kV ac.

Quite early in the development of HVDC cables, it was apparent that the cable dielectric would behave very differently under dc electric stress as compared to ac. Under dc, the insulation does not suffer from corona discharges in voids or from the dielectric tan δ loss. The electric stress distribution under dc, however, is controlled by the resistance of the cable dielectric, which is nonlinear with temperature; that is, the electric stress distribution across the dielectric will change with the temperature difference across the dielectric. Moreover, under dc stress the cable dielectric will accumulate significant space charge, which may prove detrimental if the applied electric stress is suddenly reversed. In designing HVDC cables, therefore, special attention has to be paid to the effect of load cycle and polarity reversal, as well as the impact of lightning or internally generated overvoltages. Comprehensive literature exists on these aspects and the following sections discuss the underlying physical concepts [2–20].

The power industry experience with extruded polymeric dielectrics for HVAC cables is very extensive and practical designs exist for system voltages to 500 kV. For HVDC cables, however, the space-charge accumulation phenomena for polymeric dielectrics is quite different than for layered oil-paper insulation, and the physical processes are not fully understood. The space-charge decay (dissipation) time constants are usually much longer for polymeric insulation, such as crosslinked polyethylene (XLPE). Research and development (R&D) work is underway to promote the use of XLPE for HVDC cables [21–35].

It is quite interesting to note that both oil-paper and polymeric dielectric HVDC cables have been in use for high-voltage testing equipment and for scientific R&D applications for voltages up to 1 MV. In such applications the current requirements are quite low (under a few milliamps usually), and therefore the difficulties of space-charge accumulation, polarity reversal, and changes in electric stress distribution with temperature do not present significant operational limitations.

As expected, in many areas of cable design, installation, manufacturing, and testing procedures, there is considerable overlap between HVAC and HVDC cable technology. Reference should, therefore, be made to other chapters in this book. The following sections emphasize the differences and highlight the specific technical considerations for HVDC cables.

14.2 ELECTRICAL BEHAVIOR OF DC CABLES

Under static or quasi-static voltage conditions, the electrical performance of a dc cable is determined by the resistance of its various components, that is, the resistance of the conductor, insulation, sheath, and the protective armor. Since some of these resistance parameters are nonlinear, the current rating of the cable and the electric stress across the dielectric/insulation are a complex function of the temperature across the cable insulation and the cable current. The temperature across the insulation, of course, depends upon the thermal resistance of the soil and the various components of the cable.

Paper insulated, self-contained fluid-filled (SCFF) cables are the most commonly used design as shown in Table 14.1. An analytical treatment of the electric stress distribution, under load conditions, for a typical SCFF dc cable is presented to illustrate its electrical performance. Figures 14.1, 14.2, and 14.3 show typical cross sections of such cables and a simplified equivalent circuit.

For designing a dc cable, there are two basic considerations: load that can be carried by a given cable and cable required for a given load. Of course, the cable has to operate within the specified limits of temperature, electric stress and other practical requirements for a given application. As mentioned before, the electric stress distribution is a complex function of applied voltage, cable current, and temperature difference across the dielectric.

14.2.1 Stress Distribution and Maximum Current

Assume

$$\nabla \cdot (\sigma \, \nabla V) = 0 \tag{14.1}$$

and

$$\sigma = \sigma_0 \exp(\alpha\theta + bE) \tag{14.2}$$

where V is the applied voltage, θ the temperature at radius r, α a temperature constant, b a stress constant, σ_0 the insulation base conductivity and, $E = dV/dr$.

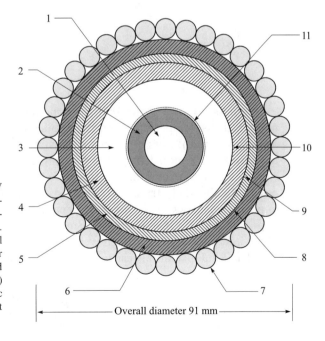

Figure 14.1 Schematic of submarine 300-kV dc oil-filled impregnated paper cable. Self-supporting copper conductor 400 mm^2 cross section with outer diameter of 32.4 mm. Insulation thickness 11.4 mm and overall cable diameter 91 mm. (1) Oil duct, (2) copper conductor, (3) oil-paper insulation, (4) lead sheath, (5) plastic sheath, (6) jute serving, (7) steel armor, (8) antiterodo tape, (9) metallic reinforcement, (10) screen, (11) screen. (Not to scale.) (Courtesy Pirelli Cables).

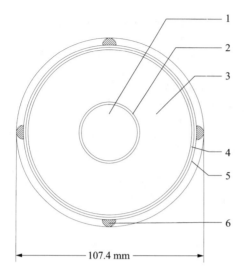

Figure 14.2 Schematic of ±600-kV pipe-type cable. Copper concentric conductor 1000 mm², 40.6 mm diameter. Oil-impregnated paper insulation 29.5 mm thickness. Overall diameter with skid wires 107.4 mm: (1) conductor, (2) conductor shielding, (3) oil paper insulation, (4) insulation shield, (5) moisture seal, (6) steel skid wires [14].

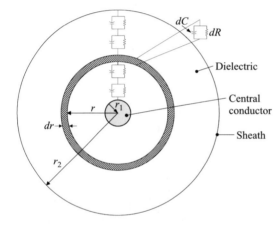

Figure 14.3 Equivalent circuit of a dc cable.

Equation (14.2) shows that the conductivity of the cable insulation is dependent upon both the temperature and the applied electric stress. Since, unlike ac cables, the dielectric constant does not control the stress distribution in the insulation, changes in the insulation conductivity, due to temperature rise under load, affect the internal electric stress distribution. Considerable research has been done on the behavior of practical cable dielectrics, and Figs. 14.4 and 14.5 show experimental data for fluid-impregnated paper [2–7, 9–14]. It should be noted that the typical values for α are approximately 0.1 per degree Celsius and 0.13 per degree Celsius for impregnated paper and polyethylene, respectively.

The total insulation resistance R is given as

$$R = \int_{r_1}^{r_2} \frac{d_r}{2\pi r \sigma} \tag{14.3}$$

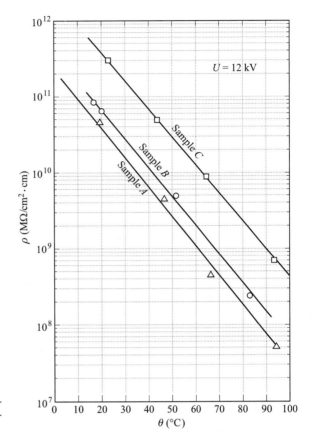

Figure 14.4 Temperature variation of resistivity of oil-impregnated paper for three different cables at dc voltage of 12 kV [7].

where r_2 and r_1 are the external and internal radii of the insulation. Then

$$I_0 = \frac{V}{\int_{r_1}^{r_2} dr/2\pi r\sigma} \tag{14.4}$$

where I_0 is the insulation loss current per unit length. Then the temperature at radius r is given as

$$\theta_r = \theta_2 + \theta(r \text{ to } r_2) = \theta_2 + \Delta\theta$$

The thermal resistance T_i is given as

$$T_i = \int_{r_2}^{r} \frac{dr\lambda}{2\pi r} = \frac{\lambda}{2\pi} \ln \frac{r_2}{r}$$

where λ is the thermal resistivity of insulation. Let

$$\mu = \frac{r_2}{r}$$

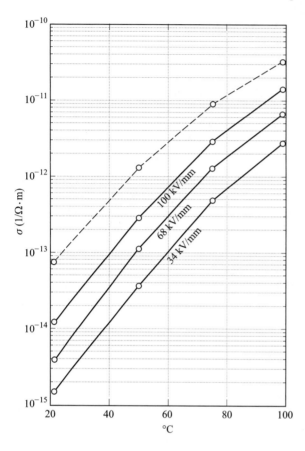

Figure 14.5. Temperature variation of conductivity of oil-impregnated paper dielectric for different electric fields. Dotted line represents the conductivity of the impregnating fluid [7].

Then

$$\theta_r = \theta_2 + \frac{\Delta\theta}{\ln\mu}\ln\frac{r_2}{r}$$

$$= \theta_2 + \frac{\beta}{\alpha}\ln\frac{r_2}{r} \tag{14.5}$$

where

$$\beta = \frac{\alpha\Delta\theta}{\ln\mu} \tag{14.6}$$

Then from Eq. (14.2)

$$\sigma_r = \sigma_0\exp\left(\alpha\theta_2 + \beta\ln\frac{r_2}{r} + bE\right)$$

$$= \sigma_0\left(\frac{r_2}{r}\right)^{\beta}\exp(\alpha\theta_2 + bE) \tag{14.7}$$

The dielectric loss current is constant:

$$I_0 = 2\pi r \sigma_r E_r$$
$$= E_r \times 2\pi r_2^\beta r^{1-\beta} \sigma_0 \exp(\alpha\theta_2 + bE) \tag{14.8}$$

From Eq. (14.4)

$$I_0 = \frac{V_0}{\displaystyle\int_{r_1}^{r_2} \frac{dr}{2\pi r_2^\beta r^{1-\beta} \sigma_0 \exp(\alpha\theta_2 + bE)}}$$

Then stress distribution is given as

$$
\begin{aligned}
E_r &= \frac{V_0}{2\pi r_2^\beta r^{1-\beta} \sigma_0 \exp(\alpha\theta_2 + bE) \displaystyle\int_{r_1}^{r_2} \frac{dr}{2\pi r_2^\beta r^{1-\beta} \sigma_0 (\alpha\theta_2 + bE_r)}} \\
&= \frac{V_0/r^{1-\beta} e^{bE}}{\displaystyle\int_{r_1}^{r_2} dr/r^{1-\beta} e^{bE_r}} \\
&= \frac{V_0 r^{\beta-1} e^{-bE}}{\displaystyle\int_{r_1}^{r_2} e^{-bE} r^{\beta-1}\, dr}
\end{aligned}
\tag{14.9}
$$

As a first approximation to solve Eq. (14.9) for stress distribution, assume

$$\sigma = \sigma_0 e^{\alpha\theta} E^\gamma \tag{14.10}$$

where

$$\gamma = \frac{V_0}{r_2 - r_1} b$$

Then Eq. (14.1) can be expressed in cylindrical coordinates, with field distribution constant in the Z and Φ directions,

$$\frac{dV}{dr}\frac{d\sigma}{dr} + \frac{\sigma}{r}\frac{dV}{dr} + \sigma\frac{d^2V}{dr^2} = 0 \tag{14.11}$$

From Eq. (14.10),

$$
\begin{aligned}
\frac{d\sigma}{dr} &= \sigma_0 \left[e^{\alpha\theta} E^\gamma \alpha \frac{d\theta}{dr} + \gamma e^{\alpha\theta} E^{\gamma-1} \frac{dE}{dr} \right] \\
&= \sigma_0 \left[e^{\alpha\theta} E^\gamma \alpha \frac{d\theta}{dr} + \gamma e^{\alpha\theta} E^\gamma \times \frac{1}{E} \frac{d^2V}{dr^2} \right] \\
&= \sigma \left[\alpha \frac{d\theta}{dr} + \gamma \left(\frac{dV}{dr} \right)^{-1} \frac{d^2V}{dr^2} \right]
\end{aligned}
\tag{14.12}
$$

Then Eq. (14.11) becomes

$$\frac{dV}{dr}\left[\alpha\frac{d\theta}{dr}+\gamma\left(\frac{dV}{dr}\right)^{-1}\frac{d^2V}{dr^2}\right]\sigma+\frac{\sigma}{r}\frac{dV}{dr}+\sigma\frac{d^2V}{dr^2}=0$$

or

$$\frac{dV}{dr}\left[\alpha\frac{d\theta}{dr}+\frac{1}{r}\right]+\frac{d^2V}{dr^2}(1+\gamma)=0$$

From Eq. (14.5)

$$\frac{d\theta}{dr}=-\frac{\beta}{\alpha}\times\frac{1}{r}$$

Then

$$\frac{1}{r}\frac{dV}{dr}(1-\beta)+\frac{d^2V}{dr^2}(1+\gamma)=0 \qquad (14.13)$$

Let

$$\sigma-1=\frac{\beta-1}{\gamma+1}$$

Then

$$\sigma=-\frac{\beta+\gamma}{\gamma+1} \qquad (14.14)$$

and

$$\frac{1}{r}(\sigma-1)\frac{dV}{dr}=\frac{d^2V}{dr^2}$$

Then

$$\frac{dV}{dr}=(k_1r)^{\sigma-1}$$

and

$$V_0=\frac{k_1^{\sigma-1}}{\sigma}r^{\sigma}+k_2$$
$$=k_3r^{\sigma}+k_2$$

where k_1, k_2, and k_3 are constants.

The boundary conditions are

$$V = \begin{cases} 0 & \text{when } r = r_2 \\ V_0 & \text{when } r = r_1 \end{cases}$$

Then

$$0 = k_3 r_2^\sigma + k_2 \qquad V_0 = k_3 r_1^\sigma + k_2 \qquad k_2 = -k_3 r_2^\sigma \qquad k_3 = \frac{V_0}{r_1^\sigma - r_2^\sigma}$$

Then

$$V_r = V_0 \frac{r_2^\sigma - r^\sigma}{r_2^\sigma - r_1^\sigma}$$

Then by differentiation

$$-\frac{dV}{dr} = E = \frac{V_0 \sigma r^{\sigma-1}}{r_2^\sigma - r_1^\sigma} \tag{14.15}$$

which is a first approximation to the stress distribution. To calculate the stress distribution, it is useful to express E_r in terms of E_c; and E_c is the stress at r_c where the ac distribution equals the dc distribution. From Eq. (14.8)

$$I_0 = \begin{vmatrix} 2\pi r_2^\beta r^{1-\beta} \sigma_0 e^{\alpha\theta_2} e^{bE_r} \times E_r \\ 2\pi r_2^\beta r_c^{1-\beta} \sigma_0 e^{\alpha\theta_2} e^{bE_c} \times E_c \end{vmatrix}$$

By equating these two expressions

$$r^{1-\beta} e^{bE_r} E_r = r_c^{1-\beta} e^{bE_c} \times E_c$$

Then

$$E_r = E_c \left(\frac{r_c}{r}\right)^{1-\beta} e^{b(E_c - E_r)} \tag{14.16}$$

In order to calculate E_c and r_c, the first approximation for E_c is used:

$$E_c = \frac{V_0 \sigma r_c^{\sigma-1}}{r_2^\sigma - r_1^\sigma} = \frac{V_0}{r_c \ln \mu}$$

$$\frac{\sigma r_c^\sigma}{(\mu^\sigma - 1) r_1^\sigma} = \frac{1}{\ln \mu}$$

and

$$r_c = r_1 \left[\frac{\mu^\sigma - 1}{\sigma \ln \mu} \right]^{1/\sigma} \qquad (14.17)$$

When the load is not given but the maximum allowable stress E_{max} is stated, the load may be calculated as follows. From Eq. (14.6),

$$\beta = \frac{\alpha \, \Delta\theta}{\ln \mu}$$

But

$$\Delta\theta = WT_i$$

where the thermal resistance of the insulation T_i

$$T_i = \frac{\lambda}{2\pi} \ln \mu \text{ and } W = I^2 R_c$$

Then

$$\beta = \frac{\alpha \lambda I^2}{2\pi} R_c \qquad (14.18)$$

where I is the load current and R_c is the per unit resistance of the conductor.

$$R_c = \frac{\rho}{r_1^2} \qquad (14.19)$$

Under maximum load conditions, the maximum stress occurs under the sheath

$$E_{r_2} = E_m = \frac{V_0 r_2^{\sigma-1}}{r_2^\sigma (1 - \mu^{-\sigma})} = \frac{V_0}{r_2 (1 - \mu^{-\sigma})}$$

and

$$V_0 = \frac{E_m r_2}{\sigma} (1 - \mu^{-\sigma})$$

Maximum power is given as

$$P = V_0 I$$

But from Eqs. (14.18) and (14.19)

$$I = \sqrt{\frac{2\pi\beta}{\alpha\rho}} r_1 = K\sqrt{\beta} r_1$$

Let

$$K = \sqrt{\frac{2\pi}{\alpha\rho}} \qquad P = KE_m r_1 r_2 \frac{\sqrt{\beta}}{\sigma}(1 - \mu^{-\sigma})$$

Then maximum power occurs when

$$\frac{dP}{d\sigma} = 0$$

But

$$\beta = \sigma(1+\gamma) - \gamma \qquad \frac{d\beta}{d\sigma} = 1 + \gamma$$

Then

$$P = K_0 \frac{\sqrt{\beta}}{\sigma}(1 - \mu^{-\sigma})$$

$$\frac{dP}{d\sigma} = \frac{1+\gamma}{2\sqrt{\beta}\sigma}(1 - \mu^{-\sigma}) = \frac{\sqrt{\beta}}{\sigma^2}(1 - \mu^{-\sigma}) + \frac{\sqrt{\beta}}{\sigma}(\ln\mu \ \mu^{-\sigma}) = 0$$

This may be rationalized as

$$(\gamma+1)\left(\frac{\beta}{\gamma+\beta} - \frac{1}{2}\right)(1 - \mu^{-\sigma}) = \beta\mu^{-\sigma}\ln\mu \qquad (14.20)$$

This may be solved numerically for β and consequently the load and load current.

References [2–7] and [9–11] give more details of the analytical approaches to the calculation of electrical characteristic of dc cables under load. Figures 14.4 and 14.5 show the typical variation of dielectric conductivity with temperature and stress, and Fig. 14.6 shows stress distribution in the dielectric for different operating conditions for one cable design.

14.2.2 Direct-Current Cable Design: Numerical Example

Specifications
(a) A 300-kV submarine cable to carry 30 MW of power per pole.
(b) Maximum electric stress not to exceed 30 kV/mm.
(c) Construction to be of oil-impregnated paper solid-type, and the maximum conductor temperature to be below 55°C.

Design Assumptions. For 30 MW and 300 kV, the maximum load current is 1000 A. Cable is buried 500 mm in soil with a thermal resistivity of 0.8 K-m/W. Let the basic impulse level (BIL) be $2.5V_0$, that is, 750 kV, and let the maximum electrical stress at BIL be 90 kV/mm. Assume a conductor size of 630 mm^2 (1250 MCM or kilo circular mils). This is equivalent to conductor diameter of 33 mm or a radius of 16.5 mm. As a conservative step, assume the surge voltage of 750 kV adds to the dc pre-stress of $V_0/2$,

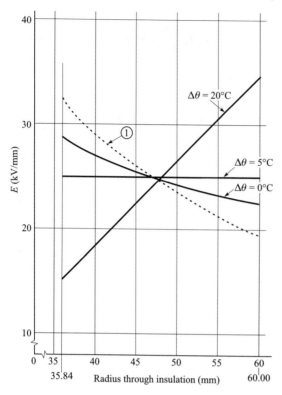

Figure 14.6. Electric stress variation in oil-paper dielectric of ±600 kV dc cable for different temperature drop across dielectric. Curve 1 shows the capacitive stress distribution [13].

that is, let the initial maximum stress be $(750 + 150)$kV, or 900 kV. Initially, in a coaxial geometry, the maximum stress may be assumed to occur at the conductor surface, that is, ignoring any stress inversion under load. Therefore,

$$\frac{900 \times 10^3}{16.5 \times \ln \mu} = 90 \, \text{kV}$$

or

$$\ln \mu = 0.606$$

or

$$\mu = \frac{r_2}{r_1} = 1.833$$

For the above value of the radial ratio r_2/r_1 and assuming a construction similar to Fig. 14.1, the following cable dimensions can be determined:

Radius of the insulation $r_2 = 30.5$ mm.

Insulation thickness $= 14$ mm.

Insulation diameter $= 61$ mm.

Shield diameter = 62 mm.

With a sheath thickness of 3 mm,

Sheath diameter = 68mm.

With a jacket thickness of 4 mm,

Jacket diameter = 76 mm.

Bedding diameter = 79 mm, assuming a bedding thickness of 1.5 mm.

Assume a 6-mm armor; the diameter over the armor = 91 mm.

With serving of 1.5 mm, the overall diameter = 94 mm.

Following the procedure of International Electrotechnical Commission (IEC) 287, [20] and as discussed in Chapter 11 on cable installation design, thermal resistances are determined as

$$T_i = \frac{6.5}{2\pi} \ln \frac{62}{33} = 0.053 \text{ K-m/W} \quad \text{(insulation)}$$

$$T_j = \frac{3.5}{2\pi} \ln \frac{76}{68} = 0.062 \text{ K-m/W} \quad \text{(jacket)}$$

$$T_b = \frac{6.5}{2\pi} \ln \frac{79}{76} = 0.040 \text{ K-m/W} \quad \text{(bedding)}$$

$$T_s = \frac{6.5}{2\pi} \ln \frac{94}{91} = 0.034 \text{ K-m/W} \quad \text{(serving)}$$

$$T_g = \frac{0.80}{\pi} \ln \frac{4 \times 500}{94} = 0.389 \text{ K-m/W} \quad \text{(ground)}$$

Therefore the sheath to ambient thermal resistance is 0.525 K-m/W, the insulation thermal resistance is 0.530 K-m/W, and the total thermal resistance is 1.055 K-m/W.

For copper conductor resistance at 50°C is given as

$$R_c = \frac{10.371}{1.25} \times \frac{285}{255} \times 3.2808 = 30.85 \, \mu\Omega\text{m}$$

Conductor loss W for 1000 A is given as

$$W = 30.85 \text{ W/m}$$

Temperature drop acros the insulation is determined as

$$\Delta\theta = 30.85 \times 0.53 = 16.4°\text{C}$$

For this type of insulation

$$\alpha = 0.1 \qquad b = 0.03$$

Average stress = 12 kV/mm.

To calculate stress distribution

$$\gamma = 0.03 \times \frac{21}{\ln \mu} = 0.2$$

Then

$$\beta = \frac{\alpha \, \Delta\theta}{\ln \mu} = 2.7 \qquad \sigma = \frac{2.9}{1.2} = 2.42$$

Then for the load condition

$$E_r = \frac{300 \times 2.42 r^{1.42}}{30.5^{2.42} - 16.5^{2.42}} = 0.24 r^{1.42}$$

Then E at the conductor is 13 kV/mm, and E at the sheath is 30 kV/mm. For 750 kV, $E = 750/16.5 \times 0.606 = 75$ kV/mm. The maximum surge stress is $75 + 13 = 88$ kV/mm. For no-load conditions

$$\beta = 0 \qquad \sigma = 0.167 \qquad E = 290 r^{-0.833}$$

Then E at the conductor is 28 kV/mm, and E at the sheath is 168 kV/mm.

The maximum temperature rise of the conductor above ambient is $30.85 \times 1.055 = 32.5°C$. Therefore, the maximum ground ambient temperature is $55 - 32.5 = 22.5°C$. Hence this design is adequate.

14.3 TRANSIENT ELECTRIC STRESSES ON HVDC CABLES

Equation (14.2) shows that the conductivity of a solid cable dielectric is dependent upon both the temperature and the applied electric stress. If a steady dc stress is applied, without any load current, it takes a relatively long time for the stress distribution to stabilize. This behavior is to be expected and is quite different from the case with ac applied voltage. In one of the very early experiments on a 400-kV solid-type paper insulated dc cable, it took over 90 min for the stress distribution to reach steady state [2]. As a direct consequence of the internal conduction mechanism inside a solid dielectric, space-charge accumulation occurs and, a cable (like a capacitor) will take a long time to return to an uncharged state after the applied stress is removed, even if the cable is short-circuited. Since a dc cable is normally a part of an ac power network, it may be subjected to transient overvoltages of either polarity for a variety of operational conditions, for example:

(a) Polarity reversal of dc voltage due to changed direction of power transport.
(b) Valve misfire in an HVDC converter giving rise to voltage oscillations that may reach twice the normal dc voltage.
(c) Switching an HVDC converter on to an open line can cause surges up to 1.5 times the rated voltage.
(d) Lightning strokes to the overhead lines associated with the cable installation. The actual overvoltage will depend upon the protection level and may involve a reverse polarity surge on the cable.

The dielectric strength of fluid-impregnated paper under combined dc and surge voltage has been investigated by several researchers [11, 15, 16]. The principal mechanisms that cause the lowering of dielectric strength are thought to be: (a) gas bubble formation in butt gap spaces at conductor–paper dielectric interface and (b) bulk space-charge polarization of the dielectric under dc voltage stress. The reduction in breakdown strength can be substantial, for example, in one investigation from 130 to 60 kV/mm when reverse polarity lightning impulses were applied to samples prestressed with dc stress of 75 kV/mm [15]. The number of surge applications and the interval between surges also affect the observed reduction in strength. Figure 14.7 shows data from [16] for a laboratory model with cylindrical geometry, and Fig. 14.8 shows data for a polymeric cable [22]. A detailed analysis of stress distribution under transient conditions is given in [11], which also gives experimental results for the Norway/Denmark (Skagerrak) solid-type 250-kV submarine cable under polarity reversal conditions. It has also been suggested that the dc cable dielectric may also experience a combined ac and dc voltage stress due to faulty HVDC valve operation in the terminal equipment. Figure 14.9 shows the results of an experiment for such a case [4].

As will be discussed in Section 14.7 the current industry practice allows for cable testing under combined dc and surge voltages, including reversal of polarity.

14.4 DESIGN OF HVDC CABLES

High-voltage direct-current transmission has been a growing area of application in electrical utilities worldwide. New technological innovations in HVDC converters

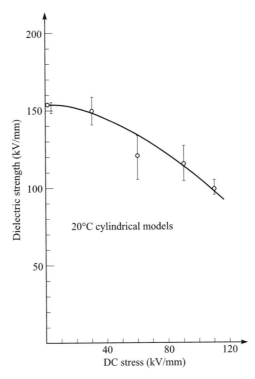

Figure 14.7 Effect of dc prestress on impulse breakdown strength of oil-impregnated samples in cylindrical geometry at room temperature [16].

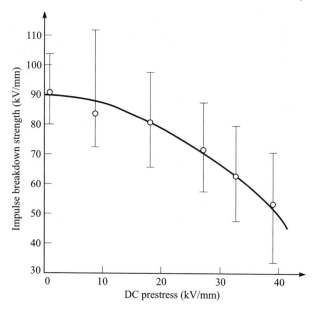

Figure 14.8 Effect of dc pre-stress on impulse breakdown strength of XLPE cable samples at room temperature [22].

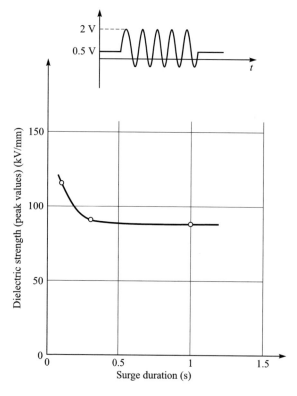

Figure 14.9 Effect of oscillating power frequency surge on dielectric strength of oil-impregnated paper as function of surge duration. One hundred shots at each level increasing in steps of 3 kV/mm. Paper 0.08 mm thick and density 0.8 g/cm³, impregnated with dodecyl benzene [4].

and substation equipment have markedly lowered costs, and HVDC schemes are often economically competitive. New developments in SCFF and high-pressure fluid-filled (HPFF) dc cable technology for voltages up to ±600kV have further enhanced the use of HVDC transmission. Table 14.1 shows only the transmission schemes utilizing cable; the list of land based schemes is considerably longer.

In economic/technical schemes involving cables, it is necessary to establish the required cable dimensions with good accuracy. In general, the design is determined by the limiting values of the temperature across and the electric stress within the dielectric. As shown in Section 14.2, the steady-state and transient stress in fluid-impregnated paper insulation involve more complex calculations for dc than for ac working conditions, and iterative design procedures are necessary.

Figure 14.1 shows that the overall construction of a dc cable is very similar to ac, and the extensive industry experience in the design, selection of material, manufacturing, and handling of ac cables is also relevant to dc. Although for ac cables, synthetic insulating materials, such as XLPE and ethylene-propylene-rubber (EPR) dominate, for dc the insulation of choice is still fluid-impregnated paper. Considerable R&D activity has been reported on the use of XLPE for dc cables, but to date there are no practical installations [21–35]. The principal technical hurdle is the lack of understanding of electrical performance under prolonged dc stress and the influence of internal space charge under polarity reversal conditions. The space-charge accumulation, its location and movement under combined dc and surge voltages is poorly understood. The discussion below is, therefore, limited to impregnated paper dielectric. In Section 14.5, however, a summary of the current development experience with the use of XLPE for dc cables is provided.

The basic constraints and the design procedure for a fluid-impregnated paper cable are outlined below.

14.4.1 Parameter Constraints

(a) The long-term dc stress of fluid-impregnated paper at normal operating temperatures is approximately 30–40 kV/mm. The acceptable transient stress level is, however, between 85 and 100 kV/mm.

(b) The maximum outer surface temperature for a stable backfill is typically between 50 and 60°C, and the maximum conductor temperature to avoid significant damage to insulation is typically 85°C.

(c) At very high operating voltages, in excess of ±1000 kV, thermal instability of insulation may occur even at operating temperatures [19].

(d) Also, for very high voltage designs, the bending radius of the cable presents processing, reeling, and shipping difficulties.

14.4.2 Outline of Design Procedure

Because of the interdependence of the resultant electric stress distribution on load current and the temperature drop across the dielectric, it is not possible to develop a linear step-by-step design procedure. An iterative methodology has to be adopted, as outlined below.

Step 1. For the given specifications of voltages (steady state and transient) and load, an initial conductor size and loss is selected for the maximum conductor temperature.

Step 2. With a given conductor size, an insulation thickness is determined for the limiting values of steady-state and transient electric stresses. An initial approximation of equating the maximum, transient stress to the stress at the outer diameter of the insulation is often used to start the iterative calculation.

Step 3. Knowing the insulation thickness and the thermal characteristics of the soil and the other cable materials, the actual conductor and cable/soil interface temperatures are calculated.

Step 4. If any of the above temperatures are higher than the specified limits, a higher conductor size is selected and the above steps recalculated. If, in the first calculation, no temperature limits are exceeded, than a smaller conductor size is selected and the calculation repeated.

Step 5. The design calculation is complete when one temperature is allowed without the other being exceeded.

It should be noted that the IEC Standard 287 [20] for ampacity calculations is almost always followed for the detailed design. Also, the other industry standards should also be taken into account for a complete installation design, as noted in Chapters 4–7 and 11.

14.5 SELECTION OF MATERIALS

In Chapter 2, a general description of cable materials has been given. Also, since many of the HVDC cable installations are in marine environment, the design of submarine cables discussed in Chapter 13 is highly relevant.

Both copper and aluminum conductors are used. However, since ac effects are not very important, the usual conductor shapes are stranded or keystone; the latter shape is necessary for oil-filled cables. The major difference in materials arises for the main electrical insulation/dielectric. All the present installations use oil-impregnated paper. The paper used in dc cables is of higher density than for ac and is of high resistivity. The loss factor tan δ is, of course, not a major consideration. The paper thickness is varied to control the stress reversal phenomena in dc cables. It is not unusual to start with a lower thickness paper near the conductor, moving to a thicker paper in the middle of the insulation buildup, and then using a thinner paper near the outer sheath. This technique considerably improves the high-voltage performance of the cable under polarity reversal conditions. Insulation taping procedure is, therefore, more complex. It is essential to keep the humidity in the paper to a minimum. For deep-sea applications, the mechanical strength of the conductor becomes a major limiting parameter. Table 14.2 shows the range of parameters for the different types of impregnated paper dc cables for marine applications.

For over 40 years, cable designers have been very interested in using polyethylene extruded cable technology for dc cables. Although there has been considerable improvement in extruded cable technology for ac, the principal long-term life limitation is due to the formation of trees in the insulation. It was initially thought that since dc cables do not have alternating voltage stress corona in voids leading to "tree" formation would not be a major concern. However, there is evidence to suggest that discharge

TABLE 14.2 Typical Oil-Paper dc Cables for Submarine Applications

Cable Type	Voltage (kV)	Maximum Conductor Cross Section (mm^2)	Maximum Length (km)	Maximum Sea Depth (m)	Maximum Transmissible Power per Pair (two cables)
Paper-solid	±150	1000	250	600	400
Paper-solid	±250	1000	200	600	600
Oil-filled S.C.	±300	2000	120	1000	1200
Oil-filled S.C.	±600	2000	80	800	2000

channel propagation may occur by a comparatively few discharges under the high stress conditions in dc cables. There is considerable data now available on the possible use of polyethylene for dc cables [3, 21–35]. For a discussion of current research see Section 14.8.

In polymeric materials, under a prolonged dc, stress charge injection occurs at the electrodes and the stress distribution, as discussed in Sections 14.3 and 14.4, is further distorted by the presence of space charge. Under surge polarity reversal conditions, the dielectric breakdown strength is considerably reduced [22]. Figures 14.7–14.9 show the experimental results. In addition to the dependence on the dc prestress, the breakdown strength will also decrease with temperature of the dielectric.

In an earlier attempt to minimize the effect of space charges, two methods were proposed [23]: (a) to suppress charge injection from the electrodes and (b) to neutralize charge injected into the dielectric.

Both methods require modification of the XLPE compound. Charge trapping by polar radicals added to XLPE and formulations that reduce the density of traps in XLPE have been tried. Experimental results for a 250-kV XLPE cable, with and without additives, showed that space-charge effects contributed to failures, in addition to other known causes, such as voids and protrusions. Research work in this area is continuing, and XLPE with charged polymers and inorganic conducting filters as additives are being studied [25–35].

14.6 DIRECT-CURRENT CABLE ACCESSORIES

14.6.1 Background

Direct-current transmission lines are usually used for the transfer of unidirectional power over long distances. In most connections, the routes cross major water barriers, which entails the use of long submarine cables. Since submarine cables are manufactured in long continuous lengths, the requirement for cable accessories has been low in volume. Thus, the R&D effort has been limited by commercial constraints, and the accessories have been developed on the basis of contractual requirement. In consequence, the basis for design has been the modification of ac cable accessories [36–40]. The basic principles discussed in Chapter 10 are valid for dc designs as well, except that the surface creepage length is increased and the insulation design stress must take into account the polarity reversal conditions. The basic accessory requirements for all types of cable constructions are, joints in the following categories:

(a) Flexible factory joints

(b) Flexible field joints

(c) Rigid field joints

A brief description and the utilization of these devices follows:

(a) *Flexible factory joint*: These are employed when the cable shipping lengths exceed the factory impregnation capacity and two or more production lengths are joined. These joints are flexible and do not exceed the overall diameter of the cable core and consequently are very long (20 ft or more in some cases). The shipping length is completed by armoring over the joints in one continuous length. These joints must be capable of withstanding the same mechanical forces as the cable, for example, forces generated during the laying operation.

(b) *Flexible field joint*: These are required in the event of cable damage or for joining two or more shipping lengths in the field. The joints are similar to the factory joints with the added requirement that the cable armor has to be carefully reconstituted over the joint. As in the case of factory joints, the joint must withstand the same mechanical forces during cable laying operations as the cable.

(c) *Rigid field joints*: These joints are used for land cable installation and submarine cable repair when the joint is not subjected to mechanical forces generated by the tensioning equipment used in laying. These joints are much shorter than flexible joints and require considerably less time for field construction.

The *outdoor terminations* on the cross-channel, United Kingdom–France, connection were of 275-kV ac standard design [40]. In the case of other projects, such as the 300-kV Georgia Strait connection in British Columbia, Canada, standard ac termination designs with extended porcelains have been employed. The basic requirement for dc terminations is an extra long surface creepage path due to pollution deposition in industrial and coastal areas.

The solid type [paper insulated lead-covered (PILC)] cable has a severe maximum conductor temperature limitation of 55°C, whereas the SCFF cable is limited to 85–90°C conductor temperature range. This limitation has restricted the use of PILC-type cable mainly to submarine applications, with SCFF-type cable employed for land installations. This has led to the need for a transition joint between the two cable types.

This requirement arose during the design for the cross-channel, United Kingdom–France connection, where the long land section on the UK side was designed with SCFF cable, and the submarine section with PILC cable. Instead of developing a classical transition joint, since only four joints were required, this problem was solved by using back-to-back SF_6 immersed terminators. Description of the accessories employed for the United Kingdom–France cables are contained in [37, 38, 40], and additional details of practical joints and terminations are available in [2, 8, 12, 18].

14.6.2 Hydraulic Systems

Self-contained fluid-filled cable systems require a fluid supply and pressure control system (see Chapters 5 and 6). In the case of land cables, the supply is provided by static reservoirs distributed along the cable route. However, for long submarine cable connections, the fluid feed locations are restricted to the termination ends only. This requires large volume fluid supplies, on the order of 10,000 gal or more, which cannot practically be sourced from static reservoirs. In additions, the supply system must provide sufficient fluid to prevent ingress of water in the event of a catastrophic cable severance. This has imposed design requirements that sufficient fluid flow be provided to meet the demand due to the cooling transient as well as sufficient fluid egress at the severed end to prevent water ingress. This translates into a requirement for a normal operating pressure on the order of 300 psi (2100 kPa) gage. In the event of cable damage or severance, this pressure is maintained for a short time period (approx. one hour). The pressure is then reduced to an intermediate level for several hours until the cable reaches ambient temperature when the pressure is reduced to a minimum level, to prevent water ingress, for an indefinite period. This pressure can be maintained for several weeks until repair procedures have been completed. Since the cable installations can be in remote areas associated with unmanned substations, the pressuring units employ mechanical pumps with remote control from a control center that may be many miles from the unit. The design and manufacturer of these units has become a specialty for a very limited number of suppliers. These units pose some additional design problems not encountered with pipe-type cable pumping units. The extremely low-viscosity fluid is ineffective as a pump lubricant. Due to the high level of gas absorption of the low viscosity fluid, the fluid reservoir tank must be maintained at all times with a vacuum blanket over the fluid surface to maintain the degassified condition.

The oil supply reservoir capacity is designed to guarantee (1) automatic supply at full pressure to a damaged cable during the thermal transient caused by load interruption, (2) sufficient fluid to manually perform proper valving changes, and (3) maintain fluid supply at a minimum pressure for an extended period to permit repair procedures to be carried out [36].

14.7 TESTING OF DC CABLES

As mentioned earlier, the first major installation of a HVDC cable was in 1954. In the early stages of the HVDC cable technology, in addition to the tests carried out by the cable manufacturer, the user utility specified its own set of tests to be carried out before shipment and at the site [41, 42]. As the usage of HVDC cables in power networks expanded in the 1960s, CIGRE (Conférence Internationales des Grands Réseaux Electriques) produced a set of test recommendations, that were later expanded in a 1980 publication to cover cables up to ±600 kV [43].

High-voltage direct-current cables almost invariably form a part of a larger power transmission network. It is to be expected that the service requirements would vary with specific application. It is, therefore, not uncommon for the user utility to specify tests in addition to those recommended by CIGRE or IEC [17, 44].

In a HVDC cable, the voltage stress profile in the insulation is dependent upon the temperature difference across the dielectric, which in turn is a function of the conductor current and temperature. It is, therefore, necessary to specify not only rated current but also the maximum conductor temperature and the maximum temperature drop across the cable insulation.

Since all the present installations of HVDC cables utilize oil-paper insulation, the test specification are applicable only to such designs. Extruded solid dielectric cable for HVDC applications are in the development stage and no comprehensive guidelines exist at present.

Several of the existing international standards for oil-filled paper insulated ac cables apply to HVDC cables as well, for example, for specifying cable characteristics (Chapters 5 and 6). The specific differences are noted in IEC documents for HVDC cables [43]. For dc test voltages, the ripple content should be below 3% and the frequency of the ac test voltage in the range 49–61 Hz. For measuring dielectric loss, the ac voltage corresponding to a maximum ac gradient of 2 kV/mm should be specified. The routine tests include measurement of conductor resistance, capacitance, power factor, and the relevant mechanical tests for land-based or submarine cables.

There are two sets of tests that are specially designed to evaluate HVDC cables for service conditions:

(i) Loading cycle and polarity reversal test
(ii) Combined dc and impulse voltage test

14.7.1 Load Cycling and Polarity Reversal Test

As noted above, the electric stress profile of a HVDC cable is critically dependent upon the temperature profile across the dielectric, which in turn depends upon the load cycle history of the cable. Polarity reversal across the dielectric, under certain conditions, may be highly detrimental to the cable insulation.

The cable sample to be tested must be at least 30 m in length complete with the required accessories. Since in some installations, for example, in submarine applications, some cable designs may experience migration of insulating compounds for steep slopes and large vertical difference between the cable ends. In such circumstances, an agreed test loop configuration, between the utility and manufacturer, is used for tests.

The load cycles and the dc stress are applied in the following manner:

(i) The cable is subjected to 30 daily loading cycles, with 8 h of heating followed by 16 h of cooling. The temperature variation limits are specified.
(ii) The first 10 cycles require the application of positive $2U_0$ between the conductor and the sheath. This is followed by an 8-h period of no current or voltage stress.
(iii) The second 10 cycles require the application of negative $2U_0$ between the conductor and the sheath. Again, followed by an 8-h rest period.
(iv) The third set of 10 loading cycles has a voltage of $1.5U_0$ applied between the conductor and the sheath. The initial voltage polarity is positive and the polarity is reversed every 4 h and at least one polarity reversal should coincide with the cessation of current in the loading cycle. The time duration of polarity reversal is 2 min. Since the insulation charge delay time constant is much longer, the above duration of polarity reversal may be extended to 10 minutes.

14.7.2 Combined DC and Impulse Voltage Test

If the HVDC cable installation is likely to experience lightning or internal overvoltages, a combined test with dc and impulse voltage of opposite polarity is recommended. In this test, the cable is loaded to raise its temperature to the maximum design temperature with a margin of plus 5°C. Initially, a negative dc voltage of $2U_0$ is applied between the conductor and the sheath for 2 h. While maintaining the dc voltage, 10 positive lightning impulses, as per IEC 230, are applied. The time interval between impulse applications should not be less than 2 min. No insulation failure should occur during this test.

14.8 EMERGING TRENDS IN HVDC CABLE TECHNOLOGY

For higher voltage HVDC cables, the combined dc and impulse voltage stress as well as the impact of polarity reversal are major design limitations. In some instances, the strategy adopted is to use HVDC terminal valve controls to limit the maximum voltage stress to which the cable is subjected. Moreover, whenever feasible, ZnO surge arrestors are used to limit transient overvoltages on the cable.

Power cable industry has almost 100 years of experience, and a great deal of it with oil-paper insulation. However, for HVDC applications, the temperature dependence of the dielectric resistance presents significant limitations in design and operation. Other insulation materials have been proposed, for example, polypropylene paper and extruded XLPE. Experimental XLPE designs have been tested for voltages up to ± 250 kV [24, 35]. Advantages and disadvantages of different dielectrics have already been discussed earlier in Section 14.5. Oil-paper designs still dominate the HVDC cable technology, and 700-kV designs have been laboratory tested in Japan. Crosslinked polyethylene insulation has a much higher space-charge dissipation time constant, since the free charges are often trapped in structural discontinuities in the XLPE matrix. In Japan research is proceeding with XLPE and inorganic additives to minimize the space-charge problems [32]. Both polarized and conductive fillers have been used and considerable improvement in the dc breakdown strength of XLPE cable material is reported. There is, however, insufficient field experience with the long-term stability of such polymeric mixtures.

The power utility industry now has over 40 years of operating experience with HVDC cables for 100 kV and above. A large proportion of such cable links are operating in marine environment. In 1981, a review of industry experience, spanning some 20 years, was undertaken by CIGRE for both ac and dc submarine cables. The survey covered 38 crossings and of these 7 were HVDC cables. Of the total 140 failures reported, for both ac and dc cables, the major cause of failure (approx. 78%) was mechanical damage inflicted by maritime vessels and the related activities. Mechanical protection for submarine cables is, therefore, a major cable and installation design consideration. Use of double steel armor is not uncommon, and in at least one instance, the power utility has resorted to constructing a tunnel for a major river crossing in Canada [17]. There are other major HVDC projects in the early stages of planning and that will require long lengths of dc cable [45, 46]. Synthetic tape insulation polypropylene-laminated paper (PPLP) has been successfully used for 500-kV ac cables in Japan, and research is underway to extend its use to dc cables [47].

Another problem often encountered is that for long lengths, oil pressure submarine cables present operational difficulties. Research has been conducted to establish feasibility of nonpressurized oil-paper dc cables. The results are, however, not very promising [48, 49].

The major current research thrust in the field of HVDC cable technology is to understand the behavior of dielectric materials. Most major cable manufacturers worldwide and other research establishments are actively engaged in this endeavor [50–57]. Since mass impregnated paper still dominates the commercially available designs, it is imperative that the designers have information on the electrical conduction processes in layered oil-paper insulation. As shown in Section 14.2, an empirical relationship has been established between electrical conductivity, temperature, and the electric stress. Reference [50] is an attempt to elucidate the physical conduction processes. It is proposed that the dominant process is ionic conduction, and the nature and concentration of the ionic carriers determines both the electric stress and temperature dependence of the electrical conductivity. Reference [54] is a very comprehensive treatment of electric fields in oil-paper dielectric. Under practical operating conditions the electric stress distribution is both capacitive and resistive and is time dependent. For instance, it depends upon the history of voltage application and whether a transient, quasi-stable, or steady-state condition obtains. The authors identify seven different electric stress regimes and indicate when an instability may lead to dielectric failure.

In the case of polymeric cables, there has been progress on two fronts: to study and analyze the behavior of space charge and to formulate molecular structures better suited for HVDC [51–57]. For HVAC applications the trend was to keep impurities in XLPE to a minimum so as to reduce the incidence of treeing. For HVDC applications, however, controlled impurities are introduced to reduce both the space-charge magnitude and the discharge time constraints. Practical HVDC cable designs utilizing modified XLPE are still in the development stage. One manufacturer, however, is offering a 100-kV dc cable as part of lower power (> 30 MW) HVDC schemes for connecting isolated generation is a large power network [58].

Reference [34] provides a summary of the current research in developing an acceptable polymer dielectric for HVDC cables.

REFERENCES

[1] "AC-DC economics and alternatives," C. T. Wu, Ed., 1987 Panel Discussion Report, *IEEE Trans. Power Delivery*, Vol. PWRD-5, 1990, pp. 1956–1976.

[2] S. S. Gorodetzky, "220-400 kV Direct Current cables," CIGRE Paper 206, 1956.

[3] R. Tellier, L. Constantin, J. M. Brenac, J. M. Oudin, and J. Bele, "Recent research work and progress in the technique of high voltage dc and ac cables," CIGRE Paper 212, 1958.

[4] P. Gazzana-Priaroggia, A. Morello, and N. Palmieri," The use of oil-filled cables for high voltage dc transmission systems," CIGRE Paper 206, 1966.

[5] F. H. Buller, "Calculation of electrical stresses in D.C. cable insulation," *IEEE Trans. Power Apparatus Syst.*, Vol. PAS-86, 1967, pp. 1169–1178.

[6] J. M. Oudin, M. Fallou, and H. Thevenon, "Design and development of direct current cables," *IEEE Trans. Power Apparatus Syst.*, Vol. PAS-86, 1967, pp. 304–311.

[7] E. Occhini and G. Maschio, "Electrical characteristics of oil-impregnated paper as insulation for HV DC cables," *IEEE Trans. Power Apparatus Syst*, Vol. PAS-86, 1967, pp. 312–326.

[8] F. H. Last, P. Gazzana Priaroggia, and F. J. Miranda, "The underground DC link for the transmission of bulk power from the Thames estuary to the centre of London," *IEEE Trans. Power Apparatus Syst.*, Vol. PAS-90, 1971, pp. 1893–1901.

[9] S. C. Chu, "Design stresses and current ratings of impregnated paper insulated cables for HVDC," *IEEE Trans. Power Apparatus Syst.*, Vol. PAS-86, 1967, pp. 1029–1036.

[10] G. Maschio and E. Occhini, "High voltage direct current cables: the state of the art," CIGRE Paper 21-10, 1974.

[11] B. R. Nyberg, K. Herstad, and K. Bjorlow-Larson, "Numerical methods for calculation of electrical stresses in HVDC cables with special application to Skagerrak cable," *IEEE Trans. Power Apparatus Syst.*, Vol. PAS-94, 1975, pp. 491–497.

[12] G. Bahder, G. S. Eager Jr., G. W. Seman, F. E. Fischer, and H. Chu, "Development of ±600 kV high and medium pressure oil filled paper insulated DC power cable system," *IEEE Trans. Power Apparatus Syst.*, Vol. PAS-97, 1978, pp. 2045–2056.

[13] G. Bianchi, G. Luoni, and A. Morello, "High voltage DC cable for bulk power transmission," *IEEE Trans. Power Apparatus Syst.*, Vol. PAS-99, 1980, pp. 2311–2317.

[14] E. M. Allam and A. L. McKean, "Laboratory development of ±600 kV DC pipe cable system," *IEEE Trans. Power Apparatus Syst.*, Vol. PAS-100, 1981, pp. 1219–1224.

[15] P. Gazzana Priaroggia, G. Mashio, and A. Palmieri, "Surge performance of impregnated paper insulation for HVDC cables," IEE (UK) Conference Record on HVDC Systems, London (UK), 1966, paper 68, pp. 325–333.

[16] W. G. Lawson, P. K. Padgham, and P. Metra, "The effect of polarity reversals on the dielectric strength of oil impregnated paper insulation for HVDC cables," *IEEE Trans. Power Apparatus Syst.*, Vol. PAS-97, 1978, pp. 884–892.

[17] "Evaluation of HVDC cables for the St. Lawrence crossing of Hydro Quebec 500kV DC line," *IEEE Trans. Power Delivery*, Vol. PWRD-7, 1992, D. Couderc, N. G. Trinh, M. Bélec, M. Chaaban, J. Leduc and Y. Beauséjour, (a) Part I: Dielectric and accelerated aging tests on prototypes," pp. 1034–1042. (b) N. G. Trinh, D. Couderc, P. Faucher, M. Chaaban, M. Bélec, and J. Leduc, "Part II: Cable testing facility for dielectric and accelerated aging," pp. 1043–1049. (c) M. Chabaan, J. Leduc, D. Couderc, N. G. Trinh, and M. Bélec, "Part III: Thermal behaviour," pp. 609–613.

[18] N. Bell, C. Picard, C. Royer, K. Isaka, and M. Nakaura, "Conception, fabrication and installation of DC oil-filled cables for the underwater crossing of the St. Lawrence river," CIGRÉ Paper 21-301, 1994.

[19] F. G. Garcia, "Design procedures for high-voltage direct-current transmission cables," *IEEE Trans. Power Delivery*, Vol. PWRD-3, 1988, pp. 425–433.

[20] International Electrotechnical Commission, "Calculation of continuous current ratings of cables—100 percent load factor," Standard IEC 287, 1982.

[21] W. G. Hawley, R. S. Broady, and J. H. Mason, "Polyethylene as a possible HVDC cable insulant," Paper 70, IEE (UK) Conference Record on HVDC Systems, 1966, pp. 338–339.

[22] Y. Sakamoto, H. Fukagawa, T. Shikama, K. Kimura, and H. Takehama, "The effect of DC prestress on opposite polarity impulse breakdown strength in XLPE cables," IEEE PES Summer Meeting, Portland, Ore., Paper A76-461-4, July 1976.

[23] T. Tanaka, "Charge injection by voltage application into polymer dielectrics—new proposal for dc cable polymer insulation system," IEEE PES Summer Meeting Portland, Ore., Paper A 76 464-8, July 1976.

[24] H. Fukagawa, H. Miyauchi, Y. Yamada, S. Yoshida, and N. Ando, "Insulation properties of 250 kV DC XLPE cables," *IEEE Trans. Power Apparatus Syst.*, Vol. PAS-100, 1981, pp. 3175–3183.

[25] B. M. Weedy and D. Chu, "HVDC extruded cables—parameters for determination of stress," *IEEE Trans. Power Apparatus Syst.*, Vol. PAS-103, 1984, pp. 662–672.

[26] B. M. Weedy, "Thermal stability in HV extruded dc cables," Proceeding of VIth Intern. Symposium on HV Engineering, New Orleans (USA), 1989, pp. 27–33.

[27] S. Ogata, Y. Mayekawa, K. Terashima, R. Okiayi, S. Yoshida, H. Yamanouchi, and S. Yokoya, "Study on the dielectric characteristics of DC XLPE cables," *IEEE Trans. Power Delivery*, Vol. PWRD-5, 1990, pp. 1239–1247.

[28] N. Yoshifuji, T. Niwa, T. Tanahashi, and H. Miyata, "Development of new polymer insulating materials for HVDC cables," *IEEE Trans. Power Delivery*, Vol. PWRD-7, 1992, pp. 1053–1059.

[29] T. Suzuki, T. Niwa, S. Yoshida, T. Takahashi, and M. Hatada, "New insulating materials for HVDC cables," Proceedings of the 3rd Intern. Conf. on Conduction and Breakdown in Solid Dielectrics, IEEE Pub. 89CH 2726, 1989, pp. 442–447.

[30] H. The-Giam, M. Farkh, M. Pays, B. Dalle, C. Simon, J. Berdaka, and D. Roy, "DC dielectric behaviour of polythene under hydrostatic pressure," *Rev. Gen. De Electricite*, No. 11, Dec. 1989, pp. 19–22.

[31] R. Hegerberg, O. Lillevik, and G. Balog, "Space charge distribution in multi-layer dielectrics," Proceedings of the Nordic Insulation Symposium, NORD-IS 90, Stockholm, 1990, pp. 9.5/1-4.

[32] J. Tanaka and D. Damon, "Space charge in polyethylene ionomer blends for DC cable insulation," Electric Power Research Institute (USA), Report EL-6977, October 1990.

[33] S. Mahdavi, Y. Zhang, C. Alquie, and J. Lewiner, "Determination of space charge distribution in polyethylene samples subjected to 120 kV DC voltage," *IEEE Trans. Electrical Insulation*, Vol. EI-26, 1991, pp. 57–62.

[34] M. S. Khalil, "International R&D trends and problems of HVDC Cables with polymeric insulation," *IEEE Electrical Insulation Mag.*, Vol. 13, No. 6, 1997, pp. 35–47.

[35] K. Terashima, H. Suzuki, M. Hara, and K. Watanabe, "Research and development of ±250 kV DC XLPE cables," *IEEE Trans. On Power Delivery* Vol. PWRD-13, January 1998, pp. 7–16.

[36] P. Gazzana Priaroggia, J. H. Pracioneri, S. W. Margolin, "The Long Island Sound submarine cable interconnection," *IEEE Trans. Power Apparatus Syst.*, Vol. PAS-90, 1971, pp. 1863–1873.

[37] C. A. Arkell, B. Gregory, and J. E. Hawkes, "Design and testing of 270 kV dc SCOF cable for the land section of the UK/France connection," *IEEE Trans. Power Apparatus Syst.*, Vol. PAS-103 , 1984, pp. 3204–3210.

[38] B. M. Weedy and S. A. M. Shekata, "2 GW Britain/France HVDC link transition joint— steady state stress," *IEEE Trans. Power Delivery*, Vol. PWRD-1, 1986, pp. 861–866.

[39] B. M. Weedy and S. A. M. Shekata, "2 GW Britain/France HVDC link transition joint— transient stresses," *IEEE Trans. Power Delivery*, Vol. PWRD-1, 1986, pp. 867–872.

[40] C. A. Arkell, E. H. Ball, K. J. H. Hocke, N. W. Waterhouse, and J. B. Yates, "Design and installation of the UK part of the 270 kV DC cable connection between England and France including reliability aspects," CIGRE Paper 21-02, 1986.

[41] E. L. Davey, "Testing of high voltage dc cables," Conference Record IEE (UK) on HVDC systems, London, 1966, Paper 71, pp. 340–342.

[42] G. Luoni, E. Occhini, and B. Parmigiani, "Long-term tests on a ±600 kV DC cable system," *IEEE Trans. Power Apparatus Syst.*," Vol. PAS-100, 1981, pp. 174–183.

[43] D. J. Skipper, "Recommendations for tests on power transmission DC cables for a rated voltage up to 600 kV," *Electra* No. 72, 1980, pp. 105–114.

[44] N. G. Trinh and D. Conderc, "Analysis of test circuits for evaluating HVDC cables," Conference Record IEEE Intern. Symposium on Electrical Insulation, Baltimore (USA), June 1992, pp. 283–286.

[45] E. Guonason, J. Henje, P. Sheperd, and D. Valenza, "A 500 MW HVDC submarine cable link: Iceland–UK–Continental Europe," Conference Record IEE (UK) Third Intern. Conference on Power Cables and Accessories 10 kV–500 kV, London, 1993, pp. 220–224.

[46] U. Arnaud, G. Bazzi, and D. Valenza, "Proposal for a commercial interconnection among the Hawaiian Islands based on the results of the Hawaii deep water cable program," *IEEE Trans. Power Delivery*, Vol. PWRD-7, 1992, pp. 1661–1666.

[47] A. Fujimori, T. Tanaka, H. Takashima, T. Imajo, R. Hata, T. Tanabe, S. Yoshida, and T. Kakihana, "Development of 500 kV DC PPLP insulated oil-filled submarine cable" *IEEE Trans. Power Delivery,* Vol. PWRD-11, 1996, pp. 43–50.

[48] E. M. Allam and A. L. McKean, "Development of 500/600 kV solid type non-pressurized oil-paper cable," Electrical Power Research Institute (USA), Report TR-100621, July 1992.

[49] P. Gazzana Priaroggia, P. Metra, and G. Miramonti, "Research on the breakdown under type test of non-pressurized paper insulated HVDC cables," European Transactions on Electric Power engineering/ETEP, Vol. 5, 1995, pp. 63–65.

[50] M. J. P. Jeroense and F. H. Krueger, "Electrical conduction in HVDC mass impregnated paper cable," *IEEE Trans. Dielectrics Electrical Insulation,"* Vol. DEIS-2, 1995, pp. 718–723.

[51] R. Coelho, B. Aladenize, and F. Guillaumond, "Charge build up in lossy dielectrics with induced inhomogeneities," *IEEE Trans. Dielectrics Electrical Insulation,* Vol. DEIS-4, 1997, pp. 477–486.

[52] R. Coelho, "Charges in non-homogeneous dielectrics," IEEE Conf. on Electrical Insulation and Dielectric Phenomena, Minneapolis, USA, October 1997, pp. 1–10.

[53] G. Katsuta, T. Itaya, T. Nakatsuka, H. Miyata, T. Takahashi, and T. Niwar, "DC and impulse treeing characteristics in insulating materials for HVDC cables," IEEE Vth Intern. Conf. on Properties and Applications of Dielectric Materials," Seoul, South Korea, May 1997, pp. 422–425.

[54] M. J. P. Jeroense and P. H. F. Morshuis, "Electrical fields in HVDC paper insulated cables," *IEEE Trans. Dielectrics Electrical Insulation,* Vol. DEIS-5, 1998, pp. 225–236.

[55] M. Ieda and Y. Suzuoki, "Space charge and solid insulating materials: in pursuit of space charge control by molecular design," *IEEE Electrical Insulation Mag.,* Vol. 13, No. 6, 1997, pp. 10-17.

[56] G. Platbrood, G. Geerts, P. Couneson, and X. Delrée, "Space charge ageing of XLPE cables," CIGRE 1998 Session, Paper 15-205.

[57] T. Takada, T. Tanaka, N. Hozumi, and T. Mizutaui, " New direct observation techniques for electric charge behaviour in insulating materials and its application to power cables," CIGRE 1998 Session, Paper 15-303.

[58] L. Weimers, "HVDC Light: a new technology for a better environment," *IEEE Power Engineering Review,* August, 1998, pp. 19–20.

Chapter 15

TELEPHONE CABLES

R. Bartnikas

15.1 HISTORICAL BACKGROUND

The first underground telephone wire or cable installation was made in 1882 and consisted of wires insulated with gutta-percha and natural rubber; the wires were plowed underground for 5 miles along a railway line [1]. At the turn of the century, there was already considerable public pressure for telephone transmission lines to be placed underground in urban areas. The extent of the situation as regards to safety and aesthetics, which were the causes of this public concern, is illustrated well by the common street scenes prevailing in New York City in 1890: poles were attaining heights of 90 ft with as many as 30 cross-arms carrying 300 wires [1]. It was largely an enacted ordinance law of the same city, requiring all overhead transmission and distribution structures to be replaced by underground installations, that was responsible for the initiation of serious efforts to install all transmission and distribution lines underground in densely populated city sectors.

The earliest telephone cables consisted of insulated conductors, used as single-wire grounded lines [2]. With increased cable lengths, difficulties arose as a result of induction effects between different adjacent conductors and due to extraneous sources. Use was then made of the so-called anti-induction cables that merely consisted of individually covered insulated conductors. Major improvements were obtained with the introduction of a lead sheath with twisted pairs in 1887 that minimized coupling or cross talk between adjacent wire circuits. A further reduction in cross talk was achieved in 1889 by Carty in New York with the connection of capacitors at frequent intervals between the wires of two circuits so as to balance the capacitances between the wires of one circuit and those of the other [2].

Voice frequency transmission over long distances was greatly improved by the introduction of inductive loading coils along the cable length, permitting the use of smaller diameter conductors or wires. In 1906, a cable using the loading-coil principle was put in operation between New York and Philadelphia, spanning a distance of 145 km; the cable consisted of 1.29- and 1.63-mm-diameter conductors twisted into pairs [2]. Telephone cable transmission was carried out over increased distances as cable test methods and manufacturing techniques were improved to control the capacitative unbalance of the cables. Another important development occurred in 1908 with the completion of the phantom circuit work in the Bell System. It was shown to be possible to derive three circuits from four wires and thereby obtain greater economy for long-distance telephone cables [2]. This required a change in construction of the telephone cable, whereby two-pair circuits were combined to form a unit of four wires termed a *quad*. The first cable of this type was manufactured by Western Electric in

1910, with a successful link being established between Washington and Boston over a distance of 730 km. In the United States during the 1920s, the telephone cable construction underwent only minor changes such as the use of staggered twists to minimize further the cross talk, reduced copper wire diameters for greater economy and smaller cable size, and, finally, changes of the lead sheathing to those of tin–lead and antimony–lead composition [1]. In Canada, a change was made from a 3% tin–lead composition to a 1% antimony–lead composition in 1916.

In the United States, the gradual replacement of the early conventional paper ribbon wound-wire insulated cable began in the 1930s when paper pulp insulated cables were introduced. However, in Canada paper pulp cables did not enter production until the 1960s. In fact, in Canada, the unusual situation arose in that the PIC (plastic insulated) cable appeared on the market before the paper pulp cable. In that respect, polyethylene (PE) insulated wires were almost concurrently introduced with PE insulated conductors in the power cable field in the 1950s. The PE insulated wires offered better electrical characteristics and greater moisture resistance than either the paper ribbon or pulp cables possibly could. The use of the standard lead sheath was also replaced by either ALPETH (corrugated aluminum shield and polyethylene jacket) and STALPETH (corrugated steel and aluminum shields and polyethylene jacket). It is, however, most noteworthy that to this present day, in spite of all the material improvements in sheathing construction, lead is still the preferred sheathing material for submarine telephone cables.

Telephone cables can be broadly divided into exchange area and toll area cables [3]. The former category comprises cables used for transmission from switching centers to subscribers or between the switching centers themselves. These cables are further subdivided into trunk feeder, distribution, and video cables. Trunk cables interconnect switching centers; when the twisted-pair cable construction was employed for this purpose, cables having more than 900 pairs with paper ribbon or paper pulp insulation were used; for PIC-type cables the number of pairs was generally less than 900. Feeder cables connect switching centers to distribution areas and are similar in construction to trunk-type cables. Distribution cables interconnect with service drop wire connections from the subscribers, and for this purpose PIC-type cables were principally employed in the past. Video cables, as the name implies, are used for picture signal links; in conventional copper conductor twisted-pair cables, the shielded video pairs are generally incorporated in the overall cable construction. The category of toll area cables includes all cables used for long-distance intercity transmission. In the past, these consisted of conventional multipair cables operating at carrier frequencies, and therefore, PIC cable construction was used. They now have been replaced entirely by fiber-optic cables. Quadded cables were also employed, but these were merely old or replacement cables in what would now be considered as antiquated systems. Finally, composite cables consisting essentially of PIC-type pairs containing coaxial cables may also be used.

With the discovery of PE in 1933, coaxial cables were developed in 1935 to serve as long-distance carrier cables providing comparatively excellent transmission characteristics. Their design and development received a great impetus during World War II, because of their importance in radar-related applications. There are various types of coaxial cable designs, such as those using solid or foamed PE insulation; air-dielectric coaxial cables consist of conductors held in place by either PE disc spacers positioned at regular intervals along the cable or spiral coaxial spacers. In the 1970s air-dielectric coaxial cables, having low attenuation losses and using pulse code modulation techniques with time division multiplexing, dominated the long-distance transmission field. However, the air-dielectric coaxial cables, very much the same as the PIC-type cables, yielded eventually their position in the long-haul application area to the high-capacity, low-attenuation optical fiber cables.

Over the last two decades, telephone cable networks have undergone enormous changes as a result of the rapid rate of conversion from analog to digital signal transmission in order to accommodate the increased traffic and bandwidth requirements for voice, video, and data transmission. Great efforts were made to apply advanced digital techniques to existing in situ cables to enhance their performance characteristics and thus prolong their useful service life, thereby averting their high replacement costs. Although in North America all major urban centers are now linked via wide-band optical fiber cables, the digital techniques have been remarkably successful thus far in extracting improved performance from existing copper conductor cables to meet the necessary bandwidth requirements for a multiplicity of applications. For the telephone subscriber loop end, the limited bandwidth requirements for voice signal transmission and the usual digital data transmission applications may still be met for the most part by the use of only the conventional twisted copper pair and coaxial telephone cables. Thus existing telephone cable systems generally comprise cables of various vintages ranging from twisted-pair-type cables to the most recently manufactured optical fiber cables. It is the intent of this chapter to discuss the transmission characteristics of both copper conductor and optical fiber telephone cables that are currently in use and to provide a description of their design, construction, and manufacture.

15.2 TRANSMISSION PARAMETERS OF COPPER CONDUCTOR TELEPHONE CABLES

The principal parameters that define the behavior of a communication transmission line are the attenuation, signal distortion or error rate, and crosstalk. The attenuation of a transmission line is normally expressed in decibels in terms of its power input P_i, and power output, P_o,

$$\alpha(\text{dB}) = 10 \log \frac{P_i}{P_o} \tag{15.1}$$

With electrical communication lines the power loss along the line is very substantial. For example in a 2000-m line, the transmitted power P_o may equal only 0.1% of the input power P_i. Note that if, in order to compensate for this loss, amplifiers are utilized such that $P_o > P_i$ then the calculated attenuation is negative or, alternatively, there is a gain.

Signal distortion occurs along the lines because the incremental inductance and capacitance of the transmission line tend to vary with frequency. As a consequence, the various frequency components in the transmitted signal undergo different attenuation and experience a difference in their respective time delays. Elimination of signal distortion would require that the I^2R loss in the conductor be equal to the insulation loss V^2G; in such circumstances, the characteristic impedance of the line Z_0 would become a real quantity and be given by [4]

$$Z_0 = \sqrt{\frac{L}{C}} = \sqrt{\frac{R}{G}} \tag{15.2a}$$

where L, C, R, and G denote respectively the incremental inductance, capacitance, conductor resistance, and insulation conductance per unit length of the line. From Eq. (15.2a) follows the condition for a distortionless line:

$$\frac{R}{L} = \frac{G}{C} \tag{15.2b}$$

But in practical communication cables, the close spacing of conductors results in high values of C and low values of L so that the condition stated in Eq. (15.2b) cannot be satisfied. However, both the distortion and attenuation of a communication cable may be substantially reduced if the cable is designed such that $R \ll \omega L$ and $G \ll \omega C$. In such circumstances, the magnitude of the characteristic impedance Z_0 may still be maintained independent of frequency, though it now has a small angle that is frequency dependent [4],

$$Z_0 \simeq \sqrt{\frac{L}{C}} \left/ \frac{1}{2} \left(\frac{G}{\omega C} - \frac{R}{\omega L} \right) \right. \tag{15.3}$$

and the corresponding attenuation in nepers per meter is

$$\alpha \simeq \frac{1}{2} \left(G|Z_0| + \frac{R}{|Z_0|} \right) \tag{15.4}$$

Note that 1 neper $= 8.686$ dB. Because of the high insulation resistance, the conductance G is virtually equal to zero, so that Eq. (15.4) becomes

$$\alpha \simeq \frac{1}{2} \frac{R}{|Z_0|} \tag{15.5}$$

and the phase angle constant β and the velocity of propagation v, which is by definition equal to ω/β, are approximately given by

$$\beta \simeq \omega\sqrt{LC} \tag{15.6}$$

and

$$v \simeq \frac{1}{\sqrt{LC}} \tag{15.7}$$

It can be perceived from Eqs. (15.3) and (15.4) that signal distortion and attenuation may be decreased by increasing the magnitude of the characteristic impedance $|Z_0|$ via an increase in L and reduction in C. The compact construction of communication cables prevents a reduction in C, because this necessarily requires larger spacing between the individual conductors. However, in the earlier days the value of L was increased by the use of loading coils, i.e., iron core inductors inserted in the copper conductors at regular intervals. However, with the introduction of carrier frequencies and later with the use of digital transmission, these coils had to be removed to permit upgrading of the lines. Attenuation was compensated by the insertion of repeaters or amplifiers along the lines. At carrier frequencies a typical twisted-pair telephone cable has an attenuation of the order of 15 dB (130 nepers) per 1000 m such that for a cable of 2 km length, the transmitted power at the receiving end without any amplification in accordance with Eq. (15.1) is only 0.1% of the power at the sending end.

From the foregoing discussion, it is apparent that distortion associated with attenuation may occur as some frequencies in a cable are attenuated more than others. For example, in a single telephone wire pair, the higher frequencies become attenuated more than the lower ones. Another form of signal distortion is that of envelope delay distortion. This type of distortion arises when the propagation time or the velocity of propagation changes with frequency, causing certain frequency components within the signal to appear at the output of a cable before the arrival of others. Thus in terms of Eq. (15.6), the phase shift β is a nonlinear function of frequency, resulting in an output signal that is distorted with respect to the input signal. Envelope delay is by definition equal to the derivative of the phase shift β with respect to the radial frequency ω [5],

$$[\text{Envelope delay}] = \frac{d\beta}{d\omega} \tag{15.8}$$

and the envelope delay distortion itself is numerically equal to the difference between the envelope delay at a given frequency and that at another frequency.

The vast majority of metallic conductor telephone cables are of the balanced twisted-pair construction. Since the common telephone cable consists of a multiplicity of twisted pairs enclosed within an overall external shield, the shield for an individual twisted pair is thus that formed by the adjacent twisted pairs. A particular pair will be perfectly balanced with respect to this shield provided its two twisted conductors have equal series impedances and admittances to the shield to prevent any induced voltage in the shield. If this balance condition is not met, then the effective shield will in turn introduce interference voltages in the twisted pair itself. The purpose of the twist in the two-wire pair is, therefore, to balance its geometry with respect to the shield. The upper frequency limit over short distances of the twisted-pair construction in an analog system appears to be in the range of 10 MHz. In contradistinction, a coaxial cable, in which the outer shield conductor is directly involved in the transmission of the signal, behaves quite differently. At voice frequencies, over which skin effect is negligible, extraneous interference becomes part of the signal in the outer conductor. However, at very high frequencies, the skin effect causes the signal to be transmitted along the inner surface of the external shield-type conductor, while the interference or noise signal remains isolated on the outer surface of the shield of the coaxial cable. The high propagation velocity and constant characteristic impedance typical of coaxial cables make them suitable for wide-bandwidth applications up to 1 GHz or even higher with well-designed cables.

In the past, when telephone transmission was carried out using analog techniques at carrier frequencies, coupling between wire pairs in multipair cables was manifest by audible voice conversation or unintelligible noise in the voice channel under use. This form of audible extraneous interference, which was induced by signals transmitted in another channel, was termed cross talk. Noise may also be introduced as a result of inductive interference from adjacent power lines, which may be either electrically induced because of voltage surges or magnetically induced due to harmonic voltage components. Switching surges on power lines may result in impulse-type noise, which may adversely influence the error rate on digital data transmission circuits. Inductive interference from power lines is effectively reduced by grounding of the cable metallic shield at regular intervals. Cross talk in cables is minimized by reducing the unbalance, which involves twisting of the individual wire pairs.

To illustrate the characteristics of cross talk let us consider two twisted-pair circuits, namely a disturbed and a disturbing circuit. By definition the cross talk measured at the near or transmitting end is equal to the ratio of powers of the disturbed to the disturbing circuit, P_n/P_0, given by [6]

$$\frac{P_n}{P_0} = k_n f^{3/2}[1 - \exp{-2(\alpha_n + \alpha_0)\ell}] \tag{15.9}$$

where k_n is a near end cross-talk constant, f is the transmission or carrier frequency, α_n and α_0 are the attenuation factors in the disturbed and disturbing circuits, respectively, and ℓ represents their length. Thus, cross talk at the near end increases markedly with the transmission frequency. A similar expression for cross talk at the far end shows that the cross talk increases with both the transmission frequency and the length of the cable.

Thermal noise, which is a measure of electron agitation above the absolute temperature, determines the utmost sensitivity level at which a telephone cable may be operated [7]. The thermal noise energy content in joules at an absolute temperature T is given by kT, where k is the Boltzmann constant and is equal to 1.3803×10^{-23} J/K^{-1}. In derived units, it is expressed as power per cycle or bandwidth in Watts per hertz,

$$P_t = kT \tag{15.10}$$

Hence, the maximum theoretical sensitivity at room temperature would be 3.76×10^{-21} W/Hz

15.3 DIGITAL TRANSMISSION

The standard voice telephone circuit requires a one-pair circuit to transmit the signal and another pair to receive the signal; i.e., a total of four wires are necessary for a two-way, or duplex, telephone conversation. In order to reduce the number of pairs required per telephone conversation and to permit more than one conversation channel per one four-wire circuit, an analog technique has now been in use for more than a half century whereby a much higher carrier frequency is mixed or modulated by the standard telephone voice frequency confined within the spectrum of 300–3400 Hz. This analog technique is commonly referred to as frequency division multiplexing (FDM). In this arrangement, 12 standard voice channels having a nominal 4 kHz bandwidth are mixed with corresponding carrier frequencies extending from 64 to 108 kHz, as delineated in Figure 15.1. With this arrangement lower side bands are employed such that, for example, channel 12 occupies the frequency between 60 and 64 kHz; use is made of

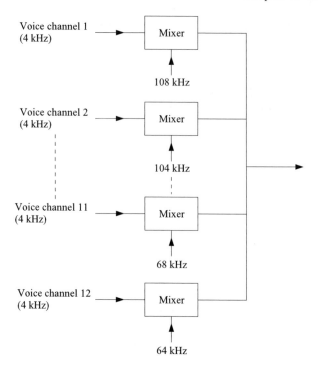

Figure 15.1 Standard CCITT group representation of analog FDM.

single side band suppressed carrier modulation [6]. An International Telegraph and Telephone Consultative Committee (CCITT) supergroup is also used. It consists of five standard CCITT groups, representing a total of 60 voice channels with carrier frequencies of 420, 468, 516, 564, and 612 kHz for the five supergroups.

At the present many of the analog carrier systems have been either already replaced or are in the process of being replaced by digital carrier systems utilizing the time division multiplexing (TDM) technique [8]. The introduction of digital carrier transmission techniques began in the late 1960s, with the increased usage of solid-state integrated circuits [9]. Digital techniques were able to deal more effectively with the cross talk, signal distortion, and electrical noise problems as opposed to analog transmission. With analog transmission as the signal is degraded along the cable, it cannot be reconstructed to its original form because the analog repeaters simply amplify the distortions introduced together with the main portion of the signal. In contradistinction, a digital regenerative repeater detects, reshapes, and amplifies the signal, restoring it to its original square-pulse form. Thus, digital techniques permit the transmission of signals over very long cable lengths without significant pulse form degradation, thereby greatly improving the quality of the transmitted voice, data, and video information over that of an equivalent analog system. Of course, this should not be interpreted as a licence for less quality control in the production of telephone cables. Electrical transmission characteristics of cables are equally important with digital trans-mission systems, since telephone cables with lower attenuation, cross talk, and signal distortion require a smaller number of regenerative digital repeaters. Cross talk in telephone cables may still adversely affect the digital transmission system by augmenting the error rate. Although with TDM, an error rate of one error in 10^5 bits over the cable length does not degrade significantly the intelligibility of the information received [5], error rates approaching 1 bit in 10^2 render the received information completely unintelligible.

With digital transmission, the signals in the cable are of binary form; i.e., each signal consists of 1's and 0's and may be transmitted in a neutral or bipolar mode. The transmitted signal is in the form of binary digits or bits, where the bit is a 1 or 0 (a square pulse or an absence of a square pulse). Consequently any degradation in the pulse shape will not alter the content of the transmitted information, since the pulse form, even though distorted following its traversal over a long cable length, may be regenerated into a square pulse unless the distortion and attenuation is so severe that an originally transmitted square pulse is detected as an absence of a pulse. Nevertheless, the use of digital procedures imposes additional performance demands on a transmission system in that the transmission of analog information over a digital system requires a much larger bandwidth compared to an equivalent analogue transmission system. For example, for an analog voice channel of 4 kHz, the required bandwidth is 64 kHz (i.e., 16×4 kHz) [5]. In effect, sampling of the analog voice signal is carried out at the rate of 8 kHz (4000×2 Hz) [6]; since each sample is accorded an 8-bit word, the voice data acquisition rate is equal to 64 kbits/s (8×8 kilobits per second) [8]. Video and other data transmission requires much higher sampling rates and correspondingly greater bandwidths. Evidently the telephone cables designed for digital transmission applications must have adequate frequency spectrum characteristics.

To transmit an analog signal in a digital mode, pulse code modulation (PCM) is employed [4]. The method is essentially based on the sampling theorem, which states that if a signal is sampled at the rate of twice or higher than its frequency, then the sampled signal will contain all the information of the original signal such that its form can be readily reconstructed by means of a low-pass filter [10–12]. The required signal processing steps for PCM are those of sampling, quantization and coding; these are schematically illustrated in Figure 15.2. Here three analog signals from three different voice channels are depicted as being sampled for transmission over a single path. Each voice channel of a nominal bandwidth of 4 kHz must be sampled at a rate of 8×10^3 samples per second. The samples from the three voice channels are multiplexed or interleaved with respect to time in the form of a series of amplitude modulated pulses. The pulse-amplitude-modulated (PAM) pattern is divided into frame segments, each representing three pulse amplitude samples from each of the three voice channels; the recurrence rate of the frames is fixed at $8 \times 10^3 \, \mathrm{s}^{-1}$ or $125 \, \mu s$. Each PAM frame is quantized by relating the PAM pulse amplitudes to a binary quantizing scale [13]. Encoding of the signal is carried out by converting the discrete pulse amplitude samples into a code word of binary digits.

The number of quantum steps in the quantizing mode must be sufficiently large to provide the characters or meanings required in practice. Thus a four-level binary code of 2^4, which could provide only 16 characters or meanings, would not even suffice for the 26-letter alphabet [5]. Current PCM systems employ either seventh- or eighth-level binary codes, i.e., 2^7 (128) or 2^8 (256) quantum steps, respectively. But it is estimated that 2048 (211) quantum steps are required to ensure acceptable sound fidelity [5]. To obviate this difficulty, either nonuniform quantizing steps or compression and expansion (i.e., companding) prior and following the quantizer are employed. If the eight-level binary code (2^8) is utilized, then each pulse amplitude sample may be converted

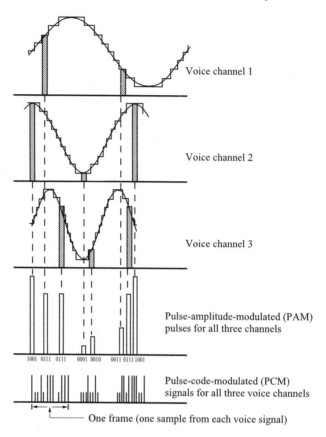

Voice channel 1

Voice channel 2

Voice channel 3

Pulse-amplitude-modulated (PAM)
pulses for all three channels

1001 0111 0111   0001 0010   0011 0111 1001

Pulse-code-modulated (PCM)
signals for all three voice channels

|◄─┬─►|
    └──── One frame (one sample from each voice signal)

Figure 15.2 Schematic sequence of TDM steps using PCM [13].

into a code word of 8 binary bits. Thus the 8 bits of the code word may denote 2^8 pulse amplitude values. In the case of a 36-voice-channel transmission system, each PCM frame would comprise 36×8 bits plus 1 framing bit, which is used for synchronization; this would involve a total of 269 bits, corresponding to a terminal bit rate of $269 \times 8 \times 10^3$) or 2.152×10^6 bits per second. Since the information rate is related to the bandwidth requirements of the transmission system, the desirability of operating with a lower level binary code is evident as it permits the use of telephone cables having lower bandwidth transmission characteristics. In practice: most TDM systems designed for voice transmission operate with 24 voice channels [5, 8, 14]. In such circumstance each frame contains a sequence of twenty-four 8-bit words, so that the overall terminal bit rate, including the framing bit, is equal to $[(24 \times 8 + 1]8 \times 10^3$, or 1.544×10^6 bits per second.

If we consider for simplicity that we are dealing again with three voice channels, then the entire PCM/TDM transmission and receiving system may be represented schematically in Figure 15.3. Note that this figure portrays the system only in the transmission direction over a single pair of wires in a telephone cable, so that the return direction of transmission would be the reverse of that depicted in Figure 15.3 involving an additional pair of wires. That is, a typical PCM/TDM system, which comprises 24

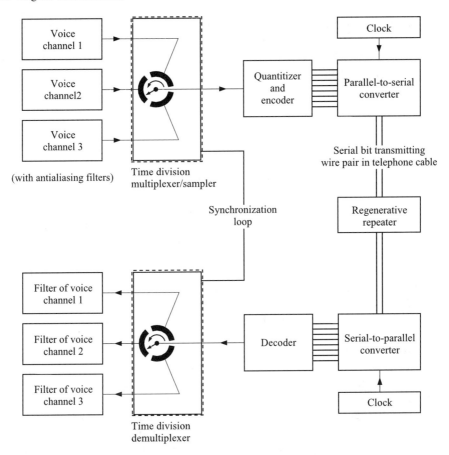

Figure 15.3 Schematic of PCM/TDM transmission system (after [14]).

voice channels, requires two wire pairs or four wires in a telephone cable. In Europe, the four-wire system carries normally 30 channels, and occasionally 32 [5]. Figure 15.3 may be best followed by making reference to Figure 15.2, which delineates the sampling steps for a simple three-voice-channel system, indicating the PAM and resulting PCM/TDM patterns. From Figure 15.3 it can be perceived that the PAM signal emerging from the TDM sampler is quantized into a PCM/TDM signal. The quantized and encoded 8-bit word signals are transmitted in parallel via an eight-wire-pair ribbon cable to a parallel-to-serial converter before entering the wire pair of a telephone transmission cable. Repeaters along the telephone cable, placed usually at 4500–6000-ft intervals, amplify, retime, and reconstitute the signal and prevent cumulative degeneration of the signal. Thus the signal is regenerated both with respect to amplitude and phase. The bandwidth of the regenerative repeaters, f_0, is given by [15]

$$f_0 = 0.3 f_b \qquad (15.11)$$

where f_b is the information bit rate; when long-haul-type coaxial cables are used, its value may extend from 224 to 1100 Mbits/s [15, 16]. The effect of bandwidth limitation is manifest by a reduction in the pulse duration of the received signals.

Antialiasing filters are indicated at the input of the TDM/sampler input to prevent the admission of noise or other spurious high-frequency components in excess of the nominal 4-kHz voice frequency [7]. These low-pass filters are designed to reject the alias frequency f_a, which is given by [14]

$$f_a = f_{in} - f_{samp} \qquad\qquad (15.12)$$

where f_{in} represents the frequency or frequencies in the analog signal that are above the sampling rate frequency f_{samp}. The presence of alias frequencies will lead to serious quantizing distortion in the sampled system.

At the receiving end the transmitted digital signals undergo the reverse digital-to-analog conversion process. The digital signals are now applied to a serial-to-parallel converter in tandem with a decoder and time division demultiplexer. Thereafter they emerge as analog voice signals at the respective channels where they enter low-pass filters to remove spurious modulation products introduced by the demultiplexer [7]. The synchronization loop between the time division multiplexer and demultiplexers prevents overlap between the signals of the three respective voice channels [14].

Most of the transmission volume over telephone cables is related to voice signals, though a substantial portion of ~15% involves data transmission [5]. For example, the data sent from a computer consists of a series of square pulses in the binary code. For this form of data to be transmitted along telephone lines, which are designed and intended for voice signal transmission, requires the use of terminal equipment. The latter is frequently a modem (a modulation-demodulation device), and the speed with which the data can be transmitted along the telephone cable is very much contingent upon the design characteristics of the modem. In many cases the modem may simply transform the binary square-wave signal from a computer into a modulated sine wave, having a frequency or frequencies within the standard nominal voice frequency band of 4 kHz. Figure 15.4 depicts schematically a data transmission link via a telephone cable between two PC computers [13]. In the portrayed form of modulation, the modulated sinusoidal signals' length in time corresponds to the width of the binary pulses. The transmission of the data may be carried out in either an asynchronous or a synchronous mode. If a given number of digital characters are forwarded in a continuous stream, then the transmission is synchronous and the receiving terminal must be maintained in

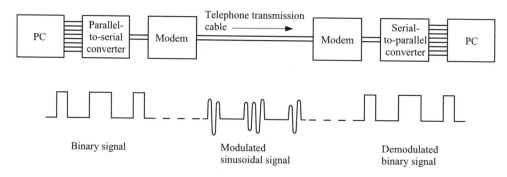

Figure 15.4 Schematic representation of data signal transmission between two modems via twisted-pair telephone line (after [13]).

phase with the transmitting terminal. When asynchronous transmission is employed, the characters are sent individually and the commencement of the transmission is indicated by a 0 as a start signal and a stop signal of 1 denotes the termination of the data transmission. The transmitted character or word is thus defined between the 0 start and 1 stop parity bit. The speed of data transmission is more rapid with synchronous transmission because no stop and start bits are required between the characters [13]. In addition, greater modulation rates may be employed.

15.4 CHARACTERISTICS OF METALLIC CONDUCTOR TELEPHONE CABLES

The characteristics of telephone cables influence the waveform of the transmitted signals and determine the amplification, pulse shaping, modulation, and equalization equipment that must be used for effective signal transmission. In this section we shall consider metallic conductor conventional twisted-pair telephone cables; the next section will deal with coaxial telephone cables.

15.4.1 Twisted-Wire Multipair Cables

The twisted-wire multipair telephone cable is perhaps one of the most common cables to be found in service even today, notwithstanding the remarkable advances achieved with optical fiber cables, which now dominate entirely the long-distance transmission field. Although the twisted-wire multipair cable was originally designed for voice frequency transmission and very restricted bandwidth, with the assistance of digital techniques its range of operation was greatly expanded and continues to expand as additional data transmission requirements arise at the subscriber service end. Particularly over very short distances involving subscriber services with many individual channels, the twisted-pair cables have been able to meet thus far with the demand for data transmission associated with the use of PCs and facsimile in addition to the voice frequency requirements. As early as 1975, single twisted pairs were capable of carrying information rates of 1.5 Mbits/s with an analog bandwidth of 1 MHz [18]. Presently rates in excess of 10 Mbits/s over short distances are possible [19] and a rate of 155 Mbits/s with a bandwidth of less than 30 MHz is obtainable with two-pair systems in a local area network (LAN) by means of a bandwidth-efficient two-dimensional pass band encoding scheme [20]. In a high-speed digital subscriber line (HDSL), the coupling between the different wire pairs results in substantial cross talk, which increases with frequency at the high information rates; at present considerable effort is put into developing more suitable transmitter designs to overcome these cross-talk problems [21]. In view of the foregoing achievements, it would appear that wire communications using digital techniques will likely continue to dominate the subscriber service lines for at least another 10 years [22].

In the design of twisted-pair telephone cables, the electrical transmission characteristics of the multipair cable are determined by the dimensions of the diameter of the metallic conductors and the type and thickness of the electrical insulation surrounding them. This fixes the resistance R and inductance L of the metallic conductors and the conductance G and mutual capacitance C between the two wires; it is these four primary parameters that determine the transmission characteristics of the cable. In North

American multipair telephone cable designs, the mutual capacitance C is standardized at $0.083\ \mu F/mile$ ($0.052\ \mu F/km$).

A compact multipair cable design requires that the space between the contiguous pairs be maintained at a minimum. Yet, simultaneously the mutual capacitance between the individual two wires must not exceed $0.052\ \mu F/km$. With the antiquated paper ribbon insulation, this requirement was readily achieved because of the porosity of the paper and the long lay of its application across the metallic conductor such that the overall paper insulation comprised an inherent air volume of approximately 50%. Likewise, the paper pulp insulating system, which eventually replaced the paper ribbons, as a result of its significant porosity, included a sufficient volume of air that in conjunction with a dielectric constant between 6 and 10 for the solid cellulose paper structure led to an overall dielectric constant below 2.0. For wires insulated with solid PE having a dielectric constant of 2.2–2.3, the insulation thickness must be augmented in order to compensate for an otherwise increased capacitance were the same insulation thickness as that employed with the paper-based insulation utilized. With PE insulated wire multipair cables, for which a cable filling compound may be used to prevent moisture ingress along the cable, a still further increase in wire insulation thickness is required to offset the increase in the overall wire-to-wire capacitance arising from the complete exclusion of air between the twisted-pair interstices.

By definition, the mutual capacitance C is equal to the capacitance between the two wires comprising the pair with the remaining pairs grounded. Accordingly, C is given by

$$C = C_d + \frac{C_1 C_2}{C_1 + C_2} \tag{15.13}$$

where C_d is the capacitance between conductors 1 and 2 comprising the pair and C_1 and C_2 are the respective capacitances of conductors 1 and 2 to ground. An approximate method for calculating C of a multipair cable has been developed by Mead [23] and Windeler [24]. The expression derived by Mead [23] for the mutual capacitance in microfarads per kilometer is given by

$$C = \frac{0.01208\varepsilon_r'}{\log\left[\dfrac{2S}{D}\left(\dfrac{D^2 - S^2}{D^2 + S^2}\right)\right] - 0.1086\delta_{12}} \tag{15.14}$$

where ε_r' is the real relative value of the permittivity or dielectric constant, d is the diameter of the metallic conductors of the uninsulated wires, S is the interaxial spacing of the wires constituting the pair, D is the effective shield diameter, and δ_{12} denotes a series function of $d/2S$ and S/D. For small-diameter flexible conductors for which the insulation of the two wires is perfectly contiguous, the interaxial spacing S is given by $d + 2d'$ where d' represents the insulation thickness over the conductor of each of the two wires. As the gage of the metallic conductors increases, the wires forming the pair become separated and S becomes greater than $d + 2d'$. The value of S is thus a function of the diameter over the dielectric (DOD) of the wire as well as being also a function of the compression or tension to which the overall cable core is subjected. Empirical measurements indicate that as DOD increases from 0.76 to 2.0 mm, the interaxial distance increases from its initially fixed value of $d + 2d'$, when the two wires are contiguous, to $(d + 2d') + (178\,\mu m)$ [24]. Windeler [24] estimates the effective shield diameter D by assuming that the two wires forming a given pair are surrounded by an oval layer of eight wires. In such a symmetrical arrangement, which is delineated in Fig. 15.5, the center of the pair to the center of the nearest conductor is equal to the interaxial distance S, while to the two further conductor centers it is $1.5S$ and to the four remaining conductors it is $1.32S$. The mean of these eight distances yields a radius of $1.285S$; hence $D = 2(1.285S)$, and the ratio of S/D is found to be 0.389. Evidently the S/D ratio will vary with the actual configuration as opposed to the idealized configuration assumed in Fig. 15.5. The symmetrical array of the pair configuration will be contingent upon the gage of the metallic conductors, insulating medium, pair twist sequence, and compression and tension to which a multipair cable may be subjected.

The series function δ_{12} appearing in Eq. (15.14) has been determined experimentally on telephone cables with known ratios of $d/2S$ and S/D. Its variation as a function of S/D with $d/2S$ as a parameter is depicted in Fig. 15.6 [24]. As can be perceived, the rate of decrease of δ_{12} with S/D is more rapid for greater ratios of $d/2S$. The interaxial separation S may also be determined from the measured inductance (in millihenrys per kilometer), which may be expressed as

$$L = 0.9211 \log \frac{2S}{d} + L_i \mu_r \qquad (15.15)$$

where μ_r is the real relative permeability and is equal to unity for copper and aluminum. Equation (15.15) will be recognized as the expression for the inductance of two wires in free space, the last term being a constant and L_i representing the internal inductance of the wires; the expression is valid up to audio frequencies.

In order to determine the effective value of the real value of the permittivity, ε_r', Windeler [24] resorts to the values of measured mutual capacitance C on cables for which the $2S/d$ ratios have been determined by means of Eq. (15.15) in terms of the measured inductance L. Figure 15.7 portrays the experimental and calculated data obtained using the most common wires having conductor gages of 19, 22, 24, and 26 AWG (American Wire Gage), corresponding to Systéme International (SI) unit diameters of 0.91, 0.64, 0.51, and 0.40 mm, respectively. The experimental mutual capacitance data points, obtained on PE insulated wire pairs, are plotted as a function of the $d/2S$ ratio. The solid and dashed curves represent calculated data obtained using ε_r' values of 1.90, 1.80, and 1.70 for a constant S/D ratio of 0.389. It is seen that the best fit

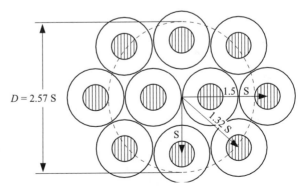

Figure 15.5 Wire pair within oval array of surrounding adjacent wires in telephone cable (after [24]).

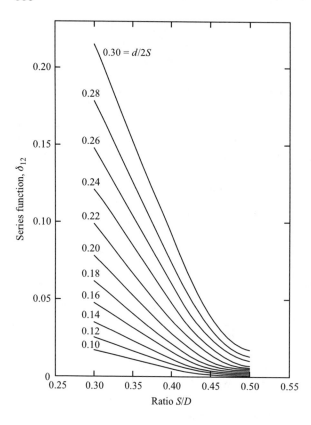

Figure 15.6 Variation of δ_{12} with ratio of inter-axial separation S to effective shield radius D for different conductor–interaxial parametric ratios (after [24]).

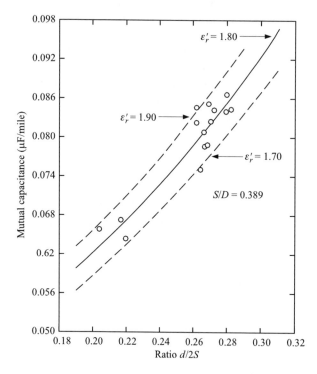

Figure 15.7 Mutual capacitance vs. ratios of $d/2S$: (○) experimental points; solid and dashed curves calculated using an S/D ratio of 0.389 (after [24]).

to the experimental data is achieved when an ε'_r value of 1.80 is employed in the case of PE insulated wire unfilled telephone cables. The values of 1.70 and 1.90 selected for ε'_r are seen to lead essentially to the upper and lower bound limits in the dielectric constant representing the equivalent PE/air-dielectric of the unfilled telephone cable.

Once the capacitance C of the wire pair is determined, the conductance G of the equivalent insulation of the wire pair follows from

$$G = \omega C \tan \delta \tag{15.16}$$

where $\tan \delta$ represents the effective value of the dissipation factor of the equivalent insulation between the two conductors forming the pair.

At low audio frequencies, the ac resistance of the copper conductors is essentially equal to the dc resistance; the dc, or low frequency, value of the resistance, R_{dc}, is given by

$$R_{dc} = \frac{d\rho}{A} \tag{15.17}$$

where ρ is the resistivity in ohm-centimeters of a conductor, having a cross-sectional area A (in square centimeters) and a diameter d (in centimeters). If one neglects the helix losses, then the low-frequency resistance of a wire pair is equal to the dc loop resistance. If the conductor diameter of the wire pair is expressed in millimeters, then the low-frequency resistance in ohms per kilometer may be stated empirically as [25]

$$R = \frac{43.9}{d^2} \tag{15.18}$$

As the frequency increases, the ac resistance will be augmented due to the appearance of the skin and proximity effects, as is apparent from Fig. 15.8, which compares the resistance of a number of different gage copper and aluminum conductor wires of filled and unfilled PE insulated wire pair cables, having a nominal mutual capacitance of 0.083 μF/mile [26].

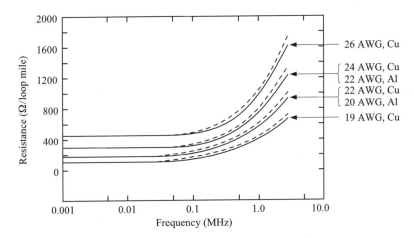

Figure 15.8 Conductor resistance versus frequency of PE insulated copper (Cu) and aluminum (Al) conductor wires: dashed curve, unfilled, dry cable; solid curve, cable filled with petroleum jelly compound (after [26]).

The appreciable increase in resistance with frequency is primarily the result of an elevated skin effect, which results from the higher current density in the proximity of the conductor surface as inductive resistance close to the surface diminishes with frequency. The skin thickness or penetration depth in millimeters over which the current is concentrated is given by

$$\delta = 505.3\sqrt{\frac{\rho}{\mu_r' f}} \qquad (15.19)$$

where ρ is the resistivity of the conductor in ohm-meters, μ_r' is the relative real value of the permeability of the conductor (equal to unity for nonmagnetic metals), and f is the frequency in megahertz. The resistance in ohms per kilometer of the δ thickness layer is equal to [3]

$$R_{ac} = \frac{0.127 \times 10^6 \sqrt{\rho f}}{d} + \frac{6.38 \times 10^6 \rho}{d^2} + \frac{0.506 \times 10^6 (S/D)^2 \sqrt{\rho_{CS} f}}{D[1 - (S/D)^4]} \qquad (15.20)$$

where ρ_{CS} is the resistivity of the cable shield in ohm-centimeters, D is the internal diameter of the equivalent wire pair shield in millimeters, and S is the axial separation of the wires in the pair in millimeters. Alternatively, both the ac resistance (R_{ac}) and internal inductance ($L_{i,ac}$) may be determined in terms of the quantity $a\sqrt{\omega\mu'/\rho}$, where a is the radius of the conductor and μ' the real value of the permeability. For this purpose data have been compiled in Table 15.1.

Returning to Fig. 15.8, it can be readily discerned that the resistance of the filled cables increases at a slightly lower rate with frequency than that of the unfilled cables. This is attributed to the greater separation between the conductors in the filled cables, which reduces the proximity effect. The aluminum conductors are observed to have a slightly higher resistance than the copper conductors of equivalent diameter. The separation between the different characteristics of the wires with different diameters but the same metallic conductors is a direct result of their difference in the base value of the dc resistance. As anticipated, the characteristics of the same conductor sizes also become shifted with respect to each other due to temperature changes, as indicated in Fig. 15.9 [27].

TABLE 15.1 The ac-to-dc Ratios of Conductor Resistance and Internal Inductance

$a\sqrt{\omega\mu'/\rho}$	R_{ac}/R_{dc}	$L_{i,ac}/L_{i,dc}$	$a\sqrt{\omega\mu'/\rho}$	R_{ac}/R_{dc}	$L_{i,ac}/L_{i,dc}$
0	1.000	1.000	10.0	3.80	0.282
1.0	1.005	0.997	20.0	7.33	0.141
2.0	1.078	0.961	30.0	10.86	0.094
3.0	1.318	0.845	40.0	14.40	0.071
4.0	1.678	0.686	50.0	17.93	0.057
5.0	2.04	0.556	60.0	21.5	0.047
6.0	2.39	0.465	80.0	28.5	0.035
7.0	2.74	0.400	90.0	31.1	0.031
8.0	3.09	0.351	100.0	35.6	0.028
9.0	3.45	0.313			

Source: From [4, 5].

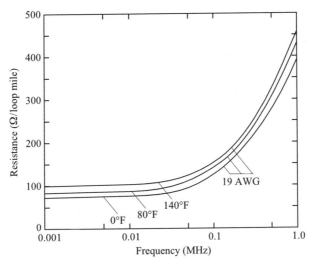

Figure 15.9 Copper conductor resistance per loop mile as a function of frequency with temperature as parameter for PE insulated 19 AWG multipair telephone cable with mutual capacitance of 0.083 μF/mile (after [27]).

Figure 15.10 portrays the inductance for PE insulated wire pairs in filled and unfilled telephone cables with a nominally designed mutual capacitance of 0.083 μF/mile as a function of frequency [26]. In order to interpret the behavior of inductance with frequency, reference should be made to Eq. (15.15). As the frequency is increased, the skin effect induces the current to concentrate increasingly toward the outer skin layer of the conductors, causing the internal flux density to diminish and thereby reduce the internal inductance L_i, which is represented by the second term of the equation. In addition, the internal inductance L_i is further decreased by the appearance of a significant proximity effect, which arises as increasingly more of the current flow becomes concentrated toward the surface of the conductors adjacent to each other. Both the skin and the proximity effects are more pronounced in the wires having larger conductor diameters.

The capacitance does not vary with frequency for the PE insulated wire pairs over the usual frequency operating régime of the multipair telephone cables, which reflects the nonpolar character of the PE molecule. For the same reason, the capacitance does not exhibit pronounced variations over the operating temperature range. This applies both to the filled and unfilled cables, though the standard deviation of the mutual capacitance for the filled cables is smaller and is of the order of 1.1 μF/mile, as compared to 1.8 μF/mile for the unfilled cables [26]. However, the exclusion of air pockets by the petroleum jelly filling compound raises the equivalent dielectric constant to 2.23 from the 1.8 value determined by Windeler [24] for the unfilled multipair PE insulated wire telephone cables.

The readily perceptible variations of the resistance and inductance of the wire pairs with frequency, allow certain deductions to be made on the characteristic impedance Z_0, of the wire pairs. The general expression of the impedance for the pair may be expressed as

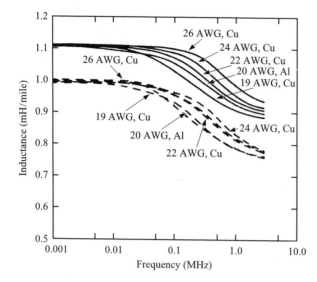

Figure 15.10 Inductance per mile as a function of frequency for multipair telephone cables with PE insulated wires with copper (Cu) and aluminum (Al) conductors at 20°C: solid lines, cable filled with jelly compound; dashed lines, unfilled cable (after [26]).

$$Z_0 = \sqrt{\frac{R + j\omega L}{G + j\omega C}} \tag{15.21}$$

Since at high frequencies $R \ll \omega L$ and $G \ll \omega L$, the characteristic impedance assumes the expression given by Eq. (15.3) for a distortionless line. Also from the relation of the propagation constant γ, given by

$$\gamma = \sqrt{(R + j\omega L)(G + j\omega C)}$$
$$= \alpha + j\beta \tag{15.22}$$

the attenuation α is approximated by Eq. (15.5) and the phase constant or phase shift β by Eq. (15.6). Under high frequencies, the resulting lossless line behavior leads to a velocity of propagation, v that is independent of frequency and is defined by Eq. (15.7). At low audio frequencies $R \gg \omega L$, but as Table 15.2 demonstrates, G is negligibly small by comparison with ωC; hence, at low frequencies, the attenuation becomes

$$\alpha = \sqrt{\tfrac{1}{2}\,\omega RC} \tag{15.23}$$

and the phase constant

$$\beta = \sqrt{\tfrac{1}{2}\omega RC} \tag{15.24}$$

and the characteristic impedance Z_0 assumes the form

$$Z_0 = \sqrt{\frac{R}{j\omega C}} \tag{15.25}$$

TABLE 15.2 Conductant G as a Function of Frequency for PE Insulated Multipair Petroleum Jelly Filled Telephone Cables

Frequency, f (MHz)	Conductance, G (μS/mile)
0.01	0.8
0.1	25.0
1.0	600
2.0	1700
3.0	3050

Source: After [26].

Thus in terms of Eqs. (15.3) and (15.25), the magnitude of the characteristic impedance, $|Z_0|$, would be expected to decrease with frequency until it attains a fixed value of $\sqrt{L/C}$ given by Eq. (15.3); its angle will also decrease with frequency, becoming zero at the frequency where $|Z_0|$ approaches asymptotically a value of $\sqrt{L/C}$. This behavior as a function of frequency is substantiated in Fig. 15.11 [26]. The magnitude of Z_0 over the higher frequencies is greater for the filled cables, since the ωL term predominates; the larger value of the inductance L contributes also to this tendency.

Figures 15.12 and 15.13 compare the attenuation as a function of frequency at room temperature for PE insulated multipair cables in the filled and unfilled states and with different degrees of moisture in the unfilled state, respectively [26, 27]. The attenuation α at low audio frequencies is governed by Eq. (15.23), which indicates dependence of α upon R and C; thus, α exhibits an increase with frequency due to the onset of the skin effect. At elevated frequencies, the value of α in unfilled cables is seen to diverge from that of the filled cables, primarily because of the higher wire pair inductance of the latter and to a lesser extent the slightly lower value of R arising from a reduced proximity effect. The difference in the attenuation characteristics between the two

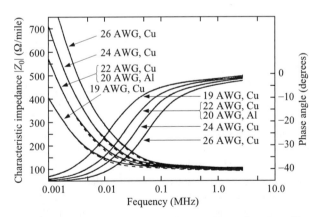

Figure 15.11 Characteristic impedance and phase angle as a function of frequency at 20°C for multipair PE insulated copper (Cu) and aluminum (Al) conductor cables, having a nominal mutual capacitance of 0.083 μF/mile: solid curves, petroleum jelly filled cable; dashed lines (shown for $|Z_0|$ only), unfilled cable (after [26]).

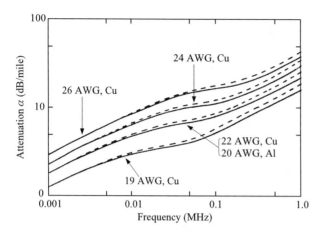

Figure 15.12 Attenuation vs. frequency of a PE insulated multipair telephone cable with a nominal mutual capacitance of 0.083 μF/mile with copper (Cu) and aluminum (Al) conductors at 20°C: solid curves, cable filled with petroleum jelly compound; dashed curves, unfilled cable (after [26]).

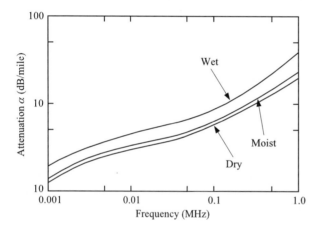

Figure 15.13 Attenuation vs. frequency of PE insulated copper conductor (19 AWG) multipair telephone cable with a nominal mutual capacitance of 0.083 μF/mile at room temperature, with moisture as a parameter (after [27]).

types of cable is sufficient to warrant consideration when carrying out design of equalizer networks. Figure 15.13 presents a comparison of the attenuation of a dry multipair cable subsequent to its immediate manufacture, followed by its complete filling with water (having a conductivity of 80 μS cm at 1 kHz at room temperature), and finally in its moist state after the removal or draining of the water for several hours while suspended in a vertical position [27]. The attenuation is seen to be augmented in proportion to the wetness of the cable; since the mutual capacitance C of the cable is increased from 0.082 to 0.186 μF/mile and the conductance G from approximately 12.5 to 140 μS/mile (at 0.1 MHz) with the cable first in a dry and then in a wet condition, respectively, the increase in attenuation with the degree of wetness is determined by the changes in the C and G parameters at low audio frequencies [27] The higher value of α results directly from the increase in C, in accordance with Eq. (15.23). While at higher frequencies the interfacial space charge polarization losses, between the PE insulation and the contiguous water film of elevated conductivity enhance the tan δ value. In accordance with Eq. (15.16), this leads to an increase in G at higher frequencies. More elevated values of G and C cause a reduction in the magnitude of the characteristic impedance $|Z_0|$, as Eq. (15.21) would infer, and as a consequence, Eq. (15.4) would in addition indicate an augmented attenuation.

It may be interesting to compare the transmission line parameters of the standard PIC cable design with those of the antecedent pulp insulated twisted-wire multipair cables. Such comparison is presented in Table 15.3 [18]. Perhaps the principal difference lies in the attenuation values at 1 MHz, which tend to be substantially higher (2–3 dB/km) for the earlier pulp insulated cables. This can be attributed to the more elevated dielectric losses in the pulp insulation associated with the polar nature of the cellulose molecules and the bound-water content in the paper structure of the pulp insulation.

Figure 15.14 illustrates the influence of temperature on PE insulated multipair copper conductor telephone cables [27]. The resistance R is significantly augmented at increased temperatures, causing a marked increase in attenuation. Over the lower frequencies, attenuation tends to vary as the square root of the frequency, while over the higher frequency regime it evinces a quasi-linear dependence. The larger the diameter of the copper conductor, the lower becomes the attenuation of the wire pair; this property has great ramifications as concerns the maximum distance over which signals may be transmitted by means of twisted-wire multipair cables. In subscriber loop networks, the wire pairs must carry two channels: one for voice and another for data transmission. The latter channel, which operates in the duplex mode for both information display and access to a data base, requires that the multipair telephone cable be capable of transmitting the data at certain rates. Figure 15.15 depicts schematically the maximum transmission distances without repeaters over which multipair cables are capable of transmitting at different data rates, with the copper conductor diameter or gage as a parameter [28]. For narrow-band digital transmission in the customer loop band, the cable insertion loss, which is principally determined by the attenuation characteristics of the multipair telephone cable, is of interest mainly at frequencies below 200 kHz [29]. In Europe, with HDSLs, a data transmission rate of 2.048 Mbits/s is employed. Recent experimentation with fast digital transmission has indicated that with suitable precoding/shaping techniques a data transmission rate of 2.048 Mbits/s could be readily achieved without repeaters over a distance of 4 km utilizing a 0.4-mm copper conductor, wire pair [30, 31]. In this particular case, modulated transmission with a center frequency of 420 kHz was utilized.

TABLE 15.3 Comparison of Transmission Line Parameters of Pulp Insulated and PIC Cables

Wire Size			Frequency (MHz)	Capacitance, C (μF/km)	Attenuation, α (dB/km)	Phase Constant, β (rad/km)	Characteristic Impedance, Z_0	
AWG	mm	Insulation					Ω	deg
22	0.64	Pulp	0.001	0.051	1.14	0.131	569	−44.1
			1.0	0.050	18.8	29.9	101	−2.8
22	0.64	PIC	0.001	0.052	1.17	0.135	578	−44.0
			1.0	0.052	16.2	31.6	99	−3.4
26	0.40	Pulp	0.001	0.049	1.74	0.202	922	−44.6
			1.0	0.048	28.3	29.6	97	−4.1
26	0.40	PIC	0.001	0.051	1.82	0.212	926	−44.6
			1.0	0.051	25.3	32.8	104	−5.0

Source: After [18].

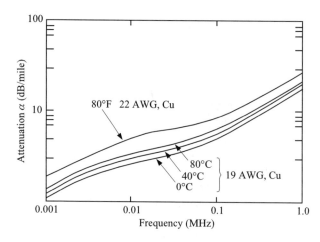

Figure 15.14 Influence of temperature upon the attenuation characteristics of PE insulated multipair telephone cables with a nominal mutual capacitance of 0.083 μF/mile (after [27]).

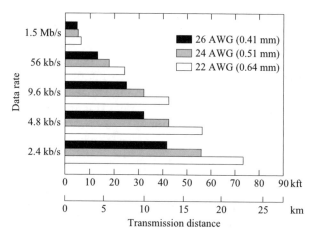

Figure 15.15 Distance constraints of nonloaded multipair telephone cables for digital data transmission without the use of repeaters (after [28]).

Figures 15.16 and 15.17 examine the effects of the petroleum filler compound and moisture, respectively, on the velocity of propagation, which is by definition equal to ω/β where β is the phase constant given by Eqs. (15.6) and (15.24) for the high- and low-frequency cases, respectively [26, 27]. Figure 15.16, which compares the filled and unfilled cables, indicates a lower velocity of propagation at carrier frequencies for the filled cables as a result of their higher inductance. The substantially higher mutual capacitance C of cables containing moisture leads to appreciably lower propagation velocities, as is evidenced in Fig. 15.17. At higher frequencies, the propagation velocity becomes less frequency dependent. Note that the velocity of propagation is commonly expressed as a percentage of the speed of light, which is equal to 2.99776×10^8 m/s, or 186,000 miles/s.

The near-end cross talk (NEXT) and the far-end cross talk (FEXT), if of a sufficiently high level, may seriously impair the transmission characteristics of multipair telephone cables. Equation (15.9) defines the general expression for NEXT; for shorter cable lengths, Eq. (15.9) reduces to [32]

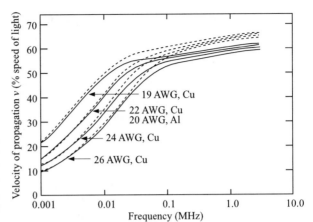

Figure 15.16 Velocity of propagation as a function of frequency for PE insulated copper (Cu) and aluminum (Al) conductor multipair telephone cables with a nominal mutual capacitance of 0.083 μF/mile at 20°C: solid curve, cable filled with petroleum jelly, dashed curve, unfilled cable (after [26]).

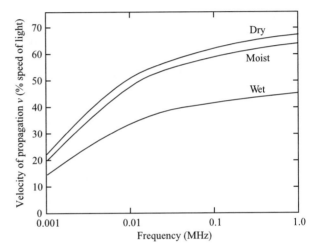

Figure 15.17 Velocity of propagation as a function of frequency of PE insulated 19 AWG copper conductor multipair telephone cable with a nominal dry-state mutual capacitance of 0.083 μF/mile, with moisture as a parameter (after [27]).

$$\frac{P_n}{P_0} = k_n f^{3/2} \tag{15.26a}$$

or alternatively

$$\text{NEXT} = 10 \log k_n + 15 \log f \tag{15.26b}$$

where, as in Eq. (15.9), k_n represents the NEXT constant; NEXT is in decibels and increases with frequency at 15 dB/decade. Similarly, FEXT for equal-level cross talk at the far end is defined by

$$\text{FEXT} = 10 \log k_f + 20 \log f + 10 \log \ell \tag{15.27}$$

where ℓ denotes the cable length and k_f is the FEXT constant. Note that the FEXT increases with frequency at the rate of 20 dB/decade. In analog transmission systems, the directions of transmission are separated in frequency, which reduces the importance of NEXT; consequently, cable cross talk at the far end becomes the design criterion, since it comprises both FEXT as well as the near-end to near-end interaction cross talk (NE-NE-IXT) [33]. With PCM transmission systems, the converse situation arises, namely that the NEXT becomes the preponderant concern because the maximum difference in the signal levels, propagating in the same direction on two different wire pairs, may be maintained at an acceptably small value [34]. It is estimated that cross talk noise coupled between adjacent wire pairs connecting the subscriber to the central office constitutes perhaps one of the most serious impairments on the subscriber loop [35].

The circuits delineated in Fig. 15.18 may be employed to determine NEXT and FEXT in metallic conductor telephone cables [27]. In the measurement procedure followed, both ends of the disturbed wire pair as well as the end of the disturbing pair are terminated with a resistance equal to their characteristic impedance. The cross talk can be determined by comparing the voltage levels in the disturbed and disturbing circuits by means of an attenuator; the latter is adjusted until an identical voltage reading on detector D is obtained as the switch position is changed from the disturbed circuit to the attenuator. The oscillator must provide a balanced output, so that all terminations may be grounded at their center taps.

It has been observed by Jachimovicz et al. [26] that the higher characteristic impedance of filled PE insulated telephone cables results in a reduction of 1 dB in the cross

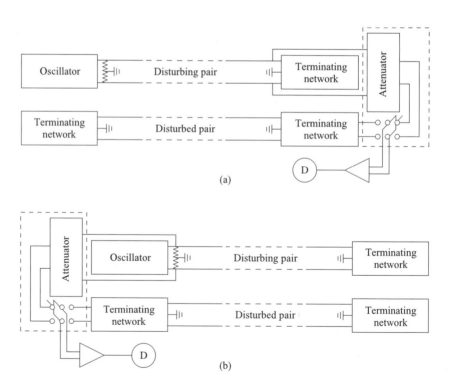

Figure 15.18 Schematic circuit diagrams for measurement of (b) NEXT and (a) FEXT coupling loss on metallic conductor telephone cables (after [27]).

talk as compared to an equivalent unfilled cable. Their data for the rms NEXT and FEXT coupling loss per test length of cable as a function of frequency are depicted in Figs. 15.19 and 15.20, respectively. They also found that FEXT is not appreciably influenced by the size of conductors within the wire gage range examined. Since FEXT depends strongly on structural unbalances of the cables, they concluded that the results point to a uniformity in the design criteria and manufacturing process used. However, NEXT is dependent upon attenuation and, consequently, insofar as wire gage is a predominant factor affecting the attenuation in terms of the conductor resistance, they were able to derive an empirical formula relating NEXT to attenuation and in turn to the conductor diameter. The empirical expression is given as

$$\text{NEXT}_1 = \text{NEXT}_2 + \log\frac{\alpha_1}{\alpha_2} \tag{15.28}$$

Thus, if the NEXT_1 results denote those obtained with a 22 AWG conductor multipair cable, the equivalent use of the attenuation α_1 for the same gage with that of α_2, corresponding to the gage of another conductor pair for which NEXT_2 is to be determined, yields the required results. As to be expected, this approach can only be employed for very long cables having identical structural unbalances. Equation (15.28) indicates that for conductor diameters less than those of 22 AWG gage NEXT would be correspondingly higher, and for those greater NEXT would be correspondingly lower.

At this juncture of the discussion, it is informative to compare the NEXT and FEXT performance characteristics of PIC cables with earlier twisted-wire multipair telephone cables in which pulp insulation was employed. Table 15.4 provides representative NEXT and FEXT data obtained on 22 AWG gage copper conductor pulp insulated cables; comparison of these NEXT and FEXT values with those of the PIC cables delineated in Figures 15.19 and 15.20 indicates the PIC cables to have improved crosstalk characteristics.

With PCM transmission systems, the use of regenerators in the presence of NEXT noise may give rise to significant error rates because of the impaired performance of the regenerator units. Hence NEXT limits the length of spacing of the regenerators, which can be used along the telephone cable route. The regenerator's susceptibility to NEXT

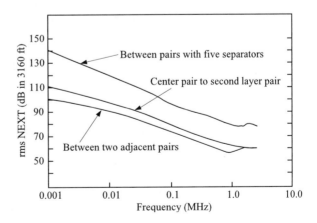

Figure 15.19 Near-end cross talk vs. frequency of 25-pair center unit of PE insulated 22 AWG copper conductor 100-pair telephone cable filled with petroleum jelly compound (after [26]).

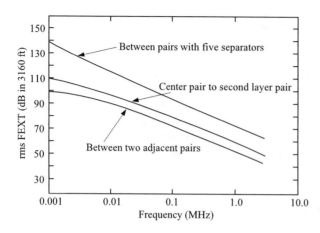

Figure 15.20 Far-end cross talk vs. frequency of a 25-pair center unit of a PE insulated 22 AWG copper conductor 100-pair telephone cable filled with petroleum jelly compound (after [26]).

TABLE 15.4 NEXT and FEXT Characteristics of Large Twisted-Wire Multipair Pulp-Insulated Telephone Cables with 100 Pair Groups at 772 kHz

	NEXT		FEXT	
Pair position	Mean (dB)	Standard Deviation (dB)	Mean (dB)	Standard Deviation (dB)
In same group	82	11	63	10
In adjacent groups	90	9	77	
In alternate groups	103	7		

Source: After [18].

noise may perhaps be best assessed in terms of the NEXT noise figure R_N, defined by [36, 37]

$$R_N = I_N - N_0 \tag{15.29}$$

where N_0 is the mean-square NEXT interferences at the regenerator decision point producing a specified error rate (e.g., 10^{-7}) and I_N is the regenerator noise amplification factor (i.e., noise amplification produced by the regenerator's equalizer). Thus the R_N factor is determined by the decision process parameters associated with intersymbol interference, deviations from the decision level thresholds, and sampling instants at the regenerator's decision point [38]. In effect, the NEXT noise figure, R_N, is equivalent to the NEXT attenuation [39]. The schematic circuit diagram for the measurement of the NEXT noise figure, R_N, is portrayed in Fig. 15.21.

Since many of the long-distance transmission links use fiber-optic cables, the volume of work associated with regenerator tests on long multipair cables has diminished greatly. The use of short lengths of multipair cables in subscriber loops and LANs in commercial buildings, where regenerators are not required, has resulted in efforts being directed to devise techniques for compensating or eliminating electronically the undesirable NEXT effects. In order to transmit data over unshielded twisted-pair

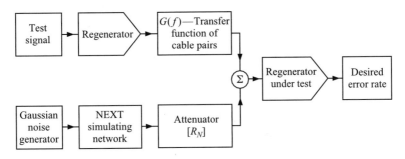

Figure 15.21 Schematic circuit diagram for measurement of NEXT noise figure to produce given error for a regenerator under test (after [37]).

(UTP) cables, which exhibit significant NEXT levels, requires cancellation of NEXT. Figure 15.22 depicts a typical arrangement for achieving NEXT cancellation [40]. The NEXT canceller follows an identical procedure employed in echo cancellation; i.e., it synthesizes a replica of the NEXT interferer, which is cancelled by subtracting the output of the canceller from that of the receiver. The input signal to the canceller is derived from the local transmitter, which consists of a series of symbols produced by the encoder of the transmitter. The output signal of the canceller is then subtracted from the received signal.

Prior to the implementation of fiber-optic cable for long-distance transmission, a considerable effort was placed into the development of lower attenuation multipair cables, which would permit a reduction in the repeater distances. This culminated in the development of expanded or foamed polypropylene copolymer insulation with superior transmission properties [41, 42]. Since such insulation contains typically 50% air, its dielectric constant is reduced to 1.53, yielding a lowered mutual capacitance of 0.0385 μF/mile. The air pore diameter in the foamed polymer is of the order of 1 mil (25.4 μm). The attenuation characteristics of an expanded polypropylene dielectric insulated 50-multipair cable are compared with a PIC equivalent multipair cable in Fig. 15.23. The attenuation of the expanded dielectric cable is significantly lower than that of the solid-dielectric insulated conductor cable because of the lower mutual capacitance and virtually zero conductance of the former, causing the line to behave more as a lossless transmission line [cf. Eqs. (15.4) and (15.5)]. The expanded insulation

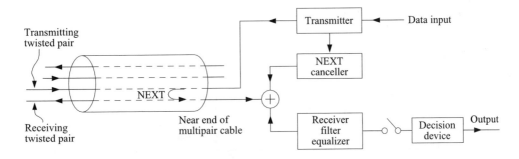

Figure 15.22 Typical arrangement for NEXT cancellation (after [40]).

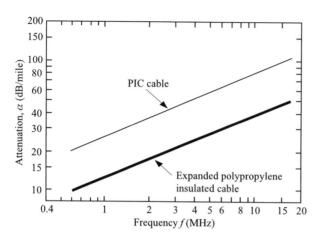

Figure 15.23 Attenuation-versus-frequency characteristics of 22 AWG conductor PIC and expanded polypropylene insulated 50-pair, 7-pair unit telephone cables at 70°F (after [42]).

thickness layer over the conductor is equal to 26.5 mils, leading to a characteristic impedance of 178 Ω at 3.15 MHz, as compared to 95 Ω for the PIC cable.

The cross-talk data at 3.15 MHz on the expanded dielectric multipair cable, which are a function of the twist length of the pairs and their geometric separation, are presented in Fig. 15.24. Shown is the curve for FEXT versus the cumulative percentage of the pairs in the 50-pair cable, i.e., the power sum of cross talk appearing in a wire pair due to the coupling of all the remaining pairs in the cable [42]. Examination of Fig. 15.24 indicates that the cable, using the expanded dielectric medium, has a cross-talk improvement of approximately 8–10 dB over the standard PIC cable. Figure 15.25 compares measurements of the capacitance unbalance obtained on expanded and solid dielectric 50-pair cables [42]. The capacitance unbalance to ground, which by definition is equal to the difference between the direct capacitance to ground of the two wires comprising a pair, constitutes a measure of the noise susceptibility of the given pair. The PIC cable is seen to contain wire pairs having a much higher percentage of pairs exceeding the specified capacitance unbalance level per given cable length.

It is generally agreed that UTP copper conductor telephone cables present a cost-effective alternative to the use of optical fiber cables for distances up to 100 m in

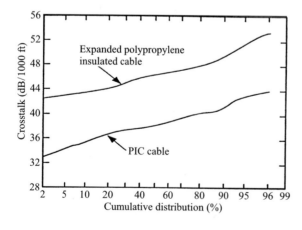

Figure 15.24 FEXT vs. cumulative distribution of 22 AWG conductor PIC and expanded polypropylene insulated 50-pair, 7-pair unit telephone cables at 70°F at a frequency of 3.15 MHz (after [42]).

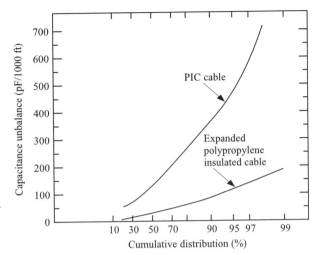

Figure 15.25 Capacitance unbalance to ground vs. cumulative distribution of 22 AWG conductor PIC and expanded polypropylene insulated 50-pair, 7-pair unit telephone cables at 1 kHz (after [42]).

LANs and up to 4 km in digital subscriber loops (DSLs) [43]. Virtually all customers in North America on DSLs and LANs both for voice and data transmission are served by an already existing and well-established UTP cable network, whose dismantling would present an unnecessary economic cost. In order to meet the demand for high-speed data transmission over the copper cable loop network, high bit-rate digital subscriber line (HDSL) and asymmetric digital subscriber line (ADSL) techniques have been developed. The former technique (HDSL) is confined to loops in a carrier serving area (CSA) and as indicated previously is capable of bidirectional transmission rates of 1.544 Mbits/s or greater. Figure 15.26, which delineates the bit rate and bandwidth of typical 24 and 26 AWG PIC cables in a NEXT-dominated environment [44], demonstrates that twisted-pair PIC cables can readily provide the necessary capacity. The transmission rate capacity is seen to decrease rapidly with the length of the telephone cable loop. The use of drop lengths (i.e., the final connection to the customer residence) increases the effective telephone cable loop. It is estimated that, on average, the addition of approximately 200 ft length to the overall loop length compensates adequately for the effect of the drop lengths [45]. The common usage of 24 AWG conductors in the drop length links minimizes impedance mismatch effects because of existing similarities in the characteristics of the loop cables. However, impedance mismatch readily arises with bridge taps on the loop cables, which are simply open-circuited pairs intentionally left in the cable for any possible future additional services. Since they essentially shunt other working pairs within the cable, a portion of a transmitted signal that may travel to the open-circuit end of the free pair will be reflected back (delayed and distorted) to the bridge tap point connection made with the working pair. Thus, not only will there develop a power loss in the main signal, but also the reflections from the end of the open pair will introduce a noise component upon the main signal that will be transmitted to the receiver; in addition, part of the reflected signal will appear as an echo at the sending end. Some additional power will be lost due to the introduction of the so-called nulls into the transfer function of the loop [45]; these nulls are shown to be empirically located at odd multiples of the frequency, f' in kilohertz,

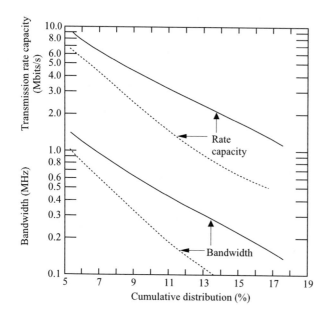

Figure 15.26 Transmission rate capacity and bandwidth of PIC telephone cables at 80 kHz with NEXT value of 57 dB: solid curve, 24 AWG copper conductor; dashed curve, 26 AWG copper conductor (after [44]).

$$f' = \frac{150}{\ell} \tag{15.30}$$

where ℓ is the length in kilofeet of the twisted pair from its open-circuited end to the point of the bridged tap connection. Theoretically f' is defined as the ratio of the phase velocity to the wavelength at the frequency f'.

With ADSL techniques, it is envisaged to transmit video signals via subscriber loops also at 1.5 Mbits/s rates within a CSA at distances up to 4.0 km [43]. However, as already mentioned previously, implementation of very high rate ADSL (VADSL) at 51.84 Mbits/s is foreseen over distances up to ~300 m. In addition, transmissions at 100 and 155.52 Mbits/s are being studied by a number of standardization committees. The UTP cables intended for some of these applications are categorized as either voice or data grade cables, and their characteristics at 16 MHz are specified in Table 15.5 [46, 47]. The higher signal attenuation together with the lower NEXT attenuation necessarily relegates the UTP-3 type cables into the voice grade category, thereby requiring NEXT cancellation procedures if these cables are to be employed for the higher 1.5-Mbits/s data transmission rates. In contrast, the substantially improved characteristics

TABLE 15.5   UTP Cable Parameters at 16 MHz

Cable Type	Signal Attenuation (dB/100 m)	NEXT Attenuation (dB)	Characteristic Impedance (Ω)
UTP-3	13.15	$\geq$23	$100 \pm 15\%$
UTP-4	8.85	$\geq$38	$100 \pm 15\%$
UTP-5	8.20	$\geq$44	$100 \pm 15\%$

Source: From [46, 47].

of UTP-4 and UTP-5 type cables render them directly suitable for high-data-rate transmission applications.

The improved characteristics of the latter cables are obtained in part by optimizing the pair twist arrangements and decreasing the number of pairs per cable, i.e., reducing the number of users. Although cross talk diminishes with twisting, the number of twists per unit length varies from pair-to-pair because no two adjacent pairs may have the same number of twists per unit length. Additional data on the different UTP cable types are provided in Table 15.6. Here the propagation loss is expressed empirically as a function of frequency. It can be perceived that the propagation loss in data grade cables varies principally as the square of the frequency, while in contrast the voice grade cable (UTP-3) exhibits substantial direct dependence upon frequency in addition to the normal square root dependence.

There have been many different electronic transmission schemes developed to increase the existing data rate transmission capacity of various grades of UTP cables [43]. Neither space nor the scope of this chapter permits us to consider these approaches. It will suffice to say that schemes involve different modulation techniques that center on the equalization of the received signal in order to remedy changes in the attenuation of the wire pairs, on reducing the effects of cross talk on signal detection, and on controlling the power spectral density of the transmitted signals. All these efforts should ensure continued relevance and importance of metallic conductor cables in the subscriber loop in terms of their cost effectiveness. A case in point concerns a recent study [47] describing a multilevel partial-response class IV (PRIV) transmission system developed for use in conjunction with category UTP-5 cables, which permits transmission rates in excess of 155.52 Mbits/s for LANs. Figure 15.27 represents schematically the allowable transmission data rate-distance scenarios for a UTP-5 cable. The dark area bounded by curves (a), (b), and (c) defines the confined region within which data grade UTP-5 cables may operate without the need of NEXT cancellation. For radiation reduction requirements, for curve (a) the spectral power of the PRIV signal above 30 MHz must remain at 20% of the total power, in curve (b) under the worst case NEXT interference from an adjacent wire pair where the symbol error probability as controlled by filters does not exceed 10^{-10}, and curve (c) delimits the maximum attenuation, which must be compensated by adaptive analog equalization.

15.5 ELECTRICAL CHARACTERISTICS OF COAXIAL CABLES

Following the development of PE in the mid-1930s, it became possible to manufacture low-dielectric-loss individually shielded concentric conductor cables. Such coaxial

TABLE 15.6 Worst-Case Propagation Loss[a] on UTP Cables at 20°C

Cable Type	Propagation Loss, $L_p(f)$ dB/100 m)	Increase of $L_p(f)/°C$	Frequency range (MHz)
UTP-3	$2.320\sqrt{f} + 0.238f + 0.000/\sqrt{f}$	1.2%	$0.772 \leq f \leq 16$
UTP-4	$2.050\sqrt{f} + 0.043f + 0.057/\sqrt{f}$	0.3%	$0.772 \leq f \leq 20$
UTP-5	$1.967\sqrt{f} + 0.023f + 0.050/\sqrt{f}$	0.3%	$0.772 \leq f \leq 100$

[a] From specification TIA/EIA-568-A.
Source: From [40, 46].

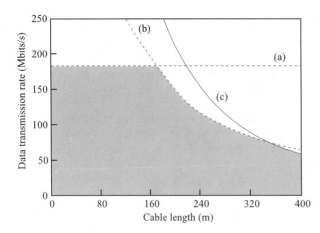

Figure 15.27 Data transmission rate vs. cable length scenarios for partial-response class transmission over UTP-5 type cables (after [47]).

cables exhibit excellent high-frequency characteristics but, as a consequence of their electrically unbalanced configuration, perform poorly at low frequencies because their outer conductor or shield is a relatively ineffective barrier to cross talk at low frequencies. For this reason the low-loss air-dielectric-type coaxial cables, that are used principally for telephone cable applications have been confined to a frequency régime between approximately 1.5 and 220 MHz [18, 48]. As Table 15.7 attests, much of the early long-haul traffic in the 1970s both in Europe and North America was carried via air-dielectric coaxial cables [18, 48–53]. In North America this involved distances up to 4000 miles; however, as has already been mentioned, at the present, the long-haul traffic has become increasingly dominated by optical fiber cables. It is interesting to note from Table 15.7 that air-dielectric coaxial cables may be used as trunk line cable in metropolitan areas because of their broadband characteristics and high-data-rate capabilities; they have become the transmission medium of choice for many LAN applications. Coaxial cables with insulations utilizing foam PE or polystyrene are employed principally for community antenna TV systems (CATV), while solid PE dielectric coaxial cables are extensively used for short-length connections for various electronic applications. However, for short-length uses between radio and TV transmitters and receivers as well as interconnections of frequency carrier equipment the air-dielectric cables are often preferred. There has been an increase in the demand for expanded polymer dielectric coaxial cables for CATV applications primarily because the large bandwidth of coaxial cables (~1 GHz) has not been fully exploited thus far. Also, the recent marked rise in the use of wireless telephony has further augmented the demand for coaxial cable systems for the associated wiring requirements in the home. A very common use of coaxial cables occurs in so-called hybrid systems, where the signal provided to a servicing area by an optical fiber cable is subsequently distributed via a coaxial cable and thence a drop cable (wire pair) to the home.

The coaxial air dielectric cables employed in North America have inner copper conductor diameters of 104 mils (2.64 mm). The inner conductor is supported by PE discs spaced 1 in. (25.4 mm) apart and having a thickness of 0.085 in. (2.16 mm) and an outside diameter of 0.361 in. (9.17 mm). Over the discs is applied a 0.012-in.

TABLE 15.7 Digital Air-Dielectric-Type Coaxial Cable Systems by Location and Date Rate (ca. 1975)

Parameter	Europe					United States	Canada	Japan	
	8.448 (Mbits/s)	34.368 (Mbits/s)	120.00 (Mbits/s)	139.264 (Mbits/s)	139.264 (Mbits/s)	274.176 (Mbits/s)	274.176 (Mbits/s)	97.728 (Mbits/s)	400.352 (Mbits/s)
Diameter (mm) Inside conductor/Outside cable	$\frac{0.7}{2.9}$	$\frac{0.7}{2.9}$	$\frac{1.2}{4.4}$	$\frac{1.2}{4.4}$	$\frac{1.2}{4.4}$	$\frac{2.6}{9.5}$	$\frac{2.6}{9.5}$	$\frac{1.2}{4.4}$	$\frac{2.6}{9.5}$
Repetition type	Regenerator	Regenerator	Regenerator	Hybrid	Regenerator	Regenerator	Regenerator	Regenerator	Regenerator
Maximum digital repeater spacing, km	4.1	2.05	2.1	80	2.05	1.74	1.9	1.6	1.6
Maximum analog repeater spacing, km	NA	NA	NA	NA	NA	NA	NA	NA	NA
Line code	HDB3	MS43	4B3T	Class IV (7 levels)	MS43	Binary	B3ZS	AMI + scrambler	AMI + scrambler
Power supply Current, mA	50	100	50	100			870	250	550
Voltage, V	±400	±400	±250	±400			±1500	±350	±1500
Power feeding span, km	80	80	30	80	100		208	24	10
Design error rate	2.5×10^{-10} per km	10^{-10} per km	2×10^{-7} for 2500 km	10^{-10} per km	10^{-10} per km		10^{-7} for 2500 km	10^{-8} for 200 km	10^{-7} for 2500 km
Application type	Short- and medium-haul trunks	Short- and medium-haul trunks	Intercity trunks	Intercity trunks	Intercity trunks	Metropolitan areas	Intercity trunks	Intercity trunks	Intercity trunks
Implementation stage	In service	Experimental link	Experimental link	Experimental link	Experimental link	In service	In service	Experimental link	Experimental link

Source: After [49].

(0.305-mm) outer cylindrical copper conductor with interlocking serrated edges along a longitudinal butt seam [52]. Thus the outside copper conductor diameter has an approximate nominal value of 0.375 in. (9.5 mm); for this reason, this cable is frequently referred to as the 0.375-in. or 9.5-mm air-dielectric coaxial cable. It is interesting to note from Table 15.7 that all the air-dielectric coaxial cables in Europe are substantially smaller. In more current designs of this type of coaxial cable, a PE pipe may be used to enclose the PE spacers with the inner conductor.

If we consider the condition of a low loss line defined by Eqs. (15.3)–(15.7), the inner and outer conductor diameters of the coaxial cable determine its characteristic impedance magnitude $|Z_0|$, which may in turn be adjusted in order to minimize the attenuation loss. Since the capacitance C and inductance L of a coaxial cable are given by [4]

$$C = \frac{\varepsilon'/\varepsilon_0}{(17.98 \times 10^9)\ln(r_2/r_1)} \qquad \text{F/m} \qquad (15.30)$$

and

$$L = (2 \times 10^{-7})\ln\frac{r_2}{r_1} \qquad \text{H/m} \qquad (15.31)$$

where r_1 and r_2 are respectively, the inner and outer conductor radii, the magnitude of $|Z_0|$ for an air dielectric line is simply given by $\sqrt{L/C}$, such that

$$|Z_0| = 60\ln\frac{r_2}{r_1} \qquad (15.32)$$

Taking into account skin effect, the ac resistance of the coaxial line may be expressed as

$$R_{ac} = \frac{\sqrt{f\mu\rho}}{2r_1\sqrt{\pi}} \qquad (15.33a)$$

where ρ and μ are, respectively, the resistivity and permeability of the copper conductors. Alternatively, Eq. (15.33) may be stated as [4]

$$R_{ac} = \sqrt{10^{-7}f\rho}\left(\frac{1}{r_1} + \frac{1}{r_2}\right) \qquad (15.33b)$$

Since the dielectric is air, the conductance G of the coaxial line reduces to zero and the attenuation constant a is given by Eq. (15.5); substitution of Eqs. (15.32) and (15.33b) into Eq. (15.5) and rearrangement gives [4]

$$\alpha = \frac{\sqrt{10^{-7}f\rho}}{2(60r_2)}\left[\frac{(r_2/r_1) + 1}{\ln(r_2/r_1)}\right] \qquad (15.34)$$

When the term in the parentheses in Eq. (15.34) is differentiated with respect to r_2/r_1 and equated to zero, the solution of the resulting equation yields the ratio r_2/r_1 for which the value of $|Z_0|$ yields a minimum in the attenuation α. For this condition $r_2/r_1 = 3.6$, which corresponds to $|Z_0| = 77\,\Omega$.

For the air-dielectric coaxial telephone cable previously considered, the ratio r_2/r_1 corresponds to the theoretical value for minimum attenuation. The actual measured characteristic impedance of the air-dielectric cables is normally approximately $75\,\Omega$. In fact, the characteristic impedance for most cables in use ranges from 50 to $75\,\Omega$. The CATV cables with foam polymer dielectrics, for which low attenuation loss is an important requirement, have $75\,\Omega$ nominal characteristic impedance. In solid PE dielectric cables used for electronic measurements, where flexibility is a necessary property for the short lengths employed, a characteristic impedance of $50\,\Omega$ is standard. The inclusion of a solid dielectric such as PE, for which the value of the relative real permittivity or dielectric constant is approximately between 2.16 and 2.23, ipso facto reduces the characteristic impedance to a value below $75\,\Omega$ as is apparent from the relation

$$|Z_0| = \frac{60}{\sqrt{\epsilon'/\epsilon_0}}\ln\frac{r_2}{r_1} \tag{15.35}$$

Assuming again the optimum ratio r_2/r_1 of 3.6, substituting in Eq. (15.35) the range of ϵ'/ϵ_0 values between 2.16 and 2.23 yields a $|Z_0|$ value of the order of $52\,\Omega$. The value of the dielectric constant or relative real permittivity of the foam PE or polystyrene insulation is to a considerable degree dependent upon the air content and bubble size of the foam insulation. In general, foam-insulation-type cables for CATV applications have characteristic impedances approaching $75\,\Omega$. Table 15.8 compares the attenuation characteristics of a number of typical solid-, foam-, and air-dielectric coaxial cables. It can be perceived that while the attenuation in solid-dielectric cables is substantially higher than in air-dielectric/PE disk cables, the attenuation in some foam-dielectric cables is either lower or comparable to that of the air-dielectric/PE disk cables.

Since in telephone applications it is primarily the air-dielectric/PE disk [0.375 in. (9.5 mm)] coaxial cable that has seen extensive use, we shall confine our discussion to this type of cable which is characterized by a propagation velocity of 95% of that of the speed of light. Its impedance decreases only slightly with frequency to approximately $75\,\Omega$ at 100 MHz from the optimum value of $77\,\Omega$, corresponding to the requirement of the lowest attenuation criterion. Above 1.5 MHz its attenuation in decibels per mile as a function of the frequency f and temperature T may be stated empirically as [48]

TABLE 15.8 Attenuation Characteristics of Selected Coaxial Cables

Cable Type	Nominal Diameter of Dielectric (mm)	Dielectric	Attenuation (dB/km) 3 MHz	216 MHz
Bell System toll cable	9.5	Air/PE disk	4.13	35.5
CATV	12.7	Polystyrene foam	—	35.1
CATV	19.0	Polystyrene foam	—	23.6
CATV	25.4	Polystyrene foam	—	19.4
RG 216/U	7.24	Solid PE	11	115
RG 164/U	17.3	Solid PE	4	56

Source: From [18].

$$20 \log |H(f, T)| \cong A\left(1 + \frac{0.0062}{\sqrt{f}}\right)\sqrt{f} + Bf + (T - T_0)D\sqrt{f} \qquad (15.36)$$

where the constants A and B are 3.9002 dB/mi.-MHz$^{1/2}$ and 0.0047 dB/mi.-MHz, respectively, $T_0 = 55°F$ and T is the ambient temperature in degrees Fahrenheit, and $D = 0.0043$ dB/mi.-MHz$^{1/2}$ °F. Here the first term denotes the loss arising from the resistive losses in the inner and outer copper conductors; as a result of the skin effect above 1.5 MHz, this loss varies as the square root of the frequency, $\sqrt{f}$. The second term increases directly with f as a consequence of the dielectric losses in the polystyrene spacer disks. The third term is associated with the resistivity changes of the copper conductors, which is a temperature function and is proportional to $\sqrt{f}$. Figures 15.28 and 15.29 delineate respectively the variation of the attenuation with respect to frequency at room temperature as a function of frequency but with temperature as a parameter.

From Figure 15.28 it is apparent that the deployment of a long-haul 0.375-in. (9.5-mm) air-dielectric PE spacer cable over a total distance of 4000 miles would present a total accumulated loss of 1.32×10^5 dB at 70 MHz [48]. This loss is readily compensated by the placement of repeaters at 1-mile (1.6-km) intervals along the entire coaxial cable length, with fixed equalizers employed to match the variations in the loss characteristics of the different adjoining cable lengths as well as any possible variation in the interrepeater distances. However, variations in the attenuation loss arising from temperature changes in the cable require the implementation of adjustable equalizers. The changes in attenuation for a $\pm 20°F$ swing in the cable exceed by one to two orders of magnitude any other time-variant effects in the coaxial transmission system [48]. Preequalization and postequalization methods are employed to limit the attenuation changes as a consequence of the temperature effects to $\sim \pm 2.5$ dB over 7-mile (11.2-km) cable segments [54]. In the foregoing scheme, a preequalizer in a preceding regulating repeater and a postequalizer in a subsequent regulating repeater compensate, each in turn, for half of the attenuation loss change along the intervening cable segment. The regulating repeaters are designed to sense the cable temperature induced attenuation changes to within ± 0.05 dB within a 1.5–70-MHz frequency band. The

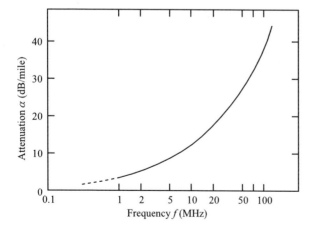

Figure 15.28 Attenuation vs. frequency at room temperature for 1-mile length of 0.375-in. (9.5-mm) air-dielectric PE spacer coaxial cable (after [48]).

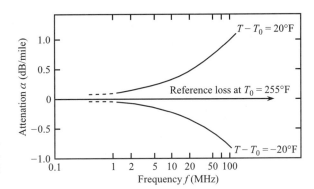

Figure 15.29 Effect of temperature on attenuation characteristics for 1-mile length of 0.375-in. (9.5-mm) air-dielectric PE spacer coaxial cable (after [48]).

regulating repeaters themselves are characterized by an intrinsic temperature coefficient, which is typically ± 0.02 dB per repeater section [48]. The latter is normally compensated continually by means of dynamic equalizers at the receiving end of each power supply span, whose used lengths do not exceed 75 miles [55].

Impedance discontinuities along the coaxial cable will give rise to signal reflections, which in telephone cables are often referred to as echoes. A high echo return loss in decibels in a coaxial cable is desirable, since it infers a low reflected signal amplitude in comparison to that of the incident signal. Echo measurements on the 0.375-in. (9.5-mm) cable are performed using a 250-ns raised-cosine pulse, which contains frequencies up to 4 MHz [52]. Figure 15.30 illustrates echo distribution results obtained on a total equivalent length of 2000 miles of a 0.375-in. (9.5-mm) coaxial cable indicating an echo magnitude average of 65.6 dB. Another method, which is particularly suited for the detection of periodic impedance changes along coaxial cables, is the so-called structural return loss (SRL) measurement in decibels whereby the input signal to the cable is swept between 8 and 220 MHz with the reflections being monitored [52]. The frequency f in

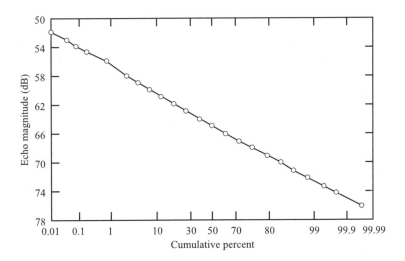

Figure 15.30 Distribution of worst echo magnitude in 0.375-in. (9.5-mm) air-dielectric PE spacer coaxial cables, representing overall length of 2000 miles (after [52]).

megahertz at which a maximum in the magnitude of reflected signal occurs, corresponding to a particular periodic irregularity along the cable, is given by

$$f = \frac{cW(1 - TU)}{2\sqrt{\varepsilon_r'}S} \tag{15.37}$$

where c is the velocity of light in space in feet per second, S is the spacing between the irregularities in feet, TU is the standing take-up (equal to 1.5% in a multiline coaxial cable containing 20 coaxial lines), and W is given by

$$W = \left[1 - \frac{L_i}{2L_e}\right] \tag{15.37a}$$

where L_i denotes the internal inductance and L_e the inductance between lines. For example, with 36-in. stranding lay irregularities, the half-wavelength frequency is found to be 156 MHz [52]. The half-wavelength at the frequency of measurement equal the distances between the periodic irregularities; some of these irregularities are cyclic and may relate to periodic extrusion variations of the polymer extruder in the case of solid or foamed insulation coaxial cables.

When coaxial cables are installed in the field, reflections may occur from impedance irregularities created by splices and taps or tees. Since the impedance of taps, including the associated electronics, is much higher than the 75 Ω nominal impedance of the coaxial cable, reflections from taps are ordinarily negligible [56]. In order to minimize the effect of taps, Ethernet and IEEE specifications require a loading less than 4 pF in parallel with 50 Ω and 6 pF in parallel with 100 Ω respectively. The taps must be located on 2.5-m markings and cannot exceed 100 per 500-m length of the 0.375-in. (9.5-mm) coaxial cable. However, the return loss resulting from cable splices may be appreciable [52]. It should be borne in mind that the return loss is not identical to the structural return loss discussed above. By definition the return loss is equal to the difference in decibels between the incident signal and reflected signal amplitudes. Figure 15.31 depicts the return loss of a field splice in a 0.375-in. (9.5-mm) coaxial cable as a function of frequency. The calculated results are based on the measured resistance and reactance of the splice, yielding a total calculated return loss of 65 dB at 20 MHz [52].

In coaxial cables it is the far end cross talk, FEXT, that constitutes the determining constraint for system operation. Figure 15.32 shows the FEXT of the inner and outer layer coaxials within an overall cable containing a total of twenty 0.375-in. (9.5-mm) coaxial cables. The cross talk results were obtained on a 2070-ft cable length on a reel and on a 2-mile cable length installed between two repeater points [52]. The differences between laboratory and field test data are attributed to splicing and relaxation of cables following unwinding and installation of the cable.

Table 15.9 gives the initial service dates and capacities of the Bell system long-haul coaxial transmission cable systems L1, L3, L4 and L5 [48]. The reduction in repeater spacing and the increase in the number of crossage channels with each new coaxial cable system is most remarkable. Table 15.10 compares the transmission capabilities of twisted-wire multipair and coaxial cables [18].

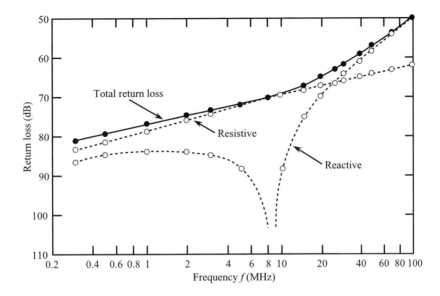

Figure 15.31 Return loss vs. frequency due to field splice in 0.375-in. (9.5-mm) air-dielectric PE spacer coaxial cable (after [52]).

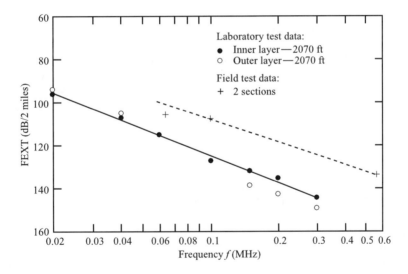

Figure 15.32 Far-end cross talk between adjacent 0.375-in. (9.5-mm) air-dielectric PE spacer coaxial lines in cable containing 20 coaxial lines (after [52]).

In Table 15.10, the twisted-pair shield cable refers to interposed shielding between bundles of twisted-wire circuits transmitting signals in opposite directions, i.e., shielding isolation between the transmitting and receiving wire pair circuits. From the comparison it is apparent that foam insulation in twisted-wire-pair cables as well as shielding improves substantially the transmission characteristics of twisted-pair cables, though these ameliorations still fall appreciably below the performance capabilities of air-dielectric PE spacers and foam PE coaxial cables. This is clearly in evidence from

TABLE 15.9   Bell Coaxial Cable System Capacities and Service Dates for System Designations L1–L5

Cable Characteristics	L1	L3	L4	L5
Beginning of commercial service	1941	1953	1967	1974
Two-way 4-kHz channels per coaxial pair	600	1860	3600	10 800
Nominal repeater spacing, miles	8	4	2	1
Number of coaxial pairs	4	6	10	11
Working pairs	3	5	9	10
Protection pairs	1	1	1	1
Total two-way 4-kHz message channels	1800	9300	32 400	108 000

Source: From [48].

Fig. 15.33, which compares the attenuation characteristics of the same cable groups. Note that in the special high-frequency twisted-wire-pair cable group are included foam insulated cables with improved staggered twisting sequences and shielding between transmitting and receiving pairs.

15.6 METALLIC CONDUCTOR TELEPHONE CABLE DESIGN AND MANUFACTURE

15.6.1 Twisted-Wire Multipair Telephone Cables

As will be recalled from our preceding discussions on the electrical characteristics of twisted-wire multipair telephone cables, each pair is designed to function as a balanced transmission line. Each pair is formed by twisting two identically insulated wires, but of different colors, in order to maintain an acceptable identification code.

Different twist patterns are employed to reduce cross talk; a total of 25 different twist lengths are utilized to ensure that adjacent pairs have different twist lengths. Commonly the twist lengths range from ~2 in. (5.1 cm) to 4.5 in. (11.4 cm). Balance with respect to the shield entails that the two wires constituting the twisted pair must have equal admittances to the shield and equal series impedances. Thus the form of the twist must be such that the geometry of the twisted-wire pair is symmetrical or balanced with respect to the equivalent shield formed by other surrounding pairs. Series impedance variations arise primarily from slight differences in the resistances of the two wire conductors. The two components constituting a capacitance unbalance are the pair-to-pair unbalance, i.e., the difference in capacitance of each conductor of the pair to ground. Even for reasonably well balanced pairs, the unbalance effects become more pronounced at higher frequencies, and it is for this reason that twisted-wire multipair cables have a low upper frequency limit, often less than 4 MHz, though with some specialized designs this limit may be substantially increased.

15.6.1.1 Paper Ribbon Twisted-Wire Multipair Telephone Cables

Although paper ribbon insulated twisted-wire multipair cables are no longer being manufactured, some remnants of these cables are still in service. In this antiquated design, in order to maintain the capacitances at a low value, the papers are applied

TABLE 15.10 Comparison of Twisted-Pair and Coaxial Cables for Broadband Services

Cable	Number of Lines per Full-Size Cable	Preferred Frequency Range (MHz)	Transmission Method	Total One-way Base Bandwidth		Cable Area per Unit BS (mm²/MHz)	Repeater Spacing (km)	Transmission Capability
				Per Line (MHz)	Per Conduit (GHz)			
Standard telephone cable, 0.4-mm pulp insulated	3600	0–1.5	Baseband	1	3.5	2	1.8	Customer line (Picturephone service)
			Digital (T1 carrier)	0.1	0.3	20	1.2	Trunk telephone
Bell system twisted-pair low-capacitance cable, 0.64-mm foam insulated	300	0–6	Digital (T2 carrier)	0.4	0.1	60	4.5	Trunk telephone
Twisted-pair shielded cable foam insulated	320	0–10	Baseband	5	1.5	2.5	2.5–3	Customer line, TV
Bell system coaxial cable, 9.5-mm air-dielectric PE spacers	22	3–300	VSB-AM	200	4	1.5	1.6	Customer line, TV

Source: After [18].

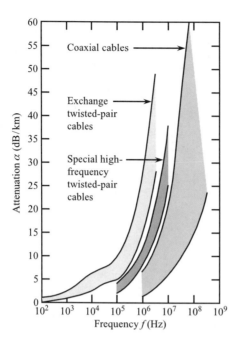

Figure 15.33 Attenuation characteristics for twisted-wire multipair and coaxial telephone cables (after [18]).

very loosely onto the wires, their prime function being to offer a mechanical separation as well as electrical insulation between the wires. The papers may be either folded loosely around the wire so that the fold runs parallel along the length of the wire or lapped spirally with an overlap, maintaining still a loose wrap. Occasionally two paper layers are applied using both application procedures. The separately insulated wires are paired together forming separate speaking circuits, with the lengths of the lay or twist of the wires of the various pairs being different to reduce interference due to coupling. The layers of such strands are assembled, with each adjacent layer being reversed in direction, until one obtains the total number of pairs required for the particular cable design. It is again to be emphasized that in cables used at voice frequencies, inductive coupling is a comparatively minor factor producing cross talk as compared to capacitive coupling caused by the capacitance unbalance of circuits. It is primarily the latter effect that is intended to be minimized by conductor twisting. In the quad construction, four insulated wires are laid up without pairing; a number of these quads are then laid up into a cable. Although some quadded cables are still in service, none are being produced at the present. Their earlier introduction arose from the implemented use of three circuits for every two pairs of wires. The third circuit, often termed the *phantom circuit*, has each pair acting as a separate conductor, so that the two pairs must be twisted together to prevent cross talk between the adjacent phantom circuits. The twisting length of the adjacent set of pairs is also different. In quadded cable terminology, the two pairs of a quad are known as the *sides*, while the term *pair* refers to those pairs outside the particular quad under consideration. It can be appreciated that the quad construction requires a stringent electrical balance of the cable components; in addition, special equipment is required. A serious disadvantage of quadded cables is that they are not suitable for use at carrier frequencies.

A typical paper ribbon insulated telephone cable is depicted in Fig. 15.34, indicating the layer construction and the lead sheath for moisture and mechanical protection.

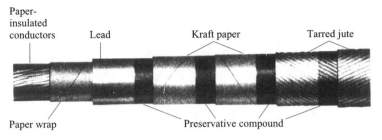

Figure 15.34 Typical paper-ribbon insulated-wire telephone cable, ca. 1958. (Courtesy of Northern-Telecome Ltd.)

All early twisted-wire multipair cables were essentially of the concentric layer construction, each layer being laid in a reverse direction to that of its preceding layer. Figure 15.35 shows a schematic diagram of a concentric layer cable of 152 pairs. The letters indicate the colors of the insulation of the two wires constituting the pair: GO (green-orange), BO (blue-orange), and RO (red-orange) The tracer pair in the middle of the cable is orange-white (OW). The shown numerals on the diagram, following the insulation color letters, denote the total number of pairs of each color combination in the respective layer. The color code is of paramount importance for the identification of pairs for testing, connecting, and splicing operations. The sequence of color within a layer is green-orange and alternately blue-orange. Also there is one red-orange or orange-white twisted-wire pair in the outside layers and in layers having an odd number of pairs.

The requirement for large numbers of pairs in telephone cables has resulted in the use of *units*, which consist normally of 25, 50, or 100 pairs. A number of these units are laid in the form of a cable that may contain a very large number of pairs. Figure 15.36 portrays a schematic cross section of a cable having 556 pairs, with 3 units forming the inner core, surrounded by 7 units that in turn are enclosed by 12 units. The letters indicate again the color code of the wire insulation: GW (green-white), RW (red-white), and BW (blue-white); whereas the letter T designates the trace pair, whose colors in this particular case are red-blue. Again, in the designation employed, the numerals following the insulation colors denote the total number of pairs within each unit including the tracer pair T wherever indicated. The unit binder colors are the same for all units

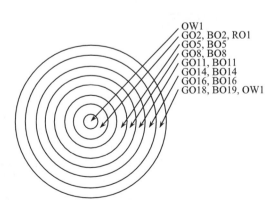

Figure 15.35 Concentric layer schematic of 152-pair concentric layer cable.

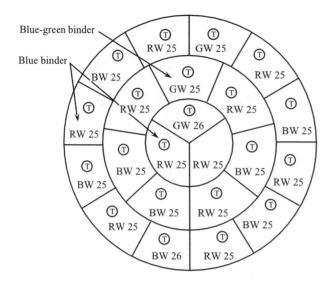

Blue-green binder

Blue binder

BW 25

RW 25   GW 25

GW 25

RW 25

RW 25

GW 26

RW 25

RW 25

BW 25

RW 25

BW 25

BW 25

RW 25   RW 25   BW 25

BW 25

RW 25

RW 25   BW 25

RW 25

BW 25

BW 25

RW 25

BW 26   RW 25

Figure 15.36 Unit array in 556-pair telephone cable.

contained within a given layer. The units, containing 26 pairs, are distinguished by a white thread in the binder. Whether the ribbon insulated wire cable is of the concentric layer or unit construction, a core wrap is used to maintain a cylindrical shape of the core. Generally two or more layers of paper are wrapped over the conductor assembly of the paper ribbon insulated cables. The thickness of the core wrap is determined by dielectric strength requirements. The core wrap tape thickness is usually 5.7 or 10 mils. The paper insulated cable cores are dried under heat and vacuum prior to sheath application. The purpose of the cable sheath is to act as a moisture barrier and provide protection from damage during normal handling and installation. Evidently, it must be sufficiently flexible to permit bending and resist corrosion in the installed environment. Early ribbon paper insulated cables always employed lead sheaths, which were subsequently replaced with lead alloys of tin, antimony, or cadmium. Lead alloy sheaths are heavy and expensive, and for this reason great effort went into sheath development following World War II. Lighter sheaths were developed for aerial installations that would permit lighter pole lines, suspension strand, and guying; also for underground installations, this permitted longer pulling-in lengths. The newly developed lightweight sheaths consist of metallic and plastic materials and are preferred to lead sheaths in most applications. However, where high strength is required, the lead alloy sheath provides without doubt the best reliability. Figure 15.34 depicts a traditional lead sheath design of an earlier paper ribbon insulated telephone cable used in underground installations requiring a lead sheath with corrosion protection but without steel reinforcement.

The antecedent paper ribbon–air dielectric system, if properly dried and maintained in a dry condition, is characterized by low mutual capacitance and loss. Another very important feature of the paper ribbon insulating system is that in the case of submarine cable failures a swelling of paper ribbon insulation at the point of water ingress would prevent free water transmission along the cable, thus temporarily isolating the fault. For these reasons, paper ribbon insulated multipair cables with lead alloy sheaths and armor were preferred for submarine application. Figure 15.37 shows

Figure 15.37 Early paper ribbon submarine telephone cable with lead sheath and double-wire armor, ca. 1958. (Courtesy of Northern Telecom Ltd.)

a profile view of a typical paper insulated submarine cable with a lead sheath reinforced with two layers of steel armor wire. Note the compact concentric construction where the twisted-wire pair layers are separated by intervening paper tapes.

15.6.2 Paper Pulp Twisted-Wire Multipair Telephone Cables

The decided advantage of paper pulp insulated wires is that the space required per pair is significantly reduced, allowing a large number of pairs per unit cross section of the telephone cable. Also the splicing procedures are greatly facilitated as compared to the paper ribbon insulated wire cables. The AWG wire sizes of annealed copper used in pulp type cables are typically 22, 24, and 26 as compared to the 19, 22, 24, and 26 AWG sizes in paper ribbon cables. In the pulp insulation process, the paper pulp is applied onto the wire by means of a cylinder mold and then the wire is passed through press rolls; subsequently the excess pulp is polished off by means of high-speed rotating polishers to provide a smooth circular insulated conductor. Prior to take-up on reels, the pulp insulated wire is passed through a drying furnace under the application of heat and vacuum. It should be emphasized that the pulp paper insulation process is economical only when large production volumes are involved, since start-ups and stops in the process are highly wasteful.

The pulp insulated conductors are twisted into pairs in a manner to minimize cross talk, and the pairs in turn are stranded into even count units, containing usually 25 pairs up to a maximum of 100 pairs. The cable core is subsequently formed by stranding the units together, as in the case of paper ribbon insulated cables; a number of spare pairs are usually stranded in along with the units. Each unit has one direction and length of stranding lay for all pairs. The pairs are stranded into units in tandem, with the subsequent cabling of all units into the final cable core in one continuous operation. Again, spiral wrappings of paper are employed to enclose the cable core prior to drying. There are basically two methods that may be employed to dry the cable cores [1]. In one process, the cable core is admitted into an oven heated to 270°F and dried under vacuum for 12 h. At the end of this drying cycle, dry air is introduced to break the vacuum. In an alternative method, the rate of drying is accelerated by tying all conductors together and passing a current through them to raise the cable rapidly to the required drying temperature in the air evacuated oven. The amount of water tolerated in telephone cables is about 4%, which is relatively high when compared to oil-impregnated-paper power cables. However, the 4% level is sufficiently adequate to maintain the low-voltage electrical characteristics at an acceptable level. The dried pulp insulation has a real relative permittivity or dielectric constant of 1.9, with an effective dielectric constant of 1.5 between the two wires of a pair.

Since paper pulp is hygroscopic, the dried pulp insulated twisted wires must be protected against moisture ingress by a sealed hermetic sheath. Frequently, added protection against moisture is provided by pressurizing the cable from within by dry air or nitrogen. The sheaths used on paper pulp insulated cables may be STALPETH, PASP, LEPETH, or simply lead (containing 1% antimony). In the STALPETH sheath, a corrugated aluminum tape is applied longitudinally with a gap followed by a corrugated tin-coated steel tape applied longitudinally over the aluminum tape; the tinned-steel tape is soldered at the overlap in order to ensure a hermetic seal. A layer of preservative compound is subsequently applied over which an external PE jacket is extruded. This construction is especially suited for ducts. In the PASP sheath, the cable core is enclosed in a PE jacket over which a standard STALPETH sheath is applied. The PASP sheath is normally used for telephone cables that are to be buried directly in ground. Finally, the LEPETH sheath construction consists of a PE inner jacket over which rests an extruded lead sheath, with a paper barrier normally separating both jackets. This type of sheath is often used for submarine cable applications.

Figure 15.38 depicts a typical pulp insulated wire telephone cable, with an ALPETH sheath applied over the paper tape enclosing the cable core. All units shown comprise 25 pairs, a number that appears standard for most present constructions. Here the wire gage is 24 AWG and the nominal value of the mutual capacitance at 5°C measured at 1000 Hz is 52 nF/km. A colored binder can be discerned to spiral over each unit, its purpose being to identify the unit position within the cable. A common color code is used within the units to identify the 25 pair groups. The cable is always maintained under nitrogen gas pressure, following the drying cycle and while in service. The shown cable construction may be enlarged to accommodate as many as 4200 pairs. The portrayed cable type is usually found in underground ducts and is used to provide large pair count exchange feeder pairs and interoffice trunks. With carrier systems, pulp insulated cables were used for toll applications.

15.6.3 Plastic Insulated Cables

When plastic insulated twisted-wire multipair (PIC) telephone cables were first introduced in the early 1950s, their high initial cost as compared to paper cables discouraged their use except for applications such as defense and military-related equipment where high reliability was required. Today with lower plastic material costs, they are extensively used because of their low cost, ready access terminals, lower installation and maintenance expenses, and excellent electrical characteristics and stability. In the early plastic insulated telephone cables manufactured in Canada, beginning in 1952, both low- and high-density PEs were used for the wire insulation. However, lately

Figure 15.38 Typical pulp insulated wire telephone cable with ALPETH sheath. (Courtesy of Northern Telecom Ltd.)

increased usage has been made of polypropylene insulation. Often all PIC-type cables are simply referred to as polyolefin insulated cables, since PE is a polymer of ethylene and polypropylene is a polymer of propylene; i.e., polyolefins comprise polymers produced either from ethylene or propylene. The main advantages of polyolefin insulated cables are associated with their low-dielectric-loss characteristics, which render them ideal for use at carrier frequencies or pulse code modulation where comparable attenuation losses in paper or pulp insulated cables are relatively high. Although the transmission characteristics of polyolefin insulated wires themselves are not affected by moisture or water, the mutual capacitance between pairs is increased with water ingress into the cable. In addition, the occurrence of pinhole defects in thin layers ($\sim$0.2 mm) of polyolefin insulation creates channel sites into which the water may migrate; this gives rise to electrical short circuits to ground, thereby resulting in noise generation and eventually loss in conductor continuity due to electrolytic corrosion maintained by the dc power supply. Accordingly, in wet environments it is common practice to use filled-core PIC cables that are essentially PIC cables that have a paraffinic hydrocarbon compound (petroleum jelly based) filled core to prevent moisture ingress; the petroleum jelly compound matches the low loss and dielectric constant of the polyolefin insulation, which varies between 2.28 and 2.33. The unfilled-core PIC cables are thus simply referred to as air core cables. The filled-core cables are further subdivided into compound-filled solid insulation and compound-filled foam-skin insulation. The former consist of solid polyolefin insulated wire cables, while the latter comprise wires insulated by an inner layer of foam polyolefin covered by an outer skin of solid pigmented polyolefin. The foam or cellular structure polyolefin insulation, introduced in 1968, is obtained by use of blowing agents that are activated during the insulation extrusion process. The foam dielectric has a lower dielectric constant ($\sim$1.50–1.75), thus allowing to decrease effectively the size of the cable; the purpose of the skin polyolefin layer is to prevent the hydrocarbon compound from penetrating into the foam dielectric. Table 15.11 provides additional data on various polyolefin extrusion compounds that may be employed to insulate telephone wires and cables. The high-tensile-strength, and medium-density compound 3 characterized by extremely low dielectric losses is typical of present-day PIC cable insulation. The low-density compound 2 is less susceptible to melt fracture and is thus suitable for high extrusion speeds. The low-density compound 1 exhibits the highest dielectric loss; it is typical of present-day loop-up wire insulation employed for entry to customer services. The foam or expanded insulation compound 4 is a high-density polyolefin that contains chemical blowing agents and provides 50% expansion. As mentioned previously, when the highly expanded insulation is exposed to petroleum jelly in filled PIC cables, it must be protected by an impervious skin of another polymer compound. Compound 3 may be used to provide this type of skin cover. Note that as the polyolefin compounds are nonpolar, the dielectric constant increases directly with the density of the polymer. It is thus interesting to observe that the foam compound 4 in its solid form has a $\tan\delta$ value in excess of that of compound 5; this is perhaps due to the presence of more lossy expansion inducing or controlling additives. However, in its expanded form, the presence of the air phase would be expected to lower its $\tan\delta$ value considerably below that of the solid compound 5.

The usual PIC–air core construction consists of 19, 22, 24, or 26 AWG gage conductors of annealed copper insulated with color-coded polyolefin and twisted into pairs to reduce cross talk (using up to 25 different twist lengths). The length of the

TABLE 15.11 Typical Electrical and Physical Properties of Some Commercial PE Insulation Compounds for Metallic Conductor Multipair Telephone Cables

Property and Test Method	Low-Density		Medium-Density	High-Density	
	Compound 1	Compound 2	Compound, 3	Compound 4 (50% foam)	Compound 5 (35% foam)
Melt index (ASTM D1238), g/10 min	0.20	0.65	0.33	0.80	0.80
Density at 20°C (ASTM D1505), g/cm^3					
In solid form	0.921	0.921	0.930	0.945	0.945
In expanded form	—	—	—	0.450	—
Tensile strength (ASTM D638), psi	2.2×10^3	2.3×10^3	2.8×10^3	2.9×10^3	3.2×10^3
Elongation (ASTM D638), %					
In solid form	600	700	650	600	600
In expanded form	—	—	—	350	350
Brittleness temperature (ASTM D746), °C	−90	<−100	<−100	<−100	<−100
Dielectric constant at 1 MHz (ASTM D1531)					
In solid form	2.29	2.29	2.30	2.33	2.32
In expanded form	—	—	—	—	—
Dissipation factor at 1 MHz (ASTM D1531), in solid form	1×10^{-4}	7×10^{-5}	7×10^{-5}	3×10^{-4}	7×10^{-5}

Note: Compounds 1, 2, 3, 4, and 5 refer to Union Carbide compounds DFDB-6005, DFDA-7540, DHDA-8880, DGDA-3485, and DGDK-3364. All data obtained with ASTM test procedures.
(Courtesy of Union Carbide Corp.)

twists ranges generally from 4 to 15 cm: reduced twist lengths lead to increased costs due to longer manufacturing times, while increased twist lengths require more time to identify the two sides of an unraveled pair for field splicing purposes. Ten basic colors are used, and these are combined to form 20 of the colors employed in a unit. Each unit is enclosed within a binder of proper color code to facilitate identification of each pair in the cable. After the units are stranded together, the cable core is enclosed in a nonhygroscopic tape applied longitudinally using an overlap.

The filled-core solid-insulated-type cables have essentially the same construction as the air core cables except that the stranded core is impregnated with the water-resistant paraffinic hydrocarbon jelly compound that is chemically compatible, thus inducing no measurable degradation effects in the polyolefin insulation. Again a nonhygroscopic tape (that is, coated with the compound) is applied longitudinally over the core. In the filled-core foam-skin-insulated-type cables, the construction is identical to that of the air core cable, with the exception that 26 AWG gage wires are not used. The petroleum jelly impregnation process is the same as that for the filled-core solid-insulated-type cable.

Expanded or foam insulation diminishes the space per twisted-wire pair requirement and reduces the cable material costs in comparison to solid polyolefin insulated twisted-wire pairs. However, foam insulation does not quite attain the space per pair achieved by pulp insulated wire pairs. The wire-to-wire dielectric strength of foam insulation is substantially less than that of solid polyolefin insulation, which is usually an order of magnitude greater than that of paper pulp insulation.

Figure 15.39 portrays a typical single-unit air core PIC cable in which the cable core has a SEALPETH sheath consisting of a carbon-loaded PE jacket applied over a nonhygroscopic core wrap. Common conductor AWG sizes employed in the cable are 19 (0.9 mm), 22 (0.63 mm), 24 (0.5 mm), or 26 (0.4 mm); the average value of mutual capacitance is of the order of 53 μF/km. The shown construction is particularly suited for aerial installations. The SEALPETH sheath construction provides good dielectric strength characteristics and, therefore, constitutes an effective protection against lightning.

Various sheathing arrangements may be used in conjunction with PIC cables, depending upon the application. It will be recalled that the prime function of a sheath is to prevent moisture ingress and at the same time conform with the electrical, chemical, and mechanical requirements. Electrically, the sheath must be able to provide an effective shield against inductive interference and against lightning. Chemical resistance must be provided against possible exposure to acids and bases in the environment. Mechanical strength must be sufficient to protect the cable against damage during installation and while in service. In the 1950s, the lead sheath used on paper insulated cables was gradually replaced by STALPETH (steel-aluminum-PE) sheaths, which consist of a corrugated steel tape covering with a 0.13 mm thickness applied over an aluminum tape overlap and soldered along the longitudinal

Figure 15.39 PIC telephone cable with SEALPETH sheath. (Courtesy of Northern Telecom Ltd.)

seam. Following an application of a flooding compound, a PE jacket is extruded over the corrugated steel tape. Initially, the STALPETH jacket has been principally employed with pulp insulated twisted-wire pair cables in duct installations. Since the seam is generally not free of soldering defects, the STALPETH sheath cannot be assumed to be entirely hermetic. With the introduction of the PIC cables also appearing in the 1950s, various ALPETH (aluminum-PE) sheaths were introduced. This tendency toward light sheaths continued in the new generation of sealed sheaths in which the coated aluminum tape was bonded to itself and to the underside of the outer carbon-black-loaded PE jacket. The SEALPETH sheath of the type depicted in Fig. 15.39 consists of a 200-μm aluminum tape shield coated with a 50-μm-thick polymer film and is primarily intended for aerial use for air core PIC cables. It may also be used in direct burial or duct applications, but only with filled-core PIC cables. For aerial applications an ALPETH sheath may be used as an alternative; in the ALPETH sheath, the 0.2-mm layer of aluminum is normally corrugated and never polymer coated. When air core and filled-core PIC cables are directly buried or placed in ducts, a SEALPAP sheath is utilized. It consists of an inner jacket of carbon-loaded PE extruded over the core wrap. In this sheath, a noncorrugated aluminum tape shield of 200 μm (8 mils) thickness, coated on each side with a 50-μm (2-mil) polymer film, is applied longitudinally over the core wrap and heat sealed at the overlap. A carbon-black-loaded, high-molecular-weight PE jacket is subsequently extruded over the aluminum layer. Often a PAP sheath is used as an alternative to the SEALPAP sheath on air core cables. The former differs from the latter in that the aluminum shield is normally corrugated but not coated with the polymer film.

For direct burial applications, where increased mechanical protection is required, PASP-type sheaths are employed. The PASP sheath consists of an inner jacket of carbon-loaded PE extruded over the core wrap, followed in sequence by corrugated-aluminum (200-μm) and tinned-steel (150-μm) tapes applied longitudinally. The former is applied with a gap and the latter is soldered at the overlap. A thermoplastic flooding compound is applied over the tinned-steel tape, and this is followed by an extruded carbon-loaded high-molecular-weight PE jacket. The choice of the PE jacket is contingent upon the toughness and abrasion resistance required by the particular application of the telephone cable. Since the abrasion resistance of PE increases with its density, a high-abrasion-resistant jacket necessitates the use of high-density PE. As Table 15.12 indicates, there is a range of different density PE jacket compounds available to suit the application. The medium-density compound has a sufficiently high tensile strength and deformation resistance to render it suitable for jackets with buried cables. Recently considerable interest has developed in low-density PE jackets (low-density compound A) that retain reasonably good flexibility at low temperatures and certain acceptable stress cracking resistance. A good compromise between low-temperature flexibility requirements and abrasion resistance may be obtained with medium-density PEs (e.g., medium-density compound D), which performs adequately well on both optical fiber and conventional metallic conductor cables.

Various protective steel armor coverings are available for metallic conductor multipair cables to provide increased mechanical strength under adverse conditions. The protective steel armor also assists in reducing the effects of low-frequency induction from nearby power lines and cables. For aerial applications, steel tape is an integral part of the sheath armor construction; in addition, tarred jute is used for protection

TABLE 15.12 Typical Electrical and Physical Properties of Some Commercially Available PE Jacketing Compounds for Metallic Conductor Multipair and Optical Fiber Telephone Cables

Property and Test Method	Low-Density Compound		Medium-Density Compound		High-Density Compound	
	A	B	C	D	E	F
Melt index (ASTM D1238), g/10 min	0.55	0.21	0.65	0.65	0.75	0.22
Density at 20°C (ASTM D1505), g/cm³	0.932	0.931	0.945	0.941	0.954	0.956
Tensile strength (ASTM D638), psi	2.35×10^3	2.20×10^3	3.90×10^3	4.00×10^3	3.00×10^3	4.00×10^3
Elongation (ASTM D638), %	800	800	800	800	800	800
Brittleness temperature (ASTM D746), °C	<-100	-85	<-100	<-100	<-100	<-100
Carbon black content (ASTM D1603), %	2.6	2.6	2.6	2.6	2.6	2.6
Abrasion resistance, mg/100 rev	23	30	17	18	16	14
Dielectric constant at 1 MHz (ASTM D1531)	2.48	2.48	2.54	2.51	2.55	2.55
Dissipation factor at 1 MHz (ASTM D1531)	3.0×10^{-4}	3.0×10^{-4}	3.0×10^{-4}	3.0×10^{-4}	3.0×10^{-4}	3.0×10^{-4}
Dielectric strength, 125 mils thickness (ASTM D149), V/mil	500	500	500	500	500	500

Note: Compounds A, B, C, D, E, and F refer respectively to Union Carbide Compounds DFDD-6059, DFDD-0588, DHDA-6548, DHDA-8864, DGDA-6318, and DGDA-8479. Compounds D and E have been optimized for optical fiber cable applications. (Courtesy of Union Carbide Corp.)

against corrosion. With underground installations, the types of required armors are subdivided into three classes: light steel wire, single steel wire, and double steel wire armor categories. In light wire armor systems either a lead or a PAP sheath can be used in conjunction with tarred jute and galvanized steel wire armor (cf. Fig. 15.40). For longer submarine crossings, the single steel wire armor system is used; it consists of a lead sheath and a larger steel wire gage (cf. Chapter 16). For use in deep water under strong tides and heavy ice formations, a heavy armor construction consisting of double steel wire layers is used and is of the type depicted in Fig. 15.37.

Some remarks should be made concerning telephone cable sizes. The outside diameter sizes of telephone cables are limited by the duct sizes in existence. In the United States the duct diameter is fixed at 3.25 in., whereas in Canada it is 3.5 in. so that in the latter instance somewhat larger cables may be accommodated. For a duct diameter of 3.25 in., the largest diameter lead sheathed PIC cable that may be pulled into the duct without causing damage is considered to be 2.6 in. diameter. With pressures toward increased number of pairs in telephone cables and to reduce the amount of copper used, there is developing a trend toward the use of smaller diameter wires. The decrease in wire diameters resulted in a corresponding increase in the resistance per unit length of the wires employed, which had to be offset by the use of repeaters, loading coils (for noncarrier systems only), and improved telephone sets. This resulted primarily in the deployment of the 22, 24, and 26 AWG wire gages, although there still appears to be a limited demand for the 19 AWG gage wire pairs. Figure 15.41 illustrates a quasi-exponential growth in the maximum number of pairs in telephone cables over the first part of the century, with wire gage as the parameter. Whereas the graph indicates the maximum number of pairs per cable, having 26 AWG wires, to be 2727 in 1955, it should be pointed out that in 1982 cables having more than 4000 pairs were being manufactured. In retrospect, it is interesting to make the observation that the enormous increase in the number of twisted-wire pairs in conventional telephone cables was at that time perhaps only an inconspicuous harbinger foretelling that conventional multi-pair telephone cables were rapidly approaching their capacity limits for long-haul transmission.

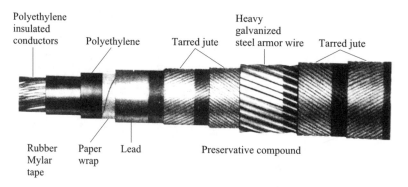

Figure 15.40 Single steel wire armored PIC cable, ca. 1958. (Courtesy of Northern
Telecom Ltd.)

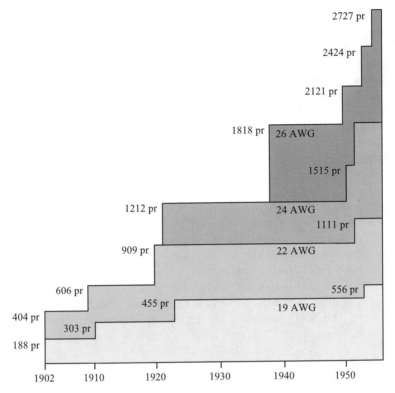

Figure 15.41 Pair capacities of telephone cables manufactured in first half of twentieth century, ca. 1955. (Courtesy of Northern Telecom Ltd., Lachine, Québec.)

15.6.4 Electrical Tests of Twisted-Wire Multipair Telephone Cables

Many of the tests and measured electrical quantities were discussed in Sections 15.2–15.4 and will therefore not be repeated here. However, a few general observations must be made. Since twisted-wire multipair telephone cables contain numerous pairs, the electrical tests are carried out by means of automated/computerized equipment and frequently on a sampling basis. The tests for shorts, open circuits, and crosses between a conductor of one pair and another pair, which in the earlier paper ribbon twisted-wire multipair cables were performed manually by means of battery power buzzers, are still required on pulp insulated cables. The pulp insulated cables must undergo a wire-to-wire ac dielectric strength test and must withstand an ac peak voltage of 0.5 kV with the exception of 19 AWG wires, for which the specified withstand voltage is 0.7 kV [57]. The wire-to-wire dielectric strength of PIC cables is evaluated using automated ac test sets, for which the required withstand voltage is specified as a function of the wire gage and varies from 2.4 kV for 26 AWG wires to 5 kV for 19 AWG wires [57].

A cable core to sheath voltage test is required to ensure that voltage surges arising from adjacent electrical power system faults and lightning are not transmitted to the twisted-wire pairs. If the insulation between the core and sheath consists of paper, then the required ac withstand voltage is 1.4 kV peak [57], while for cables with an inner PE jacket the Bellcore specifications require a dc withstand voltage of 20 kV. The insulation resistance between the wires of the pairs must also be evaluated. Ordinarily, this test is performed on two pairs per reel with all other pairs grounded. The insulation resistance of PIC cables is normally very high, but for paper/pulp insulated conductors, the insulation resistance requirement represents a critical test since the insulation resistance diminishes rapidly with the moisture constant of the paper/pulp insulation. The minimum insulation resistance requirement, expressed as a product of the measured insulation resistance in ohms and the length of the cable in miles, is fixed at 500 MΩ-miles [57].

Other routine tests carried out on twisted-wire multipair telephone cables include those described in Sections 15.2–15.4, namely conductor resistance, near- and far-end cross talk, mutual capacitance and conductance (or tan δ), resistance unbalance, and the pair-to-ground, pair-to-pair and pair-to-shield capacitance unbalances. Much of the tedium in these measurements is obviated by means of computer-controlled test equipment. For example, a cross-talk test on a cable containing 100 twisted-wire pairs involves 4950 measurement step combinations [57].

As in the case of coaxial cables, return loss (RL) and structural return loss (SRL) measurements are also performed on twisted pairs. Both quantities are measures of the degree of roughness or, alternatively, the degree of smoothness in the structure of a twisted pair. The SRL compares the input impedance, Z_{in} of a given pair with its characteristic impedance Z_0 over a range of frequencies and is given by

$$\text{SRL} = -20 \log \left| \frac{Z_{in} - Z_0}{Z_{in} + Z_0} \right| \tag{15.38}$$

where for short twisted-pair lines, Z_0 may be obtained in terms of the open-circuit impedance Z_{oc} and short-circuit impedance Z_{sc}, measurements as

$$Z_0 = \sqrt{Z_{oc} Z_{sc}} \tag{15.38a}$$

When the twisted-pair line becomes longer, i.e., greater than one-eighth of the wavelength at the frequency of measurement, structural variations become significant and the short- and open-circuit measurement data substituted into Eq. (15.38a) yield in effect the input impedance Z_{in} in lieu of the characteristic impedance Z_0 [57a]. For long twisted-wire pair lines, the characteristic impedance may be determined from the relation

$$Z_0 = \frac{\alpha + j\beta}{G + j\omega C} \tag{15.38b}$$

or, alternatively, its value may be deduced from the input impedance measurements using the least-squares function techniques as outlined in ASTM standard D4566 [57a]. The input impedance, Z_{in} may be obtained in terms of the complex reflection coefficient Γ and load resistance Z_L, using the expression

$$Z_{in} = Z_L \left[\frac{1 + \Gamma}{1 - \Gamma} \right] \tag{15.38c}$$

where Z_L may be a nominal resistance of 100 Ω with a 1% tolerance [57a]. Network analyzers are commonly employed for SRL measurements; all measurements are of the balanced impedance type.

The return loss measurement is more straightforward in the sense that it does not involve the determination of the characteristic impedance Z_0. The RL measure compares the input impedance Z_{in} with a nominal impedance Z_L; it is defined by the relation

$$RL = -20 \log \left| \frac{Z_{in} - Z_L}{Z_{in} + Z_L} \right| \qquad (15.38d)$$

Here, as for the SRL measurement, Z_L usually consists of a 100-Ω resistive load [57a]. Note that there exists an implicit equivalence between the RL and SRL measures when $Z_0 \simeq Z_L$.

15.6.5 Outside Plant and Station Connection Wires and Cables

It is perhaps appropriate at this juncture of our discussion to make a few cursory comments on outside plant wires and station connection wires and cables. Many subscribers are served by aerial drop wires from poles to their premises. These consist normally of black polyvinyl chloride (PVC) insulated conductors laid in parallel, with a groove on each flat face of the insulation to permit their separation at the two end connection points. The ac dielectric strength between conductors must be better than 1.5 kV. Mechanical strength is imparted in the conductions by a concentric steel core that is covered by copper to maintain the necessary electrical conductivity. The overall diameter of the copper-steel reinforced conductor is approximately 0.98 mm. Aerial drops to multiple dwellings consist of cables of several polyethylene insulated twisted-wire pairs wrapped in Mylar tape and laid parallel to a bare galvanized steel messenger wire. The overall combination is enclosed in an extruded black PVC jacket such that the cable cross section forms a figure 8. The electrical conductors are made of 22 AWG gage solid annealed copper and the steel wire strength member has a diameter of 2.76 mm. The same design is employed even for single twisted-wire pairs in heavily treed rural areas. A minimum ac breakdown voltage equal to or greater than 1.5 kV is required between the electrical conductors and between all conductors and the steel messenger.

For buried connections between subscriber's premises and the terminal, jacketed underground cables are employed. These may consist of spiraled color coded polypropylene insulated wire pairs; again the electrical conductor has a 22 AWG gage and is made of solid annealed copper. An inner carbon-free PVC jacket is extruded over the spiraled wire pairs over which is placed a flat galvanized steel wire armor in spiral form. A final outside black (carbon-containing) PVC jacket is applied over the armor. Dielectric integrity tests involve an application of a 5-kV alternating voltage between the conductors and each conductor to shield. The cable design for distribution in rural areas is similar in that an overall black PVC jacket is extruded over a spirally applied flat galvanized steel wire armor, which directly overlays two parallel clear PE insulated 19 AWG solid annealed copper conductors. One of these conductors may be tined in order to to facilitate identification of the two insulated wires in the pair. Again the cable must withstand an alternating test voltage of 5 kV applied between the conductors and the conductors and armor (shield). For ground connections from the cable terminals and entrance protectors in the subscriber premises, PVC insulated solid annealed copper conductors are employed with 6, 10, 12, or 14 AWG gages to meet the current rating requirements. A color code is often employed to identify the gage of the ground wire.

Station connection wires and cables are primarily used for wiring purposes within customer's premises. The most common and conspicuous form of connection is that from the service entrance device (usually the protector) to the connecting block of the telephone set. Typically, the cable consists of four color-coded polypropylene insulated 22 AWG solid annealed copper conductors placed within a PVC extruded jacket having a standard chameleon gray color. The size of station connection cables varies with the services required on the customer premises and may attain outside diameters approaching 20 mm in which the total number of pairs may exceed 100. For these multipair cables, 24 AWG gage solid annealed copper conductors are utilized. In these applications fire prevention considerations may give preference to the use of semirigid PVC jackets. The cable must withstand a dielectric strength test of 2 kV ac applied between the conductors.

15.7 COAXIAL CABLE DESIGN AND CONSTRUCTION

Apart from their extensive use in military and navigational applications, coaxial cables are used in voice, data, and television toll circuits in urban centers where microwave radio relay systems have limited access. Under certain conditions multitube coaxial cables may be used for complete circuits between toll offices. In most instances, coaxial cables have greater use flexibility than video pair cables; the preference for the latter is at low frequencies where the shielding of a balanced twisted pair is better than that of a coaxial cable [58]. Most coaxial cables are designed for a characteristic impedance of either 50 or 75 Ω. Where lower attenuation is desired, the 75-Ω cables are preferred; also, for very low capacitance requirements, coaxial cables having a characteristic impedance greater than 75-Ω are used. The diameter size of coaxial cables increases with the characteristic impedance value, and large-impedance cables such as 200 Ω are significantly bulkier; to achieve lower attenuation values, however, large-diameter cables may be required. The latter is also necessary where larger voltage ratings may be imposed.

In the earlier days, stranded inner conductors were used where greater flexibility was required; however, with improved construction techniques, this practice is much less common now, since stranded conductors tend to give rise to higher losses. For low-capacitance and low-attenuation applications, the solid PE dielectric may be replaced by a foamed or expanded dielectric or an air-dielectric disc-spacer-type system. The latter construction is vulnerable to moisture effects, and consequently, the cable must be well sealed, such as by an hermetic sheath or other water protective covering. For higher voltage applications, surface leakage paths along the disc spacers and the lower breakdown strength of the air bubble inclusions of the foam dielectric render the latter two systems less desirable. Solid-dielectric insulated PE cables need less mechanical protection and have greater moisture resistance. Some typical polyolefin compounds available for coaxial cable insulation are given in Table 15.13. Some of the compounds

TABLE 15.13 Typical Electrical and Physical Properties of Some Commercial PE Insulating Compounds for Coaxial Cables

Property and Test Method	Low Density, Compound, B_1	High-Density Compound		Low-Density Compound		
		B_2	B_3	B_4	B_5	B_6
Melt index (ASTM D1238), g/10 min	2.0	8.0	8.0	0.20	0.65	0.20
Density at 23°C (ASTM D1505), g/cm^3						
In solid form	0.92	0.96	0.965	0.92	0.921	0.92
In expanded form	0.44	—	0.23			
Tensile strength (ASTM D638), psi						
In solid form	2.2×10^3	4.8×10^3	4.8×10^3	2.2×10^3	2.3×10^3	1.8×10^3
In expanded form	0.6×10^3					
Elongation (ASTM D638), %						
In solid form	600	650	650	600	700	600
In expanded form	300					
Dielectric constant at 1 MHz (ASTM D1531)						
In solid form	2.28	2.34	2.36	2.28	2.29	2.29
In expanded form	1.50					
Dissipation factor at 1 MHz (ASTM D1531)						
In solid form	2.2×10^{-4}	5.0×10^{-5}	3.0×10^{-5}	2.0×10^{-4}	7.0×10^{-5}	6.0×10^{-5}
In expanded form	1.5×10^{-4}					

Note: Compounds B_1, B_2, B_3, B_4, B_5, and B_6 refer respectively to Union Carbide compound designations DFD-4960, DGDA-5944, DGDA 6944, DFDA-6005, DFDA-7540, and DYNH/DFDA-12353 NT. (Courtesy of Union Carbide Corp.)

for solid extruded coaxial cables are identical to those that may be utilized for twisted-wire insulation. In this regard note that compound B_5 is identical to compound 2 in Table 15.11. Flexibility in coaxial cables constitutes an important feature of the compounds employed, and consequently, low-density polymers tend to be utilized, with the exception of compound B_2. However, the latter, which is characterized by extremely low dielectric losses, is generally mixed with a low-density compound to obtain optimum mechanical properties. It is interesting to note that expansion of the compound B_1 does not reduce greatly the dielectric loss; this may perhaps be again due to some added losses by the chemical blowing agents.

In the past it had been standard practice to employ chemical blowing agents in the production of expanded PE insulated coaxial cables. However, in order to reduce dielectric losses and capacitance, a larger volume expansion ($> 60\%$) is preferred; for this purpose a gas injection process is currently utilized whereby nitrogen under pressure is dissolved in molten PE [58a]. The injected gas must be distributed uniformly and incorporated in a stable form into the PE to produce small gas bubbles of uniform size and thereby avert fluctuations in the capacitance and diameter of the foamed insulation coaxial cable; this necessitates stringent control of the extrusion pressures, a requirement achieved by injecting the nitrogen gas at sonic velocity. The condition for critical pressure P_c to ensure sonic flow of the nitrogen stream is given by [58a]

$$\frac{P_c}{P_a} = \frac{2}{(C_p/C_v) + 1} \frac{C_p/C_v}{(C_p/C_v) - 1} \tag{15.39}$$

where P_a is the upstream pressure in the extruder, C_p is the specific heat of the nitrogen gas at constant pressure, and C_v denotes the specific heat of nitrogen at constant volume. For nitrogen, $C_p/C_v = 1.41$ and $P_c/P_a = 0.528$; hence, the choice of the nozzle size for a given upstream pressure determines the critical gas pressure P_c necessary to attain sonic flow. The relationship between the nozzle diameter D (in mils), regulator pressure P_R (in pounds per square inch), and line speed v (in feet per minute) for a specific expansion level is given by [58a]

$$DP_R = (6.8 \times 10^4)vA\left(1 - \frac{\rho f}{\rho S}\right) \tag{15.39a}$$

where A is the cross-sectional area in square feet and ρ_f is the density of the foam insulation and ρ_S the density of the base PE resin, both in pounds per cubic foot. The nozzle housing must be able to withstand gas injection pressures up to 10^4 psi.

Complete mixing of the nitrogen gas and polymer after the injection point requires a substantial extruder length. Extruders with length-to-diameter ratios of 30:1 to 38:1 are commonly used. The gas is normally injected at an intermediate point along the length of the extruder, or if extruders are used in tandem, the injection point is between two of the extruders. If three extruders are employed in tandem, then one extruder melts the polymer while gas addition occurs in the second extruder; the latter also acts as a heat exchanger. Lastly, the third extruder precoats the conductor prior to its contact with the foam to ensure an adequate adhesion between the conductor and the foam insulation. The tandem extrusion procedure is normally employed with foam insulations having expansions substantially above 60%. Figure 15.42(a) illus-

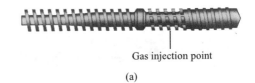

Gas injection point

(a)

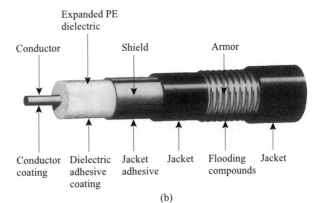

Expanded PE
dielectric

Conductor   Shield   Armor

Figure 15.42 (a) Typical screw for a single gas injection extruder for foam PE, coaxial cables. (b) Typical foam PE insulated distribution coaxial cable. (Courtesy of Union Carbide Corp. [58a].)

Conductor   Dielectric   Jacket   Jacket   Flooding   Jacket
coating   adhesive   adhesive   compounds
            coating

(b)

trates the screw of a single gas injection extruder. In the first section of the extruder the PE is melted and pumped, while in the second section, which follows the gas injection point, the molten polymer is mixed with the gas; here a blister ring with a narrow clearance prevents the gas from flowing back. Figure 15.42(b) depicts a typical foam insulated coaxial cable designed for distribution application. It is to be emphasized that in the gas injection process used on foam coaxial cables, it is common practice to employ a blend of low- and high-density PEs. For example, referring to Table 15.13 the high-density compound B_3 may be deployed in ratios of 80–60% to that of 20–40% respectively of the low-density compound B_6. In this combination, the high-density compound imparts toughness upon crystallization and its low shear viscosity permits a higher volume of gas to be incorporated; furthermore, the low-density long branched-chain polymer, because of its strain-hardening characteristics of the melt, leads to smaller and more uniform gas-containing pores [58a].

All coaxial cables must be adequately shielded; the most effective type of shield is a continuously extruded aluminum tube over the insulation. The aluminum shield construction has been extensively used on 75-Ω foam-dielectric coaxial cables for CATV applications. Where flexibility is a requirement, coaxial cables employ braided wire or tape shields. Braids are woven from plain, tinned, or silvered copper wire; for greater shielding effectiveness a double braid may be used, as depicted in the coaxial cable specimen in Fig. 15.43(a). This is a typical coaxial cable used in equipment and studio wiring for operation up to 10 MHz. The cable consists of a solid, annealed copper center conductor insulated with solid PE, over which two closely woven layers of tinned, annealed 34 AWG copper wire braids that constitute the outer conductor are applied. A slate-colored PE jacket is applied over the braid shields, yielding an overall cable diameter of 7.7 mm (0.305 in.). The cable is designed to operate with a characteristic impedance of $75 \pm 3\,\Omega$ at 2MHz. The propagation velocity relative to that in free space at 10 MHz is 66%.

Figure 15.43 (a) Flexible solid PE insulated interior wiring coaxial cable with double-braided shields (Courtesy of Northern Telecom Ltd.)

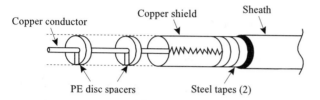

Copper conductor

Copper shield

Sheath

PE disc spacers

Steel tapes (2)

Figure 15.43 (b) Typical air-dielectric coaxial cable.

The propagation velocity relative to that in free space may be increased to 95% at 100 MHz if air-dielectric coaxial cables are employed. Such cables, commonly referred to as 0.375-in. (9.5-mm) coaxial cables in our previous discussions and operated generally between 5 and 300 MHz with a nominal impedance of 75 Ω, are utilized for broadband transmission as trunk or main feeder cables in CATV. In their composite cable form they have found extensive application in the antecedent high-capacity analog systems as well as in the latter date long-haul, high-capacity, 274.176-Mbits/s digital telephone transmission systems. They are also employed as connections to the outputs of multiplex terminals to transmitting sites of radio systems. A typical construction schematic of an air-dielectric 0.375-in. coaxial cable system is depicted in Fig. 15.43(b). The cable consists of a 0.1-in. diameter solid inner copper conductor centered by PE insulating discs of 0.361-in. outer diameter spaced at equal intervals of approximately 1-in. (2.54 mm) along the cable length. The outer conductor is formed with a 0.012-in. thick copper tape, having corrugated edges, that is applied longitudinally over the disc insulated center conductor and formed into a tube such that the corrugated edges are butted together. Two 0.006-in. thick steel tapes are applied spirally in the same direction over the copper tape to provide both mechanical strength and electromagnetic shielding [52]. Depending upon the application, PASP or SEALPETH sheaths may be utilized; in some cases, a PVC jacket may be applied over the SEALPETH sheath. Polyvinyl chloride is considered to be flame retardant and exhibits good chemical resistance when exposed to hydrocarbon liquids. The outside diameter dimensions range from approximately 1.4 to 1.8 cm.

High-capacity transmission requirements necessitated the development of composite coaxial cables, comprising up to 22 coaxial cables, or 11 coaxial pairs (one coaxial line for transmission and the other for receiving the signals) [18]. Included within the composite coaxial cables were units of twisted pairs employed for the transmission of alarm and control signals as well as voice frequency signals. The earlier designs contained pulp insulated twisted-wire pairs, but these were subsequently replaced with polyolefin insulated PIC-type twisted-wire pairs. The composite coaxial cables employed various sheaths, depending upon their service environment. Table 15.9 illustrates a number of possible arrangements of composite coaxial cables, containing different numbers of individual coaxial cables and the associated twisted-wire pairs. It is readily perceived that the outside diameter of the composite coaxial cable varies slightly with the nature of the cable shield utilized.

It is to be observed with the sheaths given in Table 15.14 that the Lepeth PJ sheath constitutes a truly hermetic sheath in that it includes a lead sheath [58]. The Lepeth PJ sheath consists of a PE jacket extruded over the cable core, a paper heat barrier, a lead sheath, an asphalt flooding compound, and an outer PE jacket. The purpose of the inner PE layer is to provide a higher dielectric strength insulation interface between the core of the cable and the lead sheath, while the outside PE jacket over the lead sheath protects the lead against electrolytic corrosion. It is to be emphasized that the principal differences between the Lepeth and Lepeth PJ sheaths rest in that the former omits the outside PE jacket protection. When added mechanical protection is required, a wire armor may be applied over the Lepeth PJ sheath. Further, in regard to sheath construction, it is appropriate to emphasize the underlying difference between the PASP and STALPETH sheaths. The construction of the two sheaths is identical, with the major exception that the PASP sheath contains an inner PE jacket separating the cable core from the normally contiguous aluminum tape in the STALPETH sheath. The resulting improvement in the dielectric strength between the cable core and the shield renders the PASP sheath particularly suitable for aerial and duct cable installations in high-incidence lightning areas. In instances where greater mechanical protection is necessary, a SEALPASP sheath is utilized in lieu of the PASP sheath. A flat polymer-coated aluminum tape, which is overlapped and sealed to the inner surface of the inner PE jacket, constitutes the distinguishing feature of the SEALPASP sheath (as the term "seal" emphasizes). The remainder portion of the jacket consists of a corrugated aluminum tape over which is applied an overlapped and soldered corrugated steel tape followed by a flooding compound and thence the outer PE jacket.

Table 15.15 illustrates a variety of some low-, medium-, and high-density PE jacket compounds commercially available for coaxial cables, depending upon the cable flexibility and other mechanical requirements. It should be borne in mind that wherever greater jacket toughness is required the higher density compounds must be selected. Note that the substantial improvement in the dielectric constant and $\tan \delta$ values of compounds C_5 and C_6 follows directly from the exclusion of the carbon additive in these compounds vis-à-vis the compounds C_1, C_2, C_3, and C_4.

Ordinarily the PIC pair units, which are enveloped in a plastic dielectric film to enhance the dielectric strength between them and the adjacent coaxial lines, are placed

TABLE 15.14 Composite-Type NT-Z375 (0.375-in.) Coaxial Cable Diameters and Sheaths

Number of Coaxial Lines	Number of PIC pairs[a]		Nominal Cable Diameter						Normal Maximum Shipping Length	
			PASP		SEALPASP		Lepeth-PJ			
	19 AWG	22 AWG	mm	in.	mm	in.	mm	in.	m	ft
2	8	8	32	1.25	34	1.35	35	1.39	1800	5900
4	8	8	37	1.44	39	1.54	40	1.59	1640	5390
6	12	12	44	1.73	47	1.85	48	1.90	1180	3875
8	16	25	51	2.02	54	2.13	55	2.17	895	2940
10	12	12	56	2.21	59	2.32	60	2.36	745	2450
12	16	16	61	2.41	64	2.52	65	2.55	695	2285

[a]Cables contain either 19 AWG or 22 AWG pairs, but not both.
(Courtesy of Northern Telecom Ltd.)

TABLE 15.15 Typical Electrical and Physical Properties of Some Commercially Available PE Jacketing Compounds

Property and Test Method	Medium-Density Compound		High-Density Compound		Low-Density Compound, C_5	Medium-Density Compound, C_6
	C_1	C_2	C_3	C_4		
Melt index (ASTM D1238), g/10 min	0.30	0.55	0.70	0.15	0.80	0.80
Density at 23°C (ASTM D1505), g/cm³	0.93	0.93	0.95	0.96	0.92	0.94
Tensile strength (ASTM D638), psi	2.10×10^3	2.25×10^3	3.50×10^3	3.60×10^3	2.10×10^3	4.00×10^3
Elongation (ASTM D638), %	650	650	900	800	700	1000
Brittleness temperature (ASTM D746), °C	−85	<−100	<−100	<−100	<−100	<−100
Environmental stress cracking (ASTM D1693), time in days	>21	>21	>21	>21	>21	>21
Dielectric constant at 1 MHz (ASTM D1531)	2.60	2.50	2.52	2.66	2.29	2.32
Dissipation factor at 1 MHz (ASTM D1531)	5×10^{-3}	3×10^{-4}	3×10^{-4}	3×10^{-4}	4×10^{-5}	5×10^{-5}
Dielectric strength, 125 mils thickness (ASTM D149) V/mil	500	500	500	500	500	500

Note: Compounds C_1, C_2, C_3, C_4, and C_5 refer respectively to Union Carbide compound deisgnations DFDA-6423, DFDB-6425, DFDB-6448, DFDB-3479, GRSN-7530, and GRSN-6549.
(Courtesy of Union Carbide Corp.)

in the center of the cable core. Thus, the usual compact cable configuration consists of PIC units at the center of the cable core surrounded by a symmetrical array of coaxial lines. The minimum bending radius of the overall cable is primarily determined by the stiffer construction of the coaxial lines and the associated sheaths. Table 15.16 gives the values of the minimum bending radii as a function of the number of coaxial lines within the composite cable and the type of sheath employed.

15.8 VIDEO PAIR CABLE DESIGN AND CONSTRUCTION

Video pair cables are primarily utilized to transmit voice and video signals in urban areas, such as in central broadcasting studios and transmission sites. They are commonly employed in the transmission of signals from studios or other remote areas to transmitters for closed circuit television and microwave entrance links. Additional applications include data transmission and satellite system earth stations. Video pair cables are designed to operate between a few hertz to 10 MHz, so that low cross talk, low attenuation, and shielding effectiveness are of prime importance. Figure 15.44 displays typical single-pair and multipair composite video cables. The pairs consist of 16 AWG annealed copper wire conductors insulated with expanded (cellular) PE that are twisted together into a pair. The video pair is then covered with a belt of extruded, expanded (cellular) PE to provide added strength to ensure stable electrical characteristics, notwithstanding the imposed flexing during its handling and use. The shown video pair shield consists of two layers of copper tape, with the first layer being applied longitudinally having an overlap and the second layer applied as a helical spiral also with an overlap. Video pair cables are essentially designed to behave as balanced, broadband, shielded transmission lines. Composite video pair cables are formed by cabling together video cables with twisted-wire pair units [cf. Fig. 15.44(b)]. In such cables paper tapes are wrapped helically upon each unit and video pair. A concentric lay-up is employed with the twisted-wire pair units situated at the center of the cable core with the video pairs forming the peripheral array. The commonly applied sheaths are Lepeth, Lepeth-PJ, PASP, or STALPETH, depending upon the nature of the environment.

Where increased flexibility is required, the two-layer copper tape shield of the expanded PE insulated video pair cable, depicted in Fig. 15.44(a), is replaced by a flexible double-braid tinned copper shield as shown in Fig. 15.45. In addition, in the double-braid shield video pair cable construction, solid PE insulation is employed in

TABLE 15.16 Minimum Bending Radii of Composite Coaxial Cables

	Minimum Bending Radius					
Number of Coaxial Lines	PASP Sheath		SEALPASP Sheath		Lepeth-PJ Sheath	
	mm	in.	mm	in.	mm	in.
2	255	10	280	11	280	11
4	305	12	330	13	330	13
6	355	14	380	15	380	15
8	405	16	430	17	430	17
10	455	18	485	19	485	19
12	485	19	510	20	535	21

[a]The minimum dynamic bending radii are 30% higher.
(Courtesy of Northern Telecom Ltd.)

(a)   (b)

Figure 15.44 Video pair cables: (a) Single-video-pair cable; (b) combined multipair video cable. (Courtesy of Northern Telecom Ltd., Lachine Québec.)

lieu of the expanded PE in the less flexible cable. As a consequence, the transmission characteristics of the more flexible cable are compromised resulting in a slightly elevated attenuation over the operational frequency range as evidenced in Fig. 15.46.

A considerable growth in the use of video pair cables is anticipated as a result of the introduction of a rapid self-configuring serial bus (IEEE standard 1394), which represents a viable alternative for the parallel port based cable connections of typical home PC systems. The configuration of the cable, to be used in conjunction with the IEEE 1394 serial bus, contains two video pairs and two power supply cables, as delineated in Fig. 15.47. The shield of the overall composite cable is isolated from the shields of the two video pair cables. The power provided to the computer, printer, and other peripherals is supplied via the same cable by the two power wires. One of the video pairs transmits two-way non-return-to-zero data, while the other carries a strobe signal that undergoes change of state for the condition when two consecutive non-return-to-zero data bits are identical [59]. For this application short cable lengths are employed to minimize signal distortion.

15.9 OPTICAL FIBER TELEPHONE CABLES

The principle of guiding light via highly reflecting hollow tubes or glass fiber bundles has been known for many decades, but it was only with the development of 20 dB/km loss silica fibers [60] coupled together with the availability of lasers as a coherent

Figure 15.45 Flexible double-braid shield video pair cable. (Courtesy of Northern Telecom Ltd.)

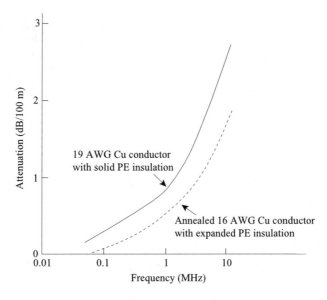

Figure 15.46 Attenuation characteristics of solid PE double-shield braid and expanded PE two-layer copper tape video pair cable. (Courtesy of Northern Telecom Ltd.)

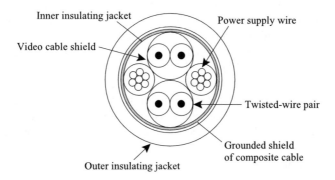

Figure 15.47 Composite video pair interconnecting cable for IEEE standard 1394 self-configuring serial bus [59].

radiation source and photodiode detectors that transmission of voice and data signals along optical fiber telephone cables was realized in the 1970s. The first trial run of 650-m-long optical fiber cable was installed in Atlanta in 1976 [61]. During this trial it was established that optical fiber cables could operate successfully as trunk cables with an information rate capacity of 44.7 Mbits/s and a maximum repeater spacing of 10.9 km; thus optical cables were demonstrated to be viably competitive with copper conductor cables on long-distance transmission routes. Shortly thereafter followed other successful trial runs of experimental optical fiber cables as well as permanent installations in the United States [61] as well as elsewhere [62–64]. Although the attenuation of the first multimode optical fiber cables operating at a wavelength of 0.82 μm was still rather elevated in the order of approximately 5 dB/km [65], their bandwidth capabilities were substantially in excess of that of coaxial cables at their utmost limit of 1 GHz at that time. Improved single-mode optical fiber cable designs in the late 1980s led to bandwidths approaching 10 GHz with data transmission rates of ~250 Mbits/s at an operational wavelength of 1.55 μm with attenuation losses reduced appreciably down to 0.2 dB/km [66]. It is currently estimated that with the development of erbium-doped optical amplifiers, bandwidths of 100 GHz will be readily attainable [67]. Optical fiber cables have several decisive advantages over metallic conductor cables, which are primarily associated with their high-bandwidth capacity, low attenuation, and nonmetallic nature, which renders them electrically immune to lightning strikes, electromagnetic interference, and dangerous fault currents in the vicinity of electrical power lines. On the other hand, optical fiber cables must be well protected mechanically and shielded from moisture ingress to prevent water-induced fiber degradation. Since individual fibers may carry many circuits, a cable breakage or dig-in at an optical cable site may cause the interruption of an enormously large number of circuits.

15.9.1 Optical Fiber Manufacture

Low-loss silica or silicon dioxide (SiO_2), is the most common base material from which optical fibers are currently made to operate within the infrared wavelength region extending from 0.80 to 1.55 μm. The light wave guiding characteristics of the optical fiber are determined by the index of refraction of the inner core of the fiber, which is usually enhanced by the use of phosphorus pentoxide (P_2O_5) and germanium oxide (GeO_2) dopants, and that of the lower index of refraction of the skin or cladding, which envelops the inner core and whose index of refraction is usually diminished with respect to the core by the addition of chlorine, fluorine, or boron oxide, (B_2O_3). In this manner the index of refraction of the core is adjusted with respect to the cladding to provide the optimum angle of reflection of the transmitted light signals within the optical fiber (cf. Chapter 17).

The optical fibers must be drawn from a preformed slab of the material. There are a number of methods available for producing the preformed slabs [68], but principally only three methods have been commercialized on a large scale: the modified chemical vapor deposition (MCVD) process [68], the outside chemical vapor deposition (OCVD) process [69], and the vapor phase axial deposition (VAD) process [70]. A schematic diagram depicting the MCVD technique is given in Fig. 15.48, where highly pure con-

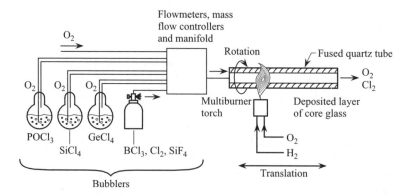

Figure 15.48 Schematic diagram of preform preparation using MCVD process (after [68]).

stituents are made to react within a silica tube [71]. Layers of SiO_2 are formed from a reaction between O_2 and $SiCl_4$ and deposited in layer form within the tube. Additional mixing occurs with oxygen in the manifold, and subsequently the reacted mixture flows into the substrate tube. A mobile oxygen-hydrogen multiburner-torch promotes further reaction along the fused quartz tube and thereby ensures a uniform deposition of the SiO_2 film layers. Thus the cladding material layers are formed first and core layers subsequently; as the cladding material layers are being formed, dopants of chlorine, fluorine, or boron oxide are produced by the admission of suitable reactants. Similarly for the core material, $GeCl_4$ and $POCl_3$ are reacted with oxygen to form the P_2O_5 and GeO_2 dopants. The reaction rate is controlled by the torch temperature, which is usually 1800°C; however, when the formation of the cladding and core layers is complete, the preform is formed by raising the temperature, thereby causing the tube to collapse into a solid rod.

The efficiency of the MCVD process is enhanced by an increase in the deposition rate [68], which is a function of the temperature gradient. It may be expressed as [72]

$$(\text{Eff})_{\text{dep}} = 1 - \frac{T_e}{T_r} \tag{15.40}$$

where T_e represents the temperature at which the gas stream of the reactants reaches equilibrium with the wall surroundings downstream from the hot zone of temperature T_r. The length ℓ over which deposition occurs is given by [68]

$$\ell = \frac{F}{\alpha} \tag{15.41}$$

where F denotes the flow rate of the reactants and α the diffusivity. The nature of Eq. (15.41) indicates that high flow rates lead to longer deposition lengths and that the deposits assume a tapered profile. Oxide deposition rates as high as 1 g/min have been achieved with reactant flow rates of 8 g/min; the taper effect prevailing at high flow rates may be reduced by water cooling of the support tube downstream from the hot zone [68]. High deposition rates ($\sim$2.5 g/min) may also be achieved at atmospheric pressure with rf plasma deposition techniques, whereby the resulting efficiency of deposition is also enhanced ($\sim$75%) due to the small ratios of T_e/T_r [cf. Eq. (15.40)] or temperature gradients within the plasma [73].

The VAD method for the fabrication of silica preforms represents a continuous fabrication process in the axial direction and consists of a tensile machine, which is depicted in Figure 15.49 [74]. The $GeCl_4$, BBr, $POCl_3$ and $SiCl_4$ reactants are admitted into a reaction chamber via coaxial silica tubes into a central tube situated within the oxygen-hydrogen torches, in which the oxygen and hydrogen themselves are being allowed to enter through outer tubes. A porous preform is grown unto the surface of a starting silica glass rod by the deposition of SiO_2 particles, which are synthesized in a vapor phase reaction process. A carbon ring heater is employed to produce zone melting and render the preform transparent. The continuous upward pulling of the preform in the axial direction permits the manufacture of long transparent preforms with large diameters ($\sim$20–30 mm). The original porous preform diameters range from 50 to 70 mm.

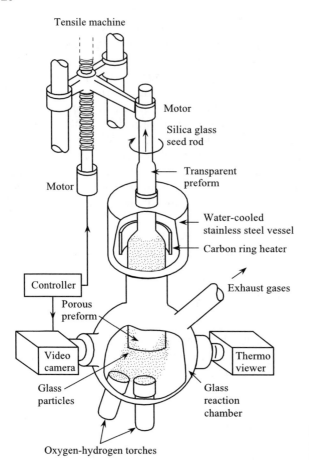

Tensile machine

Motor

Silica glass seed rod

Transparent preform

Motor

Water-cooled stainless steel vessel

Carbon ring heater

Controller

Exhaust gases

Porous preform

Video camera

Thermo viewer

Glass particles

Glass reaction chamber

Oxygen-hydrogen torches

Figure 15.49 Schematic diagram of VAD apparatus for fiber preform fabrication (after [74]).

The refractive index profile in the VAD process is controlled by the spatial distribution of the dopant concentration in the flame [74]. Since the different dopant-reactant mix ratios are injected into the oxygen-hydrogen flame from the torch nozzles, fine control of the refractive index profile requires the use of multiple nozzles. In addition, the surface temperature of the porous preform must be well controlled to ensure that the GeO_2 dopant formed during the reaction is incorporated into the SiO_2 in its melted amorphous glassy state and not as a low-temperature crystalline phase. Refractive index profile control also entails that the outer diameter of the porous preform be maintained constant.

The VAD process may be utilized to manufacture single-mode and graded-index multimode fibers. For single-mode fiber preforms, a porous preform, containing an outer thick cladding layer with an inner core, is fabricated such that the ratio of the cladding to the core diameter is equal to approximately 3. A thick OH ion-free silica tube is applied in the form of a jacket over the resulting preform. Oxide deposition rates of the order of 1 g/min at deposition efficiencies extending from 60 to 80% are readily achieved [74].

In the OCVD process, depicted schematically in Fig. 15.50, the fabrication steps consist of deposition of small particles of very high purity glass (often termed "soot")

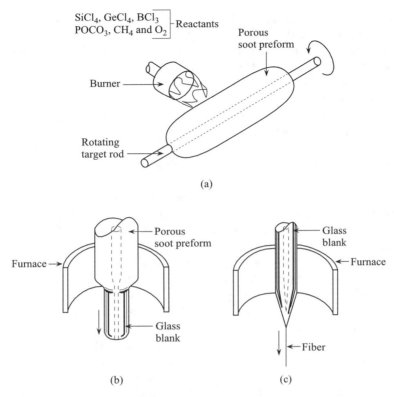

Figure 15.50 Fiber preparation sequence by OCVD process: (a) soot deposition; (b) sintering; (c) fiber drawing (after [75]).

on a target rod, followed by preform sintering and, finally, by fiber drawing from the transparent preform, termed the glass blank [75]. In this process, $SiCl_4$, BCl_3, $POCl_3$ and O_3 are made to react at high temperature in either a flame of methane and oxygen or oxygen and hydrogen to produce high-purity particles of soot, having diameters in the range from 100 to 1000 Å. These emitted particles from the flame are directed toward and deposited upon a rotating ceramic rod placed ~15 cm from the flame nozzle. The resulting preform formed upon the mandrel has a porosity of approximately 75%, with an average pore size of the order of 0.3 μm. Sequenced control of the halide vapor stream composition during the fabrication of the porous preform permits radial variation of the deposited glass layer composition and thereby the value of the index of refraction in the core and cladding sections. This procedure allows the fabrication of step-index and graded-index multimode fibers [76]. Deposition rates of ~2.0 g/min are readily achieved at an efficiency of approximately 50%.

Following its completion the porous preform is removed from the ceramic rod and is sintered at 1500°C in a helium atmosphere as it passes through a furnace and becomes a cavity-free transparent glass blank [77]. The sintering step involves a treatment with Cl_2 that removes the OH impurities from the optical fiber glass that were originally introduced by the flame combustion gases [78]. The hydroxyl ion content, which is responsible for attenuation losses at 1.39 μm, is thereby reduced to < 0.1 ppm from a level on the order of 200 ppm in the untreated porous preform [75]. Prior to actual fiber drawing the glass blank is preheated to temperatures in the range of 1800–2200°C, depending upon the glass composition, such that its viscosity is decreased to a value between 10^5 and 10^6 P [75]. As drawing of the multimode fiber takes place down to a final fiber radius of 125 μm, the center hole is eliminated. Since the presence of the center hole may lead to fracture of the sintered blanks as a result of the tensile stresses, which arise from a difference between the thermal expansion coefficients of the core and cladding glass composition, the soot sintering and fiber drawing steps may be combined. Thus the center hole in the sintered glass is maintained above the glass transition temperature until it is made to collapse during the fiber drawing step [79]. Alternatively, the central hole may be closed during the soot sintering step [80].

The mechanical strength of fibers is an important property as concerns optical fiber cable installation practices such as cable pulling forces, and permissible bending radius. Although minimum optical fiber tensile strengths of > 3.5 GPa (5×10^5 psi) may be readily achieved, contamination inclusions and surface irregularities of 0.1 μm or less may seriously reduce the tensile strength [81, 82]. The fiber strength is thus contingent upon both the defects present in the preform as well as the cleanliness and temperature stability of the drawing furnace [83]. Prior to drawing, surface defects may be removed from the blank preform by means of either etching or fire polishing.

A number of clean heating sources for fiber drawing at temperatures of 2000°C or higher are available. These include graphite resistance and induction furnaces, oxygen-hydrogen flames, high-power CO_2 lasers, and zirconia induction furnaces [81, 84]. A fiber drawing and coating system using a zirconia induction furnace is illustrated in Fig. 15.51. [81] The zirconia induction furnace is constructed with rings of yttria-stabilized zirconia, enclosing the rf susceptor. The temperature is controlled to within

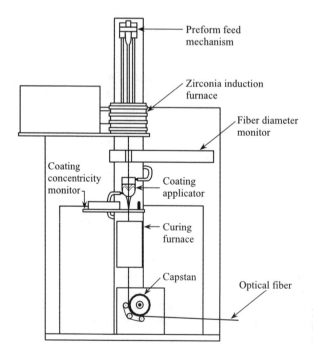

Figure 15.51 Schematic of optical fiber drawing and coating apparatus using zirconia furnace (after [81]).

±2°C by means of an optical pyrometer on the external surface of the zirconia suscep-tor. Zirconia insulation within a fused silica vessel inside a water-cooled copper con-tainer encloses the rings. The convective gas flow within the furnace is reduced by constricting the opening at the top of the furnace through which the preform enters. Reduced convection leads to greater temperature stability within the furnace, which results in improved diameter control of the drawn optical fiber [81].

Further improvements in fiber diameter control may be achieved by means of a laser forward scattering monitoring technique in terms of the diameter-dependent fringe interference patterns that permits diameter accuracy regulation to within ±0.10 μm [81, 85]. The diameter control is effected by means of a feedback loop, which involves speed regulation of the capstan drive.

The fiber drawing step also involves the application of the coating onto the fiber, as indicated in Fig. 15.51. The coating is applied to protect the fiber against abrasive damage. Its purpose is thus to protect the optical transmission characteristics of the fiber as well as impart it with additional strength. Normally polymer-type coatings are employed and care is taken that during their application the optical fiber is not damaged and the polymer characteristics are such that the polymer solidifies prior to reaching the capstan, which requires rapid crosslinking reactions as well as either thermally or ultraviolet activated curing mechanisms [81, 86]. Considerable care must be exercised to eliminate any cavity formations in the applied polymer coating, which would inevitably lead to an increase in optical transmission losses of the drawn fiber. The geometry of the coating applicator (cf. Fig. 15.51) plays also an important role in ensuring homogeneity of the polymer coating film. Particular attention is given to prevent contact between coating dies and the drawn fiber; this is accomplished by either monitoring the concentricity of the coating by means of laser scattering measurements [87] in order to allow appropriate positioning of the coating applicator or the use of a compliant coating applicator [81]. Compliant coating materials, such as silicone rubber, are normally employed to avert imparting microbending losses into the optical fibers by cushioning them against externally imposed mechanical stresses during cabling and installation [88]. The compliant silicone coatings must frequently be protected them-selves against abrasion by mechanically strong nylon jackets [81].

15.9.2 Transmission Parameters of Optical Fibers

Although this section deals with the parameters determining the transmission characteristics of the different types of optical fibers and is thus both complementary and supplementary in content to that considered on the introductory aspects on the subject of optical fibers, it does not delve into the introductory details of the transmis-sion characteristics in terms of geometric optics as is done in Chapter 17. Readers may thus familiarize themselves with that aspect of the subject by reference to Chapter 17.

The silica optical fibers currently in use are of the single- or multimode type. In the latter the signal is propagated in a number of different modes, each of which travels over a different path length with the result that each mode reaches the receiving end of the fiber at a different time, whereas in a single-mode, or monomode, fiber, wherein the core diameter exceeds only by several integers the wavelength of the transmitted light signal, the propagation occurs at a single mode along the shortest path, i.e., the center or axis of the cylindrical fiber. It is thus the much larger diameter of the core, usually 50 μm, as compared to 8–10 μm of the single-mode fibers, that gives rise to multimode transmission. The cladding material surrounding the core of the fiber, whose index of refraction is less than that of the core, imparts the guiding characteristics to the optical fiber. Since the propagation of the various modes over different path lengths in the multimode fiber results in distortion of the transmitted pulse, the index of refraction is normally graded, using the doping techniques discussed in the previous sections. With refractive index grading, it is possible to increase the propagation velocity over the longer signal transmission paths and decrease it over the shorter paths, so that the transmitted light signal of each mode arrives at the output end of the fiber at the same time. For this purpose, it is common practice to employ a parabolic index-grading distribution. Multimode optical fibers, whose index of refraction is graded, are referred to as graded-index multimode optical fibers, as distinguished from the step-index-type multimode fibers wherein the index of refraction decreases abruptly in the form of a step at the core-cladding interface.

Single-mode fibers are essentially step-index-type fibers, which may be categorized as either dispersion-shifted or dispersion-flattened fibers. The former are designed such that their minimum chromatic dispersion, which occurs at a wavelength of 1.31 μm is displaced to 1.55 μm, which is the wavelength of lowest attenuation for silica-based fibers. In this fiber structure, the cladding is homogeneous and has a fixed index of refraction throughout. While in the depressed-cladding configuration, the cladding consists of two layers with the outer undoped silica layer having a higher index of refraction than that which is contiguous with the core of the fiber.

Figure 15.52 compares two representative designs of multimode and single-mode fibers, with the former having the usual parabolic graded-index distribution [89]. The typical numerical values given for the multimode graded-index fiber refer to a core composition of GeO_2-P_2O_5-SiO_2 with a parabolically graded fiber of $\Delta n = 0.018$, where

$$\Delta n = n_1 - n_2 \tag{15.42}$$

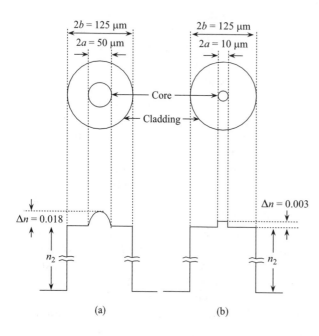

(a)   (b)

Figure 15.52 Typical cross sections of (a) multimode graded-index and, (b) single-mode optical fibers and associated index of refraction profiles (after [89]).

where n_1 is the refractive index of the core and n_2 that of the cladding; the numerical aperture (NA) of the multimode fiber is defined by

$$NA = \sqrt{n_1^2 - n_2^2} \qquad (15.43)$$

For the particular multimode graded-index fiber specified in Fig. 15.52, NA = 0.23 and the number of guided modes, N, at the transmission wavelength $\lambda = 1.3\,\mu m$ would be 192 [89]. Note that in actual practice, optical fibers produced by the MCVD process will be characterized by a crevicelike depression in the refractive-index profile at the center of the core, as evident in Fig. 15.53, due to depletion of the dopants at the center of the preform during the tube collapse step [90, 91].

The criterion of a step-index fiber to operate in a single mode can be defined in terms of the normalized frequency, or f_n number, as [89]

$$f_n = \frac{2\pi a}{\lambda_0} \sqrt{n_1^2 - n_2^2} \qquad (15.44)$$

such that $f_n < 2.405$. Here λ_0 is the wavelength in free space and a is the core radius. The cut-off wavelengths λ_c for the single-mode fiber configuration in Fig. 15.52 above which higher modes cannot propagate is 1.225 μm. Depressions in the index of refraction of the type appearing at the axis of a graded-index core, as depicted in Fig. 15.53, increase the value of f_n defined by Eq. (15.44) and decrease λ_c [92]. Since generally, $n_2 \simeq n_1$, Eq. (15.44) is frequently expressed as [93]

$$f_n = \frac{2\pi a}{\lambda_0} n_1 \sqrt{2\Delta} \qquad (15.45)$$

where Δ is the fractional refractive index difference given by

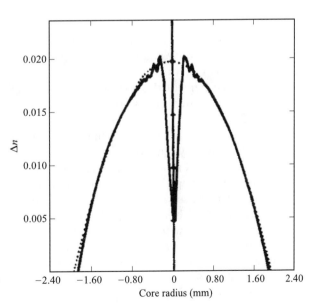

Figure 15.53 Index of refraction profile of graded-index preform (after Kaminow [90]).

$$\Delta = \frac{n_1^2 - n_2^2}{2n_1^2} \tag{15.46}$$

If cylindrical waveguides are considered, then the number of possible propagating modes in terms of the core radius a is [94]

$$N_m = 2a \left(\frac{2\pi}{\lambda_0}\right)^2 (n_1^2 - n_2^2) \tag{15.47}$$

and using Eq. (15.43) yields

$$N_m = 2 \left(\frac{2\pi a}{\lambda_0}\right)^2 (NA)^2 \tag{15.48}$$

i.e., the number of modes increases with the square of the diameter of the fiber. Letting $N_m \leq 1$ gives the condition of single-mode propagation as [95]

$$2 \left(\frac{2\pi a}{\lambda_0}\right)^2 (NA)^2 \leq 1 \tag{15.49}$$

This condition requires a core diameter $2a$ of

$$2a \leq \frac{\lambda_0}{(2\pi)^{1/2} \, NA} \tag{15.50}$$

or

$$2a \simeq \frac{0.22\lambda_0}{NA} \tag{15.51}$$

In practice, single-mode transmission fibers have core diameters only a few times larger than the wavelength of the transmitted light signal. The typical single-mode fiber for transmission wavelengths of 1.3 and 1.55 μm has a core diameter between 8 and 10 μm. The numerical apertures for these single-mode fibers fall between 0.029 and 0.042, as compared to multimode fibers, whose NA values are commonly of the order of 0.22.

Polarization effects within single-mode fibers may be of concern in coherent optical communication systems where the polarization signal of the local oscillator must match the polarization of the light signal at the output of the optical fiber [96]. The rotation of the polarization plane in the single-mode optical fibers is caused by thermal and mechanical stresses as well as geometric irregularities such as cable curvature, bends, and twists [97]. These strain-induced variations of polarization along the fiber tend to be unpredictable with time, as illustrated in Fig. 15.54.

Attempts to introduce increased birefringence into the single-mode optical fibers in order to stabilize the polarization effects to prevent modal degeneration have generally led to enhanced losses [98]. However, since the swings in polarization with time are in general not great, effective polarization controllers have been devised whereby the

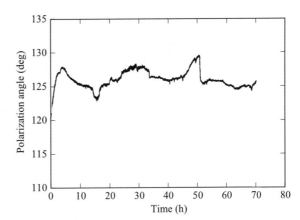

Figure 15.54 Polarization angle variation in directly buried acrylate coated single-mode optical fiber telephone cable (after [96]).

polarization of the local optical signal is matched to that of the received signal [99]. In the case of multimode fibers, the polarization effects are less important because of the prevalent use of photodiode detectors [89].

Much more serious effects in optical fibers than polarization are those due to pulse broadening and dispersion, because they are inversely related to the bandwidth and information-carrying capacity of fibers [100–102]. Pulse broadening and dispersion occur with increasing transmission distance over the optical fiber; in extreme cases, the pulse widening of two adjacent originally narrow pulses can lead to an overlap whereby the two pulses, whose original time spacing is still retained, may not be resolved as two discrete separate pulses. This situation is illustrated in Fig. 15.55. It is well to distinguish between pulse dispersion effects and pulse spreading in multimode fibers. The former are associated with material and waveguide dispersion, whereas the latter, also referred to as intermodal dispersion, results from group delay differences among modes, i.e., a difference in the travel times of each mode as it propagates along its particular travel path length and is thus dependent upon the refractive-index profile of the fiber.

Material pulse dispersion occurs both in single- and multimode fibers and is caused by atomic resonance effects, which in glass fibers appear over the infrared frequency régime and manifest themselves as a wavelength dependence of the index refraction. Waveguide dispersion is characteristic of only single-mode fibers in which light pulse spectral components propagate at different velocities whose magnitudes are determined by the material and geometry of the optical fiber waveguide. Frequently the term chromatic, or intramodal, dispersion is used when jointly referring to material and waveguide dispersion. Note that pulse dispersion phenomena are wavelength

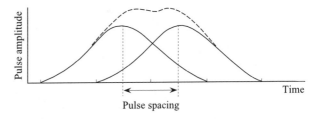

Figure 15.55 Pulse superposition of two adjacent light pulses caused by pulse broadening and dispersion effects.

dependent, with the exception of the pulse intermodal dispersion. The units of dispersion are picoseconds of pulse broadening per kilometer per nanometer of transmission source spectral width (ps/km.nm).

The material dispersion $M(\lambda)$ may be expressed as [103, 104]

$$M(\lambda) = \frac{\lambda}{c} \frac{d^2 n_1}{d\lambda^2} \tag{15.52}$$

where λ is the transmitted wavelength, c is the velocity of light in free space, and $d^2 n_1 / d\lambda^2$ is the second derivative of the refractive index of the fiber material. The wavelength λ_{01}, at which pulse broadening $M(\lambda)$ reduces to zero is dependent upon the concentration of germania (GeO_2) [103]. This property of the multimode fiber is illustrated in Fig. 15.56, from which it is apparent that material dispersion in multimode fibers can be greatly reduced by operating light transmission sources at wavelengths in close proximity of λ_{01} of the multimode fiber at which $M(\lambda) = 0$ [103]. Material dispersion arises from the refractive-index dependence upon frequency or wavelength, which is caused by electronic and atomic absorption resonance effects in the material, with the dopants playing an important role.

The dependence of the refractive index is often expressed in terms of the Sellmeier equation, which is based upon the classical resonance absorption expressions for the permittivity or index of refraction. This equation may be stated as [106, 107]

$$n^2(\lambda) - 1 = \sum_{i=1}^{n} \frac{A_i \lambda^2}{\lambda_v - B_i^2} \tag{15.53}$$

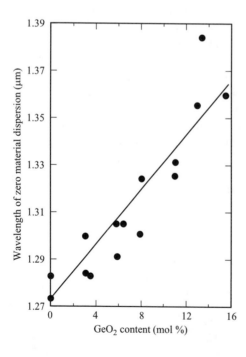

Figure 15.56 Effect of germania concentration on material dispersion in silica multimode fibers [103, 105].

where $n(\lambda)$ represents the refractive index of the core or cladding materials, λ_ν, is the wavelength related to the resonance frequency ($\nu = 2\pi c/\omega$), and constants A_i and B_i depend upon the harmonic oscillator strengths in the material. The latter are determined empirically by fitting the calculated results to the measured dispersion curves and depend significantly on the type and concentration of the dopant employed. The dependence of material dispersion upon wavelength for borosilicate (13 mol % B_2O_3–87 mol % SiO_2) and germanium borosilicate (10 mol % GeO_2–4 mol % B_2O_3–86 mol % SiO_2) is illustrated in Fig. 15.57 [108–110]. The wavelength for zero material distortion is seen to occur at approximately 1.3 μm, which is near the value of 1.27 for pure silica (SiO_2).

Pulse broadening due to material dispersion, described by Eq. (15.52), increases directly with the length of the optical fiber line, ℓ, as well as with the spectral width of the transmission source, $\delta\lambda$. Thus, Eq. (15.52) may be expressed as [100]

$$t_m = \delta\lambda_s \frac{\lambda\ell}{c}\frac{d^2 n_1}{d\lambda^2} \tag{15.54}$$

The empirical form of Eq. (15.54) takes into account the fact that the pulse width t_m in time units increases with the optical fiber length ℓ and the spectral width $\delta\lambda_S$ of the transmitting source. Since the practical units of pulse broadening are in picoseconds per kilometer (ps/km) per spectral width of source in nanometers, Eq. (15.54) is rearranged in the form

$$\frac{t_m}{\ell(\delta\lambda_S)} = \frac{\lambda}{c}\frac{d^2 n_1}{d\lambda^2} \tag{15.55}$$

The use of light-emitting diodes (LEDs) sources with broad spectral widths of 40–50 nm with multimode optical fibers as opposed to the relatively narrower $\delta\lambda_S$ values of lasers (usually equal to or less than 2 nm) requires material dispersion to be taken into account in multimode optical fiber transmission systems. In LED systems a material dispersion value of ~ 100 ps/km-nm may be considered to be typical of germania-doped silica multimode fibers [103, 104].

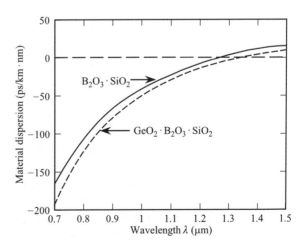

Figure 15.57 Material dispersion of borosilicate and germanium borosilicate glass of silica fibers [108–110].

Intermodal pulse broadening is the predominant dispersion process in multimode optical fibers. The group delay spread, arising from differences in group velocities of the various modes, leads to an intermodal pulse broadening of [100]

$$t_n = \frac{\ell n_g \Delta}{2c} f(i) \tag{15.56}$$

where n_g is a group index constant and $f(i)$ represents an empirical function of i, which is a profile parameter such that at $i = 2.25$ a minimum value of $f(i)$ also results in a minimum of the intermodal pulse broadening t_n. The number of disposable constants in Eq. (15.56) diminishes its usefulness. However, the dependence of t_n upon Δ indicates that with suitable refractive-index profile control, intermodal pulse broadening may be significantly reduced. The so-called power law is often employed to illustrate the effectiveness of graded-index profiles in reducing the intermodal pulse broadening; it can be expressed as [103]

$$n^2(r) = n_{0r}^2 \left[1 - 2\Delta \left(\frac{r}{a} \right)^g \right] \tag{15.57}$$

where a is the radius of the core, n_{0r} is the maximum refractive index at the center of the core, $n(r)$ is the refractive index at radius r, with, $r \leq a$ and g is the power law profile parameter. Least modal pulse broadening is obtained at $g = 2$ [104]; it can be demonstrated that the optimum power law refractive index profile leads to an rms pulse width of [111–113]

$$\frac{t_{rms}}{\ell} \approx 0.14\Delta^2 \; \mu s/km \tag{15.58}$$

for which the group index N_{0r} value of 1.46 is assumed; the group index N_1 given by [114]

$$N_1 = n_1 - \lambda \frac{dn_1}{d\lambda} \tag{15.59}$$

is a parameter that appears in the profile dispersion expression associated with the optimum value of the exponent g determination [101, 112]. The corresponding maximum achievable bit rate of the multimode optical fiber is given in terms of Eq. (15.58) as [113]

$$B = \frac{\ell}{4 t_{rms}} \tag{15.60}$$

or

$$B = 1.8\Delta^{-2} \; Mb\text{-}km/s \tag{15.60a}$$

If $\Delta = 0.0135$ is substituted in Eq. (15.60a), then the maximum bit rate achievable becomes equal to 10 Gb-km/s. Note that in optical fibers the fiber bandwidth is expressed as a frequency-distance product; Figure 15.58 depicts the theoretical bandwidth of a multimode optical fiber for $\Delta = 0.0135$ and a core-cladding refractive index

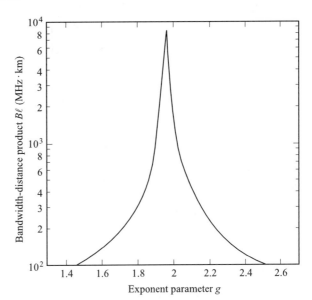

Figure 15.58 Bandwidth-distance product vs. actual value of exponent parameter g, for multimode optical fiber with refractive index difference $\Delta n = 0.02$ and $\Delta = 0.0135$ [113].

difference $\Delta n = 0.02$ is a function of the power law exponent g [113]. The bandwidth characteristic of the multimode fiber in Fig. 15.58 is idealized in the sense that the exponent g is not constant along the fiber length because exact index grading in accordance with the power law is generally not attainable in practice. It can be perceived from the nature of the bandwidth characteristic in Fig. 15.58 that a relatively small deviation in the value of g from that at the bandwidth maximum causes a very pronounced decrement in the bandwidth of the multimode fiber. Since minor deformations in the graded profile of multimode fibers may influence adversely its bandwidth, production of large-bandwidth multimode fibers require stringent manufacturing quality control methods [115]. The maximum bandwidth at the optimum value of g_0 of a graded-index multimode fiber exceeds that of a step-index fiber with same value of Δ by a factor of $10/\Delta$. With profile synthesis employing several dopants, the spectral width of the peaked bandwidth at g_0 may be broadened [116]. Figure 15.59 portrays the optimum bandwidth exponent value, g_0, as a function of the wavelength with the type of dopant in the glass fiber as a parameter. The measured characteristics for the GeO_2-SiO_2 and P_2O_5-SiO_2 binary compound fibers show that the phosphorus-doped fiber results in a g_0 value that varies substantially less with the wavelength λ than that of a comparatively doped germanium fiber. An assumed concentration of 11.6 mol% phosphorus and zero percent germanium at the center of the core ($r = 0$) and 1 mol% germanium and zero phosphorus at the maximum core radius ($r = a$) predicts for the tertiary combination of P_2O_3-GeO_2-SiO_2 a flat g_0-versus-λ characteristic with $g_0 = 1.92$ and $dg_0/d\lambda = 0$ at $\lambda = 0.8\,\mu m$ [113, 116]. Various schemes of index grading are possible for the increase of bandwidth of multimode fibers [118, 119]; however, highly accurate control of refractive-index variation across the cable core is difficult to achieve in practice.

Bandwidth improvement in multimode fibers can arise from attenuation effects introduced due to geometric variations along the optical waveguide [110]. As the higher order modes are attenuated, pulse spreading due to these modes is eliminated; hence,

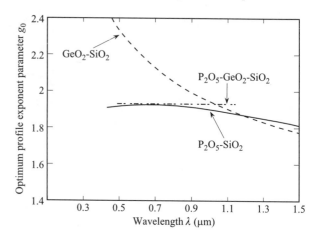

Figure 15.59 Optimum profile exponent parameter, g_0 vs. wavelength, λ of silica SiO_2 multimode fibers with type of dopant as parameter [113, 116, 117].

pulse spreading due to the higher modes diminishes with the length of the fiber, ℓ. It has been established that the bandwidth of a fiber is a function of $\ell^{-\gamma}$, with complete mode mixing occurring at $\gamma = 0.5$ [110].

Nonfundamental modes of singly clad multimode fibers may have similar dispersion characteristics as multiclad single-mode fibers, which can exhibit more than one zero dispersion wavelength [107, 114]. This particular behavior of singly clad multimode fibers is achievable because they, as multiclad single-mode optical fibers, are characterized by a cut-off wavelength. The existence of several dispersion-free wavelengths can be illustrated by considering the expressions for the *modal second-order dispersion* β_2 and the *material second order dispersion* β_{2m}. These are given by [107, 120]

$$\beta_2 = \frac{\lambda^3}{(2\pi c)^2} \left(\lambda \frac{d^2 \beta}{d\lambda^2} + 2 \frac{d\beta}{d\lambda} \right) \tag{15.61}$$

and

$$\beta_{2m} = \frac{\lambda^3}{2\pi c^2} \frac{d^2 n_0(\lambda)}{d\lambda^2} \tag{15.61a}$$

Note that β_2 is a function of the propagation constant β and its derivatives with respect to the wavelength λ while β_{2m} is obtained by noting that $\beta = 2\pi n_0/\lambda$. The second-order material dispersion β_{2m} is calculated by means of Eq. (15.53), for which the constants A_i and B_i given in Table 15.17 refer to a step-index multimode fiber with a composition as indicated [107, 108].

Figure 15.60 depicts the second-order modal dispersion together with the second-order material dispersion of a step-index multimode optical fiber with a core radius of 6.7 μm, whose A_i and B_i coefficients are given in Table 15.17. The material second-order dispersion can be perceived to disappear in the proximity of $\lambda = 1.42\,\mu m$, whereas the second-order modal dispersion is observed to become zero at approximately $\lambda = 1.1\,\mu m$ and $\lambda = 1.35\,\mu m$. The displacement of the second-order modal dispersion curve vis-à-vis the second-order material dispersion is ascribed to the negative

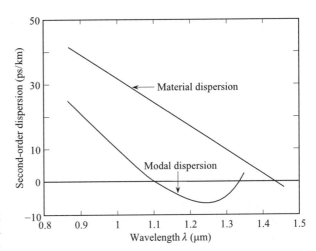

Figure 15.60 Second-order material and second-order modal dispersions of step-index multimode fiber with core radius 6.7 μm [107].

contribution of the waveguide dispersion to the second order material dispersion. The modal dispersion behavior of singly clad multimode fibers, which is characterized by two modal zero dispersion wavelengths and near-zero dispersion wavelengths over a substantial range of wavelengths, may permit to extend the transmission capabilities of such multimode optical fibers. It is conceivable that with selective modal excitation, wavelength division multiplexing (WDM) in multimode optical fibers may thus be achievable [121].

In single-mode fibers, the zero-dispersion wavelength is determined by the material and waveguide dispersions; since the slopes of these two characteristics are of opposite sign, suitable adjustment of the optical fiber parameters can produce a displacement of the zero-dispersion wavelength λ_0'. This is routinely accomplished by altering the refractive-index profile; one such arrangement is that of the so-called depressed cladding profile, obtained with doubly clad single-mode fiber, also frequently referred to as the W-fiber. The depressed cladding profile is portrayed in Fig. 15.61.

If $n_1 > n_3 > n_2$ and $(n_1 - n_3)/n_1 = (n_3 - n_2)/n_1 = 0.01$, then, for a core diameter ($2a$) of 7.2 μm and an inner cladding thickness (d) of 1.08μm, the resultant dispersion between the wavelengths of 1.35 and 1.67 μm is equal to less than ± 1 ps/km-nm [89,

TABLE 15.17 Sellmeier Coefficients A_i and B_i for Multimode Optical Fiber[a]

Coefficients A_i and B_i^{b}	Core	Cladding
A_1	0.72454395	0.70724622
A_2	0.42710828	0.39412626
A_3	0.82103399	0.63301929
B_1	0.8697693×10^{-8}	$0.80478054 \times 10^{-7}$
B_2	$0.11195191 \times 10^{-6}$	$0.10925792 \times 10^{-6}$
B_3	0.1084654×10^4	$0.78908063 \times 10^{-5}$

[a] Core with 13% mole concentration of GeO_2 in SiO_2; cladding with 13% mole concentration of B_2O_3 in SiO_2 [107, 108].
[b] Cf. Eq. (15.53).

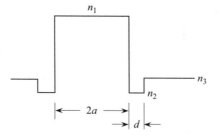

Figure 15.61 Depressed-cladding refractive-index profile of double clad single-mode W-type fiber.

122]. Figure 15.62 shows that the low resultant dispersion is a direct outcome of cancellation between the waveguide and material dispersion contributions. It must be emphasized that to achieve the desired behavior, the diameter of the core must be rigidly controlled as variations in $2a$ will cause significant changes in the value of the minimum-dispersion wavelengths [123].

The overall flattened dispersion characteristic in Fig. 15.62, which leads to very low dispersion values over an extended wavelength range, permits the efficient use of a large number of optical channels, thereby in effect creating a large-bandwidth fiber. As pointed out, it is desirable to shift the zero-dispersion wavelength λ_0'' to $1.55\,\mu\text{m}$, at which the silica fibers also exhibit their lowest attenuation loss; this may be achieved with a depressed cladding refractive-index profile for which the refractive index of the core has a triangular distribution of the type portrayed in Fig. 15.63. Also with multi-clad monomode fibers having a triangular core refractive-index profile, it is readily possible to achieve a nearly zero dispersion flattened characteristic with zero dispersion occurring at $1.55\,\mu\text{m}$. Such a refractive-index profile with four claddings is depicted in Fig. 15.64.

The overall dispersion in monomode optical fibers not only is contingent upon the material and waveguide dispersion intrinsic to the optical fiber itself, but also is depen-

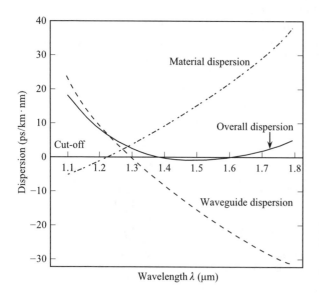

Figure 15.62 Dispersion vs. wavelength of monomode fiber of depressed-cladding profile represented in Fig. 15.63 with $(n_1 - n_3)/n_1 = (n_3 - n_2)/n_1 = 0.001$, $d = 0.3a$, and $2a = 7.2\,\mu\text{m}$ [89, 122].

Figure 15.63 Depressed cladding with triangular-core refractive-index profile of doubly clad monomode fiber [124].

Figure 15.64 Depressed cladding with triangular-core refractive-index profile of multiclad monomode fiber having total of four claddings [124].

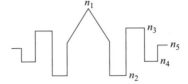

dent upon the rms half-width of the pulse, $\delta\lambda$, of the single-mode injection laser source. The minimum pulse broadening occurring at the zero-dispersion wavelength λ_0'' for a monomode fiber may be expressed as [124, 125]

$$t_s = \frac{\ell\delta\lambda^2}{8}\frac{d^2n}{dt^2} \tag{15.63}$$

The bit rate or bandwidth-distance product $B\ell$ in gigabits per second-kilometer may be written in an empirical form as [124, 126]

$$B\ell \leq \frac{341}{\delta\lambda(d^2n/d\lambda^2)} \tag{15.64}$$

and the corresponding $B\ell$ value for systems operating at or in the vicinity of the zero-dispersion wavelength is augmented to

$$B\ell \leq \frac{11,207}{\delta\lambda^2(d^2n/d\lambda^2)} \tag{15.65}$$

Fluctuations in the signal as a result of laser mode partitioning at the receiving end exert an adverse effect on the $B\ell$ product; this effect, which causes line width broadening of a single longitudinal mode, normally referred to as chirping [127], is dealt with by introducing a factor K to take into account the fluctuation of the power not present in the dominant mode of the laser [124]. When the K parameter of the laser source is greater than 0.1, the bandwidth of the monomode fiber becomes compromised so that Eqs. (15.64) and (15.65) must be modified accordingly as [124]

$$B\ell \leq \frac{341}{\delta\lambda(d^2n/d\lambda^2)\sqrt{K}} \tag{15.66}$$

and

$$B\ell \leq \frac{1173}{\delta\lambda^2(d^2n/d\lambda^2)\sqrt{K}} \tag{15.67}$$

When single-mode lasers with very narrow spectral pulse widths are employed such that $\delta\lambda \leq (\lambda^2/c)B$, Eqs. (15.64) and (15.66) must be adjusted by the substitution of $(\lambda^2/c)B$ for $\delta\lambda$ [124].

Figure 15.65 portrays the product of bandwidth or bit rate and distance as a function of the transmission wavelength with the laser pulse spectral rms half width $\delta\lambda$ as a parameter for a single-mode optical fiber, whose zero-dispersion wavelength λ_0 is centered at 1.3 μm [124]. The bandwidth-distance product is seen to exhibit a very pronounced dependence upon the $\delta\lambda$ value; i.e., a small increase in $\delta\lambda$ causes a marked reduction in the bandwidth-distance product of the fiber.

The majority of the monomode optical fibers in present transmission systems operate at the minimum-dispersion wavelength of 1.3 μm. At the wavelength of 1.3 μm, the transmission capacity of those systems is principally limited by the electronic repeater bandwidth and the attenuation intrinsic to the silica fibers themselves. However, to shift the operation of the monomode fibers to the attenuation minimum at 1.55 μm would require to reduce the pulse dispersion in the monomode optical fibers, which is rather elevated and is of the order of 17 ps/km-nm [128, 129]. In an effort to circumvent the dispersion that would be introduced at 1.55 μm, several dispersion compensation techniques have been proposed. One possible approach involves spectral inversion, whereby at the midpoint of a given fiber link the signal spectrum is inverted, thus making the dispersion in the remaining portion of the fiber behave in a reverse manner and cause compression of the broadened signal emerging from the first half of the fiber [130]. However, this constitutes an awkward solution as it necessitates the use of nonlinear components such as the insertion of sizable lengths of dispersion-shifted optical fibers. Dispersion compensation may alternatively be achieved by utilizing linear components such as optical fibers with opposite dispersion [131]. But with linear dispersion compensation, the required fiber lengths are much longer, attaining overall lengths that may approach a total of nearly half the length of the original fiber for which the compensation is to be effected. Dispersion compensation may also be obtained using electrical techniques. These relatively more sophisticated approaches

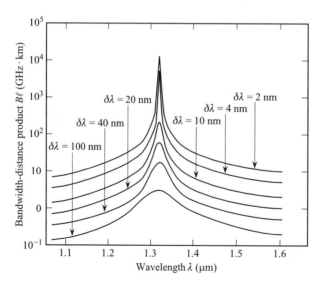

Figure 15.65 Effect of spectral line width $\delta\lambda$ of laser source upon bandwidth-distance characteristics of monomode fiber with zero-dispersion wavelength λ_0' of 1.3 μm (after [124]).

involve predistorted signal synthesis [132] and in-line amplified dispersion supported transmission [133] methods. The former method, in which a frequency sweep generator is employed to control the bias current of a laser diode whereby the signal to be transmitted is injected by means of an absorption modulator, is capable of improving the dispersion by 20%. More recently a three-level modulation scheme has been devised to counteract dispersion [128]. In this electrical procedure use is made of the interaction between fiber dispersion and its nonlinearities. The pulse shape and the peak power are adjusted in a three-level modulation scheme; the shape of the pulse is controlled in the electrical domain with the result that the dispersion of the fiber may be compensated electrically. The procedure involves the calculation of the power penalty in terms of the so-called eye diagrams, where the received eye opening is a function of the attenuation, dispersion, and self-phase modulation. All three quantities are related via the generalized nonlinear Schrödinger equation to the complex field envelope, which is a function of longitudinal fiber coordinate and the relative time frame moving with the pulse at the group velocity [128].

As new developments occur in the area of optical amplifiers, the attenuation loss in optical fibers is becoming less critical. However, there are many multimode and single-mode fibers in use at the wavelengths of 0.8 and 1.3 μm where amplification must be performed by electronic repeaters. It is thus necessary to consider the attenuation characteristics of these fibers in more detail in order to appreciate more fully their transmission capacity limitations in terms of their attenuation behavior.

Optical fibers are operated within the wavelength range extending from approximately 0.7 to 1.8 μm, which overlaps the ultraviolet spectrum and the tail end of the electron resonance absorption region at the lower wavelengths, whereas the upper wavelength range encroaches on the tail end of the infrared atomic resonance absorption region. Since multimode fibers operate at the wavelength of 0.8 μm and step-index monomode fibers at the zero-dispersion wavelength of 1.3 μm and dispersion-shifted monomode fibers at the wavelength of least attenuation of 1.55 μm, their signal attenuation behavior at these specific wavelengths is of particular importance. Although atomic resonance absorption induced attenuation at 1.55 μm constitutes a predominant loss contribution, it is essentially negligible at 1.3 μm; while electron resonance absorption effects are also insignificantly small at 1.3 μm, they may lead to attenuation values as high as 0.8–1.0 dB/km at 0.8 μm. Thus in multimode fibers the overall attenuation ($\sim$3.5 dB/km) is principally caused by Rayleigh scattering and electron resonance absorption, with the former exerting the prevailing influence. Rayleigh scattering arises from density fluctuations (variations in the refractive index) in the lattice of the amorphous glass material of the optical fiber, whereby the transmitted light signal undergoes loss of energy as a portion of its light energy is scattered in various directions. Since the variation of the index of refraction within the glass structure is on a microscopic scale, Rayleigh scattering is more pronounced at the lower wavelengths, with its magnitude varying inversely as λ^4. Consequently, its intensity diminishes rapidly with wavelength, which accounts in part for the much lower losses in the monomode fibers that are operated either at 1.3 or 1.55 μm. As the intensity of Rayleigh scattering is enhanced by the dopant concentration in the fibers, the higher attenuation observed with multimode fibers must in part be attributed to their greater dopant levels.

Contaminants such as transition metals (copper and iron) as well as water may contribute markedly to absorption-induced attenuation losses at particular wavelengths. Over the transmission wavelength spectrum, the transition metal contribution to resonance absorption is usually widely dispersed, while the peroxide radical (OH) induced losses exhibit three well-defined resonance absorption peaks. The most intense and detrimental peak, which may readily result in an attenuation of $\sim$3.5 dB/km, appears at approximately 1.39 μm. Two minor OH absorption peaks, which occur at approximately 0.96 and 1.26 μm, augment the attenuation only by about 0.1 dB/km.

Figure 15.66 depicts the effect of dopant on the attenuation characteristics of silica cores, having different dopants but with identical borosilicate claddings ($B_2O_3 \cdot S_iO_2$) [134]. For comparison purposes, the attenuation characteristic of a single-mode fiber is also shown [135]. It can be seen that with the multimode fibers, the germania dopant (GeO_2) yields the lowest attenuation ($\sim$ 9.4 dB/km) both at 0.8 and 1.55 μm. Whereas at 1.3 μm an attenuation of $\sim$0.6 dB/km is typical for both the GeO_2- and P_2O_5-doped silica fibers. The combination of GeO_2 and B_2O_3 dopants is perceived to lead to excessively high losses for transmission wavelengths greater than 1.3 μm and, consequently, the B_2O_3 dopant is rarely used. The three multimode fibers and the monomode fiber are seen to exhibit a very pronounced hydroxyl (OH^-) resonance absorption peak at 1.39 μm. The monomode fiber, in particular, is observed to manifest an unusually high OH^- absorption maximum, notwithstanding its low attenuation of $\sim$0.2 dB/km at 1.55 μm. The presence of the OH^- radicals is undesirable because of their adverse influence on the attenuation characteristics of the fiber at the longer wavelengths. As already mentioned previously, the most common sources of OH^- contamination are associated with hydrogen in the starting materials, such as trichlorosilane, air leaks into the deposition atmosphere, carrier gas, and direct OH^- radical diffusion from the substrate tubes [68]. Various techniques are employed to eliminate hydrogen by means of reaction with chlorine gas.

The cladding used on monomode fibers must be characterized by low atomic resonance absorption losses at the higher wavelengths (i.e., 1.55 μm), with the consequence that dopants such as B_2O_3, which exhibit high absorption losses, cannot be utilized in claddings. In addition, the cladding over the single-mode fibers must be of sufficient thickness, usually at least five times the core diameter to reduce the ingress of OH^- via the diffusion mechanism, as well as for waveguide considerations [68, 136]. The single-mode fiber given in Fig. 15.66 has a GeO_2-SiO_2 core but employs an SiO_2

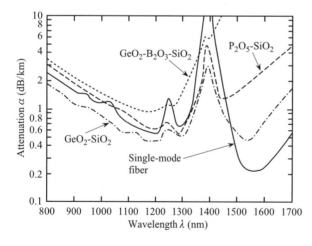

Figure 15.66 Attenuation characteristics of single-mode fiber and three multimode fibers with different dopants in core [68, 134, 135].

cladding to reduce the atomic resonance absorption at 1.55 μm in order to attain a low attenuation level of 0.2 dB/km. The more refractory SiO_2 is deposited by means of a pressure-controlling device to assist in the deposition of an adequately thick cladding layer [136].

The MCVD process has been used to prepare low OH^- radical content graded-index multimode GeO_2-P_2O_5-SiO_2 fibers with an NA of 0.23 and 50 μm core diameter. Figure 15.67 indicates the spread in attenuation values over a 41-km length of such fibers [103, 137]. The OH^- content in terms of the absorption maxima at 1.39 μm is of the order of 15 ppb.

Table 15.18 compares the attenuation characteristics of multimode graded-index fibers as a function of the type of dopants in the core and the cladding materials, manufactured using the outside vapor deposition (OCVD) process [69]. It is interesting to observe that the lowest attenuation values result with fibers having GeO_2-P_2O_5-SiO_2 cores and SiO_2 claddings.

As Figure 15.68 demonstrates, it is readily feasible with dispersion-shifted single-mode fibers to achieve attenuation values as low as 0.2 dB/km at the operating wavelength of 1.55 μm. Figure 15.68 is shown here for illustrative purposes, since it depicts proportionately the various components of the contributing loss mechanisms to the overall attenuation curve of the fiber [138]. However, it should be pointed out that it evinces a relatively large OH^- loss peak at 1.39 μm, which with currently manufactured single-mode fibers may be readily reduced to $\sim$ 1 dB/km. The principal and second major OH^- loss peaks, which occur at 1.39 and 1.25 μm, respectively, influence appreciably the attenuation values of dispersion-shifted monomode fibers at 1.55 μm and multimode fibers at the zero-dispersion wavelength λ_0' of 1.3 μm. Hydrogen diffusion into the optical fibers has been found to result in increased losses over the longer wavelengths [139–142]. Also, graded-index multimode cables, with GeO_2-P_2O_5 doped cores, having protective silicone and nylon coatings were found to exhibit higher OH^- loss peaks following a two-year operating period, as substantiated by the results in Figure 15.69 [143]. The observed effect, which was attributed to hydrogen ingress into the core material, was found to be substantially less pronounced with fibers having reduced P_2O_5 dopant concentrations. However, in another study, the converse was observed in that the presence of P_2O_5 dopant was found to exert a beneficial influence in reducing the hydrogen diffusion induced losses [143]. Lastly a study involving single-

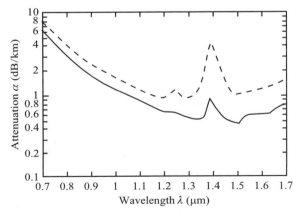

Figure 15.67 Typical attenuation range spread as function of wavelength for multimode GeO_2-P_2O_5-SiO_2 graded-index fibers [103, 137].

TABLE 15.18 Attenuation Characteristics of Selected Multimode Graded-Index Fibers Manufactured Using OCVD Process

Parameter	A–D	B–E	A–D	C–D
Core diameter, μm	50	50	100	100
Numerical aperture	0.2	0.2	0.3	0.3
Cladding diameter, μm	125	125	140	140
Bandwidth, GHz-km	>1.0	>1.0	>1.200	>0.002
OH concentration, ppm	<0.1	<0.1	<0.1	<0.1
Attenuation, dB/km				
At 0.82 μm	3.0	2.5	4.0	4.0
At 1.20 μm	1.0	0.6	2.0	1.5
At 1.30 μm	3.0	0.7	3.0	1.5
At 1.50 μm	20.0	0.8	—	2.0
At 1.60 μm	—	0.3	—	2.5

Note: Cores: A, GeO_2-B_2O_3-SiO_2; B, GeO_2-P_2O_5-SiO_2; C, GeO_2-SiO_2. Claddings: D, B_2O_3-SiO_2; E, SiO_2.
Source: From [69].

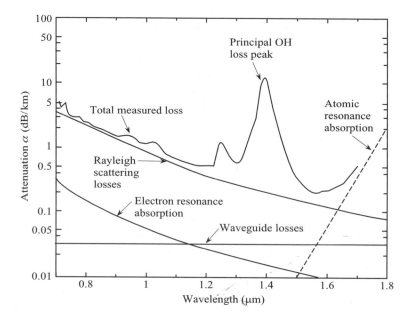

Figure 15.68 Measured attenuation characteristic with estimated contributions of Rayleigh scattering, electron and atomic resonance absorption, and waveguide losses as shown for single-mode GeO_2-SiO_2 fiber of 9.4 μm core diameter and $\Delta n = 0.0028$ [89, 135].

mode optical fibers exposed to hydrogen indicated an increase in the attenuation of 0.02 dB/km at 20°C after 25 years of operation [144].

The attenuation data given in Fig. 15.69 is ordinarily obtained in situ by means of an optical time-domain reflectometer (OTDR), which launches wide high-power light pulses of 50 ns duration into the optical fibers via a directional coupler. Since Fresnel reflection is negligible, the essentially Rayleigh backscattered light is recorded with

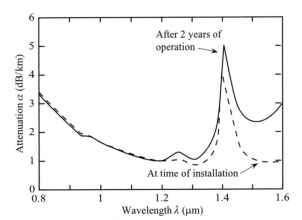

Figure 15.69 Effect of aging upon attenuation characteristics of GeO_2-P_2O_5 doped-core graded-index multimode optical fiber cable [143].

respect to the length of the fiber by means of a sensitive receiver [145, 146]. The OTDR unit is rendered more flexible in that markers superposed upon the attenuation trace may be employed to locate more precisely any attenuation faults along the optical line. The accuracy of fault location may be further augmented by comparing measurement data obtained from both ends of the fiber. The use of OTDR permitted the authors of [143] to obtain the inserted attenuation loss of each splice employed in their optical cable system and thus determine the stability with time at the operating wavelength of 1.3 μm; their results are presented in Table 15.19. The data indicate that the splice losses are between 0.10 and 0.12 dB and exhibit little change with time.

A dielectric waveguide such as an optical fiber will dissipate power by radiation when its axis becomes curved [147–151]. For a curved rectangular dielectric slab, the radiation loss at the bend or curvature of the axis is given by [148]

$$2\alpha = \frac{\Delta P}{P} \tag{15.68}$$

where ΔP is the power outflow at the bend and P is the power transmitted along the straight portion of the dielectric waveguide; the factor of 2 appears in Eq. (15.68) because α represents the amplitude loss coefficient of the guided wave. There have been a number of complex derivations of the power loss in optical fibers due to

TABLE 15.19 Attenuation of Splices Used on GeO_2-P_2O_5-SiO_2 Graded-Index Multimode Optical Fiber Cables at 1.3 μm

Splice Number	Attenuation at Time of Installation (dB)	Attenuation after 2 years of Operation (dB)	Change in Attenuation (dB)
1	0.11	0.07	−0.04
2	0.10	0.12	0.02
3	0.12	0.10	−0.02
4	0.10	0.11	0.01

Source: From [143].

bends, but the most recent one [151] indicates that the bend loss may be estimated in terms of the bend radius and the parameters of the fiber comprising the core and cladding radii, the core-cladding refractive index difference, and the refractive index of the coating. It is thus evident that any defects that render the optical fiber asymmetric will essentially result in microbends. Evidently, external mechanical forces may cause damage to the fiber and result in bends, unless the fiber is adequately protected.

Bends in the fiber will enhance the waveguide loss contribution to the overall attenuation of the fiber. Figures 15.70 and 15.71 delineate the attenuation behavior due to the macrobending or half-turns of two single-mode fibers having cut-off wavelengths of 0.85 and 0.62 μm respectively [151]; the bend loss characteristics were obtained by means of a spectrophotometer with a resolution capability of 0.01 dB using a pulsed light source [152]. The attenuation due to severe bends is seen to be significant and to become more pronounced at higher wavelengths. When strong coupling takes place between the propagating field and the backward radiating field, the observed bend loss oscillations disappear [153]. It should be emphasized that the magnitude of the bend loss is substantially influenced by the refractive index of the coating, n_{c3}, whose wavelength dependence is determined by the curing or crosslinking level of the polymer material employed [151].

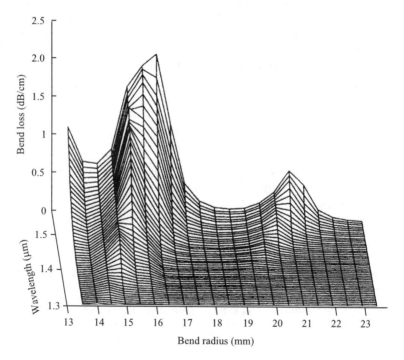

Figure 15.70 Measured attenuation of half-turn bend of single-mode fiber of 2.4-μm-diameter core of refractive-index difference, $\Delta n = 6.5$, NA $= 0.137$, $\lambda_c = 0.859\mu$m, and coating refractive index $n_{c3} = 1.5$ (after [151]).

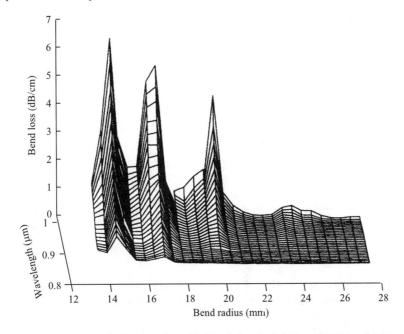

Figure 15.71 Measured attenuation of half-turn bend of single-mode fiber of 2.64-μm-diameter core of refractive-index difference $\Delta n = 2.79$, NA $= 0.09$, $\lambda_c = 0.621\mu$m, and coating refractive index $n_{c3} = 1.475$ (after [151]).

15.9.3 Construction and Design of Optical Fiber Cables

In the construction and manufacture of optical fiber cables, a pliant polymer coating is extruded over the cladded core of the individual fibers to prevent the development of microbends in the fibers during the fiber twisting and cabling operation. In this respect optical fibers differ greatly from metallic conductor communication cables, whose characteristics are substantially less influenced by cabling operations. The padding of a soft coating protects the fiber against externally applied mechanical forces. Lateral loads on the fiber due to surface roughness and compression loads during manufacture caused by constriction due to cooling of extruding compounds or from thermal contraction of the cable when subjected to low temperatures may introduce microbending faults [154]. Since the thermal coefficient of expansion for polymers is substantially larger than that for glass, the looseness of most conventional optical fiber cable designs, which aim at reducing the coupling between the various stresses within the cable, does provide some necessary protection for a fiber within a tube, which may buckle as a result of low-temperature-induced contraction.

An effective cable construction must protect the fiber against moisture permeation, not only to avert the OH$^-$ radical effects upon the attenuation characteristics of the fiber, but also because silica fibers exposed to moisture undergo static fatigue, which causes crack or fissure growth in the glass structure. Figure 15.72 indicates that the strain of silica fibers at the mechanical fracture stress decreases with the fissure size, while Fig. 15.73 further demonstrates that the time to failure of the silica fibers

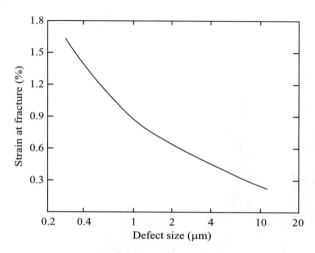

Figure 15.72 Dependence of strain value at failure stress of silica fibers upon size of defects [154].

diminishes markedly with a decrease in the strain capability of the fiber following its exposure to moisture [154]. It is thus apparent that to ensure the extended life of the fibers, stringent manufacturing quality controls are required to minimize the incidence of defects and their magnitude and to provide a dry operational environment within the cables. Since all polymers transmit moisture at various degrees depending upon their molecular structure, the use of jelly petroleum filling compounds in the interior of the cables will at best only retard the eventual ingress of moisture. A true hermetic sheath would thus require a metallic covering. However, some improvement in the static fatigue resistance of silica fibers has been achieved by the use of special fiber coatings

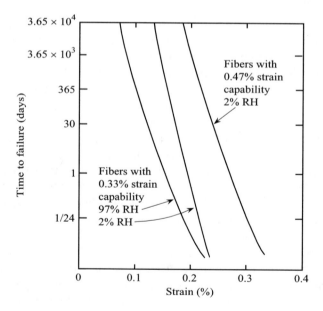

Figure 15.73 Time to failure as function of strain with strain capability of fiber and relative humidity (RH) as parameters [154].

[155, 156]. These coatings consist usually of ultraviolet cross-linkable resins followed by a thin layer of amorphous carbon that acts as a relatively impervious cover to moisture ingress. Nevertheless, it is interesting to observe that tests on complete cable systems indicate that hermetic-coated fibers exhibit somewhat higher attenuation losses; it is interesting that the splices employed in these cable systems on both hermetic- and nonhermetic-coated fibers do not reveal any particular trend (cf. Table 15.20) [156].

To ensure decoupling of the fibers from the mechanical forces, which are externally applied upon the cable during its manufacture, installation in the field, and subsequent service, pliant buffer materials must be employed in the cable construction. The individual fibers may have a tight buffering of an extruded polymer, as for example polyethylene; alternatively, a polymer buffer tube may be employed to protect a loose helically coiled fiber within the tube. The individual fibers may also be arranged in a row to form a ribbon or placed in V-shaped grooves. These four variations in design are depicted in Fig. 15.74 [157]. Dielectric polymers such as polyethylene, polyvinyl chloride, polyester, and polyamide, which have suitable thermal expansion coefficients, Young's moduli, and lateral pressure behavior, constitute the most common type of buffer materials in use.

In general terms, the optical fiber cables may be categorized into three generic designs: layered, bundled, and ribbon [154]. Typical layered and bundled cable cross sections are portrayed in Fig. 15.75. The layered design is commonly used for low-fiber-count cables ($\sim$20) and consists usually of a single layer of 6 to 8 fibers, resulting in an overall diameter of approximately 7 mm. The bundled design comprises layers of stranded optical fiber bundles. Each fiber in a bundle has an applied buffering polymer layer of 1 mm outer diameter. The individual fibers are packaged into a unit or bundle, usually consisting of between 2 and 8 fibers, and the units are then stranded together either by themselves or with strength members around a common central strength member. In certain cables, the central or some of the outer strength members may consist of copper pairs. Bundled cables with outer diameters of 15–25 mm may contain up to 40 fibers [154]. Both layered and bundled cables have the disadvantages of requiring a fiber-by-fiber splicing procedure. Although bundled cables may be designed for large numbers of fibers, they are not considered to be space efficient.

TABLE 15.20 Attenuation in Hermetically and Nonhermetically Coated Single-Mode Fiber Cable Systems

Fiber Type	Fiber Cable Length (km)	Number of Joints	Fiber Coating Type	Average Splice Loss for Joints (dB)	Optical Loss of Link for Cables (dB/km)
Dispersion			Hermetic	0.065	0.24
shifted	129.1	45 splices,	Nonhermetic	0.066	0.23
single mode		1 connector	Nonhermetic	0.072	0.23
(1.55 μm)			Nonhermetic	0.050	0.23
Normal			Hermetic	0.069	0.38
single-mode	44.9	16 splices	Nonhermetic	0.054	0.36
(1.30 μm)			Nonhermetic	0.044	0.36
			Nonhermetic	0.040	0.35

Source: From [156].

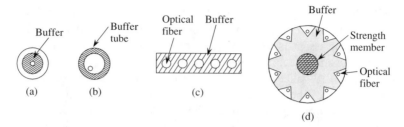

Figure 15.74 Buffering of optical fibers: (a) tight buffering of single fiber; (b) loose
buffering of single fiber; (c) buffered ribbon of 5 fibers; and (d) V-
groove array of 10 fibers [157].

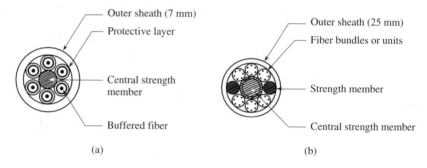

Figure 15.75 Cross sections of layered (a) and bundled (b) optical fiber cables [154].

In the so-called ribbon design, the optical fibers are packed by means of a polymer coating into ribbons containing either 5–10 fibers (European approach) or 6–12 fibers (North American approach). The diameter of the fibers is usually equal to 0.25 mm; the ribbons are covered at the top and bottom by protective tapes [154]. The individual ribbons are stacked on top of each other, with the resulting multifiber stack being situated in the center of the manufactured cable, as depicted in Fig. 15.76. [124] This leads to a compact and space-efficient design of a multifiber cable, e.g., a cable with 144 fibers is characterized by an outside diameter of only 12 mm. The ribbon-type cable is amenable to mass-splicing techniques, thereby greatly diminishing splicing inherent costs.

Frequently, composite optical fiber cables may also contain metallic conductor twisted pairs for control purposes; this practice was particularly prevalent with the earlier generation of optical fiber cables. Such a composite cable is shown in Fig. 15.77.

Additional cross-sectional designs for various long-haul optical fiber cables are presented in Chapter 17 and, consequently, will not be repeated here. For short-distance subscriber loops, the configuration of the optical fiber cables tends to approximate that of twisted copper pair cables [157]. In subscriber loop applications, each copper conductor twisted pair is generally replaced with an optical fiber, resulting in relatively large outside diameter cables, containing many optical fibers. Figure 15.78 demonstrates that the ribbon construction may accommodate more fibers per given outside cable diameter than either that of the loose buffering or V-groove constructions.

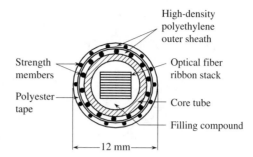

Figure 15.76 Cross section of multifiber ribbon-type cable [124].

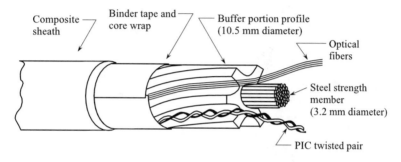

Figure 15.77 Early optical fiber cable with accommodation for single PIC twisted pair in buffer slot. (Courtesy of Northern Telecom Ltd.)

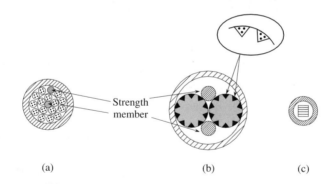

Figure 15.78 Relative outside cable diameter sizes of cables containing 60 optical fibers using (a) loose, (b) V-groove, and (c) ribbon buffer arrangements [157].

The ribbon buffer construction represents a popular design for both long- and short-distance applications. It must be emphasized that even with the ribbon-type construction there are particular configurations of ribbon stacks with which one can achieve greater reductions in the outside diameter of cables; this is illustrated in Fig. 15.79, which considers two possible designs for a ribbon-type cable, comprising a total of 1000 optical fibers. Evidently, examination of Fig. 15.79 indicates that the design with 10 fibers per ribbon leads to the smaller cable core diameter of 37 mm.

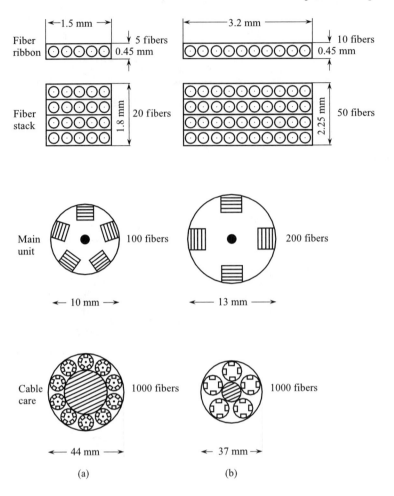

Figure 15.79 Comparison of cable core designs containing 1000 fibers using (a) 5 and
(b) 10 fibers per ribbon (after [157]).

   The overall outside jackets for optical fiber cables are generally similar to the
extruded polymer jackets, which are applied on metallic conductor telephone cables.
The optical fiber cable may contain either outer strength members, over which the
polymeric jacket is extruded or, alternatively, central strength members over which
the optical fiber bundle units may be stranded. In the latter situation, the stranding
procedure would be similar to that of twisted copper conductor pair cables where the
central member consists of a multipair unit in lieu of the strength member. However,
the external strength member construction presents the decided advantage of additional
protection to the fibers when they are subjected to bending stresses as well as greater
immunity to abrasion and damage due to cuts [158]. The binder tape or core wrap of an
optical fiber cable usually consists of polyester tapes. An overlapping aluminum sheath
may be applied under the PE jacket to provide added moisture protection to filled
optical fiber cables; however, although the filling compound and noncontinuous aluminum sheath will retard moisture ingress, they will not render the cable completely
hermetically sealed.

To guard against rodent damage, 125-μm-thick stainless steel tape armor may be applied. However, use of steel armor or aluminum sheathing may weaken the cable to lightning-induced voltage surges along the metallic covering. These surges may produce pinholes in the adjacent dielectric material layers as a result of discharges. Also the cables may be subjected to electromagnetic compression forces that may lead to crushing damage as a result of the magnetic forces created by the lightning at or in the vicinity of the cables [139].

The intentionally loose construction of the optical fiber cables imposes stringent requirements on the manufacturing process. The small amount of frictional coupling between the fibers and the core supporting structure results in differences in length between the various component parts of the cable. Since the coefficient of thermal expansion of the optical fibers is low compared to the plastic and strength components, allowances must be made for the elongation of cables. The cable construction design must attempt to minimize temperature-related length changes, which may result in axial compressive strains. A common effect observed with cables with polymeric extruded materials is that of shrinkback, which involves strain recovery phenomena. Since metals have a lower expansion coefficient than polymers, as strength members they perform a critical function in counteracting the relatively high elongation of polymeric materials employed in the optical fiber cable [154]. Also, their use in the cable provides added tensile strength because of their higher Young's modulus as compared to polymers such as Kevlar and fiberglass.

15.9.4 Fiber Cable Installation: Splices and Connectors

Installation procedures for optical fibers follow closely those practiced with metallic conductor multi-twisted-pair telephone cables. Similar to other telephone cables, optical fiber telephone cables may be directly buried, suspended aerially, placed in ducts, or installed in buildings along service passages or ducts. Optical fiber cables weigh less than equivalent copper twisted pair cables and, consequently, are easier to manipulate. This permits larger pulling lengths for the optical fiber cables, thereby reducing the number of splices required and, hence, decreasing the splice-associated costs as well as eliminating the attenuation that would be introduced by each of the omitted splices.

When optical fiber telephone cables are to be pulled into ducts by means of a rope attached to the pulling eye, which is in turn secured to the strength members of the optical cable, North American practice specifies that the tension force on the cable should not exceed 2700 N [154, 159]. However, European practice allows higher tension; for example, the permissible pulling force on a 200-fiber cable is set at 5000 N, which by comparison is twice that on a 200-twisted-pair metallic conductor cable (2500 N) [157]. In contrast, a 500-twisted-pair cable with copper conductor diameters of 0.4 mm, whose outside diameter of 46 mm and weight exceed that of a 500-optical-fiber cable of 33 mm outside diameter, has a permissible tensile force of 5000 N. The pulling forces on the optical cable may vary greatly depending upon the configuration of the ducts. For instance, pulling forces as low as 450 N were observed over a 1-km distance along a straight duct as compared to 2700 N over a relatively short but more tortuous stretch of 200 m [157]. Large tension forces encountered over short distances are frequently the result of snaking or twists in the duct geometry. The radii of twists in the ducts should not be smaller than 20 times that of the external diameter of the optical fiber telephone cable, which corresponds to the value of the minimum permissible bending radius. Careful monitoring of the tensile force by means of various tension-sensitive devices such as dynamometers, line tensiometers, or remote sensing pullers will assist in averting axial tensile stress damage to the optical fiber cable.

Optical fiber telephone cables are often installed into existing twisted copper wire pair telephone cable ducts, with diameters of ∼100 mm. This is accomplished by inserting several 25-mm-diameter subducts in the larger duct in order to enhance the versatility of the existing duct system [159]. Thus subsequent installations or removals of optical fiber cables may be carried out without disturbing the remaining in situ cables.

The relatively longer lengths of optical fiber telephone cables require additional precaution to obviate the development of kinks during their installation process. It is common practice to employ a procedure normally termed center pulling, whereby the cable reel is transported to some intermediate point along the projected cable route at a position, preferably half of the total cable length, where a section of the cable is pulled into the duct [159]. The remaining portion of the cable is subsequently unreeled and arranged in a figure 8 and thence pulled into the oppositely facing duct of the cable route. Continuous cable lengths of 3 km may be readily handled with this procedure.

In the 25-mm-diameter conduits of high-density polyethylene (HDPE) into which long optical fiber telephone cables must be pulled, any marked tortuosity in the polymeric pipe or duct will present appreciable surface resistance to the entering cable. If this situation arises, great pulling forces will be required, with the result that capstans may be needed approximately every 200 m. In such circumstances use may be made of the viscous flow air principle, involving the deployment of a simple air compressor to inject the optical fiber cable along with a strong air flow into the ducts and allow the viscous air stream to carry the cable along until it emerges at the other extreme end of the duct [160]. This principle of operation is similar to that commonly employed to induce small flexible individual optical fibers to flow into conduits of empty cables installed in buildings [161]. For the cable blowing operation a compressor of 8 bars capacity may be sufficiently adequate to blow in an 800-m-long cable into a HDPE conduit at a speed of 60 m/min; lengths up to 2 km may be installed in conduits using several of these compressors in series. A certain degree of stiffness in the optical fiber cables, rather than being detrimental, is found to be advantageous. Oils may be used to reduce the frictional resistance along the interior of the polymeric conduits.

Optical fiber telephone cables may also be directly buried. Their installation may be carried out using trenching or by means of a cable-laying plow, with the cable being directed over a deflector into the conduit of the plow blade. It is common to use standard metallic conductor cable-laying equipment whereby several of the appreciably lighter optical fiber cables may be placed simultaneously down to a depth of 1.5 m [159]. Protection against rodent attack would require a metallic armor protection, having a thickness greater than 125 μm.

Two main accessories that must be used in conjunction with optical fiber cables are those of connectors and splices. The former is employed in cases where the optical fibers must be periodically connected and disconnected, i.e., a connector must function as a demountable devise, whereas splices entail a permanent connection of the fibers in cases where the continuous lengths of the cables are insufficient to span the entire distance between the transmitting and receiving points and where certain practical limitations prevent the installation of a single cable of sufficient length.

The splice is essentially a device with application in the field, and we have already in previous discussions noted that splices connecting optical fibers may add substantial attenuation losses to the overall optical fiber link. A low-loss connection requires accurate alignment of the two fiber cores to be joined. Losses in the contact may arise from a mismatch in the type of the fibers joined, i.e., differences in the core diameter, core-cladding eccentricity, numerical aperture NA, and refractive index profile [162]. Additional contact losses will be introduced due to extrinsic effects such as core axial tilt, Fresnel reflections, fiber end angle, separation of core ends, and transverse offset between the adjoining fiber cores. Figure 15.80 illustrates the effect of fiber misalignments due to end separation, axial tilt, and transverse offset of a multimode fiber, with measurements carried out using an LED source [162, 163]. It can be perceived that the attenuation loss is most susceptible to the transverse offset. A mismatch in the fiber characteristics such as the NA, the cores, and refractive-index profiles could also produce large losses, e.g., transmission from a high NA of 0.18 to a low NA of 0.16 could result in a 1-dB loss [162].

Splicing of both single-mode and multimode fibers is normally carried out by means of either fusion or adhesive bonding into an alignment member [162]. The latter is a mechanical splicing method whereby a number of fibers (usually 12) are inserted between precision-etched silicon chips to form a well-defined linear array [158, 164], as depicted in Fig. 15.81. Bridging chips with spring-loaded chips ensure a retainable alignment of the adjoining fibers, whose ends are prepared by grinding followed by polishing. In the field where the stringent precision procedure implemented in the factory cannot be employed, the refractive indices of the clipped-together arrays are matched and the attenuation of each optical fiber joint is verified by means of an OTDR. Mechanical mass splicing may also be performed using a grooved cylindrical PE structure.

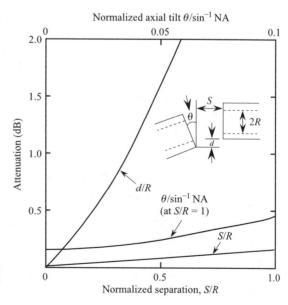

Figure 15.80 Influence of misalignment of two adjoining multimode fibers upon attenuation of joint [162, 163].

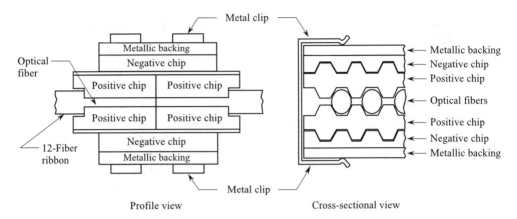

Figure 15.81 Mechanical mass splicing technique employing precision-etched silicon chips (after [158, 164]).

It is to be emphasized that irrespective of the splicing procedure followed, the principal splicing steps involve the removal of the primary coating and cutting of the fiber, preparing the fiber by scoring and breaking, then grinding and polishing, aligning the fibers and ensuring the retention of the alignment [157]. Particular attention must be given to the protection of fusion splices by creating a stress-free environment [139, 165].

Fusion splicing techniques are more suitable for the splicing of single fibers and are, consequently, more labor intensive. They are thus more popular with cables having a smaller number of fibers, such as interoffice trunk cables. The fusion splicing technique has the advantage in that it ensures a splice of the same dimensions as the original fiber, with the fusion time and temperature suitably adjusted to minimize any core deformation of the fiber [158]. Self-alignment of the fibers is assisted by the surface tension force associated with the molten fiber ends at the arc fusion point. In the fusion technique, the two fibers to be spliced are aligned and clamped in a V-groove, and a physical contact is made between them before the creation of the arc (cf. Fig. 15.82). Ribbons containing a number of fibers may also be spliced, using the fusion technique, with the proviso that great care is exercised to align accurately the V-groove array. Attenuation values for fusion ribbon splice joints as low as 0.045 dB have been achieved; this compares to a value of 0.034 dB for single fiber-to-fiber splice joints [157]. It is noteworthy that, already in 1983, well-controlled laboratory experiments had demonstrated that with single-fiber splicing techniques, employing identical fibers, attenuation values as low as 0.02 dB were attainable.

The attenuation of the splice is greatly contingent upon the outer diameter tolerances and core/cladding eccentricity. If the former is taken as $\pm 1\,\mu$m and the latter as $\leq 0.5\,\mu$m, the attenuation of fibers spliced in a ribbon type construction may be anticipated to be in the order of 0.1 dB [157, 166]. Splicing requirements for single-mode fibers are more stringent than for multimode fibers, because accurate core centering demands greater care with a reduction of core diameter. Figure 15.83 demonstrates that attenuation losses for monomode fibers increase very markedly with offset; the effect is

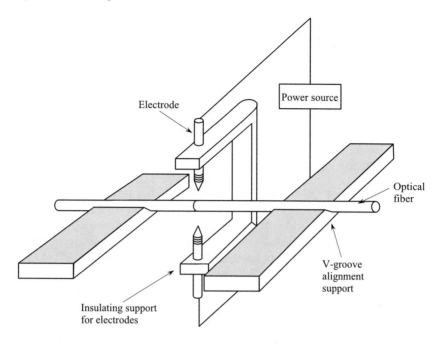

Figure 15.82 Fusion splicing system [158].

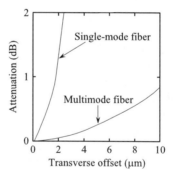

Figure 15.83 Influence of transverse offset upon attenuation of joints of spliced multi-mode and single-mode fibers (after [158]).

seen to be much less pronounced in multimode fibers [158]. In addition, it should be observed that optical glass fibers whose ends are clamped in joints or terminations may experience lateral offset effects when soldered into ferrules or epoxy bonded into capillaries [167]. The stresses associated with these offsets can lead to mechanical instability and degrade the optical performance of the splices.

The accuracy of single-mode optical fiber alignment may be simply verified by measuring the attenuation loss of a splice in situ. Power is injected into the fiber at a point preceding the splice by coupling the light input via a 4-mm-radius bend of the fiber; the received signal intensity at another light coupling point after the splice, where the fiber is again bent to a radius of 4 mm as portrayed in Fig. 15.84, may then be compared to that at the light injection point. Optimum alignment of the fibers corresponds to the situation where the adjustment of the axial center of one core with respect

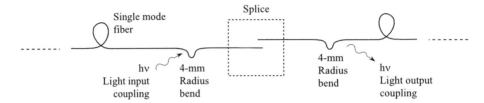

Figure 15.84 Splice alignment technique employing local power injection and detection in single-mode optical fiber [168].

to the other core of the single-mode fiber results in a minimum difference between the input and output powers at the two respective light signal injection and detection points. The highly accurate alignment requirement of monomode optical glass fibers may be circumvented in the case of short-distance applications on building premises by the use of large-core graded-index optical polymeric fibers [169, 170]. These polymeric fibers, which are of the poly-methyl-methacrylate (PMMA) type, are capable of 3 GHz bandwidths and may be used in conjunction with inexpensive injection-molded plastic connectors.

In comparison to splices, connectors have the added constraint in that they must be capable of establishing repeatedly low-loss contacts between optical fibers each time that a reconnection is made, usually at a receiving or transmitting equipment point. There have been many designs developed for both multimode and monomode optical fibers. A common design for multimode fibers interposes a ferrule or plug structure between the alignment mechanism and the optical fiber, with accurate centering of the fiber within well-controlled ferrule dimensions [162]. Figure 15.85 illustrates a filled-epoxy molded plug connector that is transfer molded upon an optical fiber in which the core concentricity with respect to the tapered end surface of the plug can be maintained within a few micrometers. As indicated in the figure, a biconical molded center sleeve is utilized to align the plugs. With refractive index matching on a 55-μm-core graded-index multimode fiber, an attenuation value of 0.4 dB is achieved; in the absence of index matching, this value may be degraded to as high as 0.7 dB [171].

Single-mode fiber connectors using core-centered ferrules were introduced in the 1980s [172]. The light signal emerging from the end of the core was employed to serve as a reference to achieve concentric cuts of the ferrule's cylindrical surface, which were

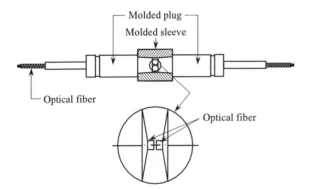

Figure 15.85 Epoxy molded plug connector with biconical molded center sleeve [162, 171].

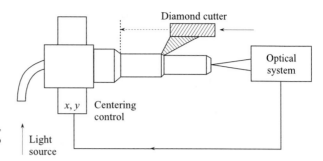

Figure 15.86 Core-centered connector ferrule, with light emission from fiber end used to achieve concentric cut (after [173]).

polished to assume a convex form in order to provide direct contact between the end of the two adjoining fibers (cf. Fig. 15.86). This procedure precluded any need for refractive index matching [172]. Connectors prepared with core-centered single-mode ferrules have low attenuation values as substantiated by the data in Fig. 15.87. Temperature changes between −20 and 70°C are found to result in attenuation changes of only 0.05 dB. There are now a significant number of connectors available commercially. In North America, the AT&T (ST contacting type) connector has a twist-on locking device similar to those employed on coaxial cables; its attenuation loss is typically below 1 dB.

For large-capacity switching systems and computer interconnections, optical multifiber connectors are employed to meet the optical parallel line transmission requirements. For this purpose, 12-optical-fiber ribbon connectors are commonly utilized and double 12-fiber (24-fiber) connectors have been recently proposed [174, 175]. A thermosetting epoxy resin is employed in a transfer-molding procedure. V-grooves cut in the metallic mold permit precise positioning of the insert pins, which form the pin holes of the connector guide and the holes for the optical fibers. The V-grooves are machined in both the upper and lower molds, as depicted in Fig. 15.88. Note that the insert pins of the two guide pin holes are determined by V-grooves of the upper and lower parts of the mold. The ferrule eccentricity error, i.e., the offset of the hole position of the optical fiber, is approximately 3 μm. The 24-fiber connector is designed for 1.3 μm wavelength multimode fibers of 50 μm core diameter. The application of the 24-fiber connector to single-mode fibers would require improvement in the eccentricity error. Figure 15.89 represents the attenuation distribution characteristic of the

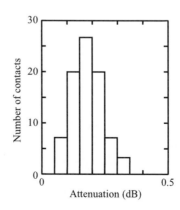

Figure 15.87 Histogram of mean attenuation values at 1.3μm for 10 insertions per pair for connector employing core-centered ferrules (after [172]).

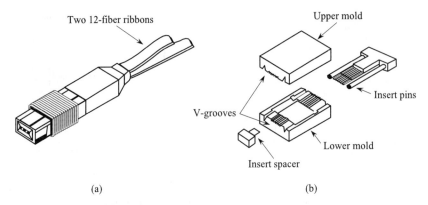

Two 12-fiber ribbons

Upper mold

V-grooves

Insert pins

Insert spacer

Lower mold

(a)   (b)

Figure 15.88 (a) A 24-fiber multimode optical connector with (b) associated transfer mold (after [174]).

multimode connector. Comparison of Figs. 15.87 and 15.89 would suggest that multifiber connectors have poorer contact attenuation performance than the connectors utilized with single fibers.

Frequently it is required to couple a fraction of the light signal out of or into an existing fiber transmission line. This may be accomplished by means of passive taps [176, 177]. Two adjacent fibers, which are fused together, constitute a passive tap (cf. Fig. 15.90). Their application is illustrated schematically in Fig. 15.91, with the added transmission and reception points as indicated. A 4×4 star coupler, portrayed in Fig. 15.92, may be constructed by a concatenation of four identical four-port couplers of the type depicted in Fig. 15.91, with the coupling ratios adjusted to the desired values (usually $1:1$) at each fused joint. It is also interesting to note that an all-fiber beam splitter arrangement may be employed to partition the transmitted beams of light of two adjacent fibers, as portrayed in Fig. 15.93.

The reliability of installed optical fiber cables must be ensured by providing rapid fiber break fault location and repair methods. Optical time-domain reflectometry (OTDR) techniques constitute reliable break location indicators, but procedures utilizing them suffer because of their limited dynamic range [178–180]. An automated technique based on bit-counting data obtained at the two ends of an optical fiber line appears to have shown promise as an effective break locator in cases where the breaks

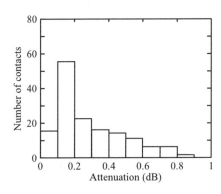

Figure 15.89 Attenuation characteristic of 24-fiber multimode optical connector (after [174]).

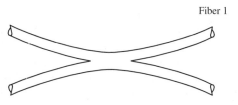

Figure 15.90 A fused fiber coupler.

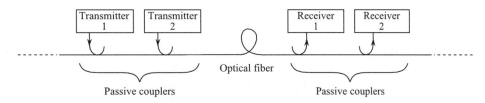

Figure 15.91 Passive couplers along optical fiber transmission line [176].

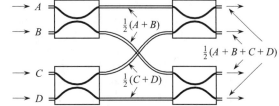

Figure 15.92 A 4×4 star coupler formed by concatenation of four four-port couplers (after [173]).

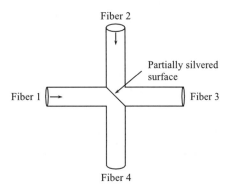

Figure 15.93 Fiber beam splitter arrangement (after [176]).

in the fiber develop rapidly [179, 180]. It has been demonstrated that the accuracy with which breaks in the optical fiber may be located is contingent upon the decay time of the transmitted light pulse following fiber failure [180]. The method is principally intended for failures caused by backhoe digging accidents, whereby the unduly high pressure forces exerted upon the buried optical cables may induce the optical fibers to fail suddenly due to tensile fracture [181]. The proposed bit-counting method is not suitable for detecting failures in fibers due to slow rate degradation processes arising from temperature fluctuations, water ingress, and splice aging. If τ is defined as the time required for the transmitted light to decay after failure has taken place and v is the group velocity of the light in the fiber, ($\sim 2 \times 10^5$ km/s), then the measure of uncertainty, Δx, in the break position along the fiber is given by [180]

$$\Delta x = \tau v \tag{15.69}$$

It is seen from the nature of Eq. (15.69) that the longer the decay time of the light intensity in comparison to a bit period following a break-type failure of the fiber, the poorer the resolution accuracy becomes with which the failure site may be located.

Rapid failure breakage mechanisms may be characterized into three distinct categories [180]. The first case concerns severe bending of fibers to less than a radius of 10 mm such that the core no longer is capable of guiding the light, which then couples into the cladding and, as a consequence, undergoes pronounced attenuation. In this case the decay time is directly related to the time required to bend the fiber through a critical radius. A relatively slow bending process results in a long value of τ. For example, should $\tau = 3\,\mu$s, then the break fault in the fiber could only be located with a resolution of 600 m. In the second possible scenario the fiber may fracture under stress suddenly before any appreciable bending can occur; in such circumstances τ may fail to exhibit any appreciable dependence upon stress. Such behavior has been found to be characteristic of single-mode fibers, which fail with breaking strengths above 0.7×10^9 N/m^2 (1×10^5 lb/in.^{2}) [180]. Figure 15.94 shows a typical light intensity–decay time curve of a coated single-mode fiber that undergoes rapid fracture when subjected to a dynamic tension of 2.7×10^9 N/m^2. The 90 to 10% decay time is seen to be 14 ns, which in terms of Eq. (15.69) indicates a fault location length resolution limit of approximately 3 m.

The third category of failures involves fiber breaks at low stress levels at which the sheared fiber ends remain in close contact with each other [180]. At the two severed fiber end interfaces, transmission losses arise due to Fresnel reflections; also, a major transmission power loss occurs as the two fiber sections develop a tilt with respect to each

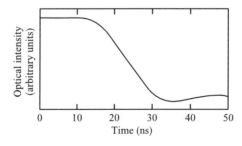

Figure 15.94 Light intensity decay characteristic of single-mode fiber of 8.3 μm cladding diameter, with rapid fracture occurring at dynamic tension of 2.7×10^9 N/m^2, with strain rate of 51 cm/min and diode laser source with wavelength of 1.3 μm (after [180].

other. The critical tilt angle θ_e at which the transmitted power decreases by a factor of $1/e$ is given by [180]

$$\theta_e = \lambda n w \qquad (15.70)$$

where λ is the wavelength of the laser source, n the index of refraction, and w the width parameter of the Gaussian field. If λ, n and w are taken as 1.3 μm, 1.45, and 4.75 μm, respectively, then Eq. (15.70) leads to a tilt angle value slightly in excess of 3°, which is sufficient to give rise to large transmission losses. As the separation of the fiber ends increases, the magnitude of the applied stress levels diminish with an attending increase of the decay time, τ. The foregoing behavior is observed to occur with single-mode fibers with breaking strengths less than 0.7×10^9 N/m² (1×10^5 lb/in.²); Fig. 15.95 depicts the decay time characteristic of a fused splice of a single-mode fiber that underwent fracture at a dynamic tension of 0.62×10^9 N/m² [180]. Fast ($\sim$25 ns) and slow ($\sim$140 ns) time constants are perceived to describe the decay of the measured light intensity. The former, which accounts for 60% of the amplitude decrement, is ascribed to the initial tilting displacement of the splice ends, whereas the latter is believed to be associated with the subsequent translational separation of the two fractured ends of the splice. It is fortuitous that in this particular case the fast time constant permits the location of the fault to within 5 m, so that the presence of the long-time delay poses no practical consequences. However, fractures occurring at dynamic tensions substantially less than 0.6×10^9 N/m² yield Δx values that are much too large to permit effective location of the fracture faults. Hence, in practice, fracture faults in optical fiber cables can only be accurately located using the light intensity decay method when the time constant describing the decay is substantially less than 100 ns.

15.9.5 Optical Fiber Transmission Systems

It is beyond the scope of this book to attempt any detailed treatment on the subject of optical fiber transmission systems. However, insofar as the transmission capabilities of optical fiber cables are greatly influenced by the characteristics of the active components, such as sources and transmitters, detectors and receivers, and amplification devices, all of which form an integral part of the overall transmission system, the subject matter must be at least addressed in a very cursory and descriptive manner.

In any transmission system, one of the main objectives is to deliver power from the source with the least possible loss. If P_{in} represents the power coupled from the source

Figure 15.95 Light intensity decay characteristic of splice of single-mode fiber of 8.3 μm core and 125 μm cladding diameters undergoing fracture at a dynamic tension of 0.62×10^9 N/m² under strain rate of 13 cm/min and diode laser source wavelength of 1.3 μm (after [180]).

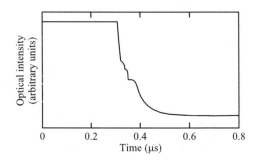

into the fiber, then the attenuation, α (in decibels per kilometer) of the optical fiber may be expressed as

$$\alpha = \frac{10 \ln(P_{in}/P_{out})}{\ell} \tag{15.71}$$

where P_{out} denotes the output power at the far end of the optical fiber, whose length ℓ is in kilometers. Equation (15.71) may also be written as

$$\frac{P_{out}}{P_{in}} = \exp\left(-\frac{\alpha\ell}{10}\right) \tag{15.72}$$

The overall power transmission capacity is thus contingent upon the attenuation of the optical fiber line as well as the efficacy of coupling between the source and the input to the optical fiber line. The light source is coupled to the input of the fiber by either placing the prepared optically flat surface of the fiber end against the surface of the light-emitting source or by means of a lens system. Assuming that the source emits light radiation from a surface A_s with an intensity of I_b watts per steradian-area within a solid angle Ω_S, the power coupled into the fiber without the use of a lens is given by [95]

$$P_{in} = \begin{cases} I_b\Omega_p A_{in} & \text{for } A_{in} < A_s \\ I_b\Omega_p A_s & \text{for } A_s < A_{in} \end{cases} \tag{15.73}$$

where Ω_p is related to the critical angle θ_p (i.e., $\Omega_p \simeq \frac{1}{4}\pi\theta_p^2$) of the fiber and A_{in} equals the surface area of the fiber. Since the total source power P_s, is defined as

$$P_s = I_b\Omega_s A_s \tag{15.74}$$

then the coupling power loss into the fiber becomes

$$\frac{P_{in}}{P_s} = \begin{cases} \dfrac{\Omega_p}{\Omega_s} \dfrac{A_{in}}{A_s} & \text{for } A_{in} < A_s \\[2ex] \dfrac{\Omega_p}{\Omega_s} & \text{for } A_s < A_{in} \end{cases} \tag{15.75}$$

It is thus evident that the coupling power loss into the fiber will increase with a decrease in the core diameter or area of the fiber and an increase in the solid angle of the light radiation angle of the source. Consequently, the coupling power loss into single-mode fibers having small-core diameters will be particularly high unless a lens is employed to focus the light power beam into the fiber core. Some degree of focusing may be achieved also with refractive index matching fluids [95].

As has been pointed out previously, the spectrum width size of the light source influences adversely the bandwidth or the transmission capacity of an optical fiber cable. The spectrum width of light emitting diode (LED) sources is usually on the order of $\sim$50 nm, while that of laser diodes (LD) is appreciably less. The spectrum of Fabry-Perot-type laser diodes consists of a number of lines over the wavelength scale with an effective width of approximately 10 nm; in contrast, the distributed feedback

(DFB) laser diode provides a single line spectrum with a spectral width of 0.5 nm. Both LED and LD sources are forward-biased devices in that they emit light radiation as a result of electron and hole recombination at the *p-n* junction. Intelligence and data are transmitted by directly modulating the output of the light-emitting sources, with the modulation being produced by variation of the current in a suitable drive circuit.

The LEDs are particularly suitable for use as light-emitting sources with multi-mode fiber cables that have been designed to carry data rates up to 150 Mbits/s over short transmission distances or repeater intervals of 20 km or less [18]. They are characterized by high-performance reliability and do not require output stabilization, because their power output exhibits negligible variation with temperature. The so-called Burrus-type LED structure, depicted in Fig. 15.96, has been commercialized and employed for operation at 0.85 and 1.3 μm [176]. The Burrus-type LED has an etched well to permit butt coupling to optical fibers and is characterized by a small area or dot of contact. Heterojunctions are formed with different compositions of $Ga_sAl_{1-x}As$, thereby creating different energy band levels that form potential barriers, which confine the injected electrons and holes within a thin active layer at the *p–n* junction.

As the electrons and holes combine, photons are emitted because of a difference in the energy level of the valence (holes) and the conduction (electrons) bands. The wavelength of the emitted light radiation is inversely proportional to the difference in the energy level E and is given by [183]

$$\lambda = \frac{ch}{E} \tag{15.76}$$

where c is the velocity of high and h represents Planck's constant. The recombination process of electrons and holes takes place in the p layer in proximity to the *p-n* junction. The purpose of the thin transparent layer above the activated layer prevents readsorption of light before it reaches the optical fiber, thereby improving coupling efficiency. The metallic dot contact at the bottom of the structure provides a small emitting area. The LEDs normally yield output powers in the range from 50 to 100 μW, though the

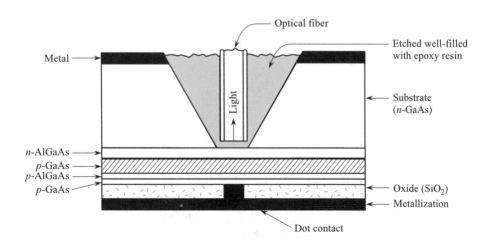

Figure 15.96 Burrus-type LED source [176, 177]

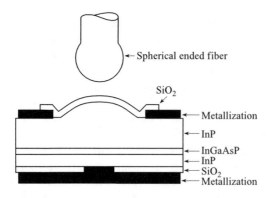

Figure 15.97 Arrangement using monolithic lens for coupling to spherically ended fiber from surface emitting InGaAsP/InP LED source [182, 185, 186].

exact power that may be launched into a multimode fibre will evidently depend upon the core size and numerical aperture [182].

The Burrus-type LED design has also been implemented with InGaAsP hetero-structures [182, 184]. Figure 15.97 shows such application wherein a monolithic lens is utilized on an InP substrate in conjunction with a spherical-ended optical fiber to achieve improved coupling [185]. Surface-emitting InGaAsP LEDs have a bandwidth of 120 nm and have been found to be highly reliable light sources for operation at 0.85 1.3, and 1.55 μm, with a projected service life expectancy of 10^9 h.

Gallium arsenide and aluminum arsenide laser diodes are extensively used for transmission at 0.85 μm over multimode fibers. For transmission at 1.3 μm, where silica fibers exhibit a minimum in dispersion, as well as for applications over the minimum of attenuation at 1.5 μm, InGaAsP laser diodes are employed. As in the case of LEDs, use is made of heterojunctions to confine the injected electrons and holes to the active layer. This is schematically illustrated in Fig. 15.98 [176, 187]. The resonant cavity of the laser is formed by the three directions of confinement. When the injected current density in the active layer exceeds a certain threshold level, the cavity commences to resonate. The thickness of the active layer must thus be sufficiently large to prevent losses in the p and n layers adjacent to the junction. Figures 15.99 and 15.100 compare the power launched by LED and LD sources into a multimode fiber having a 50-μm core and an NA of 0.2; evidently, the transmitted power with the LD source is much higher, though it becomes only a linear function of the drive current at much higher values of the latter.

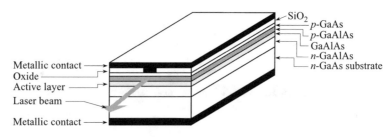

Figure 15.98 Heterojunction LD source [176, 187].

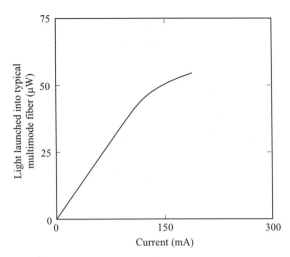

Figure 15.99 Power output of LED source vs. drive current (after [176]).

In an optical fiber transmission system, a detector converts the received optical signal into an electrical current. Photodiodes of the *p-i-n* type and avalanche photodiodes are the common detectors employed for this purpose. For every incident or absorbed photon, a *p-i-n* photodiode generates a single electron-hole pair with a maximum conversion ratio of 0.8 A/W at the wavelength of 0.8 μm [176]. In this process the quantum efficiency of the *p-i-n* photodiode is necessarily less than unity, because some photons that are generated may become reflected or the recombination of electrons and holes may not always result in a capacitive or displacement current. In comparison, the avalanche photodiode (APD) detector produces many more electron-hole pairs per photon impact, because its operation is based on ionization impact collision phenomena that produce charge carrier multiplication in the form of charge carrier avalanches under externally applied high voltages. Hence, a single photon impact may readily give rise to 100 electron-hole pairs; however, this random avalanche process does generate also some noise, which imposes an inherent sensitivity limit on avalanche photo diodes. For 0.8-μm-wavelength transmission applications, silicon-based devices are principally employed because of their high quantum efficiency,

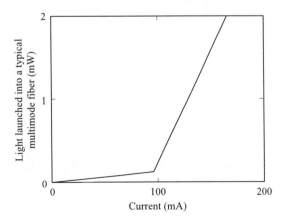

Figure 15.100 Power output of LD source vs. drive current (after [176]).

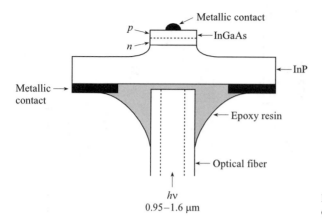

Figure 15.101 An InGaAs/InP *p-i-n* photodiode detector [182, 190].

rapid response of <1 ns, and low capacitance; whereas for transmission at longer wavelengths, Ge-, InGaAs-, and HgCdTe-based devices have higher quantum efficiencies [176].

GeAPDs have a gain-bandwidth product of approximately 60 GHz, but their dark current is relatively high (0.1–1 μA); however, their quantum efficiency is 0.8 at 1.3 μm [182, 188, 189]. An InGaAs *p-i-n* photodiode, having a quantum efficiency of 0.7 at 1.3 μm and dark current of 2–5 nA, is depicted in Fig. 15.101 [182, 190]. It is characterized by a junction capacitance of 0.5 pF at a bias of −10 V.

In order to reduce the dark current due to tunneling in avalanche photodiodes, a narrow-bandgap layer of InGaAs is utilized for absorption of the light radiation and a wide-bandgap layer of InP is employed as well for charge pair multiplication. An avalanche photodiode detector construction, which exhibits an intrinsically low dark current of 3 nA, is portrayed in Fig. 15.102.

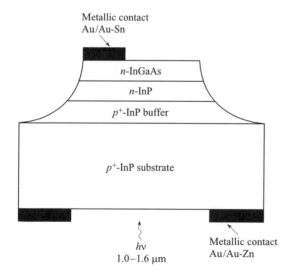

Figure 15.102 An InGaAs/InP avalanche photodiode detector [182, 191].

The transmission of optical signals over optical fiber cables is generally carried out using the techniques of time division multiplexing (TDM), code division multiplexing (CDM), or wavelength division multiplexing (WDM) [95]. As we have seen from our earlier considerations of copper wire cables, in the TDM technique each source is assigned a specific time interval during which it is permitted to transmit the information, which is in turn retrieved by a receiver with suitable time gating to the intervals of the particular source. As has been pointed out previously, since TDM is a synchronous technique, the respective transmitters and receivers require alignment between their given time intervals. In analogy to the TDM technique, the CDM technique necessitates that both sources and receivers be assigned a particular pulse code sequence. In contrast, however, the asynchronous nature of the CDM technique requires the receiver to decode the transmitted information by means of suitable correlation procedures. As the name of the WDM technique implies, the transmission is carried out over unique wavelength bands, such that the laser source must transmit with a modulation bandwidth commensurate with the specifically assigned wavelength. At the receiving end, optical filtering procedures are employed to retrieve the information transmitted at the respective wavelengths. Evidently, the WDM technique bears direct resemblance to the frequency division multiplexing (FDM) employed in the early analog transmission systems over twisted-pair copper wire telephone cables. In WDM as in FDM, the sources operate independently of each other so that the transmission is synchronous.

Optical fiber cables generally function as a one-way transmission medium. Most of the present optical fiber systems in operation are of the synchronous type and can be schematically represented as in Fig. 15.103 [176]. The signal source consists of the intelligence and data to be transmitted and a clock circuit; the emerging signal is coded to produce a new signal; i.e., the coder encodes n input bits into m output bits. The purpose of the driver circuit is to convert the coded signals from the coder into currents, which in turn modulate the output of the LED or LD optical source.

The signals from the optical sources are transmitted over the optical fiber line to the optical receiver unit, whose schematic circuit diagram is delineated in Fig. 15.104 [176]. The amplified and filtered output from the photodiode detector is applied to the phase locked loop of a clock recovery circuit (c.f. Fig. 15.103), whose output is employed to regenerate the coded binary signal. The decoder receives the regenerated signal and recovered clock information and reproduces in turn the clock and data signals representative of the original signal inputs to the optical transmission system.

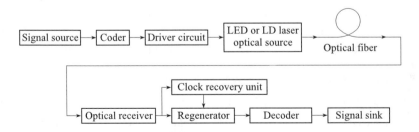

Figure 15.103 Schematic diagram of synchronous digital optical fiber point-to-point transmission system (after [176]).

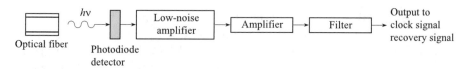

Figure 15.104 Schematic circuit diagram of optical receiver circuit (after [176]).

   With the development of optical amplifiers, WDM provides the possibility of achieving enormous transmission bandwidths of the order of ~3 THz; in the near future, data rates already on the order of 100 Gbits/s should be readily attainable. The current commercial WDM system in operation at the present has a transmission capacity of ~40 Gbits/s, which is substantially above that of the various optical fiber systems installed over the last 17 years that employ TDM in the electronic domain [66] (*cf.* Table 15.21). In analogy with FDM, the WDM scheme may be represented schematically as portrayed in Fig. 15.105. Here a number of single frequency or wavelength-emitting diode lasers (e.g., distributed feedback lasers) are coupled into a single wide-band optical fiber line. A spacing of 50–500 GHz (0.4–4 nm) is maintained between the respective input channels to ensure well-defined channel separation and to circumvent wavelength drift effects such as those caused by temperature changes in optoelectronic components [66].

   A passive star coupler is employed to distribute the multiwave signal transmitted via a wide-band optical fiber to different signal receiver facilities, where tunable optical filters are utilized to retrieve the single wavelength associated with the respective transmission channel. It should be pointed out that in order to attain the full potential of WDM to transmit at terabit-per-second speeds, the latter must be utilized in conjunction with TDM to time multiplex many of the signals either optically or electronically.

   In WDM, optical filters play an important role in determining the characteristics of the overall WDM system design. A design for a single-mode fiber multiplexer/demultiplexer using the dielectric multilayer film approach is depicted in Fig. 15.106. In this lensless filter device, a dielectric film filter is sandwiched between oblique fiber ends [172, 182]; here the light reflected by the filter is captured by another filter situated in close proximity to the joint. Although the depicted filter design has been employed for operation specifically at 1.285, 1.355, 1.480, and 1.560 μm, cascading of the filter elements may yield devices suitable for many more wavelengths [192]; however, the design is less suitable for multi-WDM channel systems with highly reduced separation in the respective wavelengths [172].

   Lensless filter designs result in compact devices with a reduced number of parts, thus rendering them more immune to temperature effects and mechanically induced vibrations. However, ball lens ferrules are frequently employed in conjunction with

TABLE 15.21 Approximate Transmission Capacities of Commercial Optical Fiber Systems [66]

Year of installation	1980	1983	1985	1986	1987	1991	1995	1996
Transmission mode	TDM	TDM	TDM	TDM	TDM	TDM	WDM	TDM
Transmission capacity (Gb/s)	0.045	0.095	0.17	0.46	1.8	2.5	40	10

Source: From [66].

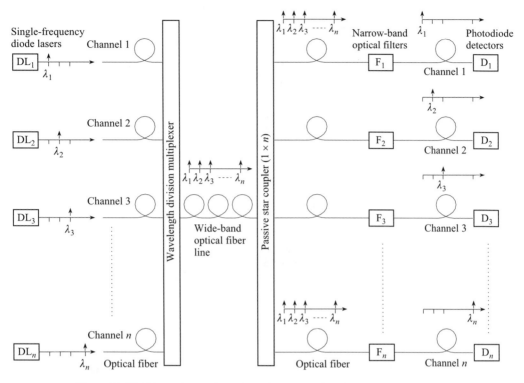

Figure 15.105 Schematic representation of WDM optical fiber transmission system (after [66]).

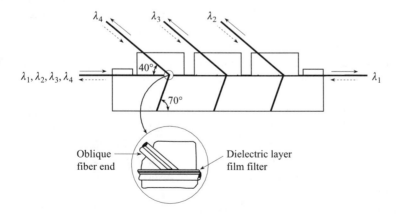

Figure 15.106 Sandwiched dielectric layer WDM multiplexer-demultiplexer design for operation at four different wavelengths [172, 192].

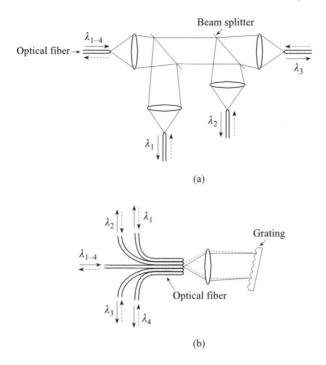

Figure 15.107 Wavelength filtering by means of lenses in conjunction with (a) beam splitters and (b) lenses with gratings (after [173]).

dispersive elements inserted between the lenses to separate the wavelengths. Figure 15.107 illustrates the use of lenses in conjunction with beam splitters and gratings to achieve separation of different wavelengths [173]. Other means available for filtering purposes include wavelength-sensitive directional couplers and Fabry-Perot etalons. The latter are particularly applicable to systems with single-wavelength laser sources, because of their high wavelength selectivity response. A Fabry-Perot etalon filter consists essentially of a wavelength-selective resonant cavity formed by two mirrors whose separation may be adjusted to tune the filter and thereby alter the wavelengths of resonance within the cavity. These filters are characterized by subnanometer pass bands and are tunable in times of the order of a few milliseconds [66].

Optical amplifiers are particularly suited to WDM transmission systems, since the optical signal may be directly amplified as opposed to electronically operated repeaters. The electronic repeaters must convert the optical signal to an electrical signal before they can retime, reshape, and amplify the received signal. This electronically readjusted and distortion corrected signal is then applied to a diode laser, which emits an optical signal identical in form and intensity to that at the input end of the optical fiber. In contrast to the rather intricate and expensive electronic repeater units, optical amplifiers are relatively simple devices: they essentially consist of a short length of silica glass fiber doped with erbium metal ions [193–195].

The operational principle of the optical amplifiers can be readily understood by considering a simplified energy diagram of the erbium-doped silica fiber, depicted in Fig. 15.108 [66]. When the optical fiber is illuminated with a pump laser at 0.98 and

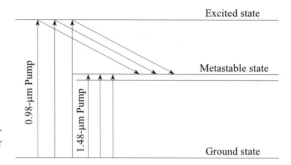

Figure 15.108 Energy diagram of erbium-doped optical fiber amplifying medium (after [66]).

1.48 μm wavelength, the erbium ions are excited into a higher energy level. The erbium ions are pumped to a higher energy state by means of an infrared red beam, which is coupled along with the light signal beam into the erbium-doped optical fiber via a three-part wavelength-selective coupler. The erbium ions are raised to an excited state by the 0.98-μm pump, from which they decay rapidly to a metastable state to which other erbium ions have been raised by the 1.48-μm pump. If undisturbed, the erbium ions will spontaneously decay to the ground state in times ranging from a few microseconds to milliseconds; during their decay process, they will emit photons in random directions (spontaneous emission) within the 1.53–1.56-μm-wavelength spectrum. However, when transmitted photons along the optical fiber within the same wavelength band (1.53–1.56 μm) arrive at the doped optical fiber, they stimulate the erbium ions to emit photons in the same direction, phase and wavelength as the transmitted light, which in effect results in an amplification of the transmitted light signal. Remarkably, the erbium ions emit light radiation within the same wavelength region over which the silica fiber exhibits maximum transparency and, therefore, minimum attenuation. The gain of erbium-doped silica fibers is ordinarily 20–30 dB, and because of their cylindrical geometry, it is not significantly affected by polarization [66, 193]. Spontaneous emission is

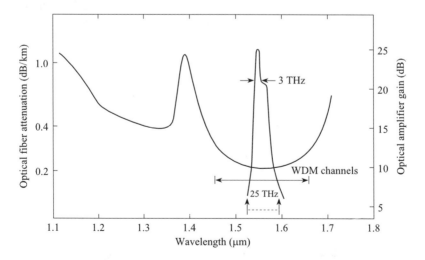

Figure 15.109 Attenuation of single-mode silica fiber and gain of erbium-doped silica fiber as function of wavelength (after [66]).

not entirely eliminated and accounts for a noise level of approximately 4–5 dB over a bandwidth, which is in excess of that of the transmitted light signal. Since the meta-stable state time of the erbium ion is substantially longer than the duration of the rapid nanosecond digital bit duration, the gain of the optical amplifier will essentially remain constant over the individual bit time even in the presence of saturation caused by intense sporadic signals [66].

The gain characteristic of the erbium-doped silica fiber is delineated in relation to the attenuation characteristic of the silica fiber itself in Fig. 15.109 [66]. The vertical arrows at the base of the gain curve indicate the WDM wavelength range proposed by the International Telecommunications Union: a reference wavelength of 1.5525 μm (193.1 THz) is indicated with an individual channel separation of 0.8 nm or 100 GHz. The nonlinearities in the gain characteristics of the erbium-doped fiber ampli-fier (EDFA) are conspicuously apparent in Fig. 15.109. Since these nonlinearities pose problems in WDM systems operated at 1.55 μm, there has been considerable effort expended to achieve gain flatness in the optical amplifier. Codoping the core of the fiber with aluminum has improved the flatness, but it has not eliminated the peak at 1.530 μm [196]. A further extension of the flatband range has been obtained with erbium-doped fluoride fiber amplifiers (EDFFAs) [197-200]. The improvement of band-width flatness attained with the EDFFA is demonstrated in Fig. 15.110, from which it can also be discerned that whereas the gain of the amplifier is enhanced with longer lengths of doped fiber, the flatness of the bandwidth is adversely affected.

The gain of optical amplifiers is susceptible to temperature variations [202]. These variations were found to be caused by changes in the absorption and emission cross sections of the optical amplifiers, which are linear functions of temperature and are dependent upon wavelength [201]. Although the EDFFA provides the best flatband characteristics, its gain is more affected by temperature-induced changes than that of either the erbium-doped aluminosilicate fiber amplifier (EDSFA) or the standard EDFA. The behavior of an EDFFA with fiber length and temperature is portrayed in Fig. 15.111, from which it is apparent that the optimum gain for the different fiber lengths is temperature dependent. Vis-à-vis Fig. 15.110, it is thus palpably evident that the gain efficiency may be seriously compromised by shortening the fiber length to attain the most desirable flatband characteristic.

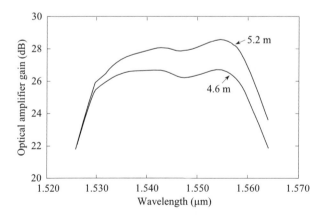

Figure 15.110 Gain vs. wavelength of 5.2- and 4.6-m long wavelength-multiplexed EDFFA (after [201]).

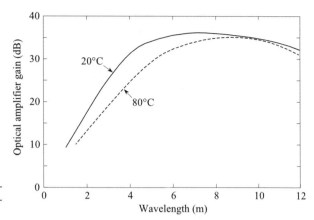

Figure 15.111 Gain vs. fiber length with temperature as parameter of EDFFA at wavelength of 1.540 μm (after [201]).

As may be anticipated, the efficiency of the optical amplifiers should exhibit a dependence upon the erbium ion concentration. There has been very little work done in this area, but recent results have been reported on a study that dealt with this question [203]. Figure 15.112 shows the quantum efficiency characteristics of aluminum codoped fiber amplifiers (EDSFAs) as a function of erbium ion concentration for the transmission wavelengths of 1.530 and 1.555 μm. A significant degradation in the efficiency can be perceived to occur with rising erbium ion concentration. Currently designed EDFAs require over 80% quantum/conversion efficiency, which in the case of the aluminum codoped fibers would correspond in terms of Fig. 15.112 to an erbium ion concentration of $<20 \times 10^{24}$m^{-3} (900 mol ppm Er^{3+} [203].

The development of optical amplifiers [204] has not only permitted the introduction of very high capacity optical fiber WDM techniques, but also made possible to contemplate the upgrading of a large volume of conventional single-mode silica fiber lines that are characterized by a dispersion minimum at the wavelength of 1.3 μm and an attenuation minimum at 1.55 μm [205–206]. In order to take advantage of optical amplification, the upgrading must be carried out at the attenuation minimum of 1.55 μm, which falls within the bandwidth of the erbium-metal-doped silica fibers.

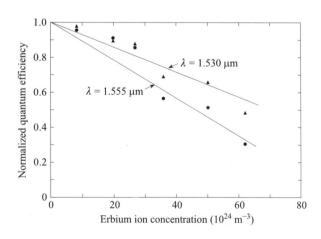

Figure 15.112 Quantum efficiency vs. erbium ion concentration with transmission wavelength as parameter of aluminum EDSFA (after [203])

TABLE 15.22 Dispersion-Limited Single-Mode Optical Fiber Lengths for Selected Fixed Data Rates Obtained by Sliding-Block Line Code Technique

Modulation	Data rate (Gbits/s)	Dispersion-Limited Distance (km)	
		Uncoded	Coded
External	8.0	172	250
	2.0	2750	4000
Direct	8.0	9	91
	2.0	1750	3250

Source: From [207].

Since most of the installed single-mode fiber cables are nondispersion shifted, the pulse dispersion at 1.55 μm becomes significantly elevated at high values of the data rate–fiber length product to cause intersymbol interference [207]. In the past the most common digital techniques that have been employed to reduce intersymbol interference have been those of maximum-likelihood sequence detection, Tomlinson-type precoding, partial-response signaling decision feedback equalization, and adaptive nonlinear cancellation [208–211]. Perhaps one of the most promising recently developed techniques is that utilizing the sliding-block line codes [207]. In this approach, the considered code prohibits the occurrence of isolated 1's, i.e., the 010 pattern in the encoded transmitted signal sequence. With the addition of suitable compensation circuitry at the receiver, the implementation of the code is found to lead to significant increases in the otherwise dispersion-limited data rates or fiber lengths, as is apparent from the results given in Table 15.22 [207].

Note that the dispersion-limited distances decrease with a rise in the data transmission rate over the optical fiber. Also the dispersion-limited distances are found to be higher when external modulation is employed. In the externally modulated procedure, an external circuit is employed to modulate the continuous-wave output of a semiconductor DFB laser; in the case of direct modulation, the drive current of the laser transmission source is modulated directly.

REFERENCES

[1] F. W. Horn, *Lee's ABC of the Telephone*, Vol. 5, *Cable Inside and Out*, Training Manual, Geneva, Illinois, 1976.

[2] W. E. Mougey, "The testing of long distance telephone during installation," *Electrical Commun.*, Vol. II, 1923, pp. 219–223.

[3] *Electrical Conductors Handbook*, Northern Telecom Limited, Lachine, Quebec, Canda, 1977.

[4] E. W. Kimbark, *Electrical Transmission of Power and Signals*, Wiley, New York, 1955.

[5] R. L. Freeman, *Telecommunication System Engineering*, Wiley, New York, 1996.

[6] D. H. Hamsher, *Communications System Engineering Handbook*, McGraw-Hill., New York, 1967.

[7] H. Nyquist, "Thermal agitation of electricity in conductors," *Phys. Rev.*, Vol. 32, 1928, pp. 110–113.

[8] J. G. Nellist, *Understanding Telecommunications and Lightwave Systems*, IEEE Press, New York, 1996.

[9] C. G. Davis, "An experimental pulse code modulation system for short haul trunks," *Bell Syst. Tech. J.*, Vol. 41, 1962, pp. 1–24.

[10] H. Nyquist, "Certain topics in telegraph transmission theory," *Trans. AIEE*, Vol. 47, 1928, pp. 617–644.

[11] C. E. Shannon, "A mathematical theory of communication—I," *Bell Syst. Tech. J.*, Vol. 27, 1948, pp. 379–423.

[12] C. E. Shannon, "A Mathematical theory of communication—II," *Bell Syst. Tech. J.*, Vol. 27, 1948, pp. 623–656.

[13] J. Martin, *Telecommunications and the Computer*, Prentice-Hall, Englewood Cliffs, NJ, 1969.

[14] G. M. Miller, *Modern Electronic Communication*, Regents/Prentice-Hall, Englewood Cliffs, NJ, 1993.

[15] H. Marko, R. Weiß, and G. Binkert, "A digital hybrid transmission system for 280 Mbits/s and 500 Mbits/s," *IEEE Trans. Commun.*, Vol. COM-23, 1975, pp. 274–282.

[16] I. Dorros, J. M. Sipress, and F.D. Waldhaner, "An experimental 224 Mb/s digital repeatered line," *Bell Syst. Tech. J.*, 1966, pp. 993–1043.

[17] P. Bylanski and D. G. W. Ingram, *Digital Transmission Systems*, Peter Peregrinus, Stevenage, United Kingdom, 1976.

[18] D. H. Hoth, "Broad-band media for urban communications," *IEEE Trans. Commun.*, Vol. COM23, 1975, pp. 121–131.

[19] J. W. Lechleider, "High bit rate digital subscriber lines: a review of HDSL progress," *IEEE J. Selected Areas Commun.*, Vol. 9, 1991, pp. 769–784.

[20] G. H. Im and J.-J. Werner, "Bandwidth-efficient digital transmission over unshielded twisted-pair wiring," *IEEE J. Selected Areas Commun.*, Vol. 13, 1995, pp. 1643–1655.

[21] G. D. Golden, J. E. Mago, and J. Salz, "Transmitter design for data transmission in the presence of a data-like interferer," *IEEE Trans. Commun.*, Vol. 43, 1995, pp. 837–850.

[22] W. Walkoe and J. J. Starr, "High bit rate digital subscriber line: a copper bridge to the network of the future," *IEEE J. Selected Areas in Commun.*, Vol. 9, 1991, pp. 765–768.

[23] S. P. Mead, "Shielded cable system," U.S. Pat. 2,086,620, April 14, 1936.

[24] A. S. Windeler, "Design of polyethylene insulated multipair telephone cable," *AIEE Trans. Commun. and Electr.*, Vol. 78, Part I, 1960, pp. 736–739.

[25] E. B. Rosa and F. W. Grover, "Formulae and tables for the calculation of mutual and self-inductance," Paper 169, *Bull. Bureau Standards*, Vol. 8, No. 1, 1912, pp. 1–237.

[26] L. Jachimowicz, J. A. Olszewski, and I. Kolodny, "Transmission properties of filled thermoplastic insulated and jacketed telephone cables at voice and carrier frequencies," *IEEE Trans. Commun.*, Vol. COM-21, 1973, pp. 203–209.

[27] G. S. Eager, L. Jachimowicz, I. Kolodny, and D. E. Robinson, "Transmission properties of polyethylene-insulated telephone cables at voice and carrier frequencies," *AIEE Trans. Commun. Electr.*, Vol. 78, 1959, pp. 618-639.

[28] T. P. Byrne, R. Coburn, H. C. Mazzoni, G. W. Aughenbaugh, and J. L. Duffany, "Positioning the subscriber loop network for digital services," *IEEE Trans. Commun.*, Vol. COM-30, 1982, pp. 2006–2011.

[29] S. Tomazic and A. Umek, "A simple formula for calculation of the power loss in digital transmission lines," *IEEE Trans. Commun.*, Vol. 40, 1992, pp. 484–486.

[30] R. F. H. Fischer, W. H. Gerstacker, and J. P. Huber, "Dynamics limited precoding, shaping and blind equalization for fast digital transmission over twisted pair lines," *IEEE J. Selected Areas Commun.*, Vol. 13, 1995, pp. 1622–1633.

[31] R. F. H. Fischer and J. B. Huber, "Comparison of precoding schemes for digital subscriber lines," *IEEE Trans. Commun.*, Vol. 45, 1997, pp. 334–343.

[32] S. D. Bradley, "Crosstalk considerations for a 48-channel PCM repeatered line," *IEEE Trans. Commun.* Vol. COM-23, 1975, pp. 722–728.

[33] D. H. Morgen, "Expected crosstalk performance of analog multichannel subscriber carrier systems," *IEEE Trans. Commun.*, Vol. COM-23, 1975, pp. 240–245.

[34] J. W. Lechleider, "Broad signal constraints for management of the spectrum in telephone loop cables," *IEEE Trans. Commun.*, Vol. COM-34, 1986, pp. 641–646.

[35] J. T. Aslanis and J. M. Cioffi, "Achievable information rates on digital subscriber loops: limiting information rates with crosstalk noise," *IEEE Trans. Commun.* Vol. 40, 1992, pp. 361–372.

[36] A. J. Gibbs, "Measurement of PCM regenerator crosstalk performance," *Electr. Lett.*, Vol. 15, 1979, pp. 82–83.

[37] G. J. Semple and A. J. Gibbs, "Assessment methods for evaluating the immunity of PCM regenerators to near end crosstalk," *IEEE Trans. Commun.*, Vol. COM-30, 1982, pp. 1791–1797.

[38] G. J. Semple, "The effect of intersymbol interference on the operation of PCM line regenerators," *Australian Telecommun. Res.*, Vol. 12, 1978, pp. 17–31.

[39] R. J. Catchpole and P. J. Dyke, "Design and assessment of primary PCM line repeaters," IEE Telecommunications Transmission Conf., London, March, 1981.

[40] G. H. Im, D. D. Marman, G. Huang, A.V. Mandgik, M. H. Nguyen, and J. J. Werner, "51.84 Mb/s 16 CAP ATM LAN Standard," *IEEE J. Selected Areas Commun.*, Vol. 13, 1995, pp. 620–632.

[41] D. E. Setzer and A. S. Windeler, "A low capacitance cable for the T2 Digital Transmission Line," 19th International Wire and Cable Symposium", Atlantic City, NJ, 1970.

[42] D. E. Setzer, "Low capacitance multipair cable for 6.3 Megabit per second transmission system," International Conference on Communications, IEEE Conf. Record 72 CHO 622-1-COM, Philadelphia, June 19–21, 1972, pp. 19/19–19/23.

[43] W. E. Stephens, H. Samuel, and G. Cherubini, "Copper wire access technologies for high performance networks," *IEEE J. Selected Areas Commun.*, Vol. 13, 1995, pp. 1537–1538.

[44] J. J. Werner, "The HDSL environment," *IEEE J. Selected Areas Commun.*, Vol. 9, 1991, pp. 785–800.

[45] S. V. Ahamed, P. L. Gruber, and J. J. Werner, "Digital subscriber line (HDSL and ADSL) capacity of the outside loop plant," *IEEE J. Selected Areas Commun.*, Vol. 13, 1995, pp. 1540–1549.

[46] Additional cable specifications for unshielded twisted pair cables, Technical System Bulletin No. 36, EIA/TIA Document, Nov. 1991.

[47] G. Cherubini, S. Ölger, and G. Ungerboeck, "Adaptive analog equalization and receiver front-end control for multi-level partial response transmission over metallic cables," *IEEE Trans. Commun.*, Vol. 6, 1996, pp. 675–685.

[48] F. C. Kelcourse and R. A. Tabox, "The design of repeated lined for long-haul coaxial systems," *IEEE Trans. Commun.*, Vol. COM-22, 1976, pp. 200–205.

[49] M. Decina and U. de Julio, "Constraints of basic transmission media on the hierarchies," IEEE International Conf. on Communications, IEEE Conf. Record 75CHO 971-2 CSBC, San Francisco, June 16-18, 1975, Vol. I, pp. 16-9/16-13.

[50] U. de Julio, B. Fabbri, L. Sacchi, F. Tosco, and G. Veglio, "An experimental system for hybrid digital transmission at 140 Mbits/s on coaxial cables," *IEEE Trans. Commun.*, Vol. COM-24, 1976, pp. 881–890.

[51] F. H. Blecker, R. C. Boyd, F. J. Hallenbeck, and F. J. Herr, "The L-4 coaxial system," *Bell Syst. Tech. J.*, Vol. 48, 1969, pp. 821-1099.

[52] G. H. Duval and L. M. Rackson, "Coaxial cable and apparatus," *Bell Syst. Tech. J.*, Vol. 48, 1969, pp. 1065–1093.

[53] J. F. Gunn, J. S. Ronne, and D. C. Weller, "Mastergroup digital transmission on modern coaxial systems," *Bell Syst. Tech. J.*, Vol. 50, 1971, pp. 501–520.

[54] E. H. Angell and M. M. Luniewicz, "Low noise ultralinear line repeaters for the L5 coaxial system," *IEEE Trans. Commun.*, Vol. COM-22, 1974, pp. 206–211.

[55] Y. S. Cho, C. R. Crue, and M. R. Giammusso, "Static and dynamic equalization of the L5 repeated line," *IEEE Trans. Commun.*, Vol. COM-22, 1974, pp. 212–217.

[56] C. S. Yen and R. D. Crawford, "Distribution and equalization of signal on coaxial cables used in 10 Mbits/s baseband local area networks," *IEEE Trans. Commun.*, Vol. COM-31, 1983, pp. 1181–1186.

[57] *Telecommunications Transmission Engineering*, Vol. 2, 3rd ed., Bellcore, 1990.

[57a] ASTM Standard Test Methods for Electrical Performance Properties of Insulations and Jackets for Telecommunication Wire and Cable, *ASTM Book of Standards*, Vol. 10, 2, ASTM, Conshohocken, PA, 1997.

[58] G. W. A. Dummer and W. T. Blackband, *Wires and R.F. Cables*, Pitman, London, 1961.

[58a] S. Maki, "High expansion insulation compounds for coaxial cables," *Kabelitems*, No. 166, Union Carbide Corporation, 1997.

[59] I. J. Wickelgren, "The facts about firewire," *IEEE Spectrum*, Vol. 34, 1997, pp. 19–25.

[60] F. P. Kapron, D. B. Beck, and R. D. Maurer, "Radiation losses in glass optical waveguides," *Appl. Phys. Lett.*, Vol. 17, 1970, pp. 423-425.

[61] J. S. Cook and O. I. Szentesi, "North American field trials and early applications in telephony," *IEEE J. Selected Areas Commun.*, Vol. SAC-1, 1983, pp. 393–397.

[62] D. R. Nicol, J. T. Harvey, V. Svovoda, F. A. Donaghy, C. H. Storey, and K. R. Ballinger, "A 2 km optical fiber communication trial," *IEEE Trans. Commun.*, Vol. COM-26, 1978, pp. 1061–1067.

[63] J. E. Midwinter and J. R. Stern, "Propagation studies of 40 km of graded index fiber installed in cable in an operational duct route," *IEEE Trans. Commun.*, Vol. COM-26, 1978, pp. 1015–1020.

[64] G. Cocito, B. Costa, S. Longoni, L. Michetti, L. Silvestri, D. Tibone, and F. Tosco, "COS2 experiment in Turin: field tests on an optical cable in ducts," *IEEE Trans. Commun.*, Vol. COM-26, 1978, pp. 1028–1036.

[65] J. A. Olszewski, G. H. Foot, and Y. Y. Huang, "Development and installation of an optical-fiber cable for communications," *IEEE Trans. Commun.*, Vol. COM-26, 1978, pp. 991–998.

[66] A. E. Willner, "Mining the optical bandwidth for a terabit per second," *IEEE Spectrum*, Vol. 34, 1997, pp. 32–41.

[67] Special Issue on Optical Networks (Papers on optical TDM networks), *IEEE J. Selected Areas Commun.*, Vol. 14, 1996, pp. 979–1052.

[68] J. B. MacChesney, "Materials and processes for preform fabrication-modified chemical vapor deposition and plasma chemical vapor deposition," *Proc. IEEE*, Vol. 68, 1980, pp. 1181–1184.

[69] P. C. Schultz, "Fabrication of optical waveguides by the outside vapor deposition process," *Proc. IEEE*, Vol. 68, 1980, pp. 1187–1190.

[70] T. Izawa and N. Inagaki, "Materials and processes for fiber preform fabrication-vapor-phase axial deposition," *Proc. IEEE*, Vol. 68, 1980, pp. 1184–1187.

[71] S. R. Nagel, J. B. MacChesney, and K. L. Walker in *Optical Fiber Communications*, T. Li, Ed., Academic, Orlando, 1985.

[72] K. L. Walker, G. M. Homsey, S. R. Nagel, and F. T. Geyling, Abstract 137, Extended Abstracts, Electrochemical Society Meeting, Pittsburgh, 1978.

[73] J. W. Fleming and P. B. O'Connor in Proc. Optical Fiber Symposium, American Ceramic. Soc., Annual Meeting, Chicago, 1980.

[74] T. Izawa and N. Inagaki, "Materials and processes for fiber preform fabrication–vapor phase axial deposition," *Proc. IEEE*, Vol. 68, 1980, pp. 1184–1187.

[75] P. C. Schultz, "Progress in optical waveguide process and materials," *Appl. Opt.*, Vol. 18, 1979, pp. 3684–3693.

[76] P. C. Schultz, "Method of forming a light focusing fiber waveguide," U.S. Pat. 3,826,560, July 30, 1976.

[77] G. W. Scherer, "Sintering of low density glasses: I, II and III," *J. Am. Ceramics. Soc.*, Vol. 60, 1977, pp. 236–246.

[78] D. R. Powers, "Method of making dry optical waveguides," U.S. 4,165,223, Aug. 21, 1979.

[79] A. C. Bailey, "Method of drawing glass optical waveguides," U.S. Pat. 4,157,906, June 12, 1979.

[80] G. W. Scherer, "Sintering inhomogeneous glasses : Applications to optical waveguides," *J. Non-Cryst. Solids*, Vol. 34, Oct. 1979.

[81] L. L. Blyler and F. V. DiMarcello, "Fiber drawing, coating and jacketing," *Proc. IEEE*, Vol. 68, 1980, pp. 1194–1198.

[82] T. J. Miller, A. C. Hart, W. I. Vroom, and M. J. Bowden, *Electronic Lett.*, Vol. 14, 1978, pp. 603–605.

[83] F. V. DiMarcello, A. C. Hart, J. C. Williams, and C. R. Kurkjian in *Fiber Optics: Advances in Research and Development*, B. Bendow and S.S. Mitra, Eds., Plenum, New York, 1979.

[84] T. Izawa and S. Subdo, *Optical Fibers: Materials and Fabrication*, Reide, Boston, 1987.

[85] D. H. Smithgall, L. S. Watkins, and R. E. Frazee, "High speed noncontact fiber-diameter measurement using forward light scattering," *Appl. Opt.*, Vol. 16, 1977, pp. 2395–2402.

[86] H. Schonhorn, C. R. Kurkjian, R. E. Jaeger, H. N. Vazirani, R. B. Albarino, and F. V. DiMarcello, "Epoxy-acrylate-coated fused silica fibers with tensile strengths > 500 psi (3.5 GN/m^2) in /km gauge lengths," *Appl. Phys. Lett.*, Vol. 29, 1976, pp. 712–714.

[87] H. M. Presky, "Geometrical uniformity of plastic coatings on optical fibers," *Bell Syst. Tech. J.*, Vol. 55, 1976, pp. 1525–1537.

[88] D. Gloge, "Optical-fiber packaging and its influence on fiber straightness and loss," *Bell Syst. Tech. J.*, Vol. 54, 1975, pp. 245-262.

[89] T. Li, "Structures, parameters and transmission properties of optical fibers," *Proc. IEEE*, Vol. 68, 1980, pp. 1175–1180.

[90] I. P. Kaminow, D. Marcuse, and H. M. Presky, "Multimode fiber bandwidth: theory and practice," *Proc. IEEE*, Vol. 68, 1980, pp. 1209–1213.

[91] D. Marcuse, *Principles of Optical Fiber Measurements*, Academic, New York, 1981.

[92] W. A. Gambling, H. Matsumura, and C. M. Ragdale, "Wave propagation in a single mode fiber with dip in refractive index," *Optical Quantum Electron*, Vol. 10, 1978, pp. 301–309.

[93] D. Marcuse, *Theory on Dielectric Optical Waveguides*, Academic, New York, 1974.

[94] H. Killen, *Fiber Optic Communications*, Prentice-Hall, Englewood Cliffs, NJ, 1991.

[95] R. M. Gagliardi and S. Karp, *Optical Communications*, Wiley, New York, 1995.

[96] G. Nicholson and D. J. Temple, "Polarization fluctuation measurements on installed single-mode optical fiber cables," *J. Lightwave Tech.*, Vol. 7, 1989, pp. 1197–1200.

[97] E. M. Frins and W. Dultz, "Rotation of the polarization plane in optical fibers," *J. Lightwave Tech.*, Vol. 15, 1997, pp. 144–146.

[98] I. P. Kaminow, J. R. Simpson, H. M. Presky, and J. B. MacChesney, "Strain birefringence in single-polarization germanosilicate optical fibers," *Electr. Lett.*, Vol. 15, 1979, pp. 677–679.

[99] I. W. Stanley, G. R. Hill, and D. W. Smith, "The application of coherent optical techniques to wide-band networks," *J. Lightwave Tech.*, Vol. LT-5, 1987, pp. 439–451.

[100] H. F. Wolf, Editor, *Handbook of Fiber Optics: Theory and Applications*, Garland STPM Press, New York, 1979.

[101] R. Olshanski and D. B. Keck, "Pulse broadening in graded-index optical fibers," *Appl. Optics*, Vol. 15, 1976, pp. 483–491.

[102] S. O. Agbo, A. H. Cherin, and B. K. Tariyal, "Lightwave" in *The Electrical Engineering Handbook*, R. Dorf, Ed., CRC/IEEE Press, New York, 1997, Chapter 42.

[103] D. G. Gloge and T. Li, "Multimode-fiber technology for digital transmission," *Proc. IEEE*, Vol. 68, 1980, pp. 1269–1275.

[104] D. Gloge, E. A. J. Marcatili, D. Marcuse, and S. D. Personick, in *Optical Fiber Telecommunications*, S. E. Miller and A. G. Chynoweth, Eds., Academic, New York, 1979.

[105] M. J. Adams, D. N. Payne, F. M. E. Sladen, and A. H. Hertog, "Wavelength-dispersive properties of glasses for optical fibers, the germania enigma," *Electr. Lett.*, Vol. 14, 1978, pp. 703–705.

[106] I. H. Malitson, "Refractive index of fused silica," *J. Optical Soc. Am.*, Vol. 55, 1965, pp. 1205.

[107] M. Badolo and P. Emplit, "Prediction of close-to-zero model dispersion over a wide range of wavelengths in singly clad multimode fibers," *J. Lightwave Tech.*, Vol. 15, 1997, pp. 121–124.

[108] J. W. Fleming, "Material and mode dispersion in $GeO_2 \cdot B_2O_3 \cdot SiO_2$ glasses," *J. Am. Ceramic. Soc.*, Vol. 59, 1976, pp. 503–507.

[109] J. W. Fleming, "Material dispersion in light guide glasses," *Electr. Lett.*, Vol. 14, 1978, p. 326.

[110] T. Li, "Optical fiber communications—the state of the art," *IEEE Trans. Commun.*, Vol. COM-26, 1978, pp. 946–955.

[111] D. Gloge and E. A. J. Marcatili, "Multimode theory of graded-core fibers," *Bell Syst. Tech. J.*, Vol. 52, 1973, pp. 1563–1578.

[112] E. A. J. Marcatili, "Model dispersion in optical fibers with arbitrary numerical aperture and profile dispersion," *Bell Syst. Tech. J.*, Vol. 56, 1977, pp. 49–63.

[113] I. P. Kaminov, D. Marcuse, and H. M. Presby, "Multimode fiber bandwidth: theory and practice," *Proc. IEEE*, Vol. 68, 1980, pp. 1209–1213.

[114] D. Gloge, I. P. Kaminov, and H. M. Presby, "Profile dispersion in multimode fibers: measurement and analysis," *Electr. Lett.*, Vol. 11, 1975, pp. 469–470.

[115] D. Marcuse and H. M. Presby, "Effects of profile deformation on fiber bandwidth," *Appl. Optics*, Vol. 19, 1980, p. 188.

[116] I. P. Kaminov and H. M. Presby, "Profile synthesis in multi component glass optical fiber," *Appl. Optics*, Vol. 16, 1977, pp. 108–112, 1977.

[117] H. M. Presby and I. P. Kaminov, "Binary silica optical fibers: Refractive index and profile measurements," *Appl. Optics*, Vol. 15, 1976, pp. 3029–3036; also see errata to above in *Appl. Optics*, Vol. 17, 1978, p. 3530.

[118] A. Weierholt, "Model dispersion of optical fibers with a composite α-profile graded index core," *Electr. Lett.*, 1979, Vol. 15, pp. 733–734.

[119] B. J. Ainslie and C. R. Day, "A review of single-mode fibers with modified dispersion characteristics," *J. Lightwave Tech.*, Vol. 4, 1986 pp. 967–979.

[120] M. Badolo and P. Emplit, "Numerical analysis of model dispersions in singly-clad multimode optical fibers," *Opt. Commun.*, Vol. 114, 1995, pp. 64–68.

[121] F. Dubois, P. Emplit, and O. Hugon, "Selective mode excitation in graded-index multimode fiber by a computer-generated optical mask," *Opt. Lett.*, Vol. 19, 1994, pp. 433–435.

[122] K. Okamoto, T. Edahiro, A. Kawana, and T. Miya, "Dispersion minimization in single-mode fibers over a wide spectral range," *Electr. Lett.*, Vol. 15, 1979, pp. 729–731.

[123] C. R. South, "Total dispersion in step-index monomode fibers, *Electr. Lett.*, Vol. 15, 1979, pp. 394–395.

[124] L. G. Cohen, "Trends in the U.S. broad band fiber optic transmission systems," *IEEE J. Selected Areas Commun.*, Vol. SAC-4, 1986, pp. 488–497.

[125] F. P. Kapron, "Maximum information capacity of fiber-optic waveguides," *Electr. Lett.*, Vol. 13, 1977, pp. 96–97.

[126] K. Ogawa, "Considerations for single mode fiber systems," *Bell Syst. Tech. J.*, Vol. 61, 1982, pp. 1919–1931.

[127] R. A. Linke, "Transient chirping in single frequency lasers: lightwave systems considerations," *Electr. Lett.*, Vol. 20, 1984, pp. 472–474.

[128] M. Schiess, "Three-level modulation scheme on standard fibers to extend the dispersion limit," *IEEE J. Selected Areas Commun.*, Vol. 13, 1995, pp. 479–484.

[129] J. P. Hamaide, P. Emplit, and J. M. Gabriagues, "Limitations in long hand 1M/DD optical fiber systems caused by chromatic dispersion and non-linear Kerr effect," *Electr. Letters*, Vol. 26, 1990, pp. 1451–1453.

[130] R. M. Jopson, A. H. Gnauk, and R. M. Derosier, "10Gb/s 360 km transmission over normal-dispersion fiber using mid system spectral dispersion," Proc. OFC/IOOC, 93PD3, San Jose, 1993.

[131] A. H. Gnauk, C. R. Gilles, L. J. Cimini, J. Stone, L. W. Stulz, S. K. Korotky, and J. J. Veselka, "8Gb/s 130 km transmission experiment using Er-doped fiber preamplifier and optical dispersion equalization," *IEEE Photon, Technol. Lett.*, Vol. 3, 1991, pp. 1147–1149.

[132] T. L. Koch and R. C. Alferness, "Dispersion compensation by active predistorted signal synthesis," *J. Lightwave Tech.*, Vol. 3, 1985, pp. 800–805.

[133] C. Kurtze and A. H. Gnauk, "Operating principle of in-line applied dispersion supported transmission," *Electr. Lett.*, Vol. 29, 1993, pp. 1969–1971.

[134] H. Osanai, T. Shioda, T. Moriyama, S. Araki, M. Horiguchi, T. Izawa and H. Takata, "Effect of dopants on transmission losses of low-OH$^-$ content optical fibers," *Electr. Lett.*, Vol. 12, 1976, pp. 549–550.

[135] T. Miya, Y. Terunuma, T. Hosaka, and T. Miyashita, "Ultimate low-loss single-mode fibre at 1.55 μm," *Electr. Lett.*, Vol. 15, 1979, pp. 106–108.

[136] T. Izawa and N. Shibata, "Optical attenuation in pure and doped fused silica in the ir wavelength region," *Appl. Phys. Lett.*, Vol. 31, 1977, pp. 33–35.

[137] S. R. Nagel and M. A. Saifi, "Effect of deposition rate on spectral loss of GeO_2-P_2O_5-SiO_2 graded index fibers," *Electr. Lett.*, Vol. 16, 1980, pp. 469–470.

[138] D. A. Pinnow, T. C. Rich, F. W. Ostermayer, and M. DiDomenico, "Fundamental optical attenuation limits in the liquid and glassy state with application to fiber optical waveguide materials," *Appl. Phys. Lett.*, Vol. 22, 1973, pp. 527–529.

[139] O. I. Szentesi, "Reliability of optical fibers, cables and splices," *IEEE J. Selected Areas Comm.*, Vol. SAC-4, 1986, pp. 1502–1508.

[140] K. J. Beales, D. M. Cooper, and J. D. Rush, "Increased attenuation in optical fibers caused by diffusion of molecular hydrogen at room temperature," *Electr. Lett.*, Vol. 19, 1983, pp. 917–919.

[141] Y. Namihira, K. Mochizuki, M. Kawazura, and Y. Iwamoto, "Effects of hydrogen diffusion on optical fiber loss increase," *Electr. Lett.*, Vol. 19, 1983, pp. 1034–1035.

[142] T. Tanifuji, M. Matsumoto, and M. Tokuda, "Wavelength dependent optical loss increase in graded-index optical fiber transmission lines," *Electr. Lett.*, Vol. 20, 1984, pp. 13–14.

[143] A. Tomita and P. J. Lemaire, "Hydrogen induced loss increases in germanium doped single mode optical fibers: long term predictions," *Electr. Lett.*, Vol. 21, 1985, pp. 71–72.

[144] N. J. Pitt and A. Marshall, "Long term loss stability of single-mode optical fibers exposed to hydrogen," *Electr. Lett.*, Vol. 20, 1984, pp. 512–514.

[145] M. K. Barnoski and S. D. Personik, "Measurements in fiber optics," *Proc. IEEE*, Vol., 66, 1978, pp. 429–441.

[146] L. G. Cohen, P. Kaiser, and C. Lin, "Experimental techniques for evaluation of fiber transmission loss and dispersion," *Proc. IEEE*, Vol. 66, 1978, pp. 1203–1209.

[147] E. A. J. Marcatili, "Bends in optical dielectric guides," *Bell Syst. Tech. J.*, Vol. 48, 1969, pp. 2103–2132.

[148] D. Marcuse, "Curvature loss formula for optical fibers," *J. Opt. Soc. Am.*, Vol. 66, 1976, pp. 216–220.

[149] H. Renner, "Bending losses of coated single-mode fibers : A simple approach," *J. Lightwave Tech.*, Vol. 10, 1992, pp. 544–551.

[150] R. Morgan, J. S. Barton, P. G. Harper and J. D. C. Jones, "Temperature dependence of bending loss in mono-mode optical fibers," *Electr. Lett.*, Vol. 26, 1990, pp. 937–939.

[151] L. Faustini and G. Martini, "Bend loss in single-mode fibers," *J. Lightwave Tech.*, Vol. 15, 1997, pp. 671–679.

[152] G. Martini and S. Donati, "Spectral attenuation measurements of optical fibers. Design of an instrument based on a pulsed light source," *J. Opt. Comun.*, Vol. 11, 1990, pp. 22–25.

[153] A. J. Harris and P. F. Castle, "Bend loss measurements on high numerical aperture single-mode fibers as a function of wavelength and bend radius," *J. Lightwave Tech.*, Vol. 4, 1986, pp. 34–40.

[154] M. I. Schwartz, P. F. Gagen, and M. R. Santana, "Fiber cable design and characterization," *Proc. IEEE*, Vol. 68, 1980, pp. 1214–1219.

[155] R. G. Huff, F. V. DiMarcello, and A. C. Hart, "Amorphous carbon hermetically coated optical fibers," Tech. Dig. Optical Fiber Commun. Conf., Paper TIG-2, 1988.

[156] Y. Katsuyama, N. Yoshizawa, and T. Yashiro, "Field evaluation result on hermetically coated optical fiber cables for practical application," *J. Lightwave Tech.*, Vol. 9, 1991, pp. 1041–1046.

[157] H. G. Haag, G. F. Hög, and P. E. Zamzow, "Optical fiber cables for subscriber loops," *J. of Lightwave Technology*, Vol. 11, 1989, pp. 1667-1674.

[158] C. M. Miller, "Optical fiber cables and splices," *J. Selected Areas Commun.*, Vol. SAC-1, 1983, pp. 533–540.

[159] O. I. Szentesi, "Field experience with fiber-optic cable installation, splicing, reliability and maintenance," *IEEE J. Selected Areas Commun.*, Vol. SAC-1, 1983, pp. 541–546.

[160] W. Griffioen, "The installation of conventional fiber-optic cables in conduits using the viscous flow of air," *J. Lightwave Tech.*, Vol. 7, 1989, pp. 297–302.

[161] S. A. Cassidy, S. Hornung, P. Yennadhiou, and M. H. Reeve, "Optical characterization of a trial link using blown fiber cable," Proc. IOOC-ECOC, Sept. 1985, pp. 395–397.

[162] J. F. Dalgleish, "Splices, connectors and power couplers for field and office use," *Proc. IEEE*, Vol. 68, 1980, pp. 1226–1231.

[163] T. C. Chu and A. R. McCormick, "Measurements of loss due to offset, end separation and angular misalignment in graded index fibers exhibited by an incoherent source," *Bell Syst. Tech. J.*, Vol. 57, 1978, pp. 595–602.

[164] C. Miller, "Fiber optical array splicing with etched silicon chips," *Bell Syst. Tech. J.*, Vol. 57, 1978, pp. 75–90.

[165] S. Stueflotten, "Protection of optical fiber arc fusion splices," *J. Opt. Commun.*, Vol. 3, 1982, pp. 19–25.

[166] J. B. Haber and J. Rogers, "Fiber in the loop," *AT&T Tech.*, Vol. 3, 1988, pp. 2–9.

[167] E. Suhir, "Stresses in a partially coated optical glass fiber subjected to the ends off-set," *J. Lightwave Tech.*, Vol. 15, 1997, pp. 2091–2094.

[168] C. M. de Blok and P. Matthijsse, "Field usable single-mode fibers splicing applying local core alignment," Proc. 33rd International Wire Cable Symposium, Reno, Nov. 13–15, 1984, pp. 193–196.

[169] Y. Koike, T. Ishigure, and E. Nihei, "High-bandwidth graded-index polymer optical fiber," *J. Lightwave Tech.*, Vol. 13, 1995, pp. 1475–1489.

[170] T. Ishigure, M. Satoh, O. Takanashi, E. Nihei, T. Nyu, S. Yamazaki, and Y. Koike, "Formation of the refractive index profile in the graded index polymer optical fiber for gigabit data transmission," *J. Lightwave Tech.*, Vol. 15, 1997, pp. 2095–2100.

[171] P. K. Runge, "Transfer molding of precision single optical fiber connectors," Conf. Record, IEEE International Conf. on Communications, 1979, p. 44.5.

[172] G. D. Khoe, J. H. F. M. vanLeest, and J. A. Luijendijk, "Single-mode fiber connector using core centered ferrules," *IEEE Quantum Electron.*, Vol. QE-18, 1982, pp. 1573–1580.

[173] G. D. Khoe and H. Lydtin, "European optical fibers and passive components: status and trends," *IEEE J. Selected Areas Commun.*, Vol. SAC-4, 1986, pp. 457–471.

[174] H. Ishida, K. Sakai, and T. Kakii, "Two dimensionally arranged 24-fiber optical connectors," *OSA Technical Digest Series*, Vol. 6, 1997, pp. 189–190.

[175] H. Kosaka, K. Kurihara, A. Venura, T. Yoshikawa, I. Ogura, T. Namai, M. Sugimoto, and K. Kasahara, "Uniform characteristics with low threshold and light efficiency for a single-transverse-mode vertical-cavity-surface-emitting laser-type device array," *IEEE Photon. Tech. Lett.*, Vol. 6, 1994, pp. 323-325.

[176] S. D. Personick, "Review of fundamentals of optical fiber systems," *IEEE J. Selected Areas Commun.*, Vol. SAC-1, 1983, pp. 373–380.

[177] C. Burrus and B. Miller, "Small-area, double heterostructure aluminum-gallium arsenide electron-luminescent diode source for optical fiber transmission lines," *Opt. Commun.*, Vol. 4, 1971, pp. 307–309.

[178] E. A. Cottrel and M. C. Brain, "Long reach single-mode fiber OTDR using 1.5 μm semiconductor laser," *Electr. Lett.*, Vol. 22, 1986, pp. 443–445.

[179] P. A. Rosher, S. C. Fenning, P. Cochrane, and A. R. Hunwicks, "An automatic optical fiber break location scheme for duplex and diplex transmission systems," *B. Telecom Tech. J.*, Vol. 6, 1988, pp. 54–59.

[180] S. Zemon, A. Budman, T. Wei, E. Eichen, and K. T. Ma, "Decay of transmitted light during fiber breaks–implications for break location," *J. Lightwave Tech.*, Vol. 12, 1994, pp. 1532–1535.

[181] P. G. Simkins and J. T. Krause, "Dynamic response of glass fibers during tensile fracture," *Proc. Royal Soc. London*, Vol. 350, 1979, pp. 253–265.

[182] T. Li, "Advances in optical fiber communications: an historical perspective," *IEEE J. Selected Areas Commun.*, Vol. SAC-1, 1983, pp. 356–372.

[183] W. van Etlen and J. vander Plaats, *Fundamentals of Optical Fiber Communications*, Prentice Hall, New York, 1991.

[184] A. G. Dentai, T. P. Lee, C. A. Burns, and E. Buechler, "Small area, high-radiance C.W. InGaAsP LEDs emitting at 1.2 and 1.3 μm," *Electr. Lett.*, Vol. 13, 1977, pp. 484–485.

[185] O. Wada, S. Yamakoshi, M. Abe, Y. Nishitani, and T. Sakwai, "High radiance InGaAsP/InP lensed LED's for optical communication systems at 1.2–1.3 μm," *IEEE J. Quantum Electron.*, Vol. QE-1, 1981, pp. 174–178.

[186] O. Wada, S. Yamakoshi, H. Hamaguchi, T. Tanaka, Y. Nishitani, and T. Sakurai, "Performance and reliability of high radiance InGaAsP DH LED's operating in the 1.15-1.5 μm wavelength region," *IEEE J. Quantum Electronics*, Vol. QE-18, 1982, pp. 368–373.

[187] O. Svelto, *Principles of Lasers*, Plenum, New York, 1989.

[188] H. Melchior and W. T. Lynch, "Signal and noise response of high speed germanium avalanche photodiodes," *IEEE Trans. Electron Devices*, Vol. ED-13, 1966, pp. 829–838.

[189] T. Mikawa, S. Kagowa, T. Kaneda, T. Sakurai, H. Ando, and O. Mikami, "A low noise $n^{+}np$ germanium avalanche photodiode," *IEEE J. Quantum Electron.*, Vol. QE-17, 1981, pp. 210–215.

[190] T. P. Lee, C. A. Burrus, and A. G. Dentai, "InGaAs/InP p-i-n photodiodes for lightwave communications at 0.95–1.65 μm wavelength," *IEEE J. Quantum Electron.*, Vol. QE-17, 1981, pp. 232–238.

[191] O. K. Kim. S. R. Forrest, W. A. Bonner, and R. G. Smith, "A high gain $In_{0.53}As$/InP avalanche photo diode with no tunnelling leakage current," *Appl. Phy. Lett.*, Vol. 39, 1981, pp. 402–404.

[192] H. F. Mahlein and A. Reichelt, "Micro-optischer Monomode-Multiplexer-Demultiplexer für vier Wellenlängenkanale," *Telcom. Rep.*, Vol. 8, 1985, pp. 91–95.

[193] E. Desurvire, *Erbium Doped Fiber Amplifiers, Principles and Applications*, Wiley, New York, 1993.

[194] B. Pedersen, M. L. Dakss, B. A. Thompson, W. J. Miniscalco, T. Wei, and L. J. Andrews, "Experimental and theoretical analysis of efficient erbium-doped fiber power amplifiers," *IEEE Photon. Technol. Lett.*, Vol. 3, 1991, pp. 1085–1087.

[195] M. Ohashi, "Design considerations for an Er^{3+}-doped fiber amplifier," *J. Lightwave Tech.*, Vol. 9, 1991, pp. 1099–1104.

[196] S. Yoshida, S. Kuwano, and K. Iwashita, "Gain flattened EDFA with high Al concentration for multistage repeated WDM transmission," *Electr. Lett.*, Vol. 31, 1995, pp. 1765–1767.

[197] D. Bayart, B. Clesca, L. Hamon, and J. L. Beylat, "Experimental investigation of the gain flatness characteristics for 1.55 μm erbium-doped fluoride fiber amplifiers," *IEEE Photon. Tech. Lett.*, Vol. 6, 1994, pp. 613–615.

[198] B. Clesca, D. Ronarc'h, D. Bayart, Y. Sorel, L. Hamon, M. Guibert, J. L. Beylat, J. F. Kerdiles, and M. Semenkoff', Gain flatness comparison between erbium-doped fluoride and silica fiber amplifiers with wavelength-multiplexed signals," *IEEE Photon. Tech. Lett.*, Vol. 6, 1994, pp. 509–512.

[199] B. Clesca, D. Beyart, L. Hamon, J. L. Beylat, C. Coeurjolly, and L. Berthelon, "Over 25 nm, 16 wavelength-multiplexed signal amplification through fluoride-based fiber amplifier cascade," *IEEE Photon. Tech. Lett.*, Vol. 6, 1994, pp. 586–588.

[200] M. Yamada, T. Kanamori, Y. Terunuma, K. Oikawa, M. Shimizu, S. Suodo, and K. Sagawa, "Fluoride based erbium-doped fiber amplifier with inherently flat gain spectrum," *IEEE Photon. Technol. Lett.*, Vol. 8, 1996, pp. 882–884.

[201] J. Kemtchou, M. Duhamel, and P. Leroy, "Gain temperature dependence of erbium-doped silica and fluoride fiber amplifiers in multichannel wavelength-multiplexed transmission systems," *J. Lightwave Tech.*, Vol. 15, 1997, pp. 2083–2090.

[202] M. Yamada, M. Shimizu, M. Okayasu, and M. Horiguchi, "Temperature dependence of signal gain in Er^{3+}-doped optical fiber amplifiers," *J. Quantum Electron.*, Vol. 28, 1992, pp. 640–649.

[203] P. Myslinski, D. Nguyen, and J. Chrostowski, "Effects of concentration on the performance of erbium-doped fiber amplifiers," *J. Lightwave Tech.*, Vol. 15, 1997, pp. 112–120.

[204] Special Issue on Optical Amplifiers, *J. Lightwave Tech.*, Vol. 9, 145–296, 1991.

[205] J. H. Winters and R. D. Gitlin, "Electrical signal processing techniques in long-haul fiber-optic systems," *IEEE Trans. Commun.*, Vol. 38, 1990, pp. 1439–1453.

[206] L. J. Gimini, L. J. Greenstairs, and A. A. M. Saleh, "Optical equalization to combat the effects of laser chirp and fiber dispersion," *J. Lightwave Tech.*, Vol. 8, 1990, pp. 649–659.

[207] N. L. Swenson and J. M. Cioffi, "Sliding-block line codes to increase dispersion-limited distance of optical fiber channels," *IEEE J. Selected Areas Commun.*, Vol. 13, 1995, pp. 485–498.

[208] E. A. Lee and D. G. Messerschmitt, *Digital Communication*, Kluwer, Norwell, 1988.

[209] G. D. Forney and M. V. Eyuboglu, "Combined equalization coding," *IEEE Comm. Mag.*, Vol. 30, 1991, pp. 25–34.

[210] J. H. Winters and S. Kasturia, "Adaptive non-linear cancellation for high speed fiber-optic systems," *J. Lightwave Tech.*, Vol. 10, 1992, pp. 971–977.

[211] S. Kasturia and J. H. Winters, "Techniques for high-speed implementation of non-linear cancellation," *IEEE J. Selected Areas Comm.*, Vol. 9, 1991, pp. 711–717.

Chapter 16

UNDERSEA COAXIAL COMMUNICATION CABLES

Richard T. Traut

16.1 INTRODUCTION

Since about 1850, undersea cables have provided important telecommunications links across the world's seas and oceans. The evolution of undersea communication cables from early designs that could carry telegraph signals to improved designs capable of carrying telephone signals over vast transoceanic distances required about a century of time. These newer systems, using improved coaxial undersea cable designs, then became the primary telecommunication links between continents for over 30 years, from the 1950s through the 1980s. These successful coaxial designs then evolved into the first successful transoceanic communication cables using optical fibers as the transmission medium. Since the late 1980s, many coaxial undersea cable systems are being supplanted by fiber-optic undersea cable systems. The considerable knowledge and experience gained over nearly 40 years of manufacture, installation, and operation of the coaxial cables were fundamental to the success of the undersea fiber-optic cable designs.

This chapter is intended to introduce the technology of undersea coaxial telecommunication cables. The general information and basic formulas provided should serve the average reader, as these coaxial systems are generally being replaced with fiber-optic systems that provide much greater transmission capacity. However, for more in-depth study, an extensive published reference list is provided.

A coaxial cable has an outer "tube" conductor that is concentric around a central conductor, with the inner and outer conductors separated by electrical insulation. The term *coaxial* means that the inner and outer conductors have the same axis.

This chapter is divided into two sections. The first, on undersea Cable telecommunications, provides an overview of the evolution of undersea telecommunication followed by a listing of the major installed transoceanic cable systems up through the first transatlantic fiber-optic system in 1988. The second section, on undersea coaxial cable design, discusses the electrical and mechanical design requirements, including an overview of some practical design limitations as they influence system performance. This section also includes practical considerations regarding manufacturing, handling, installation, and cost. Finally, a discussion of the design interrelationship between the coaxial undersea designs and the new generation of fiber-optic undersea designs is presented.

16.2 UNDERSEA CABLE TELECOMMUNICATIONS

16.2.1 History of Undersea Telecommunications Via Cable

The impetus to develop undersea cables to span seas and oceans, which are hundreds and thousands of kilometers wide, was from two major nineteenth-century inventions. The first of these was the telegraph, invented by Samuel F. B. Morse. Morse began his work on the telegraph in 1832, applied for a patent in 1837 and was granted a U.S. Patent in 1844. The second major invention was the telephone, by Alexander Graham Bell, in 1876. Following each of these momentous telecommunications inventions, there was considerable activity to install terrestrial wire systems in a number of countries. Not surprisingly, there was also great interest in finding ways to transmit the output of each of these important communication devices between continents. However, this was not easily done. The following sections will summarize these challenging, often discouraging but ultimately successful attempts to install telecommunications cables across the oceans.

Table 16.1 summarizes milestones in the installation of undersea cable systems and some other events that are related or lend historical perspective to the subject.

Prior to 1858 it was not possible to communicate rapidly across the oceans. At that point in history, the telephone and radio had not yet been invented. Communication across large bodies of water was a slow process since the fastest ships took many days to cross the Atlantic or Pacific. Of course every communication was necessarily a "round trip," and one message and reply took weeks to accomplish. Today, with the existence of instantaneous worldwide voice and data communication, such a situation may seem unbelievable.

Against the backdrop of this very slow communication over long distances, it is easy to understand how significant was the invention of the first practical electric telegraph. In 1844, Samuel F. B. Morse demonstrated the first working overhead wire telegraph system by transmitting the message "What hath God wrought" between Baltimore and Washington, D.C. Telegraphy, which uses the interruption of an electrical signal in long and short intervals to transmit words by Morse code, then developed rapidly as the preferred means for long-distance communication. The installation of noninsulated wires, using stand-off insulators to isolate the relatively low voltage telegraph electrical pulses from ground, was relatively easy to achieve in terrestrial telegraph systems. Therefore, over a number of years, vast land telegraph transmission systems were installed. However, it proved to be much more difficult to find reliable cable designs and installation techniques for telegraph transmission across seas and oceans [1–8].

16.2.1.1 Technical Challenges

Perhaps the largest technical challenge to developing viable undersea telegraph cable designs, as of 1840, was the lack of a suitable dielectric material to insulate the conductor from ground. Since this need predated the development of synthetic plastics by many decades, it was necessary to find a "natural" material that would have the right properties for the undersea application. These properties were essential from the standpoint of both cable performance and practical considerations in producing

TABLE 16.1 Historical Perspective of Undersea Telecommunication Cables[a]

Year	Event
1815	Napolean defeated at Waterloo
1821	First electric motor—Michael Faraday (England)
1823	Monroe Doctrine
1825	First railroad locomotive—George Stephenson (England)
1827	Invention of photography—Joseph Nicephore Niepce (France)
1832	**Samuel F. B. Morse begins work on telegraph**
1839	Goodyear discovers rubber vulcanization
1843	**Discovery of gutta percha—William Montgomerie et al. (Malaysia)**
1844	**Morse granted U.S. patent on telegraph**
1845	Irish potato famine
1850	**First undersea telegraph communication cable (Calais, France, to Dover, England)**
1856	Pasteurization of milk—Louis Pasteur (France)
1856	**Telegraph cable laid from St. John's, Newfoundland, to Cape Breton Island, Canada (85 miles)**
1857	First land telegraph system completed across the United States
1858	**First transatlantic telegraph cable (Ireland–Newfoundland)**
1861–1865	U.S. Civil War
1866	**First fully successful transatlantic telegraph cable (Ireland–Newfoundland)**
1876	**Telephone invented—Alexander Graham Bell**
1901	Guglielmo Marconi demonstrates first transatlantic radio signal (Newfoundland from England)
1902	**First transpacific telegraph cable (Vancouver, British Columbia–Brisbane, Australia)**
1903	First airplane—Orville Wright
1906	**Announcement of invention of audion tube (forerunner of electron tube)—Lee DeForest**
1914–1918	World War I
1917	Invention of sonar (sound navigation and ranging)—Paul Langevin (France)
1927	Start of radiotelephony
1935	Radar system patent—Robert Watson-Watt (Scotland)
1937	**After discovery of polyethylene, first quantities available for cable trials**
1939–1945	World War II
1948	**Invention of transistor (John Bardeen, Walter A. Brattain, William B. Shockley)**
1956	**First transatlantic telephone cable—TAT-1 (Scotland–Nova Scotia)**
1956	**First Pacific telephone cable (Port Angeles, Washington–Ketchikan, Alaska)**
1963	**First U.S.–Europe transatlantic telephone cable—TAT-3 (Tuckerton, N.J.–Cornwall, England)**
1963	**First transpacific telephone cable system (British Columbia–Hawaii–Fiji–New Zealand–Australia)**
1964	**First U.S.–Japan telephone cable system—HAW-2/TPO-1 United Stated–Hawaii–Midway–Wake–Guam–Japan)**
1965	**World first commercial communication satellite—Intelsat 1**
1988	**First transatlantic fiber-optic cable system—TAT-8 (United States–England and France)**

[a]Bold face indicates events closely related to undersea cable development.

the long lengths necessary for undersea systems. These properties included the ability to be applied around the center conductor by a continuous process that would yield an insulation of sufficiently high quality to perform over many years without failure. It must certainly have been evident to many of those directly involved in the planning, engineering, and installation of these early undersea cable systems that failure of a transoceanic cable would be extremely costly and very difficult, if not impossible, to repair. Therefore, the lack of a suitable existing cable dielectric at that time was a very significant problem [9–14].

The important technical requirements for this insulation included good dielectric strength, low dielectric constant, and relatively high insulation resistance both initially after manufacture and during a long operating lifetime submerged in seawater. Of equal importance were the mechanical properties of the insulation, including compressive modulus, water absorption, elongation properties and the ability to maintain geometry during the high stresses that occur during installation in deep water. These properties, are also among those required in modern undersea communication cables. However, in the early days the available selection of natural materials was small and inadequate.

In 1843, a medical doctor named William Montgomerie from Great Britain was serving on the staff of the Governor of Singapore. He noticed the Malayan natives using a rubberlike substance obtained from the gum of a tree known as gutta percha or gutta taban. He was inspired to write a long paper on the properties of this material. He had noticed that the material was rubberlike at normal temperatures but became plastic when immersed in hot water. In that state, he observed, it could be molded into any shape. He sent his paper and samples of the material to the Royal Society of Arts in London. The following year, several events led to gutta percha becoming an important insulation material for undersea cable. The use of gutta percha as an electrical insulation was first suggested by Michael Faraday. Others designed equipment to extrude the material, and this led to the manufacture of the first cable in 1848. Successful undersea testing of this 2-mile length led to talk of schemes to use gutta percha insulated cables for transoceanic communications. Typically, gutta percha telegraph cables were wrapped with hemp bedding and covered with round steel armor wires applied helically.

In 1850, the first successful undersea telegraph cable was installed between Calais, France, and Dover, England. This 46-km-long cable was insulated with gutta percha but was made without external armor protection. This cable was successful at first but was soon damaged by an anchor. The next year, another cross-channel cable was made that was then overarmored with protective steel wires by a wire rope manufacturer. The success of this cable led to a series of orders. By 1864, an amalgamation between the Gutta Percha Company and the Telegraph Construction and Maintenance Company was also successful and led to better cables and advancement of the manufacturing processes. Gutta percha remained as the preferred undersea telegraph cable insulation for more than 80 years, until the advent of polyethylene in the late 1930s.

In order to ascertain the integrity of the gutta percha insulated long-length cables, before committing the cables to shiploading and ocean installation, a new manufacturing practice was established in the 1850s. The cables were coiled in stationary tanks followed by flooding of the tanks with water. This permitted the water to serve as an electrical ground. The center conductor was then impressed with a direct-current (dc) voltage to measure effectively the insulation resistance and ensure that the insulation was intact. Without this, a tremendous gamble would have been taken. The consequences of not ensuring insulation integrity at the factory stage were either high cost of repair after postinstallation test or worst case, if not recoverable for repair, total inoperability of the cable system.

The center conductor and outer armor were not really problems for the manufacture of these early transoceanic telegraph cables. The preferred center conductor material was copper, usually a solid rod, due to its high conductivity, availability, and favorable mechanical properties. Techniques to join these copper conductors, such as welding and brazing, were used to create the very long continuous lengths necessary.

Likewise, the use of outer armoring using round steel wires was readily available and well-established due to experience with the production of steel wire rope.

The use of dc for transoceanic telegraph cables was viable from the standpoint of not having excessive electrical losses in the signal. This required a stable dielectric with reasonably high insulation resistance, which gutta percha provided. The volume resistivity of gutta percha is greater than 1×10^{15} Ω-cm^3. The loss of signal power radially through the cable insulation is very small even for a long transoceanic crossing. For example, for illustration of this low loss, let us consider a telegraph cable with a 0.432-cm-diameter solid copper conductor insulated with a 1.041-cm-diameter gutta percha layer [see Chapter 3, Eq. (3.38)]. Let us assume that the continuous system length is 2400 km. The insulation resistance of this cable system would exceed 580,000 Ω. Using the case where the sending end voltage is 200 V dc and the receiving end circuit is open, the only electrical loss occurring would be the loss of current through the insulation. With this current equal to the voltage divided by the insulation resistance (200 V/ 580,000 Ω), the dielectric loss current is less than 345 μA. The total power loss radially through the insulation is equal to $I_i^2 R_i < (0.000345)^2 \times 580,000 = 0.0691$ W, a very small loss for a very long cable length. From this example, however, it can readily be seen how important the integrity of the insulation is over the entire system length. Also, the sending-end voltage level selected influences the insulation dielectric loss proportional to $(E_2/E_1)^2$. In the preceding example, if the voltage was increased to 400 V dc, this doubling would result in the dielectric loss quadrupling to 0.2759 W. From this example it can be seen that the dielectric loss using dc is very low. Also, the reactive power loss is insignificant, unlike the situation of high capacitive "charging current" that would occur if the cable was alternating current (ac) powered.

The primary electrical loss in a dc-powered cable is the product of the conductor current squared times the dc conductor resistance ($I_c^2 R_c$). Since the value of the signal current and the conductor resistance in a telegraph cable could both be quite small, this loss could be accepted. For example, using the above illustration further, it can be seen that selection of the "right" sending dc voltage for a particular undersea telegraph system could be workable without benefit of signal amplification along the route. In the 200-V dc example with a 2400-km route length, assume that the telegraph "repeater" at the receiving end operated well at 100 V with 35 mA current. Over the 2400-km length the 0.432-cm-diameter copper conductor would have a dc resistance of < 2830 Ω at a typical ocean bottom temperature of about 3°C. The use of higher source voltage could also be used to attain the desired signal strength at the receiving end. In this example, the 0.035 A current and the 2830 Ω conductor dc resistance would produce a power loss of 3.46675 W ($I_c^2 R$) and a voltage drop of 99.05 V dc ($I_c R$). The most important performance characteristic of these cable systems was the ability to receive the signal with sufficient power and on/off potential difference to accurately reproduce the sending-end signals.

The success of laying a transoceanic cable is greatly influenced by three important factors: (1) the adequacy of the cable design, (2) the knowledge of the bottom profile, and (3) the technique of laying, including adequacy of the laying ship's equipment, experience of the crew, and the ability to lay the cable on the prescribed route.

The cable design must be capable of being laid in very deep water, up to a 6-km depth. The design must be compatible with the ship's equipment and must be capable of withstanding the rigors of laying without damage. For example, the cable components must have adequate shear coupling between layers to distribute properly the tensile loads during laying. The design must be sufficiently robust not to be damaged by handling or passage through cable fleeting drums, tensioning wheels, or linear engines.

Obtaining accurate knowledge of the cable route bottom profile was a much more daunting project in the nineteenth century than it is today. Prior to World War I and the development of sonar, water depths and bottom profiles were primarily obtained by dragging or probing a weighted, length-marked wire along the proposed route(s). This was a time-consuming, tedious, and less thorough technique than the sophisticated side-scan sonar, bottom profiling, and depth ranging used now. Then, as now, it was important to perform proposed route surveys prior to the actual installation run. One can imagine the slow acquisition of depth and profile information by the early weighted wire method and the associated problems of maintaining ship position and course at the slow speeds required. In addition, the early technique could easily miss important bottom profile features (i.e., a sunken ship) that would be important to avoid in the cable route. Modern cable-laying ships are also equipped with very sophisticated real-time data acquisition during the installation, which allows real-time close coordination of the cable payout speed with the ship speed as a function of the actual bottom profile and water depth. Given the knowledge of how important this information is to successful cable laying, it can be understood how difficult it was to properly lay deep-water cables prior to sonar.

The design of the ship itself is, of course, an important factor in successful cable laying. Reference [15] gives a fascinating history of the many cable-laying ships that have been deployed up to 1968. Since then a number of new state-of-the art cable ships have been built specifically for installing and maintaining the vast worldwide undersea cable network. The early cable layers did not have the advantage of modern ships with sonar bottom profiling and electronic navigation. Yet through the use of the best available methods they persisted to lay many significant undersea cables.

16.2.1.2 First Undersea Telegraph Cables

When Cyrus W. Field, an American capitalist, made a proposal to lay an undersea telegraph cable from America to Ireland, he met with ridicule. Nevertheless, he proceeded to form a company that, in 1856, laid an 85-mile-long cable from St. John's, Newfoundland, to Cape Breton Island. This was to be the first link in the first transatlantic line. Cyrus Field then formed a company in England to finance the first transatlantic line. In 1857, the laying of the first attempt was begun from the coast of Ireland. Unfortunately, the cable broke when about 640 km out from the start. In 1858, another attempt was made but was abandoned after the cable broke three times. But just a month later another attempt was made, and the first transatlantic cable laying was completed on August 5, 1858, with a length of 3219 km from Ireland to Newfoundland. The cable system worked and Queen Victoria and President Buchanan exchanged congratulations. However, the signal degraded over time and by October 20, 1858, had ceased to function.

A new transatlantic cable laying was attempted in 1865, but the cable broke halfway across. In 1866 a new cable was successfully laid from Ireland to Newfoundland, and this cable was operated for a number of years. This influenced further laying of telegraph cable until, by the end of the century, virtually every major city in the world was linked by telegraph. An example of a typical undersea telegraph cable design is shown in Fig. 16.1.

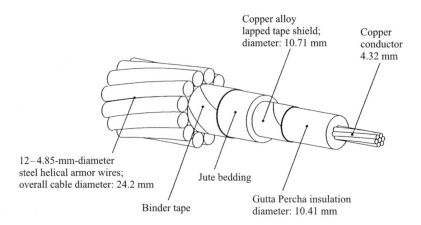

Figure 16.1 Typical undersea telegraph cable.

As difficult as it had been to achieve success crossing the Atlantic, the increased experience and improvements in laying techniques no doubt contributed to greater success with later installations. In the Pacific, where lengths are longer and the water is generally deeper, the first transpacific telegraph cable was laid in 1902, connecting Vancouver, British Columbia, with Brisbane, Australia, a distance of 12,900 km.

16.2.1.3 First Undersea Telephone Cables

In 1927, overseas radio telephony was introduced. In the ensuing years there were a number of improvements made to radio telephony technology that improved both the quality and reliability of the service. This technology provided the first-ever ability to routinely transmit the human voice instantaneously overseas. The primary drawbacks to this service were the limited number of voice channels and the disturbances in the transmission quality caused by the earth's atmosphere. It was also a quite expensive communication service. However, despite the cost, variable transmission quality, and frequent unavailability of voice channels, many people used the service. From this popularity, it was foreseen that good, relatively low-cost transoceanic telephone service had the potential to be quite profitable. Based upon the overall good experience with undersea telegraph cable systems, it was envisioned that a suitably designed undersea telephone cable would be viable. However, there were technical obstacles to overcome [16–25]. A satisfactory voice signal was much more difficult to transmit than a telegraph signal. Also, efforts were initiated in the 1920s to develop improved cable insulations.

A forerunner of truly long-haul undersea telephone cables was a cable system installed between Florida and Cuba in 1921. This preceded the invention of polyethylene and used gutta percha as the insulation and was designed in such a way as to provide increased internal inductance to counter some of the capacitance. Gutta percha had proved to be an excellent insulation for undersea telegraph cables, but for use on voice frequency it had characteristics that caused greatly increased electrical losses. These losses were even greater at carrier frequencies, and the need was recognized for a new insulation with much improved dielectric properties.

An important characteristic of undersea telephone cables is their capacitance. Cables made with gutta percha insulation had, relative to later generation polyethylene insulated cable designs, high capacitance and capacitance that was more frequency dependent. In order to balance out some of this capacitance, it was necessary to introduce additional inductance into telephone cables. In long undersea cable systems this could be achieved, to an extent, by designing the cable to have inherently more inductance in a *continuous* design approach by designing to enhance inductance evenly along the cable length. This was primarily achieved by including either small helical wires or gapped helical copper alloy tapes directly over the center conductor (often a solid copper rod wire) and under the insulation. While this approach was effective in providing some increased inductance evenly along the cable length, and hence balancing out some of the capacitance, there were certainly design limitations physically on how much inductance could be introduced through this technique. Another design approach was through the use of *lump-loaded* cable system designs, which functioned by insertion of inductive load coils periodically within the cable spans. These inductive coils were installed, typically, at the cable factory prior to ship loading. Use of this system design approach allowed a greater amount of inductance to be installed within the system in order to enable balancing out of capacitance to a more ideal system design level. In general, the higher the telephone signal frequency, the more important it was to inductively balance out the inherent capacitance of the telephone cable insulation. Later introduction of synthetic insulation, particularly polyethylene, resulted in significantly lower capacitance and less frequency-dependent change in cable capacitance over the frequency bands of interest. This improvement had the significant effect of greatly reducing the need to introduce compensating inductance.

The invention of polyethylene in the 1930s provided an insulation with electrical and mechanical properties that favored development of undersea telephone cables. Early versions of polyethylene did not contain plasticizers and tended to be prone to cracking. To correct this mechanical drawback and make polyethylene suitable for extrusion on cable conductors, a small amount of such materials as polyisobutylene or butyl rubber (such as 5%) were compounded with the polyethylene. This practice was followed satisfactorily for many years, until improved polyethylene plasticizers were discovered.

Although steady progress was made in the development of polyethylene cable insulation from 1933 onward, it was not until 1956 that the first transatlantic telephone cable system was installed. While this delay was partly due to the diversion of cable insulation research and development (R&D) efforts toward World War II defense efforts, progress was also limited by the progress of electronics inventions. It was recognized, based on the development of terrestrial telephone cable systems, that long undersea telephone cable systems would need to incorporate periodic electronic devices designed to amplify and condition transmission frequency signals. These units, termed *repeaters*, needed to include components that performed amplification and signal conditioning. In an undersea system, failure of such components would necessitate very costly ship mobilization for repair as well as the high cost of lost service and revenues during the failure and repair. A line diagram of an SD repeater is shown in Fig. 16.2. Because of this very high potential cost associated with undersea cable component failure and repair, great attention was given to the highest grade state-of-the-art component R&D.

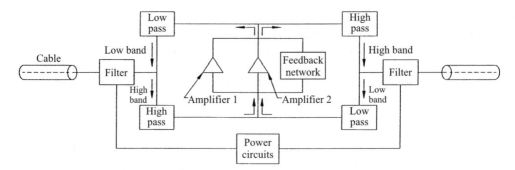

Figure 16.2 Basic configuration of an SD repeater unit (after [16]).

In addition to the fundamental performance of the components, a high level of attention was paid to determining the long-term reliability of these critical electronic components. In addition, system designers to a large extent gave importance to building redundancy into the electronics designs within these repeaters. Although top-of-the-line electronic components, with proven high reliability, added significantly to the repeater cost, this cost was well justified as being a small fraction of the cost to "fix" a failed component in a transoceanic system.

The introduction of the vacuum tube in 1906 (see Table 16.1) ushered in the age of electronics. Over the ensuing years the vacuum tube technology enabled the development of radio, radar, television, long-distance telecommunication devices, and many other important electronic devices. Although vacuum tube electronics were an important part of the ability to build extensive terrestrial telephone cable systems, their use in undersea telephone cable systems required very high attention to long-term mechanical and electrical reliability. However, in 1948, with the invention of the transistor, the future potential for transoceanic telephone cable systems was greatly enhanced. This solid-state device revolutionized the industry. As compared to the vacuum tube, the transistor provided greater reliability, smaller size, lighter weight, lower cost, greater application flexibility, and much more rugged mechanical properties. Nevertheless, it took a number of years of development, system design, testing and reliability evaluation before transistors were ready to be included in transoceanic cable systems. This included the development of repeaters that needed to be as small in size and weight as possible and compatible with the handling and deep-water installation of undersea telephone cable.

In 1956 the first transatlantic telephone cable system was installed, linking Canada and the United States with the United Kingdom. This system was named TAT-1 (for Trans-Atlantic Telephone #1). The cable design used was known as the SB design and appeared as shown in Fig. 16.3. The design features of the SB cable included a central solid 3.35-mm-diameter conductor, three 0.368-mm-thick gapped helical copper alloy tapes, polyethylene extruded to 15.7-mm diameter, six 0.404-mm-thick copper return tapes, a lapped 0.076-mm-thick copper tape (to serve as protection against teredo worms and act as an outer conductor component), a synthetic binder tape, jute serving, helical 2.18-mm-diameter high-tensile-strength steel wires, and two serving layers of impregnated jute and tar. This was the deep-water cable designed to be installed in water depths as deep as 6000 m. There were other overarmored versions of this design

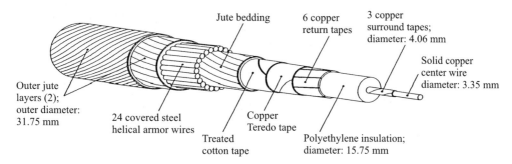

Jute bedding

6 copper
return tapes

3 copper
surround tapes;
diameter: 4.06 mm

Solid copper
center wire
diameter: 3.35 mm

Outer jute
layers (2);
outer diameter:
31.75 mm

24 covered steel
helical armor wires

Copper
Teredo tape

Treated
cotton tape

Polyethylene insulation;
diameter: 15.75 mm

Figure 16.3 External strength undersea telephone cable.

with varying amounts of armoring to protect the cable in the more hazardous environment of shallow water, as in the sections crossing the continental shelves. The commissioning of TAT-1 marked the first time that communications via voice had ever been carried on a transoceanic cable. This system had a 36-circuit capacity. One of these circuits was reserved for telegraph and low-band width use.

A second transatlantic telephone system, TAT-2, was put into service in 1959 connecting the United States with France. This cable system was of the same cable design as TAT-1 and provided 36 circuits.

During the 1950s development of an improved undersea telephone cable design was initiated [26]. By 1960 the older SB design, with external strength member, had been used successfully in two transatlantic systems, a cable transoceanic system from California to Hawaii (HAW-1), and a number of other systems. The new design (SD) would dramatically improve several cable characteristics. The concept of this design was to place high-strength steel wires inside the central conductor [26]. Over this multi-layer stranding of, typically 41 in number, extra-high-strength steel wires, a continuous longitudinal copper tape was trimmed to a precise width, continuously formed, welded continuously using tungsten-inert gas (TIG) welding, reduced in diameter within a reducing mill, and swaged to a precise diameter. A descriptive drawing showing the continuously welded copper tube after welding and after swaging is shown in Fig. 16.4. The final stage of this continuous process not only provided a precise diameter over the copper but also formed the welded copper tube in such a way as to force copper into the outer interstices of the steel wire package, thereby assuring high shear coupling between

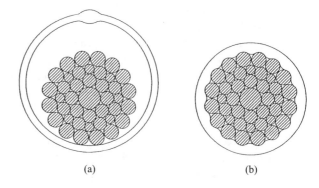

Figure 16.4 SD central strength design cross sections: (a) after stranding and copper tape forming and welding. (b) After reducing and swaging to a diameter of 8.38 mm.

(a)

(b)

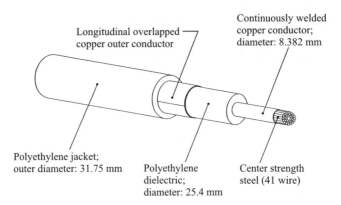

Figure 16.5 Center strength SD undersea coaxial communication cable.

the steel strength member and the copper tube. This design approach rapidly became the preferred state-of-the-art approach for deep-water undersea cable. A good description of this welding process is contained in [27]. Figure 16.5 shows the configuration of the SD design.

The center strength undersea design was developed for the SD, SF, and SG coaxial undersea telephone cable designs [28–30]. Over the copper tube, polyethylene was extruded and then, in the next operation, it was circumferentially shaved to a precise diameter. The shaving process was necessary because extrusion alone could not provide the very close diameter tolerances needed to achieve the extremely uniform diameter. The very uniform diameter was necessary to optimize signal transmission and minimize echoes and distortion. Following shaving, the next and final manufacturing operation for SD, SF, and SG cables was the forming of a longitudinal copper tape applied with a small overlap followed in tandem by extrusion of an outer polyethylene jacket.

The S-series design designation was unique to this type of cable produced in the United States. In Europe, a nearly identical design was developed concurrently. The primary design difference between designs was that the U.S. design used a copper outer conductor longitudinal tape whereas the European design used an aluminum outer conductor longitudinal tape. Both design approaches enclosed the outer conductor within an extruded polyethylene outer jacket. The four operations of (1) stranding and tubing, (2) extrusion, (3) shaving, and (4) outer conductor and jacket provided an excellent cable for transoceanic telephone signal transmission. The SD design, for example, made it preferable to the earlier SB cable for several important reasons. These included lighter weight, higher strength-to-weight ratio, ability to install safely in greater than 6000-m water depths, lower cost, improved handleability, and much lower torque and twisting under tension. This latter property was especially important because of the tendency of the earlier external strength member cable designs to throw kinks under certain installation conditions due to their relatively high torque and relatively high rotational properties. The SD design also possessed significantly improved electrical characteristics, resulting in attenuation that was on the order of two-thirds of the attenuation of SB, as well as the designed ability to carry many more communication channels. The lower attenuation also resulted in greater spacing capability between repeaters, but because of the higher top frequency of SD (1.0 MHz vs.

164 KHz for SB), repeater spacing was actually reduced. However, SD systems only required one cable for two-way transmission, whereas SB systems required one cable for each direction. The design of the next-generation undersea coaxial cable after SD was SF, which is shown in Fig. 16.6.

The new center strength (often called "armorless" or "lightweight") deep-sea communication cable designs did not have any external steel wire armoring over the vast majority of the transoceanic system length. Such cables were purposefully designed for the benign ocean bed environment that typically exists in ocean depths greater than 900 m. Here, in the depth range of 900–6000 m or more, these cables are not subject to damage by anchors, trawlers, abrasion from cable movement in water currents, or damage from marine life or other hazards that exist in shallower waters. Deep-sea communication cables are installed with virtually zero residual tensile load when in place on the ocean bottom. This is to promote conformance of the cable to the bottom profile along the preselected cable route and to avoid catenary suspensions. In this deep ocean environment the cable is completely screened from ultraviolet light. The ocean pressure is high but constant, there is insignificant water current and the temperature is constant in the 3–6°C range. Under this constant and nonhazardous ambient condition, the cable, repeaters, equalizers, and system splices are in a favorable environment for highly stable and reliable long-term operation. These systems typically have design lifetimes of 20 years or more. Where external armoring is needed, in depths of 0–900 m typically, varying levels of external steel armoring are applied to the lightweight cable sections. In addition, many times, in particularly hazardous locations, i.e., where anchor and fishing activity level is significant, the shore ends of transoceanic systems are trenched in or plowed in to depths of several meters below the bottom. The varying levels of external armoring, in some systems, typically progress from shore seaward as follows: (1) rock armor (heavy steel armor layer applied at a high helical angle over double-armored lightweight cable), (2) double-armored cable, (3) single-armored cable, and (4) lightweight deep-water cable. It can be appreciated that it is very important to the systems designers, in order to ensure long-term total system reliability, to have accurate knowledge of the route bottom, subbottom, and fishing/anchor activities along the entire route with emphasis on the continental approaches.

The first transoceanic system using the new center strength design approach was CANTAT-1 installed from Scotland to Newfoundland in 1961. This system had 80 telephone circuits, each channel being 3 kHz bandwidth in the 60–300- and

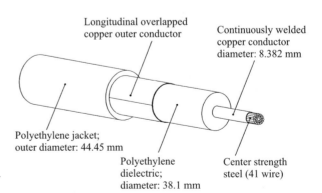

Figure 16.6 Center strength SF undersea coaxial communication cable.

360–608-kHz operating range. In 1962 the first SD system was installed from Florida to Jamaica to Panama. In 1963 the first transatlantic SD system, TAT-3, was installed from Cornwall, England, to Tuckerton, NJ, United States. This system had 138 telephone circuits, each with 3-kHz bandwidth. TAT-3 was 6683 km in length and had 183 rigid two-way repeaters. In 1964, additional SD systems were installed, including HAW-2 from California to Hawaii, TP-1 from Hawaii to Japan, and St. T-1 from Florida to St. Thomas, Virgin Islands. (See Table 16.2.)

16.2.1.4 Installed Transoceanic Cable Systems

The advent of the first commercial communication satellite, *Intelsat 1*, in 1965 made the use of intercontinental satellite communications a reality. This important technology field has important inherent capabilities, such as visual geographical and infrared mapping of the Earth, navigational positioning, meteorology, and so forth, which would not be possible without satellites. Satellites are also an important means of communications transmission and possess the important capability of sending voice and data communication to and from remote locations anywhere on Earth, and beyond, in a "wireless" mode. At one time, some foresaw that satellites might someday replace the need for telecommunications cables. However, other forecasters could see that both technologies would be needed in a future of ever-increasing use of telecommunications. In fact, both satellites and cables are very important technologies existing side by side, and neither has supplanted the need for the other.

Table 16.3 is a partial listing of transoceanic telephone cable systems. See Table 16.2 for some additional information on coaxial undersea telephone cable systems and Chapter 17 for more information on lightwave fiber-optic (e.g., SL) transoceanic cable systems, starting with TAT-8 in 1988.

Telecommunication via cable provides some especially attractive attributes. Anyone who has done much transoceanic telephoning will recognize the truly instantaneous voice transmission that occurs when talking via cable, which sounds as if the other person is next door. Satellite voice transmissions exhibit a distinct delay in the signal transmission and receipt. Cable communications are relatively immune from any atmospheric disturbances as compared to satellites. Security is much higher with cables since tapping in is relatively difficult with coaxial cable systems and virtually impossible

TABLE 16.2   Selected Information on Undersea Telephone Cable Systems

Analog System	Cable Type	System Name	Year in Service	Voice Frequency Channels 3 kHz	Voice Frequency Channels 4 kHz	Top Frequency (MHz)	Diameter of Dielectric (mm)	Repeater Information Type	Repeater Information Spacing (km)	Repeater Information Active Device
SA	1	—	1950	—	24	0.108	11.684	Flexible	74.1	E-tube
SB	1	TAT-1	1956	48	36	0.164	15.748	Flexible	66.7	E-tube
SD	1	TAT-3	1963	148	100	1.0	25.4	Rigid	37.1	E-tube
SF	2	TAT-5	1970	845	640	6.0	38.1	Rigid	18.5	Transistor
SG	2	TAT-6	1976	4200	3150	30.0	43.2	Rigid	9.5	Transistor

Cable type: 1, external strength design; 2, center strength design.
Source: After [16].

TABLE 16.3 Transoceanic Telephone Cable Systems

Year Installed	System Name	From	To	Type	Length (km)	Repeaters (number)
1956	TAT-1	Scotland	Newfoundland	SB	7210 × 2	51 × 2
1959	TAT-2	France	Newfoundland	SB	8354 × 2	57 × 2
1960	HAW-1	California	Hawaii	SB		
1961	CANTAT-1	United Kingdom	Canada	SD (UK)	3837	90 + 20
1963	TAT-3	United States (NJ)	United Kingdom	SD	6683	183
1963	COMPAC	Vancouver, BC	Sydney, Australia	SD (UK)		
1964	HAW-2	California	Hawaii	SD		
1964	TP-1	Hawaii	Japan	SD		
1965	TAT-4	United States (NJ)	France	SD	6683	
1970	TAT-5	United States (RI)	Spain	SF	6500	332
1974	HAW-3	California	Hawaii	SF		
1974	CANTAT-2	United Kingdom	Canada		5012	490
1975	TP-2	Hawaii	Japan	SF		
1976	TAT-6	United States	France	SG		
1983	TAT-7	United States	United Kingdom	SG		
1988	TAT-1	United States	United Kingdom, France	SL		

Source: After [16, 24].

with the new lightwave cable systems. Reliability of modern cables and satellites both tend to be very high and both have acceptable error rates and overall quality of transmission. It is known that the development of communication satellites had a very positive effect in spurring along the development of fiber-optic telecommunications cables and systems.

The deep-water design of the last undersea center strength coaxial cable, known as SG, is shown in Fig. 16.7.

The development of repeaters [16, 31–44] and equalizers was essential to the feasibility of transoceanic telephone cable systems. Repeaters had the primary purpose of amplifying the transmitted analog signals. In the early systems it was found that, despite careful engineering and testing of the integrated systems prior to cable laying, there were changes to the transmitted signals after installation on the seabed. This phenomenon may have been due, at least in part, to the fact that the cable during

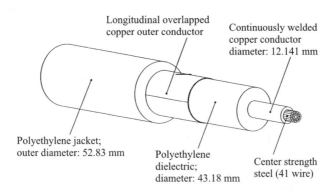

Figure 16.7 Center strength SG undersea coaxial communication cable. [16, 31–44].

installation is stored coiled in large holding tanks at usual ambient ship temperatures but then after installation on the ocean bed is essentially straight, at much lower temperatures (i.e., 3°C) and much higher hydrostatic pressure. These changes could also have been due to subtle changes in the repeater component characteristics at ocean bottom conditions. As a result, this phenomenon was given considerable attention, and it was determined that *equalizers* were needed and that they were needed approximately every 200 nautical miles (371 km). The function of the equalizers was to correct the signal distortions periodically along the transoceanic route. Equalizers were essential to high-quality transmission. Equalizers appeared outwardly like a repeater. For example, as preassembled SD systems (including repeaters and equalizers) were being installed, they were continuously monitored for transmission quality. As the ship proceeded to continuously lay the cable, these monitored signals were adjusted manually, and a final adjustment was made just before the equalizer suddenly went from the completely still mode to moving rapidly with the cable. These equalizer adjustments were important to the proper functioning of the system. These units were often called ocean block equalizers (OBEs). Adjustable equalizers were employed for both SD-type and SF-type cable systems, requiring manual equalizer adjustment shipboard during the lay. With the development of SG systems, a new equalizer design also allowed remote adjustment of the unit from the shore end. This allowed any signal loss over time to be remotely adjusted.

From Table 16.2 it can be seen that each new generation of undersea coaxial telephone cable had a significant increase in voice frequency (VF) channel capacity. The SD systems had a threefold increase in VF channels over SB systems but required repeater spacing about 55% of SB. However, SD and subsequent undersea design generations had the capability to carry two-way transmission over a single cable whereas SB systems required one cable for each direction. The SF systems had about a sixfold capacity increase compared to SD but required repeater spacing of half that of SD. Then, SG systems provided about a fivefold VF channel capacity increase over SF but again required repeater spacing to be reduced to about half of that required for SF. Comparing the last generation of analog communication cables, SG, to the first generation, SB, there was an increase of about 87 times in VF channel capacity as well as needing only one cable rather than two. In comparing SG to SD, there was an increase of about 30-fold in VF channel capacity but about a 300% increase in the number of repeaters needed for a transoceanic system. In addition, each new generation of center strength cable, while having lower loss and higher VF capacity, also required the cable to have larger overall diameter. For example, comparing diameters for deep-water armorless cables, SF [outer diameter (OD) about 44.5 mm] was about 40% greater diameter than SD (OD about 31.8 mm). In turn, SG (OD about 52.8 mm) was about 19% greater than SF and about 66% greater than SD in diameter. The primary drawback of the larger size was the effect on the ship's tank capacity, from a volume standpoint. Although the weight per unit length in air was greater for each new generation, the weight in seawater was not significantly greater because the added material was polyethylene, which has a specific gravity of about 91% that of seawater. Due to this, the permissible recovery depths were not significantly altered with the newer high-capacity designs. The larger cables, of course, were more expensive because of added materials and associated manufacturing costs. Overall, the increased costs of the newer system cables and repeaters were cost-justified based on the vastly increased VF capacity.

When the last transatlantic coaxial cable SG system (TAT-7) was installed in 1983, study had already been undertaken on another new generation of analog cables [16]. This cable type would have been known as SH and would have had capability of transmitting 16,000 3-kHz-spaced channels and a top frequency of 125 MHz. However, the repeater spacing would have been about 4.6 km, or about half that of SG. In addition, there were some formidable transmission problems to overcome. By 1983, intensive multiyear R&D programs to develop fiber-optic transoceanic cables were well underway. For example, in the United States an 8-year R&D program to develop SL lightwave cable was beyond the 50% completion point in 1983. Due to the enormous potential of this new transmission medium technology, next-generation transoceanic telephone cable system development was primarily concentrated on fiber optics from about 1976 until 1988, when the first fiber-optic SL transatlantic system (TAT-8) was placed in service. It is of interest to note that the era of telegraph transoceanic cables lasted about 100 years and the era of analog coaxial transoceanic cables prevailed for over 30 years after that. Overlapping this coaxial era, the era of lightwave transmission really gained impetus in 1970, when it was first practically demonstrated that light could be transmitted through a glass fiber with attenuation of about 20 dB/km, which is about 100 times the decibel attenuation of the standard optical fibers of the 1990s.

16.3 UNDERSEA COAXIAL CABLE DESIGN

The first transoceanic coaxial telephone cables used a central conductor of copper, polyethylene dielectric, outer copper conductor (free-flooding), and outer strength steel armor wires, such as the SB design. The next generations of coaxial undersea cables were dramatically different in design, incorporating the cable strength in the center in the form of a densely packed stranding of high-strength steel wires enclosed tightly within a continuous copper tube. The advantages of these designs, such as SD, SF, and SG, over the earlier designs were significant from both the electrical signal transmission standpoint and the mechanical design standpoint. The following section will discuss these characteristics and the design requirements for these cables (see Tables 16.4 and 16.5).

16.3.1 Design Requirements: Electrical

In the design of a coaxial communication cable, the primary focus is on minimizing signal attenuation and on attempting to reduce signal distortion attributable to the cable. These design objectives were especially important in the design of transoceanic coaxial telephone cables. The attenuation is primarily influenced by the characteristics of the inner conductor, the dielectric, and the outer conductor. (See 15.5 for discussion of the design of coaxial cables.)

16.3.1.1 Power Supply to System Repeaters

Undersea coaxial communication cables are also "power" cables due to the need to provide power to the repeaters. Each repeater has a requirement for a certain wattage in order to operate the electronic components within each repeater housing. Unlike terrestrial communication cable systems, where repeaters can be powered from a source

TABLE 16.4 Properties of SB and SD Undersea Coaxial Communication Cables

Property	SB by Cable Type				SD by List Number				
	A	B	D	H	1[b]	2[c]	3[a]	4[a]	5[d]
Center conductor diameter									
mm	4.06	4.06	4.06	4.06	8.382	8.382	8.382	8.382	8.382
in.	0.160	0.160	0.160	0.160	0.330	0.330	0.330	0.330	0.330
Jacket over outer conductor?	No	No	No	No	Yes	Yes	Yes	Yes	Yes
Diameter over dielectric									
mm	15.748	15.748	15.748	15.748	25.4	25.4	25.4	25.4	25.4
in.	0.620	0.620	0.620	0.620	1.000	1.000	1.000	1.000	1.000
Overall outer diameter									
mm	46.48	36.32	31.75	31.75	31.75	38.10	46.99	53.34	68.58
in.	1.83	1.43	1.25	1.25	1.25	1.50	1.85	2.10	2.70
Weight in air									
kg/m	5.610	3.125	1.711	1.711	1.284	2.024	4.390	5.908	10.745
lb/ft	3.77	2.10	1.15	1.15	0.863	1.36	2.95	3.97	7.22
Weight in seawater									
kg/m	4.152	2.083	0.923	0.923	0.472	0.848	2.872	3.869	7.545
lb/ft	2.79	1.40	0.62	0.62	0.317	0.570	1.93	2.60	5.07
Specific gravity in seawater	3.847	3.000	2.17	2.17	1.580	1.721	2.892	2.898	3.538
Breaking strength									
kgf	23,133	10,432	12,247	14,515	8,845	—	22,679	22,679	45,359
lbf	51,000	23,000	27,000	32,000	19,500	—	50,000	50,000	100,000
Cable modulus									
km	5.566	5.00	13.25	15.73	18.75	10.43	7.89	5.86	6.01
nautical miles	3.00	2.699	7.15	8.479	10.106	5.62	4.256	3.159	3.240
Predicted hydrodynamic constant (degree-knots)	83	67	48	48	35	42	69		
Transverse sinking speed (U), knots	1.45	1.17	0.84	0.84	0.61	0.73	1.20		

Note: Standard (cable-laying) nautical mile = 6087 ft = 1.8553 km.
[a]Copper central conductor, single external armor.
[b]Steel/copper central conductor, no external armor.
[c]Steel/copper central conductor, single external armor.
[d]Copper central conductor, double external armor.
Source: After [45].

independent of the cable, undersea cable systems must receive their repeater power from within the cable. This was done by imposing a dc voltage, between the inner and outer conductors, from the shore terminal stations. Each transoceanic cable system was designed for a specific constant dc voltage applied to the cable at these stations. The necessary voltage would depend upon the cable design and characteristics, particularly the inner conductor dc resistance and total cable system length, as well as upon the power requirements of each system repeater.

An extremely important component within the undersea cable system is the voltage surge protection device. Failure to design adequately this feature into the system would likely result in extensive damage to system electronics. Such a result would likely render the system inoperable and would require an extremely expensive repair.

The applied dc potential was typically on the order of several thousand volts. Often the systems were designed to have a positive dc potential applied on one continental station and a negative dc potential of equal and opposite polarity voltage level applied

799

TABLE 16.5 Properties of SF and SG Undersea Coaxial Communication Cables

Property	SF					SG				
	1[a]	3[b]	4[b]	5[c]	6[b]	1[a]	2[d]	3[e]	4[b]	5[c]
Center conductor diameter										
mm	8.382	8.382	5.588	5.588	8.382	12.141	12.141	12.141	7.163	7.163
in.	0.330	0.330	0.220	0.220	0.330	0.478	0.478	0.478	0.282	0.282
Jacket over outer conductor?	yes	yes	yes	yes	yes	yes	yes	yes	yes	yes
Diameter over dielectric										
mm	38.10	38.10	25.40	25.40	38.10	43.18	43.18	43.18	25.40	25.40
in.	1.500	1.500	1.000	1.000	1.500	1.700	1.700	1.700	1.000	1.000
Overall outer diameter										
mm	44.45	67.31	67.31	86.61	86.61	52.83	77.22	97.28	71.63	91.44
in.	1.750	2.65	2.65	3.41	3.41	2.08	3.04	3.83	2.82	3.60
Weight in air										
kg/m	2.083	10.313	10.328	20.894	20.834	3.084	12.902	25.373	11.280	22.575
lb/ft	1.40	6.93	6.94	14.04	14.00	2.072	8.67	17.05	7.58	15.17
Weight in seawater										
kg/m	0.491	6.660	6.667	14.837	14.777	0.841	8.081	17.754	7.128	15.819
lb/ft	0.33	4.475	4.480	9.97	9.93	0.565	5.43	11.93	4.79	10.63
Specific gravity in seawater	1.31	2.82	2.82	3.45	3.44	1.375	2.68	3.33	2.72	3.34
Breaking strength										
kgf	8,391	36,287	36,287	77,110	77,110	16,780	72,570	172,360	63,500	149,680
lbf	18,500	80,000	80,000	170,000	170,000	37,000	160,000	380,000	140,000	330,000
Cable modulus										
km	17.08	5.45	5.44	5.20	5.22	19.96	8.98	9.71	8.91	9.46
nm	9.21	2.93	2.93	2.80	2.81	10.75	4.84	5.23	4.80	5.10
Predicted hydrodynamic constant, degree-knots	29.7	85.4	85.45	112.37	112.15	34.25	87.83	115	85.65	112.93
Transverse sinking speed (U), knots	0.520	1.491	1.491	1.961	1.957	0.598	1.533	2.024	1.495	1.971

[a] Steel/copper central conductor, no external armor.
[b] Copper central conductor, single external armor.
[c] Copper central conductor, double external armor.
[d] Steel/copper central conductor, single external armor.
[e] Steel/copper central conductor, double external armor.
Source: After [45].

at the other shore station. In this way about half of the system repeaters actually received their power from each terminal station.

The actual voltage appearing across the dielectric at any point was largely a proportional function of the distance of that point from the terminal station to the mid-ocean zero voltage "null" point. In a dc power cable, virtually all of the electrical power loss occurs as the result of $I^2 R$ losses due to the inner conductor dc resistance. There is no significant loss of electrical power from radial loss current through the dielectric due to the very high ($> 10^{17}$ Ω-cm) resistivity of communication cable grade polyethylene. The only other "loss" occurring to the dc power is the amount of power loss, or voltage drop, associated with each repeater. The high degree of nearly linear voltage drop that occurs along the inner conductor from the shore power source to the midocean zero-voltage point allows the system designer to predict quite accurately the actual inner conductor voltage appearing at each repeater.

As an example, perhaps oversimplified, take the case of a 6700-km-long system extending from the United States to the United Kingdom. For example purposes only, assume that the applied voltage at the U.S. end is +4000 V and the applied voltage at the UK end is −4000 V dc and that the repeaters are evenly spaced. The point where +2000 V exists on the inner conductor will be about one-fourth of the total system distance from the U.S. station, or about 1675 km. The zero-voltage null point would be midway in the system, at about 3350 km from each shore. In this example the voltage drop along the inner conductor is about 4000/3350 = 1.1940 V/km. It is evident that the zero-voltage null point needs to occur about midway in the cable span connecting the last positive dc polarity powered repeater to the last negative dc polarity powered repeater in order to ensure the right amount of power is available to each repeater. It is also evident that the voltage magnitude at one or both terminal stations could be adjusted to change the location of the midspan zero-voltage null point if needed.

The transoceanic cable system environment is particularly favorable to the stable operation of the dc power transmission to the repeaters. The cable ambient temperature is not subject to the diurnal temperature changes that always occur in outdoor terrestrial cable systems. Most of the cable system is at constant deep ocean bottom temperatures of about 3°C. Therefore, the cable and repeater electronics are naturally within a virtually unchanging thermal environment. Once powered, the small amount of heat given off by the repeater quickly reaches a state of thermal equilibrium because of the vast heat sink effect of the unchanging temperature water ambient. Likewise, when the cable is first installed, it quickly (typically within 2 h of reaching its permanent location) reaches equilibrium temperature with the ambient. An example of this would be of a cable temperature onboard the cable ship in the 20–35°C range, cooling rapidly to equilibrium on the ocean bottom at about a 3°C ambient temperature. Most deep-water bottom temperatures are in the 3–5°C range with very small variation over the system lifetime. The next most significant sections of the cable system are from deep water up the continental shelves where stable bottom temperatures in the 3–10°C range are typical. The I^2R losses occurring along the inner conductor are constant in magnitude and not significant enough, due to the very low amperage dc carried, to result in any measurable heating of the inner conductor. This fact is also favorable to a very stable condition. The very low amperage dc results in ≪1°C inner conductor temperature rise.

The great advantage of the use of dc powering of transoceanic cables is the efficiency of power transmission. The most significant loss is the I^2R inner conductor loss. The power level supplied at a terminal station is defined as the product EI, where E is the supplied dc voltage and I is the inner conductor current. For example, if $E =$ 4000 V and $I = 1$ A, the power level is equal to 4000 W, or 4 kW. To illustrate the great advantage of using higher transmission voltage, look at how a lower transmission voltage would influence conductor power loss. If, for some reason, the voltage was changed to 2000 V, in order to carry the same 4000 W of power, the conductor current would double to 2 A. Since the value of R, conductor dc resistance, is a constant (heating of the cable inner conductor due to 2 A current would be insignificant), it can be seen that the I^2R loss increases to $40 \times R$ compared to the $1 \times R$ loss in the first case. This quadrupling of loss with halving of the transmission voltage obviously is less efficient. For a given system power requirement it may generally be stated that inner conductor loss reduces by a factor of 4 with a doubling of the system voltage. The general equation that describes this effect on inner conductor loss as a function of change of dc transmission voltage is

$$\Delta(I_2)^2 R = \left(\frac{1}{(E_2/E_1)}\right) I^2 (I_1)^2 R \tag{16.1}$$

where E_1 and I_1 are first-case voltage and current and E_2 and I_2 are second-case voltage and current with R the constant conductor dc resistance.

The cable system designer must select a transmission voltage level that supplies the required power without jeopardizing the system reliability. System voltage level must take into account that the entire length of the system electrical insulation must operate without insulation failure over the system lifetime. As a result of this concern, system voltages are conservatively selected to be well within insulation electrical stress levels that are known to have operated reliably without failure. Vast experience has been gained over the last 40+ years using dc powered transoceanic telephone cable systems. Polyethylene has proven to be an extremely reliable dielectric for this application at the dc stress levels employed. In the manufacture of undersea cable, an extremely high level of attention is given to controlling the integrity of the undersea insulation system. This attention extends to the supply of specified polyethylene compound for the cable insulation extrusion and specially manufactured polyethylene compound used in injection molding of the splice housing insulation. Both the cable insulation process and the splicing and molding processes are tightly controlled under well-established procedures employing thoroughly qualified operators. In addition, there is continuous monitoring of the cable insulation after extrusion to ensure quality. The splice molding machines must first be subjected to qualification procedures and are controlled by a calibration program that includes periodic requalification. Following molding, the molded polyethylene is thoroughly inspected by x-ray and quality analysis. Due to the extreme importance of transoceanic system reliability and the high cost to repair a fault, the measures that are taken to control both the cable quality and the molded splice quality are unparalleled in the cable industry. Following cable manufacture, the way in which the cable is handled for storage, system integration, shiploading, and system installation is designed to minimize risk of damage to the cable. In addition, a series of electrical measurements to ensure polyethylene integrity are carried out. These include insulation resistance measurements and a high-voltage dc withstand test of the dielectric system. As a result of the many steps taken to ensure insulation integrity of the cable and molded components, the probability of any significant defect passing through the entire process chain without detection is virtually zero.

The dc power flows axially through both the enclosing copper tube and the central strength steel wires in proportion to the dc resistance of each component. Although the steel has a much larger cross-sectional area than the copper, its conductivity is much lower than copper. Consequently, the dc power, in a typical coaxial transoceanic cable design, flows about 80% through the copper and about 20% through the steel. This is in contrast to the high-frequency communication transmission signal that flows only through the copper.

16.3.1.2 Transmission Signal Multiplexing

The high-frequency transmission signals are superimposed on the inner conductor with the outer conductor carrying the return signal. There are many design considerations for the cable design related to creating a cable that has the right transmission properties. One of the important developments of the twentieth century has been the invention of polyethylene and the subsequent development of high-purity electrical telecommunication cable insulation grade polyethylene. This, along with parallel development of cable designs and manufacturing processes, made the realization of transoceanic transmission of telephone voice channels via polyethylene insulated coaxial undersea cables possible. Of course, the development of terminal station and undersea cable repeater electronics was equally important to the success of this endeavor.

The human voice frequency range is from 300 to 3400 Hz. This approximate 3100-Hz wide frequency band would require a separate conductor for each additional voice channel if it were not for the invention of a process called multiplexing. Multiplexing electronically performs the process of shifting the original voice frequency channel to a higher frequency [46]. Many incoming voice channels are handled simultaneously, each assigned to a different frequency band of about 3–4 kHz width. Obviously, if many voice channels are to be "stacked up," the total bandwidth of the carrier signal must be wider than the sum of all of the voice channel frequency widths. For analog transmission the dominant multiplexing technique is frequency division multiplexing (FDM), whereby each channel occupies a different portion of the frequency spectrum. Another method is time division multiplexing (TDM), which uses a technique of separating channels by splitting time into narrow slices. Time division multiplexing is particularly appropriate for digital carrier systems such as the lightwave systems, and many analog carrier systems that used FDM have been replaced by digital carrier systems using TDM. However, FDM was the dominant technique for some 50 years for analog communication systems such as transoceanic telephone cables.

Frequency division multiplexing acts as a voice channel "combiner" or "gatherer" at the sending end. At the receiving end, demultiplexers perform the reverse function by separating the voice channels from the carrier signal and returning them to natural voice frequency. This process may be repeated a number of times in the course of the communication channel signal reaching its destination. With FDM, each channel amplitude modulates a different carrier frequency. For example, twelve 4-kHz voice channels make up a group with a frequency range from 60 to 108 kHz. For example, TAT-1 (reference Table 16.2) had a top frequency of 164 kHz and could carry 36 voice channels of 4 kHz width or 48 voice channels of 3 kHz width. The entire bandwidth cannot be used because it is necessary to use small portions for frequency separation between channels. For a more detailed discussion of the FDM and TDM transmission techniques the reader is referred to Chapter 15.

16.3.1.3 Undersea Coaxial Cable Design
for Communications Transmission

In an ideal transoceanic coaxial cable system there would need to be a cable having no loss of signal strength (attenuation) and no distortion or misalignment of the voice channel signals. If such a cable were possible, there would be no need for any undersea electronics such as repeaters and equalizers. The signals would be just as strong and undistorted at the receiving end as they were at the sending station. However, returning to reality, such a system is not within the realm of possibility. Therefore, cable systems are designed to approach the ideal as much as is possible with materials, designs, and system principals that reduce and/or control signal attenuation and distortion to acceptable levels for reliable voice transmission.

Attenuation is most often referred to in decibels per unit length. The opposite of attenuation is gain, also referenced as decibels. The bel is defined in terms of two powers:

$$\text{Bel} = \log\left(\frac{P_o}{P_i}\right) \tag{16.2}$$

The decibel is defined as

$$\text{Decibel} = 10\log\left(\frac{P_o}{P_i}\right) = \text{dB} \tag{16.3}$$

where P_i is power in and P_o is power out.

In the Naperian system a different unit is used for attenuation. This is the neper [sometimes called the transmission unit (TU)]. The formula for the neper is

$$\text{Neper} = 0.5\ln\left(\frac{P_o}{P_i}\right) \tag{16.4}$$

To convert decibels to nepers, divide decibels by 8.686.

By convention, system gain is referred to as positive decibels (+) and system loss is referred to as negative decibels (−). A value such as −10 dB might also be commonly referred to as 10 dB down. Alternatively, if the attenuation in dB is taken as a positive quantity, then gain becomes a negative quantity. In such convention, the power ratios are written inversely as P_i/P_o in Eqs. 16.2–16.4. Attenuation in a coaxial undersea cable was generally expressed as decibels per nautical mile (dB/nm). (See Table 16.6.) The abbreviation nm for nautical miles is unfortunate in that the transmission wavelength

TABLE 16.6 Power Ratio Relationship to Decibel Level

P_o/P_i Power Ratio	Decibels	Signal Effect
0.00001	−50	Attenuation
0.0001	−40	
0.001	−30	
0.01	−20	
0.1	−10	
1	0	Neither gain nor attenuation
10	+10	Gain
100	+20	
1000	+30	
10,000	+40	
100,000	+50	

of optical fibers is frequently expressed in nanometers or as nm in its abbreviated unit nomenclature.

An example of how attenuation might be expressed in an actual system follows. An SF List 1 deep-sea cable has a top frequency of 6 MHz and an attenuation at that frequency of −4.020 dB/nm. For a 10-nautical-mile (10 × 6087-ft) span between repeaters, the total cable attenuation is −40.2 dB/10 nm. This indicates that the power ratio for that 10-nm span is 0.0000955. Another way of expressing this is that the attenuated signal at the last end has only about 1/10,471 of the signal power it had leaving the starting end repeater. This type of attenuation or loss is typical of undersea systems and points up the critical importance of repeater gain to system operation. Even though the signal is greatly reduced in power at the end of the span, the quality of the signal is perfectly adequate. The reconstitution and amplification of the signal are repeated many times in a transoceanic cable system. For example (see Table 16.3) the TAT-5 SF system between the United States and Spain, a distance of about 6500 km, had some 332 repeaters. A voice signal carried on TAT-5 thus had to go through some 333 amplifications (332 undersea repeaters plus 1 at the shore receiving station). Again, this illustration may make it easier to realize how critical the absolute reliability of each repeater is to the system operation.

Four primary cable design parameters are controlled by the cable geometry and by the materials used in the cable. These primary parameters control the secondary parameters. It is the secondary parameters that are of greatest concern to the cable system designer. These secondary parameters are attenuation, characteristic impedance, phase shift, and velocity.

The four primary parameters are resistance R and inductance L of the current-carrying circuit, and capacitance C and conductance G of the dielectric circuit. The terms R and L are series parameters acting axially along the cable; C and G are shunt parameters acting radially along the cable dielectric. These primary parameters will be discussed before proceeding to discuss the secondary parameters.

The series parameters resistance and inductance, are more influenced by frequency than are the shunt parameters capacitance and conductance. This is due to the phenomenon known as skin effect, which will be discussed. Resistance and inductance also vary with temperature and, to a lesser extent, with pressure. The proximity of other conductors, armor wire, and sea water also affect the resistance and inductance.

The resistance of the inner conductor and of the outer conductor are very important. The equation for dc resistance is

$$R_{dc} = \frac{\rho \ell}{A} \tag{16.5}$$

where R_{dc} is the dc resistance in ohms per unit length, ℓ is the unit length in feet, ρ is the volume resistivity of the conductor [for 100% conductivity copper at 20°C (68°F) this is equal to 10.371 Ω-circular mil/ft, and A is the conductor area in circular mils [the circular mil area is found by converting the cross-sectional area of the conductor into an equivalent area round rod with diameter measured in thousandths of an inch (mils); this diameter in mils is squared to yield the circular mil area]. For copper conductivity other than 100% IACS (International Annealed Copper Standard) the numerator of Eq. (16.5) is multiplied by the factor [100/(conductivity in % IACS)] such that R_{dc} is given by

$$R_{dc} = \frac{[100/(\text{conductivity in \% IACS})]\,\rho\ell}{A} \tag{16.6}$$

For example, for 102.1% conductivity, the factor would be 0.979432.

The dc resistance of copper of a given conductivity at 20°C can be corrected for a different temperature by the formula:

$$R_2 = R_1[1 + a_1(t_2 - t_1)] \tag{16.7}$$

where R_2 is the resistance at temperature t_2 (in degrees Celsius), R_1 is the resistance at temperature t_1 [20°C (68°F)] and a_1 is the temperature coefficient of resistance at t_1 (20°C, which is equal to 0.00393/°C).

The temperature coefficient of the dc resistance of copper at any other reference temperature may be determined from the following formula, which depends upon the fact that the temperature coefficient is proportional to conductivity for copper:

$$a_x = \frac{0.0407}{\rho_x} \tag{16.8}$$

where the subscript x refers to the temperature of interest and ρ_x is expressed in Ω-cmil/ft.

In purely dc resistance calculations, as for the inner conductor of designs such as SD, SF, and SG, which comprise both copper and steel components, the dc resistance of the steel must be accounted for to derive the total dc resistance of the conductor. The volume resistivity for the particular steel must be known but then can be used in Eq. (16.6) to derive the numerator correction factor for Eq. (16.5). The copper dc resistance and the steel dc resistance add together as conductors in parallel:

$$R_{total} = \frac{R_{copper} \times R_{steel}}{R_{copper} + R_{steel}} = \Omega/\text{unit length} \tag{16.9}$$

The steel dc resistance is only important electrically in terms of the total dc resistance that determines the cable voltage drop for the supply of dc power to the repeaters. For the superimposed carrier transmission frequency the steel does not carry any of the signal. This is due to the skin effect.

The ac resistance is determined based on the phenomenon known as skin effect. This causes the conductor current at a frequency to travel closer to the conductor surface as frequency increases. As frequency increases, the current travels closer to the outside surface of the inner conductor and closer to the inside surface of the outer conductor in a coaxial cable. The ratio of ac-to-dc resistance is termed the skin effect ratio. First, the factor X is determined:

$$X = 0.0277\left(\frac{\mu f}{R}\right)^{0.5} \tag{16.10}$$

where μ is the permeability of the conductor (for copper it is equal to 1.0), f is the frequency in cycles per second (hertz), and R is the dc resistance in Ω per 1000 ft at 20°C.

By reference to Table 16.7, the skin effect ratios for resistance and inductance can be determined. Then the ac resistance is determined by multiplying the resistance skin effect ratio by the dc resistance.

The formula for the skin effect depth of penetration is

$$\delta = 1.980 \left(\frac{\rho}{f} \right)^{0.5}$$

(16.11)

TABLE 16.7 Skin Effect Ratios

X	R_{ac}/R_{dc}	L_{ac}/L_{dc}	X	R_{ac}/R_{dc}	L_{ac}/L_{dc}	X	R_{ac}/R_{dc}	L_{ac}/L_{dc}
0.0	1.000	1.000	3.9	1.641	0.702	11.5	4.327	0.245
0.1	1.000	1.000	4.0	1.678	0.686	12.0	4.504	0.235
0.2	1.000	1.000	4.1	1.715	0.671	12.5	4.680	0.226
0.3	1.000	1.000	4.2	1.752	0.657	13.0	4.856	0.217
0.4	1.000	1.000	4.3	1.789	0.643	13.5	5.033	0.209
0.5	1.000	1.000	4.4	1.826	0.629	14.0	5.209	0.202
0.6	1.001	1.000	4.5	1.863	0.616	14.5	5.386	0.195
0.7	1.001	0.999	4.6	1.899	0.603	15.0	5.562	0.188
0.8	1.002	0.999	4.7	1.935	0.590	16.0	5.915	0.176
0.9	1.003	0.998	4.8	1.971	0.579	17.0	6.268	0.166
1.0	1.005	0.997	4.9	2.007	0.567	18.0	6.621	0.157
1.1	1.008	0.996	5.0	2.043	0.556	19.0	6.974	0.149
1.2	1.011	0.995	5.2	2.114	0.535	20.0	7.328	0.141
1.3	1.015	0.993	5.4	2.184	0.516	21.0	7.681	0.135
1.4	1.020	0.990	5.6	2.254	0.498	22.0	8.034	0.128
1.5	1.026	0.987	5.8	2.324	0.481	23.0	8.387	0.123
1.6	1.033	0.983	6.0	2.394	0.465	24.0	8.741	0.118
1.7	1.042	0.979	6.2	2.463	0.451	25.0	9.094	0.113
1.8	1.052	0.973	6.4	2.533	0.437	26.0	9.447	0.109
1.9	1.065	0.968	6.6	2.603	0.424	28.0	10.134	0.101
2.0	1.078	0.961	6.8	2.673	0.412	30.0	10.861	0.094
2.1	1.094	0.953	7.0	2.743	0.400	32.0	11.568	0.088
2.2	1.111	0.945	7.2	2.813	0.389	34.0	12.275	0.083
2.3	1.133	0.935	7.4	2.884	0.379	36.0	12.982	0.079
2.4	1.152	0.925	7.6	2.954	0.369	38.0	13.689	0.074
2.5	1.175	0.913	7.8	3.024	0.360	40.0	14.395	0.071
2.6	1.201	0.901	8.0	3.094	0.351	42.0	15.102	0.067
2.7	1.228	0.888	8.2	3.165	0.343	44.0	15.809	0.064
2.8	1.256	0.875	8.4	3.235	0.335	46.0	16.516	0.061
2.9	1.286	0.860	8.6	3.306	0.327	48.0	17.223	0.059
3.0	1.318	0.845	8.8	3.376	0.320	50.0	17.930	0.057
3.1	1.351	0.830	9.0	3.446	0.313	60.0	21.465	0.047
3.2	1.385	0.814	9.2	3.517	0.306	70.0	25.001	0.040
3.3	1.420	0.798	9.4	3.587	0.299	80.0	28.536	0.035
3.4	1.456	0.782	9.6	3.658	0.293	90.0	32.071	0.031
3.5	1.492	0.766	9.8	3.728	0.287	100.0	35.607	0.028
3.6	1.529	0.749	10.0	3.799	0.282			
3.7	1.566	0.733	10.5	3.975	0.268			
3.8	1.603	0.717	11.0	4.151	0.256			

Source: After [21]

where δ is the skin effect depth of penetration in inches, ρ is the volume resistivity of the conductor, and f is the frequency in hertz.

From inspection of Table 16.7 it can be seen that skin effect does not become an important factor until X values of 0.6 or higher are reached [see Eq. (16.10)]. In undersea coaxial communication cable systems, due to the frequencies used, skin effect is always an important factor. The skin effect causes the current to flow in a smaller and smaller portion of the conductor cross section with each frequency increase, hence the significant increase in ac resistance with increased frequency.

In transmission calculations the total series resistance of the inner and outer conductors, sometimes termed the "loop" resistance, is used.

The other series parameter used in calculations is inductance. For a coaxial cable the total inductance is the sum of the internal inductance of the center conductor (L_i), the space inductance (L_s), and the internal inductance of the outer concentric (or return) conductor (L_o). The internal inductance is derived from the formula

$$L_i = \frac{0.0152\mu(L_{ac}/L_{dc})}{1000} = \text{henrys/1000 ft} \tag{16.12}$$

The inductance skin effect ratio L_{ac}/L_{dc} can be obtained from Table 16.7. The formula for the space inductance of a coaxial cable is

$$L_s = \frac{0.1404 \log(r_2/r_1)}{1000} = \text{henrys/1000 ft} \tag{16.13}$$

where r_1 is the radius of the inner conductor and r_2 is the inner radius of the outer conductor.

The internal inductance of the outer conductor is given by the following approximate formula:

$$L_o = \frac{(1.2)(10^4)}{\omega}\left(\frac{\mu_1 f}{R_2 R_3 \pi^9}\right)^{0.5}\left(\frac{\sinh\frac{x_1}{2}\cos\frac{x_1}{2} - \cosh\frac{x_1}{2}\sin\frac{x_1}{2}}{\cosh x_1 - \cos x_1}\right) \tag{16.14}$$

$$= \text{henrys/1000 ft}$$

where

$$x_1 = \delta_t(2\sigma\mu_1\omega)^{0.5}$$
$$= 0.5436\delta_t(f)^{0.5} \quad \text{for 100\% conductivity copper at 20°C} \tag{16.15}$$

and δ_t is the thickness of the outer conductor (tube) in inches, μ_1 is the permeability (for air or copper, $4\pi \times 10^9$ H/cm, σ is the conductivity in mhos per centimeter (for copper, 5.8×10^5), f is the frequency in hertz, $\omega = 2\pi f$, r_2 is the inner radius of outer conductor in inches, and r_3 is the outer radius of outer conductor in inches. This approximation is quite accurate if $r_3/r_2 \leq 1.25$. The internal inductance of the outer conductor L_o is negligible for nonmagnetic materials and, in most cases, may be ignored. The total inductance of a coaxial cable is given as

$$L_{\text{total}} = L_s + L_i + L_o = \text{henrys}/1000 \text{ feet} \tag{16.16}$$

The shunt parameters of capacitance and conductance will now be discussed. The electrostatic capacitance of an insulated conductor is

$$C = \frac{0.00736\varepsilon}{\log(D/d)} = \text{microfarads}/1000 \text{ ft} \tag{16.17}$$

where ε is the dielectric constant of the insulation material (also called specific inductive capacitance), D is the diameter over the insulation (dielectric) in inches, and d is the diameter over the conductor in inches.

The typical dielectric constant for undersea coaxial grade polyethylene is 2.282. The capacitance of polyethylene is not influenced by frequency but is influenced by geometric changes of D/d. Such changes most often occur due to temperature changes and the resulting slight radial expansion or contraction of the polyethylene. For this reason, undersea communication coaxial cable is typically stabilized to 10°C in special tanking facilities prior to factory measurements of the transmission properties of each length (such as capacitance, attenuation, group delay, and characteristic impedance). Polyethylene has a relatively high thermal expansion coefficient when compared to the other major cable components (copper, steel, and, in some designs, aluminum).

The last of the four primary cable electrical design parameters is conductance. Conductance is the reciprocal of insulation resistance and is defined as the dielectric loss from leakage current radially (shunt characteristic) through the insulation per unit of cable length. Since polyethylene has a volume resistivity that is extremely high ($>10^{17}$ Ω-cm^3), with dc voltage and low-frequency voltage, the loss is extremely small and generally considered negligible. At higher frequencies the conductance of polyethylene can be reasonably estimated by the following formula:

$$G = 2\pi f C \tan \delta = \text{mhos}/1000 \text{ ft} \tag{16.18}$$

where C is the capacitance in farads/1000 ft., and $\tan \delta$ is the dissipation factor of the insulation. Undersea coaxial communication cable polyethylene typically has a dissipation factor equal to about 0.00012.

The following section will discuss the important secondary parameters that depend upon the primary parameters of resistance, inductance, capacitance, and conductance. These secondary parameters are very important in cable system design. They are characteristic impedance, attenuation, phase shift, and velocity of propagation.

The characteristic impedance is a complex quantity and is expressed as

$$Z_0 = \left(\frac{R + j\omega L}{G + j\omega C}\right)^{0.5} = \text{ohms} \tag{16.19}$$

where R is the series resistance (in ohms per unit length), L is the series inductance (henrys per unit length), G is the shunt conductance (mhos per unit length), C is the shunt capacitance (farads per unit length), and $\omega = 2\pi f$, where f is the frequency of interest in hertz.

One extremely accurate method of measuring characteristic impedance is described [47]. The characteristic impedance is the impedance that, if used for a sending impedance and terminating impedance together with a cable of that impedance, will cause the line to act as an infinite line. A shorter formula to calculate approximately the characteristic impedance is

$$Z_0 = \frac{138}{(\varepsilon)^{0.5}} \log \frac{D}{d} = \text{ohms} \tag{16.20}$$

where ε is the insulation dielectric constant. Yet another equivalent formula for characteristic impedance is used extensively in the actual measurement of Z_0:

$$Z_0 = (Z_{oc} Z_{sc})^{0.5} = \text{ohms} \tag{16.21}$$

where Z_{oc} is the measured open-circuit impedance, and Z_{sc} is the measured short-circuit impedance.

The propagation constant is a complex quantity that represents the change in magnitude and phase in a signal wave as it passes along the transmission line. It is expressed as

$$\gamma = [R + j\omega L)(G + j\omega C)]^{0.5} = \alpha + j\beta/\text{unit length} \tag{16.22}$$

where α is the attenuation in nepers per unit length [see Eq. (16.4)] and β is the phase shift in radians per unit length. Attenuation indicates the change in signal power as the signal passes through the transmission line. [See Eqs. (16.3) and (16.4) and Table 16.6.] The phase shift is the imaginary part of the propagation constant and is related to the amount of time required for the signal to be propagated through the length of transmission line at the frequency of interest.

The velocity of propagation is the speed at which a signal travels through the transmission line and is expressed as

$$v = \frac{\omega}{\beta} = \text{velocity of propagation in unit length per second} \tag{16.23}$$

A primary practical concern in production of undersea coaxial communication cable systems was the occurrence of signal distortions or discontinuities caused by minor discrete changes in the cable geometry during production, handling, system integration, and installation. Seemingly small discrete localized changes in, for example, the effective local insulation diameter D created localized changes in the effective characteristic impedance and could distort the signal and/or set up reflections [48]. Some further discussion of this will be given in the next section.

16.3.1.4 System Requirements versus Practical Design Limitations

Transoceanic cable system manufacturers used stringent procedures to ensure that the as-built cable systems would have transmission properties that matched the design requirements as closely as possible. Measures taken by such manufacturers for this purpose might be viewed as extreme by manufacturers of other less critical cable structures. Such measures included very stringent quality control of incoming materials, tightly toleranced components, narrow manufacturing tolerances, high levels of quality assurance activity, special processes to ensure the exacting and uniform control of the extruded dielectric diameter and concentricity, and extensive measuring of key transmission properties during the stages of manufacture and after system integration.

Parameters that do not change their effect on attenuation over the frequency range are said to have "cable shape," meaning that they can be compensated by adjusting the cable length. An example of a parameter with cable shape is the dielectric constant. Parameters that do not have cable shape cause attenuation variation with frequency. Examples of parameters that do not have cable shape are dissipation factor and conductor thicknesses. The effects of these parameters on attenuation cannot be compensated by adjusting the cable length. One of the features of the electronics within ocean block equalizers was their designed frequency characteristic that could compensate for variations in dissipation factor. (See Table 16.8.)

In manufacture of transoceanic coaxial cable systems, testing included pulse echo measurements in order to identify and control signal reflections caused by discrete variations in cable properties. Such variations had several causes, such as midspan splices, slight air gaps at the outer conductor overlap, and localized variations in geometry. The pulse echo test was performed with a time-domain reflectometer (TDR) and functioned by sending out discrete pulses and receiving and recording the reflections. The interpretation and analysis of these pulse echo results were an important part of the cable-making process and an art of its own. The skilled interpreter could quite accurately pinpoint the causes and precise lengthwise locations within the cable. Most echoes were well-understood as to cause and most were also tolerable without degrading system performance. Often, it was noted that echoes would disappear or appear as the result of subsequent manufacturing operations such as turning and coiling the length into another tank. This measurement and analysis procedure for reflections enabled the interpreters of the echo traces to discriminate between minor echoes having no serious effect on system performance and more serious echoes that, in some cases, were solved by removing a discrete section from the cable.

TABLE 16.8 Cable Parameter Tolerances and Effects on Attenuation

Parameter	Dimension	Tolerance	Effect on Attenuation (%)	
			At 100 kHz	At 1 MHz
Inner conductor diameter	0.330 inch	±0.001 inch	±0.04	±0.04
Diameter over dielectric	1.000 inch	±0.001 inch	±0.12	±0.12
Thickness of inner conductor	0.023 inch	±0.0005 inch	±0.10	±0.00
Thickness of outer conductor	0.010 inch	±0.0002 inch	±0.18	±0.01
Conductivity of inner conductor	99.1% IACS	±0.3%	±0.11	±0.11
Conductivity of outer conductor	100.6% IACS	±0.3%	±0.05	±0.03
Dielectric constant	2.282	±0.005	±0.11	±0.11
Dissipation factor	0.00012	±0.00002	±0.07	±0.21
Totals				
Algebraic			±0.78	±0.63
Root sum square			±0.30	±0.29

Note: Parameters are for type SD coaxial undersea cable.
Source: After [34].

16.3.2 Design Requirements: Mechanical

16.3.2.1 Fundamental Mechanical Requirements

The cable must have strength greater than it would need to have to avoid breaking during installation or recovery. The tension that occurs during recovery is typically on the order of twice the installation tension. Although the probability of any particular cable section ever needing to be recovered or repaired is extremely low, the entire cable must be designed to be capable of recovery in deep-water system depths. Such depths are often in the 5–6-km range and in some transoceanic crossings may range up to about 8 km. In the case of recovery and repair in deep water the tension on the recovered cable during the repair includes the static weight of the cable in seawater as well as additional dynamic tension due to the ship motions constantly occurring during the period of time involved.

In deep-sea coaxial cables it is desirable to have as high a cable modulus as practically possible. Cable modulus is defined as the length of cable (without repeaters) that could be suspended vertically in static ocean water without exceeding the cable breaking strength. This is a common measure of strength to weight in seawater used in the industry. Tables 16.4 and 16.5 give the breaking strengths and cable moduli of undersea coaxial cables. The deep-water designs shown in these tables are the SB Type D and H and the SD, SF, and SG List 1 designs. It can be seen that the cable modulus of these deep-water designs ranges from 13.25 to 19.96 km. This gives an indication of the strength margin that exists in these designs when installed or recovered in the deepest water.

The acceptable level of maximum tension on deep-water coaxial cables of the center strength design (such as SD, SF, and SG) is typically no higher than about 70% of the cable-breaking strength. This "safety margin" is maintained for primarily two important reasons: (1) to maintain safety during installation and recovery/repair and (2) to avoid permanently changing the cable length by placing the steel strength members beyond their elastic limit. This latter condition is also referred to as exceeding the yield stress. If cable tension is so high as to exceed the yield point for the steel, the cable elongation then enters an undefined inelastic region, which results in a permanent added length. Increasing the tension further and further will eventually result in breaking the cable. The tension and steel stress region lying between the yield point tension and the breaking point tension is avoided in cable laying/recovery practice. Cable that has been subjected to this level of tension will have permanently altered radial and axial dimensions. These changes will also alter the electrical transmission properties.

All cables and wire ropes exhibit nonlinear permanent elongation response when subjected to operating tensions not exceeding their rated working load (always below the yield stress of steel strength members). This permanent elongation remains after taking the cable to a tension within its operating tension range and returning to zero tension. This permanent elongation within the elastic operating range is also referred to as residual strain or constructional stretch. In a solid rod of steel at a "lay angle" of zero degrees from the axial and truly not exceeding its tensile yield stress, this residual strain would be equal to zero. In cables and wire ropes composed of helical steel wires, the amount of residual strain resulting will vary depending upon (1) the steel wires angle(s) of application, (2) the radial "tightness" and wire tension control during manufacture, and (3) inherent (often tiny) gaps occurring between the layers. In practice, in order to minimize this residual strain, the helical angles are kept low and the tightness of the manufacture is well controlled. This results in residual strain of such center strength designs as SD, SF, and SG being very low, typically 0.1–0.15%, which is comparable to the lowest residual strain design wire ropes. This property is desirable in order not to change significantly the signal transmission properties of the coaxial cable.

Another important mechanical property of undersea coaxial cables is the amount of torque generated when the cable is placed under varying levels of tension. The torque characteristic as well as the torsional stiffness (resistance to rotation) is very important in the installation and recovery operations. Cables that generate high torque and/or have low torsional stiffness, are prone to handling difficulties. Sometimes, cables with that undesirable combination of properties have been known to "throw a loop" and therefore result in the need for a repair. Experienced cable-laying ship staff are well aware of the difficulties encountered in laying such cables. Certainly one of the primary advantages of going from the early undersea coaxial external strength designs (i.e.: SB type) to the center strength designs (i.e., SD, SF, SG) was the dramatic reduction in torque.

Yet another important mechanical property of undersea coaxial cable is the ability to withstand permanently very high hydrostatic pressures without significant change in geometry or detrimental effects on the signal transmission properties. Table 16.9 shows the hydrostatic seawater pressures encountered at various ocean depths. The primary property of importance in withstanding the high deep-water pressures without detriment to cable performance is the compressive modulus of the cable materials. Both copper and steel have very high compressive modulus and polyethylene, which makes up the majority of the cross-sectional area of transoceanic coaxial cables, has a high enough compressive modulus to prevent any significant change in radial dimensions due to high hydrostatic pressure. When the cable is installed, it is paid out behind the cable-laying ship and may take hours to reach the bottom. Therefore, it is gradually and uniformly subjected to increasing pressure and is, at the same time, cooling down from its initial temperature in the ship's tank to ocean bottom temperatures in the 3–5°C range. This does produce change in the D/d ratio which in turn changes the capacitance and the conductance while the temperature decrease changes the resistance and inductance.

TABLE 16.9 Ocean Pressure at Deep-Water Transoceanic Cable Depths

Water Depth		Hydrostatic Ocean Pressure (psi)
1000 m	3281 ft	1458
2000 m	6562 ft	2916
3000 m	9842 ft	4375
4000 m	13,123 ft	5833
5000 m	16,404 ft	7291
6000 m	19,685 ft	8749
7000 m	22,966 ft	10,208
8000 m	26,246 ft	11,666

The net effect of cooling and pressurization is a small change in the signal transmission properties once installed on the deep-water ocean floor. Through long experience this phenomenon is well-understood and predictable. The inner conductor is so designed that increasing hydrostatic pressure has virtually no effect on its diameter. The high ocean pressure is, if anything, beneficial to the dielectric properties of the polyethylene. Any tiny gaps that may have developed between the outside of the dielectric and the inside of the outer conductor are virtually certain to be completely closed under high hydrostatic pressure. Therefore the occurrence of pulse echoes in the manufacturing processes can often be expected to reduce significantly under high pressure.

16.3.2.2 Evolution from External to Center Strength Designs

As discussed earlier, the R&D effort initiated in the 1950s to improve upon the early transoceanic telephone cable designs was very successful in developing the SD center strength design. This new design had many advantages. Some comparisons between the deep-water versions of the early external strength member SB (Type D) design properties and those of the comparable SD (List 1) deep-water design are of interest (see Table 16.4). First, both have the same diameter of 31.75 mm but the weight in air of the SD design is 75% of that of the SB design. While this weight reduction is favorable to shiploading and implies lower cable material cost, an even more important advantage of SD List 1 was that its weight in seawater was just 51.1% of that of SB Type D. This fact was, of course, very favorable to installation and recovery tension requirements in deep water and the cost and complexity of ship equipment to control that tension. As a result of this lighter weight in seawater, SD List 1 had a cable modulus that was 41.5% higher than SB Type D and 19.2% higher than the stronger SB deep-water design (SB Type H). In spite of this higher operating strength-to-weight ratio, SD List 1 had only 72.2% of the breaking strength of SB Type D. This reflects the lower need for steel in the SD design and resultant lower cable material costs. Table 16.2 shows that SD provided about 2.8 times as many VF channels as compared to SB. In addition, SD permitted two-way channels over the same cable whereas SB required one cable for each direction. The SD systems did require about 80% more repeaters *per cable*, but because only one cable was needed versus two with SB, the total number of repeaters per system was reduced by about 10% using SD.

Significant advantages of the new designs were dramatically reduced torque under tension, significantly greater resistance to twisting (torsional stiffness), significantly easier handling, and improved cable characteristics during laying. Overall, the center strength system cable designs proved to be less costly to manufacture, had significantly higher channel capacity, and had superior mechanical and handling properties.

16.3.2.3 Protection From Damage

The early external strength SB designs, once armored, were well protected from damage during handling, installation, service, and recovery. Table 16.4 provides further information on the armored versions of the SB design, which were used according to the known hazards such as fishing draggers, anchors, and ocean bottom conditions in relatively shallow continental shelf waters less than 1 km in depth. Notably, the SB designs did not include an extruded plastic jacket over the outer conductor and were designed to be free-flooding to seawater. These designs were subject to long-term corrosion effects on the outer conductor due to their direct contact with seawater.

The central strength designs for deep water required no external steel armor but did always include a high-density polyethylene jacket. This jacket served as a binder, as protection for the outer conductor during handling, installation, and recovery, as a reliable means of transferring shear load during installation and recovery, and as a reliable means of protecting the outer copper conductor from seawater and corrosion. In addition, as indicated in Tables 16.4 and 16.5, SD, SF, and SG also had armored version designs for hazardous continental shelf conditions.

With the development of improved equipment and techniques for trenching or plowing-in undersea cables, more and more armored near-shore sections of transoceanic cable systems were buried below the ocean floor, wherever practicable, as this technology progressed [49–51]. Fishing activities are still the number one enemy of undersea cables, along with anchors.

16.3.2.4 Practical Considerations Regarding Manufacturing, Installation, and Recovery

The SD, SF, and SG designs were less costly to manufacture than the earlier SB designs. They were also significantly more efficient to process, and as a result of the improved design features and development of improved manufacturing techniques, they were able to be produced to more stringent requirements and tolerances.

An interesting characteristic of transoceanic cables is their transverse sinking speed. Such cables are deployed during a transoceanic cable-laying operation at ship speeds typically in the 5–10-knot range (1 knot = 1 nautical mile per hour = 6076 feet per hour = 1852 meters per hour) while paying out cable at a carefully engineered and controlled rate as a function of water depth, bottom profile, and ship speed. The interesting point here is that during this continuous laying process in deep water, the cable paid out at any instant in time does not reach the ocean floor for a long time. The cable (and repeaters if applicable) is literally in a sinking suspension in the ocean for miles behind the ship. The cable is suspended basically parallel to the ocean surface, with a small down angle, and its rate of sinking is determined by its specific gravity, diameter, surface condition and flow properties of the seawater. The transverse sinking speeds for SB Type D, SD List 1, SF List 1, and SG List 1 deep-water cable designs are summarized in Table 16.10. For example, an SD List 1 cable that is being laid in 6000 m of water would take about $6000/(0.61 \times 1852) = 5.31$ h to reach the bottom.

TABLE 16.10   Comparison of Undersea Coaxial Communication Cable Sinking Speed Factors

Property	SB Type D	Deep-Water Cable Design		
		SD List 1	SF List 1	SG List 1
Overall outer diameter, in.	1.25	1.25	1.75	2.08
Weight in seawater, lbs/ft	0.620	0.472	0.330	0.565
Specific gravity in seawater	2.17	1.58	1.31	1.375
Predicted hydrodynamic constant, degree-knots	48	35	29.7	34.25
Transverse sinking speed (U) knots	0.84	0.61	0.52	0.598

It is beyond the scope of this text to attempt to cover the very interesting and complex technology involved with laying these cable systems from specially designed ships. Early laying attempts with transoceanic telegraph cables in the mid-1800s often ended disastrously by breaking or seriously damaging the cable at sea. This not only was due to the lack of experience and detailed understanding of the art and science of laying deep-water cable, but also was seriously handicapped by the state-of-the-art equipment and ship-positioning capability. The invention of sonar, of tremendous importance to this complex procedure, did not occur until 1917 and was not useful to cable laying until many years later.

Reference [45] presents detailed treatment of the art and science of laying cable. A casual review of this document will convince one of the many important considerations involved in this technology and will provide insight into the procedures and techniques required. Also, in [34] there is a very interesting detailed description of the background, design requirements, and techniques employed to develop and build the C.S. Long Lines, one of the major cable-laying ships of the last 40 years. Also, particularly interesting technical writings discuss cable-laying aspects in particular. Many of the listed references contain information related to undersea cable installation. Finally, for a very interesting, detailed, and complete (up to 1968) work on the extensive history of cable-laying ships, the book by Haigh [15] is recommended.

16.3.3 Influence on Design of Undersea Digital Fiber-Optic Cables

In 1858 the first transatlantic telegraph cable was laid, and it was not until 1956 that the first coaxial transatlantic telephone cable was installed, a span of 98 years. Then, it took only 32 years, to 1988, for the first transatlantic fiberoptic cable to be placed in service. During the approximately 40-year period preceding 1988, the vast experience gained with transoceanic analog coaxial cable systems was extremely valuable input to the successful development of transoceanic digital fiber-optic cable designs.

In the design of the SL cable [54], it was recognized early that many of the important design features inherent in the center strength coaxial cable designs such as SD, SF, and SG were also important to successful fiber-optic transoceanic designs. Features such as low residual strain, high strength-to-weight ratio in seawater, high hydrodynamic pressure resistance, handleability, ease of installation, and established and well-proven manufacturing techniques were of key importance to the design of this new generation of cables. The vast experience gained with the coaxial analog cables was exceedingly important to the successful development and commercialization of completely successful transoceanic digital fiber-optic telephone cable systems starting with TAT-8 in 1988.

REFERENCES

[1] C. C., Adley, "The electric telegraph, its history, theory and present applications," *Proc. Inst. Commun. Engr.*, Vol. 11, 1852, pp. 299–329.

[2] J. W. Brett, "On the submarine telegraph," *Proc. Roy. Inst.*, Vol. 2, 1857, pp. 394–403.

[3] J. W. Brett, *On the Origin and Progress of the Oceanic Telegraph*, 1858.

[4] G. Seward, *The Trans-atlantic Submarine Cable*, 1878.

[5] *Personal Recollections of Werner Von Siemens, 1893*, Asher, London.

[6] A. E. Foster, P. G. Ledger, and A. Rosen, "The continuously-loaded submarine telegraph cable," *J. IEE.*, Vol. 67, 1929, pp. 475–506.

[7] J. D. Scott, *Siemens Brothers 1858–1958*, Wiedenfeld & Nicholson, London.

[8] W. H. Russell, *The Atlantic Telegraph (1865)*, David & Charles, 1972.

[9] H. F. Wilson, "Gutta percha and its use in the 19th century," *Trans. Plastics Inst.*, Vol. 17, No. 30, 1949, pp. 28–38.

[10] H. F. Wilson, A. L. Mayers, and R. C. Mildner, "Polyethylene insulated communication cables," AIEE Symposium on Polyethylene, Cleveland, OH, 1951.

[11] M. W. Perrin, "The story of polythene," *Research*, Vol. 6, March 1953, pp. 111–117.

[12] E. Baguley, "Polyethylene submarine telephone cables," *Trans. Plastics Inst.*, Vol. 32, 1964, p. J118.

[13] G. E. Conklin, "Reduction of dielectric loss of polyethylene," *J. Appl. Phys.*, Vol. 35, 1964, 3228.

[14] E. T. Mottram, "Submarine telephone cables," *IEEE Spectrum*, Vol. 2, No. 5, 1965, pp. 96–103.

[15] K. R. Haigh, *Cableships and Submarine Cables*, 1968.

[16] R. D. Ehrbar, "Undersea cables for telephony," in *Undersea Lightwave Communications*, IEEE Press, 1986, Chapter 1, pp. 3–22.

[17] W. A. J. Omeara, "Submarine cables for long-distance telephone circuits. *J. IEE*, Vol. 46, 1911, pp. 309–427.

[18] O. E. Buckley, "The future of transoceanic telephony," *J. IEE*, Vol. 89, Part 1, 1941, pp. 454–461.

[19] C. S. Lawton and L. H. Hutchins, "A non-armoured submarine telephone cable," *Trans. AIEE*, Vol. 72, Part 1, 1953, p. 153.

[20] C. S. Lawton, "More about non-armoured submarine cable," *Trans. AIEE*, Vol. 78, Part 1, 1959, p. 367.

[21] "Manual for submarine cables," Simplex Wire and Cable Company, 1960.

[22] D. P. J. Retief, "The South Africa–Portugal submarine telephone cable, (SAT-1) (360 speech channels)," *Trans. S. Afr. Inst. Electr. Eng.*, Vol. 61, 1970, pp. 413–435.

[23] J. H. H. Merriman, "'Looking ahead to the next century," *Post Office Telecommun. J.*, Vol. 28, No. 1, 1976, pp. 22–25.

[24] C. C. Barnes, *Submarine Telecommunication and Power Cables*, IEE, 1977.

[25] H. M. Brinser, outline and notes on "Underwater communication cables," 1982.

[26] R. J. Halsey, "A proposed new telephone cable between the UK and Canada," *Post Office Electrical Eng. J.*, Vol. 50, 1957, p. 104.

[27] "Gas metal arc welding," in *American Welding Society Welding Handbook*, 8th ed., Vol. 2, Chapter 4, 1991.

[28] I. Welber, "The s.f. submarine cable system (720 two-way channels using transistors and a large diameter single coaxial cable)," *Bell Lab. Record*, Vol. 45, No. 5, 1967, pp. 139–143.

[29] A. W. Lebert and J. Kreutzberg, "Ocean cable for the s.f. system," *Bell Labs Record*, Vol. 45, No. 10, 1967, pp. 320–323.

[30] J. D. Bishop, J. L. Thomas, and C. A. Van Roesgen, "Terminals and power for the s.f. submarine cable system," *Bell Lab. Record*, Vol. 46, No. 5, 1968, pp. 155–160.

[31] J. J. Gilbert, "Telephone cables with submerged repeaters," *Bell Syst. Tech. J.*, Vol. 30, 1951, pp. 65–87.

[32] *Bell Syst. Tech. J.*, Vol. 36, No. 1, Jan. 1957.

[33] O. R. Bates and R. A. Brockbank, "Anglo-Canadian transatlantic telephone cable (CANTAT): laying the North Atlantic line," *Post Office Electrical Engr. J.*, Vol. 56, 1963, p. 99.

[34] *Bell Syst. Tech. J.*, Vol. 43, No. 4, Part 1, July 1964.

[35] "Repeaters and equalizers for the SD submarine cable system," *Bell Syst. Tech. J.*, July 1964.

[36] *Bell Syst. Tech. J.*, Vol. 48, No. 6, pp. 1853–1864, July–August 1969.

[37] *Bell Syst. Tech. J.*, Vol. 49, No. 5, May–June 1970.

[38] "Repeater and equalizer design," *Bell Syst. Tech. J.*, May–June 1970.

[39] Bell Telephone Laboratories, "Physical design of electronic systems," Vol. IV, in *Design Process*, Prentice-Hall, Englewood Cliffs, NJ, 1972.

[40] J. V. Milos and P. A. Yeisley, "Manufacturing aluminum castings and extrusions for use in SG submarine cable repeaters," *Western Electric Engr.*, July 1975.

[41] *Bell Syst. Tech. J.*, Vol. 57, No. 7, Part 1, September 1978.

[42] "SG undersea cable system: Repeater and equalizer design and manufacture," *Bell Syst. Tech. J.*, September 1978.

[43] *Bell Lab. Record*, OTC 3995, 1981 Offshore Technology Conference, Houston, TX, September 1981.

[44] "TASI-E Communication System," *IEEE Trans. Commun.*, Vol. COM-30, No. 4, April, 1982.

[45] C. E. Roden, "Submarine cable mechanics and recommended laying procedures," Bell Telephone Laboratories, December, 1964.

[46] J. G. Nellist, *Understanding Telecommunications and Lightware Systems*, IEEE Press, 1996.

[47] S. A. Schelkunoff, "The electromagnetic theory of transmission lines and cylindrical shields," *Bell Syst. Tech. J.*, October 1934, pp. 532–579.

[48] A. Judy, "Pulse echo analysis," Bell Telphone Laboratories, Atlanta, GA, undated.

[49] "Ends of undersea cables are buried to protect them against damage from vessels," *IEEE Spectrum*, Vol. 4, No. 9, 1967, p. 136.

[50] 'Tilling the ocean floor (protecting TAT cables from damage by fishing fleets)," *Post Office Telecommun. J.*, Vol. 20, No. 2, 1968, pp. 12–15.

[51] "Sea Plow IV—Digging in the newest transatlantic cable," *Bell Lab. Record*, September 1976.

[52] E. E. Zajac, "Dynamics and kinematics of the laying and recovery of submarine cable," *Bell Syst. Tech. J.*, September 1957.

[53] T. Yabuta, K. Ishihara, and Y. Negishi, "Submarine optical-fiber cable design considering low elongation under tension," *Electronics Lett*, Vol. 18, No. 22, 1982, pp. 943–944.

[54] T. C. Chu, "A method to characterize the mechanical properties of undersea cables," *Bell Syst. Tech. J.*, Vol. 62, No. 3, March, 1983.

Chapter 17

TERRESTRIAL AND UNDERWATER OPTICAL FIBER CABLES

William F. Wright

17.1 INTRODUCTION

The purpose of this chapter is to convey to the reader a fundamental understanding of optical fibers and optical fiber cables and to present a review of undersea fiber-optic communication systems as an example of this technology in service. It includes a concise explanation of single-mode fiber characteristics and an overview of several cable designs and their applications. If a deeper understanding of these topics is desired, the reader may consult other sources, including Chapter 15 as well as the references listed at the end of this chapter.

It is of considerable practical value for a cable engineer to be equally knowledgeable in both the communication and power cable fields. For example, a power cable engineer may frequently encounter optical fiber lines laid along or used in conjunction with power cables. Since optical fibers are immune to the effects of electromagnetic fields, they are often placed in close proximity to high-voltage cables; a common type of composite cable is the so-called optical ground wire (OPGW) cable, which essentially is a grounded cable containing optical fiber communication lines. Electrical power companies utilize their existing rights of way to install optical fiber cable, which creates a source of additional revenue and helps satisfy the increasing demand for telecommunication lines. Also, fiber-optic cables are often stranded into undersea power cables. Here, the optical fibers not only provide communication links, but also can provide temperature sensing along the entire length of the fiber (when used in conjunction with specialized optical measurement instrumentation). Thus, it benefits engineers associated with the electrical power industry to know the basics of optical fiber transmission and cable functionality.

Optical fibers used for contemporary communication systems are certainly microscopic marvels of modern technology, but a typical 250-μm-diameter coated fiber is not capable of field deployment without further protection. Glass optical fibers need to be protected from various environmental effects and external forces such as axial and transverse loading, moisture, bending, and temperature excursions. This is the purpose of the external optical fiber cable construction. Proper cable design is critical to provide the protection needed to ensure reliability for the fiber's installation and operating life.

What special qualities do fibers offer that make them the best choice for high-bit-rate, long-haul telecommunications? Consider the following collection of valuable attributes of optical fibers.

17.1.1 Low-Signal Loss and High Bandwidth

Since the function of an optical fiber is to guide high-bit-rate optical pulses over long distances, the qualities of low loss and high bandwidth (i.e., low-signal distortion) are the most important for a waveguide to possess. The term *bandwidth* refers to the information-carrying capacity of the optical fiber. Optical communication systems differ from radio frequency coaxial cable transmission systems only in the frequency range of the carrier wave energy. A typical optical carrier frequency is in the range of 100 THz, while a typical microwave carrier frequency is approximately 1–10 GHz [1]. This translates into an increase in information-carrying capacity of approximately 10,000 for the optical communications system. Compared to coaxial cable transmission, optical fiber communication systems carry much more data, over greater distances between repeaters, over fewer cables. This comparative reduction in transmission equipment makes optical fiber systems more reliable and less costly.

17.1.2 Immunity to Electromagnetic Interference

Optical fiber is a dielectric medium (i.e., a nonconductor of electricity). This characteristic provides optical waveguides with immunity to the effects of electromagnetic interference (EMI) from electric motors, power cables, or lightning. Thus, as a dielectric material, not only are fibers immune to EMI, but also optical fibers do not radiate electromagnetic fields. Optical fibers are also immune to EMP (electromagnetic pulse) effects, which makes them desirable for military use [2].

17.1.3 Small Size and Low Weight

Compared to copper wires, the vastly superior bandwidth and attenuation properties exhibited by optical fibers allow very few fibers to carry a large volume of data. This small size and light weight makes fiber-optic cable a desirable choice when installation space is limited, such as on board ships or in cable ducts. Fiber-optic guided missiles and lightweight tactical deployment cables are just a few of the many specialized military uses that take advantage of this size and weight benefit.

17.1.4 Security

As previously mentioned, optical fibers do not radiate any electromagnetic energy; thus to extract and detect data (i.e., pulses of light) traveling in the fiber, one must either *bore* into the light guiding core or bend the fiber to cause some of the light to leak out. These operations are very difficult to do without causing severe signal loss. Although it is possible, it is very difficult to tap into an optical fiber, which makes it a highly secure transmission medium. Additionally, special *secure* optical fiber systems can be constructed that are truly impossible to invade without detection.

17.1.5 Safety

Optical fiber, being electrically nonconductive, does not create a spark hazard or ground loops. This is a highly desirable characteristic when operating in flammable or explosive environments.

17.2 HISTORICAL PERSPECTIVE

When reflecting upon the following brief list of historical milestones, the reader should be cognizant of the fact that optical fibers are just one component among many in a telecommunication system. Fundamentally, any communication system contains the message origin (e.g., voice, data), a transmitter, a signal transmission medium, and finally, a receiver [2]. The following incomplete list of selected historical milestones shows how people have always been interested in improving their communications technology:

(i) The Greeks of the eighth century BC devised a system of communicating through fire signals.

(ii) In 1838, Samuel Morse invented the telegraph.

(iii) In 1858, the first submarine telegraph cable was installed between North America and Great Britain [3].

(iv) In 1870, John Tyndall, an English scientist, demonstrated the principle of *total internal reflection* (to be discussed in greater depth subsequently in this chapter) by launching a beam of light through a water stream as it poured out the side of a tank [4].

(v) In 1878, the first telephone system was installed in New Haven, Connecticut [2]. (Bell invented the telephone in 1876.)

(vi) In 1880, Alexander Graham Bell invented the photophone, a device that used modulated sunlight transmitted a short distance through the atmosphere to a photosensitive selenium receiver. This was the first example of an instrumented atmospheric optical link [4].

(vii) In 1895, Guglielmo Marconi demonstrated the first radio transmission [2]. Overseas radio telephony was introduced much later in 1927 [3].

(viii) In 1956, the first voice quality transatlantic telephone submarine cable system, TAT-1, went into service. TAT-1 carried a maximum of 36 telephone conversations and was taken out of service in 1979, after exceeding its 20-year design life [3].

(ix) In 1960, the laser was invented. This was important because a laser emits coherent monochromatic light of sufficiently high intensity to be considered as a potential emitter for optical communications. Just two years later, in 1962, the first semiconductor laser was constructed [2]. A laser has a data transmission capacity approximately equal to 10 million television channels.

(x) In 1966, the team of K. C. Kao and G. A. Hockman, and another independent researcher named A. Werts, issued research papers stating that low-loss optical fibers could be achieved by controlling the level of impurities in the chemicals and glass used to make the fiber.

(xi)) In 1970, scientists at Corning Glass Works fabricated the first optical fiber with an attenuation of less than 20-dB/km. This was significant because a 20 dB/km optical signal loss was comparable to copper cable transmission system loss, thus positioning optical fiber as a viable alternative transmission medium.

Since 1970, more than 10 million kilometers of optical fiber has been installed worldwide in less than two decades [2]. (Present telecommunication-grade single-mode fiber loss is about 0.2 dB/km at 1550 nm wavelength, two orders of magnitude lower than the state-of-the-art 20-dB/km loss level of 1970.) In 1988, TAT-8, the first transoceanic fiber-optic cable telecommunication system went into service, which linked the United States to France and the United Kingdom. It contained three fiber pairs (one pair served as a spare) and operated at a wavelength of 1310 nm and a transmission rate of 280 Mbits/s (this provided 20,000 voice channels per fiber pair). When fully completed, the FLAG system will span 27,300 km from Great Britain to Japan (FLAG is an acronym for Fiber Link Around the Globe) [5]. FLAG will operate at transmission rates of 5 and 10 Gbits/s, and will have a capacity equivalent to 120,000 telephone conversations. The rapid advances in lightwave telecommunication technology over the past 20 years were made possible by technological progress in many areas such as electro-optical devices (i.e., lasers and photodiodes), high-speed switches, fiber-optic couplers, connectors, cables, and many other optical devices. Figure 17.1 shows how transmission rates increased as lightwave communication systems evolved.

17.3 OPTICAL FIBER CHARACTERISTICS

17.3.1 Physical Description

An optical fiber can be imagined as consisting of two concentric cylinders of glass. The cylinder at the center, called the core, is surrounded by an outer cylinder called the cladding. Surrounding the cladding is a plastic coating to enhance handling. The core, which has a higher optical refractive index than the cladding, is the region where the majority of the light is guided. Figure 17.2 shows a depiction of a generic single-mode fiber. Not all light launched into the core will be captured. Only light rays that strike the core-cladding boundary at a sufficiently low angle will be guided in the fiber core. The *critical angle* is defined as the maximum angle that permits all light to be reflected back

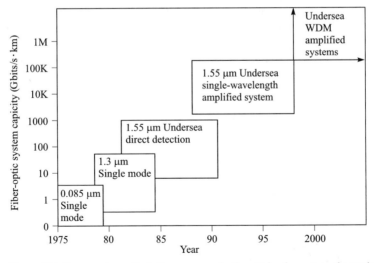

Figure 17.1 Progress in optical fiber communication technology over the period 1974–1992. Different curves show the increase in bit rate–distance product for the five generations of fiber-optic communication systems (after [1]).

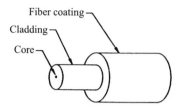

Fiber coating
Cladding
Core

Figure 17.2 Single-mode optical fibre consisting of core diameter of approximately 9 μm, cladding diameter of 125 μm and coating diameter of 250 μm.

into the core. At angles above the critical angle, light is refracted out of the core and into the cladding, where it is lost. Figure 17.3 illustrates this effect as light rays are launched into a fiber core at different angles.

Fibers can be grouped into two types, multimode and single-mode. One of the basic differentiating characteristics is that multimode fibers have a larger light-guiding core than single-mode fibers. The larger multimode fiber core offers both advantages and disadvantages, depending on the specific system requirements. This chapter will confine itself exclusively to describing only single-mode fiber, since it is the most commonly used fiber for long-haul telecommunications; an in-depth treatment of multimode fibers may be found in Chapter 15.

17.3.2 Refractive Index and Total Internal Reflection

As previously stated, in order for light to be captured and guided in the core, the light rays must intersect with the core-cladding boundary at a sufficiently low angle; otherwise, instead of being reflected back into the core, the light will refract out of the core and into the cladding. This phenomenon of light rays being reflected back into the core at each intersection of the core-cladding interface is known as *total internal reflection*. The maximum angle of incidence where total internal reflection can occur is determined by the refractive indices of the core and the cladding. The bending of light at the interface between two optically transmissive mediums is known as refraction. The index of refraction, *n*, of any optically transmissive material is defined as the ratio of the velocity of light in a vacuum, *c*, to the velocity of light in the specified optical medium, *v*. The concept of *refractive index* enables one to quantify a material's ability to refract light (i.e., a rating of its optical density). Refractive index can be expressed mathematically by

$$n = \frac{c}{v} \qquad (17.1)$$

The index of refraction of a vacuum is 1.0000; for water it is 1.3. For fused silica (the base material of optical fiber) it is 1.46 and for diamond it is 2.4. Further clarification of the concept of refractive index can be advanced by considering Fig. 17.4.

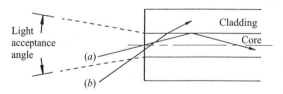

Light
acceptance
angle

(a)

(b)

Cladding

Core

Figure 17.3 Light ray paths in optical fiber: ray *a* is captured in core, but light ray *b* refracts out of the core and into cladding.

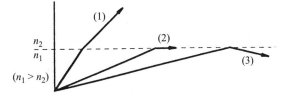

Figure 17.4 Effect of refractive index upon refraction or reflecting of light rays at optical interface.

When a ray of light travels from one medium to another, the ray will be refracted. If the refractive index of the second medium is greater than the first, the ray will be bent toward the normal; if the refractive index of the second medium is less than the first, the ray will be bent away from the normal (the *normal* is a line drawn perpendicular to the optical interface). The amount of refraction that occurs is described by Snell's law, which can be expressed mathematically as

$$n_1 \sin \theta_1 = n_2 \sin \theta_2 \qquad (17.2)$$

where n is the index of refraction and θ is the angle measured with respect to the normal. In Fig. 17.4, light ray 1 strikes the refractive index interface at an angle that results in the ray being refracted as it travels from medium 1 to medium 2. As the angle of incidence increases, depicted by ray 2, the angle of the resulting refracted light ray increases to a maximum of $90°$. The *critical angle* is the angle of incidence that produces a refracted ray that is tangent to the optical surface (i.e., a ray that is $90°$ with respect to the normal). If the angle of incidence becomes larger than the critical angle, as depicted by ray 3, the resulting action is a reflection of the light back into the original optical medium. This is the basis for the guidance of light by the core of an optical fiber.

17.3.3 Mechanical Characteristics

Glass optical fibers must possess adequate strength and flexibility to be of practical value to withstand the stresses of cable manufacturing as well as the in-service environmental conditions of a lightguide transmission system. Fortunately, fibers can be made with sufficiently small flaws and with rugged coatings that protect them from a variety of harsh environmental conditions.

Tensile strength and static fatigue are two important physical qualities of an optical fiber. A flaw-free glass optical fiber is about ten times stronger than a steel wire of equal diameter. Maximum tensile strengths of 2000 kpsi (14 GPa) have been observed in short-gauge-length glass fibers [2]. However, microscopic surface flaws in a fiber create weak points that could ultimately lead to premature breakage if they are not screened out. Fiber manufacturers are able to control material impurities and surface flaws on the glass fibers to very low levels. Fiber strength is maintained through the application of protective polymer coatings immediately as the fiber is drawn, in a process where the surface of the glass fiber is never touched until after the protective coating material is applied and cured. To ensure further reliability and protection from premature breakage, fibers are subjected to a low-level tensile loading over their entire length to establish a minimum tensile strength requirement. This is called *proof testing* the fiber. Typical proof test levels are 50–100 kpsi for terrestrial fiber and 150–200 kpsi for submarine fiber cabling applications. Static fatigue, the slow growth of existing flaws in a fiber, can be accelerated by mechanical, chemical, and other environmental conditions, resulting in weakening of the fiber. Fortunately, this phenomenon is well understood and can be controlled through the use of an appropriate cable design that suits the installation environment and provides the necessary protection for the fibers.

17.3.4 Basic Optical Performance Characteristics

As pulses of light travel down an optical fiber, both the amplitude and the width of the pulses are affected. These properties are referred to as attenuation and dispersion, respectively. Both of these phenomena are strongly dependent on the wavelength of the light pulses that are being transmitted in the optical fiber. Optical communication systems are typically designed so that the wavelength of the light source and the critical transmission properties of the optical fiber are optimized to suit the system's requirements. Figure 17.5 depicts a typical attenuation-versus-wavelength curve for a modern single-mode optical fiber. The two primary intrinsic loss mechanisms are *absorption* of the signal energy by the glass atomic structure (causing energy transformation into heat) and *scattering* caused by inhomogeneities in the density of the glass molecular structure. In silica-based optical fibers, such as those commonly used for lightwave transmission systems, Rayleigh scattering is the dominant loss mechanism in the 1.5–1.6 µm wavelength region. The Rayleigh scattering loss component can be mathematically expressed as:

$$L_R = A\lambda^{-4} \tag{17.3}$$

where A is the scattering coefficient for the specific type of optical fiber in question, with units of $(dB/km)\ (\mu m^{-4})$, and λ is the wavelength in micrometers. Scattering coefficient values, used in the preceding equation, range from 0.7 to 1.0 for typical single-mode fibers [6].

Optical fiber attenuation is expressed in units of dB/km (decibels per kilometer). This unit allows the fiber's logarithmic loss characteristic to be simplified when calculating additive losses of fiber spans and other optical components in a system (i.e., by using dBs, all loss components in a fiber-optic link simply add arithmetically). A fiber's attenuation value can be determined by using an optical source with focusing optics to launch light into the fiber and a power meter to measure the amount of power emerging

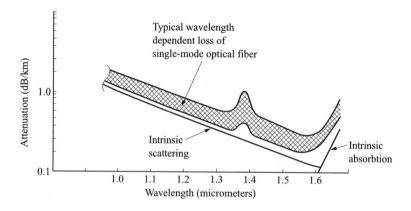

Figure 17.5 Typical spectral loss range of commercial single-mode fibers (after [6]).

from the far end of the fiber. When both the input power and output power are measured, the fiber loss is calculated using the following mathematical expression:

$$\text{Fiber loss (dB/km)} = \frac{-10\log(P_o/P_i)}{\text{length (km)}} \tag{17.4}$$

where P_o is the output power at the far end of the fiber and P_i is the power launched into the fiber at the near end.

The previously described loss measurement technique requires both fiber ends to be accessible to the tester and generates a loss value at one or more wavelengths (depending on the test system). Another loss measurement technique known as optical time domain reflectometry is widely used and offers some advantages. An instrument called an optical time domain reflectometer (OTDR) launches pulses of light into the fiber, whereupon the energy traveling through the fiber generates a small amount of backscattered light that is detected by an optical receiver located at the same end where the pulse was originally launched. Although the level of backscattered light is very low, OTDRs are capable of acquiring large amounts of data and employ signal-averaging techniques to enhance the data. The OTDR measurement results in a plot that depicts the loss characteristics of the fiber over the entire length of the fiber. Figure 17.6 shows an example of an OTDR data plot.

In addition to the previously mentioned intrinsic loss mechanisms, there are also extrinsic bending loss mechanisms. The two most commonly referred to are macrobending and microbending. A simplified distinction between these two mechanisms is the obvious dimensional difference (i.e., bend radius and period) of the bending. Microbending is characterized by a very small bending amplitude on the order of nanometers and a period of critical spatial frequency of about a millimeter [6]. Figure 17.7 portrays the wavelength dependency of macrobending and microbending for single-mode optical fiber. Notice the greater loss incurred in the 1550-nm-wavelength region compared to the 1310-nm region. The 1550-nm region, which exhibits lower optical loss, also exhibits higher bending loss sensitivity. Additional sources of extrinsic loss include radiation induced losses, hydrogen-induced losses, chemical contaminants introduced during the manufacturing of the optical fiber, and any environmental factors that have a detrimental effect on the physical path of the fiber causing additional bending losses while it is in service (c.f. Chapter 15). It is very important for cable designers to choose

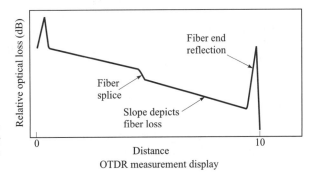

Figure 17.6 OTDR measurement display indicating relative optical loss as function of distance (km) of single-mode optical fiber with splice.

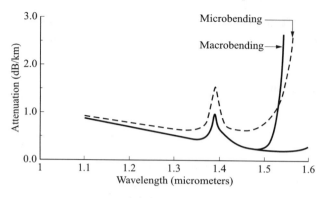

Figure 17.7 Single-mode fiber bending induced loss characteristics (after [6]).

compatible materials and create manufacturing processes that minimize the influence of microbending losses over a cable's rated temperature and tensile load range.

Unlike attenuation properties, which are mainly determined by the bulk material properties of fused silica glass (the base material of single-mode fiber), the dispersion characteristics of optical fiber can be controlled to a limited degree to achieve specific amounts of dispersion throughout specific wavelength regions. As pulses of light travel through an optical fiber, they become wider or more dispersed because the light pulses contain a range of optical wavelengths, and the propagation speeds of the various wavelength components are not equal. It is similar to a group of sprinters who start a race at the same point in time but subsequently arrive at the finish line fractions of a second apart from each other. It can be considered that the time interval between the arrival of the fastest runner and the slowest runner is the same concept as pulse broadening.

In single-mode fiber, chromatic dispersion is the dominant dispersive mechanism. The chromatic dispersion properties of the optical fiber cause broadening of the optical pulses as they travel through a fiber-optic transmission system. Chromatic dispersion consists of two components known as material dispersion and waveguide dispersion. Material dispersion is caused by the dependence of refractive index (and optical pulse propagation velocity) on wavelength. Since an optical pulse contains a group of wavelengths, each component in the group will not travel at the same speed. The second dispersive mechanism, waveguide dispersion, results from the fact that the region of the fiber that guides the optical pulses contains a range of refractive index values. As a pulse travels through the fiber, the central portion of the optical energy envelope sees a relatively high refractive index medium (because the center of the fiber's core contains the highest refractive index material), compared to the outer extremities of the envelope which travel in or near the cladding and see a relatively lower refractive index medium. It is the combination of waveguide and material dispersion effects that is largely responsible for establishing a fiber's overall dispersion profile. The total dispersion (i.e., pulse broadening) that will occur in a fiber-optic transmission system can be calculated with the following equation:

$$D(\text{ps/nm-km}) \times \Delta\lambda\,(\text{nm}) \times L\,(\text{km}) = \text{total pulse broadening (ps)} \qquad (17.5)$$

where D is the fiber's dispersion value at a specified wavelength in units of picoseconds per nanometer-kilometer, $\Delta\lambda$ is the spectral width of the optical source (the optical transmitter) in nanometers, and L is the length of the optical transmission system in kilometers [6]. For example, a system containing optical fiber with a dispersion coefficient of 2.0 ps/nm-km, an optical signal spectral width of 0.1 nm, and a length of 100 km would experience a total pulse broadening of 20 ps.

One common method of measuring single-mode fiber chromatic dispersion is done by transmitting light pulses of various wavelengths through a fiber and measuring the relative arrival times at the receiver. Traditionally, the acquired data are plotted in the form of *relative time delay* versus wavelength, and calculations yield the fiber's dispersion properties over a range of wavelengths. It is possible to keep chromatic dispersion effects well under control by employing various dispersion-compensating schemes. Dispersion compensation methods may be in the form of a specially designed fiber or an optical device that causes the transmitted optical pulses to experience an opposite amount of chromatic dispersion compared to that induced by the transmission system. When chromatic dispersion is minimized, a dispersive mechanism known as Polarization Mode Dispersion (PMD), which is very small in relation to chromatic dispersion, can become the dominant mechanism. Polarization mode dispersion is the result of very slight amounts of imperfect circularity in the optical fiber core, which causes the optical pulses traveling in the fiber to split into two roughly equal parts, with each part being of opposite polarization states. These two differently polarized groups will travel at different velocities; thus the optical pulse experiences broadening the same as previously described for chromatic dispersion.

17.4 INTRODUCTION TO FIBER-OPTIC CABLES

Fiber-optic cables have been commercially manufactured since about 1980 [6]. Early optical fiber telecommunication systems utilized multimode fibers. The best optical losses at this time were around 1 dB/km at 1300 nm, and transmission rates were typically no greater than 90 Mbits/s. In 1983, the wide acceptance of single-mode fiber to the long-haul telecommunication industry marked the beginning of a period of steady growth. Initially, systems operated at 1310 nm, where fibers exhibited losses of about 0.5 dB/km [6]. In contrast, today's optical systems are capable of transmission rates in excess of 10 Gbits/s, with optical fiber losses of less than 0.2 dB/km at 1550 nm. Today's fiber-optic cables are routinely installed in locations such as underground, aerially along poles, in buildings, onboard ships, and across the world's oceans. The nature of the installation environment determines which design of the fiber optic cable should be used.

A fiber-optic cable can be defined as a structure whose purpose is to provide the fibers with protection from detrimental levels of longitudinal and transverse stresses and provide a benign chemical and physical environment for the service life of the fibers. Additionally, the cable structure provides a convenient way to handle and organize the many tiny glass filaments contained within. What attributes would the ideal cable possess? High strength, high crush resistance, ability to withstand small radius bends, ability to withstand large temperature variations and a wide range of environmental conditions, easy fiber accessibility, and low cost would surely be on the list of desirable features. Economic and functional viability ultimately determine which cable designs will be chosen to meet the rigors of the installation process, provide the necessary protection for the service life of the fibers, and be manufacturable within acceptable cost limits. There are as many different types of fiber-optic cable as there are different installation environments. The following sections will provide an overview of some common fiber-optic cable designs and explain various performance considerations. Descriptive categories include terrestrial outside plant cable, submarine cable, and "specialized" cable designs. The latter category includes some cables designed for unusual environments or performance criteria.

17.4.1 Fiber-Optic Cable Design Criteria

A fundamental difference between fiber-optic cable and electrical power cable is that the metal conductors in a power cable carry the tensile stresses created during installation and in-service conditions, whereas fiber-optic cables contain tensile strength members integrated into the cable specifically to isolate the fibers from detrimental levels of stress applied during installation and in-service conditions. Detrimental levels of tensile stress can cause fibers to break or to weaken and make them susceptible to subsequent breaking at low stress levels.

Some of the factors that should be considered in a cable design are the in-service environment (e.g., temperatures, weather, static or dynamic cable loading), optical loss, fiber count, installation practices, bend radius, cable diameter and weight, cable repair/ replacement issues, fiber accessibility, and splicing issues. Cable engineers have created many designs to meet optical fiber protection and reliability requirements. One major design differentiation is either to couple "positively" the fibers to the cable strength members or allow the fibers to float freely in a gel-filled tube or a slot. These are referred to as "tight-buffer" and "loose-buffer" (or loose-tube) designs, respectively. Each design offers advantages and disadvantages, depending on the installation practices and in-service requirements for the cable. In loose-buffer fiber-optic cables, the fibers reside in a cavity, typically in a gel-filled plastic tube or in a slotted central member. The fiber-filled tubes or slots are often stranded around a central member, which causes the linear length of the fibers to be greater than the length of the tubes. When this type of cable sees a tensile stress, the cable will stretch (i.e., strain), but the fibers that are freely floating will experience no strain up to the point where the cable strain equals the excess fiber length. Upon experiencing further cable strain the fibers are effectively coupled to the cable and will experience the same strain as the cable. Fiber strain relief values of up to about 1% are possible [6]. Figure 17.8 depicts the concept of a loose-buffer cable.

The amount of excess fiber length is affected by the stranding pitch and the inner diameter of the loose tube. This type of cable provides a finite degree of fiber stress isolation for both elongation and compressive strains. Also, the gel-filled tube environment is conducive to very low added cabling losses. Loose-buffer fiber-optic cable is very popular in terrestrial outside plant installations. Tight-buffer cable designs are so named because the fibers are tightly coupled to the structure; thus the fibers experience the same strains as the cable. One might think that this design may not provide adequate protection for the fibers. On the contrary, tight buffer cable does offer adequate protection and finds applications in the high-reliability environment of undersea cable. Whether tight or loose buffer, the cable strength members provide a finite degree of protection to prevent the fibers from experiencing detrimental levels of stress and strain. As previously stated, fibers are proof tested to ensure that they are able to withstand specified levels of stress. Typically, a cable's tensile load rating includes a safety margin that takes into consideration short- and long-term fiber loading (short-term loading

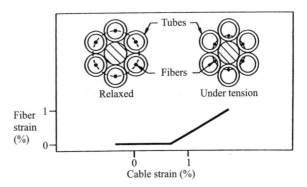

Figure 17.8 Fiber strain in loose tube cable design.

should be limited to less than 60% of the fiber-proof stress, and long-term loading should be limited to less than 20% of the fiber-proof stress [6]. Whatever the design philosophy used, the cable must demonstrate successful performance under test conditions established by industry standards.

Fiber-optic cables utilize a variety of strength members, jacketing materials, and filling compounds. Common jacketing materials include polyethylenes, polyurethanes, and poly vinyl chloride (PVC). Commonly used strength members include steel, Kevlar (a high-strength aramid yarn made by Dupont), and fiber-reinforced plastics. In general, high strength-to-weight ratio is a requirement for these strength members. Again, the selection of any specific material is influenced by many of the previously mentioned cable design factors. This writing will not expand on these details; however, it is important to state that all cable materials and designs are subjected to rigorous qualification testing.

17.4.2 Terrestrial Outside Plant Fiber-Optic Cable

The various cable types that this section includes under "terrestrial outside plant" are designs applicable to telephony aerial, direct buried, and underground duct installation. The loose-buffer types known as stranded loose-tube, slotted-core, and central-core loose-tube designs are very common. Figure 17.9 shows a typical stranded loose-tube cable. The fibers are placed in tubes filled with a gel that prevents water ingress. The other interstices between the tubes and the central member are also flooded with a water-blocking compound.

The tubes are stranded (either helically or S-Z oscillated) around a central member that is upcoated to a specified diameter. This central member provides strength and stiffness. It may be electrically conductive (i.e., steel) or it may be constructed from a dielectric material such as fiber-reinforced plastic rod. The major strength components in the cable depicted in Fig. 17.9 are high-strength yarns which are stranded over the tubes. The materials used for this are often either a Dupont Kevlar aramid yarn or a similarly functioning high-strength glass yarn. Finally, the most common material used for terrestrial outside plant cable jackets is polyethylene, mixed with carbon black to enable it to withstand exposure to sunlight. Cables of this type are appropriate for installation in an underground duct or for lashing to a support cable in an aerial installation. If an outer armor layer is added (typically corrugated steel, aluminum, or copper tape), then the cable would be appropriate for a direct buried installation.

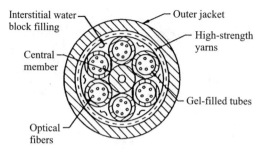

Figure 17.9 Loose-tube-type cable construction.

The metallic armor aids in surviving a lightning strike and provides protection against gnawing rodents.

A slotted-core cable design, shown in Fig. 17.10, functions the same as the loose-tube cable. The slots are often configured in an oscillated lay. As previously described, the fibers float freely in their slots and are decoupled, to some extent, from cable strains. The slotted-core cable design offers a high degree of manufacturing versatility. For this reason, *composite* cable designs (i.e., cables that include a variety of different components such as fibers and copper wires in the same cable) often utilize the slotted-core design.

The central-core cable design made by AT&T is shown in Fig. 17.11. It offers a large centrally placed tube containing fibers (in either wrapped loose bundles or fiber ribbons) and water-blocking gel. Corrugated armor, steel wires (i.e., strength members) and a polyethylene jacket complete the construction. Fiber ribbons, as shown in Fig. 17.11, are an efficient way of handling optical fibers. Fiber ribbons consist of multiple fibers placed side by side in a common plane, and all bonded together by an encapsulating material. Optical fiber ribbons offer two major advantages compared to loose fibers. First, ribbons provide a higher packing density than loose fibers (the highest fiber count cables are made with ribbon cable designs; fiber counts up to 4000 have been designed [7]). Second, fiber ribbons can be spliced with high efficiency with *mass fusion* fiber-splicing equipment.

17.4.3 Submarine Fiber-Optic Cable

Undersea fiber-optic cable systems are designed to meet some very stringent reliability requirements that are quite different from terrestrial-based optical fiber communication systems. Fiber-optic cables used for undersea lightwave communication systems (referred to as *underwater*, *undersea*, or *submarine* cables) are designed for

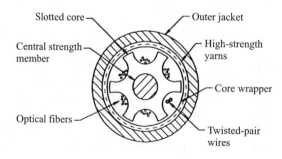

Figure 17.10 Slotted-core cable.

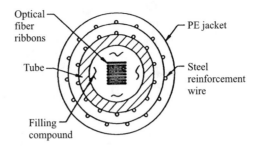

Figure 17.11 Central-core cable (containing optical fiber ribbons) with steel-reinforced crossply sheath (after [6]).

service life on the ocean floor. The following short list summarizes some basic performance requirements for submarine cables: (i) provide a high-strength structure to protect fibers from detrimental levels of elongation strain that could occur during deep-water deployment or recovery; (ii) provide a benign environment for the fibers, with respect to detrimental phenomena such as moisture, high hydrostatic pressure, and hydrogen; (iii) provide an electrical power path for repeaters (when necessary); and (iv) provide a robust exterior designed to withstand expected levels of external aggression commensurate with the cable installation environment. In comparison to terrestrial cables, submarine cables and all other components in a submarine system (e.g., electronic components and cable jointing hardware) are designed for extremely high reliability. This is necessary because when a fault occurs in a submarine cable system, a specially equipped cable ship must be sent to recover, repair, and redeploy the cable. Not only is the repair operation very costly, but while the system is out of service, operating revenues are lost. There are two major categories of submarine fiber-optic cables: *repeatered* and *non-repeatered* cables. Both types will be briefly described in this section. As the name implies, repeatered cable systems incorporate signal repeaters (or amplifiers) into the cable path. Repeatered systems, by definition, are of long length and usually traverse deep regions of the world's oceans. These cable systems typically are engineered for a service life of 25 years and a reliability factor of less than 3 ship repairs over their lifetime [3].

It is obvious that submarine cables that are deployed to the ocean bottom must have high strength and high crush resistance to withstand hydrostatic pressures of 10,000 psi. Cable strength and weight specifications are determined in part by establishing requirements for minimum cable sinking speed, maximum weight in water, and minimum and maximum strength. With the establishment of these critical performance characteristics, a *design window* can be created. Figure 17.12 depicts the concept of a cable design window.

In the deep-sea environment, a cable typically experiences high hydrostatic pressures (10,000 psi at a depth of 4000 fathoms), a constant 2°C temperature, and very little threat from other potentially damaging phenomena. However, in shallow water the potential for cable damage is considerably greater. Cable damage can be caused by fishing trawlers or anchors, and tidal currents can expose once-buried cables. To counter these real-world threats, shallow-water submarine cable installations are usually armored with heavy steel wires, and the cables are often trenched into the sea bottom. Submarine fiber-optic cable manufacturers offer a variety of armor designs to fit the appropriate environmental hazard level. Figure 17.13 depicts a repeatered submarine fiber-optic cable design that has a proven performance record of over 100,000 km of

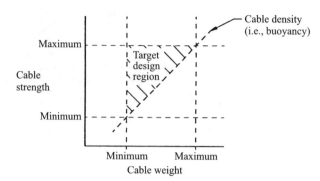

Figure 17.12 Submarine cable design window (after [8]).

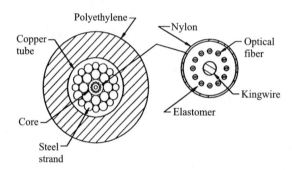

Figure 17.13 Submarine optical fiber cable [9].

worldwide cable installations [9]. At the center of the cable is a structure containing a central steel kingwire surrounded by several fibers, all embedded in a soft elastomeric matrix. This elastomer provides a benign, low-loss environment for the fibers. It is critical to the fiber's reliability that the longitudinal strain on the fibers is controlled during the manufacturing process. The philosophy of this tight-buffer design is to couple the fibers to the cable so that fiber strains can be predicted and controlled to acceptable levels. A high-strength pressure vessel is formed by a copper tube drawn down over a package of 24 helically stranded steel wires that surround the optical core. Water-blocking compound is introduced into the steel wire package to prevent water ingress, and to aid in shear coupling the fiber structure to the steel wire strength members. The continuously welded copper tube acts as a barrier to hydrogen and serves as a low-resistance electrical conductor to bring power to the repeaters. It is critical for an undersea fiber-optic cable to include a hermetic barrier to prevent hydrogen in the ocean environment from diffusing into the silica glass fibers, which would result in a permanent increase in optical loss. An outer polyethylene jacket provides both electrical insulation and abrasion resistance. Various armor designs may be applied over this cable core to provide additional protection. To reiterate, control of fiber strain is of paramount importance. Submarine cables, which experience some of the most severe stresses and strains of any cable type, are designed to preserve fiber reliability levels by proof testing fibers to an appropriate conservative level and then placing the fibers in a structure that allows the cabled fiber strain to be predictable and controlled during cable deployment and recovery operations.

Nonrepeatered submarine optical fiber cable is designed appropriately to meet a modified set of cable performance requirements. Repeatered submarine fiber-optic cable systems, by virtue of the fact that each fiber path requires many repeaters to boost the optical signal to its destination, typically contain relatively low fiber counts (i.e., two to four fiber pairs). In contrast, nonrepeatered submarine cables, which are typically added as extensions or upgrades to existing terrestrial telephone systems, contain much higher fiber counts (12–48). Nonrepeatered cables are often used for interisland links, mainland-to-island links, and coastal festoon installations. These installations are often in relatively shallow water compared to the deep-sea repeatered cable designs. Although the basic performance requirements of repeatered and nonrepeatered cables are similar, specific strength, crush resistance, and electrical power conductivity characteristics of nonrepeatered cable are modified to meet the shallow-water environment. Consequently, nonrepeatered cable designs typically contain less copper and a thinner polyethylene jacket.

The shallow-water environment of nonrepeatered submarine cable systems typically requires armoring of the cables to protect them from tidal currents, abrasive sea bottom conditions, and fishing gear. Figure 17.14 shows a *double-armor* nonrepeatered submarine fiber-optic cable. Submarine cable armor typically consists of wrapping the cable core with a textile yarn such as polypropylene to create a bedding layer. Galvanized steel armor wires are then stranded over the textile yarn bedding. The armor wires are stranded in the opposite direction of the steel strand wires in the cable core to balance the torque forces created when the cable is under tension. A flooding or slushing compound such as a pine tar–asphalt blend is applied to penetrate the bedding yarns and the armor wire interstices. Finally, an outer layer of yarn saturated with slushing compound covers the exterior. It is important that the slushing compound fully penetrates the bedding yarn fibers and wire interstices to provide long-term protection for the steel wires in the potentially harmful marine environment [10].

The critical need for high reliability has resulted in tight-buffer designs being dominant in the (deep-water) repeatered submarine market. However, in the mostly shallow water nonrepeatered cable environment, less severe levels of elongation strain permit a wider range of designs where a variety of both tight- and loose-buffer cable designs can be found. Another element of the nonrepeatered submarine cable market is that fiber break-out and jointing procedures need to be crafts-person-friendly to allow greater flexibility in deployment and repair operations where specialized cable ships and crews may not always be available. Submarine fiber-optic cable manufacturers have created a variety of tests to verify every aspect of performance. Tests such as ocean

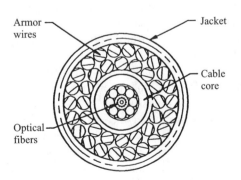

Figure 17.14 Nonrepeatered undersea fiber-optic cable with two layers of armor wire. (Courtesy of Simplex Technologies, Inc. [11])

environment simulations, tension, torque, cyclic bending, material compatibilities, accelerated material aging tests, and more are performed to ensure that high reliability and long service life are realized.

17.4.4 Specialized Fiber-Optic Cable Designs

The following limited collection of information on "specialized" fiber-optic cables provides an overview of selected cable constructions that have been developed to meet some severe performance requirements.

17.4.4.1 Optical Ground Wire Cable

The OPGW cables are installed at the top of high-voltage power transmission line towers and are designed to survive lightning strikes and short-time-duration *short-circuit* events while providing simultaneously a benign environment for the optical fibers. Originally, the ground wire consisted of steel-reinforced aluminum conductors with no communication components. Its sole purpose was to protect the power lines from lightning strikes and provide short-circuit ground paths. Eventually, coaxial or symmetrically paired cables were designed into the ground wires to provide communications. However, being electrically conductive elements, high-voltage surges on the telecommunication lines were possible [12]. In the late 1970s, when optical fiber communications became an established technology, ground wire designs naturally evolved to integrate the desirable dielectric properties and superior transmission characteristics of optical fibers into a new cable design, which became known as the optical ground wire. The OPGW has been successfully constructed in both tight- and loose-buffer designs. It is designed to meet very rigorous mechanical performance characteristics. The aerial installation environment of OPGW places it under both a permanent tensile stress and additional variable stresses from the effects of wind and ice loads as well as temperature variations. In addition, the OPGW cable must survive the effects of lightning strikes, which can vaporize the impacted regions of the outer aluminum wires. A lightning strike is a localized event, but a short-circuit condition subjects a large portion of the cable to a high current surge, resulting in resistive heating of the cable, which creates thermal elongation strain. The OPGW cable is typically constructed with aluminum alloy components or combinations of aluminum alloy, aluminum-clad steel, and steel components. Loose-buffer OPGW designs use fiber containment structures of plastic tubes, aluminum or steel tubes constructed from longitudinally welded metal tapes, or slotted-core components. In tight-buffer OPGW designs, individual fibers or fiber ribbons are embedded in a matrix material that provides shear coupling of the fiber structure to the cable. As always, the cable must provide protection from detrimental levels of fiber strain. Figure 17.15 shows a loose-buffer design and a tight-buffer design OPGW cable.

A variety of cable performance qualification tests have been created to verify critical performance characteristics of these specialized cables. An incomplete list of tests includes crush resistance, tensile stress and strain, torque, aeolian vibration, short circuit, and bending over sheaves under specified diameters and tensions. These tests are designed to simulate worst-case cable installation practices or in-service conditions (e.g., Hydro-Quebec's OPGW system is subject to 45 mm of radial ice accumulation, $-50°C$ temperatures, and very high short-circuit currents) [15]. The most important

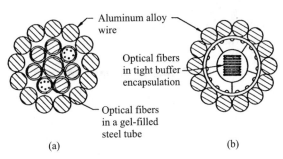

Figure 17.15 (a) Loose-buffer [13] and (b) tight-buffer [14] OPGW cable designs.

function of OPGW is to protect the high-voltage conductors from lightning strikes (and subsequential power outages). Factors that should be considered when choosing the most suitable OPGW for a specific application are the anticipated level of severity of local lightning activity, the voltage level of the electrical power conductors, and the strength of the support towers. When it comes to lightning survivability, common sense prevails; the larger the OPGW, the greater the lightning strike resistance [16].

17.4.4.2 All-Dielectric Self-Supporting Cable

Another type of cable that is commonly used by power utilities is the All Dielectric Self-Supporting (ADSS) cable. This type of cable is designed for installation on support towers that carry high-voltage overhead lines. A generic ADSS cable construction shown in Fig. 17.16 includes a standard loose-tube core and an extremely robust stress relief package. The cable strength members must be a low-weight, high-strength material in order to permit the installed cable to support itself with acceptable strain and withstand wind and ice loads. This type of cable is designed to withstand permanent tensile stress and elongation while providing adequate strain protection for the fibers. Aerially installed cables exhibit aeolian vibrations created by wind. Industry standard tests that have been developed to verify robust cable performance apply a high number of low-frequency, low-amplitude vibrations to simulate many years of in-service cable conditions. Another issue faced by ADSS cable is the deterioration of the outer jacket due to a phenomenon known as dry band arcing. In the presence of a high electric field, electrical currents may flow on the cable surface when the cable gets wet. Dry band arcing can occur when wet and dry spots coexist simultaneously on the cable, promoting localized flashovers in the dry band regions. The ADSS cables can be made to

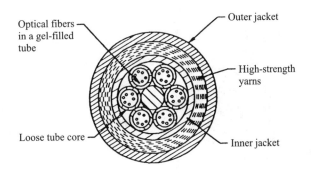

Figure 17.16 An ADSS cable.

withstand successfully the effects of dry band arcing with proper selection of the jacket material [17].

17.4.4.3 Flame-Retardant Fiber-Optic Cable

Optical fiber cables that are installed inside building plenums and risers, cable ducts, tunnels, and equipment rooms, on board ships, or in any location where people would be exposed to toxic and potentially deadly smoke from burning cables must meet special low-smoke, low-toxicity requirements. This type of cable is generally called flame-retardant fiber-optic cable. Flame-retardant cables have been successfully constructed in both tight- and loose-buffer designs. At first glance, these cables appear similar to some of the previously described outside plant cable designs, but a basic difference in design philosophy for these cables is the application of jacketing materials and other cable construction components that exhibit flame-retardant, low-smoke, and low-toxicity properties [18]. Small bend diameters and high crush resistance are desirable characteristics for these cables, since installation spaces are often cramped. Also, resistance to chemicals such as petroleum oils, gasoline, cleaning solvents, acids, and other corrosive chemicals is a consideration. A corrugated metal tape armor may be applied to protect the cable from gnawing rodents. A common flame-retardant design is one that incorporates multiple single-fiber cable units within a common outer jacket. The individually jacketed single-fiber cable units contain their own strength members which gives them good handling robustness when they are "broken out" and allows them to be easily connectorized. Figure 17.17 shows a typical flame-retardant break-out cable. Fiber counts in break-out-type flame-retardant cable typically are limited to a maximum of 24 fibers; otherwise the cable diameter becomes too large. If higher fiber count cables are required, loose-tube or slotted-core designs are usually employed.

The advantages of high-bandwidth fiber-optic communications using small-diameter, lightweight cables have also been exploited by naval communication designers. Weight and space reductions of 90% can be achieved by replacing copper cables with optical fiber cables. Shipboard fiber-optic cable systems probably represent the most stringent case of flame-retardant cable design [19]. For example, the U.S. Navy has established cable specifications for issues such as low smoke, flame retardancy, low toxicity, low acid gas generation, and resistance to hostile fluids and hostile temperatures. Also, mechanical and environmental performance requirements include thermal shock, mechanical shock and vibration, fluid immersion, salt spray, cyclic flex, shock, impact, and others.

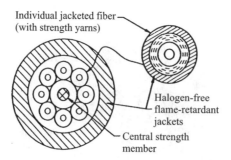

Individual jacketed fiber
(with strength yarns)

Halogen-free
flame-retardant
jackets

Central strength
member

Figure 17.17 Flame-retardant   break-out cable.

Jacketing materials used in the construction of flame-retardant cables include PVCs, polyethylenes, polyurethanes, and crosslinked polyolefins specially treated to be classified as low-smoke, halogen-free compounds. Other cable components include tapes and treated woven glass yarns for flame retardance. Cable designs include both stranded tight- and loose-buffer and slotted-core constructions. A wide range of fiber counts are possible, along with loose-fiber or ribbonized-fiber options. In North America, test specifications have been established by organizations such as Underwriters Laboratories (UL), Canadian Standards Association (CSA), the American Society for Testing and Materials (ASTM), Bellcore, and the Institute for Electrical and Electronics Engineers (IEEE), while in Europe International Standards Organization (ISO) and International Electrotechnical Commission (IEC) standards are commonly used.

17.4.4.4 Hostile Environment Fiber-Optic Cable

Fiber-optic cables can be designed to withstand exposure to corrosive chemicals, petroleum products, and high-temperature steam. These ultrarobust designs are generally referred to as "hostile environment" cables. Standard outside plant cable designs may be able to withstand some of the hostile environment conditions, but high-pressure steam lines that can expose cables to temperatures up to 140°C (284°F) would melt or degrade most jacket and tube materials [20]. Cables installed in underground ducts along with steam pipes must be able to survive exposure to steam leaks for months. In addition to special hostile environment performance requirements, these cables must also exhibit the usual collection of characteristics that promote good installation and service life. A commonly employed design philosophy that meets the many rigorous hostile environment demands is the application of a corrugated welded metal sheath over a core cable that is constructed with specially selected jacketing and fiber-tubing materials. Corrugation of the metal sheath improves flexibility and kink resistance. Figure 17.18 depicts a typical hostile environment cable design.

Several cable manufacturers offer designs that use fluoropolymer materials for the tubes and jacket and corrugated welded metallic sheaths. The application of a welded metallic sheath (aluminum or copper) is critical since it provides a moisture barrier, thereby protecting the fibers and cable core materials from the long-term effects of hydrolytic degradation. As an example, Nylon-6, which exhibits a relatively high resistance to moisture permeability when compared to other suitable polymer jacket materials, is three to four orders of magnitude less effective as a moisture barrier, compared to a welded metallic sheath [21]. The primary performance requirement for these cables

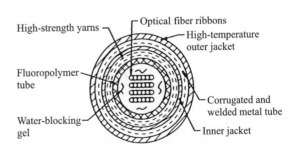

Figure 17.18 Hostile environment fiber-optic cable.

is that they maintain their physical robustness for many years, in spite of exposure to many hostile conditions (especially high pressure–high temperature steam). This is achieved through the use of properly selected materials and the application of an effective metallic sheath moisture barrier that protects the fibers from potential long-term degradation. Long-term cable reliability can also be improved by installing the maximum possible clearance or designing effective thermal insulation between the steam pipes and fiber-optic cables [22].

17.5 INTRODUCTION TO UNDERSEA FIBER-OPTIC COMMUNICATION SYSTEMS

Now that we have expounded on some of the fundamental characteristics of optical fibers and cables, we are prepared for the following brief treatise on undersea fiber-optic communication systems. The basic elements of any communication system consist of an information source, a transmitter, a transmission medium, and lastly, the receiver. The information source and/or the information recipient may be a person, a computer, or a machine. The most fundamental performance requirements for a communication system include (i) transmission of the information to the receiver in a manner that maintains *acceptable signal recognition*, (i.e., signal vs. noise) and (ii) transmission system *bandwidth* that permits an acceptable volume of information to be transmitted and received.

Mainly due to the use of computers in our technological society, the times in which we live have often been referred to as the *information age*. Modern telecommunication technologies offer us the ability to generate information and communicate wherever we are, over a global network. Technologies such as satellites, cellular telephones, microwave waveguides, radio, coaxial cable links, and optical fiber transmission systems are all in use. However, the technology that offers the best combination of low signal loss, high bandwidth, cost competitiveness, and reliability is optical fiber telecommunications technology. The design of an optical fiber transmission system is influenced by factors such as length, number of channels required, outside cable plant location (terrestrial aerial, terrestrial buried, or undersea), repair and aging concerns, compatibility and upgradability with respect to existing systems, and maintenance and repair concerns. One of the basic issues that exerts a major influence on the design of a system is whether it is a land-based (terrestrial) system or an undersea system. A comparison of terrestrial and undersea fiber-optic systems in Table 17.1 reveals several differences that influence their respective design philosophies.

Operating economics and inaccessibility of undersea fiber-optic cable communication systems create a challenging situation for system designers. In order to minimize system outages and avoid the high cost of repairs (which require specially equipped cable ships), undersea systems are engineered for very high reliability. All system components and materials in an undersea system undergo qualification testing to filter out premature failures. Also, designs incorporating critical component redundancy are used whenever feasible. The result is a high-reliability communication system that is expected to generate fewer than three repairs over an expected 25-year operation life. The focus of this writing is on repeatered undersea fiber-optic communication systems. If the reader wishes to pursue further information on terrestrial-based fiber-optic systems,

TABLE 17.1 Comparison of Terrestrial and Undersea Optical Fiber Systems

Attribute	Terrestrial	Undersea
Location and accessibility	Aerial and buried, easy accessibility	Ocean bottom, difficult accessibility
Electrical power for amplifiers	Easily accessible via separate power cable	Fiber-optic cable must also provide electrical power
Maintenance and repair	Designed for rapid fault diagnostic and modular replacement	Designed for high reliability, critical component redundancy, 25 year service life typical
Miscellaneous environmental issues	Operating temperature −40°C to +50°C, ice and windstorm damage, lightning, vandalism, rodents	Deep-ocean temperature 2°C, typical high ocean bottom pressures, fishing trawlers, increased optical attenuation due to sea bottom radiation and hydrogen.

the IEEE *Communications Magazine*, Proceedings of the IEEE, and [2] contain many excellent articles. Also reference should be made to Chapter 15.

17.5.1 Repeatered Undersea Fiber-Optic Communication System Technology

The world's first transoceanic fiber-optic cable communication system, placed in service in 1988, was TAT-8, which linked North America to Great Britain and France [6]. This was followed in 1989 by a system traversing the Pacific Ocean. These *first-generation* repeatered systems operated at a transmission rate of 280 Mbits/s per fiber pair (the equivalent of 20,000 voice channels per fiber pair), at an optical wavelength of 1310 nm, using single-mode optical fiber with minimum chromatic dispersion at 1310 nm. The repeaters in these systems (also referred to as regenerators) consisted of an optical receiver that transformed the optical signal into an electrical signal, electronic circuitry to correct the shape and timing of the digital data pulses, and finally, a semiconductor laser to retransmit the optical signal to the next repeater [3]. Regenerator-type repeaters were limited to functioning at only one wavelength and only one specified transmission rate. They could not be upgraded to higher transmission rates. One of the many high-reliability design features of these repeaters was transmitter modules that contained two lasers connected through an optical switch. In the event of a laser failure, the back-up laser could be switched in, thus maintaining system operation.

Second generation undersea fiber-optic communication systems technology continued to use regenerator-type repeaters similar to first-generation systems; however, the optical signal transmitter wavelength was changed to 1550 nm instead of 1310 nm. At a wavelength of 1550 nm, optical fiber exhibits lower attenuation and therefore allows a greater distance between repeaters. TPC-4, a system interconnecting Taiwan, the Philippines, China, and North America, installed in 1991, is an example of a second-generation undersea system. The transmission rate was increased to 560 Mbits/s per fiber pair, and repeaters were spaced about 110–120 km apart, which was a substantial improvement over first-generation system repeater spacings of about 70 km. A reduction in the number of repeaters is beneficial because it results in a reduced load on the electrical power feed equipment, fewer components, and improved system reliability.

In addition to the obvious signal transmission issues, there are many other engineering obstacles that must be overcome to manufacture, install, and operate undersea fiber-optic communication systems. Consider the critical functions of the electrical power system, the repeater supervisory system that evaluates system performance, and the role of the cable ships that carry out installation and maintenance activities.

In addition to protecting the optical waveguides, the undersea cable also conducts electric power needed to energize the in-line repeaters. The electrical power systems, which are located near the cable landing sites, are designed to supply up to 7500 V (dc.), with a highly stabilized constant current. Among the many unique features that are designed into these electrical power systems are the ability to generate low-frequency modulations that enable remote detection equipment onboard a cable ship to locate a submerged, buried cable and redundant design concepts that permit service personnel to safely deenergize and perform maintenance on selected portions of the power system, while other parts continue functioning to supply full power to the communication system [23]. The repeater supervisory system is used to monitor vital functions and remotely control some of the critical components located in the repeaters [5]. Characteristics such as *received* and *transmitted* optical power levels can be monitored to evaluate system aging. Another key function is the ability to communicate with each repeater and thereby enable system fault analysis. If a system fault occurs, each repeater is polled in succession until the fault location is verified. Also, because of critical component redundancy design, a back-up transmitter or receiver could be remotely switched into operation to replace a failed component.

It is no small task to be responsible for the deployment and maintenance of a transoceanic undersea communication system worth hundreds of millions of dollars. This awesome task is accomplished with the use of specially designed cable ships outfitted with unique equipment capable of installing, locating, recovering, and repairing the undersea cable plant. During an installation, cables placed in relatively shallow water (on the continental shelf) are buried with a remotely operated sea plow. In deep water, the cable is laid directly on the ocean bottom. The cable ships' capabilities enable cable recovery from any depth. Typically, a cable repair event consists of retrieving the damaged cable ends with a grapnel hook, splicing a short additional length of spare cable into the severed section, then reinstalling the cable. All undersea communication systems are designed with a sufficient operating margin to allow a small amount of added loss caused by a repair and still operate with acceptable signal power [24].

17.5.2 Optical Amplifier Undersea Fiber-Optic Communication System Technology

The design of *third-generation* undersea fiber-optic communication systems has taken advantage of a new device that is one of the most significant developments to impact communications technology in many years. The *erbium-doped fiber amplifier* (EDFA) is a device that creates direct optical gain (not an optical-to-electrical-to-optical conversion, as in earlier regenerator-type repeaters), operates over a broad spectrum in the 1550 nm transmission window, and is insensitive to transmission modulation rates. The initial deployment of EDFAs in undersea systems used only one wavelength, although these devices are capable of amplification over a broad wavelength region. An example of this type of system was TAT 12/13, completed in 1996, which operates at 5 Gbits/s per fiber pair. This system is a ring network configuration, with two cables spanning the Atlantic Ocean and the respective landing sites also interconnected with a separate cable. The *self-healing* ring network design is capable of rerouting communications traffic in either direction around the ring, thus enabling the system to maintain operations in the event that one point in the ring experiences a failure.

Recalling that the most fundamental performance characteristics of any communication system are signal intelligibility and transmission capacity, consider these qualities with respect to undersea fiber-optic systems. The EDFAs provide effective signal amplification, but there are other transmission requirements to be met. Signal gain is also accompanied by noise, and chromatic dispersion properties in the optical fibers cause pulse broadening, which must also be controlled in these transoceanic systems, which can be up to 9000 km in length. In an ideal world, all signal waveguides would exhibit low and uniform attenuation, all channels (i.e., optical wavelengths) would experience equal amounts of gain from noise-free amplifiers, and the communication system would be insensitive to variable signal intensity or variable states of signal polarization. However, we do not live in an ideal world; thus communication system engineers are tasked to develop solutions to the fundamental problems that impede our progress in communications technologies. The following paragraphs will address some of the more prevalent physical phenomena that effect optical fiber transmissions and the engineering approaches that have been developed to achieve high-capacity wavelength division multiplexing (WDM) undersea communication systems using erbium doped fiber amplifier chains. [Note that in all references to *signals* and *transmission*, nonreturn-to-zero (NRZ) digital pulse format is implied. Soliton pulses are not considered.]

In long-haul undersea fiber-optic communication systems, EDFAs are connected in series with fiber-optic cables (Pacific Ocean cable systems can be up to 9000 km long). In order to achieve a stable system, both the gain at each amplifier and the signal attenuation induced by the optical fiber must be controlled such that the signal arriving at each successive amplifier has an acceptable amplitude. However, since the attenuation properties throughout the many optical fibers in the cable can exhibit slight variations, uniform amplifier gain is not the correct solution. In spite of nonuniform losses in the optical fibers, stable amplified signal levels can be achieved in practice because EDFAs exhibit gain saturation [25]. Gain saturation results in *input-signal-dependent gain*. Lower level input signals experience higher gains, while higher amplitude signals experience relatively less gain (i.e., gain compression). Figure 17.19 depicts signal amplitude regulation using EDFAs with gain compression.

In addition to providing signal gain, EDFAs also generate unwanted optical noise. The level of noise generated is proportional to the amount of gain. (For example, an amplifier with a signal gain of 20 dB generates 10 times as much noise as an amplifier with a gain of 10 dB.) It is especially critical for long-haul systems to be designed to operate with a relatively low gain so that the amplified spontaneous emission (ASE) noise does not accumulate to an unacceptable level. The *signal amplitude*, compared to *noise amplitude* [i.e., signal-to-noise ratio (SNR)] is a critical system performance parameter. In practice, long transoceanic systems require short-distance amplifier spacings (40–50 km), with relatively low gain (around 8–10 dB), while shorter systems use larger amplifier gains and longer distances between amplifiers to achieve the same final SNR

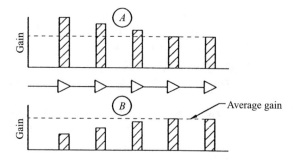

Figure 17.19 Amplifier gain compression: in *A*, a weaker than average signal receives relatively higher gain; in *B*, a stronger than average signal receives relatively less gain.

[25]. The goal of the system designer is to create a system with the correct SNR and the best reliability rating.

A very valuable feature of EDFAs is that they are capable of amplification over a broad range of wavelengths in the 1550-nm low-loss transmission region of the optical fibers. Wavelength division multiplexing systems transmit many individual wavelength channels over the wavelength region 1535–1565 nm (typical). Although the broadband gain properties of the EDFA are very useful, the gain is not uniform. A fundamental problem of nonuniform gain with respect to signal wavelength results in unacceptable variations in SNR. If left uncorrected, signals at the peak wavelength of the EDFA gain, namely 1558 nm, will incur the highest gain while signals at other wavelengths see less gain and, consequently, poorer SNR. Fortunately, there are solutions to this problem. One approach is to use a passive device called a gain equalization filter. This filter basically provides a wavelength-dependent attenuation medium whose performance characteristics are inverted with respect to the wavelength-dependent gain of the EDFA. The result is an approximate flattening of the signal gains throughout the entire transmission window [25]. Another technique, called *preemphasis*, works by applying specific levels of gain to individual channels, to compensate for the known amount of nonuniform gain in the system. Gain equalization filters and preemphasis techniques can be used separately or together to achieve uniform signal-to-noise ratios for all channels throughout the entire transmission region.

Now that signal gain issues have been discussed, we shall address the other critical parameter: *pulse broadening*. Pulse broadening resulting from chromatic dispersion, in amplified transoceanic systems is controlled by using two types of optical fibers with opposite chromatic dispersion properties. Undersea cables containing either negative-dispersion fiber or positive-dispersion fiber are connected alternately so that the signal pulse broadening is periodically returned to zero. Since the chromatic dispersion coefficient is wavelength dependent, in WDM systems only one wavelength channel can be brought back to zero total dispersion using the optical fiber in the cable. The other wavelengths requiring additional dispersion correction can be individually treated at the transmission terminals. This technique of using alternating chromatic dispersion characteristics in the fiber path to achieve a total system dispersion of zero (called dispersion mapping) is depicted in Fig. 17.20.

Chromatic dispersion is a linear phenomenon resulting from the fact that the refractive index of the optical waveguide is dependent on the wavelength of the optical pulse. However, optical fibers also exhibit a *nonlinear* behavior resulting from signal-intensity-dependent effects on the fiber's refractive index. Nonlinear refractive index

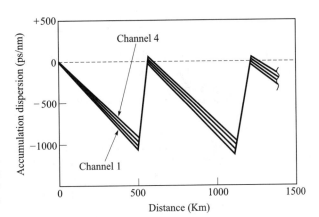

Figure 17.20 Dispersion mapping: accumulated negative chromatic dispersion is periodically counterbalanced with positive chromatic dispersion. Final dispersion adjustments for individual channels can be introduced at the terminals

effects resulting from interactions between the optical signals and the optical transmission medium can result in signal distortion and additional attenuation of the transmitted signals [25]. Transoceanic undersea amplified WDM systems overcome these detrimental phenomena by minimizing signal interaction. Signal interaction occurs when adjacent wavelength channels travel at nearly the same velocities and allow long overlap times. The chromatic dispersion properties of the optical fiber can be used to ensure that the optical signals experience sufficiently large differential velocities to keep the signal interactions short and manageable. Thus, dispersion and nonlinear effects are managed in undersea systems by creating a sufficiently large chromatic dispersion in the fiber path at all locations and alternating the use of fiber types (i.e., both negative and positive chromatic dispersion) so the final dispersion at the end of the system returns to zero [25].

Undersea fiber-optic communication technology has experienced many advancements since the first transatlantic system was installed in 1988. Over the last 10 years, system transmission capacities have increased nearly 100-fold. One of the latest published works on transmission experiments demonstrated 320 Gbits/s per fiber pair, over 7200 km, a typical transatlantic distance [26]. It seems reasonable to predict continued high demand and growth for these systems.

17.6 CONCLUDING REMARKS

This chapter was designed to provide an overview on the basic characteristics of optical fiber performance, optical fiber cable design, and undersea fiber-optic communication systems. Optical fiber communication systems offer an unsurpassed combination of reliability, data transmission capacity, and transmission quality. Through selection of a suitable cable design, almost any environment has the potential for installation of a fiber-optic cable, which will preserve the fibers' mechanical reliability and desirable low-loss optical waveguide properties.

REFERENCES

[1] G. P. Agrawal, *Fiber Optic Communication Systems*, Wiley, New York, 1992.
[2] G. Keiser, *Optical Fiber Communications*, Second Edition, McGraw-Hill, New York, 1991.

[3] P. K. Runge and P. R. Trischitta, *Undersea Lightwave Communications*, IEEE Press, New York, 1986.

[4] *Designers Guide to Fiber Optics*, AMP Incorporated, Harrisburg, PA, 1982.

[5] F. J. Denniston and P. K. Runge, "The glass necklace," *IEEE Spectrum*, Oct. 1995, Vol. 32, No. 10, p. 24.

[6] S. E. Miller and I. P. Kaminow, *Optical Fiber Telecommunications II*, Academic, San Diego, 1988.

[7] N. Okada, K. Watanabe, K. Kobayashi, and M. Miyamoto, "Ultra high-density optical fiber cable with thin coated multi-fiber ribbons for subscriber networks," International Wire & Cable Symposium, Nov. 14–17, 1994, Atlanta, Georgia, p. 28.

[8] C. Rochester, S. Barnes, P. Worthington, A. McLeod, and B. Eales, "Cable design for unrepeatered span systems," International Wire & Cable Symposium, Nov. 13–15, 1990, Reno, Nevada, p. 29.

[9] Simplex Technologies, Inc., "SL Lightweight Communication Cable," information brochure, Portsmouth, NH 03802.

[10] W. B. Wargotz, "Development of an alternate undersea armored cable protective coating compound," International Wire & Cable Symposium, Nov. 16–19, 1992, Reno, Nevada, p. 362.

[11] SL-100 Double-Armor Non-Repeatered Submarine Optical Fiber Cable, Simplex Technologies Inc., Portsmouth, NH, 03802.

[12] F. Grajewski, W. Stieb, H. Haag, and G. Hog, "Midspan jointing of optical groundwire," International Wire & Cable Symposium, Nov. 13–16, 1995, Philadelphia, PA, p. 794.

[13] H. Haag, G. Hog, U. Jansen, J. Schulte, and P. Zamzow, "Optical ground wire and all dielectric self-supporting cable: A technical comparison," International Wire and Cable Symposium, Nov. 14–17, 1994, Atlanta, GA, p. 380.

[14] Simplex Technologies Inc., "OPGW Cable (with tight buffer core)," Portsmouth, NH, 03802.

[15] E. Ghannoum, J. Chouteau, and M. Miron, "OPGW for Hydro-Quebec's telecom network," presented at the 1995 IEEE Winter Power Meeting, New York, NY, Jan. 1995.

[16] J Bonicel, O. Tatat, U. Jansen, and G. Couvrie, "Lightning strike resistance of OPGW," International Wire & Cable Symposium, Nov. 13–16, 1995, Philadelphia, PA, p. 800.

[17] O. Daneshvar, J. Hill, and X. Mann, "Development of an all dielectric self-supporting cable for use in high voltage environments," International Wire & Cable Symposium, Nov. 13–16, 1995, Philadelphia, PA, p. 763.

[18] G. Hog, J. Karlsson, N. Mariette, G. Paternostro, J. Rauchs, J. Schulte, and O. Tatat, "Flame retardant optical cables for central office, underground, or other specific applications," International Wire & Cable Symposium, Nov. 14–17, 1994, Atlanta, GA, p. 134.

[19] K. Kathiresan, J. Fluevog, S. Gentry, C. Arroyo, and L. Sherrets, "A family of non-halogen thermoplastic cables for shipboard application," International Wire & Cable Symposium, Nov. 18–21, 1991, St. Louis, MO, p. 723.

[20] K. Kathiresan, A. Panuska, J. Shea, W. Ficke, M. Santana, C. Taylor, and L. Carlton, "A fiber optic cable for hostile environments," International Wire & Cable Symposium, Nov. 16–19, 1992, Reno, NA, p. 158.

[21] M. Vyas, E. Buckland, and P. Neveaux, "Design and development of a steam resistant fiber optic cable," International Wire & Cable Symposium, Nov. 18–21, 1991, St. Louis, MO, p. 55.

[22] N. Hardwick and K. Katiresan, "Analysis of fiber optic design conditions in vicinity of steam lines—ruptured and pristine," International Wire & Cable Symposium, Nov. 16–19, 1992, Reno, NA, p. 679.

[23] R. Mortenson, B. Jackson, S. Shapiro, and W. Sirocky, "Undersea optically amplified repeatered technology, products and challenges," *AT&T Tech. J.*, Vol. 74, No. 1, 1995, p. 43.

[24] M. Kordahi, R. Gleason, and T. Chien, "Installation and maintenance technology for undersea cable systems," *AT&T Tech. J.*, Vol. 74, No. 1, 1995, p. 60.

[25] I. Kaminow and T. Koch, *Optical Fiber Telecommunications IIIA*, Academic, San Diego, 1997, p. 308.

[26] N. Bergano, C. Davidson, M. Ma, A. Pilipetskii, S. Evangelides, H. Kidorf, J. Darcie, E. Golovchenko, K. Rottwitt, P. Corbett, R. Menges, M. Mills, B. Pedersen, D. Peckham, A. Abramov, and A. Vengsarkar, "320 Gb/s WDM transmission (64 × 5 Gb/s) over 7200 km using large mode fiber spans and chirped return to zero signals," OFC-98 Post Deadline paper, Optical Fiber Communications Conference, Feb. 22–27, 1998, San Jose, CA, sponsored by the Optical Society of America, 2010 Massachusetts Avenue, Washington, DC, 20036-1023.

AUTHOR INDEX

SUBJECT INDEX

ABOUT THE EDITORS

R. Bartnikas received his early education at St. Michael's College School in Toronto, Ontario. He obtained the B.A.Sc. degree in electrical engineering from the University of Toronto in 1958 and the M.Eng. and Ph.D. degrees in electrical engineering from McGill University in 1962 and 1964, respectively.

In 1958 Dr. Bartnikas joined the Cable Development Laboratories, Northern Electric Company (now Northern Telecom), Lachine, Québec, transferring in 1963 to the Northern Electric Research and Development Laboratories (now Nortel Technologies) in Ottawa. In 1968 he joined the Institut de recherche d'Hydro-Québec and held the position of scientific director of the Materials Science Department until he was named Distinguished Senior Scientist in 1982. He is an adjunct professor in the Department of Electrical and Computer Engineering at the University of Waterloo and in the Department of Engineering Physics and Materials at École Polytechnique (Université de Montréal); he is also a visiting professor at the University of Rome at La Sapienza.

Dr. Bartnikas is the editor of the ASTM monograph/book series *Engineering Dielectrics* and two books entitled *Elements of Cable Engineering* and *Power Cable Engineering*. He is a recipient of many scientific awards, bestowed upon him in recognition for his contributions to the fields of dielectrics, electrical insulation, gaseous discharges and electrical measurements. He held the position of chairman of the ASTM Committee on Electrical and Electronic Insulating Materials from 1979 to 1985, served as president of the IEEE Dielectrics and Electrical Insulation Society, and is a member of the IEEE Energy and the IEEE Insulated Conductors Committees. Dr. Bartnikas is currently chairperson of Committee SC 15E Test Methods of the International Electrotechnical Commission (IEC). He is a registered professional engineer and is a Fellow of ASTM, the IEEE, the Institute of Physics (UK), the Royal Society of Canada (Academy of Science), and the Canadian Academy of Engineering.

K. D. Srivastava is emeritus professor in the Department of Electrical and Computer Engineering at the University of British Columbia, Vancouver, Canada. He received his early education in India graduating from the University of Roorkee with honors in electrical engineering in 1952. He received a Government of India postgraduate scholarship to study at the Royal College of Science and Technology, University of Glasgow, Scotland, where he was awarded the Ph.D. degree in 1957.

Professor Srivastava worked in several R&D laboratories in England. He was a senior research engineer at Brush Electric Co. Ltd., Loughborough and a principal scientific offer at Rutherford High Energy Laboratory, Harwell. In 1966 he immigrated to Canada. He was at the University of Waterloo, Ontario, until 1983, where he was chairman of the Electrical Engineering Department from 1972 to 1978. He then joined the University of British Columbia, where he was head of the Department of Electrical

Engineering from 1983 to 1986. He also served there as vice president from 1987 to 1994.

Professor Srivastava has been very active in international collaboration. He spent considerable time in India under the United Nations Development Program sponsorship, and in Brazil under the Canadian International Development Agency sponsorship. In recognition of his contributions to technical education, he was awarded an honorary degree by the University of Paraiba, Brazil, in 1979.

Professor Srivastava has over 45 years of professional experience in electrical insulation and high-voltage engineering and has published over 150 technical reports and papers. He coedited with Dr. Bartnikas two previous books on electric cables: *Elements of Cable Engineering* and *Power Cable Engineering*. Professor Srivastava is a registered professional engineer and is a Fellow of the IEEE, the IEE (UK) and the RSA (UK).